Hadron Structure and Lepton-Hadron Interactions

Cargèse 1977

NATO ADVANCED STUDY INSTITUTES SERIES

A series of edited volumes comprising multifaceted studies of contemporary scientific issues by some of the best scientific minds in the world, assembled in cooperation with NATO Scientific Affairs Division.

Series B: Physics

RECENT VOLUMES IN THIS SERIES

Volume 29 – Electron–Phonon Interactions and Phase Transitions
edited by Tormod Riste

Volume 30 – Many Degrees of Freedom in Field Theory
edited by L. Streit

Volume 31 – Many Degrees of Freedom in Particle Theory
edited by H. Satz

Volume 32 – Electron and Ion Spectroscopy of Solids
edited by L. Fiermans, J. Vennik, and W. Dekeyser

Volume 33 – Microscopic Structure and Dynamics of Liquids
edited by J. Dupuy and A. J. Dianoux

Volume 34 – Path Integrals and Their Applications in Quantum, Statistical, and Solid State Physics
edited by G. Papadopoulos and J. T. Devreese

Volume 35 – Correlation Functions and Quasiparticle Interactions in Condensed Matter
edited by J. Woods Halley

Volume 36 – Strongly Coupled Plasmas
edited by Gabor Kalman

Volume 37 – Coherence in Spectroscopy and Modern Physics
edited by F. T. Arecchi, R. Bonifacio, and M. O. Scully

Volume 38 – Theoretical Methods in Medium-Energy and Heavy-Ion Physics
edited by K. W. McVoy and W. A. Friedman

Volume 39 – Hadron Structure and Lepton–Hadron Interactions – *Cargèse 1977*
edited by Maurice Lévy, Jean-Louis Basdevant, David Speiser, Jacques Weyers, Raymond Gastmans, and Jean Zinn-Justin

This series is published by an international board of publishers in conjunction with NATO Scientific Affairs Division

A Life Sciences **B Physics**	Plenum Publishing Corporation London and New York
C Mathematical and Physical Sciences	D. Reidel Publishing Company Dordrecht and Boston
D Behavioral and Social Sciences	Sijthoff International Publishing Company Leiden
E Applied Sciences	Noordhoff International Publishing Leiden

Hadron Structure and Lepton-Hadron Interactions

Cargèse 1977

Edited by

Maurice Lévy and Jean-Louis Basdevant

Laboratory of Theoretical Physics and High Energies
Université Pierre et Marie Curie
Paris, France

David Speiser and Jacques Weyers

Institute of Theoretical Physics
Université Catholique de Louvain
Louvain-la-Neuve, Belgium

Raymond Gastmans

Institute of Theoretical Physics
Katholieke Universiteit Leuven
Leuven, Belgium

and

Jean Zinn-Justin

Institute of Theoretical Physics
C.E.N.
Saclay, France

PLENUM PRESS • NEW YORK AND LONDON

Published in cooperation with NATO Scientific Affairs Division

Library of Congress Cataloging in Publication Data

Cargèse Summer Institute on Hadron Structure and Lepton–Hadron Interaction, 1977.
Hadron structure and lepton–hadron interaction.

(NATO advanced study institutes series; series B, Physics; v. 39)
"Organized by the Université Pierre et Marie Curie, Paris . . . the Katholieke Universiteit te Leuven . . . and the Université catholique de Louvain."
Includes bibliographies and index.
1. Hadrons – Congresses. 2. Leptons (Nuclear physics) – Congresses. 3. Nuclear reactions – Congresses. 4. Particles (Nuclear physics) – Congresses. I. Lévy, Maurice, 1922- II. Université Pierre et Marie Curie. III. Katholieke Universiteit te Leuven. IV. Louvain. Université catholique. V. Title. VI. Series.
QC793.5.H322C37 1977 539.7'216 78-24027
ISBN-13: 978-1-4613-2885-8 e-ISBN-13: 978-1-4613-2883-4
DOI: 10.1007/ 978-1-4613-2883-4

Proceedings of the 1977 Cargèse Summer Institute held at Cargèse, France, July, 1977, sponsored in part by NATO

Softcover reprint of the hardcover 1st edition 1979
A Division of Plenum Publishing Corporation
227 West 17th Street, New York, N.Y. 10011

PREFACE

The Cargèse Summer Institute 1977 on Hadron Structure and Lepton-Hadron Interactions was organized by the Université Pierre et Marie Curie, Paris (M. LEVY and J.-L. BASDEVANT), Saclay (J. ZINN-JUSTIN), the Katholieke Universiteit te Leuven (R. GASTMANS), and the Université Catholique de Louvain (D. SPEISER and J. WEYERS) who, on the advice of NATO in 1975, had joined their efforts and worked in common. It was the 18th Summer Institute held at Cargèse and the 4th one organized by the two departments of theoretical physics at Leuven and Louvain-la-Neuve.

Again, fortune was on the side of the organizers. The experimentalists could report a great number of quite unexpected but very important events : new high mass bosons, which suggests the existence of a new quantum number, surprisingly accurate results in neutrino scattering, new electron-positron data, etc. Also, a great number of interesting theoretical ideas were presented, mostly in connection with recent developments in gauge theories, like e.g. the role of instantons, confinement of quarks, and high orders of perturbation theory.

It is a real pleasure to thank all those who have made this summer institute possible !

Thanks are due to the Scientific Committee of NATO and its President for a generous grant and especially to its scientific secretary, Dr. T. KESTER, for his constant help and encouragement.

We also thank the C.N.R.S. (France) and the N.S.F. (U.S.A.) for travel grants.

Special thanks are due to the Université de Nice for having put at our disposal the facilities of the Institut at Cargèse, and the Université Pierre et Marie Curie, Paris, the K.U. Leuven and the U.C.L. Louvain-La-Neuve for their secretarial assistance and financial support.

We wish to thank Miss M.-F. HANSELER, Chr. DETROYE and V. de Laforcade and all others from Paris, Leuven, Louvain-la-Neuve and especially Cargèse for their collaboration.

We thank Miss N. COISMAN and Mrs B. STEYAERT for typing the manuscript, and Mrs. F. de Ville, Miss M. de CROMBRUGGHE, MM M. DENIS, J. GERIS, C. LEROY and H. HAUT for their help in correcting proofs and Dr. DERHAM (Plenum Press) for his collaboration.

Finally, the financial contributions from the Brouwerijen Artois, the Kredietbank, International Business Machines of Belgium, and the Generale Bankmaatschappij are gratefully acknowledged. They helped to give this summer institute a broader international audience.

Mostly however, we would like to thank all lecturers and participants who came from over 20 countries : the willingness of the former to answer all questions and the keen interest of the latter provided the stimulus which made (we hope) this institute a success.

M. LEVY D. SPEISER

J.-L. BASDEVANT R. GASTMANS

J. ZINN-JUSTIN J. WEYERS

CONTENTS

LEPTON PAIR PRODUCTION IN HADRONIC COLLISIONS

Leon M. LEDERMAN

Columbia University, New York, N.Y. 10027

I. INTRODUCTION

This is a review of the experimental status of production of massive ($m \gtrsim 1$ GeV) lepton pairs in hadronic collisions.[1] The catch-all diagrams (Fig. 1) illustrate the principal motivations for this ever-expanding activity :

$$pp \rightarrow \ell^+\ell^- + \text{hadrons.}^2 \qquad (1a)$$

It provides a third reaction to the probe of hadronic substructure via virtual photons; the other two being

$$ep \rightarrow e + \text{hadrons}; \qquad (1b)$$

$$e^+e^- \rightarrow \text{hadrons} . \qquad (1c)$$

The relevant energy variable for (1b) is Q^2, the momentum transfer. This is limited by the energy of lepton beams $\sim$ 200 GeV at FNAL and SPS implies a $Q_{max}{}^2 \sim 400$ GeV^2.

In (1c) the total energy squared in the c.m. is limited to $\sim$ 50 GeV^2 at SPEAR/DORIS and is expected to reach $\sim$ 900 GeV^2 with the advent of PEP/PETRA in 1980.

By contrast, the effective mass of the lepton pair emerging in (1a) :

$$m^2 = 2P_+P_-(1-\cos(\theta_+ + \theta_-) = Q^2$$

is limited by the proton energies to $\sim$ 750 GeV^2 at FNAL/SPS and to 3000 GeV at ISR. With the advent of an ISABELLE of 200 on 200 GeV, m^2 can reach 160,000 GeV^2.

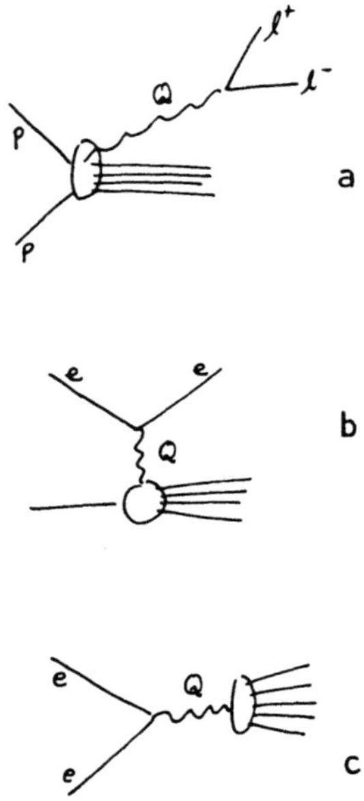

Fig. 1. Feynman diagrams : a) lepton pair production
b) deeply inelastic scattering, c) e^+e^- annihilation.

We note that hadronic substructure is probed in two ways : by the continuum distribution in dilepton masses that emerge from Fig.1a and also by resonances. These come out by themselves and may be observed as enhancements in the dilepton spectrum given sufficient running time and experimental resolution.

A second motivation has to do with the CVC argument[3] which relates reaction (1) to the corresponding weak process (Fig. 2a) in which the photon is replaced by the intermediate boson $W^\pm$ for emerging weak charged currents ($\ell^\pm\nu$) or Z° for weak neutral currents. The fact that CVC enables us to deduce the hadronic vertex part of the weak reaction from the measurements on the electromagnetic process, allows a prediction of the corresponding production of real or virtual W's.

This is in the domain of the future of high energy physics when hadronic energies $\gtrsim$ 100 GeV are available in the c.m. Furthermore, the observation of reaction (1a) near these energies permits the calibration of a large variety of esoteric particle production experiments in which the virtual photon is used as the production mode.[5-9] Usually, a purely hadronic mode is also possible, in which case reaction (1) provides a lower limit, say to production of monopole pairs,[5,6] quark pairs,[5,7] heavy lepton pairs.[5,8] Applications to intermediate boson production are also plentiful.[5,9]

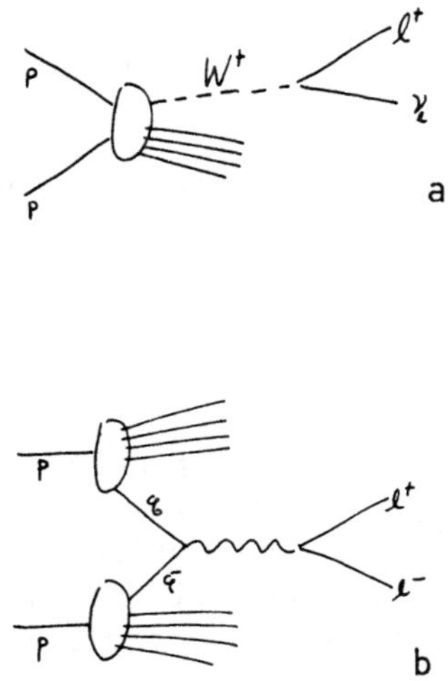

Fig. 2a. W production, compare 1a.
2b. Drell-Yan process.

II.KINEMATICS THINGS

The fundamental effective mass of the dilepton was given in Eq. (2). In addition, important variables are the dilepton Feynman variable :

$$X_F = 2P_{//}^{*}/\sqrt{s} \tag{3}$$

where $s = 2M_pE_o$; E_o = incident lab energy, and

$$p_\perp = p_+ \sin\theta_+ + p_- \sin\theta_-$$

gives the invariant transverse momentum of the dilepton in terms of measurements.

In the widely applied Drell-Yan model,[10] a quark-parton of momentum x_1p annihilates with an anti-quark of momentum x_2p (Fig.2b) to form a virtual photon of (longitudinal) momentum :

$$p(x_2 - x_1) = p_{||}^{*} \tag{4}$$

where $p = \sqrt{s}/2$.

Annihilation kinematics yield

$$x_1 x_2 s = m^2 \ . \tag{5}$$

This model will be treated in more detail in Section VI.

III. HISTORICAL MATTERS

In 1963-64, as an aftermath of the Columbia-BNL Neutrino II setup, we made a study of high transverse momentum single muons created in the beam dump.[11] The motivation was a search for $W^+ \to \mu^+\nu$. The negative result was criticized by Y. Yamaguchi[3] who noted that virtual photons would provide a background. Using the CVC argument he estimated the ratio of large transverse momentum muons from W's and from γ's of the same mass. The comment stimulated us to study the problem of detecting both muons from virtual photon decay. It quickly became clear that the variable mass could provide a true small distance probe of hadronic matter and could, if nature were so disposed, "resonate" at masses which were special. In short, we tried to dazzle the Scheduling Committee with high expectations.* In our ignorance, we ignored theoretical objections that a reaction with two protons in the initial state was just not subject to analysis, i.e. it could never be related to the inverse reaction $e^+ + e^- \to$ hadrons unless two protons were also produced there. Recall that in 1966-67, when these ideas were being discussed, "inclusive" reactions had not been invented.

The design at Columbia was with J. Christenson and two graduate students, B. Pope and G. Hicks. In the carrying out of the experiment, we were joined by P. Limon and E. Zavattini. We decided on a poor resolution, high luminosity experiment.[4]

Figure 3 illustrates the arrangement. This was a beam dump experiment, the target consisting of 1 ft of uranium in an accordion-like arrangement that permitted extrapolation to infinite density. The arrangement of steel absorber was such as to limit the transverse momentum of the emerging muons to $\gtrsim$ 500 MeV/c. The muons momentum was determined by range, the arrangement placed an effective lower limit of $P_{\parallel} \gtrsim$ 12 GeV/c on the dimuon momentum. The mass acceptance started near $\sim$1 GeV, peaked at 3 GeV, and was flat out to the kinematic limit. The data rate was accidental limited, the signal varying from 2 % of the background at $M \simeq 1$ GeV to 40 % at 5 GeV. With the smooth spill of the AGS,

---

* See Appendix I : Proposal to AGS n° 420, August 1967.

in-time subtraction of background was possible, limited only by statistical errors. It is especially interesting to note (see Fig. 4) that there were some two million events in the 29.5 GeV run, and over 50,000 in the famous "shoulder" which is now known to represent the J/ψ resonance. In the ubiquitous experimental agony as between resolution and aperture, this experiment clearly sinned in the direction of aperture. The 1974 Ting experiment,[11a] (Fig. 5) with ∿ 250 events, "sinned" in the direction of resolution, clearly, since all the events were in one bin ! I will return to a recent reanalysis of the BNL data later under the topic of "scaling" and for further tests of the Drell-Yan theory. Of course, the data we did publish stimulated a deluge of theoretical papers and when the smoke settled, only the Drell-Yan model survived. We decided to follow up this research using the newer accelerators and proposed lepton pair experiments at the ISR (1969) and at FERMILAB (1970).

The ISR experiment, a Columbia - CERN - Rockefeller collaboration ran in 1972 and failed to find any lepton pairs[12]. The alibi in this case was the appearance of a totally unexpected background which was published[13] as a discovery : high transverse momentum hadrons. Fig. 6 shows the expected yield of π°'s, and the yield found by CCR, clearly many orders of magnitude larger. The reason this was too much for a lepton pair search was in the CCR arrangement : a lead glass spectrometer with no magnetic field. Electrons were simulated by pions overlapping high p_T π°'s which generated electromagnetic showers in the glass. It is amusing that this "background" now turns out to be related to the physics of dilepton production in a deep way : via the constituent picture. We will discuss this again later.

The next stage took place at Fermilab where the lepton pair experiment was preceded by a search for single leptons - as a warm-up! Direct electrons were found[14,15] and pursued from early 1974 until January 1975. Reviews of this field abound in the literature.[2] In 1975, the Columbia-Fermilab-Stony Brook pair experiment started. Data from this experiment and other dilepton experiments appeared during 1976. See Table I.

IV. SURVEY OF DILEPTON EXPERIMENTS

A. Introduction

I found six publications related to continuum dileptons which I will list and abbreviate here :

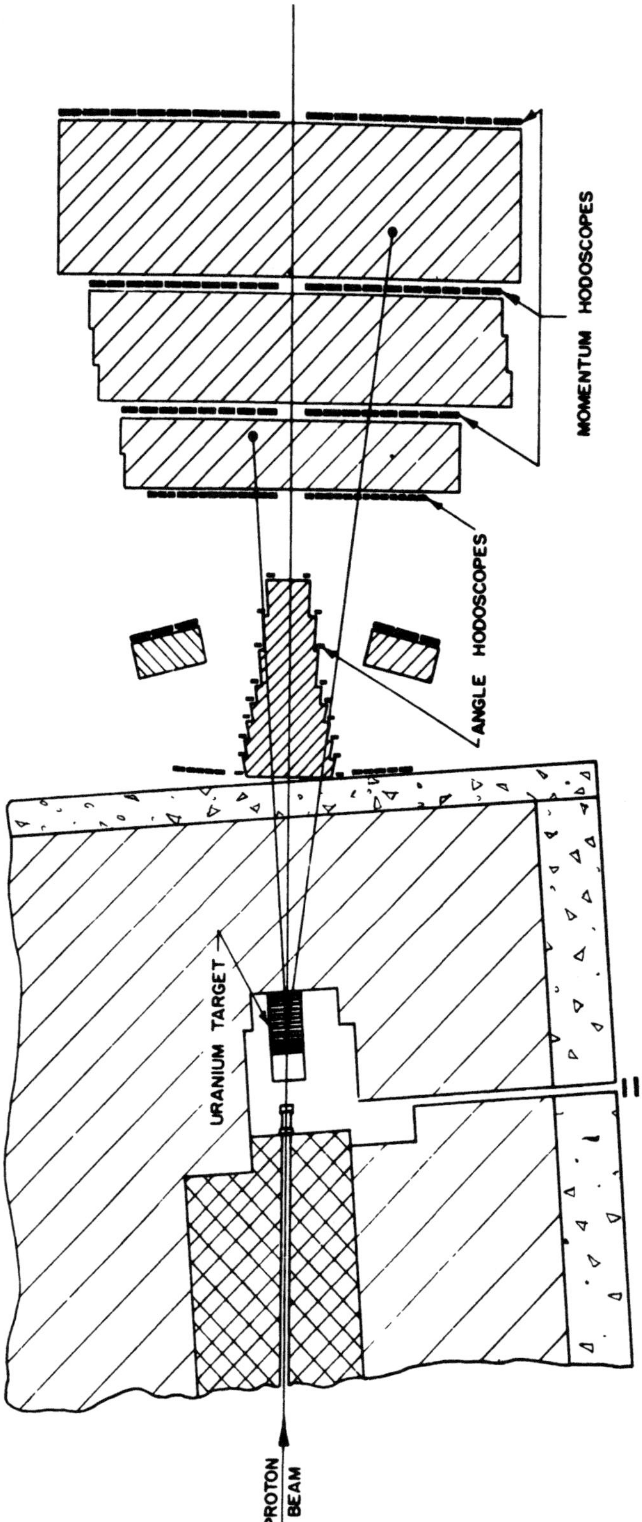

Fig. 3. Columbia-BNM dimuon apparatus.

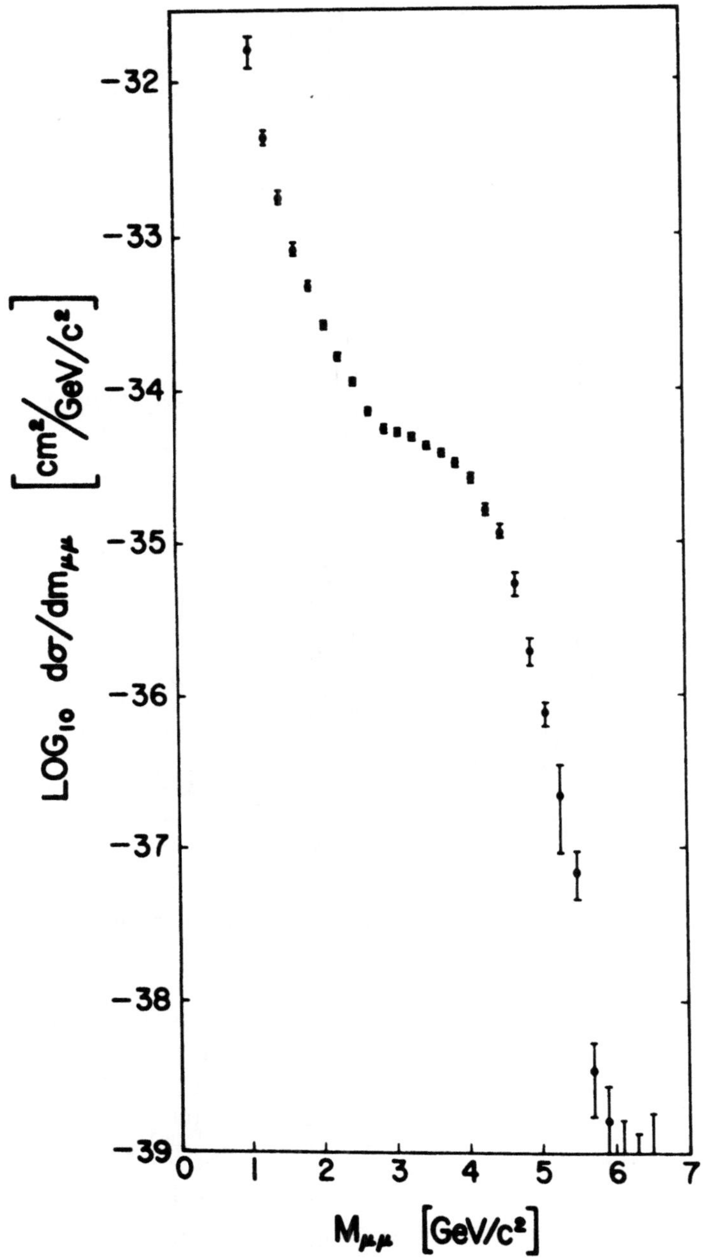

Fig.4. Columbia-BNL $\mu^+\mu^-$ data.

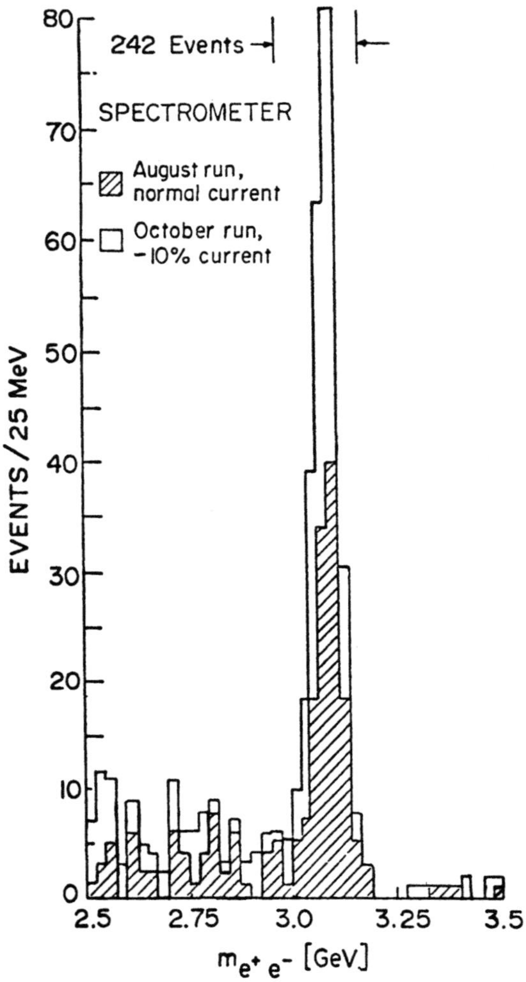

Fig.5. BNL-MIT e^+e^- data : J production.

1. Columbia-Fermilab-Stony Brook 16,17,18,19	FNAL	ee	I	300, 400 GeV
		μμ	I	400
		ee	II	200, 300, 400
		μμ	II	400
2. Columbia-Hawaii-Illinois-Fermilab[20]	FNAL	μμ		300
3. Northeastern[21]	FNAL	μμ		200
4. Chicago-Princeton[22] II	FNAL	μμ } μμ($\pi^{\pm}$ in) }		150, 250
5. Chicago-Princeton[23] I	FNAL	μμ		200, 300, 400
6. Omega Spectrometer[24]	CERN	μμ		40 GeV

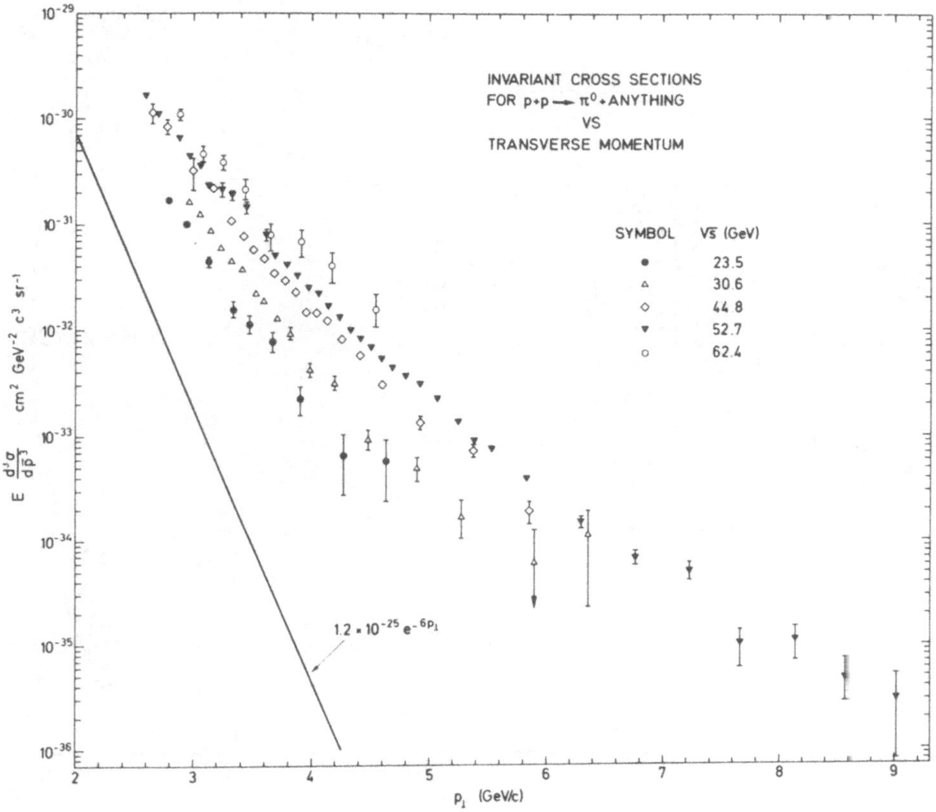

Fig. 6 : Discovery of high P_T π^o's : CCR data

1. <u>Columbia-Fermilab-Stony Brook : ee I</u> [16]. Figure 7 illustrates the layout. This is a thin target experiment (0.009 in. Be, 4 in. long) to minimize background from π^o conversions. The beam dump is 2.2 m downstream of the target, contributing about 20% to the singles rates in the various detectors. The key idea here is to have no detectors upstream of the spectrometer magnets - in fact, detectors intercept the electron trajectory only after having left the neutral beam. This permitted an incident flux of $\sim 3 \times 10^{10}$ ppp and 6×10^9 interactions/pulse - limited by singles rates in the PWC and hodoscope planes indicated. This "medium aperture" experiment subtended ± 20° around 90° in the c.m. but with a vertical aperture which limited the overall acceptance to $\sim 5 \times 10^{-4}$, the acceptance rising to this value at a mass of 5 GeV and remaining fairly flat up to $\sim$ 15 GeV. For studying the 3-6 GeV mass range, the magnet current is lowered. The other principal feature of this layout is a lead glass spec-

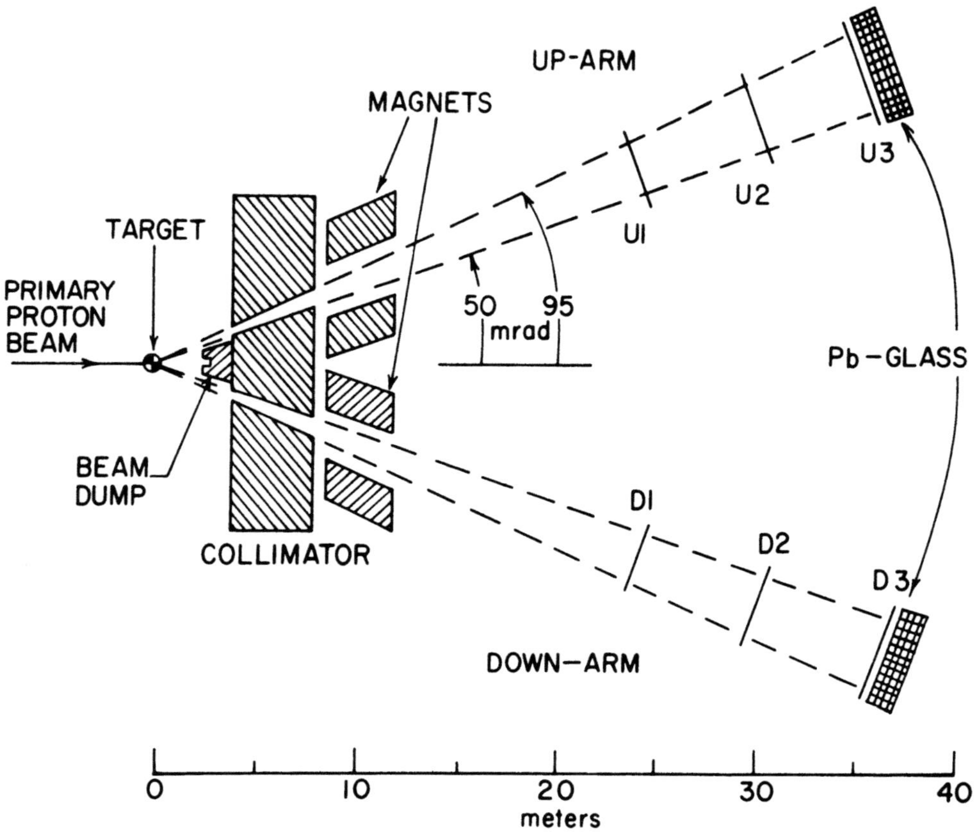

Fig. 7. CFS dilepton spectrometer.

trometer which measures the electron energy and details of electromagnetic shower development in order to select electrons out of a hadronic background that, in each arm, is $\sim 10^4$ times more numerous. In this arrangement, the single arm hadron rejection was $\sim 3 \times 10^{-4}$ leading to background free data above about 5 GeV. Accidental backgrounds dominate below $\sim$ 4 GeV and studies of the enhancement of the dilepton signal due to the insertion of converters in each arm give a measure of the correlated π° (η°) induced background. The mass resolution, although aided by the lead glass, stems mostly from the magnetic measurement and is $\Delta m/m \sim 1$ % (rms).

2. Columbia-Hawaii-Illinois-Fermilab[20]. The setup is illustrated in Fig.8 and is famous for the discovery of the charmed baryon. It is a forward spectrometer designed to look at both neutron and photon induced events. The relevant dimuon data are

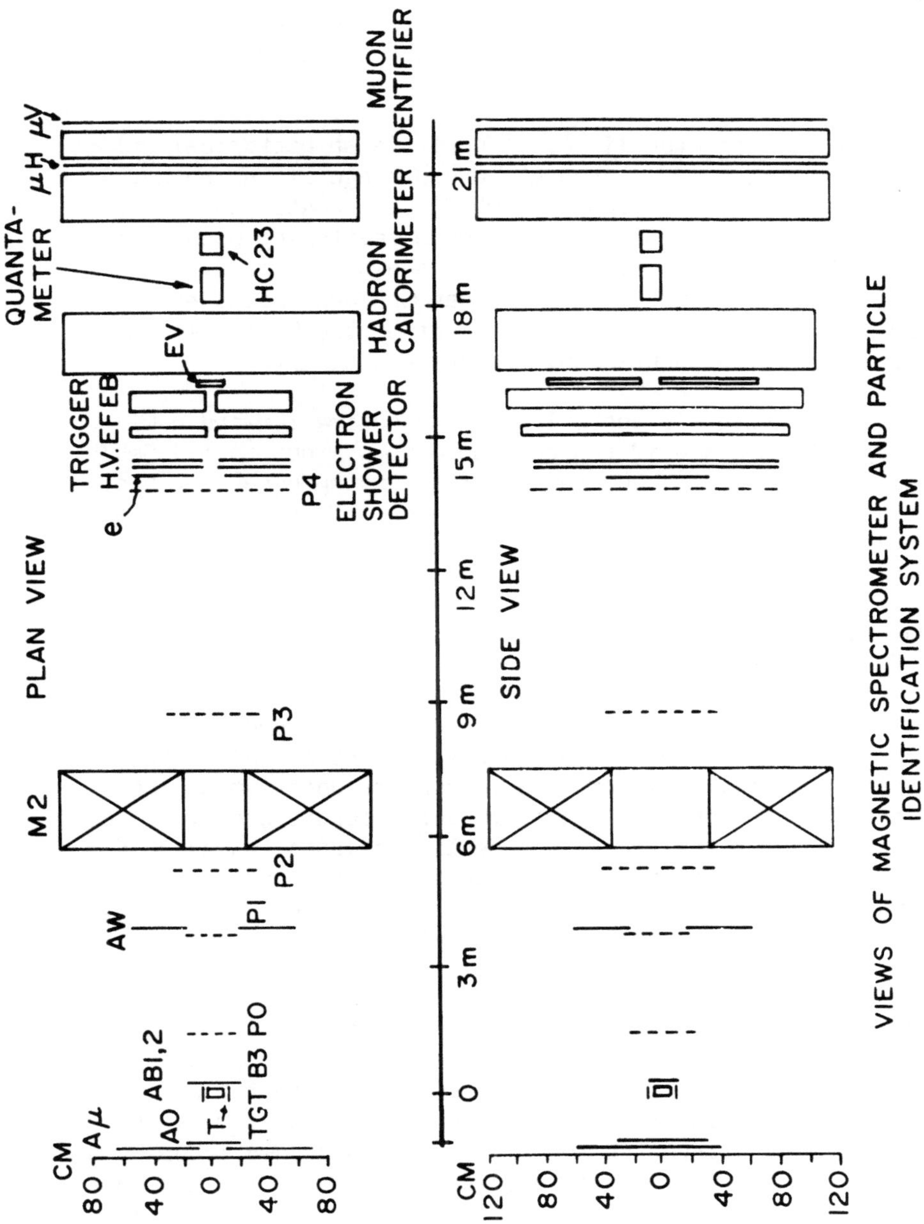

Fig. 8. CHIF apparatus.

obtained via an inclusive dimuon trigger using a spectrum of neutrons peaked at 300 MeV and tagged by a hadron calorimeter. This has a large acceptance in x_F, $p_\perp$, but because of the limited incident flux, runs out of events at masses above 5 GeV. The hadron filter imposes a cutoff in $x_F \geq 0.25$. Here again the limiting rate ($\sim$ few million interactions/pulse) is determined by the singles rates in the PWC's. The mass resolution at the J/ψ was 10 % FWHM.

3. Chicago-Princeton II [22]. (The II is an historical artifact, indicating a somewhat younger average age of the authors). This is a very large acceptance forward spectrometer based upon the Chicago cyclotron designed for a muon scattering experiment at Fermilab. The dimuon arrangement is shown in Fig. 9. This is the first dimuon experiment which has incident pions as well as protons although an earlier experiment by the Northeastern group compared J/ψ production by protons and pions.

The spectrometer is sensitive to the entire forward hemisphere ($x_F \geq 0$) and a uniform acceptance in $p_\perp$ out to $\sim$ 4 GeV/c. It is also unique in the coverage in rest system decay angle of the parent, i.e. $0 < |\cos\theta^*| < 0.8$. The incident beam is tagged by gas Cerenkov counters and was limited in the data reported here to $\sim 10^6$/pulse.

Here again, hadron filters limit the resolution to about $\sim$ 10% FWHM at the J/ψ .

4. Northeastern [21]. Mu-pairs are created at one end of a 5.6 m Fe filter. At modest beam intensity (240 GeV pions and protons in the Fermilab meson area), counters upstream of the filter serve to define the trajectory very approximately. A solid iron spectrometer after the filter deflected muons, whose outgoing trajectory is measured in spark chambers. Background is deduced from some sign pairs and is small. The acceptance in X_F and P_T is large although resolution is relatively poor.

5. Chicago-Princeton I[23]. This group has added a large solid angle detector to the magnetic spectrometer used in Fermilab Proton East to detect high $p_\perp$ hadrons. This is a small solid angle highly selective single muon detector set at 90° in the c.m. The large arm covers an interval from 60 to 125° in the c.m. with a large azimuthal coverage (-8 to 25°). This is accomplished by a series of holes dug in the shielding earth along a line displaced by 10m from the incident electron (Fig. 10). Hence, multihole spectrometer. Muons reaching a given "hole" have a minimum energy determined by the range in the earth. Data are taken as a function of the $p_\perp^s$ of the small arm setting and begins at 3.25 GeV/c. Although the mass acceptance is difficult to define, it is centered

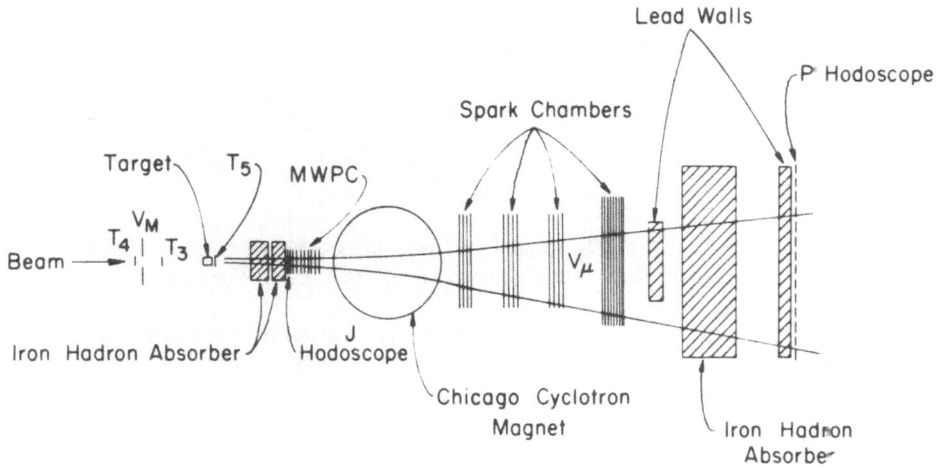

Fig.9. CP II "Chicago cyclotron" spectrometer.

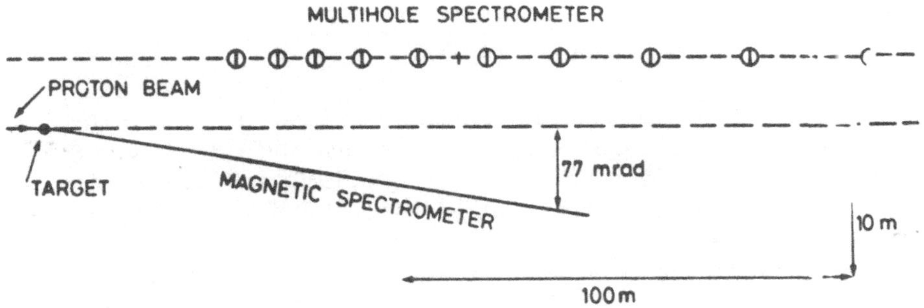

Fig. 10. Multihole spectrometer, CPI.

approximately at $m \simeq 2p_{\perp}^{s}$ and is approximately 2 GeV wide.

The evident dependence of the interpretation of the data on a production model is compensated by the high intensity which can be tolerated, leading to sensitivity to masses as high as 11 GeV/c.

6. CERN SPS Omega Spectrometer[24]. This experiment has a large acceptance ($x_p > 0$) to a fixed momentum beam of 40 GeV. Cerenkov counters permit the selection of $\pi^{\pm}$, $K^{\pm}$, p and $\bar{p}$ projectiles. The detector is an array of MWPC's inside a large magnet. Muon pairs are detected via penetration of a beam dump absorber placed just inside the magnet.

We omit here several experiments which measure only J/ψ's since this is somewhat off the main track of these lectures.

7. CFS Muons. This will be the main subject of Lecture II and details will be given there.

V. EXPERIMENTAL RESULTS OF CFS.
Proton-Nuclear Collisions

This constitutes by far the bulk of the data. We will begin by discussing only the "high mass" region defined as $m \gtrsim 5$ GeV. This has two origins : data below 4 GeV are "contaminated" by the J/ψ, ψ' and, at lower energies, the classical vector mesons ρ,φ,ω. Some continuum windows are available and we will discuss these later. Data between 4-5 GeV may also have contributions from charmonium states and, where good statistics are available as in CFS, there are significant backgrounds.

CFS Experiments

a. ee I. The dielectron mode was selected as giving the best resolution. The results of a four month run (∿ 1000 hours of data taking) are given in Fig. 11. The luminosity, i.e. the intensity of protons on target, was limited by the counting rates in the various planes of detectors. Typically, we permitted rates of the order of ∿ 5 million counts per spill of the accelerator. (This had an average effective time of ∿ 0.5 sec.) Modern scintillation counters and MWPC's do not perform well at rates much above this. The rates were generally attributed to hadrons although the individual counting rates were far higher than the $\sim 10^4$:1 for hadrons to leptons (of the same transverse momentum). Three conclusions were derived : i) a continuum signal was seen, in rough agreement with Drell-Yan predictions; ii) the average transverse

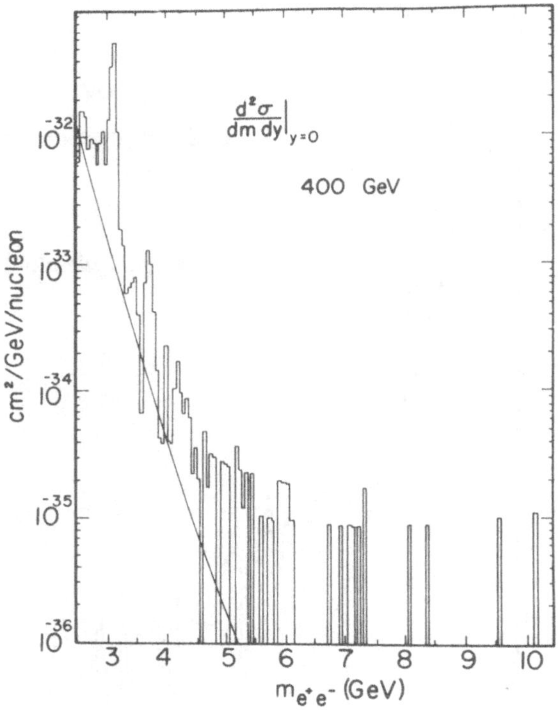

Fig. 11. ee I results.

momentum of the dielectrons was surprisingly high, $\langle p_T \rangle \sim 1.5$ GeV/c and very reminiscent of hadron production; iii) a cluster of events near 6 GeV was found to be statistically interesting - with only one chance in fifty of being a fluctuation. The interest of these high mass events and the slow rate of accumulation stimulated a dimuon version.

b. <u>μμ I.</u> To make the change, a column of about 7 m of Be was installed in each arm to filter hadrons while minimizing the perturbation of the produced leptons. Additional detectors were added to define the muon trajectory but the arrangement was very much a quick modification and far from optimized. This experiment turned on in March 1976 and initially was limited by an enormous rate in all counters. These were identified as soft muons coming from the decay of pions in the low density beryllium filter. Extensive piling up of lead and steel shielding gradually permitted a luminosity improvement and the run March-June 1976 yielded five times the electron data. See Fig. 12. This was at the expense of mass resolution which had degraded to ~ 4 % . The conclusions here were the following :

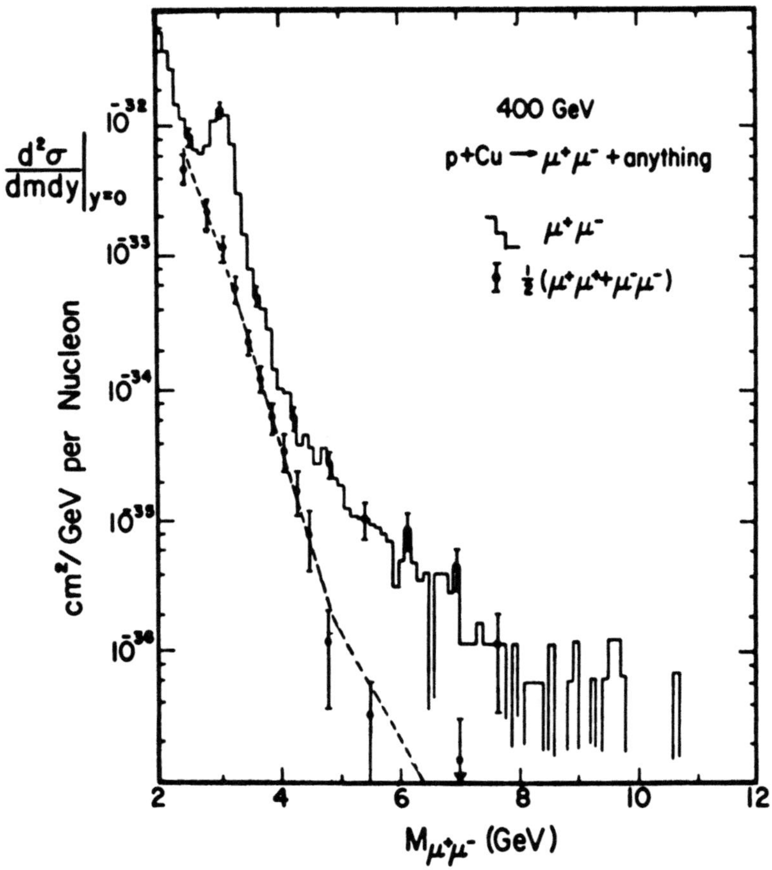

Fig. 12. μμ I results.

i) No evidence was found for a bump near 6 GeV. Although the poor resolution made this less sensitive than we had hoped.

ii) The continuum, in good agreement with ee I was also in agreement with Drell-Yan. What this actually means, we will see later.

iii) The transverse momentum was rather flat and very hadron-like. See Fig. 13a. Figure 13b summarizes the ee I and μμ I and compares them with theory.

iv) The experience gained was essentially to the design of an improved version (μμ II) which we will discuss in detail below.

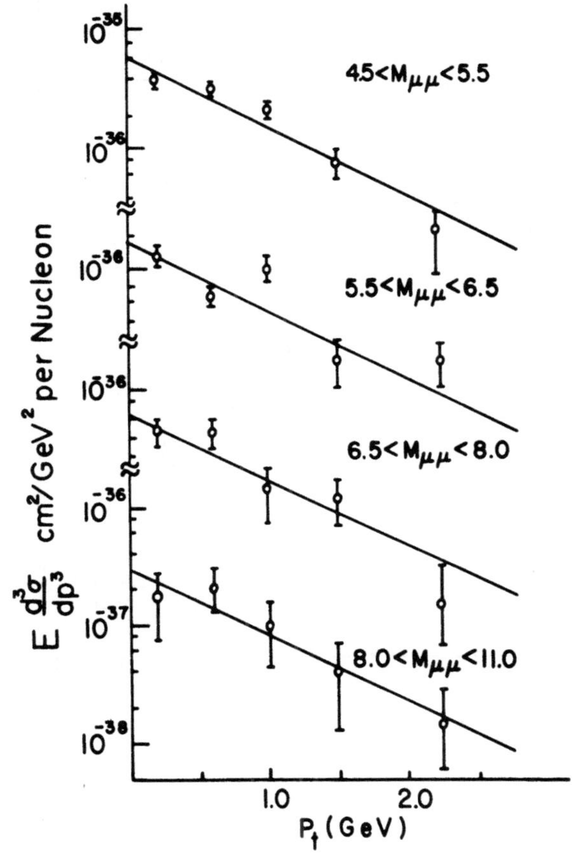

Fig. 13a μμ I high P_T data.

c. ee II. In the interval July 76 to February 77, CFS repeated the electron experiment with somewhat improved resolution. The results are still preliminary : i) the continuum was in agreement with ee I and μμ I ; ii) the combined data still showed an effect near 6 GeV although not quite so pronounced; iii) a new cluster of ∿6 events appeared, quite isolated, at 9.5 GeV. This matched a previously (and correctly!) ignored cluster in the muon data at the same mass (see Fig. 13b); iv) dihadron data were accumulated simultaneously but we will not discuss these here.

d. μμ II. (Taken from Rapporteur Talk, Budapest 1977). Table II gives the title, authors and abstract. Figure 14 presents the mass spectrum as the data taken through July 13 and is shown in order to justify the emphasis on this contribution. As can be seen from Table I, the experiment to be described extends the world collection of high mass dilepton data by two orders of magnitude.

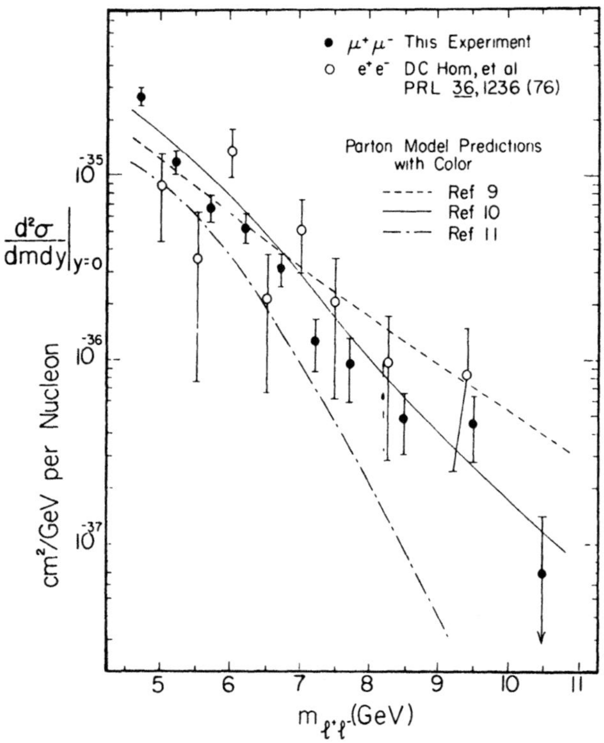

Fig. 13b Summary ee I and μμ I data.

Table I

Brief History of E-288 (Fermilab).

Run	$\mathcal{L}$		Result
e^+e^- I Sept.'75-Jan.'76 Ref. 9	∿3 x 10^9 int/spill x 0.1 % Δm/m	→	43 events ≳ 5GeV Δm/m ∿ 1 %
μμ I March-June '76 Ref. 10	∿2-3 x 10^{10} int/spill x 0.1 %	→	200 events ≳ 5GeV Δm/m ∿ 4 %
ee II Sept.'76-Feb.'77 Ref. 11	∿ 3 x 10^9 int/spill x 0.1 %	→	39 events ≳ 5GeV Δm/m ∿ 1/2 %
μμ II May 15-June '77 Ref.8	∿ 10^{11} int/spill x 1.0 %	→	9000 events>5 GeV (20 June) Δm/m ∿ 2 %

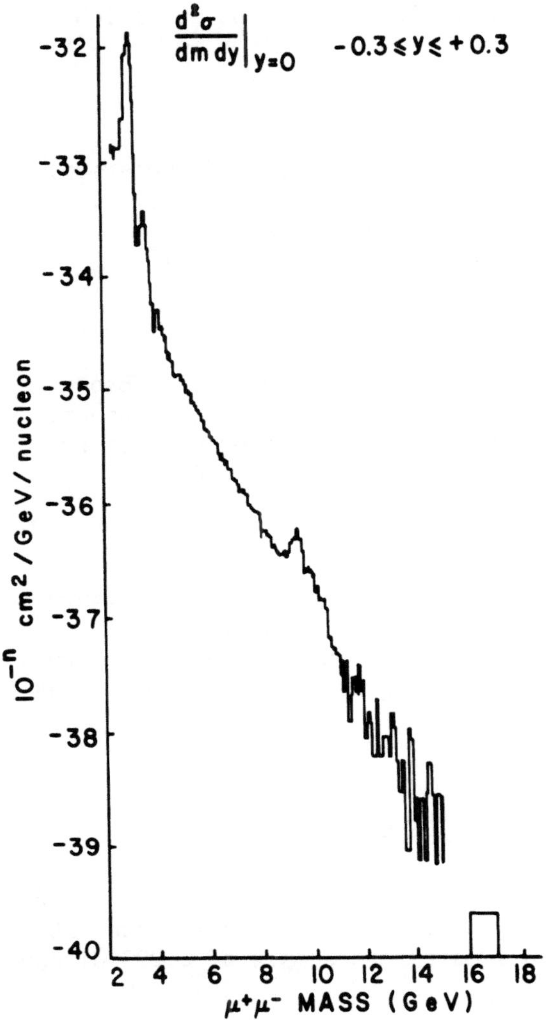

Fig. 14. Dimuon mass spectrum with 9000 events above 5 GeV. (1000 events in the ψ' peak).

Table II

OBSERVATION OF A DIMUON RESONANCE AT 9.5 GeV IN 400 GeV PROTON-NUCLEUS COLLISIONS

S.W. Herb, D. C. Hom, L. M. Lederman,
J.C. Sens,[+] H. D. Snyder, and J. K. Yoh
Columbia University
New York, New York 10027[*]

and

J. A. Appel, B. C. Brown, C. N. Brown,
W. R. Innes, K. Ueno, and T. Yamanouchi
Fermi National Accelerator Laboratory
Batavia, Illinois 60510 [++]

and

A. S. Ito, H. Jöstlein, D. M. Kaplan,
and R. D. Kephart
State University of New York at Stony Brook,
Stony Brook, Long Island, New York 11794[*]

ABSTRACT

Dimuon production is studied in 400 GeV proton-nucleus collisions. A strong enhancement is observed at 9.5 GeV mass in a sample of 9,000 dimuon events with a mass $m_{\mu+\mu-} > 5$ GeV.

---

[*] Research supported by the National Science Foundation.
[+] Permanent address : Foundation for Fundamental Research on Matter, The Netherlands.
[++] Research supported by the U.S. Energy Research and Development Administration.
Phys.Rev.Lett. August 1977

If resolution in mass and transverse momentum are included as a measure of overall sensitivity (this multiplies luminosity by a factor of at least $(M/\Delta M)^2$), the improvements are even greater.

It is relevant to recall that all our experiments were limited in data taking by the rate of single counts in the various detector planes. These are typically 5 to 50 times the rate of high energy muons that pass through one arm of the double arm spectrometer. The limitations had previously restricted the number of proton interactions per spill to $\sim 5 \times 10^9$ in dielectron running (no hadron filter) and to $\sim 3 \times 10^{10}$ int/spill in $\mu\mu$I. Note that at Fermilab, the spill is 1-2 seconds and has structure which makes the effective spill, on the average over our experience ~ 0.5 sec. Note also that in principle, 10^{13} interactions per spill are available.

Figure 15 shows the arrangement of the Columbia-Fermilab-Stony Brook (CFS) dimuon II experiment. Following the target region (described below) and a 3 meter steel shield are two air dipoles which deflect vertically. The horizontal gap between pole faces is flared out to match the angular acceptance. As in our previous exposures, magnetic deflection is in the vertical plane. Thus, momentum determinations are decoupled from the measurements of opening angle. No detectors are placed upstream of the magnets where the ambient flux is an order of magnitude higher. Thus the momentum determination must assume that the muon started from the target. The target origin is confirmed by projection, using the horizontal plane coordinates. Conventional scintillation hodoscopes and Charpak (MWPC) chambers are used.

Earlier runs taught us that the limiting singles rates did not fall off with distance to the target but depending only on the area of detector. Thus it seemed that the closer the detector is to the target, the larger the solid angle with no penalty of increased singles. The present arrangement results in three times the (vertical) solid angle, obtained by displacing all detectors toward the target. We stress that our present limitations are not imposed by laws of physics - but only of the realities of the Fermilab arrangement.

Previous exposures were dominated by the high resolution e^+e^- arrangement and had no detectors in the median plane; the east arm detected e^+, when say, these are deflected down, out of the line-of-sight to the target etc. Another factor of two was gained by symmetrizing the apparatus. Now, each arm detected both μ^+ and μ^-. A substantial advantage of this is the simultaneous recording of same sign pairs : $\mu^+\mu^-$ and $\mu^-\mu^-$ which constitute a rigorous measurement of background.

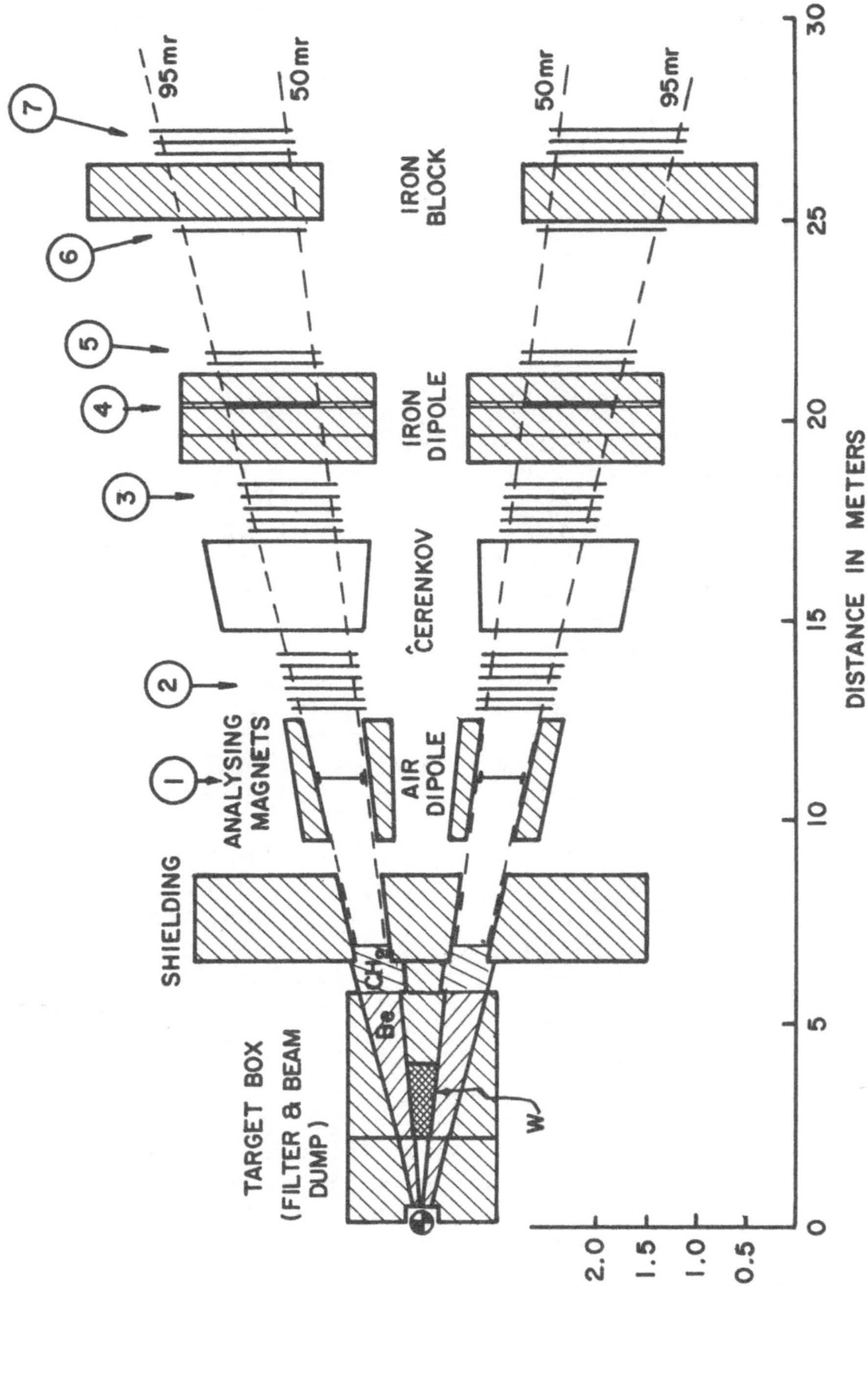

Fig. 15. μμ II apparatus.

This also improved the high mass acceptance in that particles no longer had to be deflected out of the neutral beam to be counted. The combined improvement is indicated in Fig. 16, where the mass acceptance is shown for several magnet settings. The "old $\mu\mu$ I" acceptance is also shown.

The target box (Fig. 17) provided the final improvement. Thin targets (0.7 mm wide) of Pt and Cu were used. These were 30 % of an interaction length long. The unexpended protons and forward produced secondaries were absorbed in a dense hevimet (W alloy) beam dump, 3 meters long, followed by 2 meters of steel. Particles emitted in the accepted aperture ($\pm$ 50 to 90 mrad horizontally and $\pm$12 mrad vertically) find a 7 meter long beryllium filter which is accurately fitted against the tungsten dump at small angles and steel shielding at large angles. Beryllium is everywhere at least 1 in. oversize to insure that muons do not pass though high Z material. The general scheme was used in $\mu\mu$ I but in the present arrangement, great care was taken to minimize cracks and voids. The effort was rewarded in that the flux of background π and K decay muons in the detector area was reduced by a factor of three permitting a corresponding increase in protons on target.

The first 30 cm of the filter was on a remotely controlled elevator permitting us to interchange beryllium and copper. The copper absorber reduced the rates by yet another factor of two at the expense of increased multiple coulomb scattering (MCS) and hence somewhat worse mass resolution (see below).

At the very low levels of signal being detected (we go to $\sim 10^{-13} \times \sigma_{tot}$), it is important to have overdetermination of the "muon-ness" and of the integrity of the trajectory. This was accomplished by two additions : a MWPC installed in the middle of the spectrometer dipole and a solid iron magnet downstream of the momentum chambers, the function of which is to redetermine the muon momentum. The $\sim$ 15 % precision so obtained was sufficient to remove low energy muons which, by rare combinations of scattering, appeared to have much higher momentum. A check of the effectiveness of the iron magnet was that, for those trajectories which passed the test, 96 % had a hit in the mid-magnet chamber, at the correct wire. Accidentals within $\pm$ 1 cm of the correct hit were $\lesssim$ few percent.

The mass resolution is clearly a crucial feature and Fig. 18 illustrates the contributions to this from three sources :
i) measurement errors in the 2 and 3 mm spacing MWPC's give a contribution which increases with momentum (as the bend angle decreases); ii) MCS diminishes the knowledge of the opening angle (Eq. 3) and this varies as p^{-1}; iii) finally MCS contributes to the uncertainty in the angle incident on the magnet and is independent of the muon momentum. The dotted lines show the penalty for using copper absorber.

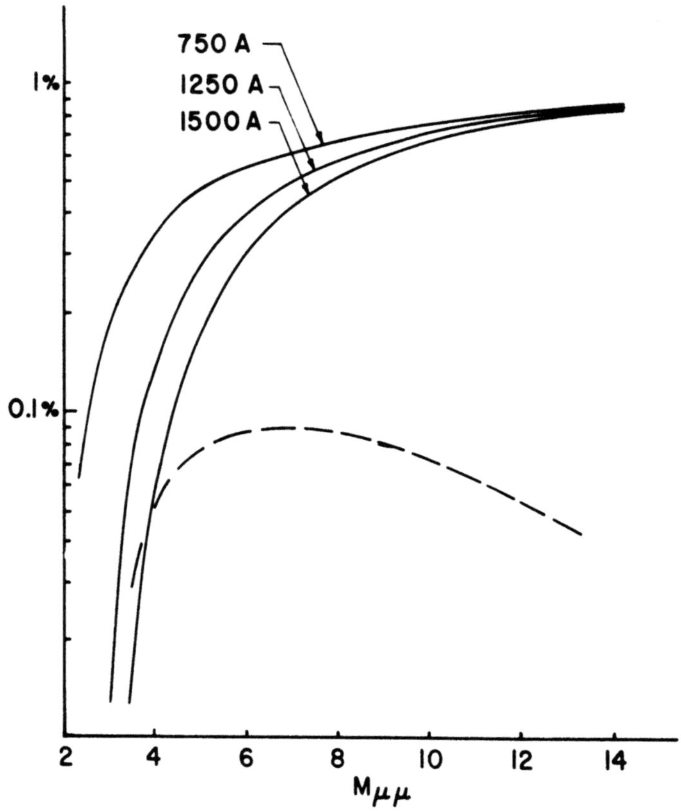

Fig. 16. Mass acceptance for various magnet currents. The dashed curve is the μμI acceptance. This acceptance is calculated with an assumed $1 + \cos^2\theta$ distribution in the helicity frame. There are ∿ 30 % effects due to other distributions.

We note that the resolution for the bulk of data taken is ∿ 2% (rms) and independent of mass for m ⩾ 6 Gev. Quantitative determinations of the resolution were carried out by detailed Monte Carlo programs (two!) checked by hand calculation and confronted with J/ψ and ψ' peaks as well as with target reconstruction data.

Results. Figure 19a shows the mass spectrum of raw data passing the loosest test :

We notice that (1) the enhancement near 9.5 GeV appears here, and (2) there is a fair amount of same sign pairs indicating background. Table 3 lists the three types of cuts applied to the data and Figs. 19b and 19c show the effects of applying the first two to the raw data. The overall cut efficiency is ∿60 % but results in a background which is ∿3 % at 5 GeV and essentially zero at higher masses. The combination of the surviving events and the Monte Carlo generated acceptance results in the

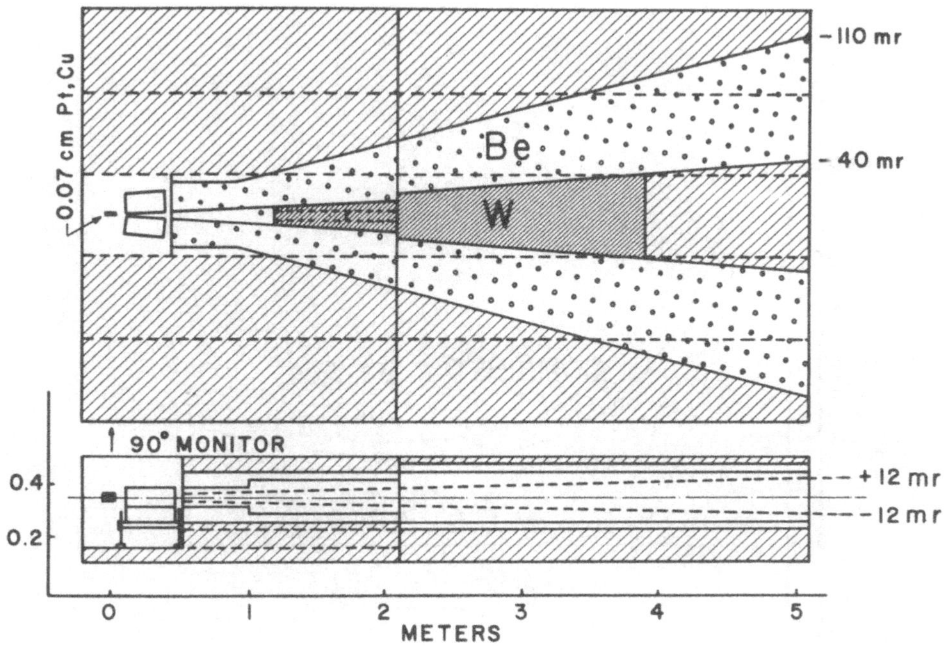

Fig. 17. Detail of target box. Note variable first 30 cm of absorber.

mass spectrum of Fig. 14. The spectrum is presented in the form :

$$B \left. \frac{d^2\sigma}{dmdy} \right|_{y=0} ,$$

where the rapidity y is used for its convenience. We note here that the acceptance in y is :

$$-0.3 \leq y \leq +0.3$$

corresponding to $|X_F| < 0.1$. We are observing muons emerging near 90° c.m. (72 - 180°) with an overall coverage of ∿2 % of the total c.m. solid angle.

In Fig. 14 the continuum dilepton yield is well fitted from 6 to 14 GeV, by a single exponential of the form, $\sim e^{-bm}$, with the slope parameter, $b \sim 1$ (see below). The fit is made with the enhancement region (9.2 - 10.6 GeV) excluded. A subtraction of the fitted continuum from the data yields the graph of Fig. 20 (data of June 20). We note that the width of the enhancement is

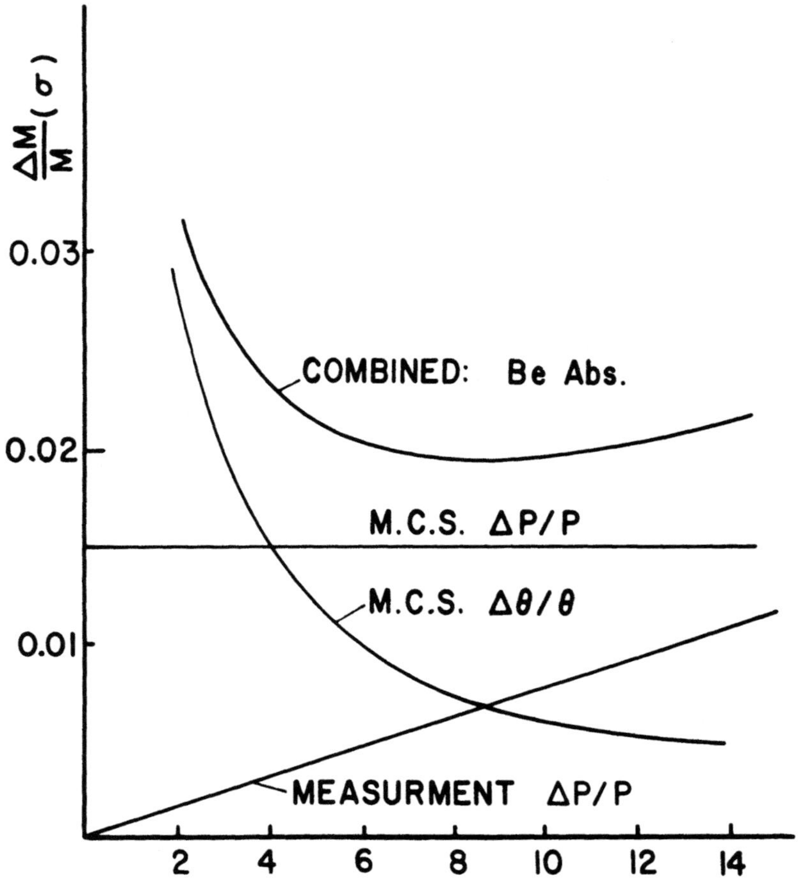

Fig. 18. Mass resolution with all Be absorber. Replacing Be by Cu increases the $\Delta\theta/\theta$ contribution only and raises the total curve to $\sim$2.1 % above 6 GeV. Analytically $\Delta m/m = 0.11/M + 0.015 + 0.0009\ M$ for Cu. Additions are in quadrature.

$\sim$ 1.2 GeV) (full width at half maximum) as compared with an experimental resolution of 500 MeV. The statistical validity of the enhancement is 440 ± 30 and hence established beyond reasonable question or uncertainties in the slope fit. There is however the possibility of systematic errors induced by the apparatus or by our analysis procedures.

Systematic Studies

We list possible sources and address each of them in what follows.

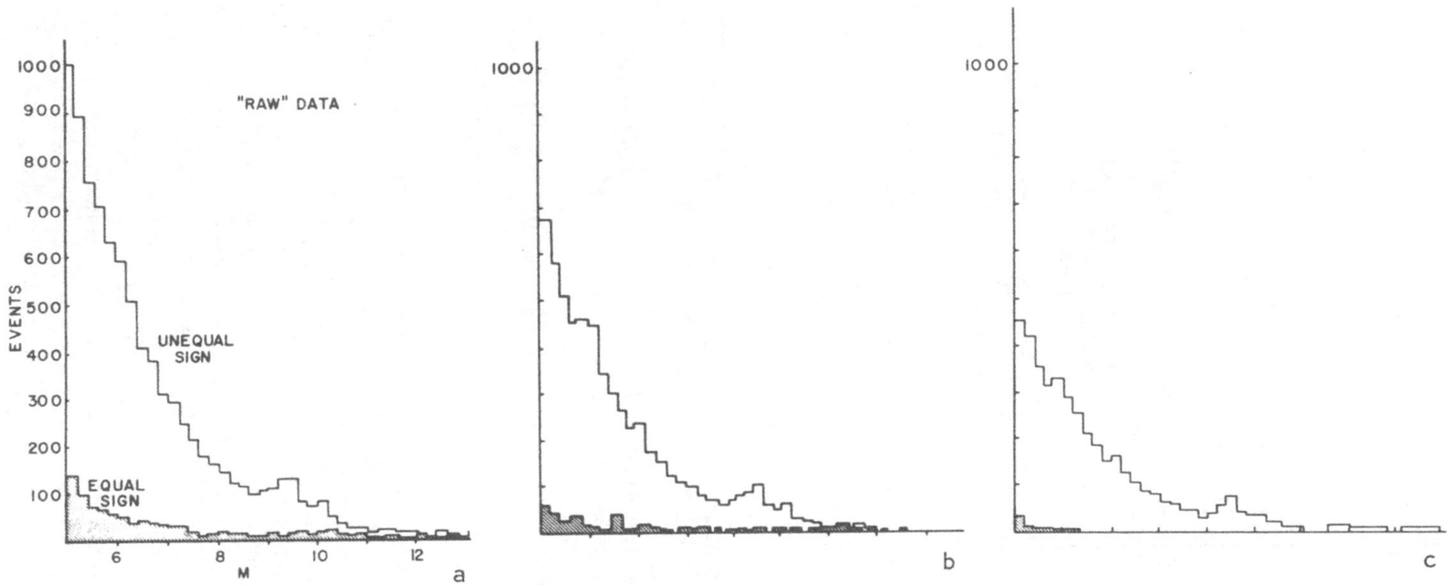

Fig. 19a Raw data, no cuts. Both opposite sign and equal sign pairs (shaded) are shown.
19b Data with fiducial cut (Table 3).
19c Data with fiducial and muon cuts.

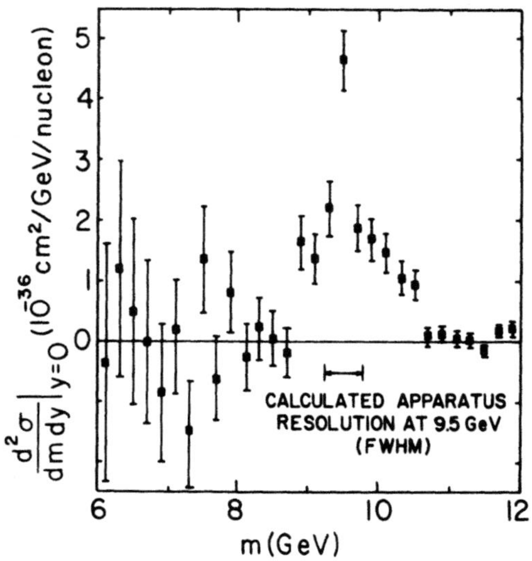

Fig. 20 Result of subtracting a fit to continuum (excluding 9.0 to 10.6 GeV) from data (June 20).

1. <u>Could the bump be an artifact induced by non-uniform response of different parts of the apparatus</u> ?

There are three varieties of tests of this hypothesis :

i) study the data under a wide variety of different run conditions,

ii) examine directly, the parts of the apparatus that are responsible for the bump data,

iii) combine muons from different events to form spurious dimuon events.

Significant data were obtained under the variety of conditions listed in Table IV. An unplanned variation was the intervention of a fire in our experimental area after one week of data taking. (We were rescued from possible corrosion of the MWPC amplifiers by a European fire salvage expert who had obtained much experience at CERN). Thus pre-fire data could be compared to post-fire data. In addition to the change of target, first 30 cm of absorber, magnet current, tests were made in which μ^+ (east arm) μ^- (west arm) pairs are compared with μ^- (east arm) μ^+ (west arm) pairs. Finally, the polarity of the magnets is reversed. Each of these changes tests for inhomogeneous response to muons in a variety of ways. The consistency tests consist in each case, of fitting the continuum, subtracting the fit and fitting the residue with a single gaussian. The fit provides the parameters of the gaussian : the central mass and the width. An

Table 3

Cuts to Raw Data.

A. Fiducial Cuts

1. Track intercepts all MWPC and counter planes.
2. Track projects through analyzing magnet.
3. Target cut $\Delta x \leq 4$ cm.

B. Muon Cuts

1. 9/10 MWPC hits and a cut on reconstructed track confidence level.
2. Target cut $\Delta x < (1.5 + 50/p)$ c.m.
3. 4/6 hits in devices downstream of iron magnet.
4. These hits be within 2σ of that projected from air mag momentum.

C. (Optional) Single Track in Each Arm.

Combined efficiency (pairs) 65%.

Table 4

Data Taken

Date	No. Hrs.	Target	Absorber	Magnet I
5/14-5/21	80	Cu	Cu	1500 A
Fire!				
5/27-6/5	100	Cu	Cu	1500 A
6/-6/12	70	Cu	Cu	1250 A
6/12-6/17	50	Pt	Be	1250 A
6/18-6/20+	75	Pt	Cu	1500 A
6/20-7/13	300	Pt	Cu	1500 A

Table 5

Fitted Results for Two Magnet Currents *

Mag Current	M_o	Γ	σB	A	b	χ^2/dof
1250 A	9.53 $\pm$0.08	1.04 $\pm$0.21	3.7 pb $\pm$0.08	2.33 μb $\pm$0.59	1.00 $\pm$0.01	48/26
1500 A	9.54 $\pm$0.06	1.20 $\pm$0.12	3.2 $\pm$0.03	1.96 $\pm$0.34	0.99 $\pm$0.03	38/27

* These are preliminary values shown for comparison. No systematic corrections applied.

example of the consistency is the effect of magnet current charge!

$$\text{Fit : } \frac{d\sigma}{dm} = Ae^{-bm} + C\,e^{-(m-m_o)^2/\Gamma^2} ,$$

where :

$$\sigma B = \int Ce^{(m-m_o)^2/\Gamma^2}\, dm$$

$$B = \text{branching ratio to } \mu^+\mu^- \quad .$$

Table V shows the results in the case of varying magnet current. We find adequate consistency although none of the fits is particularly happy with the primary hypothesis.

In summary, all running conditions gave consistent results insofar as the central mass m_o, the width Γ, and the slope parameter b . (We stress here that the experimental parameters above are an intermediate stage in the analysis and do not represent the final results).

Some results of the direct examination of the apparatus are presented in Figs. 21 and 22 where two-dimensional scatter plots are made. These (Figs. 21 a,b) are the distribution of hits in MWPC (3x,y) for muons from all masses above 5 GeV. Superposed are the hits due to masses in the bump region. A very large number of such graphs indicates nothing anomalous about the 9-10 GeV data- they show no preference for "lower left hand corners" or the like. It is obvious that this also searches for inhomogeneities in the trigger counters.

The final apparatus test is the shuffling of muons so that fictitious pairs are assembled by combining a μ^+ from one event with μ^-'s from all other events (in the $\geq$ 5 GeV sample). Thus not only tests for "luminess" in the acceptance but also tests software in the computation and histogramming of masses. Figure 23 shows the resultant mass spectrum which we find satisfactorily smooth. Note that this is not exactly the same as the accidental spectrum since presumably not all single arm muons belong to high mass pairs.

2. <u>Are the events target originated</u> ? A variety of scenarios could generate false data and conceivably a broad bump. For example, upstream sources of primary proton collisions could generate hadron showers with long flight paths, during which K, $\pi \to \mu$ decay could take place. Or, the very strong J/ψ production ($\sim$ 1000 per pulse) could load to spurious high mass levels via a scattering of each of the muons from a very energetic forward produced J. Effects of these kinds would give rise to poor reconstruction of the target and Fig. 24 presents the target reconstruction for muons from : (24a) all masses and with no target cut; (12b) the dimuon masses from 7-8.5 GeV; (24c) dimuons in the bump

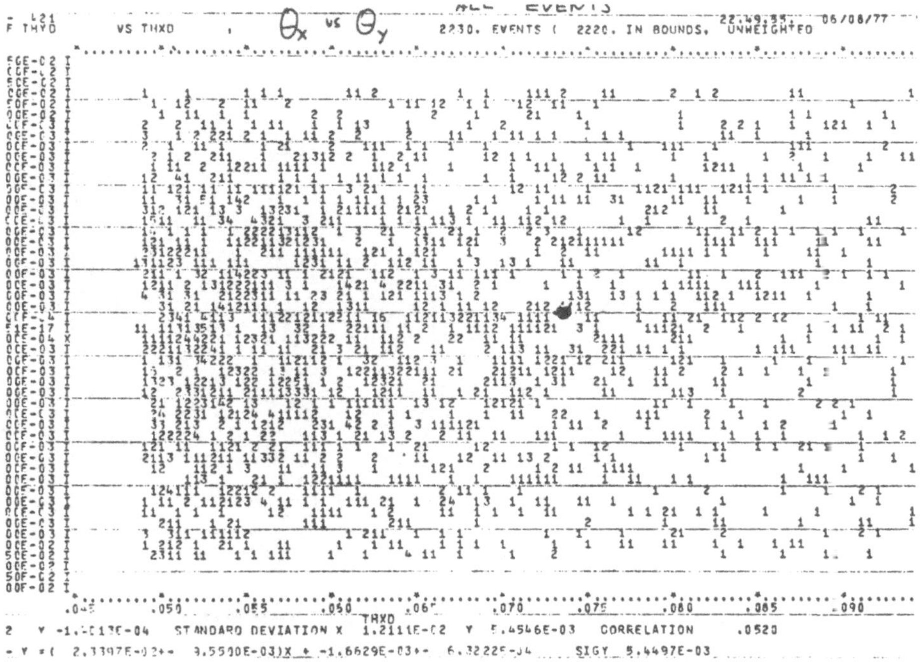

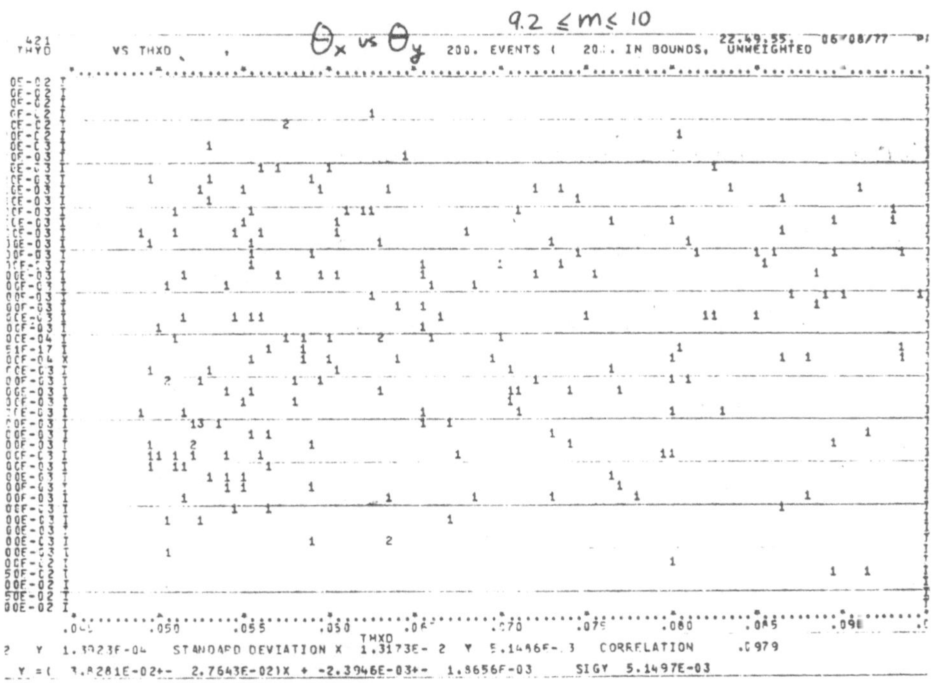

Fig. 21a Scatter plot, all masses, of horizontal angle θ_x vs. vertical angle θ_y.

21b Same as 21a but for resonance region.

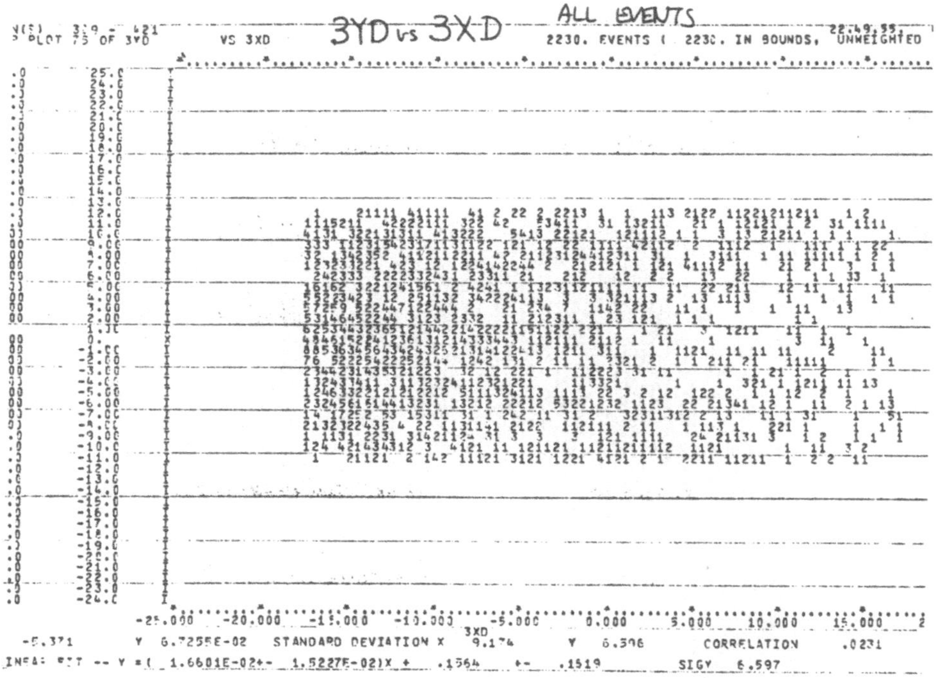

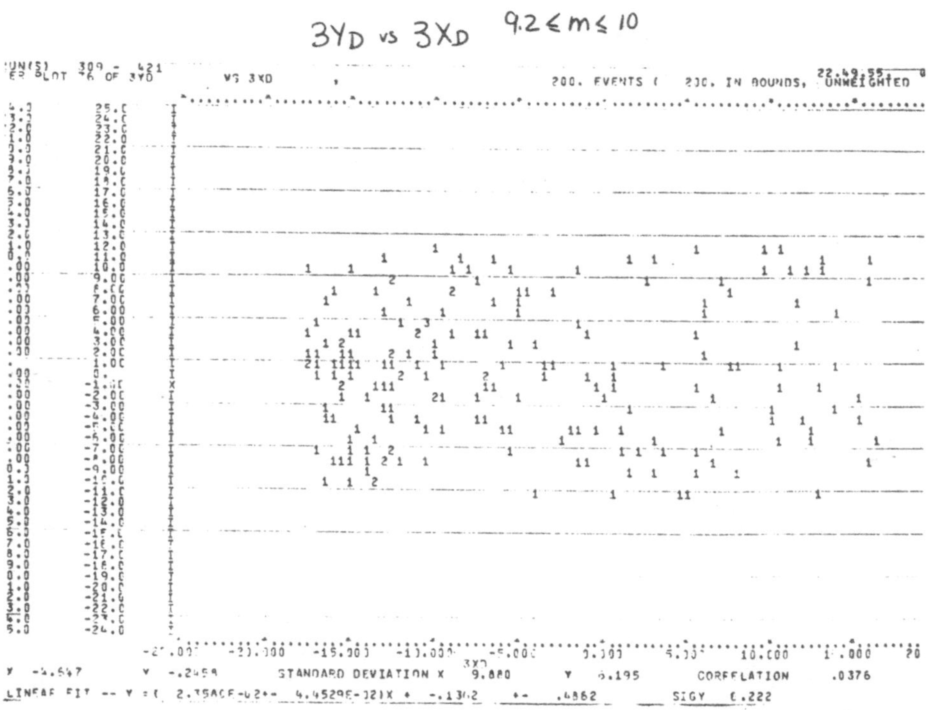

Fig. 22a Scatter plot of hits in MWPC 3x and 3y for all masses.
22b Same as 22a but for resonance region.

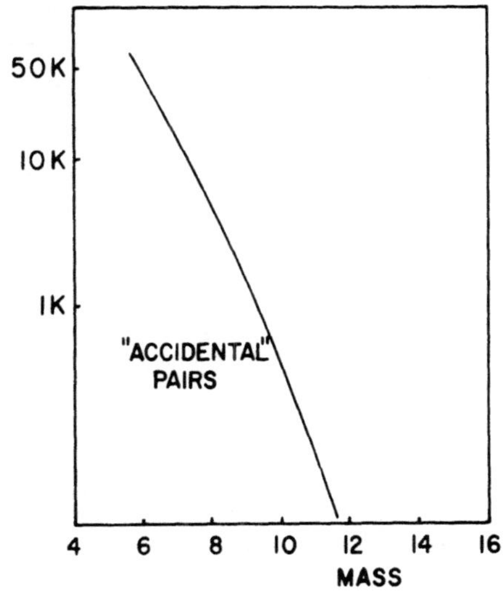

Fig. 23 Result of combining muons from different pair events.

region. Figure 24c also shows the higher mass (≳11 GeV) target picture. We note that the apparent width of the target narrows appropriately from $\sigma \sim 0.5$ in. to $\sigma \sim 0.17$ in. as the mass (hence the momentum of the muons) increases. The clear separation from nearby sources, including the beam dump is obvious, as is the fact that the bump events are normal. We note that the intrinsic width of the target is given by the projection of the length across the viewing angle and is 0.2 in. This is broadened by MCS and the Monte Carlo predictions (24c) are tested by these data. Finally, we ran with target out for long enough to have collected some 200 events above 5 GeV had we run with target in. No events were in fact collected. Thus, the events arise from the target with no evidence for anomalous trajectory.

3. Are these subtle tails to the mass resolution which could distort the mass spectrum ? This is in part tested by the target data but a very effective test is the J/ψ and ψ' particles which we collect even at high magnet currents. Figure 25 shows the 1250 A J/ψ and ψ' peaks uncorrected for acceptance. The clear separation of these peaks is a sensitive test of the resolution at low mass (see Fig. 18). The observed widths are about 15 % broader than predicted by the Monte Carlo and this is explained by the very small acceptance (Fig. 16). This means remote corners of the magnet aperture are being used. We also run the magnet current at 750 A (Fig. 16) to "spray" J/ψ over the apparatus. This permits testing the absolute energy calibration over different areas of the apparatus. It goes without saying that the 3.096 GeV

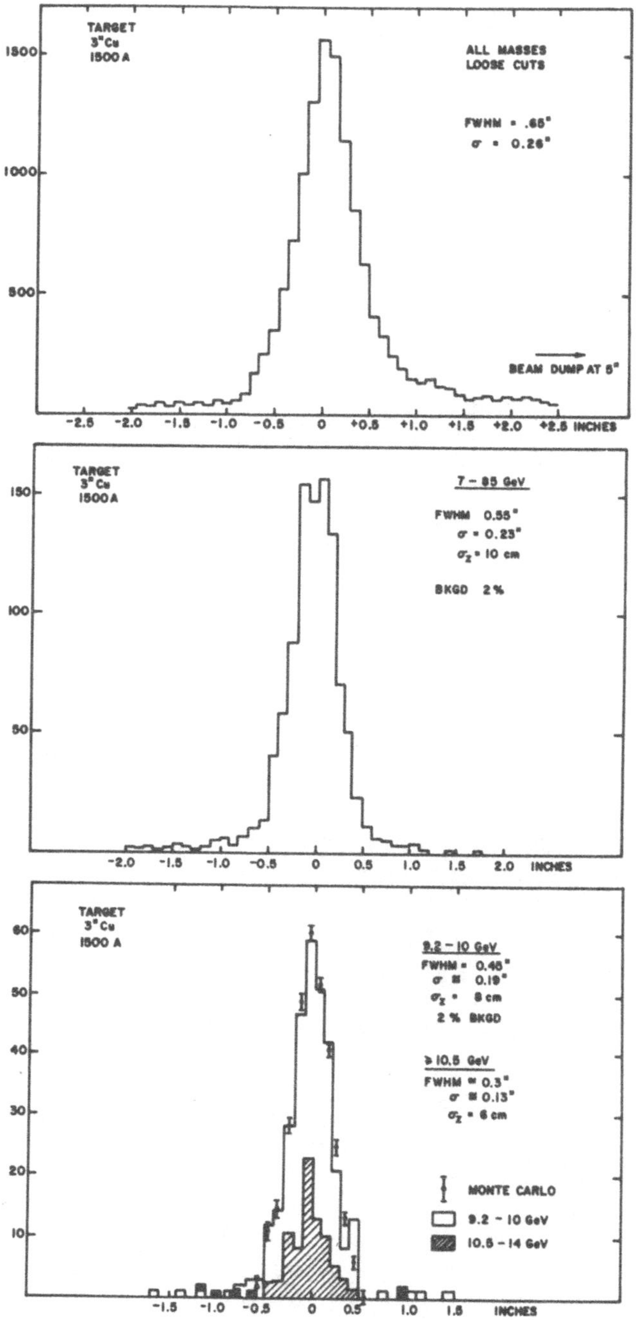

Fig. 24a Target projection for all masses with loose cuts.
24b Target projection 7-8.5 GeV masses.
24c Target projection 9.2 - 10 GeV with > 10.5 GeV (shaded).

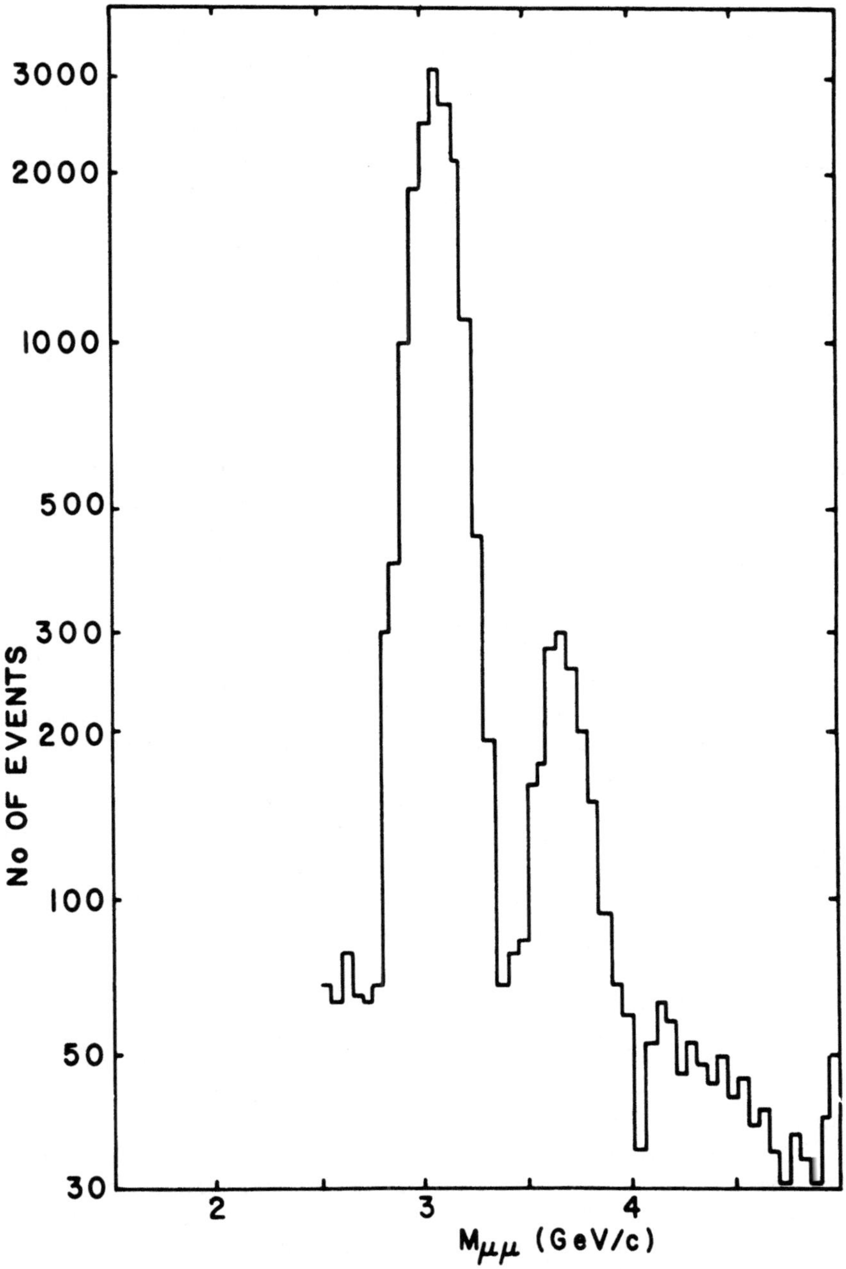

Fig. 25 J/ψ and ψ' with no acceptance correction.

J/ψ provides the primary energy calibration. However, no adjustement outside of errors is required in the magnet calibration, surveying or dE/dx correction.

4. Is the bump an artifact of the software ? Tests too numerous to detail fail to find any software culpability. An interesting one is a plot, on a very long output record, of the mass spectrum in 1 MeV bins to look for histogramming or "call 9.5" bugs. Details of the data acquisition, tape writing, reading, filter and reconstruction programs were examined and checked. The conclusion of this study leads to the conclusion that systematic effects cannot be responsible for the enhancement in the data near 9.5 GeV. Taken with the statistical results, we conclude that the effect must represent new physics.

Final results

Data have been accumulating since May 14, 1977. For the purposes of this report (given in July but written in early August), we present data as of August 13. Figure 26 shows the subtracted mass bump based upon ∿ 1000 events in the 9.2 - 10.6 region and ∿ 18,000 events above 5 GeV. The results of attempts to fit the data are given in Table VI and the following conclusions are drawn :

1. The data cannot be fit with a single wide resonance. The confidence level for the best such fit is $<< 10^{-4}$.

2. The shape of the leading edge indicates a narrow (less than the experimental resolution, i.e. $\Gamma \lesssim$ 120 MeV) resonance followed by something else.

3. The next simplest hypothesis - too narrow lines, is described in Table VI. It gives an excellent χ^2 . Clearly more lines are possible and a third peak fit is also given.

What Could It Be ?

A single broad line, if not ruled out by the data, would have run into trouble with total cross section. The absence of other broad excitations above continuum in the 5-10 GeV region also makes this hypothesis unreasonable. Other "candidates" provided by the literature include Higgs scaler meson and Z°, the neutral weak boson. The Higgs thing should, in the simplest version, couple to leptons as $(m_\mu/m_e)^2$ and hence should show essentially no electron decay. Looking back at our e^+e^- data, we see an isolated cluster of 6 events near 9.5 GeV. The muon data would predict 5 and this makes it likely that the 9.5 GeV object (which we have named Upsilon : Υ(9.5) decays universally.

The Z° is expected at m ∿ 70 GeV. However, if it showed up prematurely, one might expect interference effects with the

Table 6

Fitted Results to Subtracted Data

Υ	M	9.41 ± 0.02 GeV	9.40 ± 0.02 GeV
	$B\left.\frac{d\sigma}{dy}\right\|_0$	0.19 ± 0.01 Pb	0.18 ± 0.01 nb
Υ'	M	10.06 ± 0.04	9.99 ± 0.05 GeV
	$B\left.\frac{d\sigma}{dy}\right\|_0$	0.070± 0.007 Pb	0.067± 0.009 nb
Υ''	M	-	10.41 ± 0.12 GeV
	$B\left.\frac{d\sigma}{dy}\right\|_0$		0.016± 0.008 Pb
	χ^2	18.7/18	12.0/16

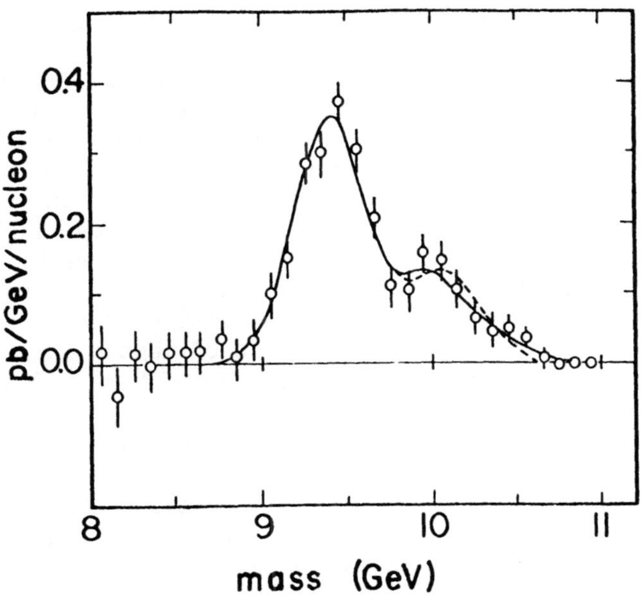

Fig. 26 Excess of the data over the continuum fit of Eq.(16) Errors shown are statistical only. The solid curve is the 3-peak fit; the dashed curve is the 2-peak fit. Data as of 15 August 1977.

electromagnetic background. In particular (with no serious theoretical thought), a C-violating asymmetry of the form : $\{<\mu^+> - <\mu^->\}/\{<\mu^+> + <\mu^->\}$ might appear if the width is sufficiently broad (the bracket represents average energy). We see no strong effect here.

The next hypothesis would be a coincidence of two vector mesons, presumably related to the ρ,φ or ω particles. This runs into the difficulty that the ρ is already 100 MeV wide and its first recurrence is much wider. Nothing is seen from 4 to 9 GeV and suddenly a narrow pair of peaks appear. It is very difficult to avoid the conclusion that here again, as with the J/ψ, a new quantum number has appeared. The livliest prospect already in the literature is the expectation of heavier quarks. Thus Eichten and Gottfried[25] have discussed the properties of bound states of "top" and "bottom" quarks, e.g. $b\bar{b}$. They predict that for $M_b \simeq 5$ GeV, there should be three bound s-states corresponding (in the charm case) to ψ, ψ'. Thus the complex structure we see (upsilon) Υ(9.4), Υ'(10.0) ... could be the s-states of "bottomonium". Detailed confirmation of this hypothesis must await increased statistics, and,hopefully, improved resolution.

VI. DRELL-YAN MODEL

A. Introduction

The literature bulges with articles on the Drell-Yan model (see Fig. 2b) with all its modifications. As a service, we list a sample of papers, some general[26-31] and some which address specific questions which have been stimulated by recent data.[32-36] The "naive" Drell-Yan theory does not take into account such recent developments as the possibility of scale breaking as is observed in ep and μp scattering, nor does it include the possibility of transverse momentum of the annihilating partons. The theory was proposed as a direct result of the Columbia-BNL data [4] as an attempt to account for the continuum distribution of lepton pairs. For a long time, agreement was poor, largely due to three effects :

(1) The Columbia-BNL "continuum" data were badly contaminated by a resolution smeared J/ψ peak.

(2) The reduction of p-uranium data to a per nucleon cross section was made via an $\sim A^{2/3}$ assumption. Although no further measurements have been made at BNL energies, data at Fermilab with similar values of Q^2/s indicate a variation closer to $A^{1.0}$.

(3) The earliest tests used unreal values of antiparton distributions (the first attempt took $q(x) = \bar{q}(x)$!). The present status is discussed below.

We summarize the motivation for studying the process as follows :

(1) The Drell-Yan model predicts the mass spectrum in terms of measurements on inelastic lepton-nucleon scattering. As such, it provides a crucial element in testing the general validity of constituent model ideas. Success of such models in ep scattering are known not to be enough.

(2) It is sensitive to the existence of the color degree of freedom and so provides a direct and almoet unique test (along with $\pi^\circ \rightarrow 2\gamma$).

(3) If and when mass distributions, angular distributions, quantum number effects (incident π, K, $\bar{p}$) etc. verify the essential ideas of the model, then the $p_\perp$ of dileptons is a direct measurement of parton transverse momenta.

(4) The continuum is a stage on which resonance perform.

(5) Finally, knowledge of the virtual photon spectrum in hadron-hadron collisions, together with the scaling features permit the design and calibration of a large number of important reactions as mentioned earlier.[5-9] Most popular is the prediction of $W^\pm$, Z° production in future accelerators. [9,28]

B. Dilepton Mass Spectrum

The general formula is :

$$\frac{d^2\sigma}{dQ^2 dX_F} = \frac{1}{3} \times \frac{4\pi\alpha^2}{3Q^4} F(\tau, X_F) \tag{6}$$

where $\sqrt{Q^2}$ is the dilepton mass (Eq. 2) and $\tau = m^2/s$. X_F is given in Eq. (3). The structure function is given in terms of the parton momentum variables (Eq. 4) as :

$$F(\tau,X_F) = \frac{x_1 x_2}{\sqrt{X_F^2+4\tau^2}} \{ \frac{4}{9} \left[u^b(x_1)\bar{u}^t(x_2) + \bar{u}^b(x_1 - u^t(x_2)\right]$$

$$+ \frac{1}{9} \left|d^b(x)\bar{d}^t(x_2) + \bar{d}^b(x_1)d^t(x_2)\right|$$

$$+ \frac{1}{9} \left[s^b(x_1)\bar{s}^t(x_2) + \bar{s}^b(x_1)s^t(x_2)\right]\} \tag{7}$$

where the superscripts b, t refer to incident beam particle and target particle. This permits evaluation of such reactions as pp,

$\bar{p}n$, etc. if the corresponding parton functions are known : u, d and s i.e. the up, down and strange quark functions. Heavier quarks are neglected. In Eq. (6) the first factor of 1/3 is due to color. Also :

$$x_{1,2} = \frac{1}{2} \left[\sqrt{X_F^2 + 4\tau^2} \pm X_F \right]. \qquad (8)$$

In the usual applications, the parton functions are written in terms of valence and sea distributions :

$$u^p(x) = u_V^p(x) + u_s^p(x) \text{ etc.}$$

and $u_s(x) = d_s(x) = s(x) = \bar{u}(x) = \bar{d}(x) = \bar{s}(x) \equiv S(x)$ (9)
in the special case of nucleon-nucleon collisions. Also : $u^p = d^N$, $u^N = d^p$. Eq. (9) is usually (not always) assumed.

In the CFS experiment, $X_F \cong 0$, $X_1 = X_2 = \sqrt{\tau}$ and the equations takes the simple form :

$$M^3 \left. \frac{d^2\sigma}{dMdy} \right|_{y=0} = \frac{8\pi\alpha^2}{81} \tau S(\sqrt{\tau}) V(\sqrt{\tau}) \qquad (10)$$

$$V(\sqrt{\tau}) = \left(\frac{3Z}{A} + 5 \right) u_V + \left(5 - \frac{3Z}{A} \right) d_V + 24S \qquad (11)$$

for a target of Z,A.

We have also transformed from X_F to the rapidity variable which has computational simplicities in the data analysis.

There exists a variety of simple functional fits for u_V, d_V and these are considered as "known". Some information on S(x) at small x comes from neutrino scattering. However the shape of the sea distribution is free and forms as varied as : $(1-x)^{3.5}/x$ and $(1-x)^{9}/x$ have been used.

The new CFS data given in Fig. 14 are statistically very precise in the mass range 5-14 GeV. At this time, all systematic corrections have not yet been evaluated. Some of these have to do with detailed checking of the Monte Carlo programs that give the mass aceeptance. These do depend somewhat on the model that is used to evaluate the acceptance, in particular, the X_F or y distribution (CFS does not have enough acceptance to determine these directly), the p_T distribution (this can be iterated) and the $\cos\theta^*$ distribution, i.e. the decay angle in the rest system of the dilepton. In the Drell-Yan model, the expected distribution is $1 + \cos^2\theta^*$ relative to the collision axis of the annihilating partons. Here again, the CFS acceptance is not large enough to determine this and there is then a residual normalization uncertainty although consistency tests with the theory are not affected. Other systematic effects relate to the nuclear size : A-dependence and effects of Fermi motion. A preliminary fit (as of August 1977) to the theory indicates qualitative agreement in absolute value (to $\lesssim$ 30 % so far). If confirmed this would prove the need for the color factor. In detail, we used the "BNL"[27] distributions :

$$
\begin{aligned}
u_V &= 1.79\ (1-x)^3(1+2.3x)/\sqrt{x} \\
d_V &= 1.107(1-x)^{3.1}/\sqrt{x} \\
S &= 0.15\ (1-x)^7/x
\end{aligned}
\tag{12}
$$

and find that the theory does not fall fast enough. This is easily corrected by choosing a faster falloff for S(X), e.g.

$$S(x) \sim (1-x)^{8-9}/x \tag{13}$$

as long as the coefficient is chosen to fit the neutrino data, this is o.k. See Fig. 27.

These statements will be sharpened as the analysis is completed in the next month or so.

Data with incident pions have been obtained by CP II and OMEGA.

Data are so far limited largely to masses below the J/ψ at Fermilab giving many low values of τ and hence uncertain applicability of the model. In fact, the data are well above the model extrapolations. Figure 28 illustrates the CP II[37] data, comparing π^-, π^+ and p on carbon. One clear result is the deviation of the low mass data from the Drell-Yan prediction. The most interesting test here is the ratio, π^+/π^-, which, according to the model, should approach 1/4 at high mass. This is easily seen since the pion is needed for its copious anti-quark supply. The charge of the π^- antiquark is 2/3, that of the π^+ is 1/3, and hence the factor of 4. This is softened by the presence of anti-quarks in the proton. The data tend to support the model but clearly need many more events especially above the J/ψ mass. Weak support for the decrease of the π^+/π^- ratio with mass is provided by the Ω spectrometer[24] where they present a ratio of 0.35 based upon 7 events above 3.5 GeV.

The CERN group's[24] π^- induced data are presented in Fig. 29 and indicate very impressive advance in dileptons via pions.

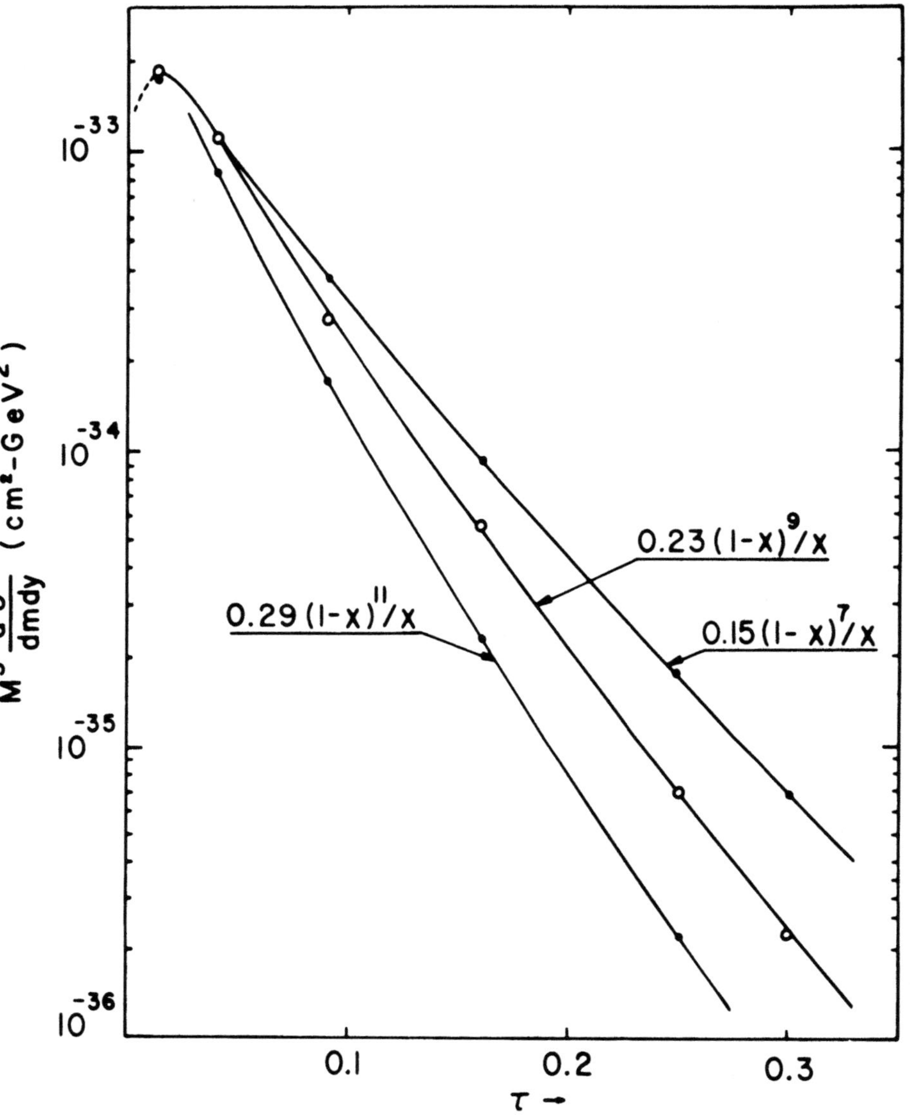

Fig. 27. Drell-Yan predictions for $X_F = 0$.

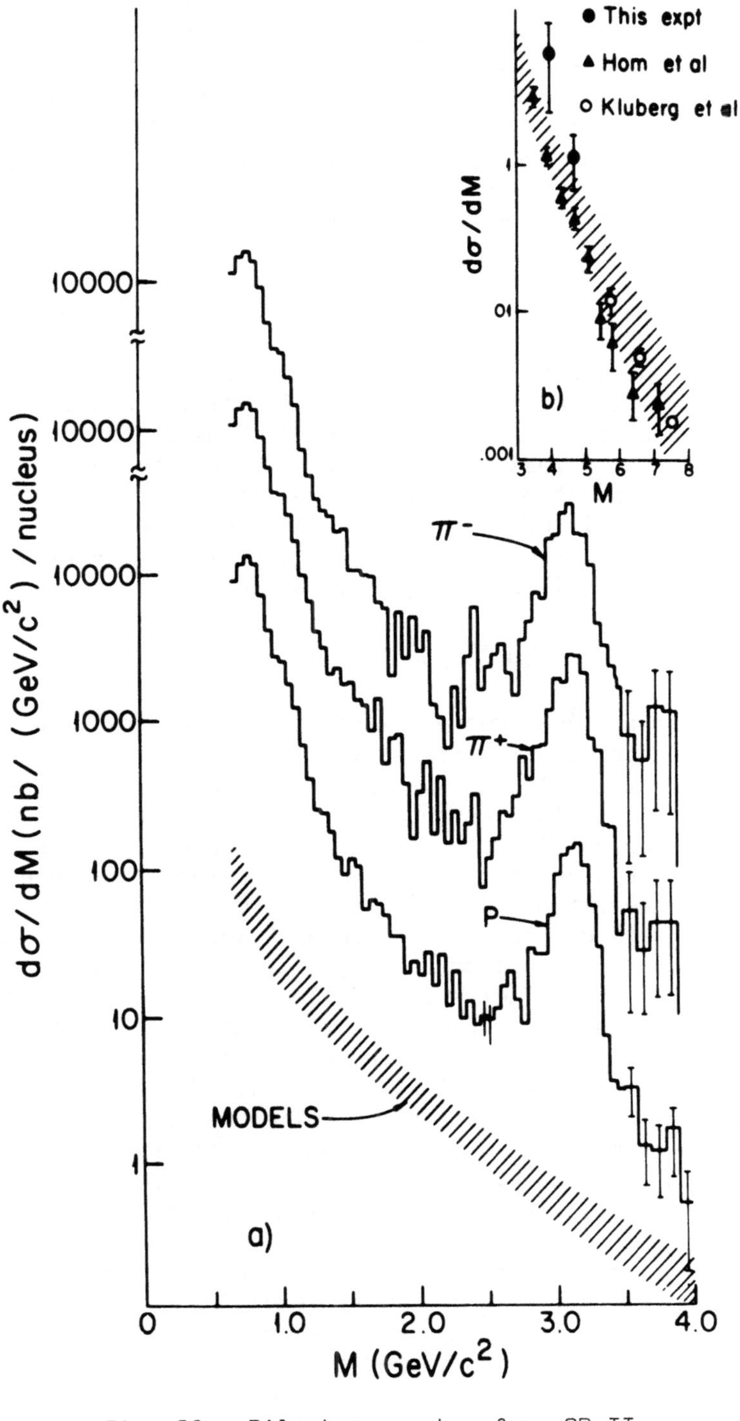

Fig. 28. Dilepton spectra from CP II.

C. X_F Distributions

This requires very large acceptance dilepton spectrometers and again wa hear essentially from CERN and CP II. Figure 30 gives the CERN results for π^- compared to Drell-Yan according to Donnachie and Landshoff.[31] Figure 31 shows analogous results from CP II. Here again the theory does reasonably well in that the X_F distribution of pions is expected to be flatter with X_F than protons, dependent as they are on the steeply falling sea.

D. P_T Distributions

Figure 32 shows the P distribution for continuum below the J/ψ from CP II. Figure 33 shows some preliminary CFS data in 1 GeV mass bins up to ∿12 GeV. Figure 34 shows a plot of the behavior of $\langle P_T \rangle$ with mass based upon μμ I high mass data. We show this because it stimulated a flood of theoretical papers. The issue is this : if Drell-Yan is really the process, as all data so far indicate, then $\langle P_T \rangle$ represents a direct observation of the $\langle k_T \rangle$ of the annihilating quarks, since

$$P_T = \frac{\sqrt{s}}{2} (X_{1\perp} + X_{2\perp}). \tag{14}$$

This requires a fundamental extension of the Drell-Yan model since the "infinite momentum frame" is now no longer viable. Figure 34 indicates a rising $\langle P_T \rangle$ with mass implying that the quark transverse momentum increases with X. The early high mass data left open the possibility that this increase continues indefinitely. It also suggested very high values for k_T. There are in general two theoretical approaches. The most popular one is based upon the work of Kogut and Susskind[38] who have reformulated the Drell-Yan model to include scaling violations as observed in ep and μp scattering. Theirs is a field theory which is scale invariant. Thus the higher the mass, the smaller the distance probed, with no cutoff. This implies a linear increase of $\langle P_T \rangle$ with mass, certainly consistent with Fig. 34. The second approach[39] follows the CIM model which has been applied to large P_T hadron physics. This uses meson-quark intermediate states to generate large $P_\perp$ virtual photons i.e. "π" + q → γ + q. Thus initial state quark k_T's are not required. For $P_T \ll M_\mu$, the theory reproduces Drell-Yan. It will be some time before data are available for the other limit $P_T \gg M_{\mu\mu}$ where definitive tests can be made. Figure 35 contains preliminary data from μμ II experiment and although the level may change somewhat before the analysis is complete, it is clear that the increase is damped at ∿ 3 GeV and is thereafter mass-independent, except for a possible bump at the Υ resonance. This result has several implications. It reassures $W^\pm$ hunters that kinematic peaks will not be washed out by high P_TW's (assuming the

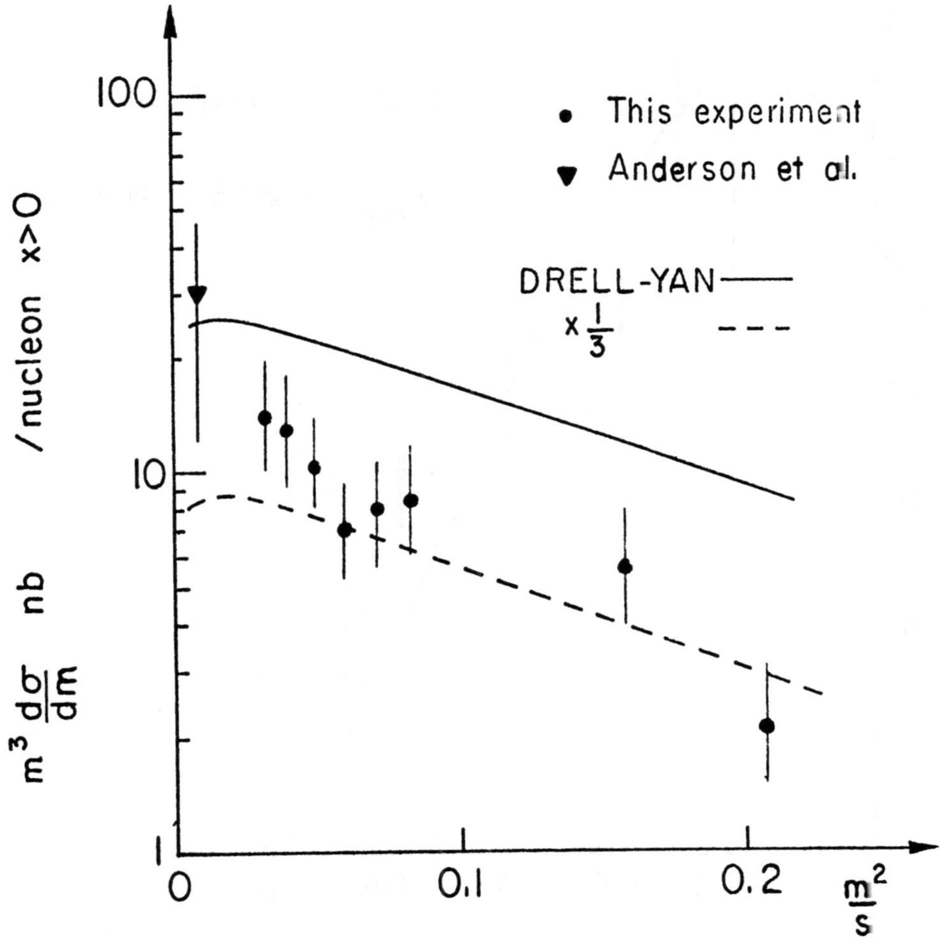

Fig. 29. π^- induced continuum from CERN OMEGA.

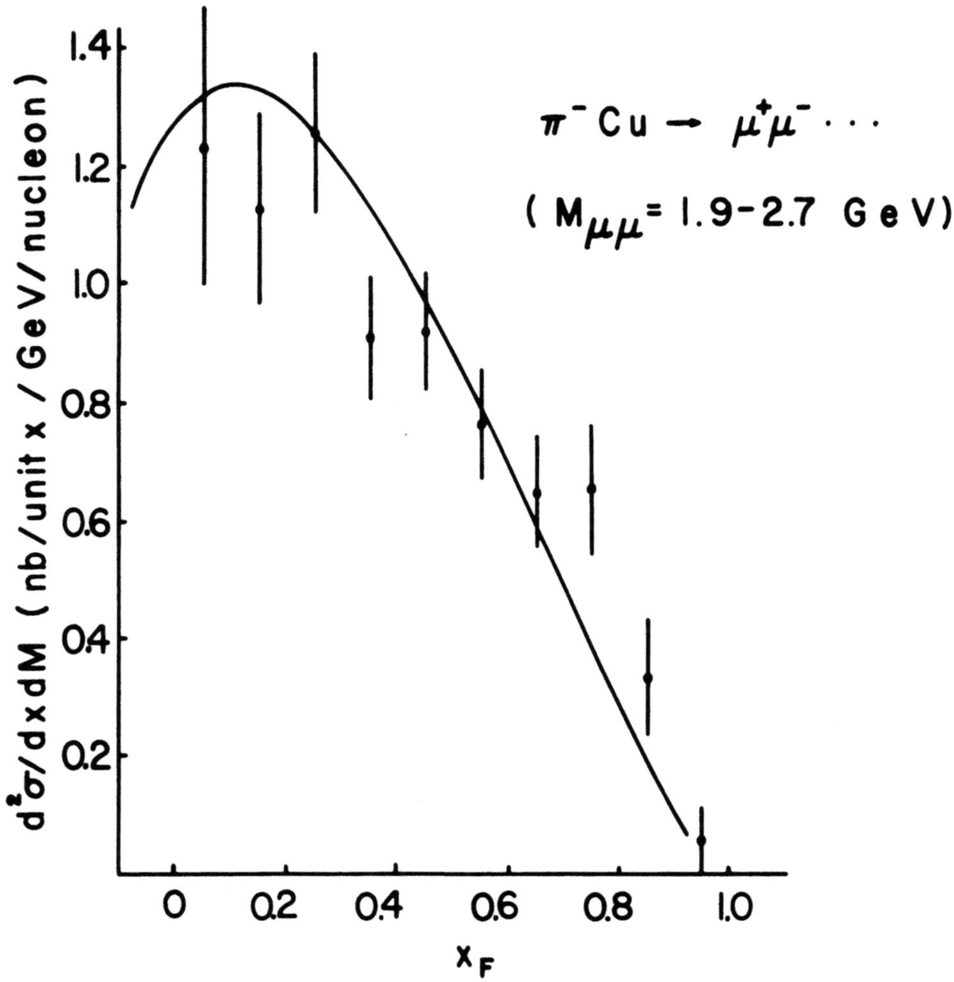

Fig. 30. X_F distribution, incident pions. CERN OMEGA.

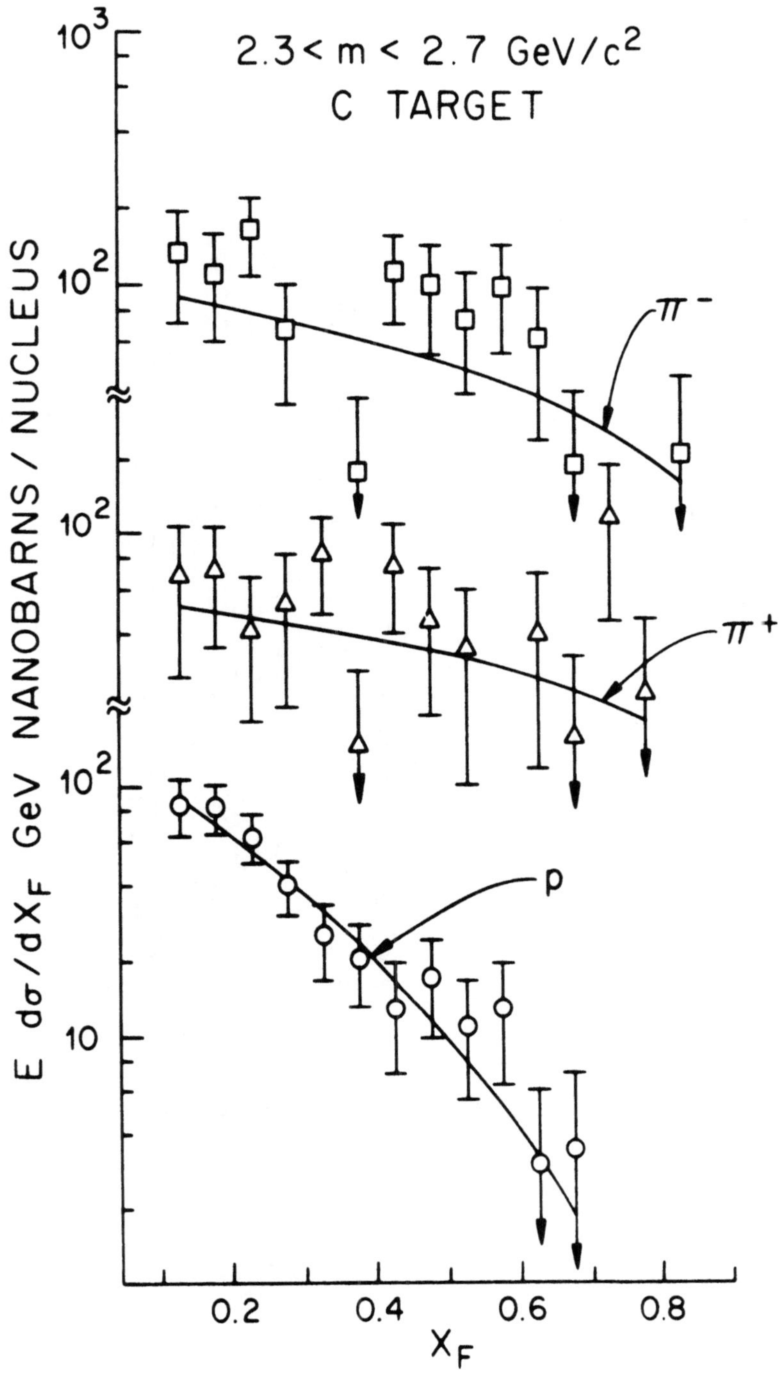

Fig. 31. X_F distribution CP II.

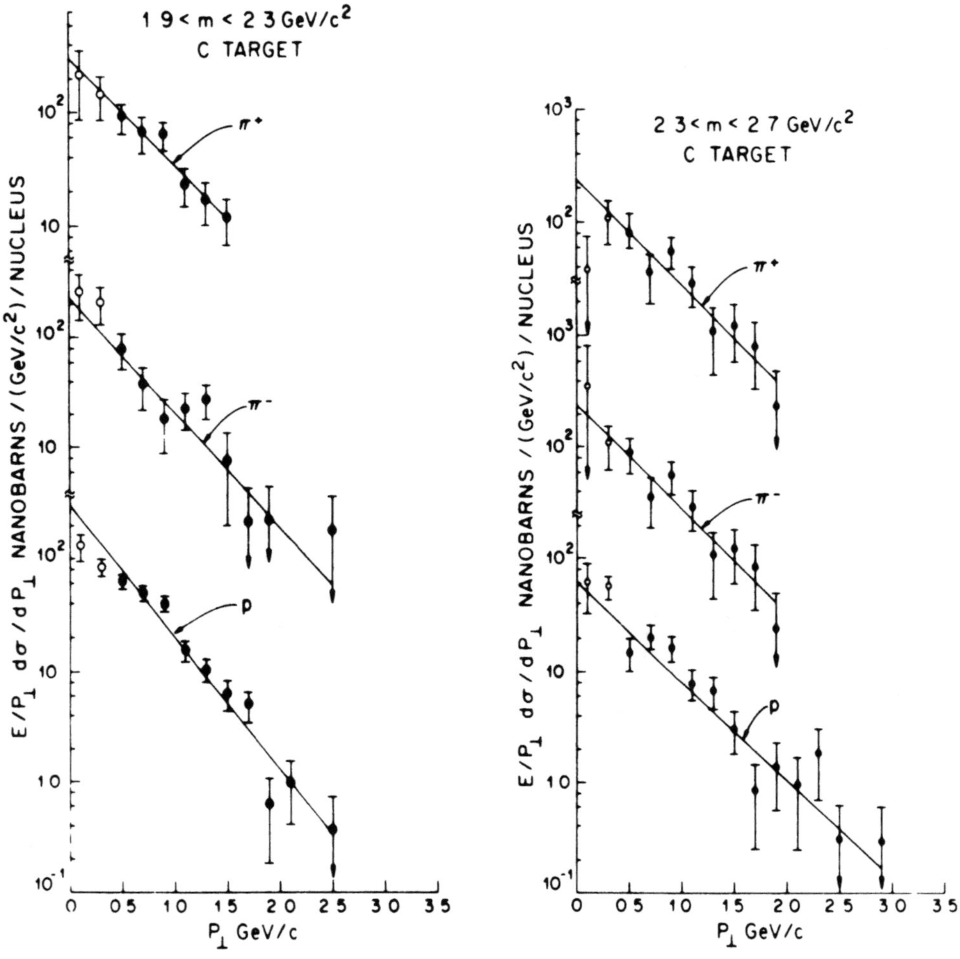

Fig. 32. P_T distributions CP II.

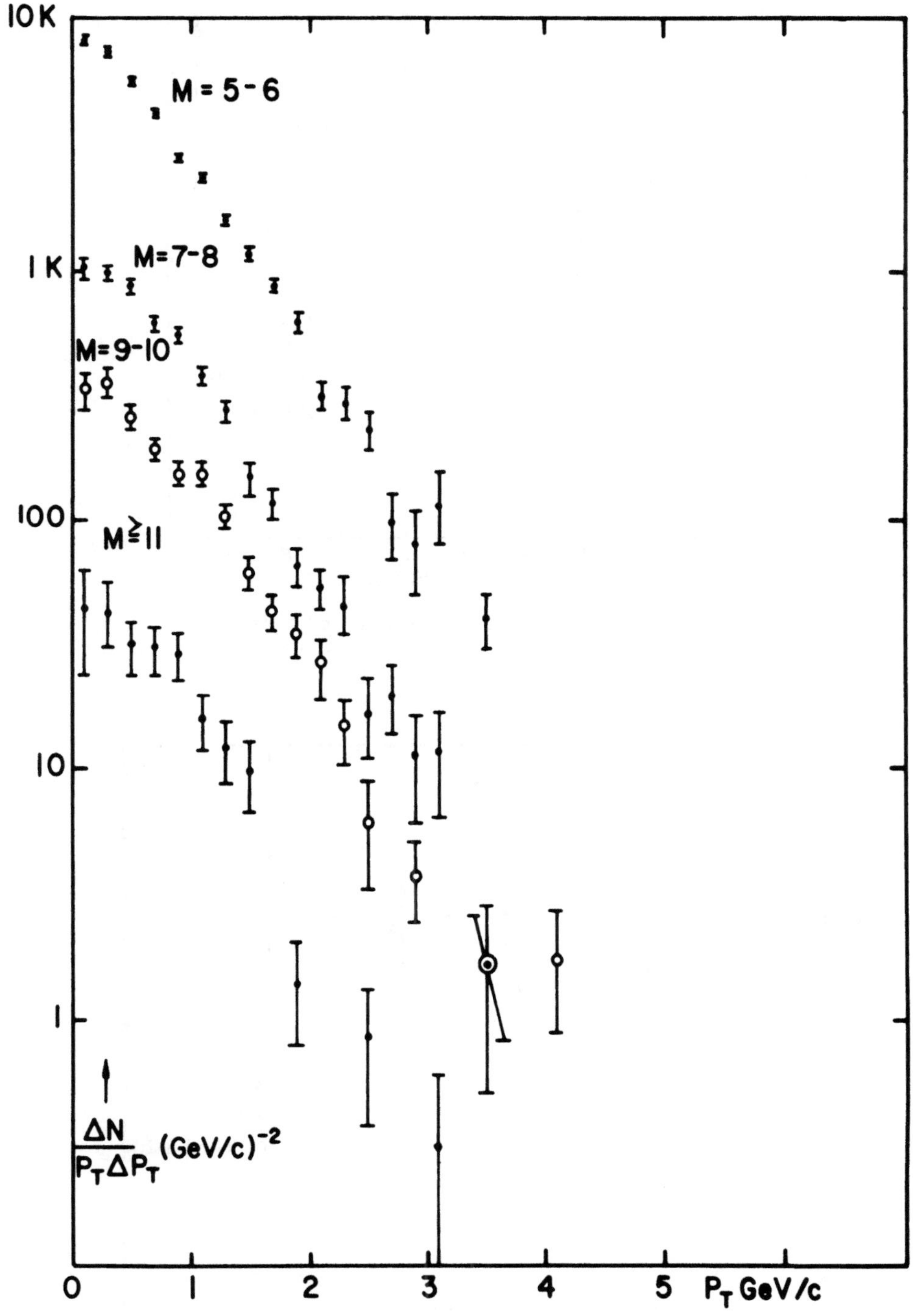

Fig. 33. Transverse momentum distribution in 1 GeV mass bins. Preliminary analysis assumes a 1 + $\cos^2\theta$ distribution in the dilepton rest system using the helicity frame.

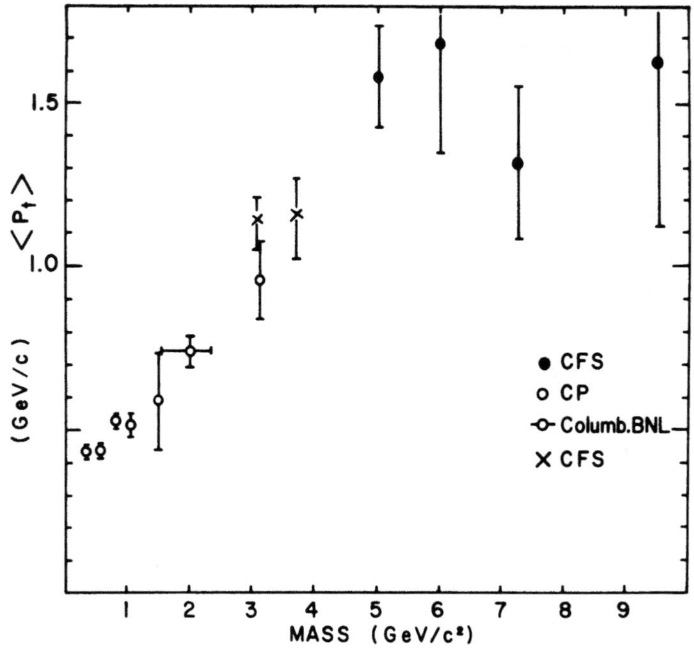

Fig. 34. Average P_T vs. mass, old data.

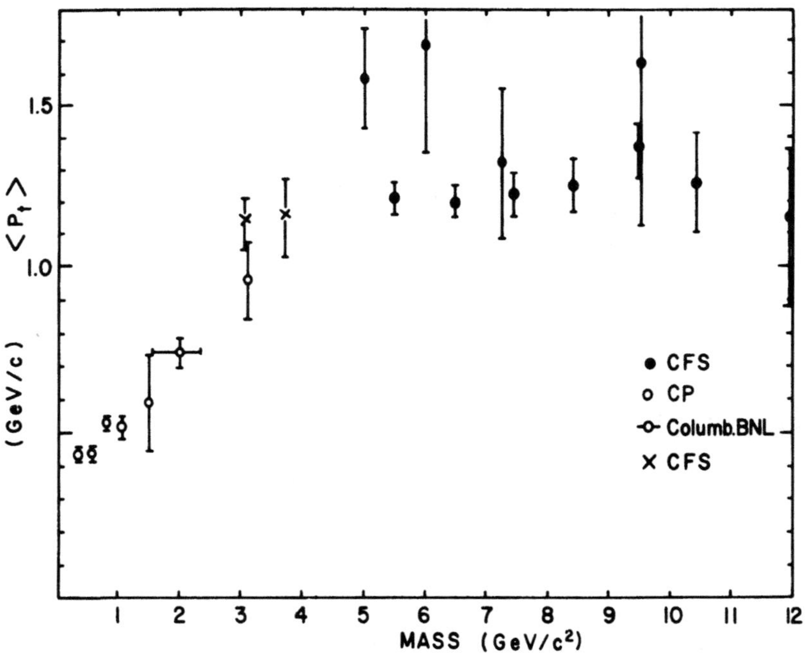

Fig. 35. Average P_T vs. mass, preliminary analysis.

trend continues). It requires some mechanism to damp the increase of $\langle P_T \rangle$ with mass. Some of the cited references[31-36] propose such mechanisms. We also rule out a popular from applied to hadron production i.e. dependence on $E_T^2 = P_T^2 + M^2$ e.g. as $\exp(-6 E_T)$. We note finally that the Duong-Van[39] scattering model predicted the experimental results with impressive accuracy. However so did a quark k_T model[36] which also gives a good account of the P_T shape. It is not clear to me how we will ever distinguish the two models. It is however exciting to recall that in high P_T hadron physics, the CCR group, as early as 1974, had reported an out-of-plane component $\langle P_{out} \rangle$ of as much as 1 GeV. A mean k_T of this order (closer to 800 MeV) would account for both the hadronic and the dileptonic processes.

E. Scaling

The final topic is the question of scaling. It had been noted soon after C-BNL that dimensional arguments[42] predicted :

$$\frac{d\sigma}{dm} = \frac{A}{m^3} F(m^2/s) \tag{15}$$

if there were no intrinsic masses comparable to $m, \sqrt{s}$. Clearly the Drell-Yan theory contains the prediction. There are three tests of this in the literature. The tests consist of multiplying data by m^3 and plotting the result against τ (or τ^{-1}) to seek a universal $F(\tau)$. Lederman and Pope[40] reanalyzed the BNL data presented earlier - subtracting the J/ψ resonance, correcting for the A-dependence and, for 3 sets of data at s = 40, 45 and 50 GeV^2, making a more informed correction for the limited aperture of the BNL experiment. The results and a comparison with μμI are given in Fig. 36. They conclude that the discrepancy of a factor of ∿ 2 - 2.5 is well within the systematic errors and, to this level, proves scaling from s = 40 to s = 750 GeV^2.

The BNL Group[27] collect a wide variety of data, mostly of m < 3 GeV but also some of the pre μμ II high mass data in a graph (Fig. 37). Clearly one can see a fairly well defined shape for "$F(\zeta)$". Finally, CP I, in a fixed apparatus, compared data at 200, 300 and 400 GeV. Corrections must be made for the change in c.m. acceptance in a fixed apparatus as s varies (from s = 380 to s = 750). Their result is in Fig. 38 and again looks very universal in spite of s-dependent cross sections which change by factors of as much as 10.

If we assume scaling in this energy range is reasonably established, one can use the precise mass spectrum of CFS μμ II to predict the dilepton yields at ISR and ISABELLE.

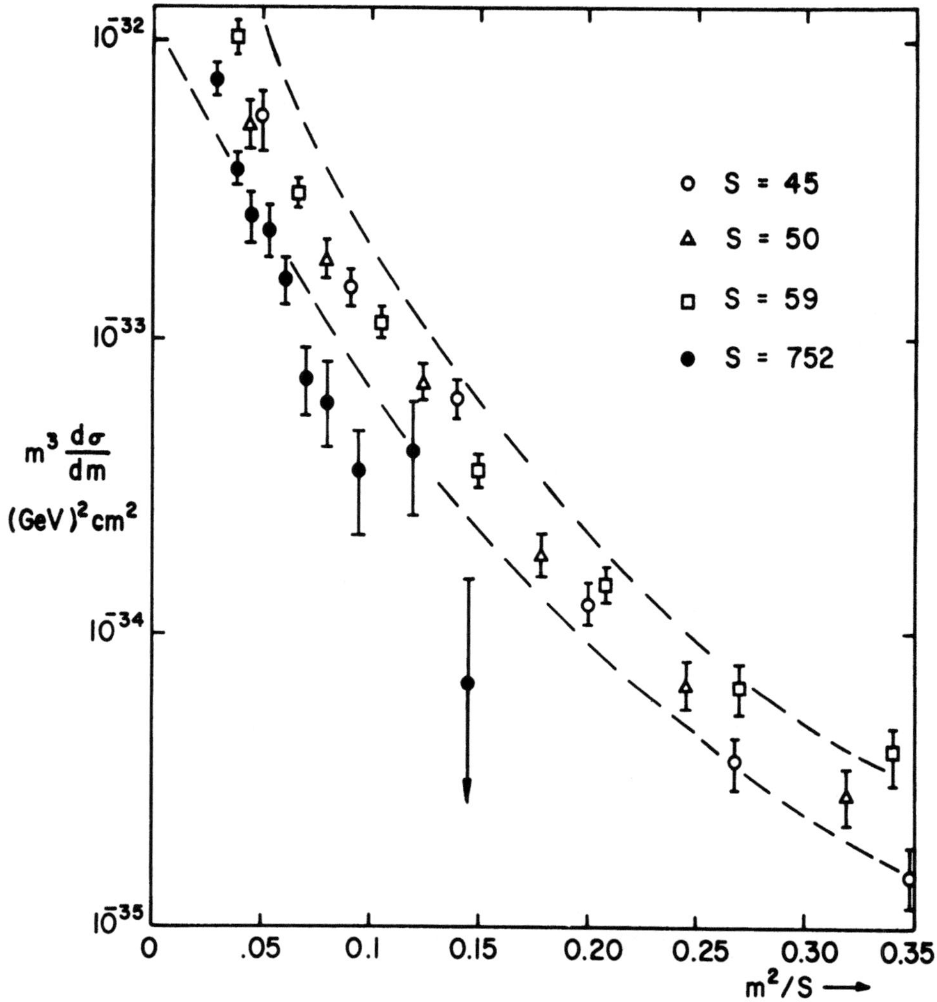

Fig. 36. Scaling test C-BNL and Fermilab

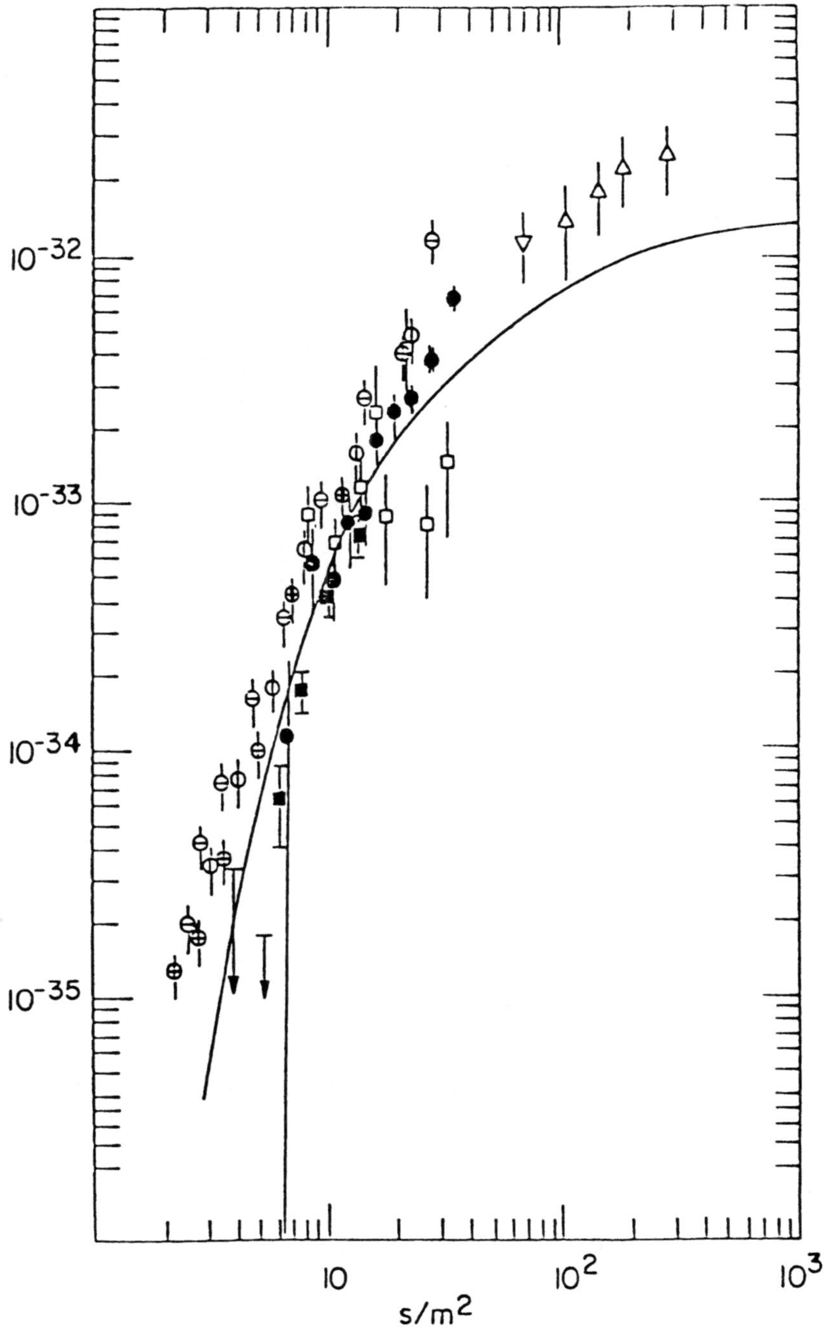

Fig. 37. Scaling test from Ref. 27.

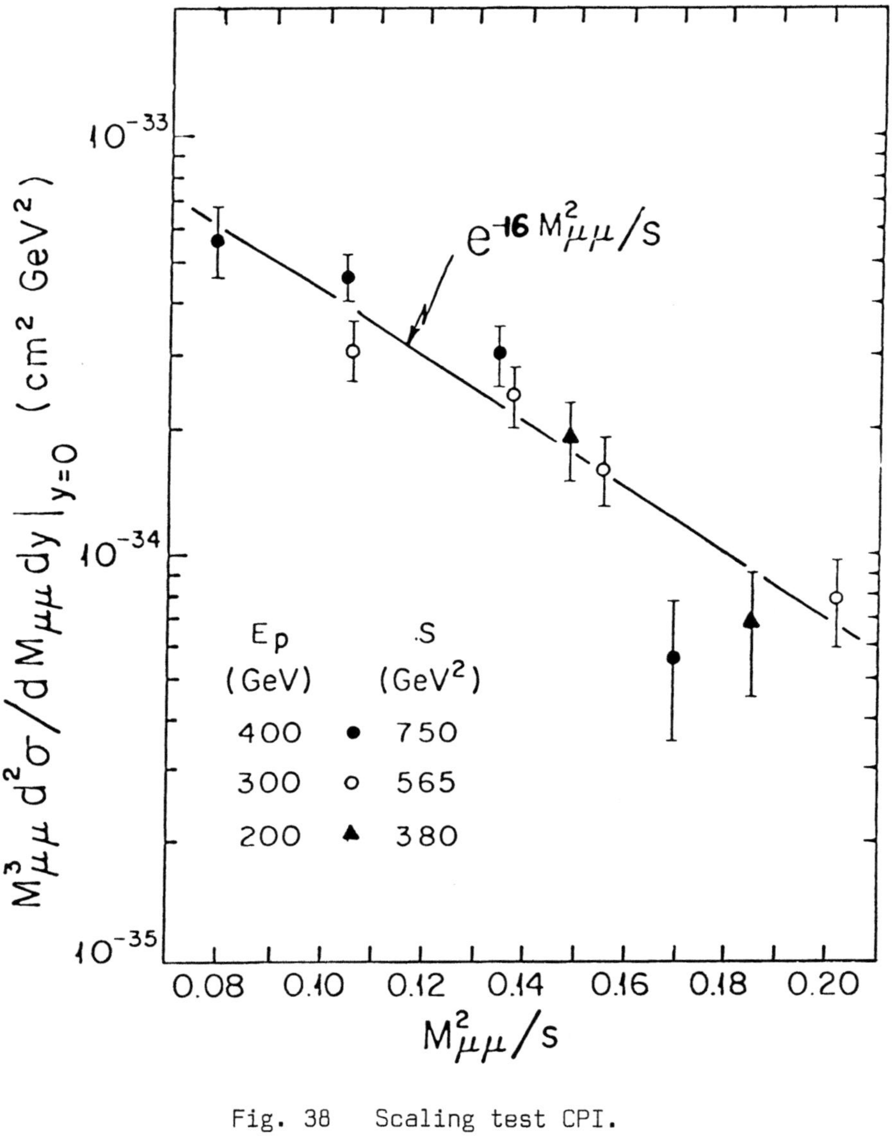

Fig. 38 Scaling test CPI.

Noting that the CFS fit is given by

$$\left.\frac{d^2\sigma}{dmdy}\right|_0 = A\ e^{-bm} \qquad (16)$$

with preliminary values :

$$A = 1.3 \times 10^{-33}\ cm^2/GeV$$

$$b \sim 1.0\ .$$

We can recast this as :

$$\left.\frac{d^2\sigma}{dmdy}\right|_0 = \frac{2.6 \times 10^{-29}}{m^3}\ \tau^{3/2}\ e^{-28\sqrt{\tau}}\ cm^2/GeV \qquad (17)$$

and compile a table of predictions :

$m_{\mu\mu}$	ISR ($\sqrt{s}$ = 60)
10 GeV	(τ = 0.028) $1.1 \times 10^{-36}\ cm^2/GeV$
20	(0.11) 1.1×10^{-38}

Confirmation of these predictions is eagerly awaited.

VII. CONCLUSIONS

The expectations of the 1967 proposal to BNL* to study lepton pairs has been fulfilled beyond reasonable optimism. What was then called a "probe of small distance structure in the hadron" has served to firmly establish the quark as a dynamical constituent. The hope that there would be resonances beyond ρ,ϕ,ω has been borne out by the J/ψ and Υ and who knows what else. The program of extending these measurements into the ISABELLE domain and into $\mu\nu$ final states looks better than ever. Cross sections are getting smaller ($\sim m^{-3}$) even at infinite energy machines and new detectors will be needed to cope with high luminosity backgrounds. I have no doubt that these too will come to pass.

---

* See Appendix I.

APPENDIX I
Proposal to the Brookhaven AGS
August 1967

Proposal for the Search for Muons Pairs in the Mass Range $\sim$ 1.5 - 6 BeV/c^2

Introduction

We propose to use the slow extracted proton beam at the AGS to search for muon pairs in the effective mass range $\sim$1.5 - 6 BeV/c^2. By employing the full energy (30 BeV) and intensity (10^{12} protons/pulse) of the AGS, we expect to observe states of mass between $\sim$1.5 - 6 BeV/c^2 having production cross section-branching ratio products (σB) as small as 10^{-38} cm^2.

The observation of muon pairs in this mass range would have several implications. They would be expected if there exist heavy vector mesons. The known resonances ρ,ω,ϕ do not, for example, account for the nucleon form factor measurements. In fact, any object that could, by virtual emission and absorption give rise to a muonic mass renormalization, would also be capable of decay into muon pairs. The role of neutral vector mesons in the hadronic electromagnetic current has been enlarged by recent theoretical considerations. The observation of the smooth background of muon pairs in this range is also of interest. It would presumably be due to real and virtual photoproduction and would have the same interest as the unsuccessful search for $p+\bar{p} \rightarrow e^+e^-$ (the result here was an upper limit of $\sim 10^{-33}$ cm^2), but with a very much wider range of energy and momentum transfer and with a sensitivity $\sim$ 2 or 3 orders of magnitude higher.

In summary, we propose a speculative exploration of the highest energy region ($E_{\mu+} + E_{\mu-} \gtrsim 15$ BeV) with great sensitivity ($\sim 10^{-38} cm^2$) but modest resolution. The detailed specification of any observed bumps would require completely different experiments; conversely the level set by this exploration would strongly influence the design of more specific proposals.

REFERENCES

1. Other reviews : J. Cronin, International School of Subnuclear Physics, Erice, July 1975.
2. Direct Lepton Reviews : L.M. Lederman, International Conference on High Energy Electron and Photon Interactions, Stanford, 1975.
   P. Piroue, Conference on Particles and Fields, BNL, October 1976; BNL 50598.
3. Y. Yamanouchi, Nuovo Cimento, 43, 193 (1966).
   L.B. Okun, Sov.J.Nucl.Phys. 3, 426 (1966).
4. Christenson, Hicks, Lederman, Limon and Zavattini, Phys.Rev.Lett. 25, 1523 (1970); Phys.rev. D8, 2016 (1973).
5. L.M. Lederman, Physics Prospects (ISABELLE), BNL 17522 (1972), p. 396; 406.
   L.M. Lederman, Esoteric Particle Production at ISR, CERN (1972);
   L.M. Lederman, D. Saxon, Nucl.Phys. B63, 313 (1973).
6. J.L. Newmeyer, J.S. Trefil, Phys.Rev.Lett. 26, 1509 (1971).
7. J. Sciulli, T. White, Phys.Rev.Lett. 26, 619 (1971).
8. L.M. Ledermann : Heavy Lepton Production in Hadron Collisions, Proceedings of the Moscow Conference on Muon-Electron Problem, Moscow (1972).
9. Palmer, Pashkos, Samios, Wang, Phys.Rev. D14, 118(1976).
   See also ref. 28.
   L.M. Lederman, B.G. Pope, Phys.Rev.Lett. 27, 765 (1971).
10. S. Drell, T.M. Yan, Phys.Rev.Lett. 25, 316 (1970).
11. R. Burns et al., Phys.Rev.Lett. 15, 830 (1965).
11.a J. Aubert et al., Phys.Rev.Lett. 33, 1404 (1974).
12. F. Busser et al., Phys.Lett. 48B, 377 (1974).
13. F. Busser et al., Phys.Lett. 46B, 471 (1973).
14. J.A. Appel et al., Phys.Rev.Lett. 33, 722 (1974).
15. J.P. Boymond et al., Phys.Rev.Lett. 33, 112 (1974).
16. D. Hom et al., Phys.Rev.Lett. 36, 1236 (1976).
17. D. Hom et al., Phys.Rev.Lett. 37, 1374 (1976).
18. D. Snyder et al., Phys.Rev.Lett.
19. S. Herb et al., Phys.Rev.Lett. 39, 252 (1977).
20. M. Binkley et al., Phys.Rev.Lett. 37, 571, 577 (1976).
21. Blanar et al., Phys.Rev.Lett. 35, 1662 (1975).
22. K. Anderson et al., Phys.Rev.Lett. 37, 799 (1976); 38, 1334(1977).
23. L. Kluberg et al., Phys.Rev.Lett. 37, 1451 (1976).
24. M.J. Corden et al., European Conference on Particle Physics, Budapest, July (1977).
25. E. Eichten, K. Gottfried, Phys.Lett. 66B, 286 (1977).
26. J. Pakvasa, D. Parashan, S.F. Tuan, Phys.Rev. D10, 2124 (1974).
27. R.F. Peierls, T.L. Trueman, L.L. Wang, BNL-22628 (1977).
28. C. Quigg, Rev.Mod.Phys. 48, 297 (1977).
29. R.D. Field, R.P. Feynman, Cal.Tech 68-565 (1976).
30. M. Duong-Van, Phys.Lett. 60B, 287 (1976).
31. A. Donnachie, P.V. Landshoff, Nucl.Phys. B112, 1234 (1976).

32. I. Hinchliffe, C. Llewellyn-Smith, Phys.Lett. 66B, 281 (1977).
33. J. Kogut, Phys.Lett. 65B, 377 (1976).
34. J. Gunion, UC, Davis 76-8 (1976).
35. J. Kogut, J. Shigemitsu, Cornell LNS Preprint (1977).
36. Kinoshita, Kinoshita, Cleymans, Peterson, Phys.Lett. 68B, 355 (1977).
37. J.G. Branson, Thesis, Princeton University COO-3072-81 (1977).
38. J. Kogut, L. Susskind, Phys.Rev. D9, 697, 3391 (1974).
39. M. Duong-Van, K.V. Vasavada, R. Blankenbecler, SLAC-PUB-1882 (1977).
40. L.M. Lederman, B.G. Pope, Phys.Lett. 66B, 486 (1977).
41. D. Antreasyan et al., CP 77-6 (1977).
42. Matveev, Muradyan, Tovkehelidze, JINR Dubna (1969).

NEW PARTICLES IN e^+e^- ANNIHILATION

B.H. WIIK

Deutsches Elektronen-Synchroton DESY
2000 Hamburg 52, Notkestrasse 85 - Federal Republic of Germany.

I. Introduction

Electron-positron annihilation is a process well suited for the study of the strong, the electromagnetic and the weak interactions. There is a simple reason for this; the annihilation diagram, which yields an intermediate state with well defined quantum numbers ($J^{PC}=1^{--}$), dominates the e^+e^- cross section at low energies and can be separated from higher order diagrams at high energies. Furthermore, the current (electromagnetic or weak) is timelike and couples directly not only to charged leptons but presumably also to quarks in a well defined way. These facts, cherished for a long time by the e^+e^- - addicts, have now come to be (almost universally) accepted by the high energy physics community. This insight has been helped by a series of stunning discoveries made in e^+e^- experiments during the past few years.

The large cross section for hadron production and its weak energy dependence, first seen at ADONE[1] and later extended to higher energies at CEA[2] and at SPEAR[3], showed that this process indeed behaved as expected in the naive quark model with point couplings. The discovery of the J/ψ particle in proton-beryllium collisions at BNL[4] and in e^+e^- annihilation at SPEAR[5] were the first hints of the existence of a new quark. The signature for a quark with a new flavour is the appearance of two new classes of hadrons; those with hidden flavour made of bound $Q\bar{Q}$ pair and those with open flavour made by combining the new quark with the old ones. Both classes[6-9] of hadrons have been observed in e^+e^- annihilation and it has also been demonstrated that the new quark predominantly decays into a strange quark as predicted[10] for charm.

Electron-positron collisions have also given convincing evidence for the existence of a new pointlike particle with a mass slightly below the mass of charmed hadrons. This particle, first observed[11] at SPEAR and later confirmed[12,13] at DORIS, has properties as expected for a heavy lepton. In these lectures I'll first acquaint you with the detectors used and then list some of the predictions made for a bound $Q\bar{Q}$ system and compare them to the experimental information available. The experimental information obtained on hadrons with open charm will be treated in section 7. The properties of the τ-particles will be treated in the last lecture. The emphasis will be on data obtained since the Tbilisi Conference.

2. Detectors.

The layouts of present e^+e^- experiments are strongly influenced by the low event rate and the slow variation with angle. A typical cross section is the point cross section :

$$\sigma(e^+e^- \to \mu^+\mu^-) = \frac{87.6\text{nb}}{s(\text{GeV})^2} .$$

At $s = 4\ \text{GeV}^2$ and a luminosity of $10^{30}\text{cm}^{-2}\text{sec}^{-1}$, the point cross section corresponds to one event per minute. Hadrons are produced with a rate a few times larger.

Such a low event rate makes it desirable to cover a large solid angle and to use loose requirements such that all genuine e^+e^- events are accepted. Demanding one or two particles emerging from the interaction region is in general sufficient to avoid flooding the data acquisition system with events from beam-gas and cosmic ray interactions. The slow variation of the cross section with production angle makes it unnecessary to cover angles very close to the beam line where the background and the contribution from two photon interaction will peak. So far, the detectors used can be grouped into three classes :

2.1. Non-magnetic Detectors.

The non-magnetic detectors are capable of measuring the direction of charged particles and photons. This allows a complete kinematical reconstruction of the final state with four or less particles. It is in general possible to identify photons, electrons and muons and to separate these particles from hadrons. Frequently, shower counters are used to measure the energy of electrons and photons. The accuracy depends on the type of shower counters used : typical values are

$$\sigma_E/E \simeq 2\%\ E^{-1/4} \text{ for NaI}$$
$$\simeq 6\%\ E^{-1/2} \text{ for lead glas}$$
$$\simeq 15\%\ E^{-1/2} \text{ for lead scintillator sandwich}$$

Detectors of this type have been used by several groups[16], Fig. 2.1.1. shows as an example the layout of the improved DESY-Heidelberg[17] detector. It consists of an inner part made of cylindrical driftchambers (CD) and scintillation counter hodoscopes (H). To improve the energy resolution photons are converted in a scintillation counter hodoscope (AC). Photon conversions are measured by the double driftchambers (CC) and the scintillation counter hodoscopes (CH) mounted behind the converters followed by NaI and lead glass shower counters. This set up is surrounded by driftchambers and an iron box 30cm thick to separate hadrons and muons.

The inner detector and the energy counters both cover an solid angle of 86% of 4π. Particles are detected in 95% of 4π. Muons are identified in 60% of 4π.

2.2. Magnetic solenoids.

These detectors cover a large solid angle with a rather homogenous magnetic field. The field is oriented either parallel (MarkI [18] and DELCO [19] at SPEAR, PLUTO [20] at DORIS or normal (MEA at ADONE) to the beam direction. The first generation of such detectors were mainly instrumented to measure the momenta and directions of charged particles. However, they can identify photons, electrons, muons and hadrons and separate low momentum pions, kaons and protons by time of flight. DELCO has also incorporated Cerenkov counters to improve the electron identification and several of the solenoid detectors now under construction for PETRA and PEP have sophisticated particle identifiers, based on dE/dx measurements or multicell Cerenkov counters.

MARK I

An exploded view of the Mark I detector is shown in Fig. 2. 2.1. The coil is 3.56m long with a diameter of 3.2m producing a magnetic field of 4 kgauss parallel to the beam axis. A particle emerging from the interaction point first traverses a thin beam tube, one of two scintillation counters surrounding the beam pipe, a pair of cylindrical proportional wire chambers, a set of four cylindrical multiwire spark chambers, one of 48 scintillation counters, the aluminium coil (about 7.5cm thick) and one of 24 shower counters, mounted between the iron yoke for the return flux and the coil. The detector subtends 65% of 4π. The momentum resolution (rms) for charged tracks is dp/p=0.015p. The iron yoke (20cm thick) serves as a hadron filter and is followed by a set of spark chambers for muon identification.

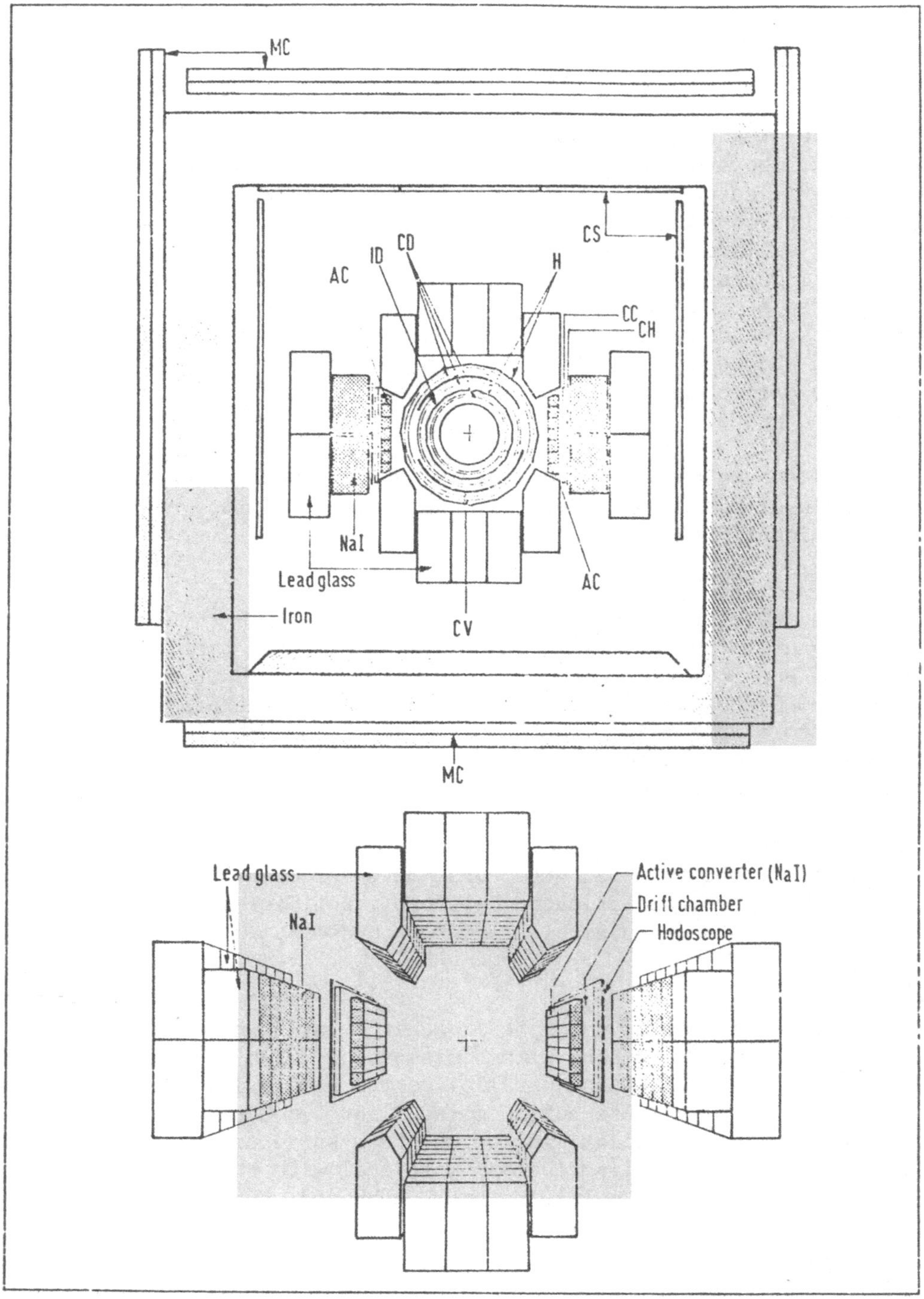

FIG. 2.1.1. : The new DESY-Heidelberg detector viewed along the beam direction.

A)

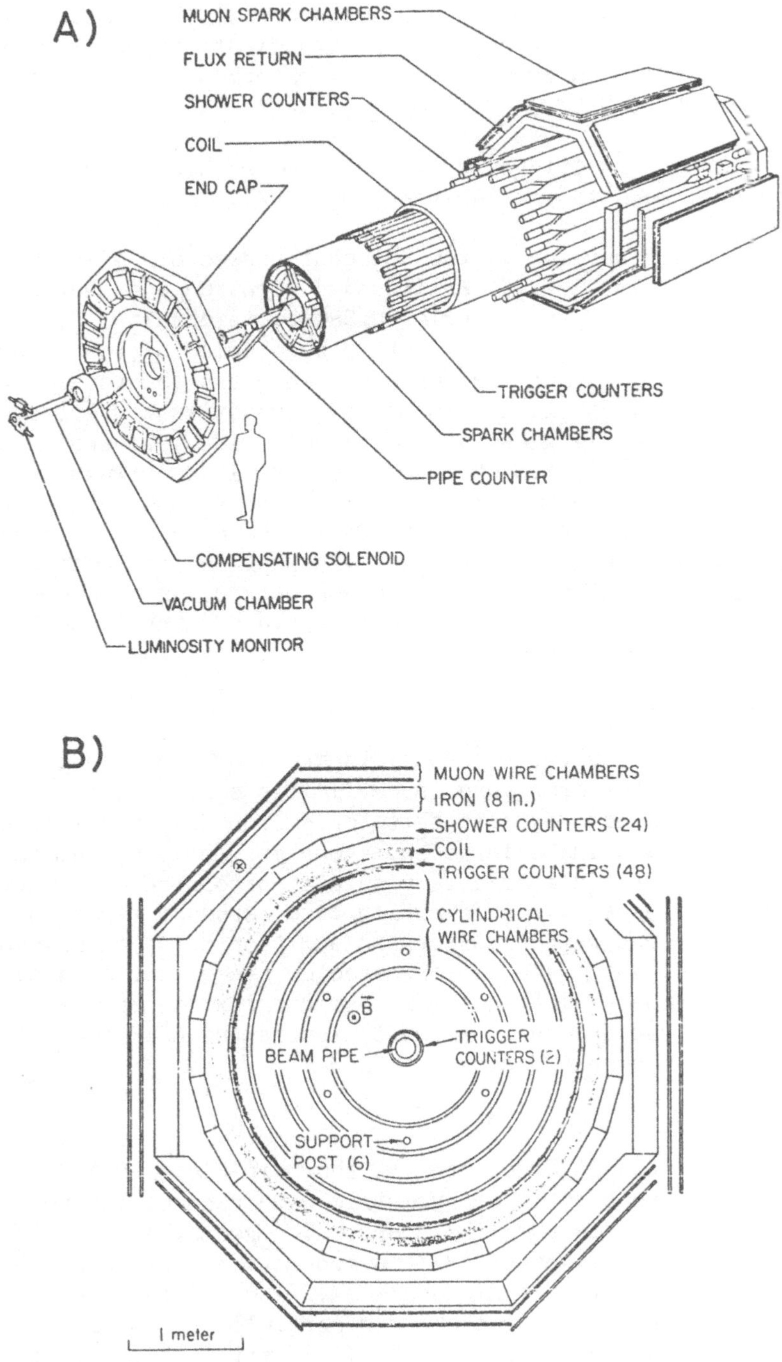

B)

FIG. 2.2.1. : A) An exploded view of the SLAC-LBL MARK 1 detector.
B) The MARK 1 detector viewed along the beam direction.

Electrons are identified by the pulseheight in the shower counters. Pions, kaons and protons are separated by a measurement of the time of flight between the two first layers of scintillation counters. An event by event separation can be made for momenta less than 0.7 GeV/c (πK) and 1.2 GeV/c (Kp).

The apparatus is triggered by two or more charged particles with transverse momenta greater than 200 MeV/c.

A more recent version of the detector viewed along the beam line is shown in Fig. 2.2.2. An improved electron and photon detection in a limited solid angle (0.69 sr) is obtained by removing a part of the magnetic flux return and replacing the shower counters by a set of scintillation and lead glass counters. The lead glass system is split in two parts - an active converter system 3.3 radiation lengths (X_0) thick followed by blocks 10.5 X_0 thick. Magnetostrictive chambers are installed between the magnet coil, the active converter and the lead glas blocks. The energy resolution is $9\%/\sqrt{E}$ with E in GeV. The muon-hadron separation is improved in a part of the solid angle by installing two concrete absorbers interleaved with magnetostrictive chambers above the detector. Muons with momenta above 900 MeV/c can be identified using this system.

2.3. Magnetic spectrometers.

This type of detector [21] consists in general of one or two magnetic spectrometer arms covering a limited solid angle in conjunction with a non magnetic detector surrounding the interaction point and covering a large solid angle. The spectrometer arms measure the particle momenta with high precision and identify the particles using the information from Cerenkov counters, shower counters, and range counters in addition to a measurement of the time of flight. The vertex detector can either be a single set of tracking chambers or a rather complex nonmagnetic detector as discussed above. We will discuss the DASP detector as a prototype of such a device.

DASP

A layout of the DASP detector is shown in Fig. 2.3.1. Two large H-magnets are positioned symmetric with respect to the interaction point and spaced 2.1 m apart. The geometric acceptance of each magnet is from 48° to 132° in production angle and ± 9° in azimuth resulting in a solid angle of 2 x 0.45 sterad for both magnets. The acceptance for a charged particle is smaller than this and depends on the momentum, the field strength and last detector plane required. The maximum field strength is 1.1 Tesla, the integrated field length 1.8 Tm. A charged particle emerging from the interaction point traverses the following detectors

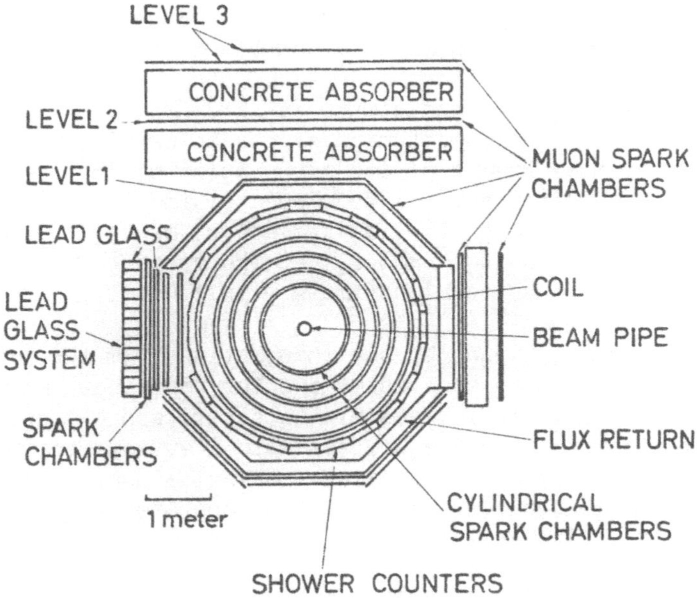

FIG. 2.2.2. : Recent version of the MARK1 detector including the lead glas wall and the improved muon identifier. The detector is viewed along the beam direction.

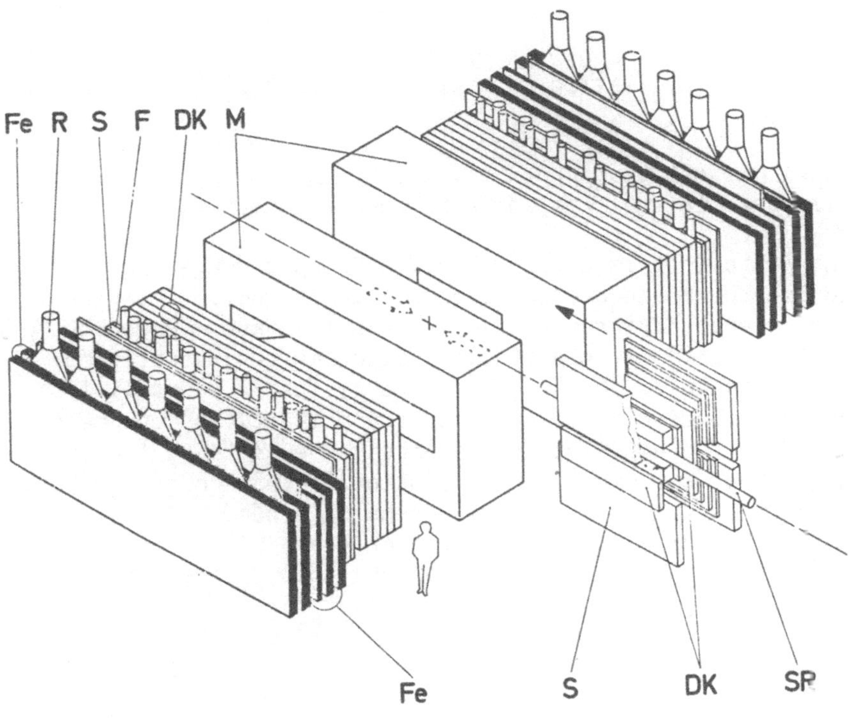

FIG. 2.3.1. : Exploded view of the DASP detector.

before reaching the magnet gap : a scintillation counter adjacent to the beam pipe, a second scintillation counter which starts the time of flight measurement, two proportional chambers, each with three signal planes a threshold gas Cerenkov counter, a third scintillation counter and a wire spark chamber with magnetostrictive readout. The momentum of a charged particle is determined from the measurement of one space point on the trajectory in front of the magnet and the knowledge of the trajectory of the particle behind the magnet. The trajectory behing the magnet is measured by 6 wire spark chambers. At the present a resolution of ± 0.8% is reached for a particle with 1.0 GeV/c momentum.

Particles are identified by the information from the time of flight system, the threshold Cerenkov counter, the shower counter and the range counter. It is feasible to separate pions and kaons for momenta less then 1.5 GeV/c and to identify protons for momenta less than 3 GeV/c by time of flight alone. Using the present Cerenkov counter, pions can be separated from kaons for momenta above 2.4 GeV/c.

Hadrons and muons are separated from electrons by the pulse-height in the shower counter and the information from the gas Cerenkov counter. The probability that a hadron is identified as an electron is 4×10^{-4}. The muons are positively identified by their range. The absorbers in the range telescope are made of iron a total of 90 cm thick, subdivided into plates of different thickness in order to allow for an optimal pion/muon separation at a chosen momentum. After each plate sufficient space for either a scintillation counter hodoscope or a spark chamber is provided.

The inner detector shown in Fig. 2.3.2. covers 70% of 4π and is located in the free space between the magnets.

The basic unit is a scintillation counter hodoscope, a sheet of lead 5mm thick and a proportional tube chamber. The tube chambers consist of three (two) signal planes and measure the position to ±5mm (±7,5mm). A particle emitted in the direction of the inner detector first traverses one of the 22 scintillation counters surrounding the beam pipe, the proportional chambers (in a part of the azimuthal acceptance), then four of the units just described, and finally a lead-scintillator shower counter. The energy resolution (rms) for photons is around 13% for an incident photon of 1 GeV. The detector has a 50% efficiency for detecting a 50 MeV photon travelling towards the detector. This efficiency increases to close to 100% for photons above 200 MeV. The detector can either be triggered by a single charged particle traversing one of the magnetic arms of the spectrometer, or by a combination of charged and neutral tracks in the inner detector.

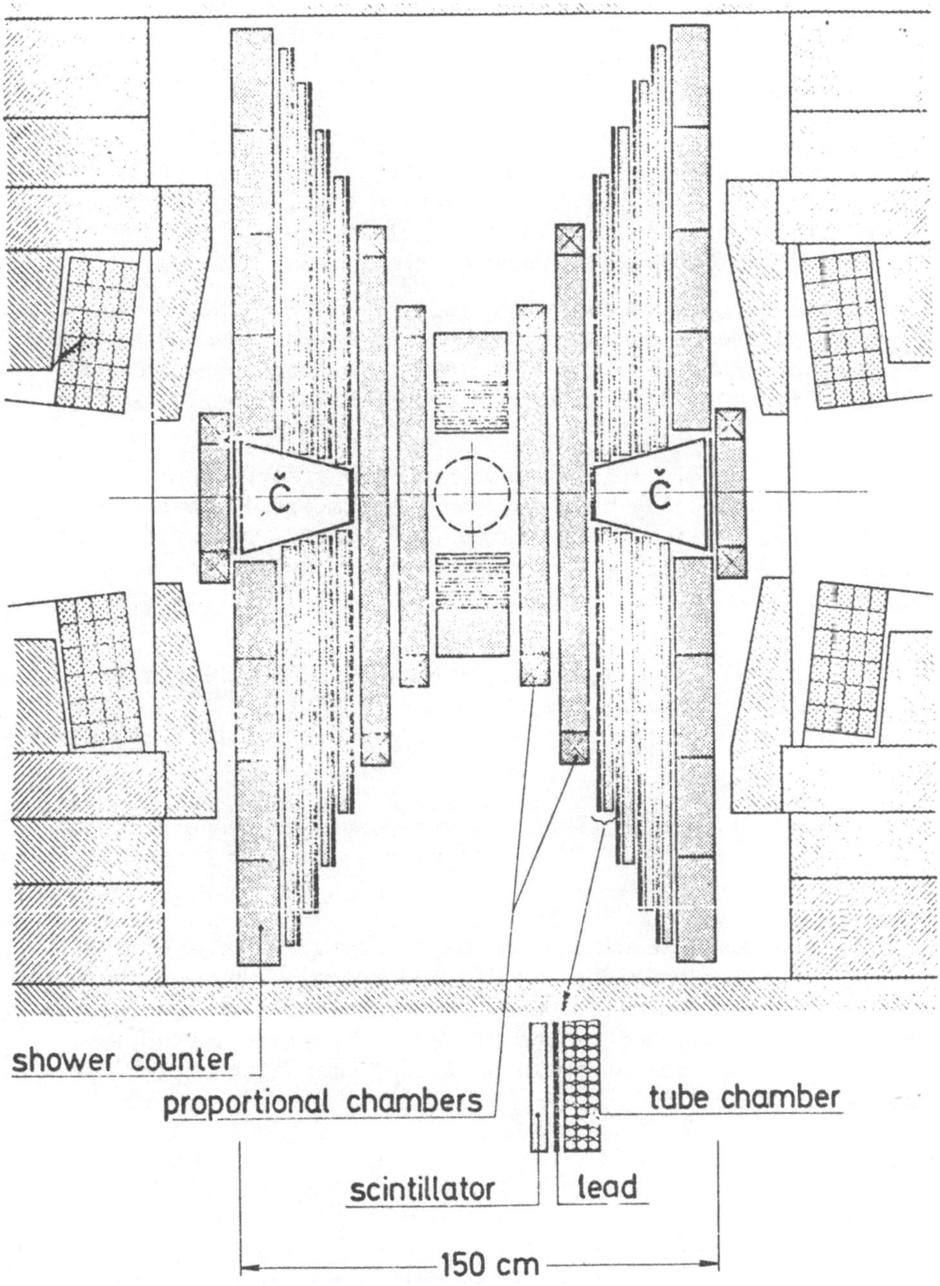

FIG. 2.3.2 : The inner detector of DASP viewed along the beam direction.

3. Properties of J/ψ and ψ'.

A large amount of data [14,15] on the properties of these resonances have been accumulated since their discovery [4,5,22] late 1974. The main facts are :

1) The resonances are hadrons. This follows from the observation that the resonances have well defined parity [23] and isospin [24]. Measurements [25] of J/ψ photoproduction in complex nuclei shows that the total cross section for J/ψ-nucleon interaction is on the order of a millibarn and hence a strong cross section.

2) The resonances are vector mesons i.e. $J^{PC}=1^{--}$. This is expected since they are produced in e^+e^- collisions and has been verified by observing [23] the interference pattern between the amplitude for $e^+e^- \to J/\psi \to \mu^+\mu^-$ and the QED amplitude for $e^+e^- \to \mu^+\mu^-$.

3) J/ψ has isospin 0. This follows from the observation [24] that J/ψ decays directly only into an odd number of pions. Since $G = -(-1)^I$ this implies that I is event 0, 2, 4. J/ψ must be an isoscalar since the direct decay $J/\psi \to p\bar{p}$ is observed [26]. The ψ' must also be an isoscalar since the decay $\psi \to \eta J/\psi$ is observed [27,28].

4) The J/ψ is predominantly an SU(3) scalar. This follows from a comparison [26,29] of $J/\psi \to \pi^{\pm}\rho^{\mp}$ and $J/\psi \to K^{\pm}K^{*\mp}$ (892) and the whole pattern [14,15] of hadronic J/ψ decays support this assignment.

All these observations are in agreement with the assumption that J/ψ and ψ' are bound $Q\bar{Q}$ states of a quark with a new flavour.

The total hadronic widths can be accounted for in terms of individual channels both for the J/ψ and the ψ'. However there is yet room for unidentified decays, in particular in the case of the ψ'. Some decay modes of the J/ψ and the ψ' are listed in tables 3.1 and 3.2, for a complete list of branching ratios see reference 15.

4. Charm.

There is now convincing evidence that the J/ψ and the ψ' are bound $Q\bar{Q}$ states where Q is quark with a new flavour. The quarks in the $Q\bar{Q}$ system must annihilate their flavour in order to decay into hadrons made of old quarks and this accounts [33] for the long lifetime observed for the J/ψ and the ψ'. It has

Table 3.1 - Decay modes of the J/ψ(3095)

Decay mode	Width(keV)	Fraction %	References
total width	69 ± 15		23
$e^+e^-(\mu^+\mu^-)$	4.8 ± 0.6	6.9 ± 0.9	23
second order electromagnetic decays	12.0 ± 2.0	17 ± 3	23
direct to hadrons	47.4 ± 14	69 ± 4	23

Table 3.2 - Decay modes of the ψ'(3684)

Decay mode	Width(keV)	Fraction %	References
total width	228 ± 56		23
$e^+e^-(\mu^+\mu^-)$	2.1 ± 0.3	0.93 ± 0.16	23
second order electromagnetic decays	6.6 ± 0.9	2.9 ± 0.4	23
direct decay to J/ψ	124 ± 30	54 ± 5	28, 30
direct decay P_c/χ	∿ 52	∿23 ± 6	31, 32
direct decay to hadrons	∿ 42	∿ 19	--

furthermore been shown that the new flavour can be identified with charm, first proposed [34] to achieve symmetry between leptons and quarks and later to explain the absence of strangeness changing neutral currents [10].

The signature for the new quark Q is the observation of two new classes of hadrons; those with hidden charm made of a bound $Q\bar{Q}$-pair and hadrons with open charm made by combining the new quark with the old ones. The $Q\bar{Q}$ states are narrow as long as they are below the threshold for the decay into states with open charm. The number and the quantum numbers of the $Q\bar{Q}$ states reflect the assumption that the quarks are fermions bound in an attractive potential and are firm predictions of such a model.

The production of hadrons with charm leads to a step in the total annihilation cross section. Hadrons with open charm decay weakly and can be identified either by their semileptonic decays or by observing very narrow states of high mass. The charm hypothesis can be identified by the preferred decay of charmed quarks into a strange quark $c \to -d.\ \sin\theta_C + s \cdot \cos\theta_C$. This leads to final states specific to the charm hypothesis like correlations between electrons (muons) and kaons and apparently exotic decays.

States with hidden and open charm have been observed confirming in detail the theoretical predictions. Here we first discuss states with hidden charm and then review the experimental knowledge on states with open charm.

4.1. Charmonium.

Charmed quarks are rather heavy and it might therefore be a good first approximation to treat the $c\bar{c}$ system as two non relativistic [35] fermions bound in an attractive potential. The general shape of the potential is determined by appealing to Quantum Chromodynamics (QCD) which perhaps is the most attractive theory of the strong interaction presently available. The quarks are coloured confined fermions which interact by emitting and absorbing coloured gluons. The gluons are massless vector particles with the quantum number of a qq pair.The strong coupling constant $\alpha_s(m^2)$ decreases with increasing mass; i.e. with smaller separation. It is therefore conjectured that the potential at small separation between the quark arises from one gluon exchange. This gives rise to a Coulombic term

$$-\frac{4}{3}\ \frac{\alpha_s(m^2)}{r}\ .$$

To prevent quarks from appearing in the final state one in general assumes an attractive potential increasing linearely with the separation between the quarks.

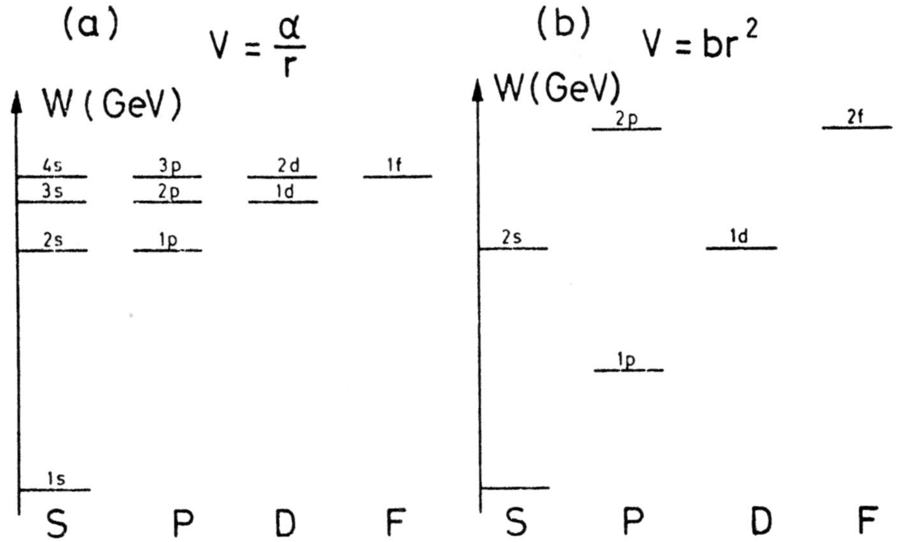

FIG. 4.1.1a,b : Level scheme predicted for two fermions bound in a rising attractive potential. a) pure Coulomb potential $V=\alpha/r$; b) harmonic oscillator potential $V=br^2$.

A potential of the form (or similar); $V = -4/3 \;\; \alpha_s(m^2)/r + ar$ has been used by many groups [36] to investigate the properties of charmonium. This potential will produce the level scheme indicated in Fig. 4.1.1c.

The levels are labeled by J^{PC} with $P = (-1)^{L+1}$ and $C = (-1)^{L+S}$. For each value of L there are two bands of radial excitations with opposite charge conjugations according to S = 0 or 1. The spectroscopic notation $^{2S+1}L_J$ is written below each set of radial excitation.

In this picture the J/ψ and the ψ' are identified with the 1^3S_1, and the 2^3S_1, respectively. The P levels split into one state with odd ($^1P_1$) and three states ($^3P_{2,1,0}$) with even charge conjugation. The first L = 2 level split into one state with even ($^2D_1$) and three states ($^3D_{3,2,1}$) with odd charge conjugation. The 3^3D_1-state has the quantum numbers of the photon and can mix with the nearby 2^3S_1 state.

It is of interest to see how sensitive the level ordering is to details of the potential, for example, in a Coulomb potential levels with the same main quantum number but different values of L are degenerate as indicated in Fig. 4.1.1a. In a

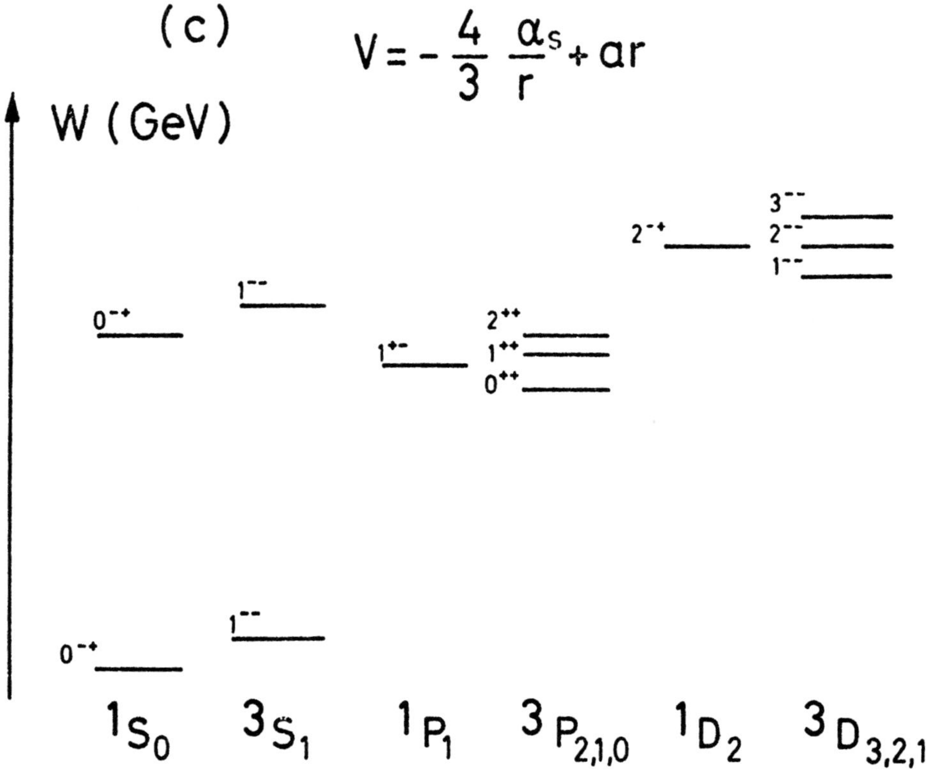

FIG. 4.1.1c : Level scheme predicted for two fermions bound in a rising attractive potential. $V=-4/3\alpha_s(m^2)/r+ar$.

harmonic oscillator potential the 2P levels would be located midway between the 1S and the 2S level and the 2S and the 3D levels would be degenerate (Fig. 4.1.1b). The general ordering of the levels in non relativistic potentials has been investigated in detail by Martin [37)] and Gross [38)]. They find that the level ordering = 1S < 1P < 2S < 1D < 2P ... is valid for rather general potentials. However, they also remark that potentials increasing faster than r^2 may invert the order of the 1D and 2S levels.

The quantum numbers of the charmonium levels are obtained with few assumptions. The level ordering is more model dependent, but the ordering shown in Fig. 4.1.1c is valid for a large class of potentials. The most model dependent predictions are those trying to reproduce the absolute widths of the $c\bar{c}$ states. The decay is a short range effect; the relevant distances are on the order of $1/m_{J/\psi}$. The $c\bar{c}$ states annihilate into massless coloured gluons which in turn materialize with nearly unit probability into normal hadrons. The rates for these transitions can be obtained [39] by analogy to similar formulas derived for positronium decays with α suitably replaced by α_s.

Some of the relevant formulas are listed below :

a) The pseudoscalar (^1S_o) levels can decay into two photons or two gluons. QCD predict for the widths :

$$\Gamma(^1S_o \to \gamma\gamma) = \frac{256\pi}{27} \alpha^2 \frac{|^1S_o(0)|^2}{M^2}$$

$$\Gamma(^1S_o \to gg \to \text{hadrons}) = \frac{32\pi}{3} \alpha_s^2 \cdot \frac{|^1S_o(0)|^2}{M^2} .$$

The vector states ($^3S_1$) can decay into a pair of leptons or into 3 gluons. The predicted widths are :

$$\Gamma(^3S_1 \to e^+e^-) = \frac{64\pi}{9} \alpha^2 \frac{|^3S_1(0)|^2}{M^2}$$

$$\Gamma(^3S_1 \to ggg \to \text{hadrons}) = \frac{160}{81} (\pi^2-9) \alpha_s^3 \frac{|^3S_1(0)|^2}{M^2}$$

b) The hadronic widths of the intermediate P-states to decay to hadrons via the gluon mechanism is proportional to the derivative of the wave function at the origin squared. These widths have been estimated [40] to

$$\Gamma_h(^3P_o) : \Gamma_h(^3P_2) : \Gamma_h(^3P_1) = 15 : 4 : 1$$

and

$$\Gamma(^3P_o) = 2.4 \text{ MeV}.$$

The $^3P_1$-state is expected to be quite narrow since it cannot decay into two massless gluons. The radiative decays $P_{2,1,0} \to \gamma 1^3S_1$ are therefore expected to be important.

The widths for the radiative M1 transitions between the vector and the pseudoscalar states can be written as

$$\Gamma(n_i{}^3S_1 \to n_f{}^1S_o+\gamma) = \frac{16}{27} \alpha \frac{k^3}{m_Q^2} |\langle n_\gamma | n_i \rangle|^2$$

This formula is valid for (kr) < 1 and assumes that the magnetic moment μ of the charmed quark is given by

$$\mu = \frac{(2/3\ e)}{m_Q} .$$

Note that the transition $2^3S_1 \rightarrow 1^1S_0 + \gamma$ is forbidden in this approximation.

The radiative decay of the ψ' into the $^3P_{2,1,0}$ states proceeds to lowest order by electric dipole radiation :

$$\Gamma(2^3S_1 \rightarrow 2^3P_{2,1,0} + \gamma) = \frac{16}{243}\ \alpha\ (2J+1)k^3 |< 2P|r|2S >|^2$$

The transition $2^3S_1 \rightarrow 2^3P_0 + \gamma$ leads to a unique angular distribution of the photon with respect to the beam axis; i.e. $d\sigma/d\Omega \sim (1 + \cos^2\theta)$. The transitions $2^3S_1 \rightarrow 2^3P_{2,1}$ do not have unique angular distributions.

The value of α_s can be extracted by taking the ratio of J/ψ → hadrons to $J/\psi \rightarrow e^+e^-$. The unknown wave function cancel in this ratio and one obtains :

$$\frac{\Gamma(1^{--} \rightarrow ggg \rightarrow \text{hadrons})}{\Gamma(1^{--} \rightarrow e^+e^-)} = \frac{5(\pi^2-9)}{18\pi} \cdot \frac{\alpha_s^3}{\alpha^2} .$$

This yield $\alpha_s \simeq 0.2$. This value is somewhat smaller than the value $\alpha_s = 0.36$ extracted from inelastic neutrino interactions.

The narrow widths of the J/ψ and the ψ' are consistent with the short range picture, however, any model which explains the OZI rule will presumably give the same answer. However, a prediction which is unique to the short range picture is the relative widths of the triplet and singlet S states. Assuming the same radial wave functions for the $^3S_1$ and the $^1S_0$ states the equations above lead to

$$\frac{\Gamma(^1S_0 \rightarrow gg \rightarrow \text{hadrons})}{\Gamma(^3S_1 \rightarrow ggg \rightarrow \text{hadrons})} = \frac{27\pi}{5(\pi^2-9)\cdot\alpha_s} \sim 100$$

The large widths of the $^1S_0$ states relative to the $^3S_1$ states is a characteristic feature of the short range gluon picture.

5. The total annihiliation cross section and 1^{--} states.

The value of R, i.e. the total cross section for $e^+e^- \rightarrow$ hadrons normalized to the point cross section, is plotted in Fig. 5.1.1 as a function of cm energy. The data [42], obtained

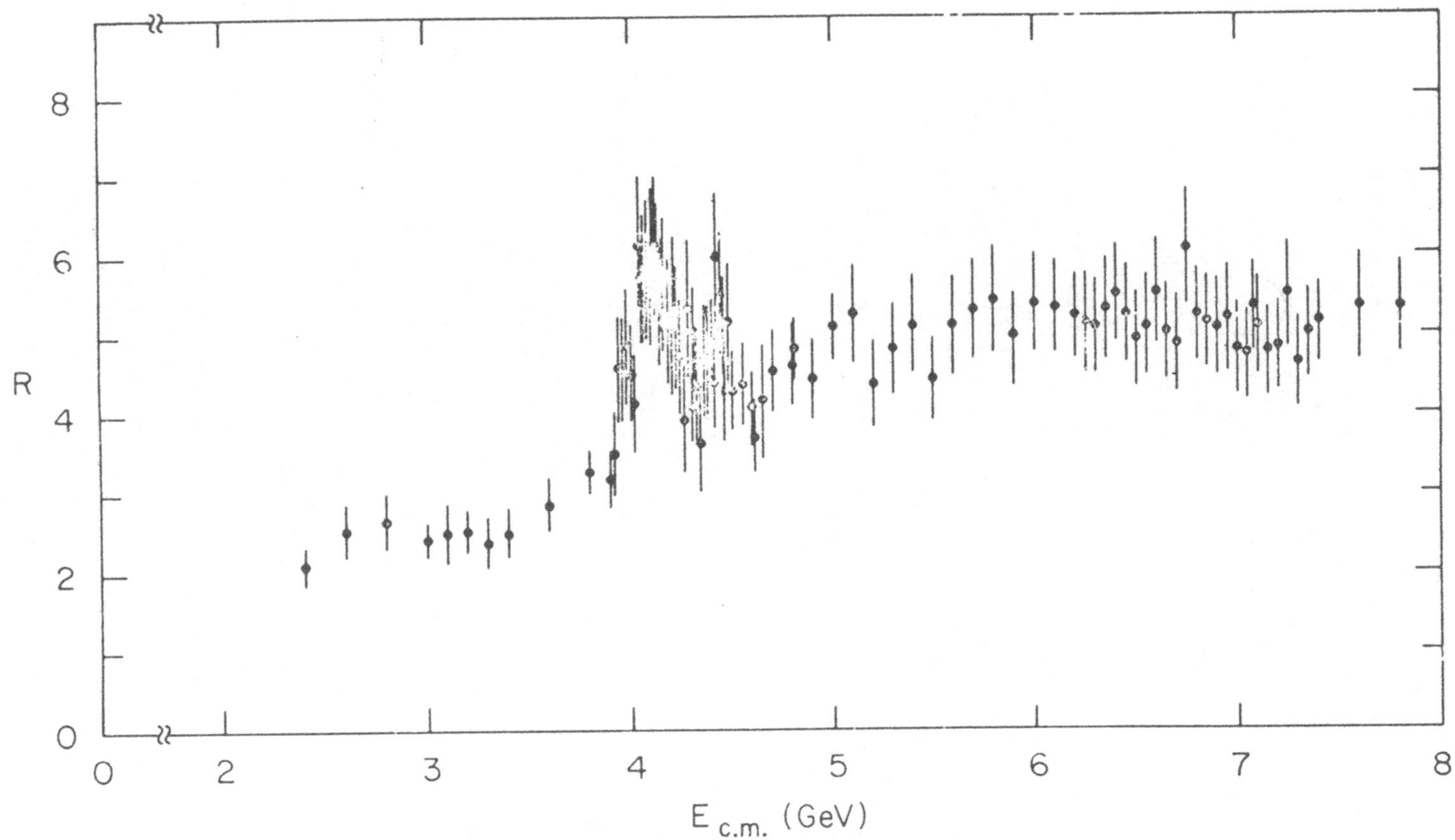

FIG. 5.1.1. : R, the total cross section for $e^+e^- \rightarrow$ hadrons normalized to the point cross section plotted as a function of energy. Data are from the SLAC-LBL Collaboration.

by the SLAC-LBL collaboration, are consistent with a constant value of R below 3.6 GeV and above 5 GeV in cm a clear step of about 2.5 units of R around 4 GeV and several resonances between 3.7 GeV and 4.5 GeV. As will be discussed later in more detail the observed step is really a superposition of two steps, one with a threshold near 3.60 GeV [13] resulting from τ pairproduction, the second with a threshold of 3.727 GeV [8] caused by production of hadrons with open charm. It is rather natural to assume that the observed 1^{--} resonances are basically $c\bar{c}$ states ($^3S_1$ or $^3D_1$) although more exotic explanations [43] have been put forward.

Several groups [44-47] have searched for narrow resonances in the energy range between 1.9 GeV and 7.4 GeV. These searches have so far yielded negative results and the upper limits are listed in table 5.1. Also the very recent discovery of the Upsilon [48] at NAL makes it rather unlikely that a new quark exists with a mass between the mass of the c-quark and 5 GeV.

The 4 GeV region has been investigated in detail by several groups. The value [*] of R obtained by SLAC-LBL [49], DELCO [19], PLUTO [50] and DASP [51], corrected for radiative effects, are plotted in Figs. 5.1.2 - 5.1.5.

All groups agree on the gross behaviour of the total cross section although there are differences, in particular in the region of the resonances between 4.03 GeV and 4.5 GeV. This is shown in Fig. 5.1.6 and 5.1.7 where the SLAC-LBL and the PLUTO data are compared to the DASP data. However, remember that all the experiments have systematic uncertainties on the order of 10% - 15% which are not included in the errors shown.

The DASP group reports that R, excluding τ pair production, increases with 2.1 ± 0.3 between 3.6 GeV and 5.0 GeV in cm. A coloured quark with charge 2/3 e will contribute 4/3 units of R at asymptotic energies. For energies near threshold QCD predicts [52] a value of ΔR larger than the simple quark model value.

We will now discuss the resonances 1^{--} resonances in more detail.

The SLAC-LBL Collaboration [49] and the DELCO Collaboration [19] at SPEAR have observed a new resonance at 3772 MeV. This is only (88 ± 3) MeV above the ψ' and 40 MeV above the threshold for $D\bar{D}$ production and below the threshold for $D\bar{D}\pi$ production. It therefore seems reasonable to assume that the large width of

*) The value of R reported by DELCO, PLUTO and the SLAC-LBL Collaboration includes τ pair production.

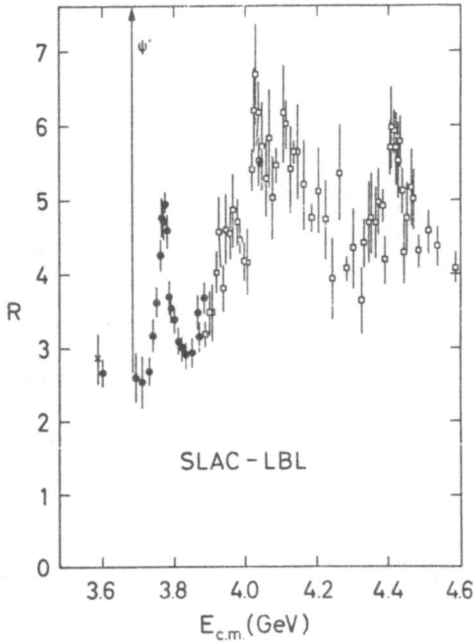

FIG. 5.1.2. : R measured by the SLAC-LBL Collaboration for energies between 3.6 GeV and 4.6 GeV. The contribution from τ production is included.

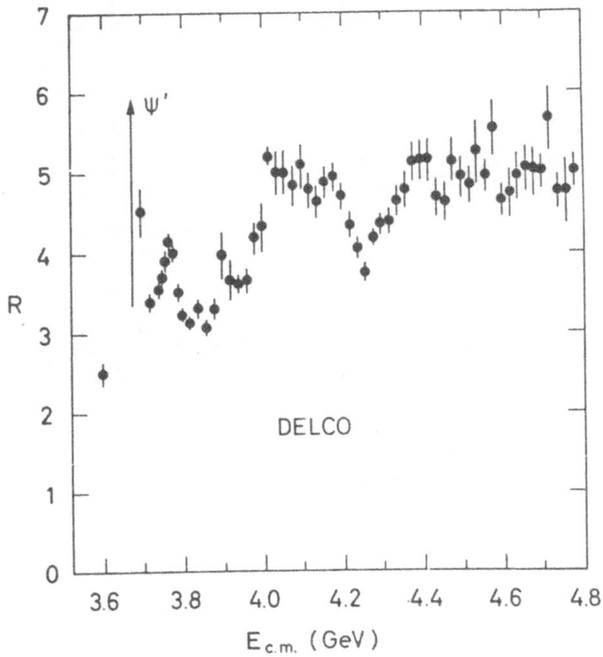

FIG. 5.1.3. : R measured by the DELCO Collaboration for cm energies between 3.6 GeV and 4.8 GeV. The contribution from τ production is included.

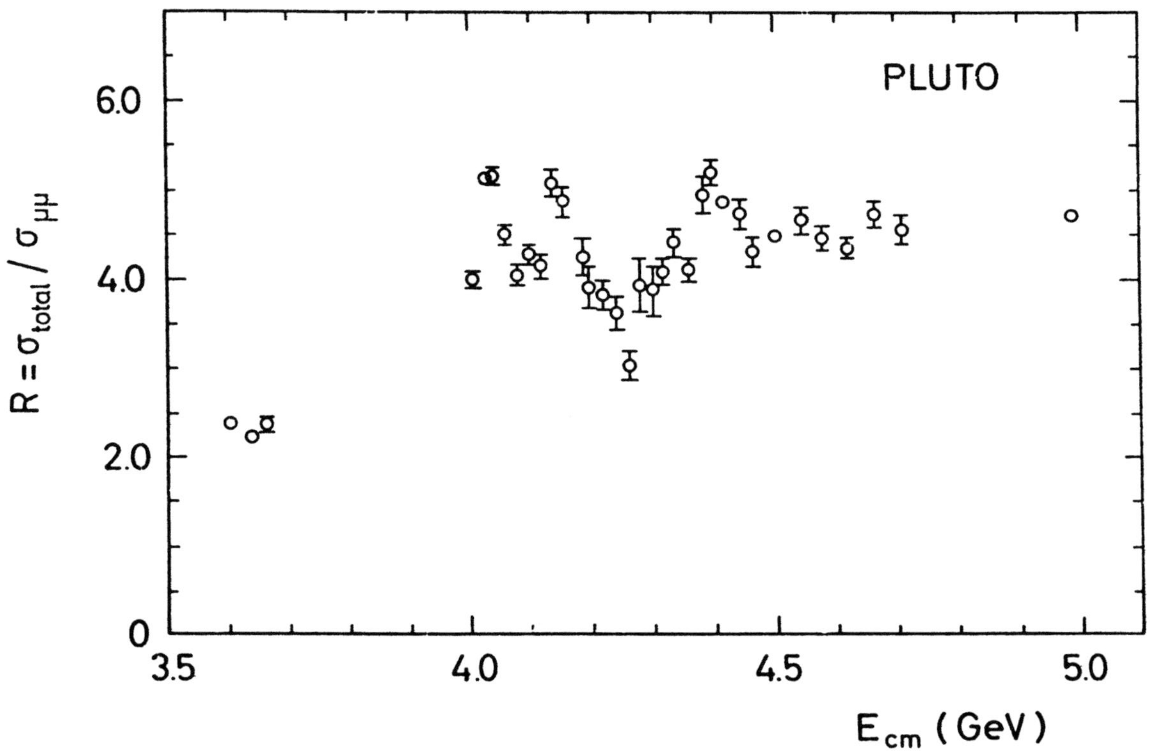

FIG. 5.1.4. : R measured by the PLUTO Collaboration. The data include the contribution from τ production.

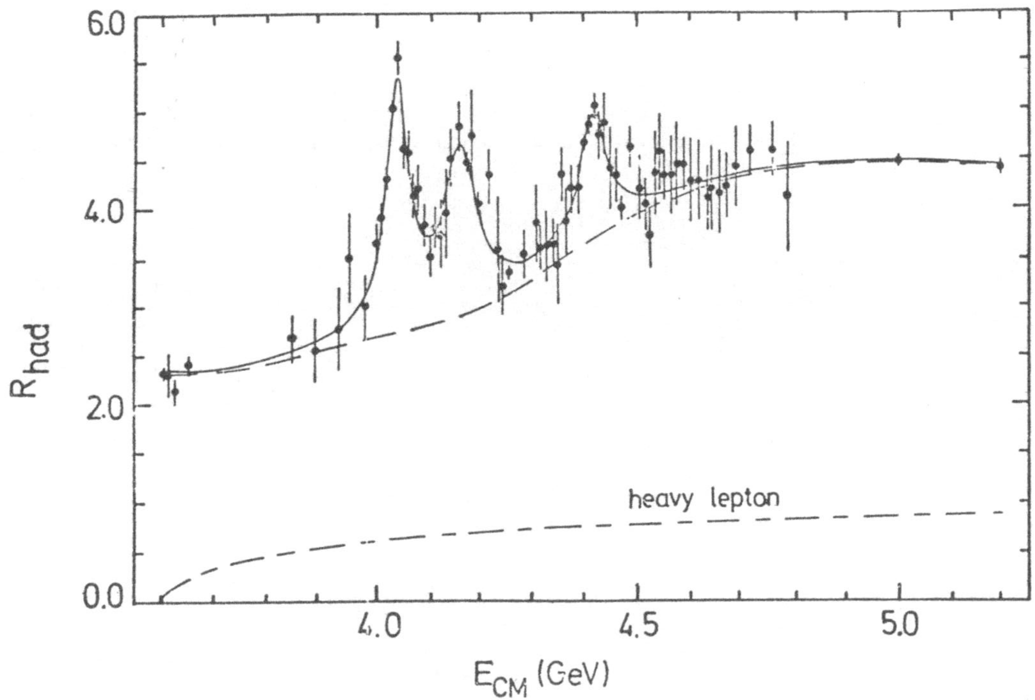

FIG. 5.1.5. : The total cross section for hadron production excluding the contribution from τ pair production normalized to the point cross section. The data are from the DASP Collaboration and the lines represent fits to the data as described in the text.

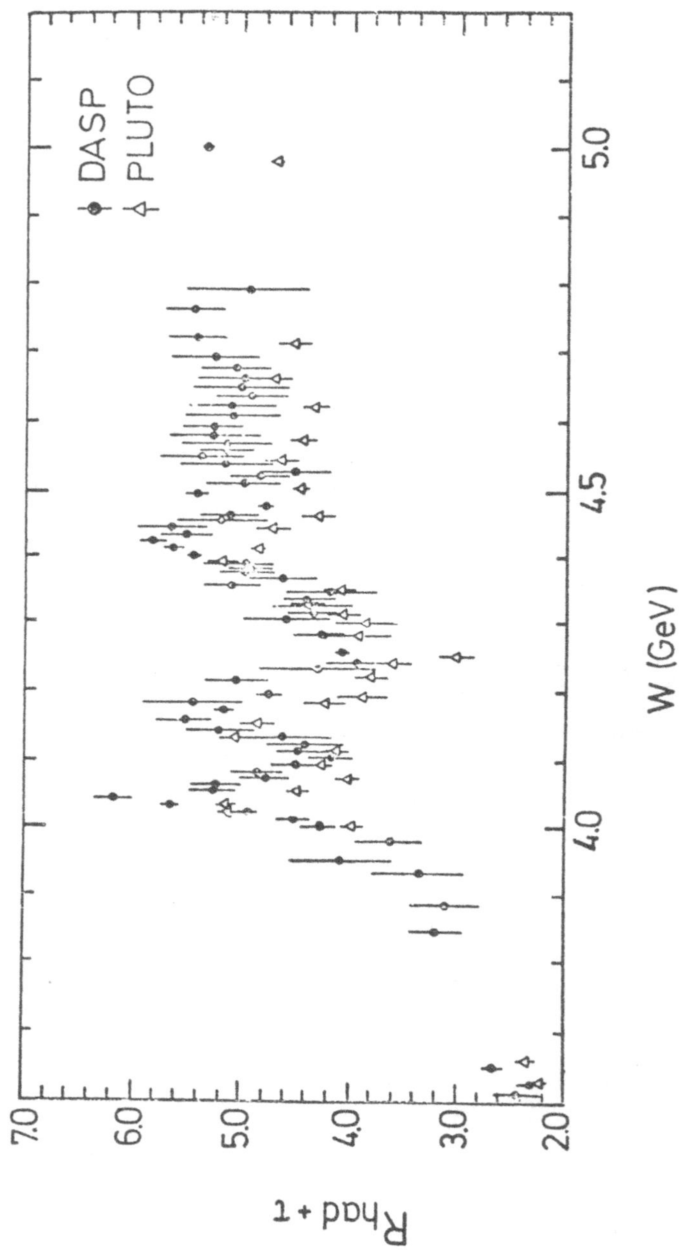

FIG. 5.1.6. : Comparison between R measured by DASP and PLUTO.

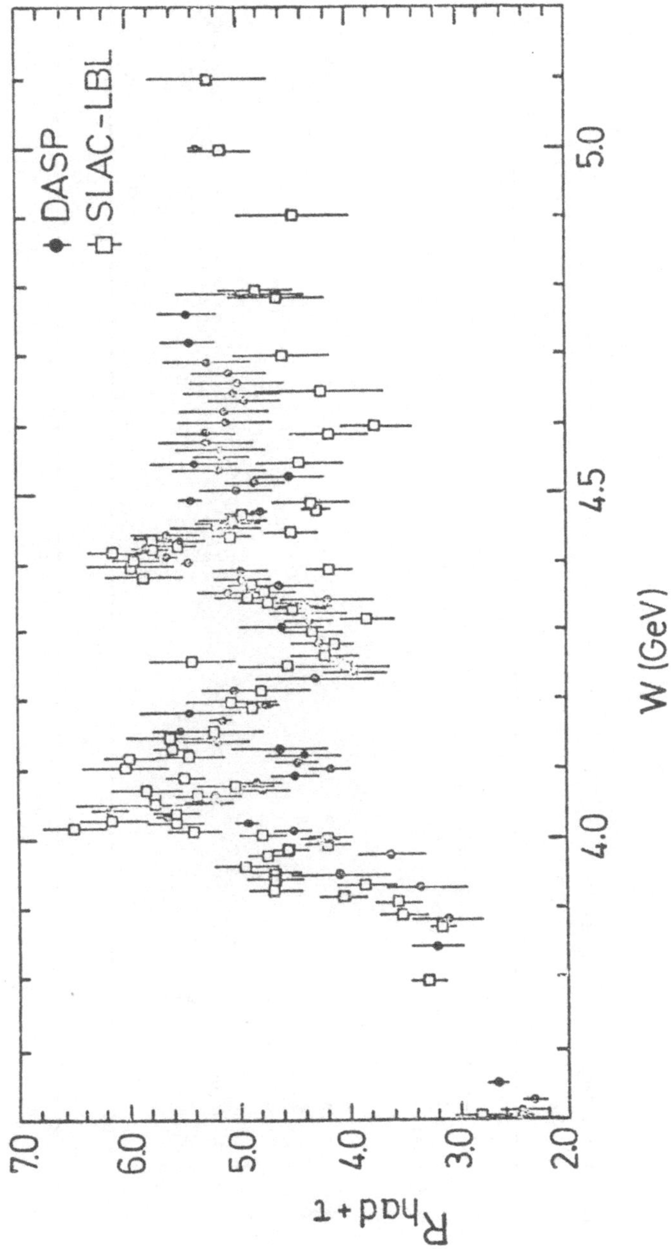

FIG. 5.1.7. : Comparison between R measured by DASP and SLAC-LBL.

Table 5.1 - Search for narrow vector states in $e^+e^- \to$ hadrons.

	Mass Range (GeV)	assumed width Γ (MeV)	$\int \sigma_{hadron} dE$ (nb-MeV)	percentage of $\int \sigma_{J/\psi} dE$	References
ADONE	1.91 - 2.2	2	< 950	< 9%	
	2.2 - 2.54	2	< 660	< 6%	44
	2.97 - 3.09	4	< 830	< 7%	
SPEAR	3.2 - 5.9	2-4	< 1000	< 9%	45
	5.9 - 7.6	2-4	< 450	< 4%	
SPEAR	5.7 - 6.4			≤ 1.2%	46,45
	7.0 - 7.4			≤ 1%	47

the ψ" (3772) relative to the width of the ψ' (3684) results from the decay $\psi' \to D\bar{D}$. This is an OZI-allowed decay and the large ψ" width is therefore a striking confirmation of the rule. The parameters extracted from the fit are listed in table 5.2.

Table 5.2 - Parameters for the new 1^{--} states.

Mass (MeV)	Γ_{tot} (MeV)	Γ_{ee} (KeV)	Reference
3095 ± 4	0.069 ± 0.015	4.8 ± 0.6	23
3684	0.228 ± 0.056	2.1 ± 0.3	23
3772	28 ± 5	0.37 ± 0.09	49
3770	24 ± 5	0.180 ± 0.06	19
4040 ± 10	52 ± 10	0.78 ± 0.1	51
4159 ± 10	78 ± 20	0.76 ± 0.2	51
4417 ± 10	66 ± 15	0.47 ± 0.1	51
4414 ± 7	33 ± 10	0.44 ± 0.14	53

Both experiments agree on the mass and the total width of the resonance, they differ however by a factor of two in the value found for Γ_{ee}. This can be traced [19] to a difference in the raw data.

The SLAC-LBL Collaboration observe a structure around 3 95 GeV. This structure has so far neither been confirmed not rejected by other groups.

There is evidence for two peaks in the 4.1 GeV region. This is clearly seen in the DASP data (Fig. 5.1.5), where the solid line is the result of a fit assuming an incoherent sum of Breit-Wigner resonances plus a smooth background term of the form

$$\sigma_B = \sigma(3.6)/s + \sum_{k=1}^{6} A_R \cdot \beta_k^3 \cdot F^2/s \ .$$

The sum is over the channels $D\bar{D}$, $D\bar{D}^*$, $D^*\bar{D}^*$, $F\bar{F}$, $F\bar{F}^*$ and $F^*\bar{F}^*$. A_k is a free parameter, β_k is the velocity of the particle k and F is a formfactor parametrized as $F = (1-s/(3.1)^2)$ with $s=(2E)^2$. The data are well represented by this Ansatz and the parameters extracted from the fit are listed in table 5.2. These parameters are insensitive to various assumption for the background.

The structure at 4.414 GeV was first observed [53] by the SLAC-LBL group and later confirmed by PLUTO and DASP although the resonance is somewhat less prominent in their data. DASP and SLAC-LBL agree well on the leptonic width of the resonance but they differ on the value found for the total width by a factor of two.

The charmonium spectrum, below threshold for charm production, is well represented by a $c\bar{c}$ bound state model. This simple approximation presumably breaks down above the threshold for charm production where virtual decays like $1^{--} \rightarrow D\bar{D} \rightarrow 1^{--}$ must be taken into account. This approach has been pursued with great success by the Cornell group [54]. The Cornell group predicted a $^3D_1$ state with mass and widths in good agreement with the values reported for the $\psi''(3772)$ well in advance of the actual discovery. It therefore seems natural to assume that the $\psi''(3772)$ indeed is the $^3D_1$ state. The mixing of the 3^3D_1 state and the nearby 3^3S_1 state is via $D\bar{D}$ intermediate state and not because of a strong tensor force in the potential. This mechanism [54,55] also gives a natural explanation for the enhancement of the decay $\psi(4.04) \rightarrow D^*\bar{D}^*$ and the relative suppression of $\psi(4.04) \rightarrow D\bar{D}$ to be discussed later.

The large width for $\psi(4.04) \rightarrow D^*\bar{D}^*$ promted various groups [43] to suggest that some of the 1^{--} resonances observed near threshold are 4 quarks states with the quark content ($c\bar{q}$ $\bar{c}q$) - "charmonium molecule". These resonances can in principle decay into $J/\psi+(q\bar{q})$ without violating the OZI-rule and hence a large J/ψ inclusive cross section is a natural consequence of these models. The PLUTO group has measured inclusive J/ψ production [56] in the mass range 4.0 GeV to 5.0 GeV. They find a total of 4 events corresponding to an inclusive J/ψ cross section on the order of 0.1% of the total cross section - i.e. a level which is consistent with a OZI-forbidden decay. Although the existence of "charmonium molecules" cannot be excluded, at present they do not seem to be required by the data.

It seems therefore natural to assume that the wave functions for the 4.04 GeV and the 4.16 state are also predominantly $c\bar{c}$ states and it is tempting to associate them with the next pair of $^3S_1$ and $^3D_1$ states. However, computations of the level schemes are not yet so advanced that they can support or reject such a conjecture. Note that the total widths of the observed 1^{--} states are all roughly within a factor of two from 3.772 GeV to 4.417 GeV. This might indicate that the high mass 1^{--} states preferentially decays into charmed mesons with a high spin.

6. Charmonium States with even charge conjugation.

The new states with even charge conjugation located below the ψ' are populated via photon emission from the ψ' and the J/ψ. The various decay modes used to search for these levels are indicated in Fig. 6.1.

6.1. The χ(2.82).

Evidence for a heavy narrow resonance with even charge conjugation has been reported [57,58] by the DASP Collaboration in the decay $J/\psi \to \gamma X \to \gamma\gamma\gamma$. Events with three photons were selected

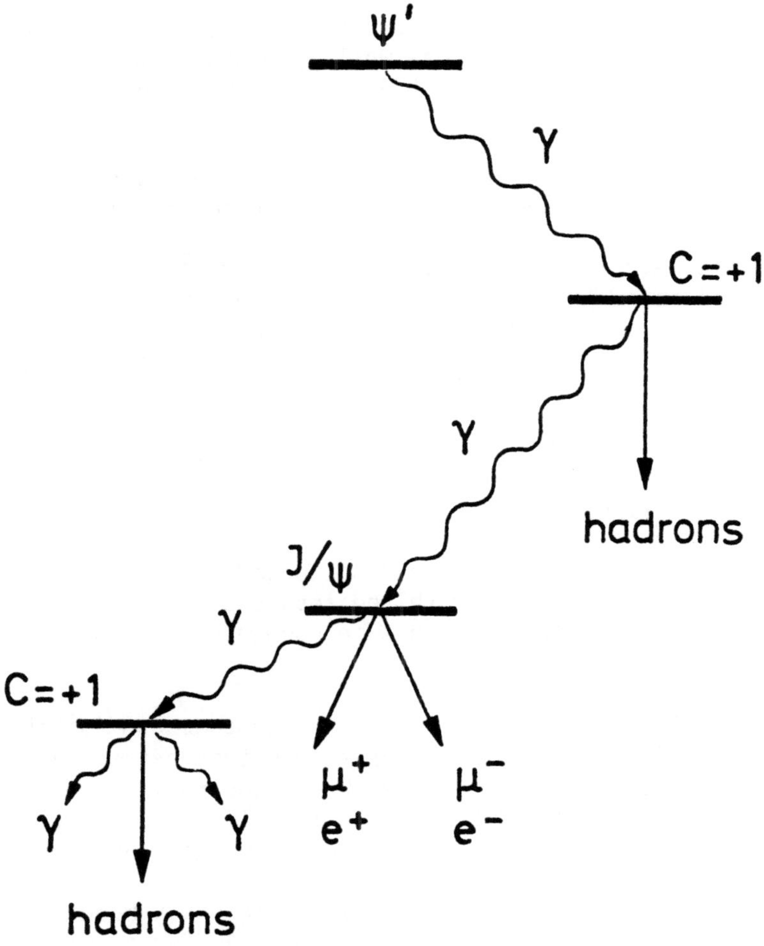

FIG. 6.1. : Possible decay modes used to search for states with even charge conjugation below the ψ' resonance.

using the DASP inner detector. For each photon the production angles were measured taking the position of the interaction point from Bhabha events observed in the same runs. A 1C fit was made and events with $\chi^2 < 2.7$ were retained. The photon energies for the remaining coplanar events were computed using energy and momentum conservation. Of the three possible $\gamma\gamma$ mass combinations only two are independent. The events are plotted in Fig. 6.1.1 and Fig. 6.1.2 as a function of the lowest and the highest mass in the event. A peak at the η mass and an indication of a peak at the η' mass is seen in the low mass plot. A peak around 2.82 GeV is clearly visible in the high mass data. The contribution of the QED reaction $e^+e^- \rightarrow 3\gamma$, shown as the dotted line in Fig. 6.1.1 and 6.1.2, was determined by a Monte Carlo computation with the matrix elements evaluated by Berends and Gastmans [59] as input. Gaussian fits to the $\eta(549)$, $\eta'(958)$ and X(282) mass peak were made simultaneously for the two distributions in order to treat the kinematical reflections properly. The fitted mass of (547.1 ± 4.2) MeV/c^2 agrees well with the known η mass. The fitted width of the η peak is (24 ± 4) MeV/c^2 in agreement with the expected resolution of 20 MeV/c^2.

In the high mass data a clear peak around 2.8 GeV/c^2 is observed superimposed on a smoothly varying background. The fitted mass is (2.82 ± 0.014) GeV/c^2 and the width is (0.040 ± 0.014) GeV/c^2. In the mass intervall (2.82 ± 0.04) GeV/c^2, 41 events are observed compared to 19 events expected from QED production and reflections of the η and the η'. This corresponds to a 5 standard deviation effect. The combined branching ratio was determined [58] to :

$$BR(J/\psi \rightarrow \gamma X) \cdot BR(X \rightarrow \gamma\gamma) = (1.4 \pm 0.4) \times 10^{-4}$$

consistent with earlier data.

Data on $e^+e^- \rightarrow 3\gamma$ have also been collected [60] for energies outside the resonance region. Both the absolute value and the mass distribution of these events are in good agreement with the QED predictions.

The DESY-Heidelberg group [17,61] has also searched for the decay $J/\psi \rightarrow \gamma X \rightarrow 3\gamma$. No peak is observed and they extract a 90% confidence upper limit on the combined branching ratio $BR(J/\psi \rightarrow \psi X) \cdot BR(X \rightarrow \gamma\gamma) < 3.2 \times 10^{-4}$. This limit is about a factor of two higher than the value measured by the DASP group.

Confirming evidence for the X(2.82) has recently been reported [62] by a group working at Serpukov using a 40 GeV π^- beam incident on a hydrogen target. They have searched for the X in the reaction $\pi^-p \rightarrow \gamma\gamma n$. Plotting the events as a function of the

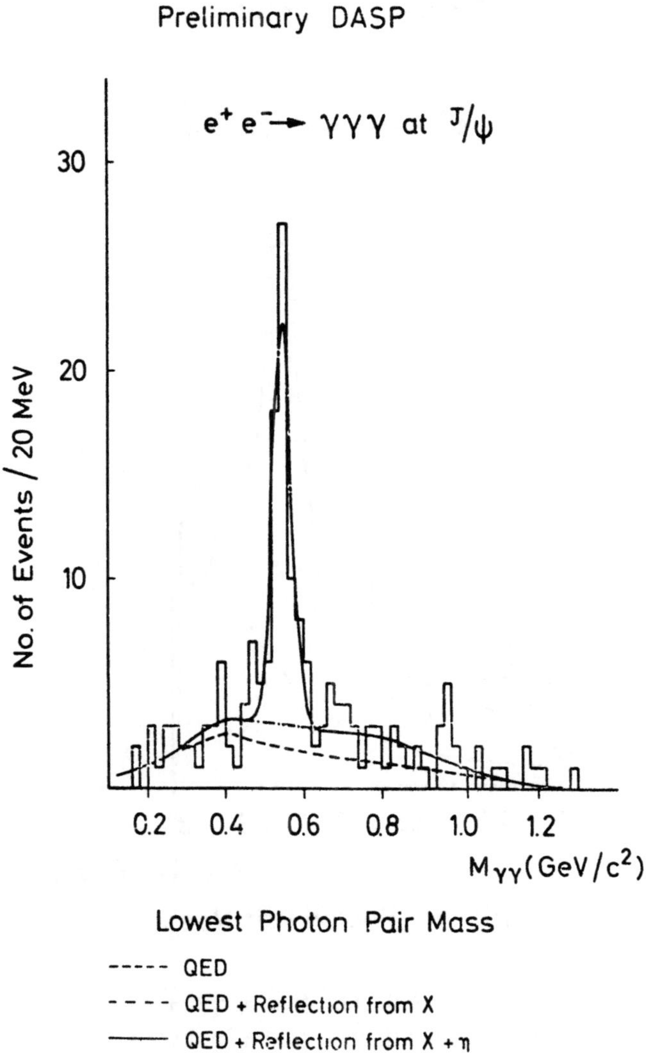

FIG. 6.1.1. : Three photon events observed at the J/ψ resonance plotted as a function of the lowest photon pair mass observed in the event.

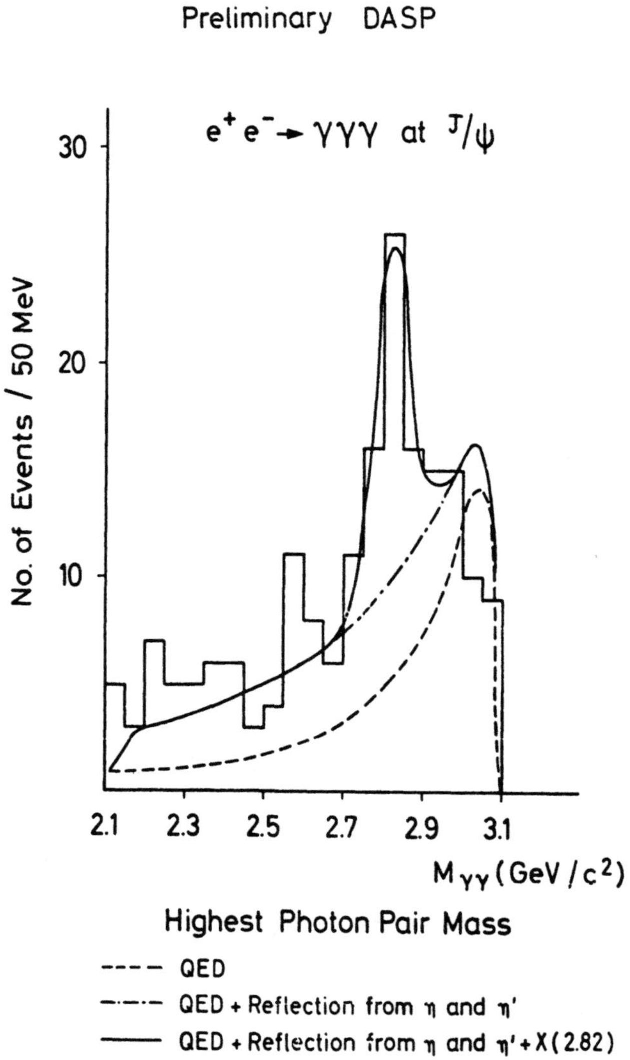

FIG. 6.1.2. : Three photon events observed at the J/ψ resonance plotted as a function of the highest photon pair mass observed in the event.

photon pair mass they observe an enhancement at (2.88 ± 0.06) GeV/c^2 with a systematic uncertainty of 0.03 GeV/c^2. The enhancement, shown in Fig. 6.1.3, has a width of 0.17 GeV/c^2 consistent with the experimental resolution. No other enhancement is seen in the $\gamma\gamma$ mass spectrum at high mass and the authors identify the peak with the X(2.82) state. The photons are emitted isotropic in the restsystem of the X as expected for a spin 0 particle.

The inclusive photon spectrum measured at the J/ψ resonance by the Maryland, Pavia, Princeton, UC-San Diego, SLAC and Stanford Collaboration [31] is plotted in Fig. 6.2.3. No significant peaks are observed and they extract a 90% confidence upper limit of BR(J/ψ → γX) < 1.7%.

We next discuss various searches for the decay ψ' → γX. None of these searches have so far yielded any evidence for the X. However, remember that singlet and triplet states with the same main quantum number n are expected to have similar wave functions in nonrelativistic charmonium models. The wave functions for 2^3S and 1^1S_0 might therefore be nearly orthogonal resulting in a strong suppression of the M1 decay ψ' → γX(2.82).

The Maryland, Pavia, Princeton, UC-San Diego, Slac and Stanford Collaboration [31] has searched for the decay ψ' → γX in the inclusive photon spectrum for the ψ'. No monochromatic photonline is observed (Fig. 6.2.3) at the appropriate energy and they report an upper limit on the branching ratio of BR(ψ' → γX) < 1% in agreement with earlier results [32] (Fig. 6.2.1) from the SLAC-LBL collaboration.

The DASP group has measured [58,60] e^+e^- → 3γ at the ψ' resonance and the events are plotted in Fig. 6.1.4 as a function of the highest mass found in the event. A total of 190 events was observed for an integrated luminosity of 1610 nb^{-1}. No statistically significant peaks are observed and both the shape and the absolute number of events are well explained by the QED process shown as the solid line. The 90% confidence upper limit on the combined branching ratio is

$$BR(\psi' \rightarrow \gamma X) \cdot BR(X \rightarrow \gamma\gamma) < 1.4 \times 10^{-4}.$$

With the results from the measurement of J/ψ → γX → $\gamma\gamma\gamma$ this can be written as

$$\frac{\Gamma(\psi' \rightarrow \gamma X)}{\Gamma(J/\psi \rightarrow \gamma X)} < 3.$$

If the X is identified with the 1^1S_0 and both transitions were allowed M1 transitions then the ratio is given by :

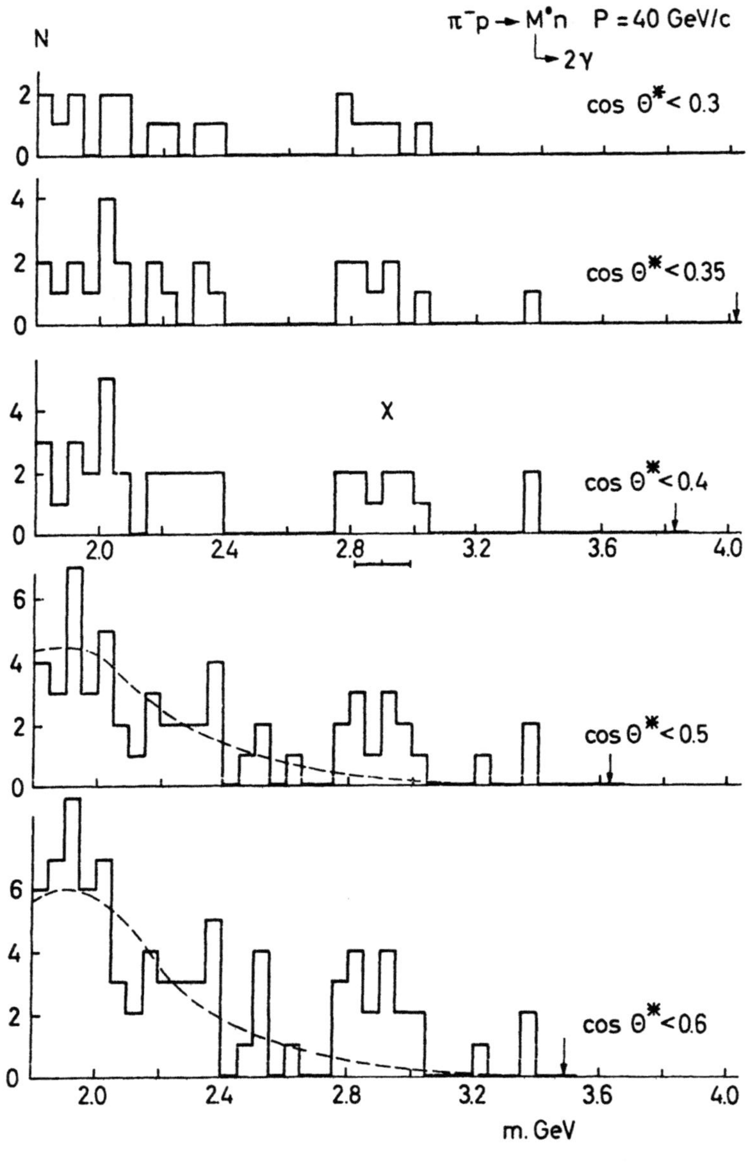

FIG. 6.1.3. : Shown are results on the reaction $\pi^- p \to \gamma\gamma n$ measured with an incident pion beam of 40 GeV at Serpukov. Plotted are the number of events versus the photon pair mass with the cosinus of the cm decay angle θ^* as a parameter. Events are detected with an efficiency of at least 0.8 for photon pair masses below the value shown by the arrow. The mass resolution (FWHM) is shown below the peak at 2.85 GeV. The background is shown by the dashed line.

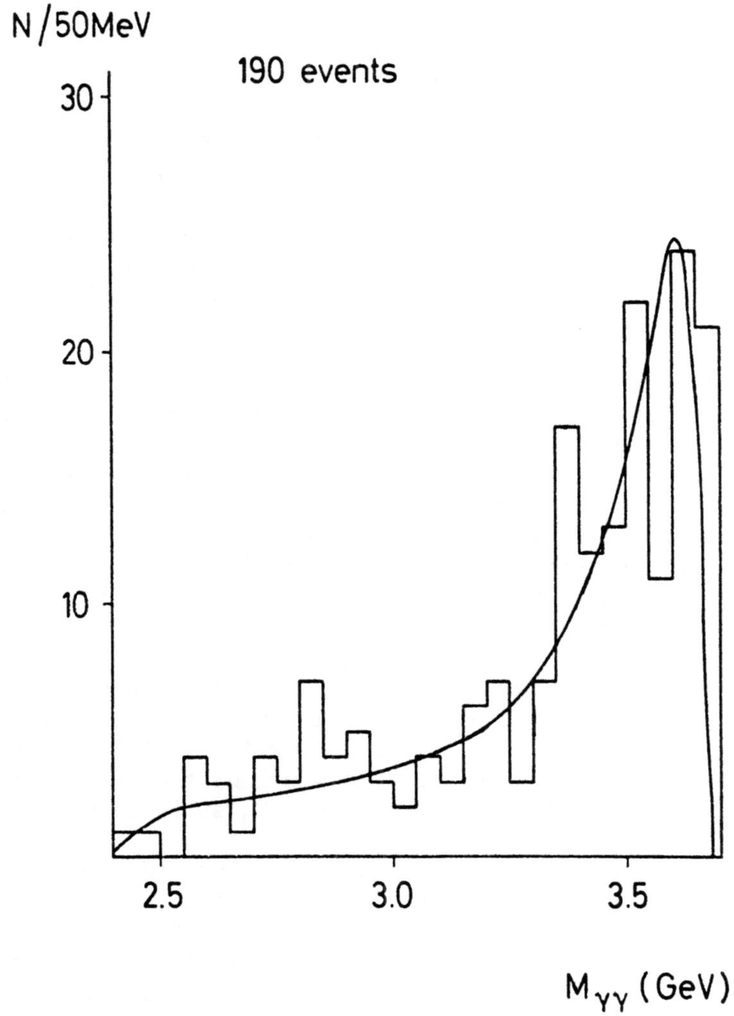

FIG. 6.1.4. : Three photon events observed at the ψ'-resonance plotted as a function of the highest pair mass observed in the event.

$$\frac{\Gamma(\psi' \to \gamma X)}{\Gamma(J/\psi \to \gamma X)} = \left(\frac{k_2}{k_1}\right)^3 \frac{|\langle\psi'|\chi\rangle|^2}{|\langle J/\psi|\chi\rangle|^2} \simeq \left(\frac{k_2}{k_1}\right)^3 \approx 25$$

where k_2 and k_1 are the corresponding photon energies. With these assumptions

$$\frac{|\langle\psi'|\chi\rangle|}{|\langle J/\psi|\chi\rangle|} < 1/8 \ .$$

The wave functions for the ψ' and the X has indeed a small overlap integral as expected in a nonrelativistic $c\bar{c}$-model.

It is clear that more data are needed in order to uniquely identify the X(2.82) with the pseudoscalar 1^1S_0 state. We will now briefly discuss if this assignment is consistent with the experimental information available.

In the simple charmonium picture the difference in mass [63] between the $^3S_1$ and the $^1S_0$ should be on the order of 30 to 60 MeV. The experimental value obtained by the DASP group is (280 ± 20) MeV. However, one should bear in mind that the potential used is by no means unique and it has been shown [64] that a large hyperfine splitting $^3S_1$ - $^1S_0$ can be obtained by adding a long range spin dependence to the potential. Also the data obtained at Serpukov give no evidence for a resonance heavier than 3.0 GeV decaying into two photons.

The upper limit on $BR(J/\psi \to \gamma X)$ and the measured value of $BR(J/\psi \to \gamma X) \cdot BR(X \to \gamma\gamma)$ yield the limit $BR(X \to \gamma\gamma) > (8 \pm 2.5)\ 10^{-3}$ compared to the predicted value of 1×10^{-3}. This discrepancy might indicate that the hadronic width of the X(2.82) is smaller than the 5 MeV predicted for the 1^1S_0 state in QCD.

A more serious problem however is the value $\Gamma(J/\psi \to \gamma X) <$ 1.2 keV. The decay $^3S_1 \to \gamma\, ^1S_0$ is a M1 transition and the value computed assuming $\mu \simeq (2/3)e/2m_Q$ and $\langle n_\gamma | n_i \rangle \sim 1$, is 28 keV, - i.e. about a factor of 25 larger than the experimental upper limit. This is serious since M1 transitions among lighter mesons are reproduced by similar computations. This has lead many people [65] to speculate that the X(2.8) is not the pseudoscalar charmonium state but rather a four quark state $c\bar{c}\ q\bar{q}$. It is further assumed that this state is below the real 1^1S_0 state and hence narrow.

Now if this is true then the decay $\psi' \to \gamma X$ and $J/\psi \to \gamma X$ are both electric dipole transitions and the ratio of the widths, assuming equal matrix elements, are

$$\frac{\Gamma(\psi' \to \gamma X)}{\Gamma(J/\psi \to \gamma X)} \simeq \left(\frac{k_2}{k_1}\right)^3 \simeq 25$$ compared to the experimental value of

$$\frac{\Gamma(\psi' \to \gamma X)}{\Gamma(J/\psi \to \gamma X)} < 3.$$

It is clearly not possible to uniquely identify the X(2.82) with the $^1S_0$ state. However, it is also fair to point out that no other candidate for the $^1S_0$ has been reported.

6.2. The P_c/χ-states.

These states have been observed (Fig. 6.1) by searching in the debris of the ψ' for the following signatures :

a) Peak in the inclusive photon spectrum.
b) The J/ψ with one monochromatic and one nearly monochromatic photon.
c) A peak in the effective mass spectrum of the hadrons in $\psi' \to \gamma$ + hadrons.

a) A first measurement of the inclusive photon spectrum at the ψ'-resonance was reported [32] by the SLAC-LBL group. This group used the solenoid as a pair-spectrometer to measure the energy of lepton pairs resulting from photons converting in the material in front of the chamber. The data plotted in Fig. 6.2.1. show a monochromatic line centered at (261 ± 10) MeV/c^2. The width is consistent with the experimental resolution and the branching ratio is (6.5 ± 2.2)% assuming the photons to have a $(1 + \cos^2\theta)$ angular distribution with respect to the beam axis. The other P_c or χ states correspond to photons with energies below 250 MeV and are masked by the rapid variation of the photon detection efficiency at low energy. Photon lines from the decay $P_c/\chi \to \gamma J/\psi$ are not observed, presumably because of the Doppler broadening.

The Maryland, Pavia, Princeton, UC-San Diego, SLAC and Stanford Collaboration determines [31] the inclusive photon spectrum by measuring the energy deposited in two sets of segmented Na(TR) crystals mounted above and below the beam pipe.
The detector is shown in Fig. 6.2.2. The energy resolution (σ) varied between 2.5% at 1500 MeV and 8% at 200 MeV. Electronic threshold effects were eliminated by demanding that the measured photon should not be a part of the trigger.
The spectra measured at the J/ψ and the ψ' resonance are plotted in Fig. 6.2.3. The spectrum measured at the ψ' resonance shows several bumps superimposed on a smoothly varying background in contrast to the structure less spectrum observed at the J/ψ resonance. The dotted line in Fig. 6.2.3 represents a Monte Carlo

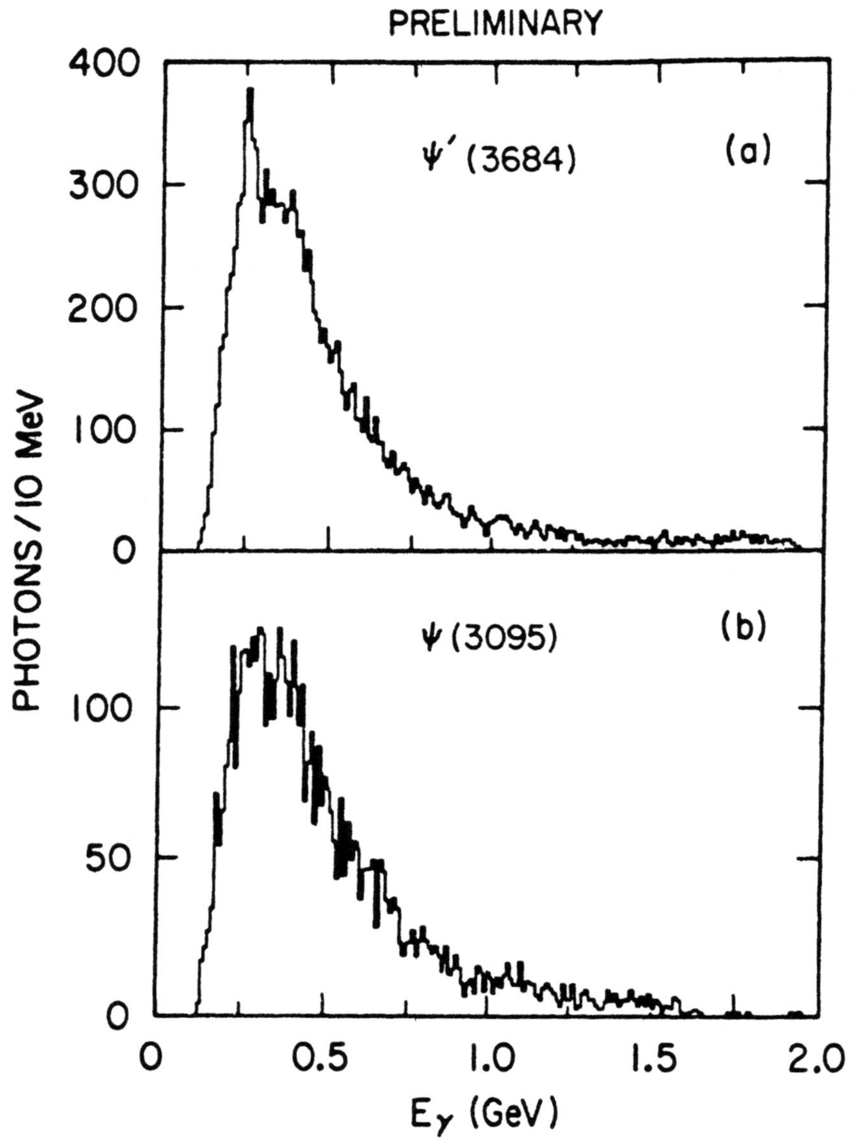

FIG. 6.2.1. : The inclusive photon distribution measured by the SLAC-LBL Collaboration at J/ψ and ψ'.

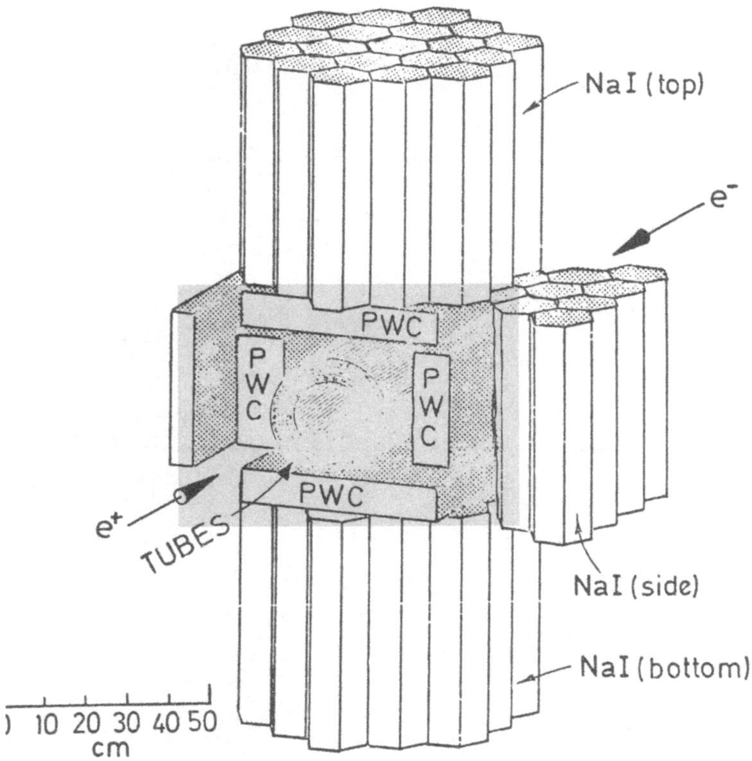

FIG. 6.2.2. : The detector used by the Maryland, Pavia, Princeton, UC San Diego, SLAC and Stanford Collaboration to measure the inclusive photon spectrum.

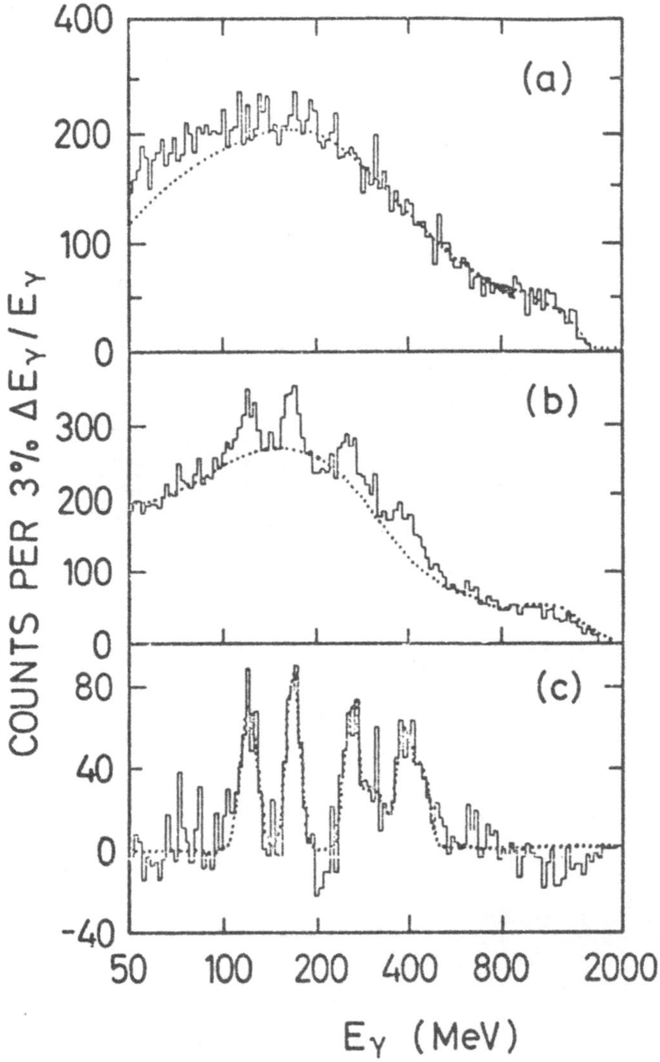

FIG. 6.2.3. : The inclusive photon distribution measured by the Maryland, Pavia, Princeton UC-San Diego, SLAC and Stanford Collaboration as a function of $\ln E_\gamma$ for :

a) the J/ψ

b) ψ'

c) shows the difference between the data and the continuum in (b).

calculation normalized to the spectrum above 150 MeV. The smooth spectrum at the ψ' was generated similarly but it includes cascade decays and the decays into the P_c/χ states, and are corrected for the discrepancy observed between the Monte Carlo calculation and the low energy photon spectrum at the J/ψ.

Three clear peaks resulting from $\psi' \to \gamma P_c/\chi$ and one Doppler smeared structure from $P_c/\chi \to \gamma J/\psi$ is seen in the spectrum. A fit was made assuming a Gaussian shape (Doppler broadened for the $P_c/\chi \to \gamma J/\psi$ decays) with the measured energy resolution of the detector as a input. The masses found for the established resonances are in good agreement with earlier results, the values obtained for the branching ratio of $\psi' \to \gamma P_c/\chi$ are listed in table 6.1. The branching ratios for $P_c/\chi \to \gamma J/\psi$ are listed in table 6.2.

Table 6.1 - P_c/χ, X states

Branching ratio for $^3S_1 \to \gamma + (C+1)$

Decay	Branching ratio %	Γ (keV)	Reference	Theory (keV)
$\psi' \to \gamma\chi(3.41)$	7.2 ± 2.0	16 ± 5	31	44
	6.5 ± 2.2	15 ± 5	32	
$\psi' \to \gamma\chi(3.45)$	< 2.5%	< 5.7	31	18
$\psi' \to \gamma P_c(3.51)$	7.1 ± 2.0	16 ± 5	31	38
$\psi' \to \gamma\chi(3.55)$	7.0 ± 2.3	16 ± 5	31	27
$\psi' \to \gamma X(2.83)$	< 1%	< 2.3	31	--
$\to \gamma X(2.83)$	< 1.1%	< 2.5	32	--
$J/\psi \to \gamma X(2.83)$	< 1.7%	< 1.2	31	28

Table 6.2 - Cascade Decay $(\psi' \rightarrow \gamma P_c/\chi)$ $(P_c/\chi \rightarrow \gamma J/\psi)$

Intermediate State	$BR(\psi' \rightarrow \gamma P_c/\chi)\cdot BR(P_c/\chi \rightarrow \gamma J/\psi)$ %	Reference
$\chi(3.41)$	3.3 ± 1.7	31
	0.3 ± 0.2	68
	0.2 ± 0.1	17
	0.2 ± 0.2	66
	(0.22 ± 0.08)[a]	
$\chi(3.45)$	< 0.61 90% CL	68
	< 0.6 90% CL	17
	0.8 ± 0.4	66
$P_c(3.51)$	5.0 ± 1.5	31
	2.1 ± 0.4	68
	3.3 ± 0.8	17
	2.4 ± 0.8	66
	(2.4 ± 0.3)[a]	
$\chi(3.56)$	2.2 ± 1.0	31
	1.6 ± 0.4	68
	2.3 ± 0.6	17
	1.0 ± 0.6	66
	(1.9 ± 0.2)[a]	

a The combined branching ratio obtained by averaging the results of references 17, 66 and 68.

b) The cascade decay.

The cascade decay : $\psi' \rightarrow \gamma P_c/\chi \rightarrow \gamma\gamma J/\psi$ has a low rate, but a very favourable signal to background ratio. Indeed the first evidence [6)] for the intermediate states was found by DASP using this method. The status as of summer 1977 is shown in Fig. 6.2.4. Plotted are the number of cascade events as a function of the low mass solution $(J/\psi\gamma)_{min}$ versus the high mass solution

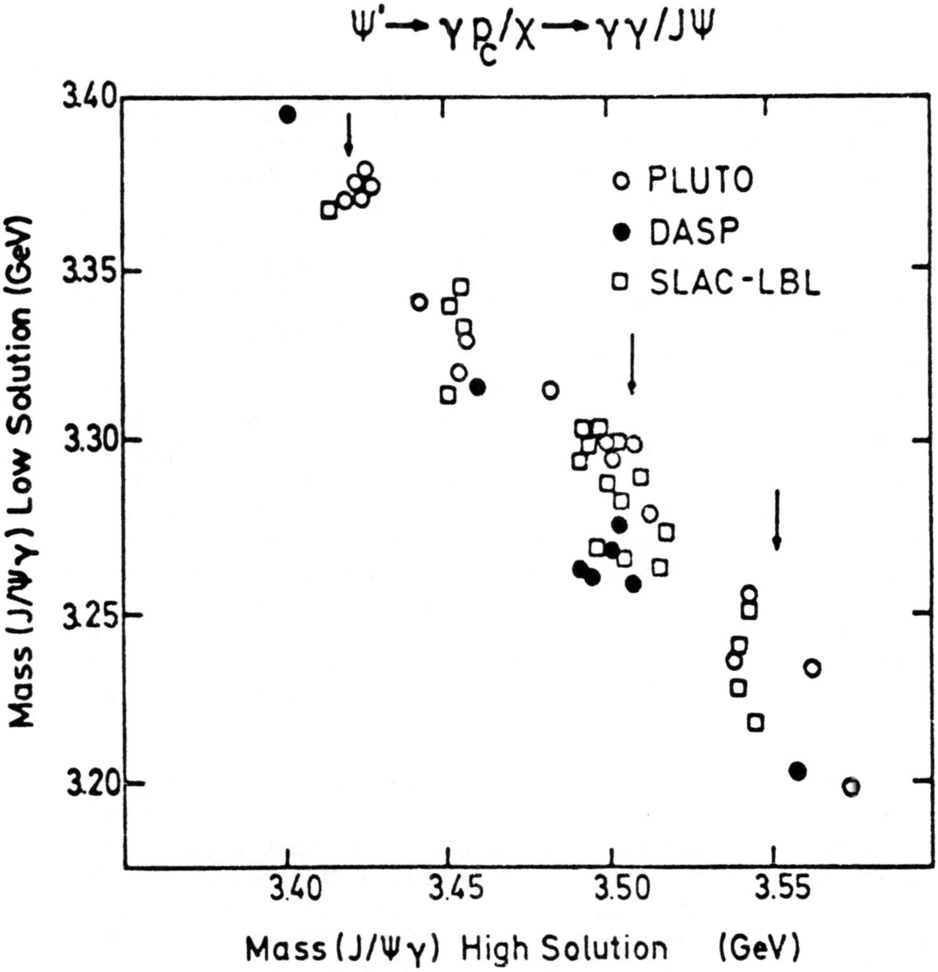

FIG. 6.2.4. : The decay $\psi' \to \gamma\gamma J/\psi$. The data are plotted as a function of the low mass solution versus the high mass solution. The plot shows the world data (DASP, PLUTO, SLAC-LBL) as of summer 1977.

$(J/\psi\gamma)_{max}$. The data are from DASP[6], SLAC-LBL [66] and PLUTO[67]. The data show clear evidence for $\chi(3.56)$, $P_c(3.51)$ and $\chi(3.41)$. In addition there is a clustering around 3.45 GeV (3.33 GeV) with 4 events from SLAC-LBL, 3 events from PLUTO (preliminary) and one event from the old DASP experiment. Although the evidence is not completely convincing it is at least a strong hint that such a resonance does exist. Preliminary results from two high statistics experiments have now become available.

The DASP group reported new data based on a fivefold increase in statistics. In this experiment [58,68] events of the type $\psi' \rightarrow \gamma\gamma\mu\mu$ are selected where the muons are measured and identified using the magnetic arms of the spectrometer and the two photons identified and their directions measured using the inner detector. A 3C-fit was made to the events assuming they all resulted from a $\gamma\gamma$-cascade via an intermediate state. The events which satisfied the fit are plotted in Fig. 6.2.5 as function of the photon pairmass versus the highest $(J/\psi\gamma)$ mass. The decay $\psi' \rightarrow \eta J/\psi \rightarrow \gamma\gamma J/\psi$ is observed with a branching ratio $BR(\psi' \rightarrow \eta J/\psi) = (4.3 \pm 0.7)\%$ in agreement with earlier data [66]. For the cascade decay only events with a photon pair mass less than 0.5 GeV/c^2 are retained. The transitions via the $\chi(3.56)$ and the $P_c(3.51)$ are clearly observed and there are two events consistent with $\psi' \rightarrow \gamma\chi(3.41) \rightarrow \gamma\gamma J/\psi$. The values of the combined branching ratios are listed in table 6.2. There is one event consistent with $\psi' \rightarrow \gamma\chi(3.45) \rightarrow \gamma\gamma J/\psi$. However the photon pairmass in this event equals the π° mass and the isospin violating decay $\psi' \rightarrow \pi^\circ J/\psi$ is expected [70] to occur with a small branching ratio. The 90% confidence upper limit on the combined branching ratio $BR(\psi' \rightarrow \gamma\chi(3.45)) \cdot BR(\chi(3.45) \rightarrow \gamma J/\psi)$ is 0.6% based on the one event observed.

The DESY-Heidelberg group [17] selected events of the type $\psi' \rightarrow \gamma\gamma\mu\mu$. The events are reconstructed from the measured directions using the constraint $m_{\mu\mu} = m_{J/\psi}$. Events are rejected if the measured and the fitted photon energies disagreed. The photon pair mass distribution, plotted in Fig. 6.2.6 gives clear evidence for the $\psi' \rightarrow \eta J/\psi$ and the combined branching ratio for the cascade decays are obtained by normalizing to this decay. The $\psi' \rightarrow \eta J/\psi$ is the excluded by a cut on the $\gamma\gamma$ mass and the remaining events are plotted in Fig. 6.2.7 as a function of the high mass solution. Two distinct peaks at a mass of respectively 3.55 GeV/c^2 and 3.51 GeV/c^2 with a 2σ enhancement at a mass of 3.41 GeV/c^2 is observed. The branching ratios are listed in table 6.2. In the data there is no evidence for the $\chi(3.45)$. The 90% confidence upper limit on the combined branching ratio is 0.6%.

It is clear that these two experiments, which have larger statistics than the old ones, find no evidence for $\chi(3.45)$. However

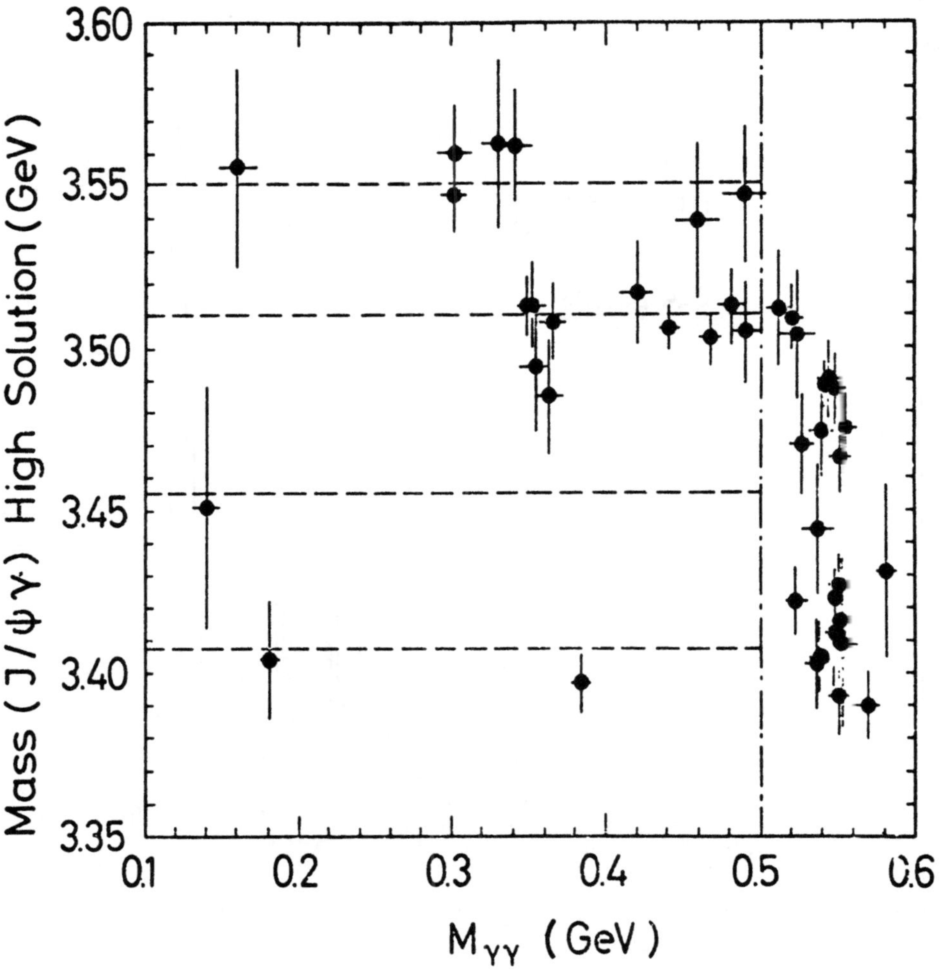

FIG. 6.2.5. : The decay $\psi' \to \gamma\gamma J/\psi$. The plot shows recent data obtained by the DASP group. The events are shown in a scatter plot of the high mass solution $(J/\psi\gamma)$ versus the photon pairmass.

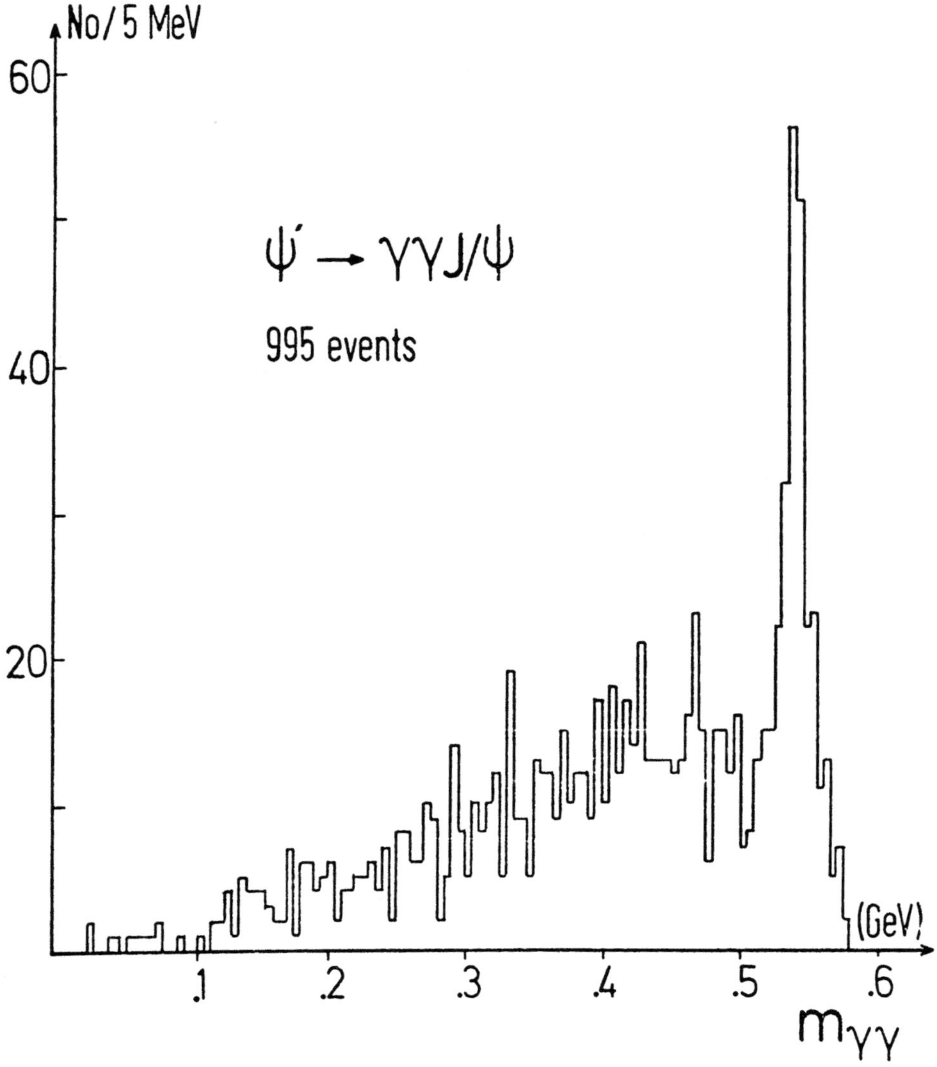

FIG. 6.2.6. : Number of events plotted versus the photon pair mass in the decay $\psi' \to \gamma\gamma J/\psi$. The data are from the DESY-Heidelberg Collaboration.

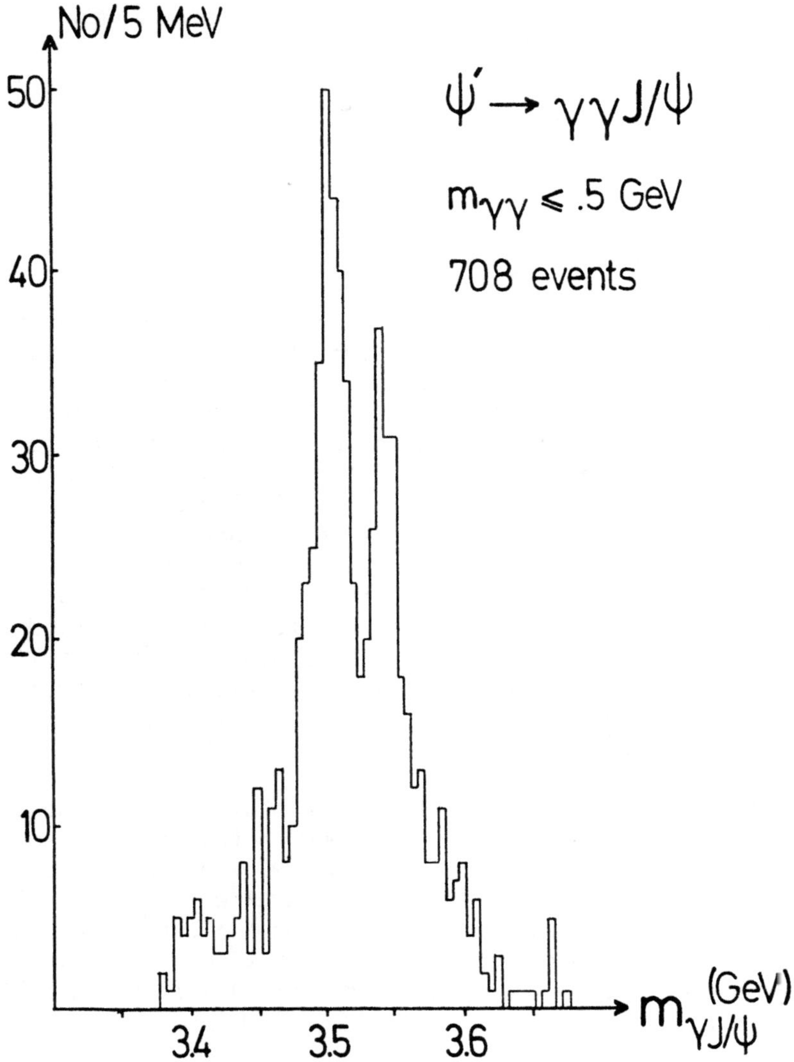

FIG. 6.2.7. : Mass distribution of the J/ψ system in the decay $\psi' \to \gamma\gamma J/\psi$ measured by the DESY-Heidelberg Collaboration. The decay $\psi' \to \eta J/\psi$ is removed from the data sample and only the highest mass solution is plotted.

note that the 90% confidence upper limits of 0.6% are still consistent with the reported [66] branching ratio of (0.8 ± 0.4)%.

c) Hadronic decays of the intermediate states.

Data on the decay $\psi' \to \gamma P_c/\chi \to \gamma$ + charged hadrons have been reported by the SLAC-LBL Collaboration. The criteria used to select these events and early evidence for the decay modes :

$$\begin{aligned} P_c/\chi &\to \pi^+\pi^-\pi^+\pi^- \\ &\to \pi^+\pi^- K^+K^- \\ &\to \pi^+\pi^-\pi^+\pi^-\pi^+\pi^- \\ &\to \pi^+\pi^-,\ K^+K^- \end{aligned}$$

can be found in reference 71.

The data are shown in Fig. 6.2.8. Plotted are the number of events in the various topologies as a function of the effective mass of the hadrons. The $2\pi^+2\pi^-$ and $\pi^+\pi^-K^+K^-$ spectra show clear evidence for three peaks centered at 3.415 GeV/c^2, 3.50 GeV/c^2 and 3.55 GeV/c^2 with an uncertainty of ± 10 MeV. The peak resulting from the direct decay of the ψ' is also seen. The two upper states are not yet resolved in the $3\pi^+3\pi^-$ mode, however, the uniform distribution of the events as a function of mass in this region indicate similar contributions from both. There is a very clear $\pi^+\pi^-$ and K^+K^- signal in the decay of the 3.415 GeV/c^2 state. A significant signal of eight events is also observed in the decay of the 3.55 GeV/c^2 state. There are a few events with a mass around 3.5 GeV/c^2 consistent with the expected background. Note that there is no evidence for the $\chi(3.45)$ in any of the decay modes. The branching ratios are listed in table 6.4.

The distribution in θ, the angle of the photon, with respect to the beam axis has been obtained by summing the various decay channels. The result is shown in Fig. 6.2.9. The angular distribution is only uniquely determined for a level with spin 0. In this case the angular distribution should be $1 + \cos^2\theta$. Higher spins could lead to a more isotropic distribution. The values of α obtained from a fit to the data of the form $1 + \alpha\cos^2\theta$ are listed in table 6.3.

Table 6.3 - Values of α

State :	$\chi(3.41)$	$P_c(3.51)$	$\chi(3.55)$
α	1.4 ± 0.4	0.1 ± 0.4	0.3 ± 0.4

The $\chi(3.41)$ is therefore consistent with J=0, whereas a J=0 assignment to the $P_c(3.51)$ and the $\chi(3.55)$ is excluded on the 2σ

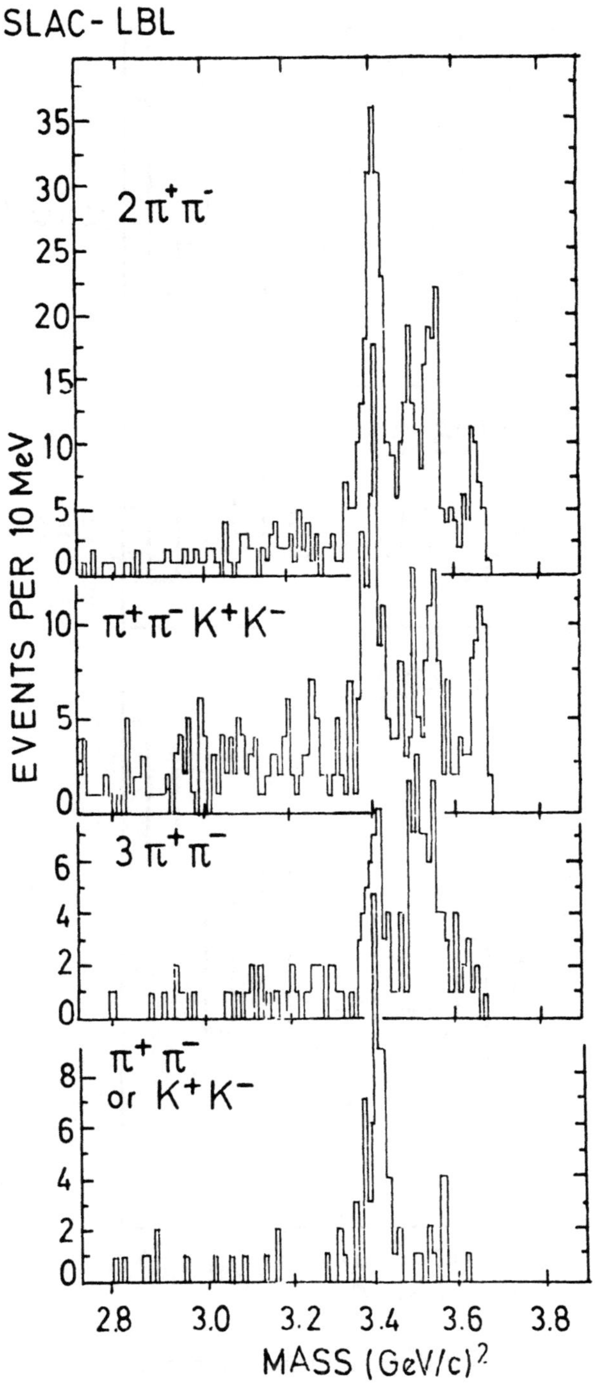

FIG. 6.2.8. : Effective mass spectra observed by the SLAC-LB_ group in the decay of the ψ' into a single pho-on and charged hadrons.

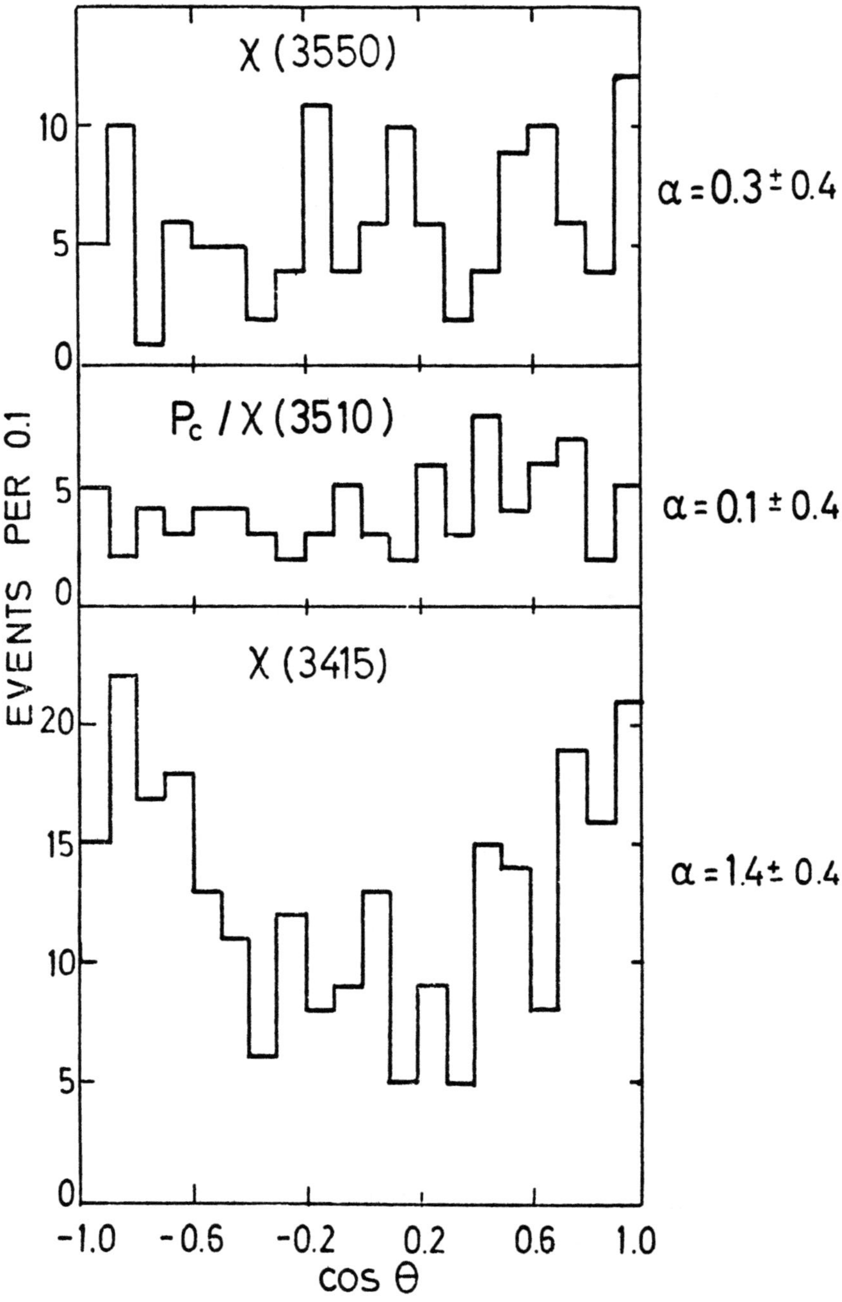

FIG. 6.2.9. : The angular distribution with respect to the beam axis of the photon from the decay $\psi' \to \gamma P_c/\chi \to \gamma$ charged hadrons. The data were obtained by the SLAC-LBL group.

level.

In the level scheme depicted in Fig. 4.1.1c there are four levels with even charge conjugation between the ψ' and the J/ψ, one pseudoscalar $^1S_0$ with $J^{PC} = 0^{-+}$ and three ^3P states with $J^{PC} = 2^{++}$, 1^{++} and 0^{++}. Are the levels found consistent with these quantum numbers ?

All the levels must have even charge conjugation since they are populated via $\psi' \rightarrow \gamma P_c/\chi$

$\chi(3.413)$ It follows from the observed decay into $\pi^+\pi^-$ and K^+K^- that the state must have natural spin-parity and be an isoscalar. The angular distribution is consistent with J=0. We can therefore safely assign this state to $J^{PC} = 0^{++}$.

$\chi(3.454)$ Needs to be confirmed.

$\chi(3.508)$ The absence of $\pi^+\pi^-$ and K^+K^- decays are consistent with the state belonging to an unnatural spin-parity sequence 0^-, 1^+, ... The angular distribution of the photon requires(2σ) $J \neq 0$, isospin must be even since the resonance decays into an even number of pions.

$\chi(3.552)$ It follows from the observed decay into $\pi^+\pi^-$ and K^-K^- that the level is an isoscalar and belongs to a natural spin-parity sequence. The angular distribution of the photon excludes $J = 0$ with 2σ.

This is all that can be deduced from the data alone. However, comparing the available information with the levels expected in the charmonium model (Fig. 4.1.1c) leads to a unique assignment using the following reasoning. The 3.413 GeV/c^2 level must be the 0^{++} state. This leaves only one natural spin-parity level - the $^3P_2(2^{++})$ level - which then must be associated with the 3.552 GeV/c^2 level. The 3.508 GeV/c^2 level has $J \neq 0$ and must therefore be the $^3P_1(1^{++})$ level. This assignment is also consistent with the fact that $P_c(3.51) \rightarrow \pi^+\pi^-$, K^+K^- has not been observed. According to Fig. 4.1.1c there is only a pseudoscalar level 0^{-+} unaccounted for and this level must then be identified with the remaining $\chi(3.45)$ level. This assignment was first suggested by Chanowitz and Gilman [72].

We will first discuss the properties of the P levels :

The decays $\psi' \rightarrow \gamma^3P_{2,1,0}$ are to lowest order all electric dipole transitions. The measured widths are listed in table 6.1 and compared to model calculations assuming E1. Note that the predicted rates are considerably larger than the measured values. If the matrix elements are independent of J then the relative rates for E1 transitions are given by $\Gamma \sim k^3(2J+1)$. In this case we expect

$$\Gamma(2^3S_1 \to \gamma\, ^3P_0) : \Gamma(2^3S_1 \to \gamma\, ^3P_1) : \Gamma(2^3S_1 \to \gamma\, ^3P_2) =$$

$$1 \quad : \quad 0.9 \quad : \quad 0.6$$

compared to the observed ratios 1 : 1.04 : 1.02. Reversing the order of the levels i.e. $^3P_0 > {}^3P_1 > {}^3P_2$ leads to a gross disagreement between the predicted and the observed rates.

In QCD the 0^{++} and the 2^{++} states can decay to lowest order via 2 gluon emission into hadrons whereas the 1^{++} state must emit at least three gluons. The total hadronic width for 0^{++} and 2^{++} is therefore proportional to α_s^2 compared to α_s^3 for the 1^{++} state. Since α_s^2 is less than one we naively expect the 1^{++} state to have a smaller hadronic width than the other two. This assertion is supported by detailed calculations [40] which find :

$$\Gamma(0^{++}) : \Gamma(2^{++}) : \Gamma(1^{++}) = 15 : 4 : 1$$

where the first ratio is the most reliable one. Experimentally we would therefore expect the 0^{++} level to have a smaller branching ratio for the radiative decay into the J/ψ than the other two and this is indeed supported by experiments.

The ratios of the hadronic widths can be estimated from the radiative transitions with the following assumptions :

1) $\Gamma_{tot} = \Gamma(^3P \to \text{hadrons}) + \Gamma(^3P \to \gamma\, ^3S_1) = \Gamma_h + \Gamma_\gamma$

2) $\Gamma(^3P \to \gamma\, ^3S_1) \sim k^3$, i.e. independent of the matrix elements.

With these assumptions we can write :

$$\Gamma_h = (\Gamma_{tot} - \Gamma_\gamma) = (\frac{1}{B_\gamma} - 1)\,\Gamma_\gamma \quad \text{where} \quad B_\gamma = \Gamma_\gamma / \Gamma_{tot}.$$

With the radiative branching ratios listed below in table 6.4 we find

$$\Gamma_h(3.41) : \Gamma_h(3.55) : \Gamma_h(3.51) = (14.2\pm6.0) : (3.8\pm1.3) : (0.5\pm0.16)$$

consistent with the prediction.

The evidence for the 3.454 GeV/c^2 level is marginal, although it seems unlikely that both SLAC-LBL and PLUTO should have a fluctuation of the background at the same mass. The value reported by SLAC-LBL for the combined branching ratio $BR(\psi' \to \gamma\chi(3.45))BR(\chi(3.45) \to \gamma J/\psi) = (0.8 \pm 0.4)\%$ is not in strong disagreement with the 90%

confidence upper limits of 0.6% found by DESY-Heidelberg and DASP. However, identifying the 3.454 GeV/c^2 level with the 2^1S_0 state causes serious problems. According to standard lore, the $^3S_1$ and $^1S_0$ states have similar wave functions while those for 2^3S_1 and 1^1S_0 should be nearly orthogonal. This indeed agrees with the observation

$$\frac{|< 2^3S_1|1^1S_0 >|^2}{|< 1^3S_1|1^1S_0 >|^2} < 1/8$$

Naively we would expect $2^1S_0 \rightarrow 1^3S_1 + \gamma$ to be negligible contradicting the fact that $\chi(3.453) \rightarrow \gamma\, ^3S_1$ is the only decay mode observed.

From the observed branching ratio for the cascade $B_1 \cdot B_2$ = $(0.8 \pm 4)\%$ and the upper limit $B_1 < 2.5\%$ we find $B(\chi(3.45)$ $\gamma J/\psi) > 0.25$.
Furthermore, using this limit and the relation

$$\Gamma(2^1S_0 \rightarrow 1^3S_1 + \gamma) \lesssim (2^3S_1 \rightarrow 1^1S_0 + \gamma) < 2.8 \text{ keV}$$

we find

$$\Gamma(2^1S_0 \rightarrow \text{all}) < 10 \text{ keV}.$$

This is, even allowing for a factor of 10 uncertainty, in strong disagreement with the standard short range picture which predicts a width of about 2 MeV for the 2^3S_0.

It is important to note that the level assignments are based on the assumption that there are only four levels with even charge conjugation between the J/ψ and the ψ'. It has recently been pointed [73] out that the large singlet-triplet splitting conjectured by identifying 1^1S_0 with the χ(2.8) could also push the $^1D_2(J^{PC} = 2^{-+})$ level below the ψ'. Then five levels with even charge conjugation are expected between the J/ψ and the ψ' and a complete assignment of the levels in a $Q\bar{Q}$ model is no longer possible. It has also been suggested that the χ(3.45) is a more exotic [65,74] resonance. It is clear that more data are needed.

Summary

The observed level scheme is shown in Fig. 6.2.10 and the various decay modes are listed in table 6.4. The number of levels and their ordering is that expected for a pair of fermions bound in a steeply rising potential. The fine structure splitting of the P-levels i.e. $^3P_2 > {}^3P_1 > {}^3P_0$ is that expected for a vector force. A pure scalar potential would reverse the order and give a level assignement in contradiction to the expe-

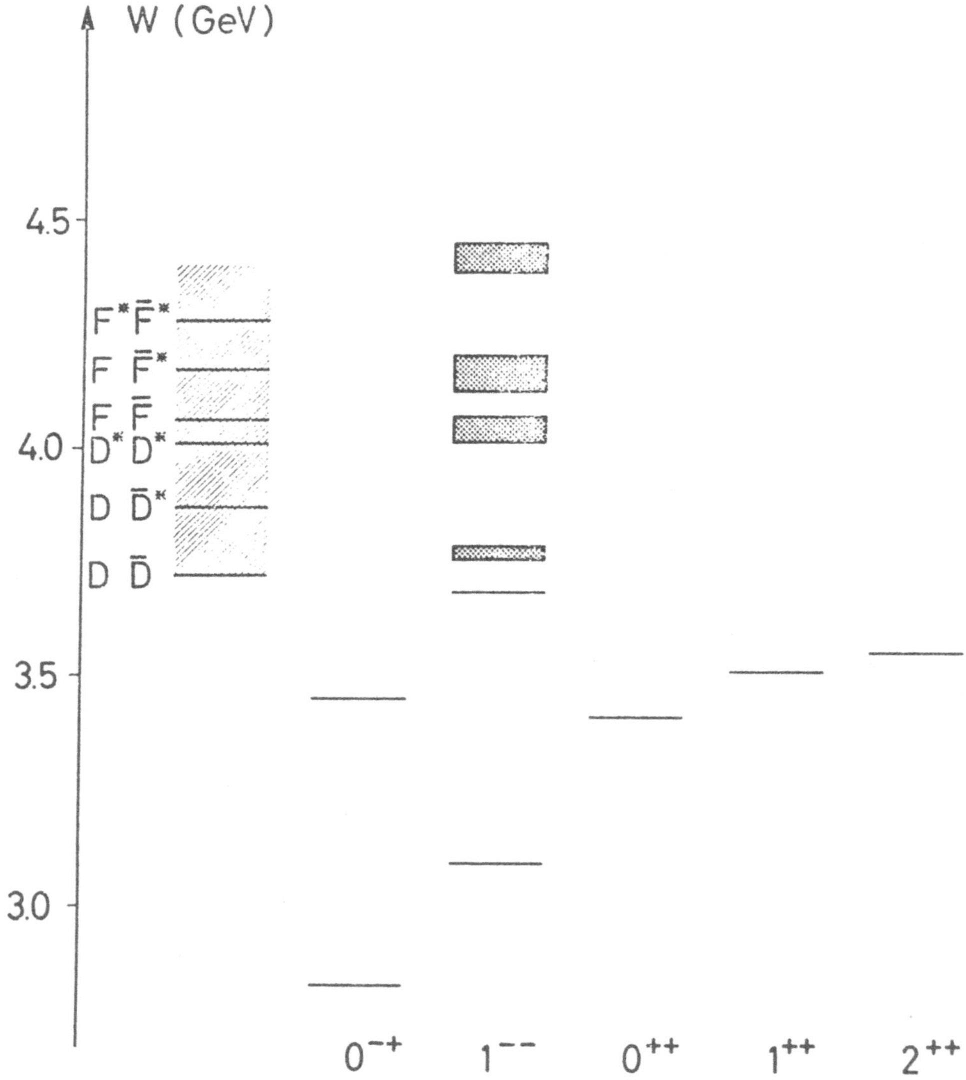

FIG. 6.2.10. : Observed charmonium levels. The thresholds for the production of mesons with open charm are indicated.

Table 6.4 - Properties of the Intermediate States.

Mass (MeV/c^2)	Decay	$BR(\psi' \to \gamma P_c/\chi)$. $BR(P_c/\chi$ decay) %	Reference	$BR(P_c/\chi$ decay)$^{1)}$ %
3413 ± 5	$\pi^+\pi^-$	0.07 ± 0.02	71	1.0 ± 0.3
	K^+K^-	0.07 ± 0.02	71	1.0 ± 0.3
	$2\pi^+2\pi^-$	0.32 ± 0.06	71	4.7 ± 0.9
	$\pi^+\pi^-K^+K^-$	0.27 ± 0.07	71	3.9 ± 1.0
	$p\bar{p}\pi^+\pi^-$	0.04 ± 0.013	71	0.6 ± 0.2
	$3\pi^+3\pi^-$	0.14 ± 0.05	71	2.0 ± 0.7
	$\gamma J/\psi$	0.22 ± 0.08	a	3.2 ± 1.2
	$\gamma\gamma$	< 0.014	58	< 0.2
	$\rho^0\pi^+\pi^-$	0.12 ± 0.04	71	1.8 ± 0.6
	$K^{0*}K^+\pi^-$	0.17 ± 0.06	71	2.5 ± 0.9
3508 ± 4	$\pi^+\pi^- + K^+K^-$	< 0.015	71	< 0.21
	$2\pi^+2\pi^-$	0.11 ± 0.04	71	1.5 ± 0.6
	$\pi^+\pi^-K^+K^-$	0.06 ± 0.03	71	0.85 ± 0.42
	$p\bar{p}\pi^+\pi^-$	0.01 ± 0.008	71	0.14 ± 0.11
	$3\pi^+3\pi^-$	0.17 ± 0.06	71	2.4 ± 0.8
	$\gamma J/\psi$	2.4 ± 0.3	a	34 ± 4
	$\gamma\gamma$	< 0.013	58	< 0.18
	$\rho^0\pi^+\pi^-$	0.026 ± 0.022	71	0.37 ± 0.31
	$K^{0*}K^+\pi^-$	0.31 ± 0.022	71	0.44 ± 0.31
3552 ± 6	$\pi^+\pi^- + K^+K^-$	0.02 ± 0.01	71	0.29 ± 0.14
	$2\pi^+2\pi^-$	0.16 ± 0.04	71	2.3 ± 0.6
	$\pi^+\pi^-K^+K^-$	0.14 ± 0.04	71	2.0 ± 0.6
	$p\bar{p}\pi^+\pi^-$	0.02 ± 0.01	71	0.29 ± 0.14
	$3\pi^+3\pi^-$	0.08 ± 0.05	71	1.1 ± 0.7
	$\gamma J/\psi$	1.9 ± 0.2	a	27 ± 3
	$\gamma\gamma$	< 0.004	58	< 0.06
	$\rho^0\pi^+\pi^-$	0.05 ± 0.03	71	0.71 ± 0.43
	$K^{0*}K^+\pi^-$	0.052 ± 0.031	71	0.74 ± 0.44

1) The values listed in table 6.1 for $\psi' \to \gamma P_c/\chi$ were used to extract the branching ratios.
a) The average values from table 6.2 were used.

---

riments. The level splittings are larger than predicted but the radiative rates are smaller by a factor of two. Indirect evidence shows that the ratio of the hadronic widths of the P states are in agreement with predictions based on QCD.

The $\chi(2.83)$ can be identified with the 1^1S_0 although there is a fair number of problems. Assigning the $\chi(3.45)$ to the 2^1S_0 state leads to disaster.

7. Mesons with Open Charm.

The existence of a fourth quark besides the familiar u, d and s quarks implies that the SU(3) nonet of 8 + 1 mesons will be replaced by a hexadecimet of 15 + 1 states as shown in Fig. 7.1 [75]. Besides ordinary SU(3) resonances with C = 0 the hexadecimet contains six states with open charm C = ± 1 and one $c\bar{c}$ state with hidden charm. The quark content and the names given to the charmed pseudoscalar mesons which are the lowest states with open charm, are listed below :

$$C = +1 \quad D^+ = c\bar{d} \quad , \quad D^o = c\bar{u} \quad , \quad F^+ = c\bar{s}$$

$$C = \;\;0 \quad \eta_c = 1^1S_o = c\bar{c}$$

$$C = -1 \quad D^- = \bar{c}d \quad , \quad \bar{D}^o = \bar{c}u \quad , \quad F^- = \bar{c}s$$

Higher mass charmed mesons will cascade by strong and/or electromagnetic decays into these states, which then decay weakly with the following decay modes :

$$D(F) \rightarrow \ell\,\bar{\nu}_\ell$$
$$\rightarrow \ell\,\bar{\nu}_\ell + \text{hadrons}$$
$$\rightarrow \text{hadrons.}$$

The leptonic decay mode is suppressed by kinematics ($J_3^D = 0 \neq J_3^e + J_3^{\bar{\nu}} = 1$). Semileptonic decays are predicted [75,76] with a branching ratio on the order of 20% and most decays (∿80%) will therefore yield only hadrons in the final state. Any new flavour will produce mixed lepton-hadron final states and show up as narrow resonances above flavour threshold. However, the flavour can be identified by the properties of the final states. The GIM mechanism (Fig. 7.2) leads to the following Cabibbo favoured decay modes :

$$D^o \rightarrow \begin{array}{l}(\ell^+ \nu_\ell)(K^- \ldots)\\ (\bar{K}\; n\pi)^o\end{array}$$

$$D^+ \rightarrow \begin{array}{l}(\ell^+ \nu_e)(\bar{K}^o \ldots)\\ (\bar{K}\; n\pi)^+\end{array}$$

$$F^+ \rightarrow \begin{array}{l}(\ell^+ \nu_\ell)(\eta,\eta', K^+K^- \ldots)\\ (\eta n\pi)^+,\ (\eta' n\pi)^+,\ (K\bar{K}n\pi)^+\end{array}$$

According to the GIM mechanism associated production of D mesons will show up as an increase in the yield of kaons, a strong correlation between leptons and kaons in multiprong final states and appearant exotic decays like $D^+ \rightarrow K^-\pi^+\pi^+$. The production of $F^\pm$ mesons should show up as an increase in the yield of particles with a large $s\bar{s}$ component in the wave function like η, η' and φ-mesons or they might predominantly decay into $K\bar{K}$ pairs.

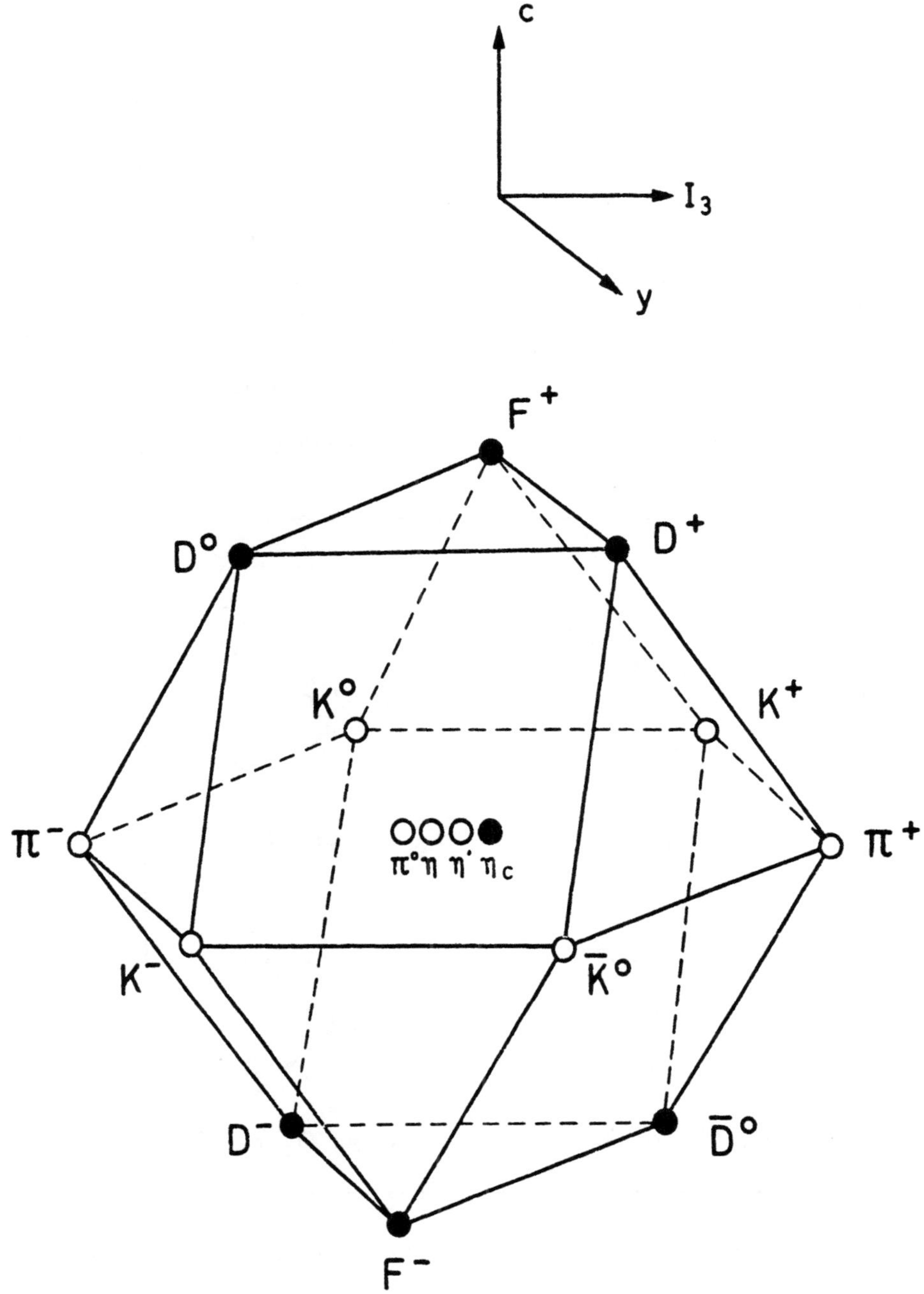

FIG. 7.1. : The hexadecimet of the pseudoscalar meson. Charm is plotted along the z-axis, Y and I_3 along respectively the y-axis and the x-axis. The π°, η and η' mesons are denoted by the open circles at the origin η_c by the black circles.

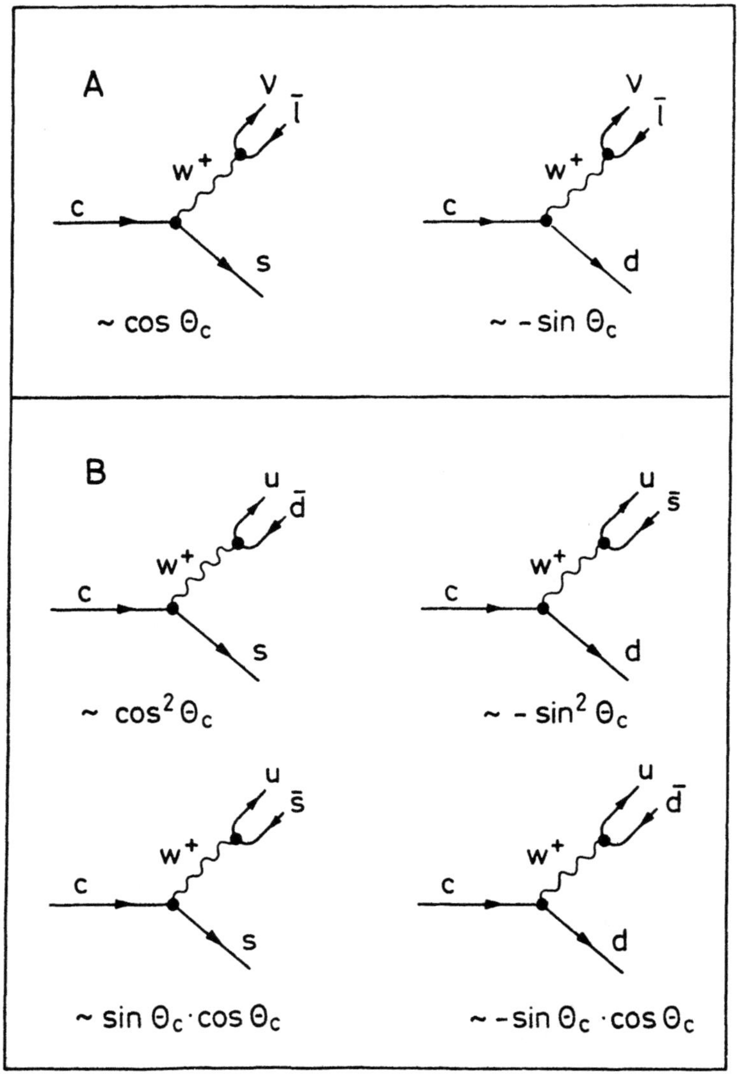

FIG. 7.2. : A) Schematic diagrams for the semileptonic decay of the charmed quark.
B) Schematic diagrams for the decay of the charmed quark into non-charmed quarks.

A strong increase in the yield of charged and neutral kaons related to the step in the total cross section at 4.0 GeV is indeed observed. This is shown, in Fig. 7.3 [77] where the ratios

$$\frac{\sigma(e^+e^- \to K^\pm X)}{\sigma(e^+e^- \to \mu^+\mu^-)}$$

measured [78] by the DASP group and

$$\frac{2\sigma(e^+e^- \to K^o_S X)}{\sigma(e^+e^- \to \mu^+\mu^-)}$$

measured by the PLUTO [79] and the SLAC-LBL [80] Collaborations are plotted as a function of cm energy. The data are in rough agreement and the energy dependence is in qualitative agreement with the energy dependence observed for the total cross section. Note that the SLAC-LBL group in general observe a higher cross section than the PLUTO group. This might be caused by losses occuring during data reduction plus the loss due to the cut off in K^o_S momentum. The SLAC-LBL groups applies a 20% correction for these effects whereas the PLUTO data have yet to be corrected. The peak in the total cross section at 4.415 GeV is only reflected in the SLAC-LBL data whereas there is no clear evidence for a peak at this mass in the DESY data. This might be due to a lack of statistics at the exact mass of the resonance.

A simple kinematical observation can be used to show that the increase in K-production is indeed due to charm production. At threshold the charmed mesons are produced at rest, hence kaons resulting from their decays must have energies less than one half of the beam energy. This is shown in Fig. 7.4, where the invariant cross section [78] $(E/4\pi p^2)\, d\sigma/dp$ is plotted as a function of kaon energy for cm energies between 3.6 GeV and 4.05 GeV. The step in kaon yield is indeed caused by particles with energy less than one half of the beam energy.

If the decay of each charmed hadron would yield either a charged or a neutral kaon then $\Delta R_{K^o} + \Delta R_{K^\pm} = 2\Delta R_C$. The step in the K-yield occuring between 3.6 GeV and 5.0 GeV in cm is around 3.3 (Fig. 7.3) compared to the value $2\Delta R_C = 4.2 \pm 0.6$ reported by the DASP group. Thus nearly every charmed hadron decays into final states containing a kaon as predicted by the GIM mechanism.

A strong correlation between electrons and neutral kaons has been observed [81] by the PLUTO group. The DASP Collaboration [58,82] finds a correlation between charged kaons and electrons in multiprong final states.

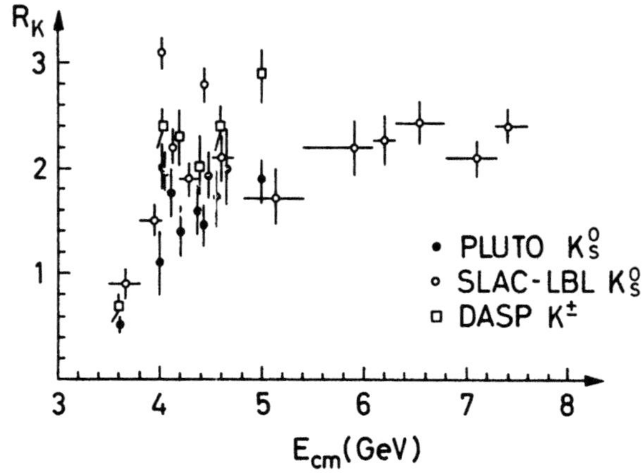

FIG. 7.3. : Inclusive K_S^0 and K production as a function of cm energy. Plotted is

$$R_{K_S^0} = \frac{2\sigma(e^+e^- \to K_S^0 X)}{\sigma(e^+e^- \to \mu^+\mu^-)} \quad \text{and} \quad R_{K^\pm} = \frac{\sigma(e^+e^- \to K^\pm X)}{\sigma(e^+e^- \to \mu^+\mu^-)}$$

as a function of cm energy.

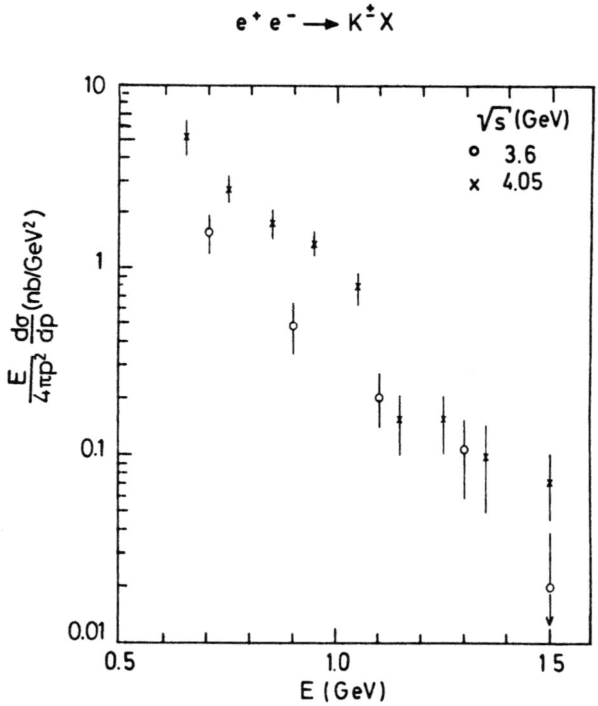

FIG. 7.4. : The invariant cross sections $E/(4\pi p^2)\, d\sigma/dp$ for the sum of K^+ and K^- production at cm energies of 3.6GeV and 4.05GeV.

Production of F-mesons are expected to show up as an increase in the yield of η's. An enhanced cross section for η production is observed [58] around 4.414 GeV and 4.15 GeV.

The SLAC-LBL group has measured [83] the yield of antiprotons and Λ's as a function of cm energy between 3.7-7.6 GeV. The data plotted in Fig. 7.5 versus energy, show the yield of antiprotons to rise rapidly between 4.5 - 5.0 GeV. This is consistent with the onset of charmed baryon production expected [84] to take place in this energy range.

7.1. Discovery of Charged and Neutral D Mesons.

The SLAC-LBL group found the first evidence [8] for charmed mesons in a sample of 29000 multihadron events collected between 3.9 GeV and 4.6 GeV and a sample of 19000 multihadron events at 4.03 GeV. The following decay modes were observed

$$D^\circ \to K^+\pi^-,\ K^+\pi^-\pi^+\pi^-$$

$$D^+ \to K^-\pi^+\pi^+$$

To observe the decay $D^\circ \to K^+\pi^-$ an effective mass for all neutral combinations of two charged tracks in 29000 multihadron events were evaluated assuming the tracks to be pions or kaons in turn. The results [8], plotted in Fig. 7.1.1a,b and c show peaks at 1.74 GeV/c^2 ($\pi\pi$), 1.87 GeV/c^2 ($K\pi$) and 1.98 GeV/c^2(KK).

These values are consistent with a single resonance decaying into one channel only with the peak in the other channels caused by assigning the wrong mass to one or both of the tracks. Better particle assignments can be obtained by using time of flight information. The time elapsed between the crossing of the e^+e^- bunches and the arrival of the produced particle at the outer ring of scintillation counters was measured with an rms resolution of 0.4 nsec. The mass of the particle is then determined from its flight time and momentum.

The results are shown in a scatter plot Fig. 7.1.2 with the square of the particle mass plotted versus momentum. The resolution obtained is not sufficient to identify pions and kaons from $D^\circ \to K^+\pi^-$ uniquely since these particles have momenta around 800 MeV/c. The following procedure was therefore adopted :
Each track was assigned a probability to be a kaon or a pion

$$P_1(\pi) \sim \exp\left(-\frac{(\tau - \tau_\pi)^2}{2\sigma^2}\right)$$

$$P_1(K) \sim \exp\left(-\frac{(\tau - \tau_K)^2}{2\sigma^2}\right)$$

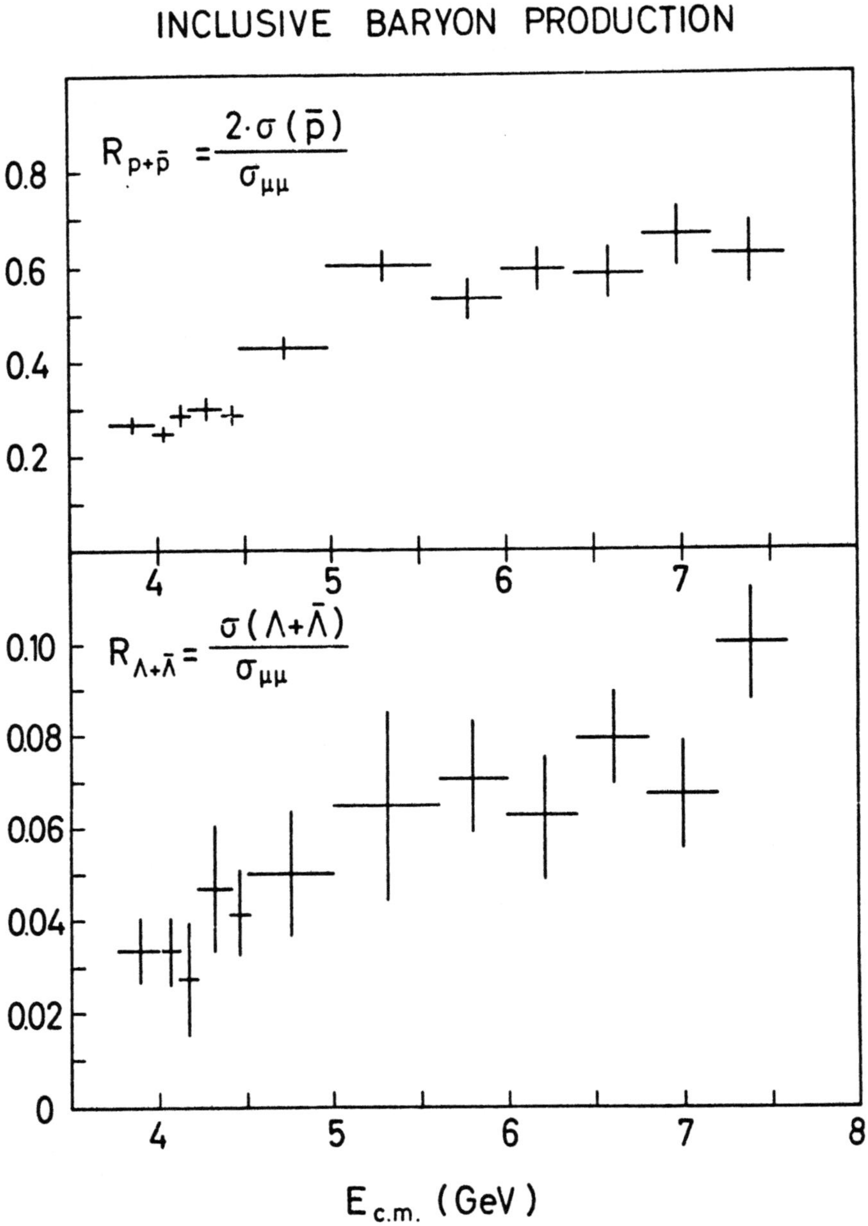

FIG. 7.5. : a) $R(p+\bar{p}) = 2\sigma(\bar{p})/\sigma_{\mu\mu}$ versus c.m. energy.
b) $R(\Lambda+\bar{\Lambda}) = \sigma(\Lambda+\bar{\Lambda})/\sigma_{\mu\mu}$ versus c.m. energy.

Here τ is the measured and τ_π or τ_K the computed time of flight assuming the track to be a pion or a kaon. Tracks with $P_1(\pi) + P_1(K) < 10^{-2}$ were rejected. Such events are presunably caused either by more than one track hitting the scintillation counter or by a proton.

The relative π-K probabilities were normalized to 1 and the weight of the two particle combinations computed as follows :

$$(P_1(\pi) + P_1(K)) \cdot (P_2(\pi) + P_2(K))$$

$$= \underbrace{P_1(\pi) \cdot P_2(\pi)}_{\pi\pi} + \underbrace{P_1(\pi) \cdot P_2(K) + P_1(K) \cdot P_2(\pi)}_{K\pi} + \underbrace{P_1(K) \cdot P_2(K)}_{KK} = 1$$

The effective mass spectra weighted in this manner are plotted in Fig. 7.1.1d, e and f. It is clear that the $K\pi$ assignment is preferred over either the $\pi\pi$ or the KK combination. In fact, the authors estimate the confidence level for the signal to arise only from $\pi^+\pi^-$ or K^+K^- to be less than 1%. They hence assume the signal results from the $K\pi$ channel only.

Evidence for the decay of the same state into $K^+\pi^-\pi^+\pi^-$ is presented in Fig. 7.1.1g, h, i. The same weighing technique was used. A clear signal is seen in the neutral K 3π system with no signal either in the $\pi^+\pi^-\pi^+\pi^-$ or the $K^+K^-\pi^+\pi^-$ combinations.

The masses and the widths of the peaks observed in the $K\pi$ and K 3π spectra are both consistent with resulting from the decay of a single resonance with a mass of (1865 ± 5) MeV/c^2. The observed width is consistent with the expected resolution.

The lowest resonance with a new quantum number must be produced in conjunction with another particle of at least the same mass. The production of new particles will therefore have a threshold at 3.73 GeV and correspondingly there should be no peaks in the recoil spectrum below 1.865 GeV/c^2.

The threshold behaviour was investigated by examining the data obtained at the ψ' and the J/ψ resonance. Here a total of 70.000 events of the type $J/\psi \rightarrow \gamma \rightarrow$ hadrons and 10000 events of the type $\psi' \rightarrow \gamma \rightarrow$ hadrons were observed. These decays will lead to final states identical to those produced by the decay of timelike photons of the same mass. No significant structure was observed either in the $K\pi$ or in the K 3π final state.

The mass spectrum recoiling against the $(K^\pm\pi^\mp)$ resonance produced at a cms energy of 4.03 GeV is shown in Fig. 7.1.3. Note the absence of events with recoil masses below 1.8 GeV and the two sharp peaks at 2.02 GeV/c^2 and around 2.15 GeV/c^2.

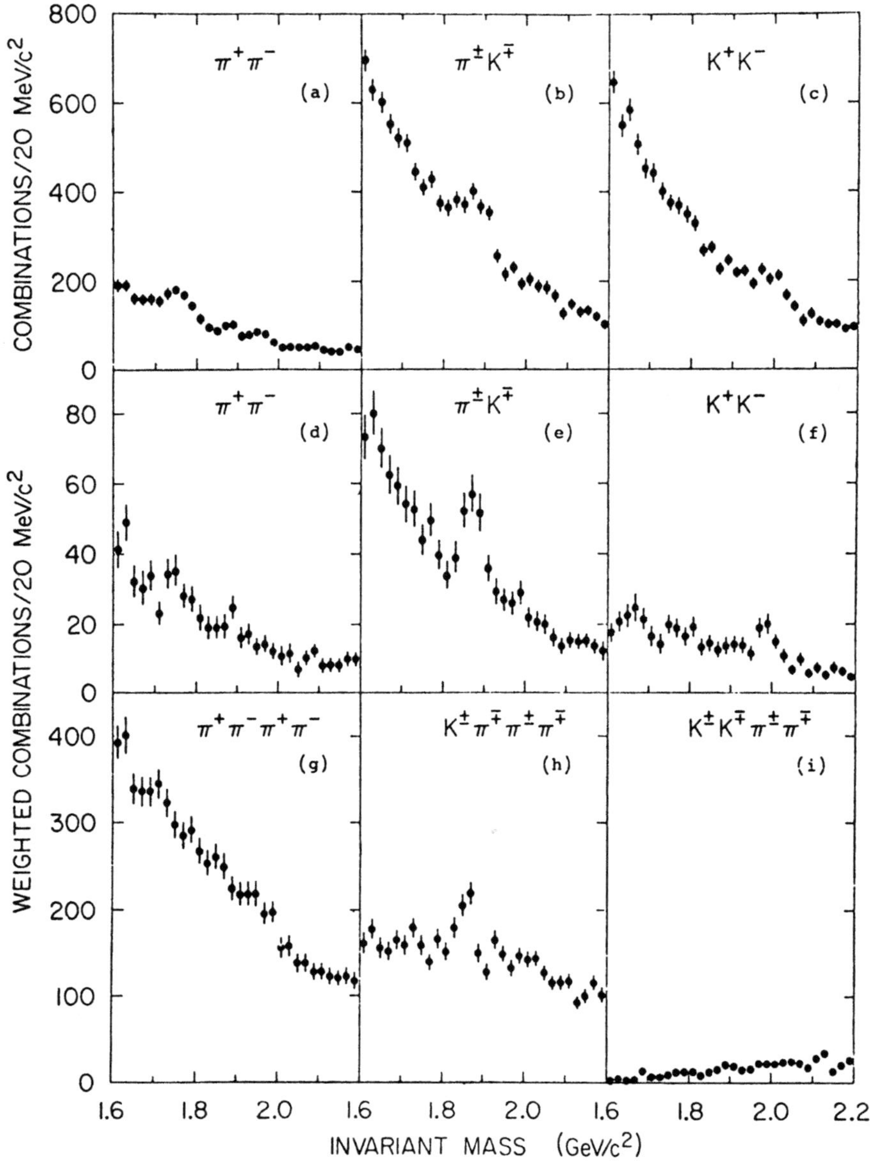

FIG. 7.1.1. : Effective mass spectra of neutral combinations of two and four charged particles.

a) $\pi^+\pi^-$ assigning pion mass to all tracks.
b) $K^\mp\pi^\pm$ assigning pion and kaon mass to all tracks.
c) K^+K^- assigning kaon mass to all tracks.
d) $\pi^+\pi^-$ weighted by the $\pi\pi$T. of F probability.
e) $K^\mp\pi^\pm$ weighted by the KπT. of F. probability.
f) K^+K^- weighted by the KKT. of F. probability.
g) $\pi^+\pi^-\pi^+\pi^-$ weighted by the $\pi\pi\pi\pi$T. of F. probability.
h) $K^+\pi^-\pi^+\pi^-$ weighted by the K$\pi\pi\pi$T. of F. probability.
i) $K^+K^-\pi^+\pi^-$ weighted by the KK$\pi\pi$T. of F. probability.

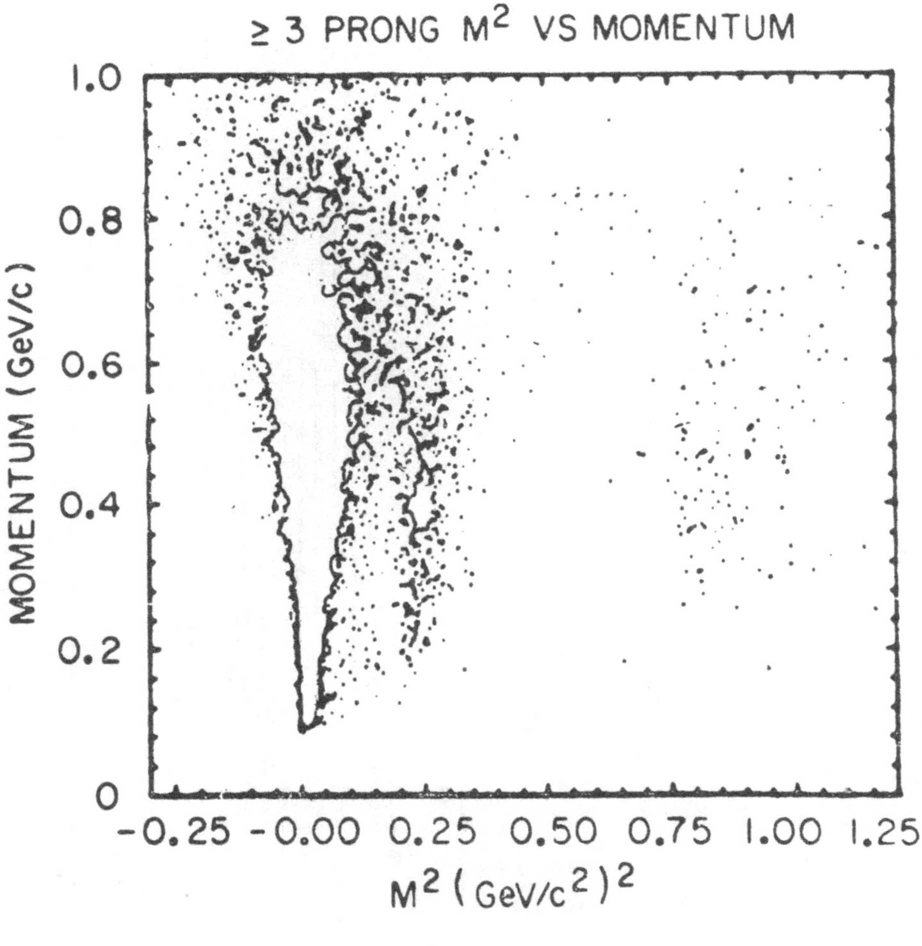

FIG. 7.1.2. : Particle mass squared derived from the time of flight as a function of particle momentum.

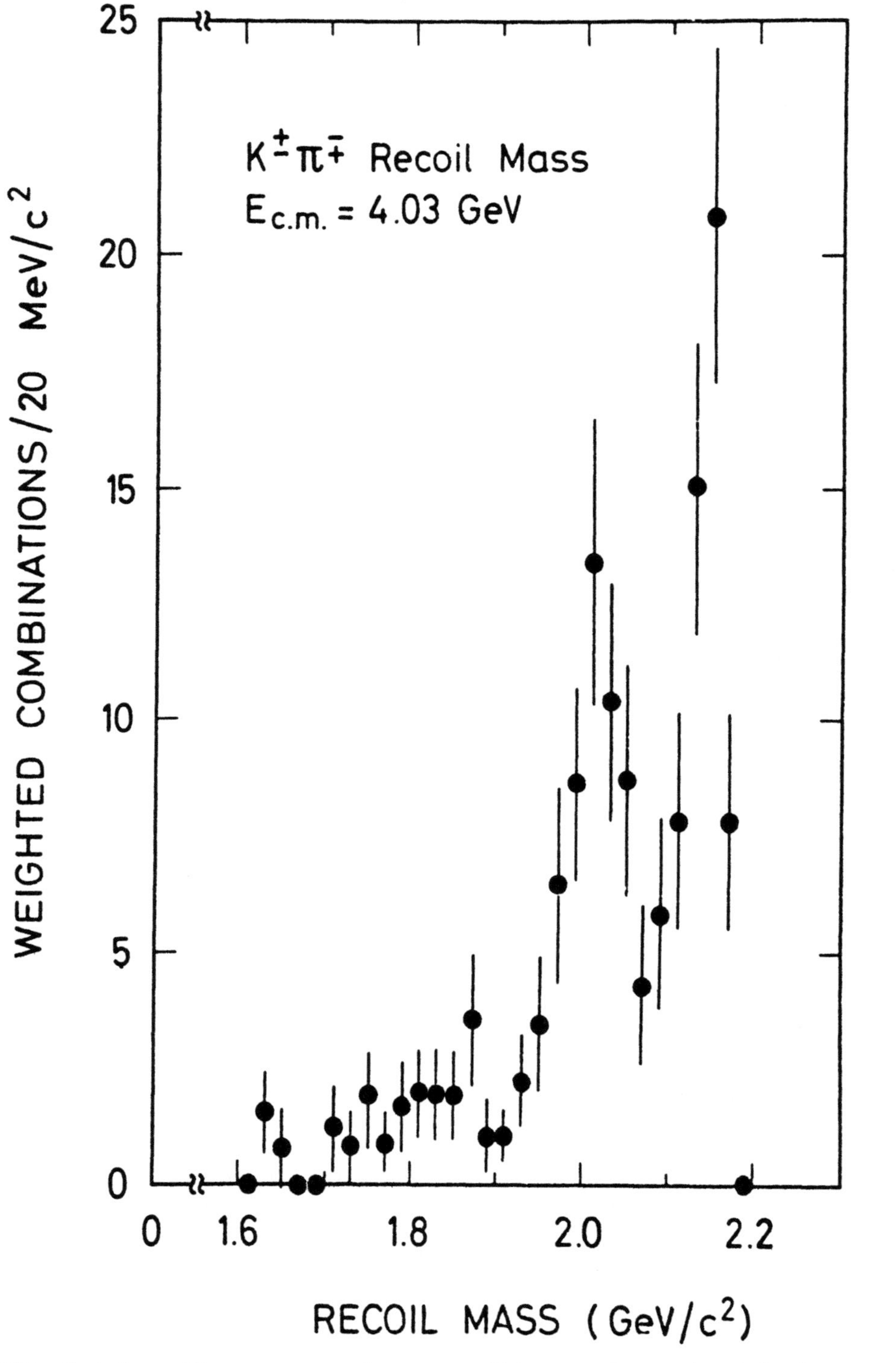

FIG. 7.1.3. : The mass spectrum recoiling against the $(K^+\pi^-)$ resonance produced at 4.03 GeV. The K system was weighted by the T. of F. probability.

This, and the appearant threshold behaviour, shows that the $K^{\pm}\pi^{\mp}$ resonance is only produced in conjunction with another heavy resonance, as predicted for a particle with a new quantum number. If energetically allowed the high mass state will cascade into the ground state by pion and/or photon emission. The sharp peak observed at 2.15 GeV/c^2 is caused by a kinematic reflection from such a cascade decay. Consider the reaction $e^+e^- \rightarrow D^{*o}\bar{D}^{*o}$ at cms energy of 4.03 GeV. Since the π^o's are nearly at rest they will, when combined with the unobserved $\bar{D}^o$ lead to a sharp structure in the mass spectrum of states recoiling against the observed D^o. This structure will appear at different masses for different cm energies.

An even more striking confirmation of the charmed quark is the observation [8] of the decay

$$D^{\mp} \rightarrow K^{\pm}\pi^{\mp}\pi^{\mp}.$$

If this final state resulted from an isospin conserving strong decay, then the initial state will have isospin 3/2 or 5/2. Such states, which cannot be described as $q\bar{q}$ pairs, are called exotic and have so far not been observed. On the other hand the GIM mechanism predicts the D^+ state to favour appearant exotic decay modes. The effective mass spectrum of $K^-\pi^+\pi^+(K^+\pi^-\pi^-)$ based on 19000 multihadron events at 4.03 GeV is plotted in Fig. 7.1.4. Pions and kaons were identified as discussed above and the mass spectrum weighted accordingly. A peak centered at (1876±10)MeV/c^2 is seen, the width is consistent with the expected resolution. There is no evidence for a resonance in the non exotic channel $K^-\pi^+\pi^-(K^+\pi^+\pi^-)$ plotted in Fig. 7.1.5. The mass-spectrum of states recoiling against the exotic resonance is plotted in Fig. 7.1.6. Only a resonance with a mass around 2.01 GeV/c^2(D^{*-}) is seen clearly.

Based on the experimental evidence discussed above it seems natural to assign the neutral resonance to the $D^o(c\bar{u})$ and the charged resonance to $D^+(c\bar{d})$.

7.2. Discovery of the F Meson.

The model also predicts that there should exist mesons with both charm and strangeness made by combining a strange quark with a charmed quark. The DASP group has reported [9,58] evidence for such a resonance.

The lowest member of the F-family should decay predominantly into a $s\bar{s}$ system leading to final states containing $K\bar{K}$, ϕ, η or η's. Copious η production (note that the η' decays yield an η 68% of the time) is therefore a strong hint of $F\bar{F}$ production.

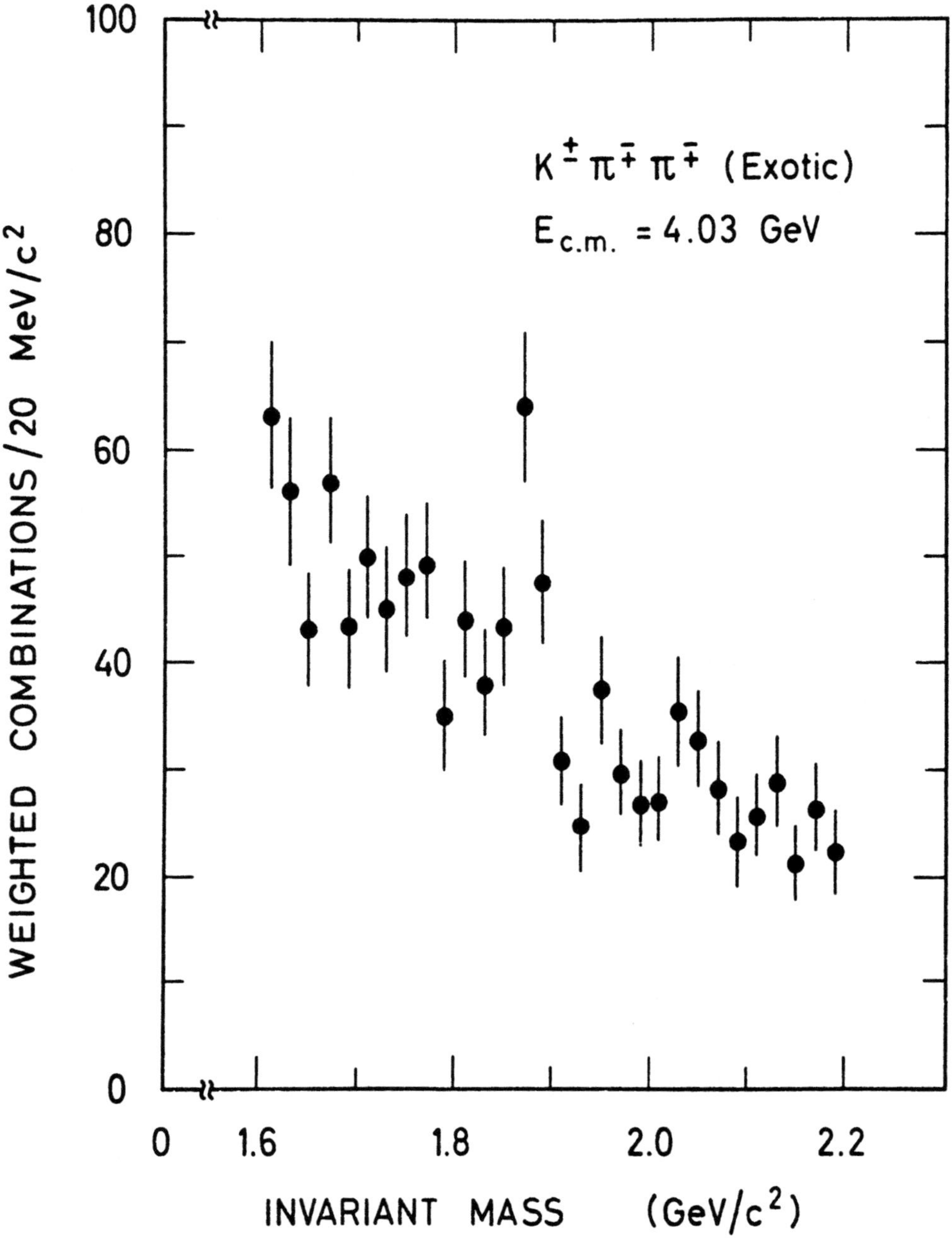

FIG. 7.1.4. : The effective mass spectrum produced at 4.03 GeV. The spectrum was weighted by the T. of F. probability.

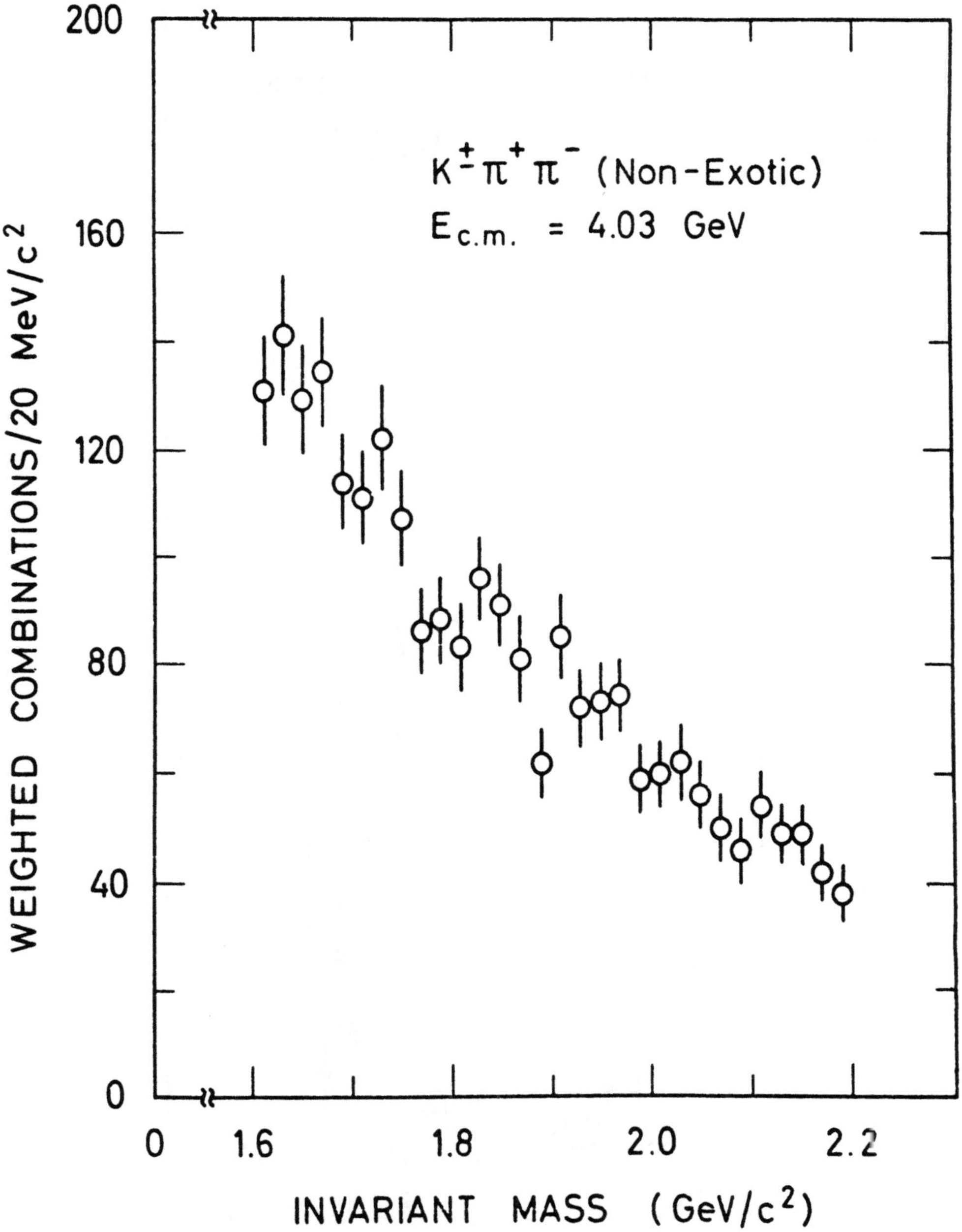

FIG. 7.1.5. : The effective mass spectrum of $K^{\pm}\pi^{+}\pi^{-}$ producec at 4.03 GeV. The spectrum was weighted by the T. of F. probability.

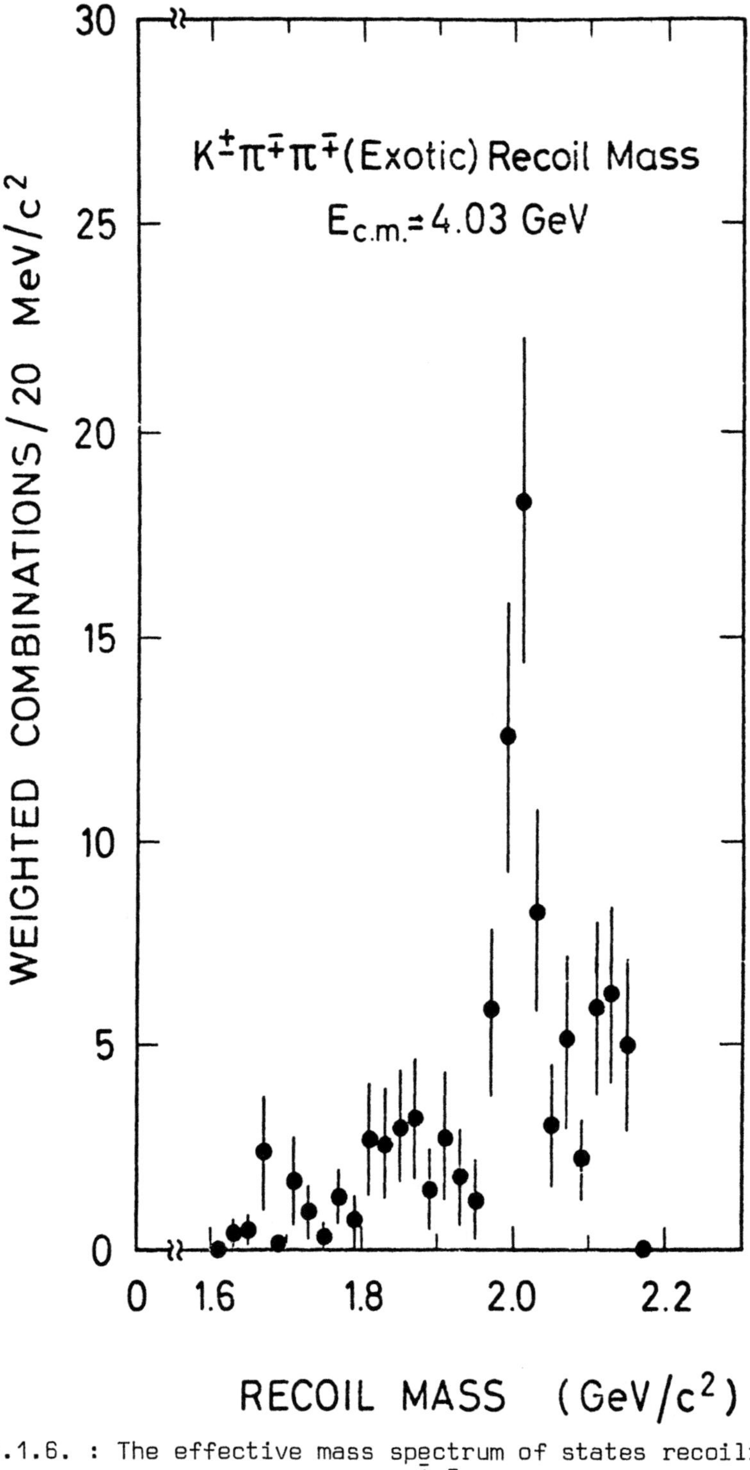

FIG. 7.1.6. : The effective mass spectrum of states recoiling against the exotic $K\pi^{\mp}\pi^{\mp}$ resonance produced at 4.03 GeV.

It is known from the SLAC-LBL experiment that the D^*-D mass difference is about one pion mass and that $D\bar{D}^*$, $D^*\bar{D}^*$ production is enhanced close to their thresholds . Assuming a similar behaviour for the F-mesons, one expects large $F\bar{F}^*$ and $F^*\bar{F}^*$ cross sections near threshold. The F^* is an isospin singlet and can therefore only decay electromagnetically via photon or π^0 emission (provided $m_{F^*} - m_F < 2m_\pi$) into the F. To enhance the signal a low energy photon was therefore required in addition to the η. The η was identified by its decay into two photons.

The measurement was carried out using the DASP inner detector. Events with two charged tracks and more than two but less than six photons were selected from the multihadron sample. At least two photons must have energies above 0/1 GeV. The vector sum of the momenta of these photons must be larger than 0.3 GeV/c but less than 1.2 GeV/c. All such photons are paired and the mass computed. The photon pairmass distribution of events containing a low energy photon is plotted in Fig. 7.2.1 for all energies except 4.4 GeV and in Fig. 7.2.2 for 4.4 GeV. The solid line represents a smooth estimate of the uncorrelated background obtained by forming the invariant mass of photons from different events having the same number of photons. The background spectrum is normalized to the data by requiring all combinations with $m_{\gamma\gamma} > 0.7$ GeV/c^2 to be ascribed to background. There is a clear η signal at 4.4 GeV, not observed at other energies. The η produced with a low energy photon at 4.4 GeV is strongly suggestive of $e^+e^- \to F\bar{F}^*(F^*\bar{F}^*)$ production at this energy.

A search was made for the GIM favoured decay mode $F \to \pi\eta$ by selecting events with the following topology :

$$e^+e^- \to \pi^\pm + \eta(\gamma\gamma) + \gamma_{low} + \text{anything}$$

The pion was measured and identified in one of the spectrometer arms and it was required to have a momentum above 0.6 GeV/c. The mass of the $\gamma\gamma$ system had to be between 0.35 GeV/c^2 and 0.65 GeV/c^2 and the low energy photon had to be below 0.2 GeV. A total of 35 candidates satisfied these criteria. The events were fitted to the reaction

1) $$e^+e^- \to F^\pm F^{*\mp} \to (\pi^\pm\eta)(F^\mp\ \gamma_{low})$$

2) $$e^+e^- \to F^{*\pm} F^{*\mp} \to (\pi^\pm\eta\gamma_{low})(X)$$

These are 2C fits because of the mass constraint $m_{\gamma\gamma}$ and the requirement that for (1) $\pi\eta$ and the missing vector must have the same mass m_F; for (2) the $\gamma_{low}F^\pm$ system and the missing vector must have the same mass m_{F^*}. A total of 25 events satisfy the criteria for (1) with a χ^2 of less than 8. These events are

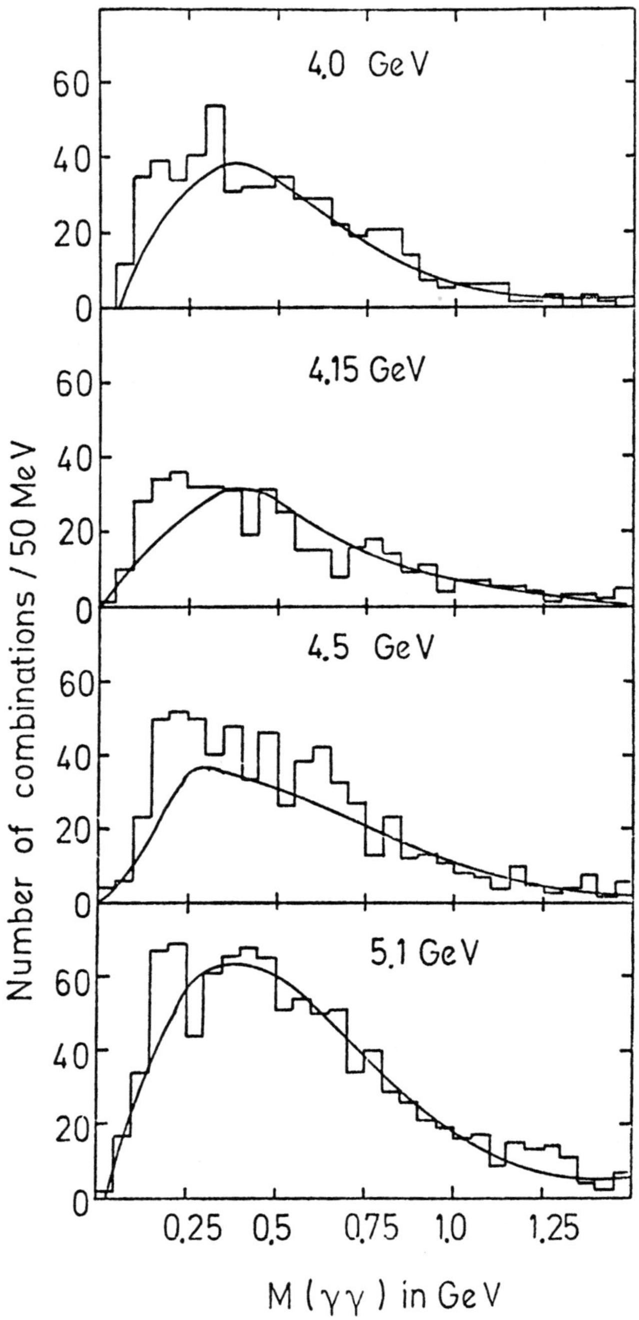

FIG. 7.2.1. : Distribution of $m_{\gamma\gamma}$ for events with a low energy photon for various values of the cm energy. The curves represent the background estimated from uncorrelated photon pairs. The background is normalized to the number of events with $m_{\gamma\gamma} > 0.7\ \mathrm{GeV/c^2}$.

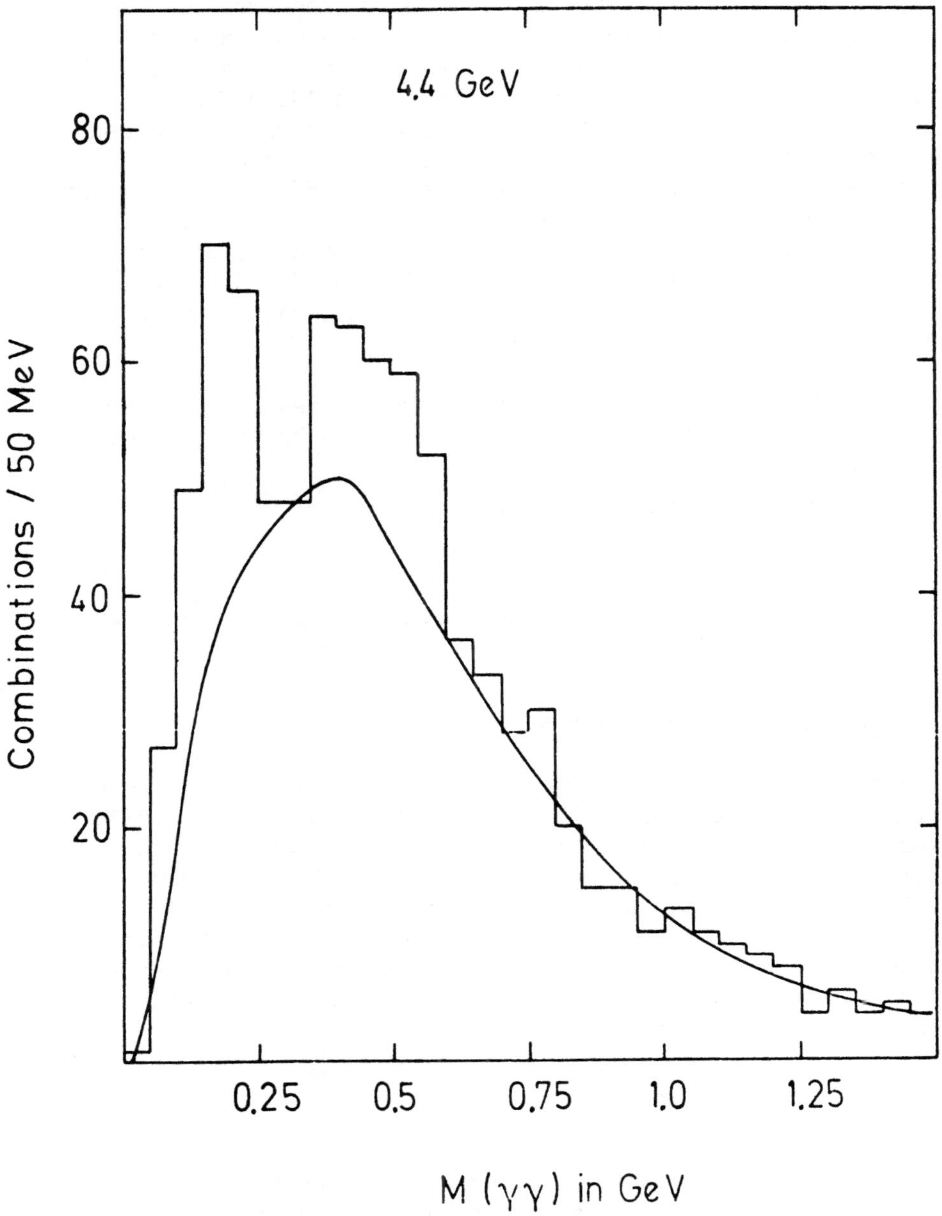

FIG. 7.2.2. : Same as above for 4.4 GeV in cm.

plotted in Fig. 7.2.3A versus Δ, the difference between the $\eta\pi$ mass computed directly from the measured momenta and directions and the value found in the fit. The events peak at $\Delta = 0$ and the 21 events with $|\Delta| < 0.25$ GeV/c^2 are retained. The events are shown in a scatter plot (Fig. 7.2.3B) of the mass difference between the recoil mass and the $\eta\pi$ mass versus the $\eta\pi$ mass. The data taken at 4.4 GeV in cm have a cluster of 6 events with a well defined $\pi\eta$ mass and a recoil mass some 100 MeV higher. This is evidence for a narrow $\pi\eta$ resonance produced in conjunction with a resonance about 100 MeV higher. It is natural to associate these resonances with the F and the F*. The background can be estimated from a similar plot (Fig. 7.2.3C) for cm energies outside the 4.414 resonance. In these data no event with a $\eta\pi$ mass compatible with the signal at 4.4 GeV was found for an integrated luminosity 5 times larger than the integrated luminosity at 4.4 GeV.

A fit to hypothesis (2) leads to similar results. An average over the values obtained by the two fits leads to the following values for the mass :

$$M_F = (2.03 \pm 0.06)\ \text{GeV}$$

$$M_{F^*} = (2.14 \pm 0.06)\ \text{GeV}$$

From the energy distribution of the low energy photon alone we find $m_{F^*} - m_F = (0.12 \pm 0.04)$ GeV.

7.3. Properties of the D Mesons.

The total width of the $\psi''(3772)$ resonance is roughly three orders of magnitude larger than the total width of the $\psi'(3684)$ although they differ only by (88 ± 3) MeV/c^2 in mass. This large difference in width is presumably due to the OZI-allowed decay $\psi''(3772) \to D\bar{D}$ forbidden by energy conservation for the ψ'. Furthermore since the resonance very likely has isospin 0 or 1 it will decay, apart from a small phase space correction, with equal widths into pairs of charged and neutral D mesons. Correcting for phase space gives :

$$\frac{\Gamma(\psi'' \to D^\circ\bar{D}^\circ)}{\Gamma(\psi'' \to \text{all})} = 0.56 \pm 0.03$$

The production cross sections for charged and neutral D mesons at the $\psi''(3772)$ resonance are therefore known and this makes it possible to determine absolute branching ratios.
The SLAC-LBL group has identified the following decay modes [85-87]

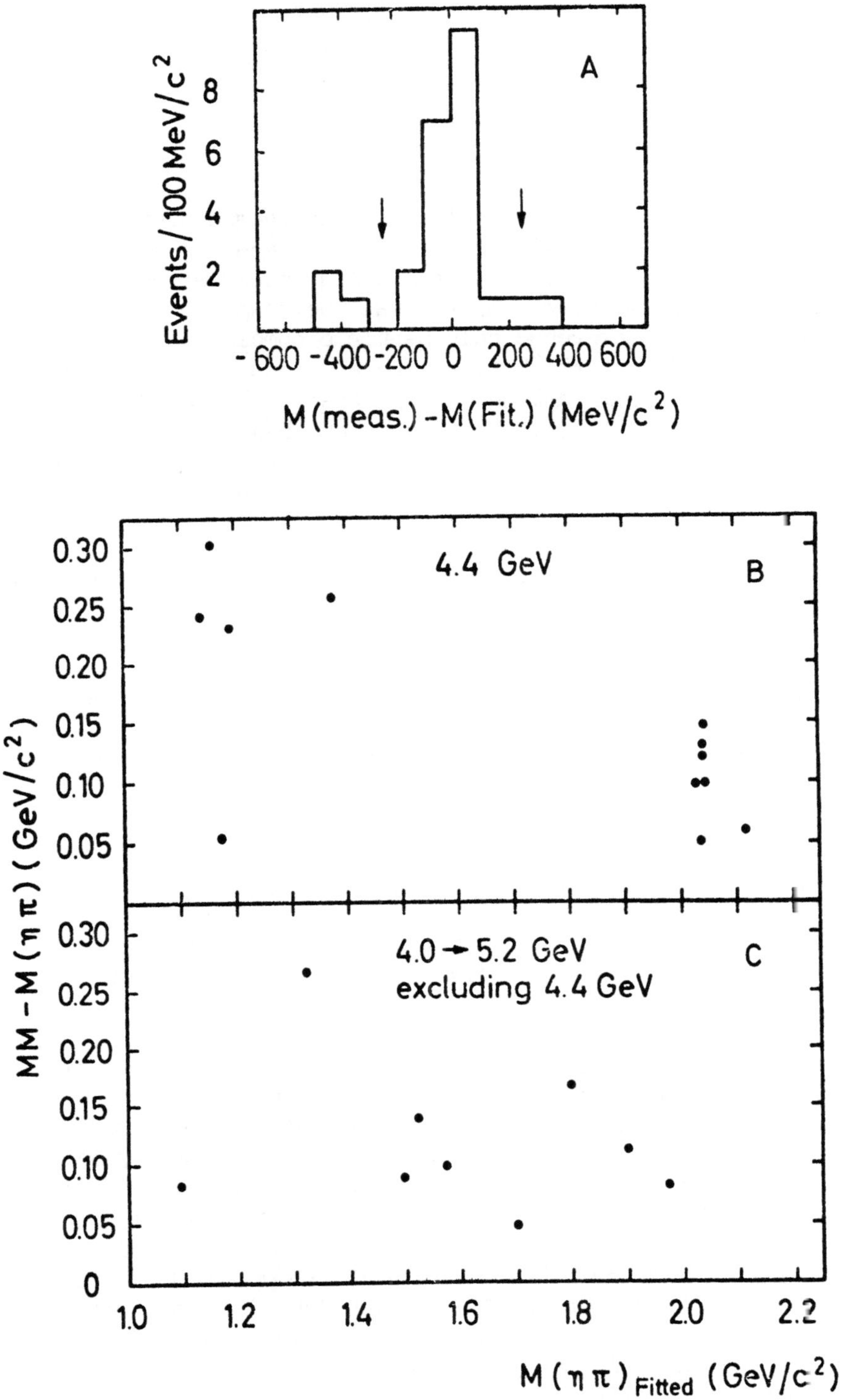

FIG. 7.2.3. : Results of a fit to the hypothesis $e^+e^- \rightarrow FF^*$ with $F^* \rightarrow \gamma F$ and $F \rightarrow \eta\pi$ as described in the text.

$D^\circ \rightarrow K^\pm\pi^\mp$ $\qquad$ $D^+ \rightarrow K^\circ_S\pi^+$

$\rightarrow K^\circ_S\pi^+\pi^-$ $\qquad$ $K^-\pi^+\pi^+$

$\rightarrow K^\mp\ \pi^\pm\pi^+\pi^-$

$\rightarrow K^-\pi^+\pi^\circ$

Charged kaons are identified as discussed above and neutral kaons are recognized by their characteristic decay $K^\circ_S \rightarrow \pi^+\pi^-$. Since the D mesons at the ψ'' mass are pairproduced without additional particles their energy must equal the beam energy. Therefore to reduce the background only particle combinations with an energy within 50 MeV of the beam energy are considered. The mass spectra for various combinations are plotted in Fig. 7.3.1 and show clear mass peaks in all channels. The absolute branching ratios are determined from these data using the known cross section for $D^\circ\bar{D}^\circ$ and D^+D^- production. The results are listed in table 7.1.

Table 7.1 - Branching Ratios for Charged and Neutral D mesons.

Decay mode	Branching ratio (%)	Reference
$D^\circ \rightarrow K^-\pi^+$	2.2 ± 0.6	85, 87
$\rightarrow \bar{K}^\circ\pi^+\pi^-$	4.0 ± 1.3	85, 87
$\rightarrow K^-\pi^+\pi^-\pi^+$	3.2 ± 1.1	85, 87
$\rightarrow K^-\pi^+\pi^\circ$	12 ± 6	86
$D^+ \rightarrow \bar{K}^\circ\pi^+$	1.5 ± 0.6	85, 87
$\rightarrow K^-\pi^+\pi^+$	3.9 ± 1.0	85, 87

In Su(3) weak decays the observed $\Delta I = 1/2$ enhancement is equal to a octet enhancement. This principle extended to SU(4) implies that the dominant part of the weak interaction should belong [88,89] to a 20 representation of SU(4). The decomposition of the 20 into SU(3) representations has an octet part with $\Delta C = 0$ and $\underline{6}$ and $\bar{\underline{6}}$ parts corresponding to transitions with $\Delta C = \pm 1$. This scheme predicts among other a small semileptonic branching ratio (discussed below) and $\Gamma(D^+ \rightarrow K^\circ_S\pi^+) = 0$. This latter decay has been observed [85] by the SLAC-LBL group with a branching ratio comparable to that observed for $D^\circ \rightarrow K^-\pi^+$. Furthermore semileptonic decays of charmed mesons do not seem to be surpressed (section 7.7).

The mass of the D mesons is determined from :

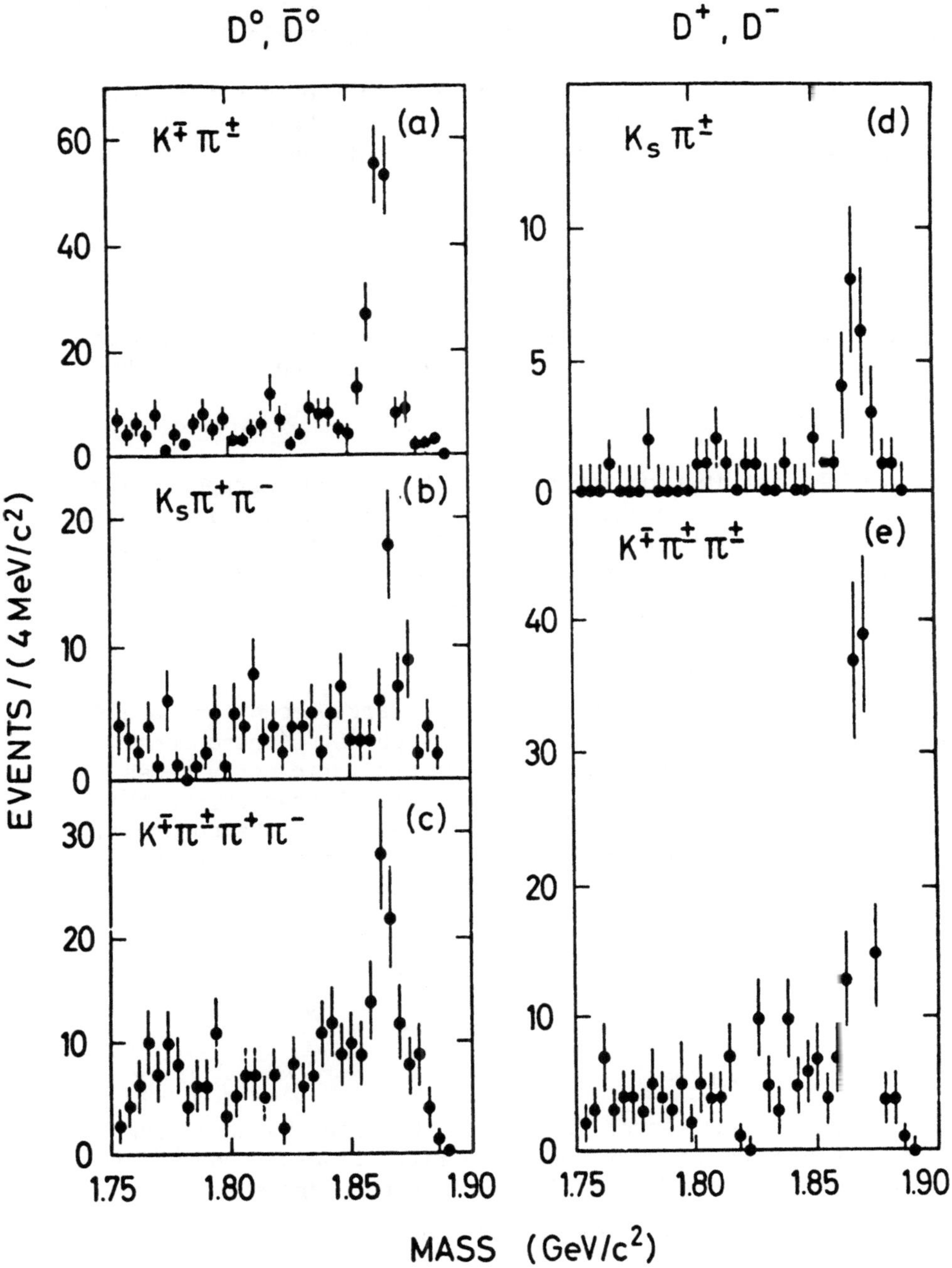

FIG. 7.3.1. : Invariant mass spectra for various decay modes of neutral and charged D-mesons.

$$M_D = \sqrt{E_D^2 - P_D^2} = \sqrt{E_B^2 - P_D^2}$$

The energy spread of the electron beam is on the other of 1 MeV and its central value can be determined very precisely. The ψ" (3772) mass is near $D\bar{D}$ threshold and the D mesons are produced with low momenta. Hence the uncertainty in the momentum does not contribute significantly to the error in the mass. A fit to the mass distributions in Fig. 7.3.2, obtained by summing the observed decay modes, yields :

$$M_{D^\circ} = 1863.3 \pm 0.9 \text{ MeV/c}^2$$

$$M_{D^+} = 1868.4 \pm 0.9 \text{ MeV/c}^2$$

In the determination of the mass difference between charged and neutral D mesons some of the systematic errors cancel and they report

$$M_{D^+} - M_{D^\circ} = 5.1 \pm 0.8 \text{ MeV/c}^2.$$

The angular distribution for neutral $(K^\pm\pi^\mp)$ and charged $(K^\pm\pi^\mp\pi^\mp)$D mesons with respect to the beam axis are plotted in Fig. 7.3.3. The observed distributions are well fit by a $\sin^2\theta$ distribution as expected for spin 0. However, higher values of the spin can in principle also lead to a $\sin^2\theta$ distribution.

7.4. Properties of the D*.

The SLAC-LBL group has collected a large amount of data [90] on charmed particle production at the 4.03 GeV/c^2 resonance. D° and D^+ are identified by their characteristic decay modes $D^\circ \to K^\pm\pi^\mp$ and $D^+ \to K^\pm\pi^\mp\pi^+$. The events are plotted versus the momentum of the D° and the D^+ in Fig. 7.4.1b,c. The spectrum, which is equivalent to a missing mass plot, is a rather complicated with several peaks and broad structures. A fit was made with the following Ansatz :

a) D and D* are produced through two body channels :

$$\begin{aligned} e^+e^- &\to D\,\bar{D} \\ &\to D\,\bar{D}^* + \bar{D}\,D^* \\ &\to D^+\bar{D}^* \end{aligned}$$

This assumption was tested for neutral D mesons by including $D^\circ\bar{D}^\circ\pi^\circ$ phase space production in the fit. It was found [90] that less than 10% of the D° signal is from such a process.

b) The following decay modes are considered :

$$D^{*\circ} \to \pi^\circ D^\circ \qquad D^{*+} \to \gamma\, D^+$$

$$D^{*\circ} \to \gamma\, D^\circ \qquad D^{*+} \to \pi^+ D^\circ$$

$$D^{*+} \to \pi^\circ D^+$$

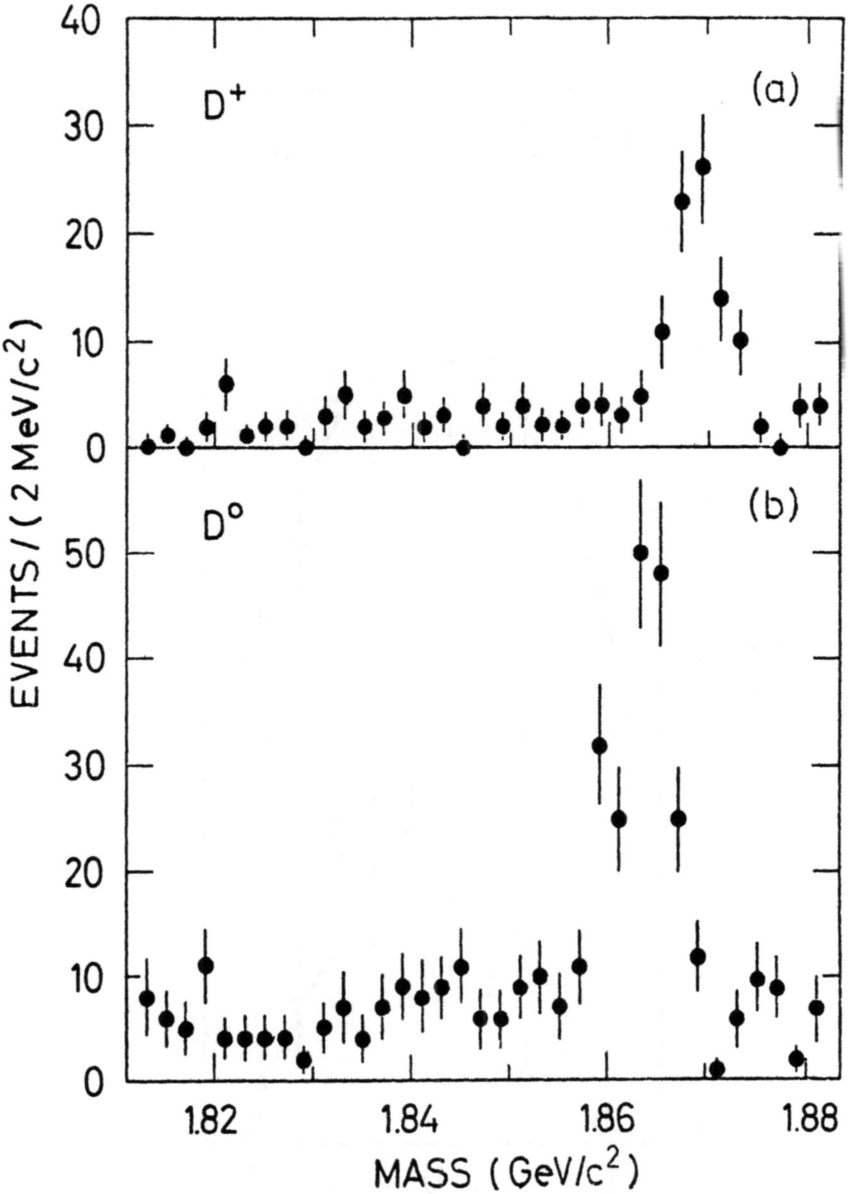

FIG. 7.3.2. : Invariant mass spectrum obtained by summing the mass spectra show in Fig. 7.3.1.

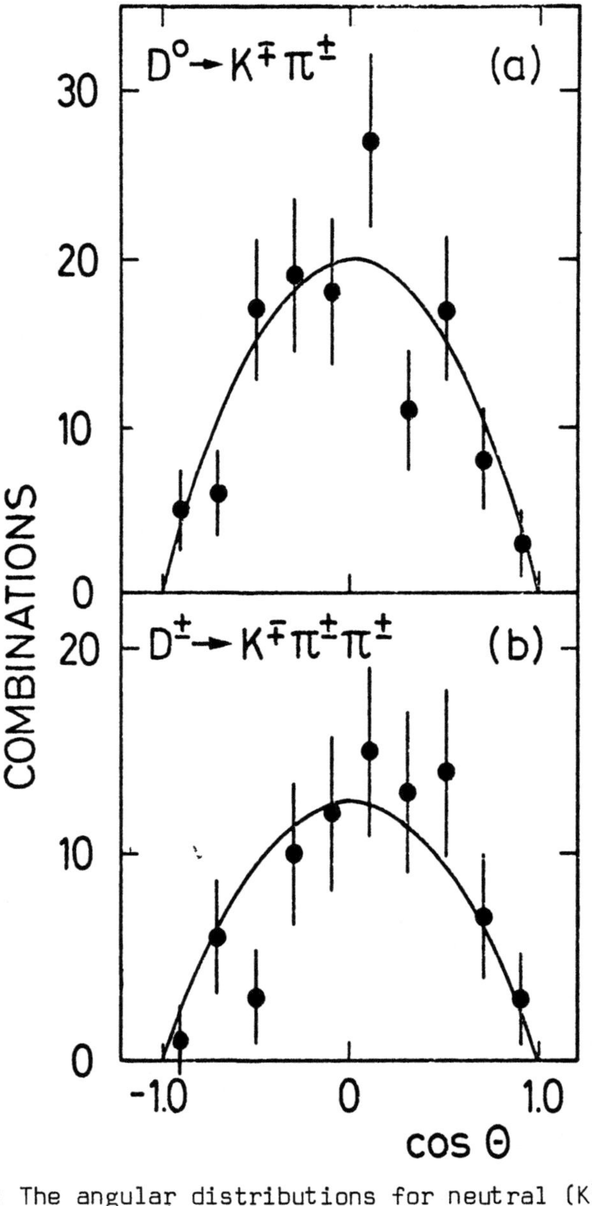

FIG. 7.3.3. : The angular distributions for neutral ($K^{\pm}\pi^{\mp}$) and charged ($K^{\pm}\pi^{\mp}\pi^{\mp}$) D mesons with respect to the beam axis. The solid curve shows a $\sin^2\theta$ distribution expected for spin 0 particles.

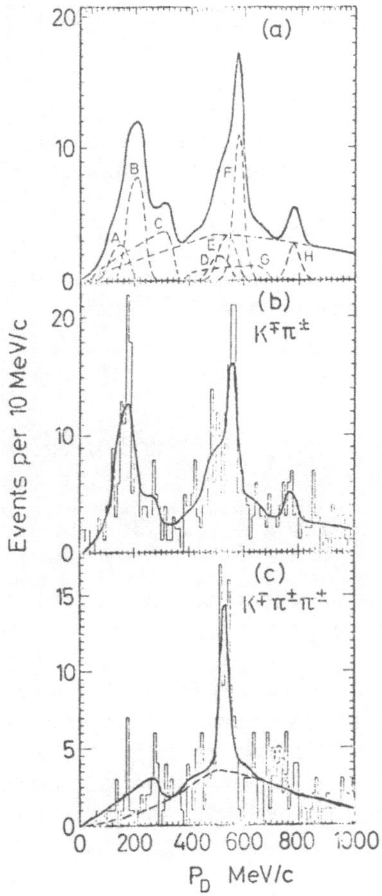

FIG. 7.4.1. : a) Example of various contributions to the D^* momentum spectrum observed at 4.03 GeV.

$e^+e^- \to D^{*+}D^{*-}$	,	$D^{*+} \to \pi^+ D^o$	(A)
$\to D^{*o}\bar{D}^{*o}$	,	$D^{*o} \to \pi^o D^o$	(B)
$\to D^{*o}\bar{D}^{*o}$	,	$D^{*o} \to \gamma D^o$	(C)
$\to D^{*+}D^-$	,	$D^{*+} \to \pi^+ D^o$	(D)
$\to D^{*o}\bar{D}^o$	,	$D^{*o} \to \pi^o D^o$	(E)
$\to \bar{D}^{*o}D^o$	,	Direct D^o	(F)
$\to D^{*o}\bar{D}^o$	,	$D^{*o} \to \gamma D^o$	(G)
$\to D^o\bar{D}^o$	,	Direct D^o	(H)

the unlabeled dashed curve corresponds to uncorrelated $K^-\pi^+$ combinations.

b) The D^o momentum spectrum observed at 4.028 GeV; The solid curve is a fit to the data.

c) The D^+ momentum spectrum observed at 4.028 GeV. The solid curve is a fit to the data. The smoothed background from nonexotic $K\pi\pi$ combinations is shown as the dotted line.

Note that the last decay mode couples charged and neutral D meson production. The corresponding decay $D^{*o} \to \pi^- D^+$ violates energy conservation.

The momentum distributions predicted for the various production and decay channels listed above are shown in Fig. 7.4.1a. Note in particular the characteristic peak at low momenta from :

$$e^+e^- \to D^{*o}\bar{D}^{*o} \quad \text{with} \quad D^{*o} \to \pi^o D.$$

This peak is clearly seen in the neutral D spectrum and it can be used to determine the mass of the D^{*o} using

$$M_{D^{*o}} = \sqrt{E^2_{D^{*o}} - P^2_{D^{*o}}} \quad .$$

Here $E_{D^{*o}} = E_B$ and, because of the small Q value, replacing $P_{D^{*o}}$ by P_{D^o} introduces an negligible error in the mass. This leads to

$$M_{D^{*o}} = 2006.0 \pm 1.5 \text{ MeV}/c^2.$$

A similar peak is not observed in the charged D spectrum because of poor statistics.

The mass of the D^{*+} is determined directly from a measurement [91] of $D^{*+} \to \pi^+ D^o$ at cm energies above 6 GeV. The observed $(D^o\pi^+ + \bar{D}^o\pi^-)$ events are plotted in Fig. 7.4.2 versus the D^{*+}-D^o mass difference in 1 MeV/c^2 bins. A narrow peak less than 2MeV (90%CL) wide is observed. A fit to the peak combined with the known mass of the neutral D meson leads to

$$m_{D^{*+}} = 2008.6 \pm 1.0 \text{ MeV}/c^2.$$

The masses of the known members of the D-family are listed in table 7.2.

The branching ratios of the decays $D^{*o} \to \gamma D^o$ and $D^{*o} \to \pi^o D^o$ are determined from Fig. 7.4.2a, b directly by fitting the relative contribution of the two decay modes to the spectrum. The fit yields $BR(D^{*o} \to \gamma D^o) = 0.45 \pm 0.15$.

The decays of $D^{*\pm}$ are extracted from the data of Fig. 7.2.1c with the following assumptions :

1) $D^* \to \pi D$ conserves isospin.
2) Phase space corrections are proportional to p^3 where p is the momentum of the D in the rest frame of the D^*.
3) The quark model prediction for $s = \Gamma(D^{*+} \to \gamma D^+) / \Gamma(D^{*o} \to \gamma D^o)$. The value of s ranges between 1/4 for $m_u/m_c = 0$ to 1/25 for $m_u/m_c = m_\rho/m_{J/\psi}$.
   The results of the fit are listed in table 7.3. The branching ratio for $D^{*+} \to \gamma D^+$ is small and varies between 0.02 and 0.10

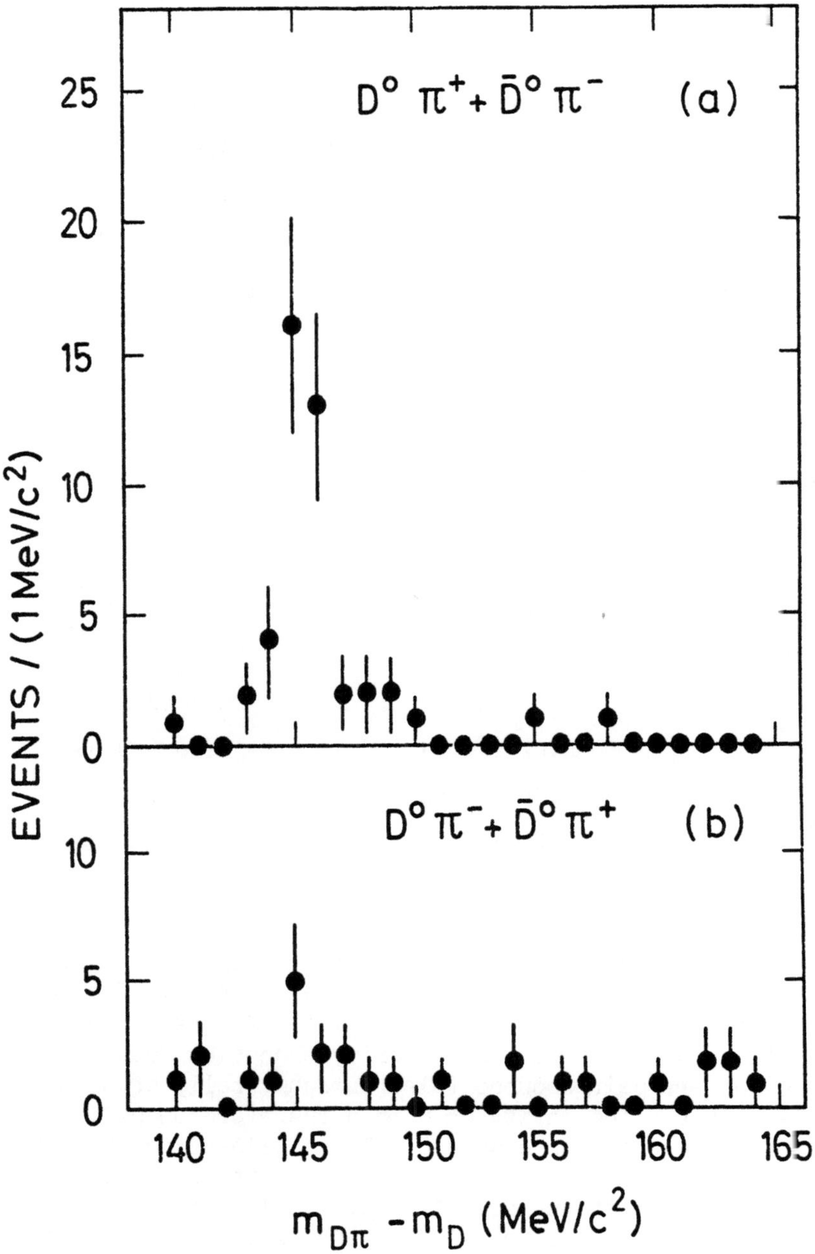

FIG. 7.4.2. : Dπ-D mass difference spectra for
a) $D^0\pi^+$ and $\bar{D}^0\pi^-$ combinations (charm conserving).
b) $\bar{D}^0\pi^+$ and $D^0\pi^-$ combinations (charm mixing).

for values of s between 1/4 and 1/25.

Table 7.2 - Masses of Neutral and Charged D Mesons.

Particle	Mass (MeV/c^2)
D^o	1863.3 ± 0.9
D^+	1866.3 ± 0.9
$D^+ - D^o$	5.0 ± 0.8
D^{*o}	2006.0 ± 1.5
D^{*+}	2008.6 ± 1.0
$D^{*+} - D^{*o}$	2.6 ± 1.8

Table 7.3 - D^* Branching Ratios.

$D^{*o} \to \gamma D^o$	0.45 ± 0.15	
$D^{*o} \to \pi^o D^o$	0.55 ± 0.15	
	s = 1/4	s = 1/25
$D^{*+} \to \gamma D^+$	0.02 ± 0.01	0.10 ± 0.05
$D^{*+} \to \pi^o D^+$	0.30 ± 0.08	0.27 ± 0.07
$D^{*+} \to \pi^+ D^o$	0.68 ± 0.08	0.63 ± 0.09

The D^o is in general considered to be a charmed, pseudoscalar state with $J^P = 0^-$ and D^{*o} to be the 1^- state. The angular distribution of the D's produced at $\psi''(3772)$ are indeed consistent with spin 0. The absolute parity cannot be determined since D mesons are pairproduced (electromagnetically or strongly) and decays weakly. The parity of D^o can therefore be defined to be negative. D^{*o} cannot have spin 0. This follows from the observed decay $D^{*o} \to \gamma D^o$. The D^{*o} must have positive parity with respect to the D^o if it has spin 1. This follows from the decay $D^{*o} \to \pi^o D^o$. A clean sample on $e^+e^- \to D^o D^{*o}$ events was obtained 92) at 4.03 GeV. The observed angular distribution of the D^o mesons is plotted in Fig. 7.4.3 and compared to various spin values. The distribution is consistent with the spin-parity

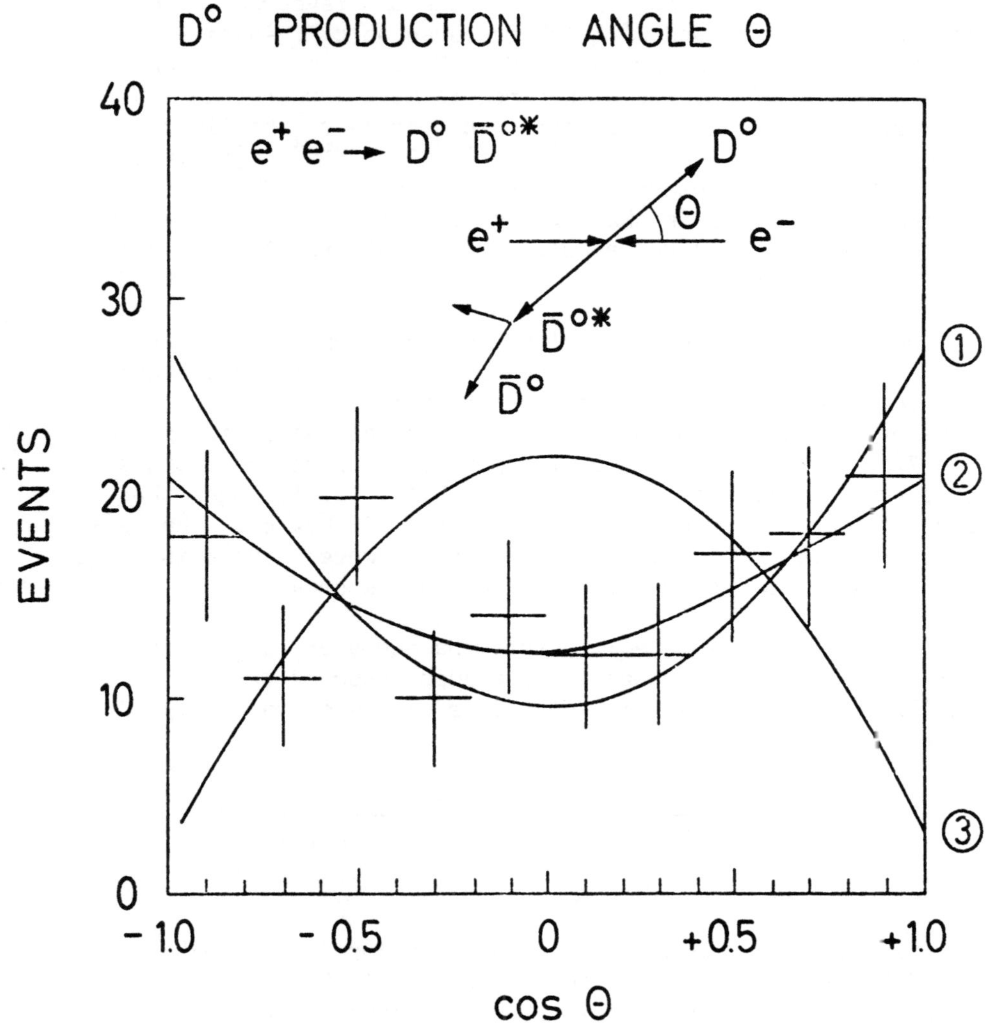

FIG. 7.4.3. : The observed angular distributions of D° produced in $e^+e^- \rightarrow D^\circ\bar{D}$ ° at 4.028 GeV. The production angle θ is defined with respect to the beam direction. Curve (1) $D^\circ=1^\pm$ $D^{*\circ}=0^\pm$ (2) $D^\circ=0^\pm$ $D^{*\circ}=1^\pm$ (3) $D^\circ=0^\pm$ $D^{*\circ}=0^\pm$

assignment assumed above. A complete discussion can be found in reference 92.

7.5. Charm production at 4.028 GeV and 4.414 GeV.

The weighted mass spectra [94] of the GIM favoured channels $K^{\pm}\pi^{\mp}$, $K_S\pi^+\pi^-$, $K^{\pm}\pi^{\mp}\pi^{\mp}$ and $K^{\pm}\pi^{\mp}\pi^+\pi^-$ measured at c.m. energies at 4.028 GeV and 4.414 GeV are plotted in Fig. 7.5.1. To reduce the background only events with a recoil mass larger than 1.8 GeV/c^2 are plotted. The weighted mass spectra of the GIM disfavoured channels $\pi^+\pi^-$, K^+K^-, $K^{\mp}\pi^+\pi^-$, $\pi^{\pm}\pi^+\pi^-$, $K^+K^-\pi^{\pm}$, $K_S^o\pi^+$ measured at 4.028 GeV resonance are plotted in Fig. 7.5.2. There are significant peaks in the $\pi^+\pi^-$ spectrum at 1.74 GeV/c^2 and in the K^+K^- spectrum of 1.98 GeV/c^2. These peaks result from the decay $D^o \to K^{\pm}\pi^{\mp}$ in which the K or the π has been misidentified.

The product of the production cross sections and the branching ratios are determined and the results are listed in table 7.4. The production cross section for charged and neutral D mesons are determined using measured branching ratios listed in table 7.1. The average value of $2R_D = \sigma(D)/\sigma_{\mu\mu}$ can be compared with the total cross sections (section 5.1). It is found that D production can account for the peak at 4.028 GeV/c^2. The data at 4.414 GeV are not entirely consistent, however, there is large cross section for D production also at this energy.

The GIM mechanism predicts that the decay of the D^o into final states of zero strangeness are suppressed by $tg^2\theta \sim 0.05$ relative to final states which contain a single K meson (Fig. 7.2). From the limits given in table 7.4 :

$$\frac{BR(D^o \to \pi^+\pi^-)}{BR(D^o \to K^{\pm}\pi^{\mp})} < 0.07 \quad \text{and} \quad \frac{BR(D^o \to K^+K^-)}{BR(D^o \to K^{\pm}\pi^{\mp})} < 0.07$$

with 90% confidence.

The relative production cross sections for $D^o\bar{D}^o$, $D^o\bar{D}^{*o} + \bar{D}^oD^{*o}$ and $D^{*o}\bar{D}^{*o}$ were extracted from a detailed fit to the D momentum distribution. All the channels above are expected to couple to the photon in a P wave. A p^3 phase space factor was therefore removed in order to facilitate a direct comparison of the couplings. This yields :

$$D^o\bar{D}^o : D^o\bar{D}^{*o} + \bar{D}^oD^{*o} : D^{*o}\bar{D}^{*o} = 0.2\pm0.1 : 4.0\pm0.8 : 128\pm40.$$

Spin counting arguments give 1 : 4 : 7. The large discrepancy provoked several groups to introduce [43] $c\bar{q}$ $q\bar{c}$ states charmonium molecules. However, this explanation is contradicted by the absence [56] of $4.028 \to J/\psi + X$ which would be a natural consequence of charmonium molecules. An alternative explanation of this small

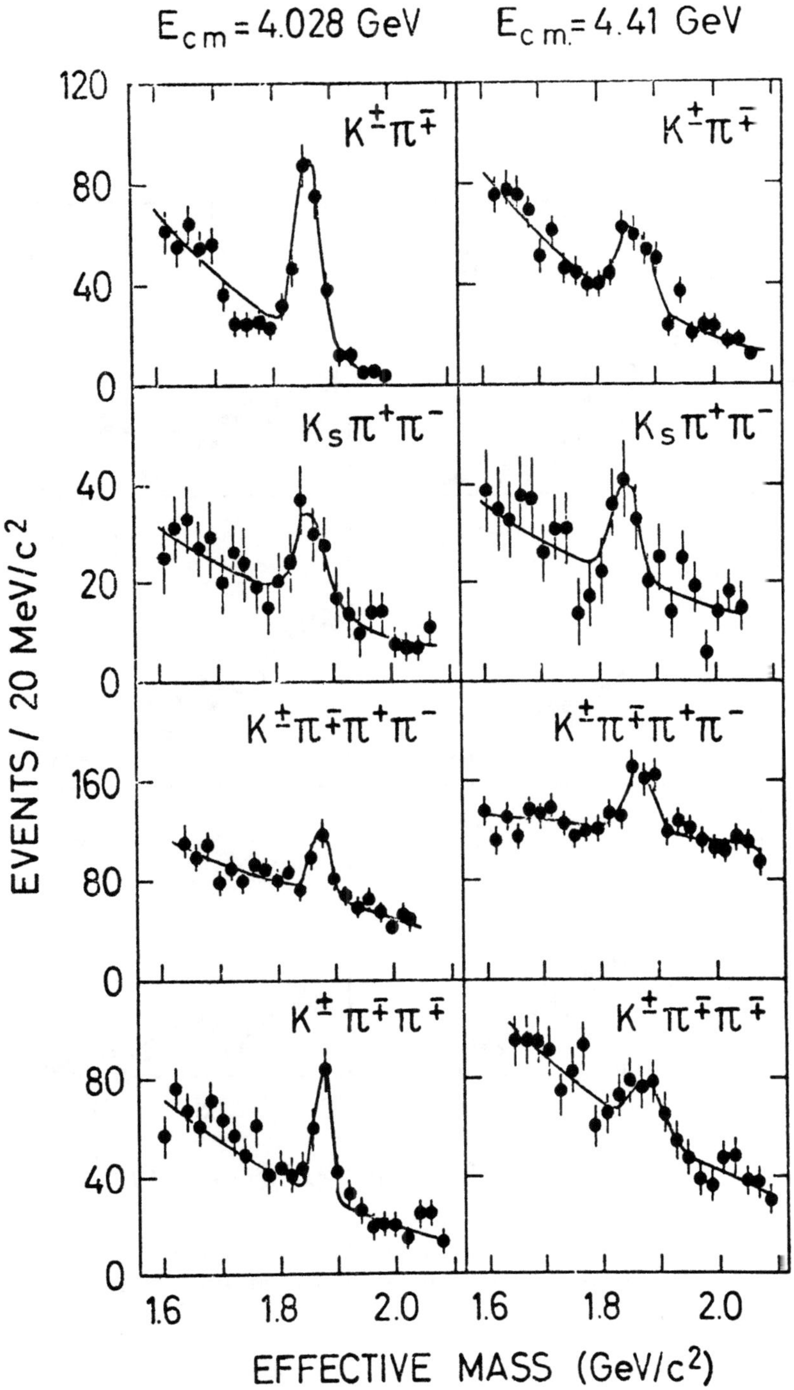

FIG. 7.5.1. : Invariant mass spectra for various GIM favoured decay modes of neutral and charged D mesons measured at 4.03 GeV and 4.41 GeV.

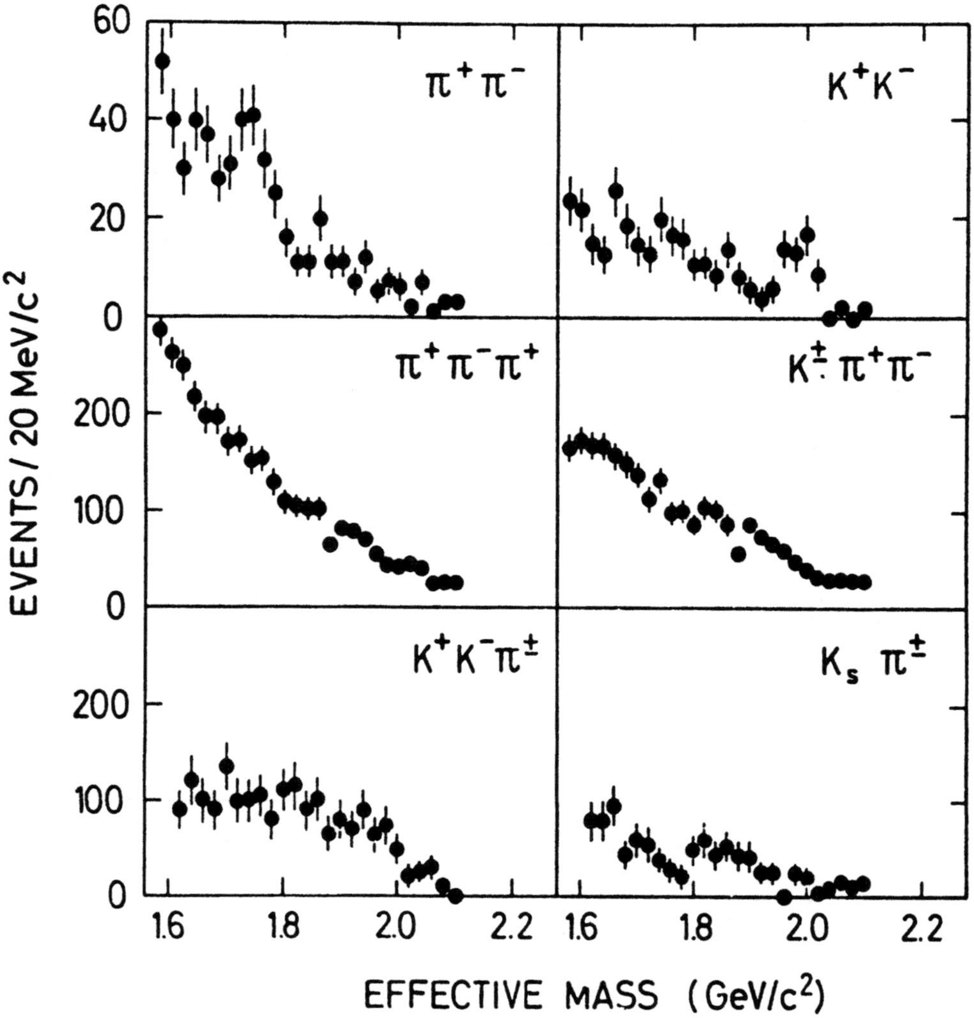

FIG. 7.5.2. : Invariant mass spectra for various GIM disfavoured decay modes of neutral and charged D mesons measured at 4.03 GeV.

$D^o\bar{D}^o$ cross section has been put forward [54,55].

7.6. $D^o - \bar{D}^o$ Mixing.

D^o and $\bar{D}^o$ differs only in the charm quantum number and can therefore be mixed by a charm changing neutral current. Such a possibility has been widely considered [75,93]. The SLAC-LBL group has searched for this mixing in two ways. The first method [91] takes advantage of the fact that the strong interaction conserves charm - i.e. the sign of the pion in $D^{*+} \to \pi^+ D^o$ ($D^{*-} \to \pi^- \bar{D}^o$) tags the charm value of the produced neutral D system. Thus a D^o-$\bar{D}^o$ mixing leads to events of the type $D^o\pi^-$ and $\bar{D}^o\pi^+$. The data are plotted in Fig. 7.4.2. Sharpening the cuts left a total of 26 events with the expected sign correlation compared to 3 events of opposite sign. Backgrounds and misidentifications are expected to yield 2 events with the wrong sign. Hence there is no evidence for a charmchanging neutral current and they find with 90% confidence that a D^o decays like a $\bar{D}^o$ less than 16% of the time.

A similar limit has been derived from a search [90] of apparent strangeness violations in events which contain a D^o meson.

7.7. Semileptonic decays of charm.

7.7.1. Origin of electron inclusive events.

The pair production in e^+e^- annihilation of new particles with large leptonic or semileptonic decay modes will lead to mixed final states containing leptons and hadrons. The observation of such final states at a level above the background expected from higher order electromagnetic interactions and semileptonic pion or kaon decays is direct evidence for the production of new particles. A new lepton or a hadron with a new flavour are but two examples of such particles. As will be discussed later there is now convincing evidence that besides charmed hadrons also a new lepton is pairproduced at cm energies above 4 GeV. However, it is still possible to assign the observed electron (muon) inclusive events to one or the other of the two competing processes.

The anticipated features, of a new lepton and a new hadron are summarized in table 7.5. The low multiplicity expected for the decay of a pair of heavy leptons into a final state with an electron (muon) plus hadrons arises as follows. In the decay $L \to \nu_L$ + Hadrons, the hadrons come from a low mass current. If the multiplicity is comparable to that from a virtual photon of the same mass, it will be small. Specific calculations support this conjecture. The leptonic decay of its partner L contributes in general only one charged track. Inelastic production i.e. $e^+e^- \to L\bar{L}$ + hadrons is negligible at these energies. The high

Table 7.4. - Production cross sections at 4.03 GeV and 4.41 GeV.

Particle	Decay Mode	Energy (GeV)	σ(D) • BR (nb)	σ(D) (nb)	$2R_D$
$D^o+\bar{D}^o$	$K^{\mp}\pi^{\pm}$	4.03	0.57 ± 0.11	25.9 ± 5	4.9 ± 1.0
$D^o+\bar{D}^o$	$\bar{K}^o\pi^+\pi^-+K^o\pi^+\pi^-$	4 03	1.09 ± 0.30	27.0 ± 7.5	5.1 ± 1.4
$D^o+\bar{D}^o$	$K^{\mp}\pi^{\pm}\pi^+\pi^-$	4.03	0.83 ± 0.27	25.9 ± 8.4	4.9 ± 1.6
D^++D^-	$K^{\mp}\pi^{\pm}\pi^{\pm}$	4.03	0.40 ± 0.10	10.3 ± 2.6	1.9 ± 0.5
$D^o+\bar{D}^o$	$K^{\mp}\pi^{\pm}$	4.41	0.30 ± 0.09	13.6 ± 4.1	3.1 ± 0.9
$D^o+\bar{D}^o$	$\bar{K}^o\pi^+\pi^-+K^o\pi^+\pi^-$	4.41	0.91 ± 0.34	22.7 ± 8.5	5.1 ± 1.9
$D^o+\bar{D}^o$	$K^{\mp}\pi^{\pm}\pi^+\pi^-$	4.41	0.91 ± 0.39	28.4 ±12.2	6.4 ± 2.8
D^++D^-	$K^{\mp}\pi^{\pm}\pi^{\pm}$	4.41	0.33 ± 0.12	8.5 ± 3.1	1.9 ± 0.7
D^++D^-	$\bar{K}^o\pi^{\pm}+K^o\pi^{\pm}$	4.03	< 0.18		
$D^o+\bar{D}^o$	$\pi^+\pi^-$	4.03	< 0.04		
$D^o+\bar{D}^o$	K^+K^-	4.03	< 0.04		
D^++D^-	$\pi^{\mp}\pi^+\pi^-$	4.03	< 0.03		
D^++D^-	$\pi^{\mp}K^+K^-$	4.03	< 0.06		
D^++D^-	$K^{\mp}\pi^+\pi^-$	4.03	< 0.02		

multiplicity expected in the decay of a pair of heavy hadrons will presumably lead to a multiplicity comparable to that observed in the decay of an ordinary hadron of the same mass - i.e. on the average of 2 to 3 charged particles plus a few photons. From the semileptonic decay of its partner we might expect one or two charged particles and a few photons. Thus a cut on the total multiplicity can be used to separate the two classes with $e^+e^- \to L\bar{L} \to e(\mu)+X$ predominantly populating the two prong class.

The momentum spectrum of the observed electron (muon) can also be used to classify inclusive lepton events. All particles in the leptonic decay $\tau \to \nu_\tau(e\bar{\nu})$ are pointlike and large values of Q^2, i.e. large electron energies are therefore not suppressed by form factors. Also at least two of the particles have negligible mass. In a semileptonic decay of a charmed particle the formfactor will disfavour large values of Q^2. The mass of the hadron system will also lead to smaller electron momenta.

We will first discuss the data on semileptonic decay of charmed particles. Information on the τ particle is presented in section 8.

7.7.2. Results on the semileptonic decays of charmed hadrons.

Semileptonic decays of charmed hadrons have been widely considered in the literature [95]. It follows directly form Fig. 7.2A that the amplitude for the favoured mode is proportional to $\cos\theta_C$ with $\Delta S = \Delta Q = \Delta C = 1$ and $\Delta I = 0$. The disfavoured mode has an amplitude proportional to $-\sin\theta_C$ with $\Delta C = \Delta Q$, $\Delta S = 0$ and $\Delta I = 1/2$. A D meson should predominantly decay into final states like $e\bar{\nu}_e K$ or $e\bar{\nu}_e K^*(892)$. Disfavoured modes like $D \to e\bar{\nu}_e \pi$ should be suppressed by $tg^2\theta_C \simeq 0.05$. The decay into $e\bar{\nu}_e K^*(1420)$ is suppressed by phase space and soft pion theorems predict that the decay $D \to e\bar{\nu}_e K(\eta\pi)$, where the $K(\eta\pi)$ system does not form a resonance, should be small.

The semileptonic branching ratio of charmed hadrons can be determined from a measurement of mixed electron hadron final states

$$e^+e^- \to C\bar{C} \to (e\ \bar{\nu}\ \ldots)\ (X).$$

The average branching ratio BR(C → eX) is obtained from

$$B(C \to eX) = \frac{\text{number of eX-events}}{2\sigma(c\bar{c})}$$

The total cross section for charmed hadron production is defined as $\sigma(e^+e^- \to c\bar{c}) = \sigma(\text{tot}) - R(3.6)\sigma(\mu\bar{\mu}) - \sigma(\tau\bar{\tau}) \cdot R(3.6)\sigma(\mu\bar{\mu})$ is the cross section for non-charmed hadron production determined below charm threshold at 3.6 GeV, $\sigma(\tau\bar{\tau})$ is the pair production cross section for a pointlike fermion.

In principle mixed muon hadron events can also be used. However, the lepton spectrum must be measured down to low momenta with a rejection factor of 10^3 against hadrons. Such a rejection factor is difficult to obtain for low energy muons.

DASP (section 2.3) has measured [96] the single electron spectrum for momenta above 200 MeV/c and cm energies between 3.6 GeV and 5.2 GeV for a total integrated luminosity of 6300 nb^{-1}. Electrons were defined as particles which gave a signal in the proper Cerenkov counter and had $\beta = 1$ ($p < 0.35$ GeV/c) or gave a large pulse height ($p > 0.35$ GeV/c) in the shower counter. A pion had a probability of 4×10^{-4} to pass these criteria. Electron pairs from Dalitz decay or pair conversion were rejected by pulse height cuts on the scintillation counters mounted before the magnet.

To reduce the background of an electromagnetic origin the event must contain at least one nonshowering track. This could be a track in a spectrometer arm identified as a hadron or a muon. A charged track traversing the inner detector could also be called nonshowering if it fired less than 1.5 tubes per layer, averaged over all layers which had at least one tube set. Events containing only two charged particles are particularly sensitive to electromagnetic background. Here we required less than 1.25 tubes per plane be activated. Test of these criteria using well defined pions showed that 95% of the pions but fewer than 5% of the electrons satisfied the tight criteria.

The background to this sample from beam gas events, Compton scattering on the material mounted in the front of the Cerenkov counter or from two photon processes was estimated and found to be small. The total background was (11.5 ± 3.5)% for the two prong sample and (15 ± 5)% for the multiprong sample. These values are in agreement with measurements done at the ψ' resonance and at 3.6 GeV.

After all cuts 60 two prong and 182 mutiprong events with an electron momentum above 0.2 GeV/c remained in this sample. The measured charged mutiplicity distributions (including the electron) for all inclusive electron events and for those with only charged tracks are plotted in Fig. 7.7.1. The distribution peaks for $n_{ch} = 2$ but it is rather wide with events up to a charged multiplicity of eleven. Both charm and τ-pairproduction are needed to explain the observed multiplicity distribution. The observed events are grouped into two prong (predominantly τ-events) and multiprong events (predominantly charm).

The lepton spectrum associated with the multiprong sample is shownin Fig. 7.7.2. The estimated background due to hadron

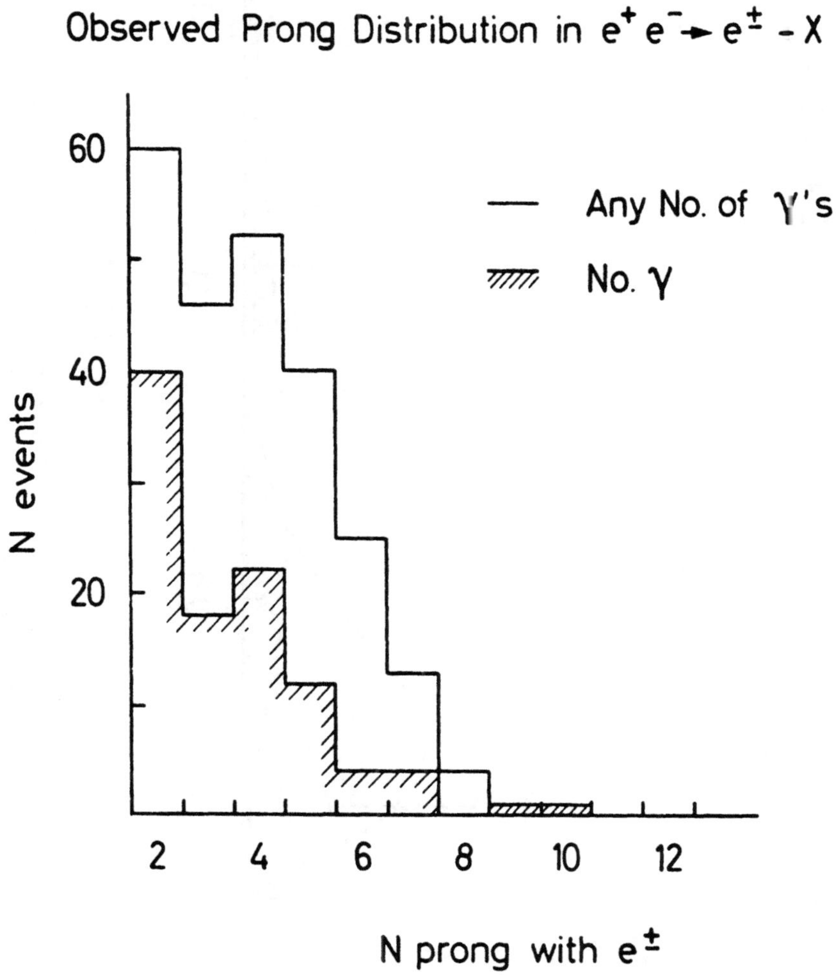

FIG. 7.7.1. : The charged track distribution observed by the DASP group for inclusive electron events. The electron is included in the prong number. The shaded distribution is for events without photons.

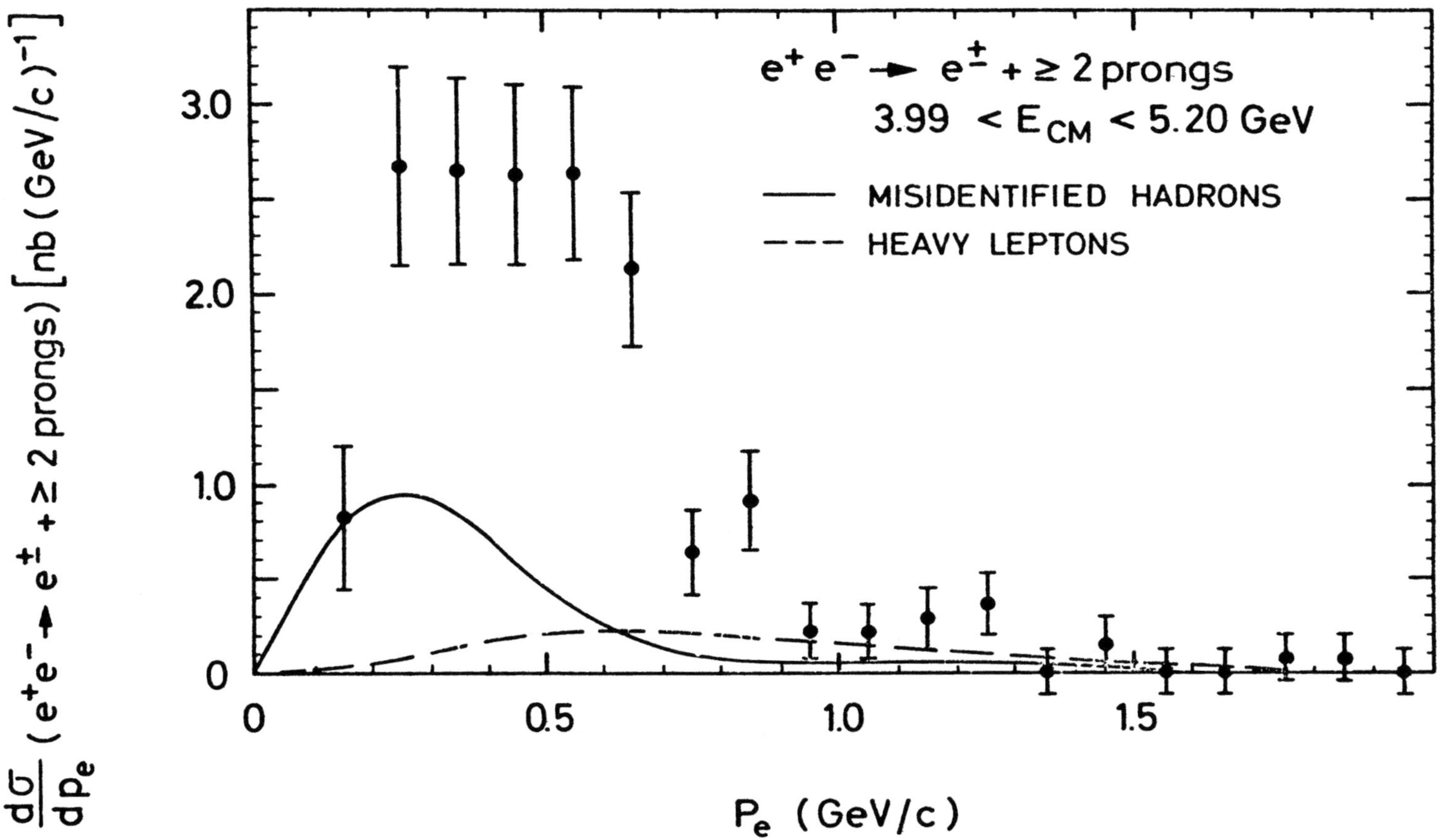

FIG. 7.7.2. : The inclusive electron momentum spectrum measured by the DASP Collaboration between 3.99 GeV and 5.2 GeV in cm for multiprong events.

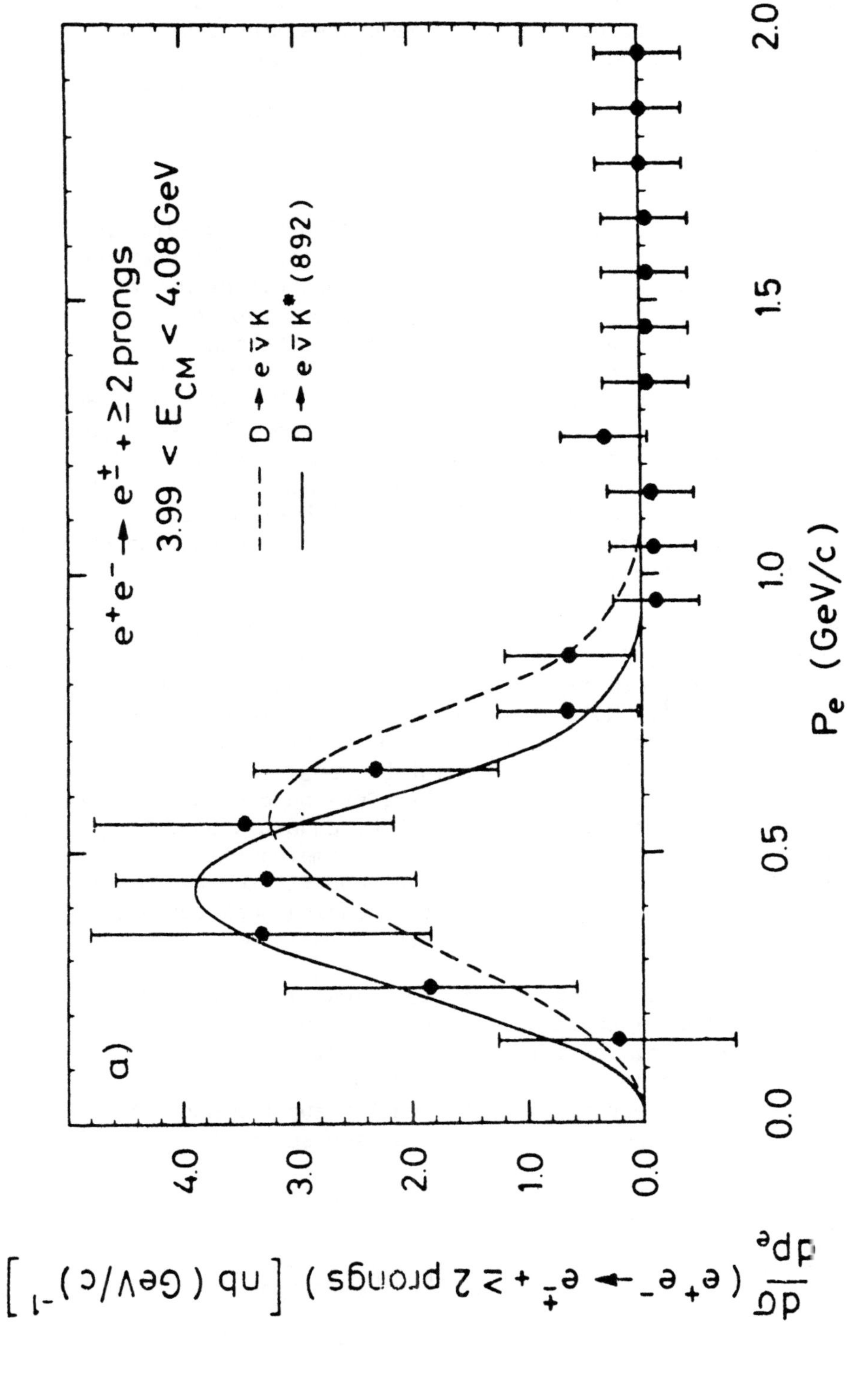

FIG. 7.7.3. : The electron momentum spectrum for D → eνX. The momentum distribution expected for D → e$\bar{\nu}_e$K and D → e$\bar{\nu}_e$K* (891) are shown by the dashed and the solid curves.

misidentification or heavy lepton production is also plotted. The background was scaled from measurements below threshold. The heavy lepton contribution was estimated assuming a τ branching ratio of 30% to decay into final states with three or more charged particles. The group finds that less than 12% of the events with $n_{ch} \geq 3$ can be explained as heavy lepton production. The simple cut on hadron multiplicity therefore yields a rather clean sample of charm decays.

The electron spectrum contains information on the semileptonic and the leptonic decay modes of lowest mass [8,10] charmed hadrons. Fig. 7.7.2 demonstrates that semileptonic decays are much more important than leptonic decays because the latter, being two body decays, would produce a peak in the electron spectrum around 1 GeV/c. This is in gross disagreement with the data which peak around an electron momentum of 0.5 GeV/c with only few events above 0.7 GeV/c. To study the observed momentum spectrum in more detail we consider the spectrum obtained for c.m. energies between 3.99 GeV and 4.08 GeV. The charm cross section in this energy region is dominated by $D\bar{D}^*$ and $D^*\bar{D}^*$ production. The spectrum, corrected for the background and the heavy lepton contribution, is shown in Fig. 7.7.3.

The spectrum in Fig. 7.7.3. was fitted to three possible channels : $D \to e\bar{\nu}_e\pi$, $D \to e\bar{\nu}_e K$ and $D \to e\bar{\nu}_e K^*(892)$. A V-A current was assumed and the form of the spectra was taken from a paper by Ali and Yang [95b]. Note that the theoretical spectra are model dependent. These fits gave a χ^2 value for 10 degrees of freedom of : 29.6 for $D \to e\bar{\nu}_e\pi$, 6.3 for $D \to e\bar{\nu}_e K$ and 2.8 for $D \to e\bar{\nu}_e K^*(892)$. The decay $D \to e\bar{\nu}_e\pi$ can therefore be excluded as the sole semileptonic decay mode of the D. The data can be fitted with either $D \to e\bar{\nu}_e K$ or $D \to e\bar{\nu}_e K^*(892)$. A combined fit attributes 35 ± 30% of the rate to $D \to e\bar{\nu}_e K$. A good fit to the spectrum can also be obtained assuming the charm changing weak current to be right handed in the decay $D \to e\bar{\nu}_e K^*(892)$.

The absolute cross section for inclusive electron production $e^+e^- \to e^\pm + X$, where X contains at least two charged tracks and any number of photons, is plotted in Fig. 7.7.4a as a function of energy. The data have been corrected for radiative effects. The background from hadron misidentification and the contribution from a heavy sequential lepton have been subtracted.

The inclusive electron cross section due to charmed particle production can be written as :

$$\sigma(e^+e^- \to e^\pm X) = \Sigma_{i,j}\sigma(e^+e^- \to C_i\bar{C}_j) \cdot (BR(C_i \to e\bar{\nu}_e X) + BR(\bar{C}_j \to e\bar{\nu}_e X))$$

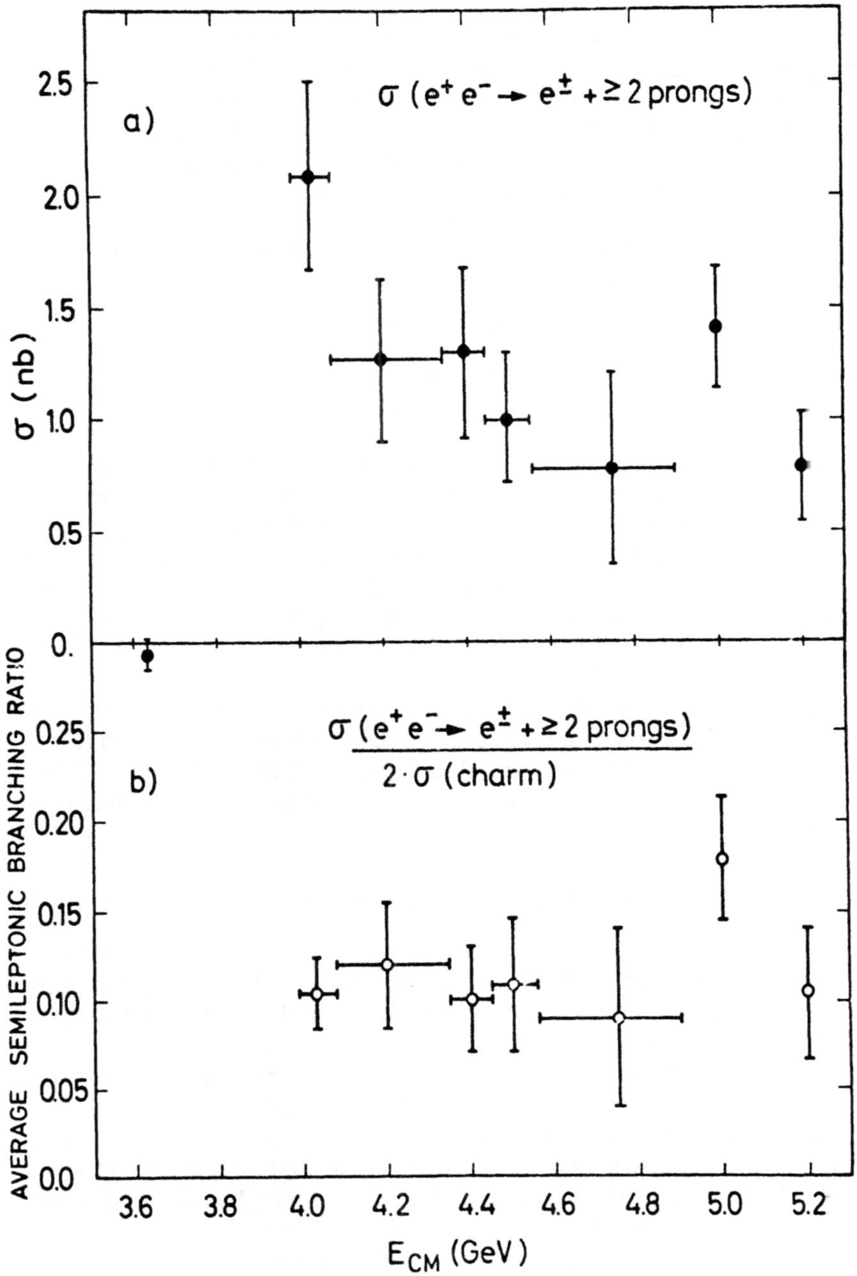

FIG. 7.7.4. : a) The cross section measured by DASP for the inclusive production of electrons plus nonshowering track plus additional charged tracks as a function of an energy.
b) The average semileptonic branching ratio for charmed hadrons as a function of energy. The error bars are statistical only.

Here $\sigma(e^+e^- \to C_i\bar{C}_j)$ denotes the effective cross section for producing the lightest charmed hadron stable against strong and electromagnetic decays. These particles might either be produced directly or result from the cascade decay of excited charmed hadrons. The cross section $\sigma(e^+e^- \to C_i\bar{C}_j)$ was obtained by subtracting the cross sections for "old" hadron production and heavy lepton production from the total cross section.

Near threshold, where only neutral and charged D production can contribute we find :

$$BR\ (D \to e + X) = 0.08 \pm 0.02$$

This should be compared to the value

$$BR\ (C \to e + X) = 0.072 \pm 0.02$$

obtained by averaging over all energies between threshold and 5.2 GeV. These values were extracted using the DASP [51] data for the total cross section and the error quoted is mainly systematic. Evaluating the branching ratio using the SPEAR data [49] on the total cross section as input leads to an average semileptonic branching ratio of 0.08 ± 0.03, compared to 0.11 ± 0.03 obtained using the PLUTO [50] total cross section.

The semileptonic branching ratio can also be determined from the fraction of inclusive electron events containing a second electron. Using this method the DASP group finds $BR(C \to e^-X) = 0.16 \pm 0.06$. Note that this value is independent of the charm cross section.

The DASP results are supported by the results from two recent experiments at SPEAR.

The DELCO group at SPEAR has measured [19] the cross section for $e^+e^- \to e^\pm + \geq 2$ hadrons ≥ 0 photons. A detailed description of the detector can be found in reference 19. Electrons are identified in 65% of 4π using an 12 module ethane filled Cerenkov counter sensitive only to electrons. (Pions have a threshold at 3.7 GeV/c). The Cerenkov counters are backed by an array of lead scintillator (2 r.l. divided in 3 layers) shower counters which cover 60% of 4π. The one-line trigger is rather loose requiring only a charged track plus a signal from two shower counters. Electron inclusive multiprong events were selected from this sample demanding that at least three charged tracks should emmerge from the interaction point and that at least two shower counters should fire. The candidate track for an electron was required to give signals in the Cerenkov and the appropriate shower counter. To reduce the background from Dalitz pairs, photon conversions and δ-rays it was required that no other track should be within an angle of 18 mrad/p(GeV) (p is the momentum of the softer hadron) with respect to the electron candidate. Hadronic events at the J/ψ resonance passed these selection criteria with an efficiency of $(3.5 \pm 0.3)\ 10^{-3}$.

The ratio

$$R_e = \frac{\sigma(e^+e^- \to e^\pm + \geq 2 \text{ charged} + \geq 0 \text{ photons})}{\sigma(e^+e^- \to \mu^+\mu^-)}$$

is plotted in Fig. 7.7.5 as a function of cm energy, the dotted curve represents the estimated background. A clear peak is seen at 3.77 GeV reflecting the decays of the ψ"(3772) into pairs of charged and neutral D mesons. The ratio mirrors the charm cross section with a rise at 4 GeV, a dip around 4.25 GeV and presumably some structure around 4.4 GeV.

Normalizing to the cross section for $e^+e^- \to \psi''(3772)$ yields $BR(D \to eX) = (11 \pm 3)\%$.

The electron momentum spectrum at the ψ"(3772) is plotted in Fig. 7.7.6. The dashed and the solid curves show the expected shapes for the decay modes $D \to e\bar{\nu}_e K$ and $D \to e\bar{\nu}_e K^*(892)$. Note that the estimated background plotted as the dot-dashed curve in Fig. 7.7.6 does not include the contribution from τ-pair-production. This contribution peaks around 800 MeV/c and extends to larger momenta.

The DELCO group has collected data at the ψ"(3772) with reduced magnetic field in order to be sensitive to electrons with very low momenta resulting from $D \to e\bar{\nu}_e K^*(1420)$. They find no evidence for this decay mode and quote

$$\frac{D \to e\bar{\nu}_e K^*(1420)}{D \to e\bar{\nu}_e + \text{anything}} < 0.1$$

The LBL-SLAC group selects [97,98] events of the type

$$e^+e^- \to e^\pm + \geq 2 \text{ charged tracks} + \geq 0 \text{ photons}$$

in the energy region between 3.7 GeV and 7.4 GeV. Electrons are identified in the lead glass wall (section 2.2) by demanding that the total energy deposited should be consistent with the momentum measured in the magnet. Furthermore the energy division between the active converters ($3.5\ X_0$) and the back blocks ($10.5\ X_0$) should be consistent with that of a showering particle and the time of flight should agree within one nanosecond with that expected for an electron.

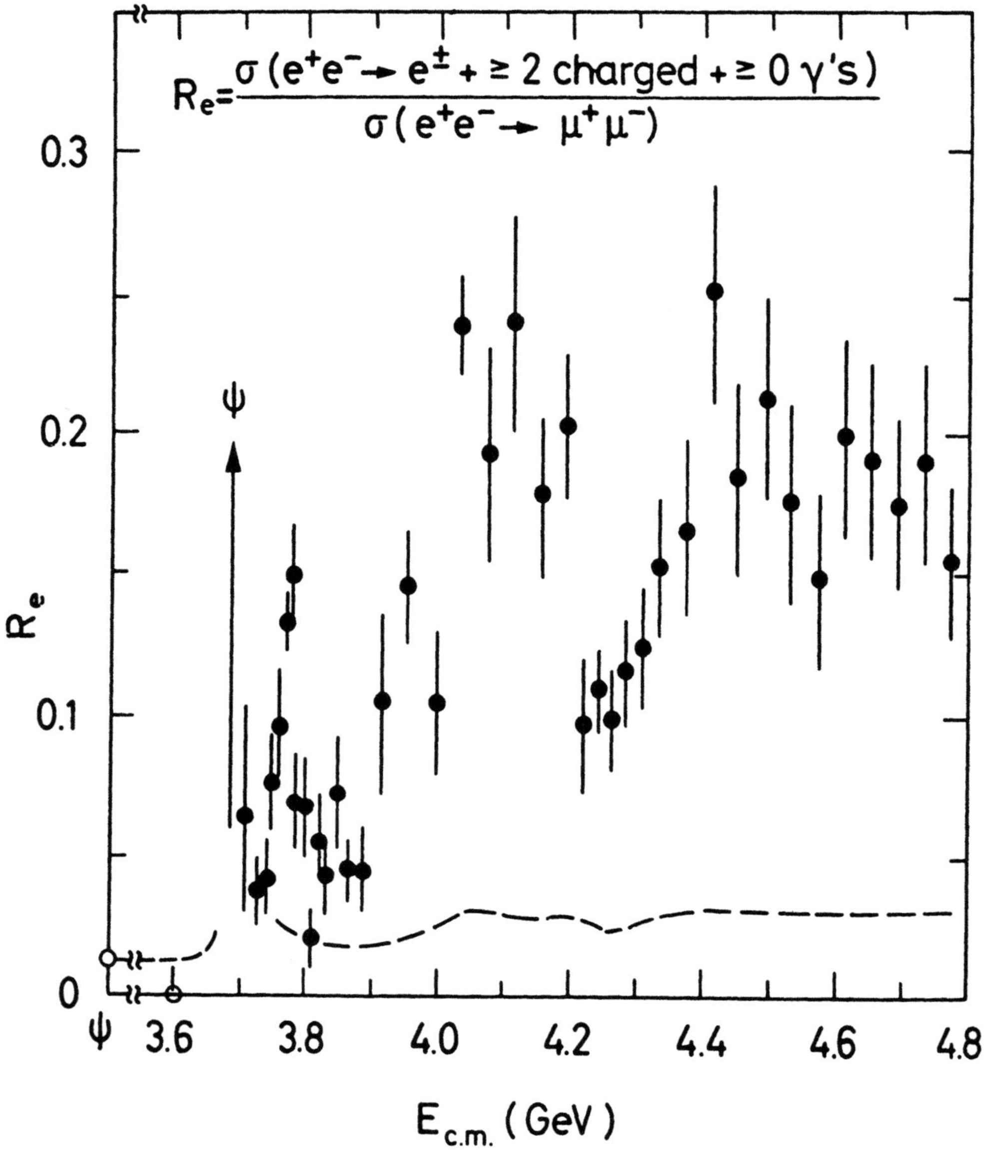

FIG. 7.7.5. : R_e, the cross section for $e^+e^- \to e^\pm + \geq 2$ prongs + ≥ 0 photons normalized to the point cross section, plotted versus the cm energy. The dotted line indicate the background. The data are from DELCO.

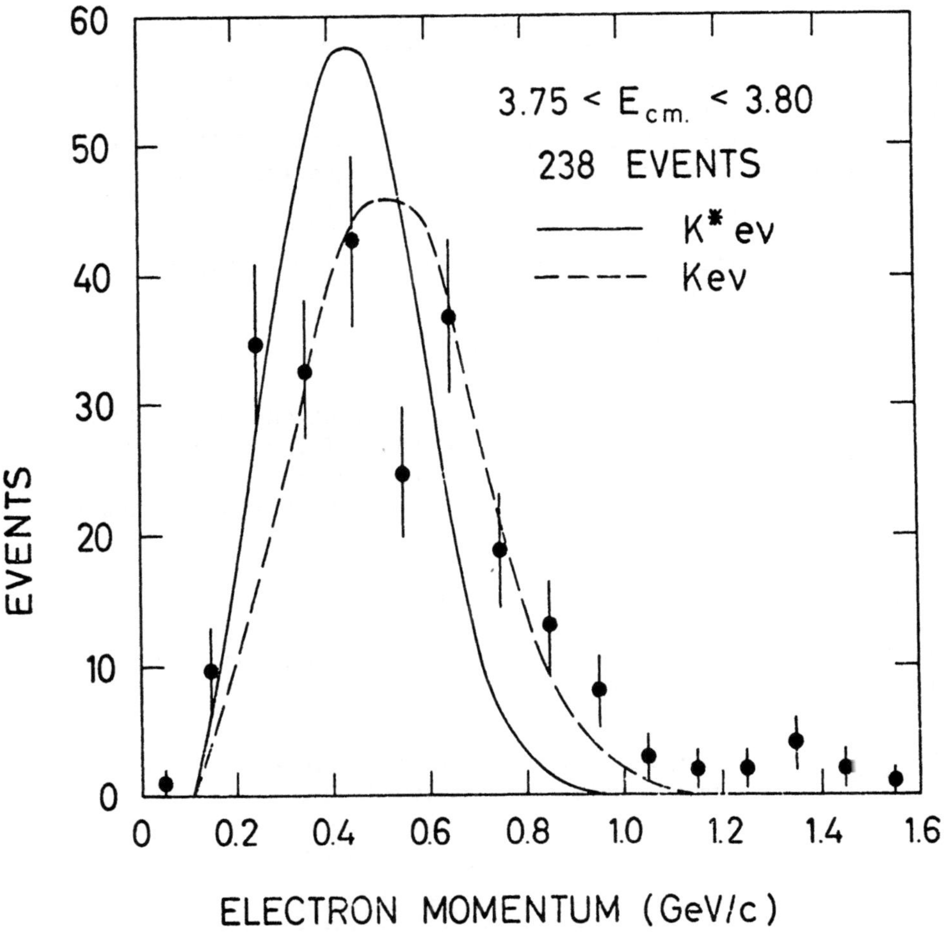

FIG. 7.7.6. : Electron momentum spectrum (measured by DELCO) for the multiprong events at the ψ''. The dashed and the solid curves show the expected shape for the decay modes $D \to e\bar{\nu}_e K$ and $D \to e\bar{\nu}_e K^*(892)$ (evaluated for V-A couplings). The dot-dashed curve gives an estimate of the background.

The probability that these criteria are satisfied by a hadron was measured at the J/ψ resonance. They find this probability to be 1.5 % for a particle of momentum 300 MeV/c decreasing to 0.4% for a particle of 1.2 GeV/c. The probability averaged over the hadron spectrum is 1.1% for momenta above 300 MeV/c.

As discussed above the resonance $\psi''(3772)$ is a clear and well defined source of D mesons. The LBL-SLAC group found a total of 61 candidates for electron inclusive multiprong events at the $\psi''(3772)$. The background, not including τ pairproduction, was estimated to contribute 25 ± 5 events. Assuming that the excess signal results from semileptonic D decays they find :

$$B(D \rightarrow eX) = (7.2 \pm 2.8)\%.$$

The momentum spectrum is plotted in Fig. 7.7.7 and compared to the spectra predicted for

$$D \rightarrow e\bar{\nu}_e\pi,\ e\bar{\nu}_e K \text{ and } e\bar{\nu}_e K^*(892).$$

The spectrum is not corrected for the contribution from τ pairproduction. The data are consistent with $e\bar{\nu}_e K$ (confidence level 33%) or $e\bar{\nu}_e K^*$(C.L. 13%) but less consistent with $D \rightarrow e\bar{\nu}_e$ (C.L. = 3%). The data are inconsistent with $D \rightarrow e\nu$ which would produce a flat electron spectrum from 810 MeV/c to 1080 MeV/c.

Data on inclusive electron multiprong events have also been collected at higher energies where not only D but also F and charmed baryon production are expected to be important. Preliminary results are listed below.

Table 7.6 - Preliminary results on inclusive electron multihadron events (Reference 97).

E_{CM} (GEV)	R_e	R_{ch}	$B(c \rightarrow eX)$
4.1 - 4.2	0.26 ± 0.08	2.1 ± 0.5	7.7 ± 3.0
4.4 - 5.7	0.23 ± 0.06	1.9 ± 0.5	7.4 ± 2.8
6.4 - 7.4	0.26 ± 0.06	1.9 ± 0.4	8.7 ± 3.2

The data on the semileptonic charm decays are summarized in table 7.7 :

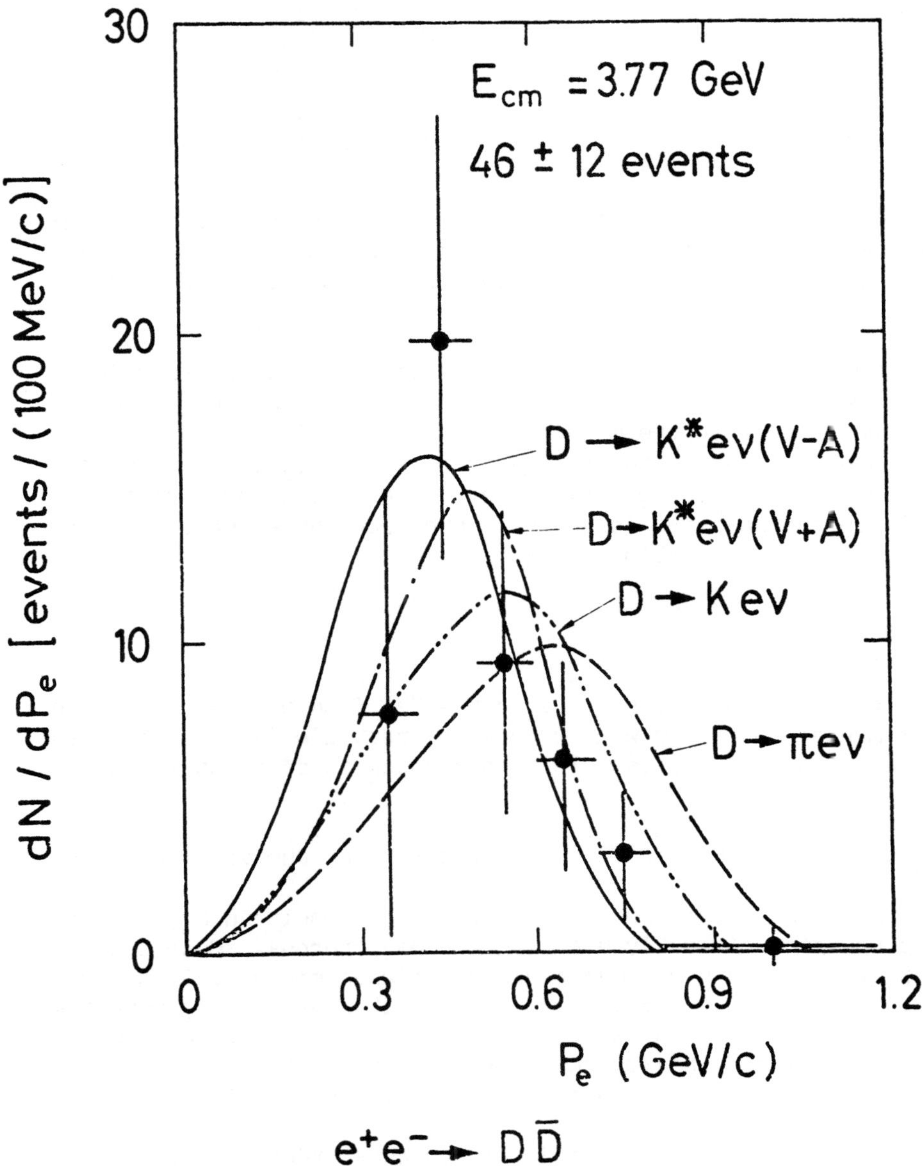

FIG. 7.7.7. : Electron momentum spectrum measured by the SLAC-LBL group for electron inclusive multiprong events. Predictions for $D \to e\bar{\nu}_e\pi$ (dotted), $e\bar{\nu}_eK$ (solid and $e\bar{\nu}_eK^*$ (892) are shown.

Table 7.7 - Semileptonic branching ratio of D mesons.

Group Reference	BR(D → eX)
DASP	8 ± 2%
DELCO	11 ± 3%
LBL-SLAC*	7.2 ± 2.8%

The quoted values are valid for a mixture of charged and neutral D mesons. In principle they can be measured separately at the ψ"(3772) resonance by selecting events with one charged or neutral D meson.

The semileptonic branching ratio is larger than the value of 4% predicted [75] from the weak decays of strange particles. This indicates that the mechanism responsible for enhancing the non leptonic channels in strange particle decays are less effective [76] for charmed particle decays. In fact if none of the available channels are selectively enhanced one expects a semileptonic branching ratio of 0.20.

8. The τ particle.

There is now convincing evidence [99] from e^+e^- annihilation experiments that a new weakly decaying particle [11] τ with a mass around 1.8 GeV/c^2 exists. This conclusion is based on measurements of :

1) $e^+e^- \rightarrow e\mu$ + nothing
2) $e^+e^- \rightarrow e(\mu)$ + minimum ionizing track + any number of photons.

Such final states can in principle be populated by associated production of weakly decaying charmed hadrons. Detailed considerations however shown that this mechanism can only account for a small fraction of (1) and (2). That a new weakly decaying particle not associated with charm indeed does exist has now been conclusively show by the DASP Collaboration which has observed τ-pairproduction below charm threshold at the ψ'.

The properties of (1) and (2) are consistent with the pairproduction and decay of a new lepton with its own quantum number. Before we discuss the data bearing on the τ-particle in more

*) including the contribution from τ pairproduction will increase the branching ratio to 7.6 ± 2.8%.

detail let me briefly remind you how a sequential lepton will appear in a e^+e^- experiment and list some of its characteristic features.

8.1. Heavy Charged Leptons.

A charged lepton is a pointlike particle with spin 1/2, charge e and no anomalous magnetic moment. Leptons are pairproduced, the cross section rises very rapidly near threshold and approaches the point cross section at high energies. These features can be used to identify a new lepton independent of its weak decays.

Decay modes of a sequential heavy lepton in standard weak interaction theory with V-A couplings are depicted in Fig. 8.1.1 A,B. The leptonic decay modes can be computed unambigously :

$$\Gamma_e \equiv \Gamma(\tau \to \nu_\tau e \bar{\nu}_e) = \frac{G^2 \cdot m_\tau^5}{192\pi^3}$$

$$\Gamma_\mu \equiv \Gamma(\tau \to \nu_\tau \mu \bar{\nu}_\mu) = \Gamma_e \cdot F(y)$$

$$G = \frac{1.02}{m_p^2} \times 10^{-5} (\mathrm{GeV}^{-2}) \qquad \text{and}$$

$$F(y) = (1 - 8y + 8y^3 - y^4 - 12y^2 \cdot \ln y).$$

$F(y=m_\mu^2/m_\tau^2)$ is a small phase space correction.

A lepton decays semihadronically as shown in Fig. 8.1.1B. Conventional theory predicts that the τ decays into final states of low multiplicity and a small ratio of kaons to pions. This is indeed reproduced in actual calculations [100] and an estimate [101] of various branching ratios is listed in table 8.1.

Some of the decay widths are rather well known

$\tau \to \nu_\tau \pi$: The pion mode is directly related to $\pi \to \bar{\nu}_\mu \mu$ as shown in Fig. 8.1.1C and D. To a good approximation :

$$\frac{\Gamma(\tau^- \to \nu_\tau \pi^-)}{\Gamma_e} = \frac{12\pi^2 f_\pi^2 \cos^2\theta_c}{m_\tau^2} \frac{1.04 m_p^2}{m_\tau^2} \quad \text{with } f_\pi = 0.137 m_p$$

$\tau \to \nu_\tau K$: This decay is directly related to $K \to \bar{\nu}_\mu \mu$

$\tau \to \nu_\tau \rho$: This decay is related via CVC to $e^+e^- \to \rho$. The width for $\tau \to \nu_\tau \rho$ can be evaluated with an accuracy of 20% using the measured values of the ρ-coupling constant.

$\tau \to \nu_\tau \cdot (n\pi)$: These decay modes are related via CVC to $e^+e^- \to (n\pi)$. (n even)

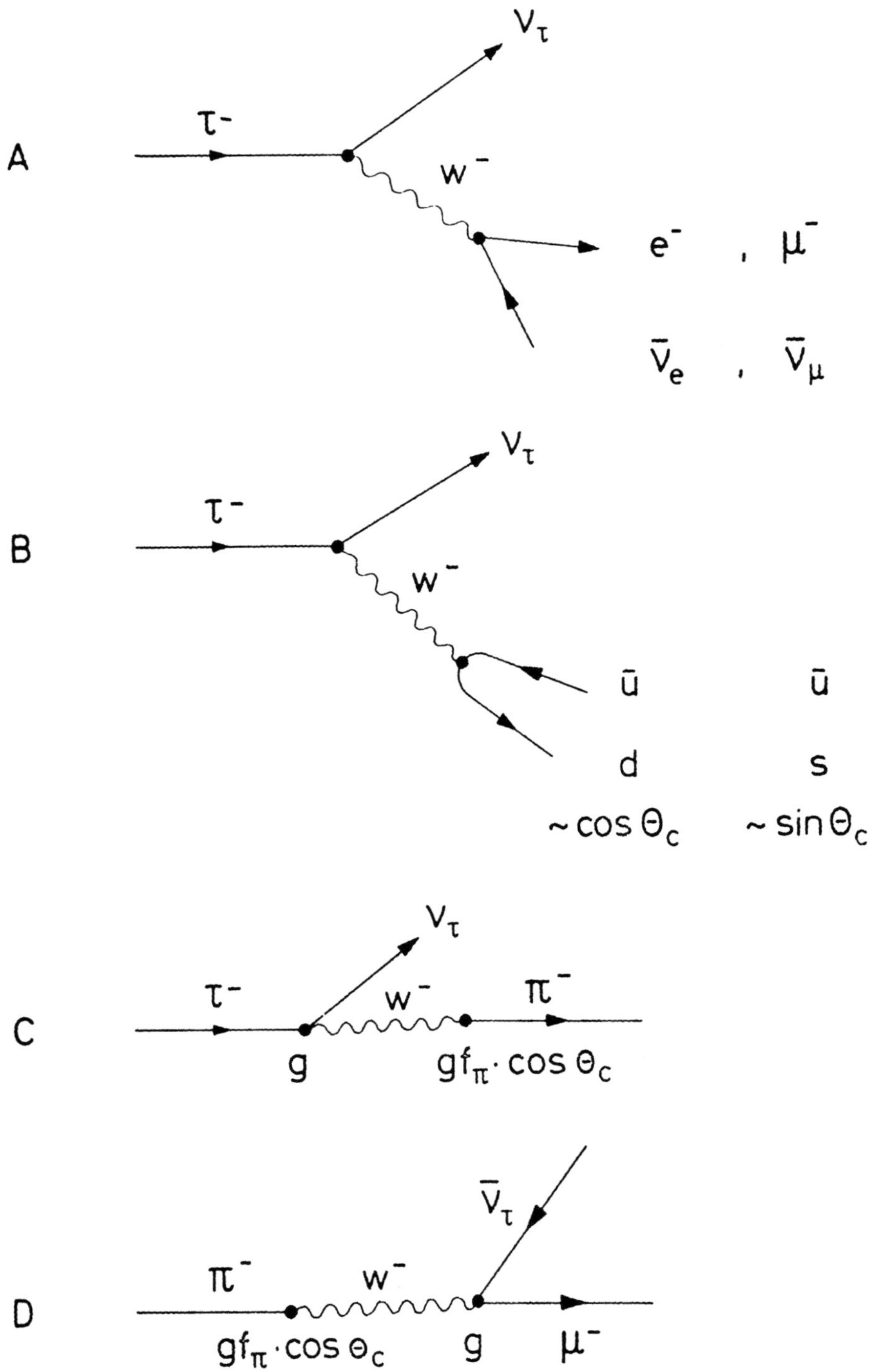

FIG. 8.1.1. : a,b) The graphs for leptonic and semihadronic decays of a heavy sequential lepton.
c,d) The relationship between $\pi^-\to\bar{\nu}_\mu\mu$ and $\tau^-\to\nu_\tau\pi^-$.

175.

Table 8.1 - Branching ratios for a sequential lepton of mass 1.81 GeV/c^2. The values are from reference 101).

Decay mode	Branching ratio	$\Gamma(\tau \to \nu_\tau X)/\Gamma(\tau \to \nu_\tau e\bar{\nu}_e)$
$\nu_\tau e\bar{\nu}_e$	0.18	1
$\nu_\tau \mu\bar{\nu}_\mu$	0.18	0.97
$\nu_\tau \pi$	0.10	0.60
$\nu_\tau \rho$	0.22	1.24
$\nu_\tau A_1$	$\sim$ 0.1	0.41
$\nu_\tau 4\pi$	0.2 (for $\nu_\tau 4\pi$ to $\nu_\tau 7\pi$)	0.44±0.13
$\nu_\tau 5\pi$		$\sim$ 0.44
$\nu_\tau 6\pi$		$\sim$ 0.11
$\nu_\tau 7\pi$		$\sim$ 0.11
$\nu_\tau K$	< 0.01	0.03
$\nu_\tau K^*(892)$	0.01	0.05
$\nu_\tau Q$	< 0.01	0.02
$\nu_\tau (Kn\pi)$ n > 3	0.01	0.07

8.2. Experimental information on the τ.

Pairproduction of new leptons will lead to mixed electron muon events via :

$$e^+e^- \to \tau\bar{\tau} \to (\nu_\tau e\bar{\nu}_e)(\bar{\nu}_\tau \bar{\mu}\nu_\mu).$$

The number of eμ-events can be written as

$$N_{e\mu} = 2\sigma(\tau\bar{\tau})\ L \cdot A_e \cdot A_\mu \cdot B_e \cdot B_\mu$$

$\sigma(\tau\bar{\tau})$, the τ-production cross section, can be written in terms of the point cross section $\sigma(\mu\bar{\mu})$ as :

$$\sigma(\tau\bar{\tau}) = \sigma(\mu\bar{\mu})\ \frac{\beta(3-\beta^2)}{2} \qquad \text{with} \quad \beta = P_\tau/E_\tau$$

and $\sigma(\mu\bar{\mu}) = 87.6\text{nb}/(2E/\text{GeV})^2$. The cross section varies rapidly near threshold and a measurement in this energy region can be used to determine the τ mass. Well above threshold the cross section is not very sensitive to the precise value of the τ mass and is given by the point cross section. L is the luminosity, A_e and A_μ the acceptances for electrons and muons. The acceptances depend on the shape of the decay lepton spectrum - i.e. on the form of the weak coupling. The difference in acceptance introduced by a V-A or a V+A coupling is small as long as the leptons are measured down to low momenta.

A measurement of the eμ-yield at high energies can therefore be used to determine $B_e \cdot B_\mu$ without knowledge of the exact mass of the lepton and the form of the weak couplings. If eμ universality is valid as expected both for sequential and ortholeptons 102) (lepton number of the τ^- = lepton number of the e^- or μ^-) then $B_e \equiv B_\mu$ and the value can be directly determined. However if the τ is a paraelectron or a paramuon (lepton number of τ^+ = lepton number of e^- or μ^-) then $B_\mu/B_e = 1/2$ or 2.

The lepton assignment can in principle be determined from a measurement of the final states :

$$e^+e^- \to \tau\bar{\tau} \to e^+e^- + \text{missing energy}$$

$$\to \tau\bar{\tau} \to \mu^+\mu^- + \text{missing energy.}$$

However, unlike the eμ channel which is rather clean, e^+e^- and the $\mu^+\mu^-$ final state events are contamined with electromagnetic events.

Information on semihadronic decays of the heavy lepton can be obtained from a measurement of inclusive electron (muon) events resulting from :

$$e^+e^- \to \tau\bar{\tau} \to (\nu_\tau \ell \bar{\nu}_e)(\bar{\nu}_\tau \cdot \text{hadrons}).$$

The rate for the inclusive events are given by

$$N_{e,h} = 2\sigma(\tau\bar{\tau}) \cdot L \cdot A_e \cdot A_h \cdot B_e \cdot B_h$$

with the nomenclature defined above.

As discussed above, τ production can be strongly enhanced relative to charm production either by selecting events with low multiplicity or by demanding the decay lepton to have momenta above 1.0 GeV. It is shown in section 7.7 that only a small fraction of the semileptonic charm decays satisfy these conditions. Most experiments reported have therefore measured the lepton two prong cross section :

$$N_{e,1p} = 2\sigma(\tau\bar{\tau}) \cdot L \cdot A_e \cdot A_{1p} \cdot B_e \cdot B_{1p}$$

with $B_{1p} = B(\tau \to \nu_\tau + 1 \text{ charged} + \geq 0 \text{ photons})$.
Electrons are sometimes excluded from the one prong modes.

8.3. Leptonic Final States.

The first evidence [11] for the production of a new pointlike particle was reported by the SLAC-LBL Collaboration at SPEAR. They observed events of the type

$$e^+e^- \to e\mu + \text{nothing}.$$

These events were selected using the following criteria :

1) Only two charged tracks with opposite charges and no photons.
2) Each track should have a momentum greater than 0.65 GeV/c.
3) One prong is identified as an electron by the pulseheight in the shower counter, the other as a muon by range. The probability of misidentifying a hadron as an electron or as a muon is respectively 18% and 20%.
4) The coplanarity angle between the planes defined by the electron and the beam direction and the muon and the beam direction must be greater than 20°.

A total of 190 events [103] with a background of 46 events was observed in the cm range from 3.8 GeV to 7.8 GeV. If these events result from charm then the leptons would be accompanied by undetected hadrons. The SLAC-LBL group made a careful evaluation of this possibility and they conclude with 90% confidence that no more than 39% of the eμ events can contain additional charged particles or photons. They conclude that the eμ events result from pairproduction of a new lepton :

$$e^+e^- \to \tau\bar{\tau} \to (\nu_\tau e\bar{\nu})(\bar{\nu}_\tau \mu \nu_\mu) \to e\mu + \text{missing energy}.$$

The observed cross section is plotted in Fig. 8.3.1 versus cm energy. The cross section varies smoothly with energy and agrees well with the energy dependence of pairproduction of a heavy lepton plotted as the solid lines in Fig. 8.3.1. Both curves evaluated for m_τ = 1.8 GeV/c^2 and 2.0 GeV/c^2, fits the data well. The momentum distribution is shown in Fig. 8.3.2. Plotted are the number of electrons and muons versus the normalized momentum variable

$$r = \frac{P - P_o}{P_{max} - P_o}$$

P_{max} is defined as the maximum momentum of the electron (muon) which is consistent with the leptonic decay of a heavy lepton of mass 1.9 GeV/c^2 with a massless neutrino. The solid curves are distributions predicted for various values of the neutrino mass assuming a V-A leptonic coupling. The dotted curve is for a V+A leptonic coupling and a massless neutrino. For m_τ=1.9GeV/c^2 and a massless τ neutrino they find that the χ^2 probability is about 0.1% for V+A compared to 50% for V-A.

The 95% confidence upper limit on the neutrino mass is m_{ν_τ} < 0.6 GeV for m_τ = 1.90 GeV/c^2 and V-A coupling. The mass of the τ was determined to (1.90 ± 0.10) GeV/c^2. This is an average over various methods and assumes V-A and a massless neutrino. The leptonic branching ratios were determined from the observed eμ cross section (Fig. 8.3.1) to

$$B_e \equiv B_\mu \equiv 0.186 \pm 0.01 \pm 0.028.$$

The first error is statistical and the second systematic. This value assumes m_{ν_τ} = 0, V-A and eμ universality.

The PLUTO Collaboration has reported [12] data on $e^+e^- \rightarrow$ eμ + missing energy with a very small contamination of other events. Electrons above 0.3 GeV/c are identified in 55% of 4π by the showers produced in traversing two radiators respectively 0.44 and 1.7 of a radiation length thick. The probability P_{he} to misidentify a hadron as an electron is determined to P_{he}= (3.5 ± 0.7)%. Muons with momenta above 1.0 GeV/c are identified in 43% of 4π as particles penetrating 68cm or iron. The probability $P_{h\mu}$ for a hadron to fake a muon was determined experimentally at the J/ψ resonance to $P_{h\mu}$ = (2.8 ± 0.7)%. PLUTO reports for cm energies between 4.0 GeV and 5.0 GeV a total of 23 eμ events with a background of 2.0 ± 0.5 events. The number and the energy dependence (Fig. 8.3.3) of the events are consistent with the heavy lepton hypothesis. They cannot be explained by charm production.

The leptonic branching ratios and the τ mass is determined from the eμ events and the data [12] on the inclusive (μ+1prong)

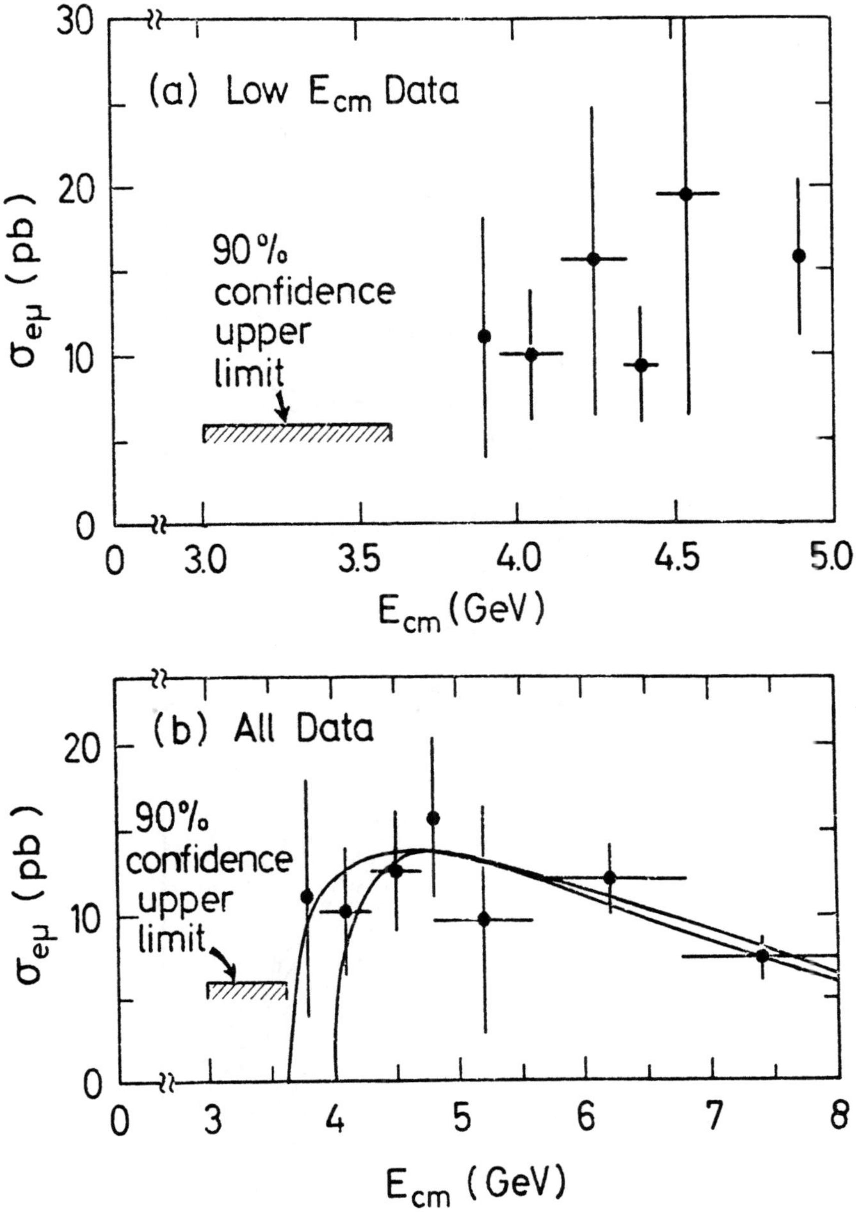

FIG. 8.3.1. : The observed eμ production cross section plotted versus cm energy. The data are from the SLAC-LBL Collaboration, only statistical errors are indicated and the horizontal lines show the energy range covered by each point. The 90% confidence upper limit on the background based on data below 3.6GeV are shown. The solid lines are predictions for a lepton of mass 1.8GeV (2.0GeV) assuming a V-A current.

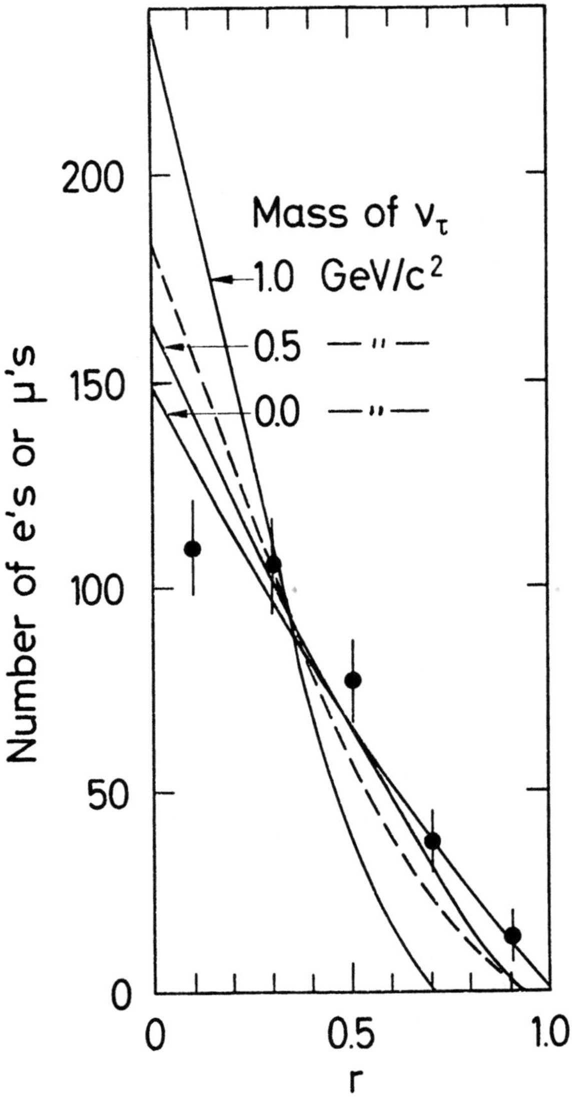

·IG. 8.3.2. : Number of electrons and muons observed in the SLAC-LBL experiment plotted versus r=P-0.65GeV/c / P_{max}-0.65GeV/c. Solid curves are for m_τ=1.9GeV/c^2 and V-A coupling with the mass of the neutrino as a parameter. The dashed curve is for V+A coupling with m_τ=1.91GeV/c^2 and a massless τ-neutrino.

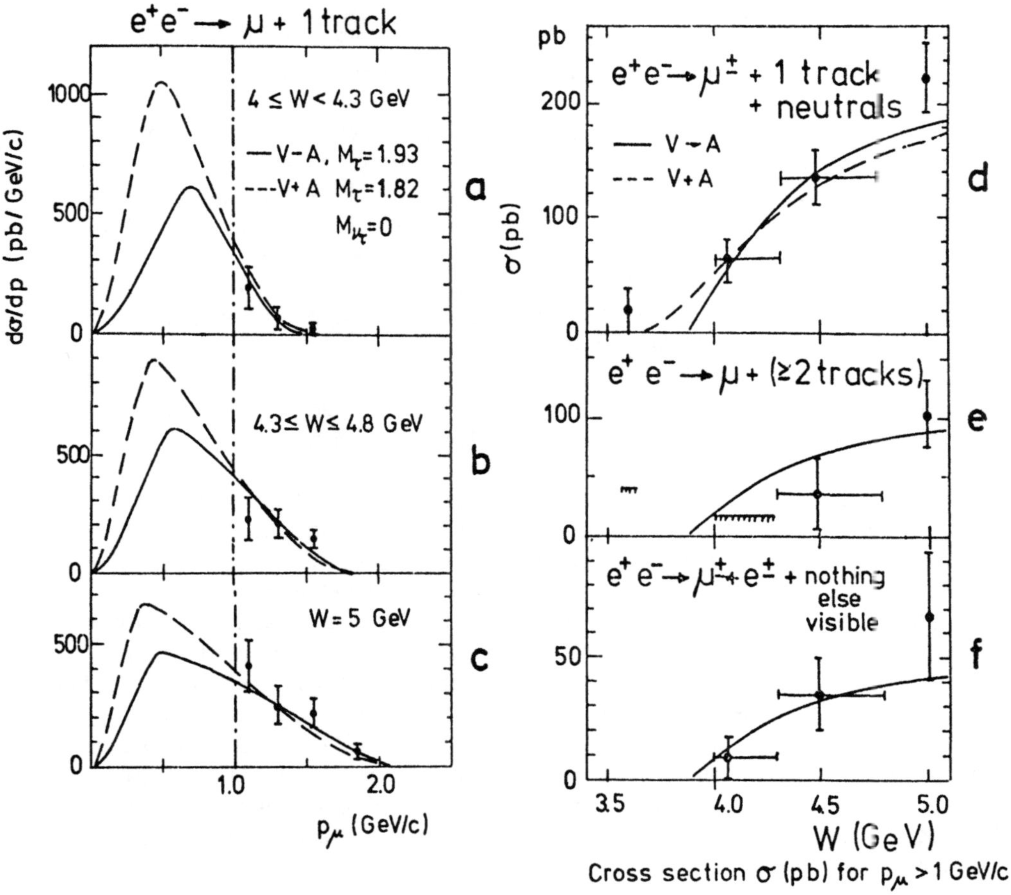

FIG. 8.3.3. : Inclusive muon data P_{μ}>1.0GeV/c measured by the PLUTO group. The data are corrected for QED-production and other sources of background.

a,b,c) The muon momentum distribution of two prong events for three different cm energies. The curves are fits to the data assuming pairproduction of a heavy lepton τ, a massless τ neutrino and V-A (solid line) or V A (dotted line) leptonic weak couplings.

d) The observed inclusive muon two prong cross section as a function of cm energy. The predicted lepton cross sections are shown.

e) The observed inclusive muon multiprong cross section as a function of energy. The solid line shows the cross section expected for a heavy lepton of mass 1.9 GeV/c^2 and a V-A current.

f) The observed eμ cross section versus energy. The solid line is the cross section expected for a heavy lepton of mass 1.9GeV/c^2 and a V-A current normalized to the data.

cross section discussed below. For V-A couplings they find $BR(\tau \to \nu_\tau \mu \bar{\nu}_\mu) = 14 \pm 3.4\%$, $BR(\tau \to \nu_\tau e \bar{\nu}_e) = (16 \pm 6\%)$ and $m_\tau = 1.93 \pm 0.05$ GeV/c^2. A V+A coupling yields $BR(\tau \to \nu_\tau \mu \bar{\nu}_\mu) = 17 \pm 5\%$, $BR(\tau \to \nu_\tau e \bar{\nu}_e) = 13 \pm 5\%$ and $m_\tau = 1.82 \pm 0.08$ GeV/c^2. The V-A fit is slightly favoured over the V+A fit.

The DASP group has measured [13] $e^+e^- \to e\mu$ + missing energy in the cm range of 4.0 GeV to 5.2 GeV. Muons of momenta greater than 0.7 GeV/c were identified in the outer detector ($P_{h\mu} = 4.2 \pm 0.8\%$) by range, electrons with momenta above 0.2 GeV/c either in the inner detector ($P_{he} = 2 \pm 0.9\%$) or the outer ($P_{he} = 4\times10^{-4}$). A total of 13 $e\mu$ events with an estimed background of 1.2 ± 0.4 events were found. Using the known production cross section and assuming $e\mu$ universality yield :
$B_e = B_\mu = 0.182 \pm 0.028 \pm 0.014$ for a V-A current and
$B_e = B_\mu = 0.206 \pm 0.033 \pm 0.015$ for a V+A current. The first error is the statistical the second the systematic.

The SLAC-LBL group has recently reported [104] new results on $e^+e^- \to e^\pm$ + a nonshowering particle + ≥ 0 photons. Data are collected for cm energies between 4.1 GeV and 7.4 GeV. The electron ($p_e \geq 0.4$ GeV/c) is identified in the lead glass wall as discussed above. The nonshowering particle must have a momentum larger than 0.65 GeV/c and it is classified as muon or hadron by range ($P_{h\mu} = 0.18$). The acoplanarity angle with respect to the beam axis should be at least 20° and the mass squared recoiling against the pair must be greater than $(0.8\ \text{GeV}/c^2)^2$ at lowest cm energy and $1.5(\text{GeV}/c^2)^2$ at the highest. They find a total of 21 $e\mu$ events with an estimated background of 0.4 events. Assuming V-A couplings $m_\tau = 1.9$ GeV/c^2, $m_{\nu_\tau} = 0$ and e universality they find $B_e = B_\mu = (22.4 \pm 5.5)\%$.

τ pairproduction will also lead to events of the type :

$$e^+e^- \to \tau\bar{\tau} \to e^+e^- + \text{missing energy}$$
$$e^+e^- \to \tau\bar{\tau} \to \mu^+\mu^- + \text{missing energy.}$$

The SLAC-LBL group [105] has selected events of this type demanding $P_e > 0.65$ GeV/c, $P_\mu > 0.65$ GeV/c. The acoplanarity angle with respect to the beam axis should be greater than 20°. After large corrections for QED and misidentified hadron events they report :

$$\frac{\text{number of ee events}}{\text{number of } e\mu \text{ events}} = 0.56 \pm 0.14 \begin{array}{l} +\ 0.16 \\ -\ 0.19 \end{array}$$

$$\frac{\text{number of } \mu\mu \text{ events}}{\text{number of } e\mu \text{ events}} = 0.70 \pm 0.15 \pm 0.19.$$

A sequential lepton or an ortholepton yield 0.5 for both ratios.

The authors [105] quote that an electron related paralepton will yield $N_{ee}/N_{e\mu} = 0.86$, $N_{\mu\mu}/N_{e\mu} = 0.29$ for the kinematical cuts used. A muon related paralepton yields the $N_{ee}/N_{e\mu} = 0.29$ and $N_{\mu\mu}/N_{e\mu} = 0.86$. The paralepton assignment is disfavoured by the data.

The Colorado-Pennsylvania-Wisconsin group [106] searched for

$$e^+e^- \rightarrow \mu^+\mu^- + \text{missing energy}$$

using the iron ball detector at SPEAR. The detector is a solid magnetized iron spectrometer with an azimuthal field of 15 kGauss and a radius of about 1m. It identifies muons above 1.3 GeV/c in nearly 90% of 4π. They select muon pairs with an acoplanarity angle with respect to the beam axis greater than 10° and demand that the system recoiling against the pair should have a mass squared greater than $4(\text{GeV}/c^2)^2$. A total of 25 $\mu^+\mu^-$ events satisfy the criteria, the estimated background is 14 events. The remaining 11 events yield a branching ratio

$$B_\mu = 0.20 \begin{array}{l} +0.10 \\ -0.08. \end{array}$$

This is evaluated assuming $m_\tau = 1.9$ GeV/c^2, V-A coupling and a massless neutrino.

8.4. Two prong inclusive electron and muon final states.

The first evidence for anomalous muon production in $e^+e^- \rightarrow \mu$ + charged track + $\geq$ photons was reported [107,106] by the Maryland, Pavia and Princeton Collaboration. They identify muons above 1.0 GeV/c by range in a single arm magnetic spectrometer. The statistics are poor, but the data are consistent [11] with the heavy lepton hypothesis based on the $e\mu$ events.

The SLAC-LBL group has measured [108] the cross section for $e^+e^- \rightarrow \mu^\pm$ + charged particle + ≥ 0 photons for cm energies between 3.9 GeV and 7.8 GeV. Muons with momenta above 0.91 GeV/c are defined by range (equivalent of 65cm of iron) in the "muon tower" (section 2.2). A hadron has approximately a 3% chance to fake a muon. Events should be acoplanar with respect to the beam axis by 20° and the mass squared recoiling against the two prongs should be greater than $1.5\ (\text{GeV}/c^2)^2$. After the background subtraction a significant signal of 103 ± 18 events remained for cm energies between 5.8 GeV and 7.8 GeV. The momentum spectrum of the muons is plotted in Fig. 8.4.1a and it can be compared to the spectrum of muons associated with multiprong events plotted in Fig. 8.4.2b. That spectrum, which mainly results from charm, is much softer than the muon spectrum of the two prong sample. A fit to the data assuming V-A couplings, BR($\tau \rightarrow \nu_\tau$ + 1 charged + ≥ 0 photons) = 0.85, $m_{\nu_\tau} = 0$ and $m_\tau = 1.9$ GeV/c^2 shown as the solid line in Fig. 8.4.1a yields $B_\mu = 0.175 \pm 0.027 \pm 0.03$. The second error shows the systematic uncertainties. Data from PLUTO [12]

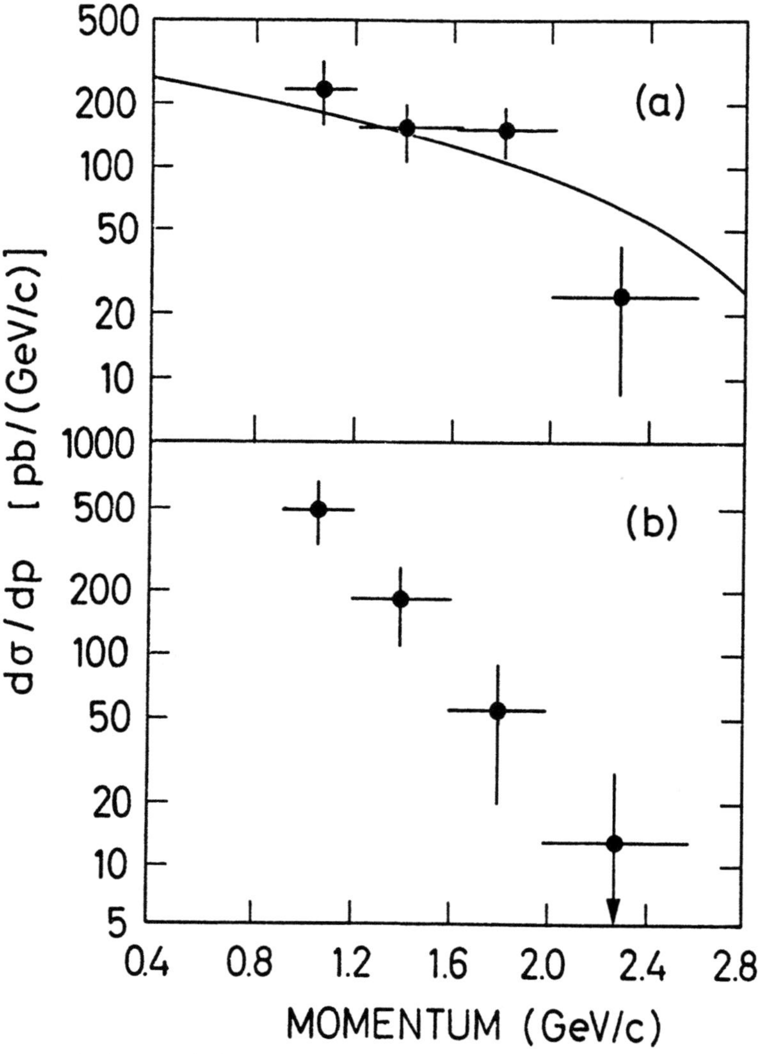

FIG. 8.4.1. : a) The momentum distribution of anomalous muon two prong events. The data were obtained by the SLAC-LBL group in the cm energy range of 5.8 to 7.8 GeV. The solid curve represents the cross section expected from heavy lepton production with the parameters listed in the text.

b) The momentum distribution of inclusive muon multiprong events.

and DASP [13] indicate that the 1 prong branching ratio might be smaller than the value assumed. Decreasing the one prong branching ratio will increase B_μ.

PLUTO has measured [12] the cross section for inclusive muon production :

$$e^+e^- \to \mu^\pm + \geq 1 \text{ charged track} + \geq 0 \text{ photons}$$

for cm energies between 3.6 GeV and 5.0 GeV. The sample is divided into two prong and multiprong events. The main background to the two prong sample results from various QED processes. This background is largely removed by demanding an acoplanarity angle of the tracks with respect to the beam axis greater than 10°, and that the missing mass squared recoiling against the observed tracks is larger than 1.4 $(\mathrm{GeV}/c^2)^2$ at 3.6 GeV and 2.7 $(\mathrm{GeV}/c^2)^2$ at 5 GeV. After all corrections a clear signal remains except at 3.6 GeV where all the events can be explained by conventional sources. The momentum distributions of the observed events are plotted in Fig. 8.3.3a,b,c. The spectra show the triangular upper end characteristic of the 3-body decay of a moving object.

The two prong cross section is plotted in Fig. 8.3.3d as a function of energy. The data indicate a threshold in the 3.6 GeV to 4.0 GeV region. The data are fitted to the heavy lepton hypothesis for V+A (dotted line) and V-A (solid line) assuming a massless neutrino. The results, plotted in Fig. 8.3.3a,b,c,d are in agreement with the heavy lepton hypothesis and yield 1.93 ± 0.05 GeV/c^2(V-A) and 1.82 ± 0.08 GeV/c^2 (V+A) for the value of the mass. The size of the cross section near threshold excludes that the muons result from the pairproduction and weak decays of a spin 0 particle. The point cross section for paiproducing a spin 0 particle is

$$\sigma(\tau\bar{\tau}) = \frac{\sigma(\mu\bar{\mu})}{4} \beta_\tau^3 .$$

Using this as an input leads to unphysical values for the leptonic branching ratios.

The branching ratio for the τ to decay into multiprong and one prong events is determined by assuming that all the events in the multiprong sample with muon momenta above 1 GeV/c result from τ pairproduction. They find BR(1 prong) = 0.70 ± 0.10 and BR(multiprong) = 0.30 ± 0.10.

The DASP Collaboration has measured [13]

$$e^+e^- \to e^\pm + \text{nonshowering track} + \geq 0 \text{ photons}$$

$$e^+e^- \to \mu^\pm + \text{nonshowering track} + \geq 0 \text{ photons}$$

for cm energies between 3.6 GeV and 5.2 GeV.

The selection criteria for the inclusive electron sample have been discussed earlier. It is important to note that electrons are measured down to 0.2 GeV/c momenta and hence the cross section extracted from these data are nearly independent whether V-A or V+A couplings are assumed.

A total of 80 events were found at cm energies above 3.9 GeV 17 events at the ψ' resonance and 1 event at 3.6 GeV. The signal at the ψ' proves conclusively that the τ particle is not associated with charm and we will discuss these data in section 8.5.

A fraction of the twoprong electron events observed above 3.9 GeV might result from associated production and semileptonic decays of charmed particles. An upper limit can be obtained by assuming that all inclusive electron events with more than two prongs are due to charm production. From the measured multiplicity distribution of these events (Fig. 7.7.1) and the known detection efficiency we estimate a total of (5 ± 2) events from this source. The direct decay of a pair of charmed hadrons into a final state with one electron and one nonshowering track is expected to contribute less than one event. The background from all other sources has been estimated to (9 ± 3) events, in agreement with (7 ± 7) events extrapolated from the 3.6 GeV data.

Candidates for muon inclusive events must have one muon track in the spectrometer, a second nonshowering track and any number of photons observed either in the inner detector or in the spectrometer arms. A charged particle was called a muon if it had a momentum greater than 1.0 GeV/c, gave no signal in the threshold Cerenkov counter, suffered an energy loss consistent with that of a minimum ionizing particle in the shower counter and penetrated at least 60 cm of iron. A total of 25 events with a background of 3.8 events was found.

After all corrections DASP observe (21 ± 5) muon inclusive events and (18.5 ± 4.6) electron inclusive events with momenta above 1.0 GeV/c. The ratio of the leptonic widths are evaluated directly from these data independent of the form of the coupling. This yields $B_\mu/B_e = 0.92 \pm 0.32$ with a systematic uncertainty of 0.07. The result is consistent with eμ universality.

The quantity $2\sigma(\tau\bar{\tau}) \cdot B_e \cdot B_{ns}$ is plotted in Fig. 8.4.2 as a function of cm energy. The data are radiatively corrected assuming a τ mass of 1.8 GeV/c^2. The data at the ψ' were corrected for the vacuum polarization enhancement. Note the rapid rise of the cross section near threshold, which is characteristic for the pairproduction of a pointlike fermion; the observed threshold and the magnitude of the cross section exclude the possibility that the τ is a pointlike particle with spin 0 or 1. This is

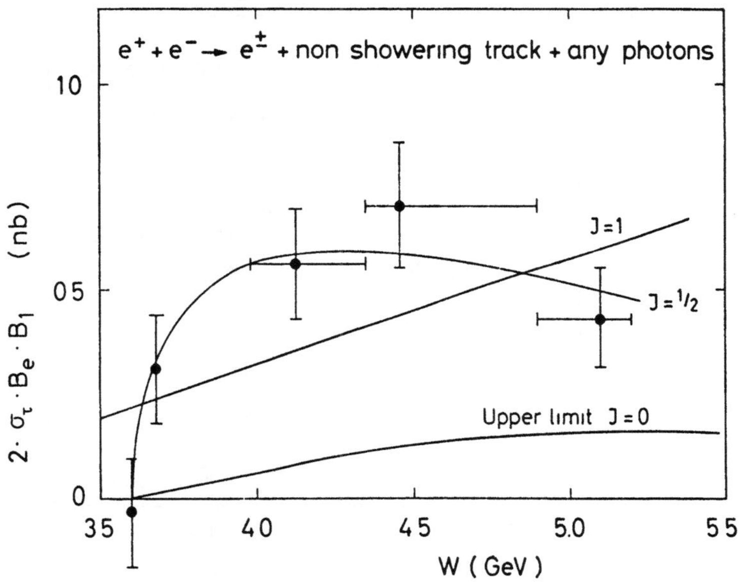

FIG. 8.4.2. : Integrated inclusive cross section for events having an identified electron, a nonshowering particle, and any number of photons as a function of cm energy. The data are from the DASP Collaboration. The solid curves show fits to the data assuming pairproduction of point particles with spin 0, 1/2 and 1.

demonstrated in Fig. 8.4.2. For spin 0 the upper limit on $2\sigma(\tau\bar{\tau})$ B_eB_{ns} was calculated with the conservative assumption that the τ has only leptonic decays and $B_e = B_\mu$. This prediction is plotted in Fig. 8.4.2 and is lower than the measured values by an order of magnitude. Assuming the mass of the τ to be well below 1.8 GeV/c^2 does not change this conclusion. For the spin 1 case the τ was assumed to have the same electromagnetic structure as a $W^\pm$ boson [109]. A fit was made to the data treating (B_eB_{ns}) and the τ mass (m_τ > 1.55 GeV) as a free parameter. The resulting curve does not fit the data. Including data obtained at higher energies [103] at SPEAR excludes J = 1.

The data are well fit by the cross section for the pairproduction of a point fermion. From the fit they determine the τ mass to be m_τ = 1.807 ± 0.020 GeV. This value, because of the data point near threshold, is more precise than earlier measurements. Since the spectrum is measured down to 0.2 GeV/c the quoted value of the mass depends very little on whether a (V-A) or a (V+A) form is taken for the current.

The fit used to evaluate the τ mass from the electron inclusive events yields $B_e \cdot B_{ns}$ = 0.086 ± 0.012. Using B_e = 0.182 ±

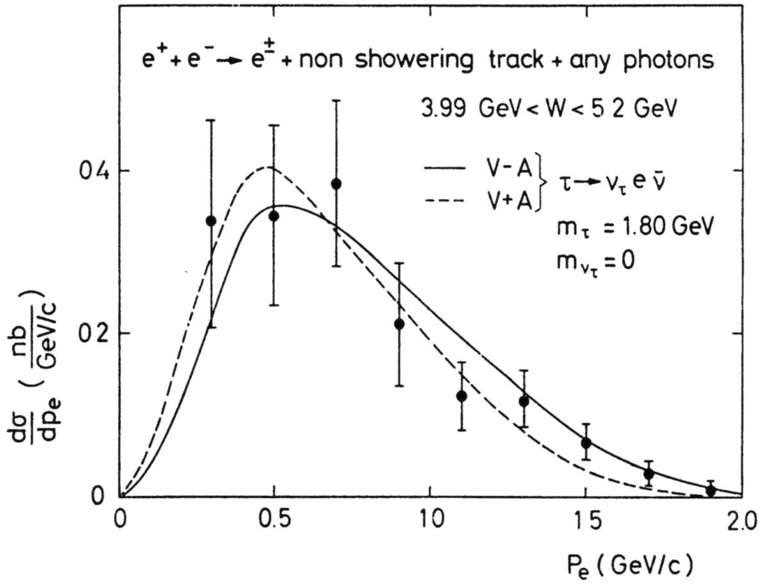

FIG. 8.4.3. : Corrected electron momentum distribution measured by DASP for events having an identified electron, a nonshowering particle, and any number of photons. (Above 1 GeV/c data having a muon instead of an electron are combined with the electron data to form a weighted mean).

0.028 the DASP group derives the branching ratio for $\tau \to \nu_\tau$ + nonshowering particle + $\geq$ 0 photons, B_{ns} = 0.47 ± 0.10. The branching ratio B_{1h} for $\tau \to \nu_\tau$ + hadron + $\geq$ 0 photons is given by $B_{1h} = B_{ns} - B_\mu = (0.29 \pm 0.11)$. The systematic errors are small compared to the statistical error. The average number of photons associated with $\tau \to \nu_\tau$ + hadron + $\geq$ 0 photons can be obtained from table 8.2 after making background corrections. Averaging the observed photon multiplicity over all two prong events in the higher energy data and correcting for the photon detection efficiency we find that a decay of the type $\tau \to \nu_\tau$ + charged hadron + any number of photons yields on the average 2.8 ± 0.7 photons.

The branching ratio B_{3h} for the τ to decay into final states with at least three charged particles can be obtained from B_{3h} = $1-B_e-B_{ns}$. (The number of electron events with p_e > 1 GeV/c and 5 or more charged tracks were found to be small). We find B_{3h} = 0.35 ± 0.11 in agreement with an earlier measurement by the PLUTO Collaboration.

The lepton momentum spectrum, obtained by combining the electron and the muon data, is plotted in Fig. 8.4.3 for cm energies between 4.0 GeV and 5.2 GeV. This spectrum extends to much higher momenta than the electron spectrum observed in the semi-leptonic decays of the charmed hadrons, reflecting the pointlike structure of the τ and the low mass of its neutrino.

The solid line shows a fit to the data assuming $m_\tau = 1.80$ GeV/c^2, a massless neutrino and a (V-A) structure of the current. The dotted line is a fit keeping the masses constant but changing the left handed V-A current into a right handed V+A current. Both fits are clearly acceptable.

Fits were also made varying the mass of the τ neutrino. The 90% confidence upper limits on the neutrino mass are $m_{\nu_\tau} <$ 0.74 GeV/c^2 for V-A and $m_{\nu_\tau} < 0.54$ GeV/c^2 for V+A.

The DELCO group has measured [19] the cross section

$$e^+e^- \rightarrow e^\pm + \text{nonshowering track} + \geq 0 \text{ photons}$$

for cm energies between 3.7 GeV and 7.4 GeV. Electrons are identified by threshold Cerenkov counters and shower counters as discussed above. In addition both tracks are required to have momenta above 0.3 GeV/c and the acoplanarity angle with respect to the beam axis must be greater than 20°. The ratio

$$\frac{\sigma(e^+e^- \rightarrow e^\pm + \text{nonshowering track} + \geq 0 \text{ photons})}{\sigma(\mu\bar{\mu})}$$

is plotted versus energy in Fig. 8.4.4. The observed cross section shows no resonances and is consistent with the cross section expected for a heavy lepton. This can be contrasted with the energy dependence plotted in Fig. 7.7.5 for $e^+e^- \rightarrow e^\pm + \geq 2$ prongs + ≥ 0 photons measured by the same group. The dotted curve in Fig. 8.4.4 shows the cross section predicted for a heavy lepton of mass 1.85 GeV/c^2 and the normalization yields B_e=0.15. However the data are still preliminary and the absolute scale is uncertain by ± 50%.

The electron momentum distribution is plotted in Fig. 8.4.5. The dotted curve shows the predictions for the leptonic decay of the τ with $m_\tau = 1.85$ GeV/c^2, $m_{\nu_\tau} = 0$ and V-A couplings. There is good agreement.

The SLAC-LBL group has measured [104] $e^+e^- \rightarrow e^\pm$ + charged + ≥ 0 photons and $e^+e^- \rightarrow e^\pm$ + missing energy for cm energies between 4.1 GeV and 7.4 GeV. The experiment and the results on $e^+e^- \rightarrow e\mu$ + missing energy is discussed above.

They have also measured the process

$$e^+e^- \rightarrow e^\pm + \text{hadron} + \geq 0 \text{ photons}.$$

A total of 31 events is observed with an estimated background of 12.1 events. The data are plotted in Fig. 8.4.6 versus r. From a fit to these events and the eμ events discussed earlier they find BR($\tau \to \nu_\tau$ + 1 hadron + $\geq$ 0 photons) = (45 ± 19)%. The fit was made assuming V-A couplings, $m_{\nu_\tau} = 0$, $m_\tau = 1.9$ GeV/c^2. The solid line shows the results of the fit.

8.5. τ-Production below charm threshold.

For some time now there has been strong evidence that a new weakly decaying particle with the properties of a sequential heavy lepton exists. This evidence can briefly be summarized as follows :

1) The cross section for $e^+e^- \to e^\pm\mu^\mp$ + missing energy is much too large to be accounted for by charm pair production followed by the semileptonic decay into $e\bar{\nu}_e$ + unseen hadrons.

2) The inclusive cross section $e^+e^- \to e^\pm$ + nonshowering track + $\geq$ 0 photons is a smooth function of energy and do not display the structure seen in the total cross section.

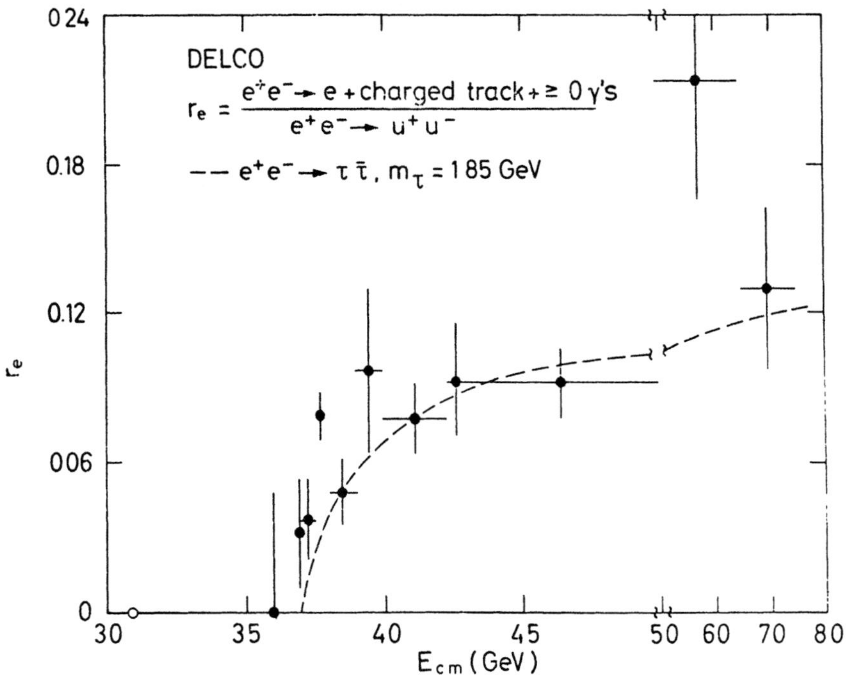

FIG. 8.8.4. : The two prong electron inclusive cross section measured by the DELCO Collaboration and normalized to the point cross section plotted as a function of cm energy. The dotted curve shows the yield expected from τ pairproduction. The absolute normalization is uncertain by ± 50%.

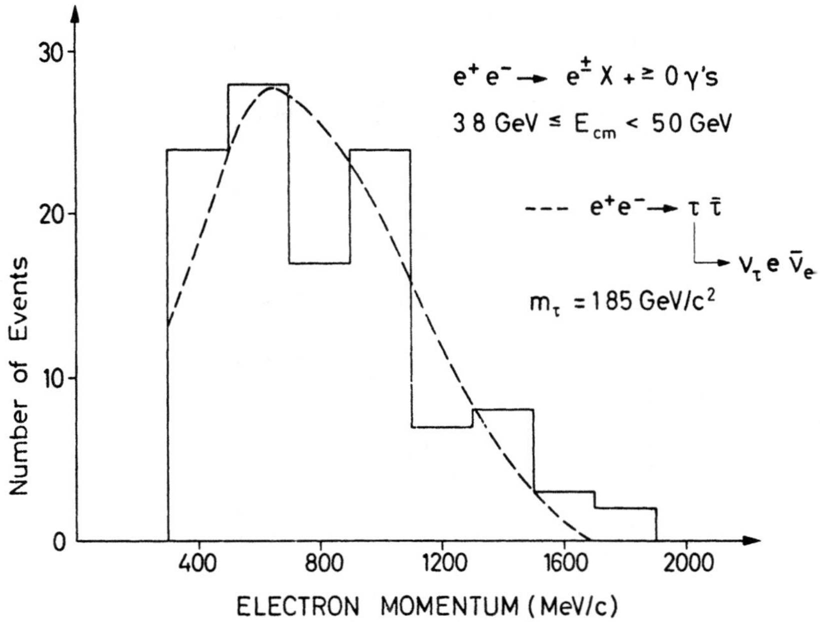

FIG. 8.4.5. : Electron momentum distribution of two-prong electron events as measured by the delco collaboration.

3) The lepton spectrum observed in the two prong class extends to high momenta and is well fit by assuming the decay of a point particle into three massless particles. It is not fit by the lepton spectrum observed in the decays of charmed particles.

4) The multiplicity distribution observed in inclusive electron (muon) events cannot be explained by charm alone.

5) The decays of the new particle yield final states with few kaons [58], contrary to the properties of the GIM current which preferentially couples to strangeness.

Recently the DASP group has observed τ pairproduction at the ψ' (3684) resonance. Since the ψ' (3684) is below threshold for associated charm production this shows conclusively that the τ is not associated with charm. We will now discuss these data in detail.

The electron momentum spectrum measured at the ψ' and plotted in Fig. 8.4.5 shows two clear clusters of events, one centered around 1.5 GeV/c and the other with momenta between 0.4 GeV/c and 0.9 GeV/c. The first cluster can be associated with the cascade decay $\psi' \to J/\psi\ X \to e^+e^-\ X$. The electrons in the second cluster have a relatively flat momentum distribution. We now

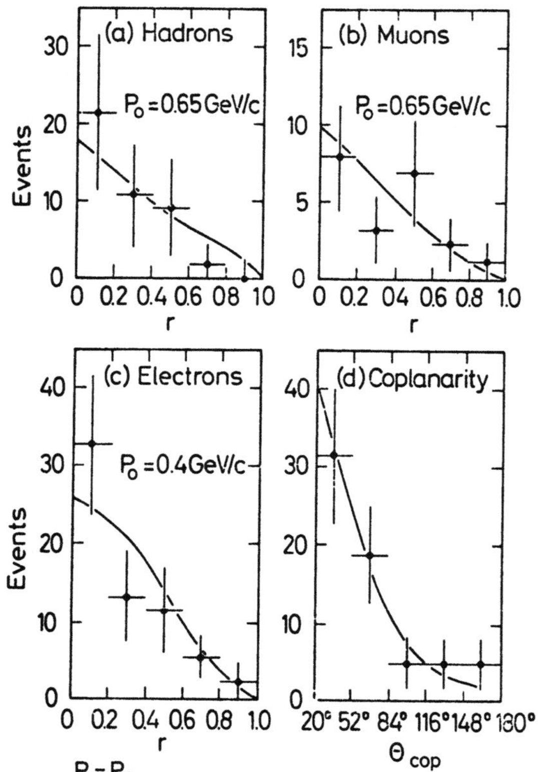

$$r = \frac{P - P_o}{P_{max} - P_o}$$

P_o = cut off momentum

P_{max} = Maximum Kinematically allowed in τ decay

FIG. 8.4.6. : The distributions for particles in inclusive electron two prong events.
a) The hadrons.
b) The muons.
c) The electrons.
d) Coplanarity distributions for all inclusive electron two-prong events. The data are from the SLAC-LBL Collaboration. The curves shown are the predictions for τ-pairproduction of a heavy lepton with m_τ = 1.9 GeV/c^2, m_{τ_ν} = 0 and V-A couplings.

consider the hypothesis that these events come from the decay $\tau \rightarrow \nu_\tau e \bar{\nu}_e$ with m_τ = 1.80 GeV/c^2 and a massless neutrino. The expected electron spectrum folded into the acceptance and normalized to nine observed events is also plotted in Fig. 8.4.5. The data are clearly consistent with the spectrum.

The background from the reaction $e^+e^- \rightarrow e^+e^- \mu^+\mu^-$ has been estimated to contribute (0.6 ± 0.2) events. An estimate using data at other energies shows that we expect less than 0.1 event from beam-gas interactions. Indeed, all the events originate within the nominal interaction volume. A twobody hadron final state can fake events of type (1) if the charged hadron traversing the magnet is misidentified as an electron. For hadrons with momenta above 0.35 GeV/c the measured probability P_{he} for this to happen is 4 x 10^{-4}. At the ψ' resonance they observe 2113 events of the type $e^+e^- \rightarrow h^\pm$ + nonshowering track + $\geq$ 0 photons where h is either a kaon or a pion traversing the magnet and the nonshowering track is observed in the inner or outer detector. This class of events therefore contributes a background of (0.84 ± 0.02) events. Dalitz decays of π° and η and photon converting in the beam pipe has been estimated using the two prong sample above. They expect a total of (0.2 ± 0.1) event from this source. We therefore estimate a background of (1.7 ± 0.3) events compared to 9 events observed. They have checked the background associated with multihadron events by searching for inclusive electron events at the J/ψ resonance. Data were recorded at the J/ψ resonance. Data were recorded at the J/ψ resonance for an integrated luminosity of 187 nb^{-1} yielding approximately 0.6 times as many multihadron events as observed at the ψ'. One event of the type $e^+e^- \rightarrow e^\pm$ + nonshowering track + $\geq$ 0 photons was found. At 3.6 GeV one inclusive electron event (1) was found. From the background rate estimated above we would expect to find 1.3 events in the data collected at the J/ψ and at 3.6 GeV compared to the two events observed. A comparison of the observed photon multiplicity is given in table 8.2 for the two prong electron inclusive events (0.4 GeV/c < p_e < 0.9 GeV/c) and for the two prong hadron inclusive events.

Table 8.2 - Photon Multiplicity.

Number of photons	0	1	2	3	4	5	6	7
$e^\pm$ + nonshowering track + $\geq$ 0 photons	4	3	1	1	0	0	0	0
$h^\pm$ + nonshowering track + $\geq$ 0 photons	207	370	440	428	312	199	99	32

The distributions are strikingly different. The electron events are accompanied by few photons as expected from τ production (table 8.1), whereas the hadron events have a large photon multiplicity.

We therefore conclude that we have observed a genuine signal for : $e^+e^- \rightarrow \tau\bar{\tau} \rightarrow (\nu_\tau e \bar{\nu}_e) + (\bar{\nu}_\tau$ + nonshowering track + $\geq$ 0 photons) at total cm energy 3.684 GeV/c. Since the ψ' is below the threshold for cham production this shows conclusively that the τ is not associated with the production of charmed mesons.

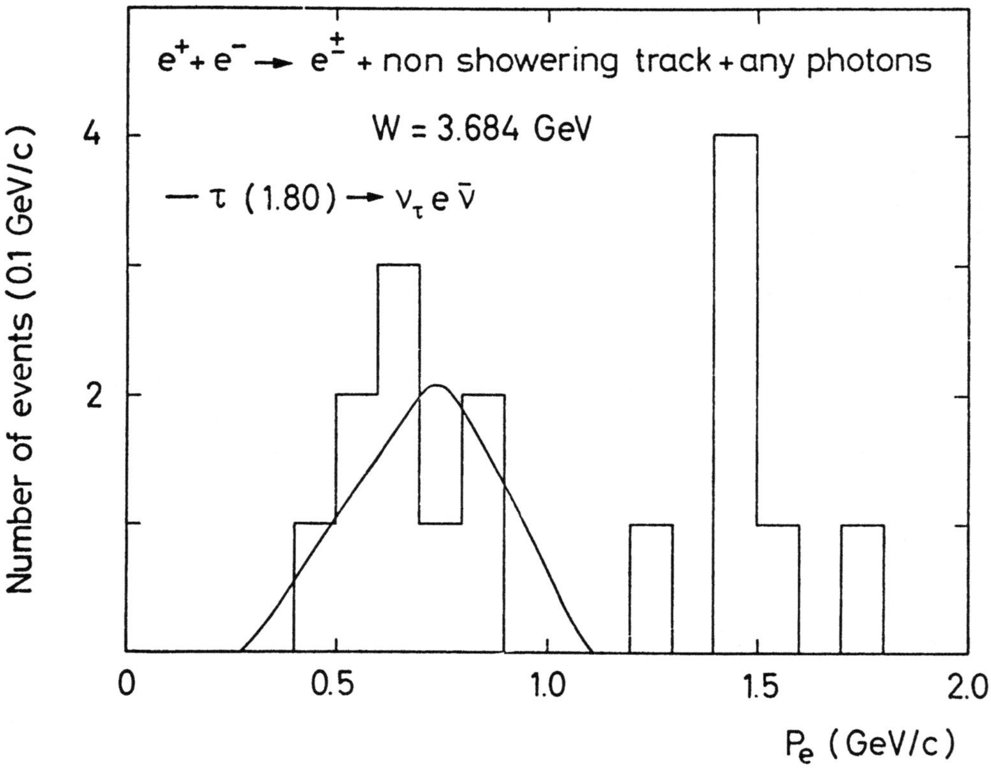

FIG. 8.5.1. : Raw electron momentum distributions observed by the DASP Collaboration at 3.684GeV. The events here are identified as electron, one non showering particle and any number of photons.

8.6. Semihadronic decays of the τ.

A characteristic feature of the standard weak interaction is that decays involving strange particles are suppressed relative to decays involving non strange final states by $tg^2\theta_C \approx 0.05$. The DASP group has determined [58] the ratio of strange to non strange particles in semihadronic τ-decays from a measurement of

$$\frac{\sigma(e^+e^- \to e^{\mp} + K^{\pm} + \geq 0 \text{ photons})}{\sigma(e^+e^- \to e^{\mp} + \pi^{\pm} + \geq 0 \text{ photons})}$$

It has been shown that the two prong cross section including one electron predominantly results from $e^+e^- \to \tau\bar{\tau} \to (\nu_\tau e\bar{\nu}_e)(\bar{\nu}_\tau$ + hadrons + $\geq$ 0 photons) with only a small contamination from charm decays. The hadrons are identified and measured using one of the spectrometer arms. The electron is identified in either the inner or the outer detector.

The DASP group finds :

$$\frac{\sigma(e^{\mp} + K^{\pm} + \geq 0 \text{ photons})}{\sigma(e^{\mp} + \pi^{\pm} + \geq 0 \text{ photons})} = 0.07 \pm 0.06.$$

Therefore on the average only 7% of all semihadronic τ-decays yield a strange particle in accordance with theory. This should be compared to multiprong events where DASP finds :

$$\frac{\sigma(e^{\mp} + K^{\pm} + \geq 1 \text{ prong} + \geq 0 \text{ photons})}{\sigma(e^{\mp} + \pi^{\pm} + \geq 1 \text{ prong} + \geq 0 \text{ photons})} = 0.24 \pm 0.05.$$

Since the charged multiplicity is on the order of 4 this is equivalent to (0.9 ± 0.18) charged kaons per multiprong event.

The decay $\tau \to \nu_\tau\pi$ is directly related to the pion lifetime as shown in Fig. 8.1.1C, D. The width $\Gamma(\tau \to \nu_\tau\pi)$ is therefore well known and can be expressed in terms of B_e as :

$$\frac{B_\pi}{B_e} = \frac{12\pi^2 f_\pi^2 \cos^2\theta_C}{m_\pi^2} = 0.60$$

for a lepton of mass 1.80 GeV.

The DASP group has searched for the $\tau \to \nu_\tau\pi$ by observing final states of $e\pi^+$ + missing energy. This final state results mainly from

$$e^+e^- \to \tau\bar{\tau} \to (\nu_\tau e\bar{\nu}_e)(\bar{\nu}_\tau\pi^+)$$

with a small contamination from other τ decays with unobserved particles or from charm production.

The pion is identified and measured with the spectrometer, the electron is identified both in the inner detector and the spectrometer. To enhance the signal relative to the background only pions with momenta above 1.0 GeV/c are considered. About 60% of all $\tau \to \nu_\tau \pi$ decay satisfy this condition. 4eπ events with $p_\pi > 1.0$ GeV/c and $p_e > 0.2$ GeV/c are observed. 10eπ events are expected from $\tau \to \nu_\tau \pi$ with $B_e = 0.18$ and $B_\pi/B_e = 0.60$ as input. The background from $\tau \to \rho\nu$ and misidentified hadrons add 2 events such that they expect to observe a total of 12 eπ events from τ decays compared to 4 events seen.
This yields $B_e B_\pi = 0.004 \pm 0.005$ compared to

$$B_e \cdot B_\pi = 0.6\ B_e^2 = 0.019 \text{ expected.}$$

The DASP data do therefore not observe the decay $\tau \to \nu_\tau \pi$ with the predicted strength. However, the statistics are too low to draw any firm conclusions.

The DASP group has also searched for events of the type

$$e^+e^- \to \tau\bar{\tau} \to \bar{\nu}_\tau K^\pm((\nu_\tau e \bar{\nu}_e) + (\nu_\tau \mu \bar{\nu}_\mu) + (\nu_\tau \pi))$$

$$= K^\pm + \text{charged track} + \text{missing energy.}$$

Only one event with $p_K > 1.0$ GeV/c was found [58]. This yields a 90% confidence upper limit of $B_K < 0.016$.

DASP has measured [58] the decay $\tau \to \nu_\tau \rho$ by selecting final states with $\pi^\pm$ + charged track + two photons. Events in which the two photons are compatible with resulting from a π° decay are retained provided that both computed photon energies are above 50 MeV. The remaining events are plotted versus $M(\pi^\pm\pi^\circ)$ in Fig. 8.6.1. Events with an identified electron are hatched. Events within the ρ-band (0.5 GeV $< M(\pi^\pm\pi^\circ) < 1.0$ GeV) are plotted versus the momenta of the $\pi^\pm\pi^\circ$-system in Fig. 8.6.2. The momentum distribution expected from the decay $\tau \to \nu_\tau \rho$ is shown as the dotted line. Note the flat distribution above 0.9 GeV/c which is characteristic for a two body decay of a moving object. The enhancement at low momenta is presumably due to multihadron events. To reduce the background, only events with a $(\pi^\pm\pi^\circ)$-momentum above 1.0 GeV/c are considered. The $\pi^\pm\pi^\circ$ mass distribution for these events are plotted in Fig. 8.6.1. These events yield

$$B_e \cdot B_\rho = 0.048 \pm 0.018 \text{ or } B_\rho = 0.27 \pm 0.10.$$

in good agreement with the theoretical predictions.

The PLUTO group has measured [109] the final state $e^+e^- \to e^\pm\pi^\mp\pi^-\pi^+$. They find, after background corrections, that these

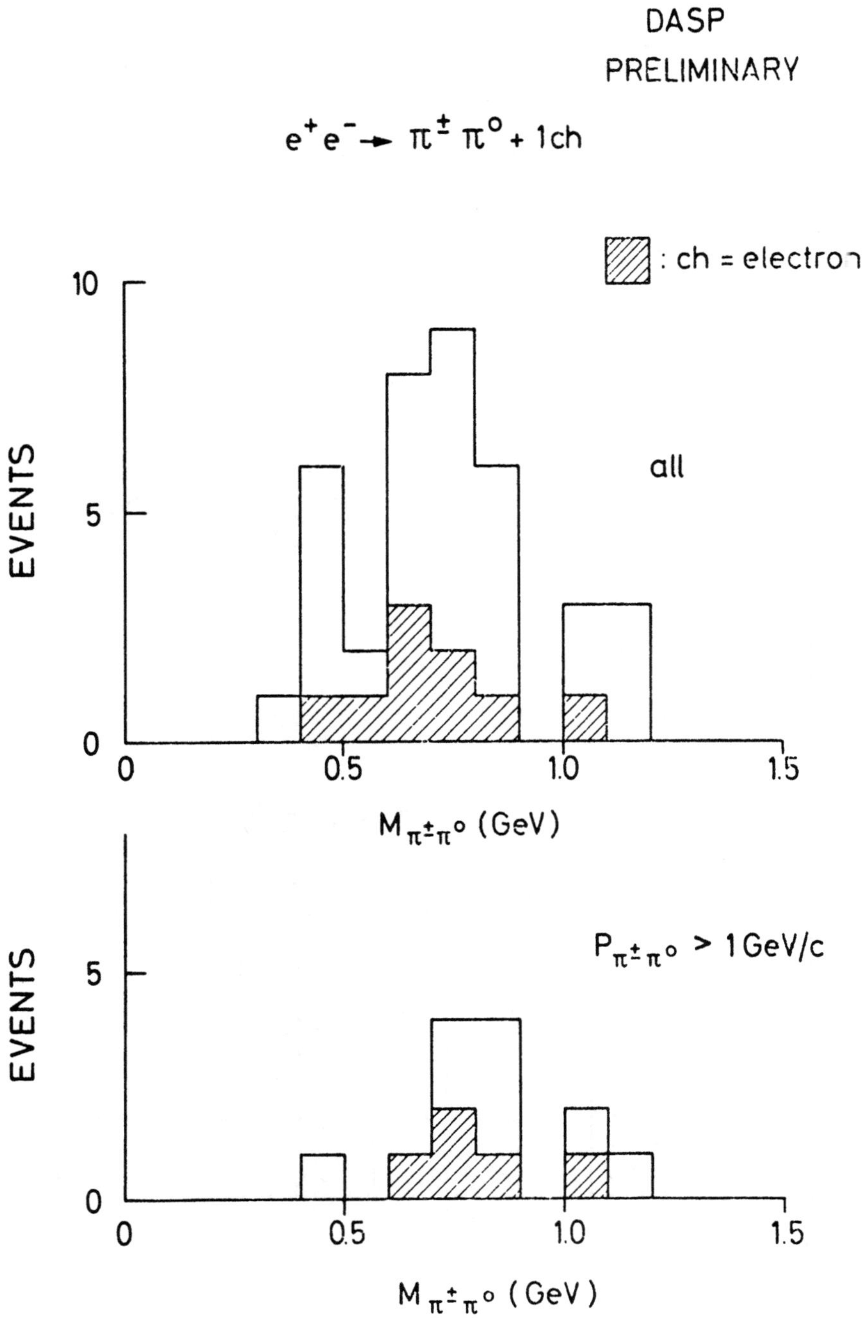

FIG. 8.6.1. : a) The distributions of $M(\pi^{\pm}\pi^{\circ})$ observed by the DASP Collaboration for events with the topology $e^+e^- \to \pi^{\pm}\pi^{\circ}$ + charged track.

b) The $M(\pi^{\pm}\pi^{\circ})$ distribution for events of the same topology as above but with $p_{\pi^{\pm}\pi^{\circ}} \to 1$ GeV/c.

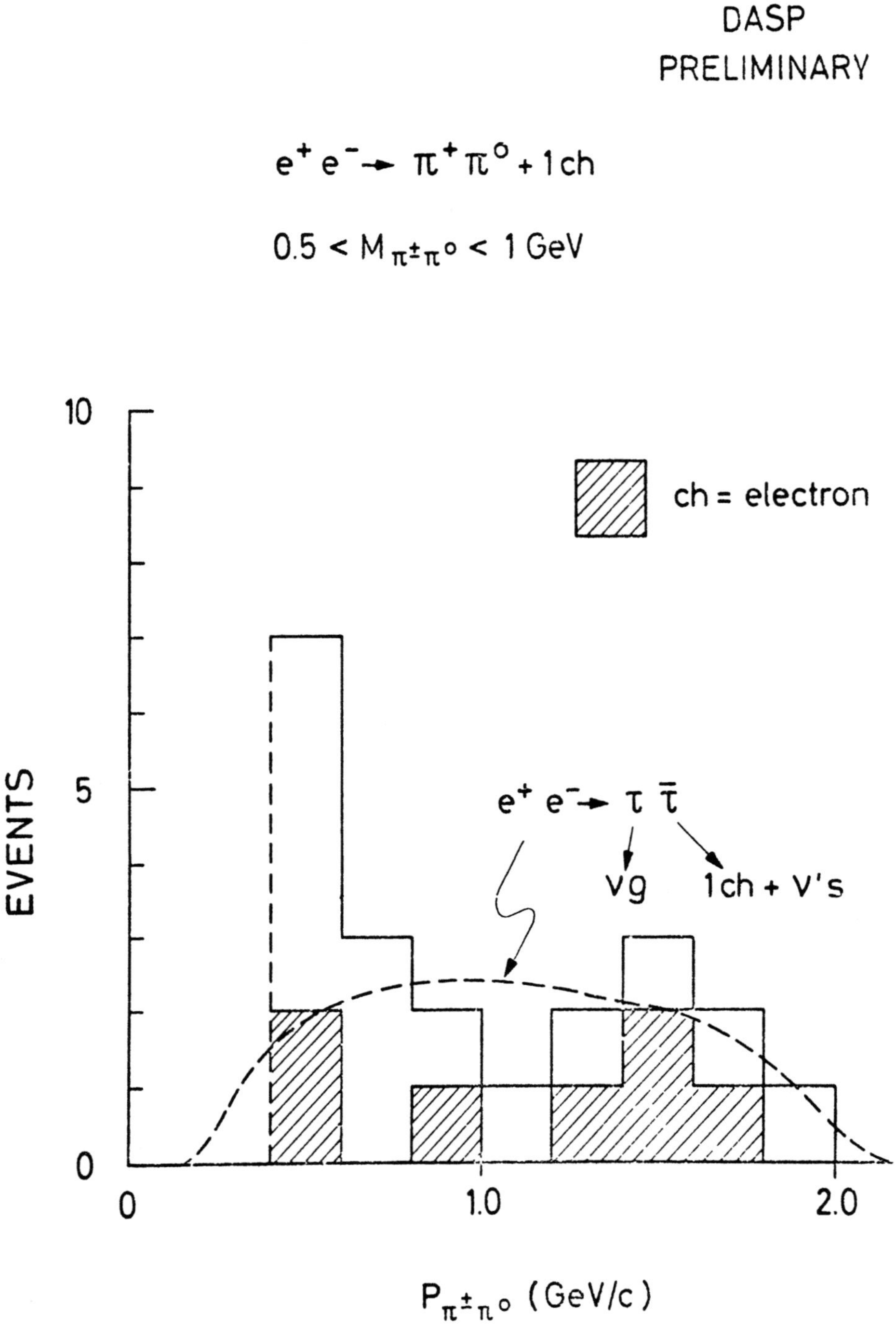

FIG. 8.6.2. : The momentum distribution of the $(\pi^{\pm}\pi^{0})$-system observed by the DASP group in $e^{+}e^{-} \to \pi^{\pm}\pi^{0}T^{\mp}$. The mass of the $\pi^{\pm}\pi^{0}$ system was between 0.5 GeV and 1.0 GeV.

events result from

$e^+e^- \to \tau\bar{\tau} \to (\nu_\tau e\bar{\nu}_e)(\bar{\nu}_\tau \rho^\circ \pi)$ where the $\rho^\circ\pi$ is consistent with $A_1 \to \rho^\circ\pi$.

They observe a total of 39 events of the type e + 3 hadrons and assume that the hadrons are all pions. From the sample of 39 events those events are selected for which the 3π-system has an invariant mass and momentum consistent with pairproduction and subsequent decay of a particle of mass $\lesssim$ 1.9 GeV/c^2. 6 events are rejected. The remaining sample of 33 events contains an estimated background of 13 ± 4 events from hadrons mislabeled as electrons and at most 3 events which had unobserved photons.

The $\pi^+\pi^-$ mass distribution for the 33 events is plotted in Fig. 8.6.3a. The background is shown by the squares. Each event contributes two mass values and the distribution can be understood as arising mainly from a ρ-signal in one combination and a broad distribution in the other. Plotting only the highest mass combination shows the ρ more clearly. 21 events have a $M(\pi^+\pi^-)$ between 0.70 - 0.84 GeV and are retained. The momentum distribution for the electrons in the remaining sample is plotted in Fig. 8.6.3b. The spectrum is hard and consistent with $\tau \to \nu_\tau e\bar{\nu}_e$ shown as the solid line but inconsistent with the charm spectrum shown as the dotted line. If the (3π) system results from the twobody decay of a moving object it should uniformly populate momenta between P_{max} and P_{min}. These limits depend on the cm energy and the mass of the 3π system. However, the quantity $U = (P_{3\pi} - P_{min})/(P_{max} - P_{min})$ is uniformly populated for any set of cm energies and mass values. The events are plotted in Fig. 8.6.3c versus U. The solid line indicates the event distribution predicted including all the experimental corrections, for a two-body decay $\tau \to \nu_\tau(3\pi)$ of a moving object.

The events can therefore safely be assigned to τ pairproduction. The 3π-mass distribution of the 21 events with at least one $\pi^+\pi^-$ combination in the ρ band is shown in Fig. 8.6.4c. The background expected from misidentified electrons is indicated by the squares. The mass histogram shows a clustering consistent with the mass and the width reported for the A_1. The non-resonant mass distribution $\tau \to \nu_\tau\rho\pi$ was computed assuming (V-A) couplings, m_τ = 1.8 GeV/c^2 and a massless τ neutrino.

This distribution is plotted as the solid line in Fig. 8.6.4 c. The χ^2 probability for this fit is 5% compared to 80% for an A_1 with mass 1.1 GeV/c^2 and width of 0.25 GeV/c^2. The resonance interpretation is thus favoured. Using B_e = 0.16 they find $B(\tau^+ \to \bar{\nu}_\tau\rho^\circ\pi^+) = 0.05 \pm 0.03$.

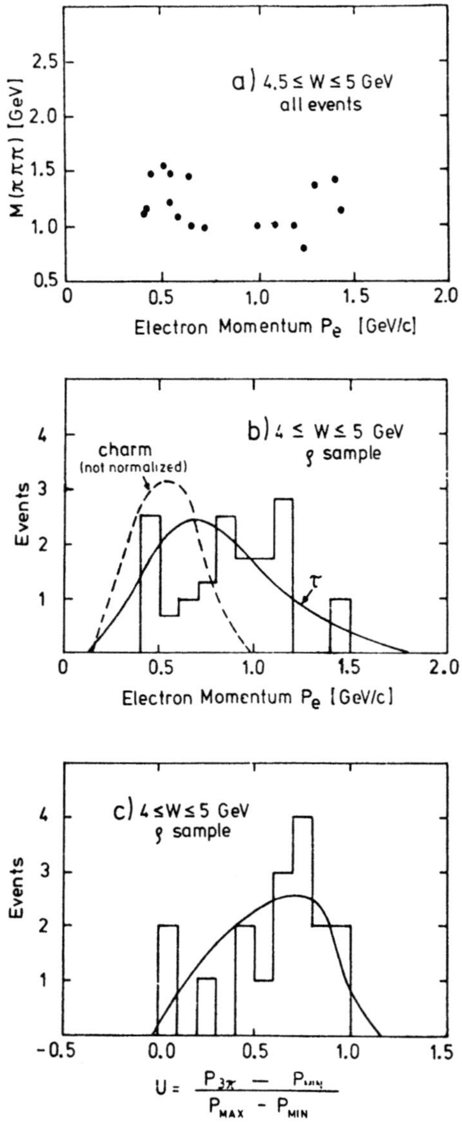

FIG. 8.6.3. : Kinematic properties of the eπππ events observed by the PLUTO Collaboration.
a) Invariant mass of the 3π-system plotted against the electron momentum.
b) The electron-momentum distribution for events from the ρ sample (see text). The solid curve is the prediction for τ-decay normalized to the data with $p_e > 0.4$ GeV/c. The dotted curve is the electron spectrum form charm decays measured by DASP. The curves are corrected for the detection efficiency.
c) Distribution of the normalized 3π-momentum (ρ-sample), compared with the expected $\tau \to \nu_\tau \pi\pi\pi$ decay distribution.

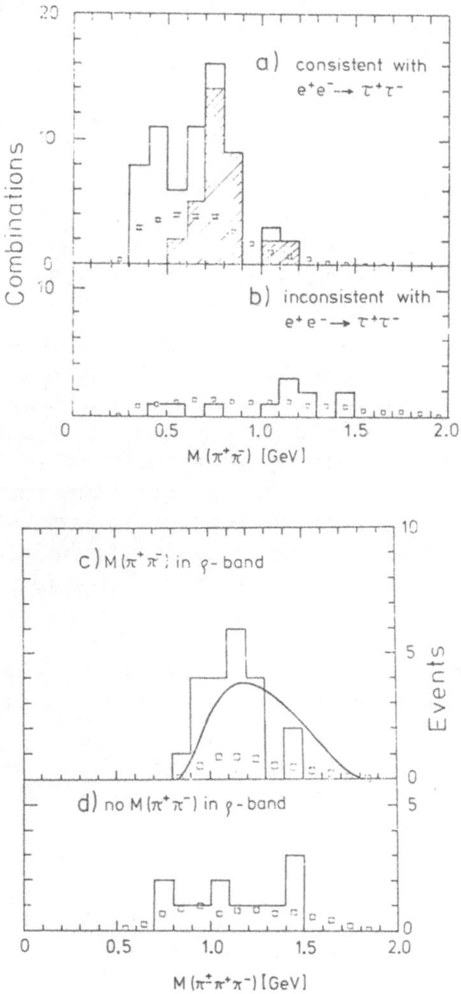

FIG. 8.6.4. : Properties of the two and three pion mass distributions in the $e\pi\pi\pi$ event sample observed by PLUTO. The squares represent the expected distributions for background from misidentified events. The solid line in c) represents the sum of the background and $\tau \to \rho^{\circ}\pi$ distribution normalized to the data.
a) $M(\pi^+\pi^-)$ for events kinematically consistent with τ pairproduction. There are two combinations per event and the shaded area represents the distribution of the higher $M(\pi^+\pi^-)$ combination.
b) $M(\pi^+\pi^-)$ for events kinematically inconsistent with τ pairproduction.
c) $M(\pi^{\pm}\pi^+\pi^-)$ distribution for events consistent with τ pairproduction and having at least one $\pi^+\pi^-$ mass in the ρ band.
d) Same as in c but with no $\pi^+\pi^-$-events in the ρ band.

The SLAC-LBL group has also reported [99] evidence for $\tau \to \nu_\tau A_1 \to \nu_\tau \rho\pi$. The mass spectrum for the 3π system is plotted in Fig. 8.6.5. The peak observed is too narrow to result from the phase space decay $\tau \to \nu\rho\pi$, plotted as the solid line.

8.7. Summary.

The τ particle, discovered by Perl and his Collaborators, is pairproduced at the ψ'. Since this is below the threshold for charm particle production it proves that the τ signal is not associated with charm. The mass of the τ is 1.81 ± 0.02 GeV/c^2. The cross section for τ production rises rapidly near threshold, it varies smoothly with energy and it reaches the value of the point cross section at high energies. This is the behaviour expected for a pointlike spin 1/2 particle with charge e - i.e. a new heavy lepton. The particle decays weakly into purely leptonic final states or into final states with hadrons plus a neutrino. The leptonic decay mode has a branching ratio around 18%, the momentum spectrum of the leptons is consistent with the decay of a pointparticle, a V-A leptonic coupling and a low mass for the τ. The data are consistent with the τ being a sequential lepton, although this is not yet proven.

The τ decays preferentially into final states of low multiplicity and a small ratio of kaons to pions. The decays $\tau \to \nu\rho$ and $\tau \to \nu\rho\pi$ where the $\rho\pi$ is consistent with the A_1 has been observed. The decay $\tau \to \nu\pi$ has not yet been observed. Some of the experimental results are summarized in table 8.3-8.5.

I am indebted to the organizers Profs. R. Gastmans, M. Lévy, D. Speiser, J. Weyers for the opportunity to give these lectures and a most enjoyable stay at Cargèse.

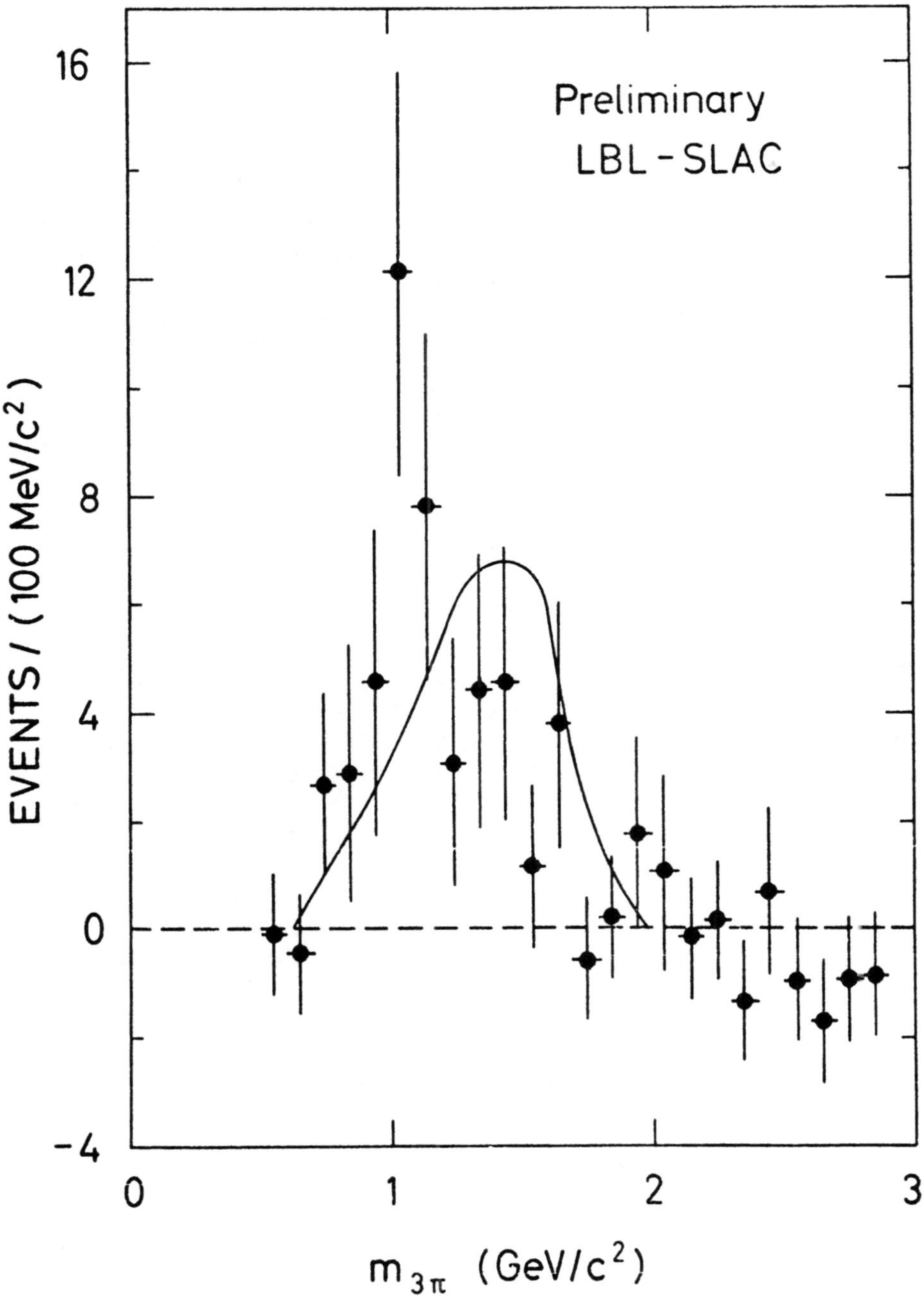

FIG. 8.6.5. : Three-particle invariant mass distribution opposite muons corrected for hadron misidentification. The solid curve is the phase space prediction for $\tau \to \nu\pi\rho$, corrected for acceptance effects. The prediction is normalized to events with a three particle effective mass between 0.7 GeV/c^2 and 1.8 GeV/c^2.

Table 8.3 - Mass of the τ

Group	Ref.	Data	Mass	Comments
SLAC-LBL	103	eμ	1.90 ± 0.10	This is an average using various methods.See ref.
PLUTO	12	μX	1.93 ± 0.05	
SLAC-LBL	99	eμ	1.80 ± 1.875	If the signal at ψ"
lead-glass wall	104	eX		(3772)results from τ-decay.
DASP	13	eX	1.807±0.020	Signal observed at ψ' (3684)

Table 8.4 - Leptonic Branching ratios.

Group	Ref.	Data	B_e or B_μ	Comments
SLAC-LBL	103	eμ	0.186±0.010±0.028	Assume $B_e = B_\mu$
SLAC-LBL	108	μX	B =0.175±0.072±0.030	Assume B_{1p}=0.85
PLUTO	12	μX	B =0.14 ± 0.034	Determine B_{1p}= 0.7±0.1 by assuming that all multiprong events with p_μ>1.0 GeV/c result from τ-decay
PLUTO	12	μe μX	B_e=0.16 ± 0.06	Using the value for B_μ from above.
SLAC-LBL lead glass	104	eμ	0.224±0.032±0.044	Assume $B_e = B_\mu$
DASP	13	eμ	0.182±0.028±0.014	Assume $B_e = B_\mu$
DELCO	19	eX	0.15	Preliminary, systematic uncertainty ± 0.50
IRON Ball	106	μμ	$0.22^{+0.07}_{-0.08}$	
Maryland, Princeton Pavia	106	μX	0.20 ± 0.10	

Table 8.5 - Hadronic Branching Ratios.

Group	Ref.	Decay Mode	BR	Comments
PLUTO	12	ν_τ+1prong+ ≥ 0 photons	0.70 ± 0.10	Assume that all muon inclusive multiprong events with $p_\mu > 1.0$ GeV/c
SLAC-LBL Lead-Glass Wall	104	- " -	0.90 ± 0.10	Assume point cross section with m_τ = 1.9 GeV/c^2.
DASP	13	- " -	0.65 ± 0.12	mass m_τ = 1.807 GeV/c^2
PLUTO	12	ν_τ+hadron+ ≥ 0 photons	0.40 ± 0.15	
SLAC-LBL	104	- " -	0.45 ± 0.19	
DASP	13	- " -	0.29 ± 0.11	
DASP	58	$\nu_\tau \pi$	0.02 ± 0.028	No clear signal observed B_e=0.18, m_τ=1.807 GeV/c^2
DASP		$\nu_\tau K$	< 0.016	90 C.L.
DASP	58	$\nu_\tau \rho$	0.27 ± 0.10	
PLUTO	110	$\nu_\tau \rho^\circ \pi^+$	0.05 ± 0.015	$\rho\pi$ spectrum consistent with resulting from A_1. Including all neutral modes gives $\Gamma(\tau \to \nu_\tau A_1)$=0.10± 0.03

REFERENCES

1. B. Bartoli et al., Nuovo Cimento 70A, 615 (1970).
Phys. Lett. 36B, 598 (1971).

G.G. Barbarino et al., Lett. Nuovo Cimento 3, 689 (1972).
G. Bacci et al., Phys. Lett. 38B, 551 (1972).
2. A. Litke et al., Phys. Rev. Lett. 30, 1189 (1973).
G. Tarnopolsky et al., Phys. Rev. Lett. 32, 432 (1974).
3. J.E. Augustin et al., Phys. Rev. Lett. 34, 764 (1975).
4. J.J. Aubert et al., Phys. Rev. Lett. 33, 1404 (1974).
5. J.E. Augustin et al., Phys. Rev. Lett. 33, 1406 (1974).
6. DASP-Collaboration, W. Braunschweig et al., Phys. Lett. 57B, 407 (1975).
7. G.J. Feldman et al., Phys. Rev. Lett. 35, 821 (1975).
8. G. Goldhaber et al., Phys. Rev. Lett. 37, 255 (1976).
I. Peruzzi et al., Phys. Rev. Lett. 37, 569 (1976).
9. DASP Collaboration, R. Brandelik et al., Phys. Rev. Lett. 70B, 132 (1977).
10. S.L. Glashow, J. Iliopoulos and L. Maiani, Phys. Rev. D2, 1285 (1970).
11. M.L. Perl et al., Phys. Rev. Lett. 35, 1489 (1975).
12. PLUTO Collaboration, J. Burmester et al., Phys. Lett. 68B, 297 (1977). and Phys. Lett 68B, 301 (1977).
13. DASP Collaboration, R. Brandelik et al., Phys. Lett. 73B, 109 (1978).
14. B.H. Wiik and G. Wolf, DESY 77/01.
To be published in the Proceedings of the Les Houches Summer School 1976.
15. G.J. Feldman and M.L. Perl, Physics Report, 336, 5 (1977).
16. For example, the detectors at ACO, the setups of the $\gamma\gamma$ and $B\bar{B}$ groups at Adone, the Stanford group, the Maryland, Pavia, Princeton, UC San Diego, SLAC and Stanford Collaboration and the Ironball experiment at SPEAR, the DESY-Heidelberg Collaboration and the Bonn-DESY-Mainz group at DORIS.
17. J. Olsson, invited paper at the 1977 International Symposium on Lepton and Photon Interactions at High Energies, Hamburg Aug. 25-31, 1977.
18. J.E. Augustin et al., Phys. Rev. Lett. 34, 233 (1973)
For the lead-glass addition see : J.M. Feller et al., SLAC-PUB-2060, LBL-6466, December 1977.
19. J. Kirkby, invited paper at the 1977 International Symposium on Lepton and Photon Interactions at High Energies, Hamburg, Aug. 25-31, 1977.
20. PLUTO Collaboration, J. Burmester et al., Phys. Lett. 66B, 395 (1977).
21. DASP Collaboration, see various publications in Phys. Lett. Maryland, Princeton and Pavia Collaboration.
M. Cavalli-Sforza et al., Phys. Rev. Lett. 36, 558 (1976).

22. G.S. Abrams et al., Phys. Rev. Lett. 33, 1453 (1974).
23. A.M. Boyarski et al., Phys. Rev. Lett. 34, 1357 (1977).
V. Lüth et al., Phys. Rev. Lett. 35, 1124 (1975).
24. B. Jean-Marie et al., Phys. Rev. Lett. 36, 241 (1976).
25. Wonyong Lee, invited paper at the 1977 International Symposium on Lepton and Photon Interactions at High Energies, Hamburg, Aug. 25-31, 1977.
26. DASP-Collaboration, W. Braunschweig et al., Phys. Lett. 57B, 297, 1975 and Phys. Lett. 63B, 487 (1976).
G. Goldhaber, LBL-Report LBL-4884 (1976).
27. W. Tannenbaum et al., Phys. Rev. Lett. 36, 402 (1976).
28. DASP Collaboration, B.H. Wiik invited paper, 1975 International Symposium on Lepton and Photon Interactions at High Energies, Stanford Univ., Aug. 21-27, 1975.
29. F. Vannucci et al., Phys. Rev. D15, 1814 (1977).
30. G.S. Abrams et al., Phys. Rev. Lett. 34, 1181 (1975).
31. C.J. Biddick et al., Phys. Rev. Lett. 38, 1324 (1977).
32. J.S. Whitaker et al., Phys. Rev. Lett. 37, 1596 (1976).
33. S. Okubo, Phys. Lett. 5, 105 (1963).
G. Zweig, CERN-Report, TH 401, 412 (1964).
J. Izuku, K. Okada and O. Shito, Proc. Theo. Physics 35, 1061 (1966).
34. Y. Hara, Phys. Rev. B134, 701 (1964).
J.D. Bjorken and S.L. Glashow, Phys. Lett. 11, 255 (1964).
35. T. Appelquist and H.D. Politzer, Phys. Rev. Lett. 34, 43 (1975).
A. De Rujula and S.L. Glashow, Phys. Rev. Lett. 34, 46 (1975)
36. T. Appelquist, A. De Rujula, H.D. Politzer and S.L. Glashow, Phys. Rev. Lett. 34, 365 (1975).
G.G. Callan, R.L. Kingsley, S.B. Treiman, F. Wilczek and A. Zee, Phys. Rev. Lett. 34, 52 (1975).
E. Eichten, K. Gottfried, T. Kinoshita, J. Kogut, K.D. Lane and T.M. Yan, Phys. Rev. Lett. 34, 369 (1975).
J. Pumplin, W. Repko and A. Sato, Phys. Rev. Lett. 35, 1538 (1975).
B.J. Harrington, S.Y. Park and A. Yildiz, Phys. Rev. Lett. 34, 168 706 (1975).
J.S. Kang and H.J. Schnitzer, Phys. Rev. D12, 841, 2791 (1975).
37. A. Martin, Phys. Lett. 67B, 74 (1976).
38. H. Grosse, Phys. Lett. 68B, 343 (1977).
39. An up to date description of the theoretical framework used to describe the new particles with a complete set of references can be found in K. Gottfried, invited paper at the 1977 International Symposium on Lepton and Photon Interactions at High Energies, Hamburg, Aug. 25-31, 1977
see also
J.D. Jackson, invited paper, European Conference on Particle Physics, Budapest, Hungary, 4-9 July, 1977.

40. R. Barbieri, R. Gatto and R. Kögerler, Phys. Lett. 60B, 183 (1976)
R. Barbieri, R. Gatto and E. Remiddi, Phys. Lett. 61B, 465 (1976).
41. O. Nachtmann, invited paper at the 1977 International Symposium on Lepton and Photon Interactions at High Energies, Hamburg, Aug. 25-31, 1977.
42. A.M. Boyarski et al., Phys. Rev. Lett. 34, 764 (1975).
43. L.B. Okun and M. Voloshin, Zh. Eksper. Theor. Fig. 23, 369 (1976).
C. Rosenzweig, Phys. Rev. Lett. 36, 697 (1976).
M. Bander et al., Phys. Rev. Lett. 36, 695 (1976).
A. De Rujula et al., Phys. Rev. Lett. 38, 317 (1977).
44. C. Bacci et al., Phys. Lett. 64B, 356 (1976).
G. Barbiellini et al., Phys. Lett. 64B, 359 (1976).
B. Esposito et al., Phys. Lett. 64B, 362 (1976).
45. A.M. Boyarski et al., Phys. Rev. Lett. 39, 762 (1975).
46. D.G. Aschman et al., Phys. Rev. Lett. 39, 124 (1975).
47. H. Lynch, invited paper at the International Conference on Production of Particles with New Quantum Numbers, Madison, Wisconsin, April 1976.
48. S.W. Herb et al., Phys. Rev. Lett. 39, 252 (1977).
49. P.A. Rapidis et al., Phys. Rev. Lett. 39, 526 (1977).
50. PLUTO Collaboration, see G. Knies, invited paper at the 1977 International Symposium on Lepton and Photon Interactions at High Energies, Hamburg, August 25-31, 1977.
51. DASP Collaboration, R. Brandelik et al., to be submitted to Phys. Lett.
52. T.W. Appelquist and H. Georgi, Phys. Rev. D8, 4000 (1973).
A Zee, Phys. Rev. D8, 4038 (1973).
53. J. Siegrist et al., Phys. Rev. Lett. 36, 700 (1976).
54. E. Eichten, K. Gottfried, T. Kinoshita, K.D. Lane and T.M. Yan, Phys. Rev. Lett. 36, 500 (1976).
K. Lane and E. Eichten, Phys. Rev. Lett. 37, 477 (1976).
55. A. LeYaouanc, L. Olivier, O. Pène and J.C. Raynat, Orsay Preprint LPTHE 77/25.
56. PLUTO Collaboration, J. Burmester et al., Phys. Lett. 68B, 283 (1977).
57. DASP Collaboration, W. Braunschweig et al., Phys. Lett. 67B, 243 (1977).
58. S. Yamada, invited paper at the 1977 International Symposium on Lepton and Photon Interactions at High Energies, Hamburg, Aug. 25-31, 1977.
59. F.A. Berends and R. Gastmans Nucl. Phys. B61, 414 (1973).
60. DASP Collaboration, W. Braunschweig et al., Phys. Lett. 67B, 249 (1977).
61. J. Heintze, invited paper, 1975 International Symposium on Lepton and Photon Interactions at High Energies, Stanford Univ., Aug. 21-22, 1975.

62. W.D. Apel et al., Phys. Lett. 72B, 500 (1978).
63. R. Barbieri, R. Gatto, R. Kögerler and Z. Kunst, Phys. Lett. 57B, 455 (1975).
A. De Rujula, International Summer Institute on Theoretical Particle Physics in Hamburg 1975, Lecture Notes in Physics 56, Springer Verlag 1976.
64. H.J. Schnitzer, Preprint Brandeis Univ., Waltham, Mass. 02159, July 1976.
65. See for example
H.J. Lipkin, Erice Lectures, Summer 1977, Fermilab Conf. 77/93-THY.
A. De Rujula and R.L. Jaffe, Proc. Conf. Exptl. Meson Spectroscopy, Northeastern University, Bosten, April 1977.
R.L. Jaffe, Phys. Rev. D15, 267 and 281 (1977).
66. W. Tannenbaum et al., Phys. Rev. Lett. 35, 1323 (1975).
67. PLUTO Collaboration, V. Blobel
Proc. of the XIIth Rencontre de Moriond, Flaine, March 1977.
68. DASP Collaboration, R. Brandelik et al., to be published in Phys. Lett.
69. W. Tannenbaum et al., Phys. Rev. Lett. 36, 402 (1976).
70. G. Segré and J. Weyers, Phys. Lett. 62B, 91 (1976).
71. F.M. Pierre, invited paper at the XVIIIth International Conference on High Energy Physics, Tbilisi, USSR, 1976.
G.H. Trilling in Proceedings of the SLAC Summer Institute on Particle Physics, Aug. 2-13, 1976 (SLAC-Report 198).
72. M.S. Chanowitz and F.J. Gilman, Phys. Lett. 63B, 178 (1976).
73. H. Harari, Phys. Lett. 64B, 169 (1976).
74. H. Krasemann and M. Krammer DESY 77/34 (June 1977).
75. M.K. Gaillard, B.W. Lee and J.L. Rosner, Rev. Mod. Phys. 47, 277 (1975).
76. J. Ellis, M.K. Gaillard and D.V. Nanopoulos, Nucl. Phys. B100, 313 (1975).
77. Taken from E. Lohrmann, invited paper at the 1977 International Symposium on Lepton and Photon Interactions at High Energies, Hamburg, August 25-31, 1977.
78. DASP Collaboration, R. Brandelik et al., Phys. Lett. 67B, 363 (1977).
79. PLUTO Collaboration, J. Burmester et al., Phys. Lett. 67B, 367 (1977).
80. V. Lüth et al., Phys. Lett. 70B, 132 (1977).
81. PLUTO Collaboration, J. Burmester et al., Phys. Lett. 64B, 369 (1976).
82. DASP Collaboration, R. Brandelik et al., Phys. Lett. 70B, 125 (1977).
83. M. Piccolo et al., Phys. Rev. Lett. 39, 1503 (1977).
84. Evidence for charmed baryon production has been reported :
E.G. Cazzoli et al., Phys. Rev. Lett. 34, 1125 (1975).
B. Knapp et al., Phys. Rev. Lett. 37, 882 (1976).
85. I. Peruzzi et al., Phys. Rev. Lett. 39, 1301 (1977).

86. D.L. Scharre et al., Phys. Rev. Lett. 40, 74 (1978).
87. G.J. Feldman, Lectures given at the Banff Summer Institute on Particles and Fields, Banff, Alberta, Canada, August 26 September 3, 1977, SLAC-PUB 2068.
The values listed for the branching ratios are from this reference. They differ from the values quoted in references 83 due to an improved radiative correction.
88. G. Altarelli, N. Cabibbo and L. Maiani, Nucl. Phys. B88, 285 (1975).
B.M. Einhorn and C. Quigg, Phys. Rev. D12, 2015 (1975).
R.L. Kingsley, F. Wilczek and A. Zee, Phys. Lett. 61B, 259 (1975).
89. J. Ellis, M.K. Gaillard and O.V. Nanopoulos, Nucl. Phys. B100, 313 (1975). The paper argues against $\underline{6}$ dominance.
90. G. Goldhaber et al., Phys. Lett. 69B, 503 (1977).
91. G.J. Feldman et al., Phys. Rev. Lett. 38, 1313 (1977).
92. H.K. Nguyen et al., Phys. Rev. Lett. 39, 262 (1977).
93. L.B. Okun, V.I. Zakharov and B.M. Pontecow, Lett. NuovoCim. 13, 218 (1975).
R.L. Kingsley, S.B. Treiman, F. Wilczek and A. Zee, Phys. Rev. D11, 1919 (1975).
A. De Rujula, H. Georgi and S.L. Glashow, Phys. Rev. Lett. 35, 69 (1975).
A. Pais and S.B. Treiman, Phys. Rev. D12, 2744 (1975).
R.L. Kingsley, Phys. Lett. 63B, 329 (1976).
M. Goldhaber and J.L. Rosner, Phys. Rev. D18, 1254 (1977).
94. M. Piccolo et al., Phys. Lett. 70B, 260 (1977).
According to reference 85 the product of branching ratio and cross section quoted should be reduced by 7.5%.
95. a) I. Hinchliffe and C.H. Llewellyn Smith, Nucl. Phys. B114, 45 (1976).
b) A. Ali and T.C. Yand, Phys. Lett. 65B, 275 (1976).
c) F. Bletzacker, H.T. Nieh and A. Soni, Phys. Rev. D16, 732 (1977).
d) R. Nabari, X.Y. Pham and W. Cotthingham, J. Phys. G : Nuclear Physics 3, 1485 (1971).
e) G.L. Kane, Phys. Lett. 70B, 272 (1977).
f) X.Y. Pham and J.M. Richard, Preprint PARIS/LPTHE 77.20.
96. DASP Collaboration, R. Brandelik et al., Phys. Lett. 70B, 387 (1977).
97. A. Barbaro-Galtieri, invited paper at the 1977 International Symposium on Lepton and Photon Interactions at High Energies Hamburg, Aug. 25-31, 1977.
98. J.M. Feller, Phys. Rev. Lett. 40, 274 (1978).
99. For a review see
M. Perl, invited paper presented at the 1977 International Symposium on Lepton and Photon Interactions at High Energies, Hamburg, August 25-31, 1977.
100. H.B. Thacker and J.J. Sakurai, Phys. Lett. 36B, 103 (1971).

100. Y.S. Tsai, Phys. Rev. D4, 2821 (1971).
J.D. Bjorken and C.H. Llewellyn-Smith, Phys. Rev. D7, 887 (1973).
K. Fujikawa and N. Kawamoto, Phys. Rev. D14, 59 (1976).
Y.I. Azimon, L.L. Frankfurt and V.A. Khoze, Leningrad Nuclear Physics Institute, Preprint 245, June 1976.
101. N. Kawamoto and A.I. Sanda DESY 78/14.
Similar results are reported also by :
F.J. Gilman and D.H. Miller, SLAC-PUB-2046, 1977
and Y.S. Tsai, private communication.
102. C.H. Llewellyn-Smith, University of Oxford, Preprint 33/76
Invited paper presented at the Royal Society Meeting on New Particles and New Quantum Number, London, 11 March, 1976.
103. M.L. Perl et al., Phys. Lett. 70B, 487 (1977).
M.L. Perl et al., Phys. Lett. 63B, 466 (1976).
104. A. Barbaro-Galtieri, Phys. Rev. Lett. 39, 1058 (1977).
105. F.B. Heile et al., SLAC-PUB-2059, December 1977.
106. Contribution to the Hamburg Conference.
see H. Sadrozinski, invited paper at the 1977 International Symposium on Lepton and Photon Interactions at High Energies Hamburg, August 25-31, 1977.
107. M. Cavalli-Sforza et al., Phys. Rev. Lett. 36, 558 (1976).
108. G.J. Feldman, Phys. Rev. Lett. 38, 117 (1977).
109. W. Alles, Ch. Boyer and A.J. Buras, CERN-TH220.
110. PLUTO Collaboration, G. Alexander et al., Phys. Lett. 73B, 99 (1978).

SOME RECENT EXPERIMENTAL RESULTS IN HIGH-ENERGY NEUTRINO PHYSICS

J. STEINBERGER

European Organization for Nuclear Research

1. INTRODUCTION

High-energy neutrino experiments can be expected to give important insight into both the structure of hadronic matter (because of the point-like nature of the collisions, or the equivalent statement, the high mass of the intermediate boson, if it exists), and also into the structure of the weak interactions. Among the more striking past results have been a) the discovery that there must be at least two neutrinos[1]; b) the discovery of neutral weak currents[2]; c) the demonstration that, at least at relative low energies, scaling and the quark parton models are resonable approximations; and d) the discovery of dileptons and their probable association with a new family of particles [3][4].

In the last two years another very important development has been the discovery of large scaling violations in the y-distribution of antineutrino charged current reactions, in the charge in this y-distribution with neutrino energy, and in the energy dependence of the ratio of charged current antineutrino to neutrino total cross-sections. These results, due mainly to the Harvard-Pennsylvania-Wisconsin-Fermilab (HPWF) collaboration at FNAL and in some measure confirmed by the California Institute of Technology (CIT) group, have been collectively called the "high-y anomaly".

In the last half year the new Super Proton Synchrotron (SPS) accelerator has been in operation, and the results collected here are the first results obtained there, although other new results, where available, are also included. The following subjects are discussed :

a) the SPS narrow band neutrino beam and CERN-Dortmund-Heidelberg-Saclay (CDHS) electronic detector;
b) charged current interactions and the "high-y anomaly";
c) neutral current interactions and the Weinberg-Salam model;
d) dilepton production and charm;
e) "wrong-sign" muons, charm-changing neutral currents and the charmed sea;
f) the x-distributions of the u, d and s quark seas.

2. THE SPS NARROW-BAND NEUTRINO BEAM AND THE CDHS DETECTOR

The neutrino experimental facilities at CERN are as follows :

a) A beam hall in which the hadron beams, produced by protons on a target, are formed. Two such beams are at present installed and can be used alternately: a "wide-band" horn focused beam and a "narrow-band" momentum selected beam. The beam hall is 120 m long from the proton target to the end.
b) An evacuated decay tunnel 300 m long.
c) A shield, made chiefly of iron, to absorb hadrons and muons.
d) Devices to monitor the absolute beam flux and the beam composition.
e) The neutrino interaction detector.

The overall layout is shown in Fig. 1.

Until now, only the narrow-band beam has been used. In this beam secondary hadrons are focused on a momentum slit and then made as nearly parallel as possible (see Fig. 2). Some of the parameters of the CERN beam are :

maximum hadron momentum :	275 GeV
acceptance, solid angle :	10 μsr
momentum resolution :	$\Delta p/p = \pm 5\%$
angular divergence :	0.2 mrad

Since the neutrinos are dominantly ($\sim$ 98 %) produced in the two-body decays $\pi \to \mu + \nu$ and $K \to \mu + \nu$, the neutrino momentum can be determined for each event from the decay kinematics (sse Fig. 3), subject to the double valued π-K ambiguity. Because of the low π-μ mass difference, the pion neutrinos are less energetic. Some data from the CDHS experiment are shown in Figs. 4 and 5 to illustrate the kinematics. The uncertainty in the neutrino energy varies with decay angle (detector radius) from $\sim$ 5% at $r = 0$ to $\sim$ 20 % for $r = r_{max} = 1.8$ m, for kaon neutrinos at a hadron beam energy of 200 GeV. The resolution is much worse for pion neutrinos. At the SPS the entire neutrino spectrum, from quite low energies due to pion neutrinos up to the highest energies from kaon neutrinos, is covered for one setting of the hadron beam energy. This can be seen in Fig. 6, where the wide-band and "effective" narrow-band spectra for the CDHS detector are shown.

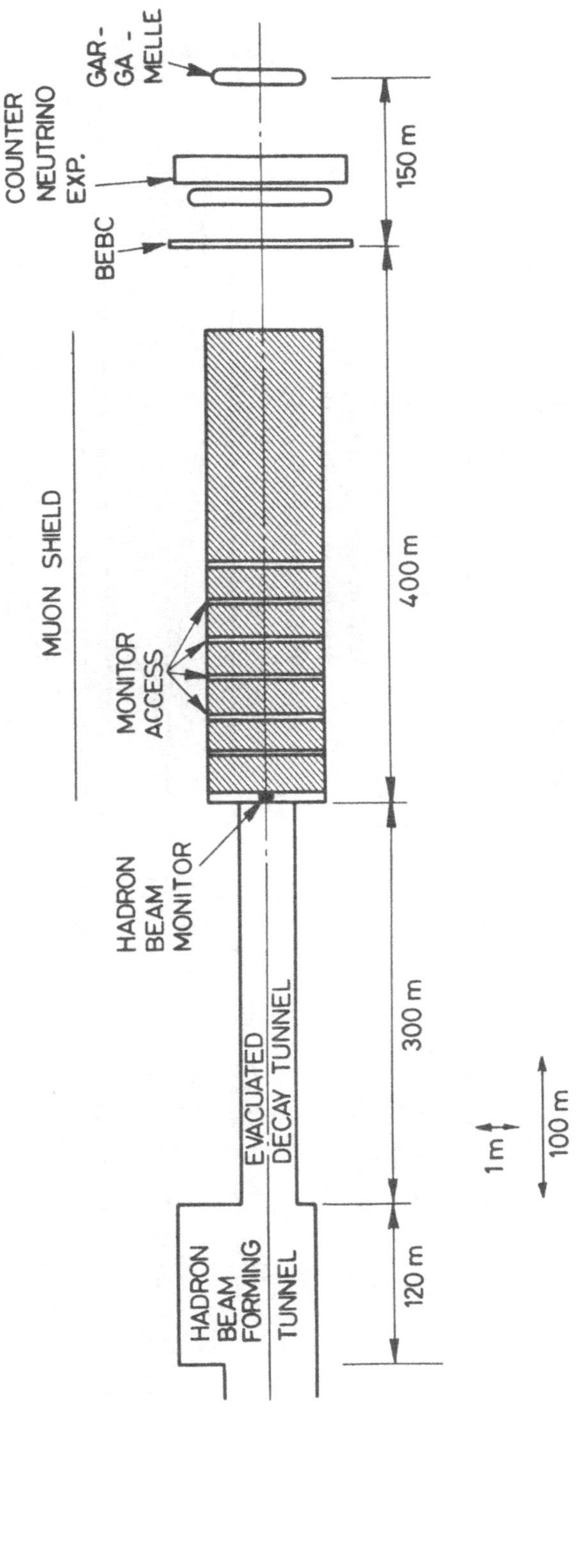

Fig.1. Over-all layout of neutrino area at SPS

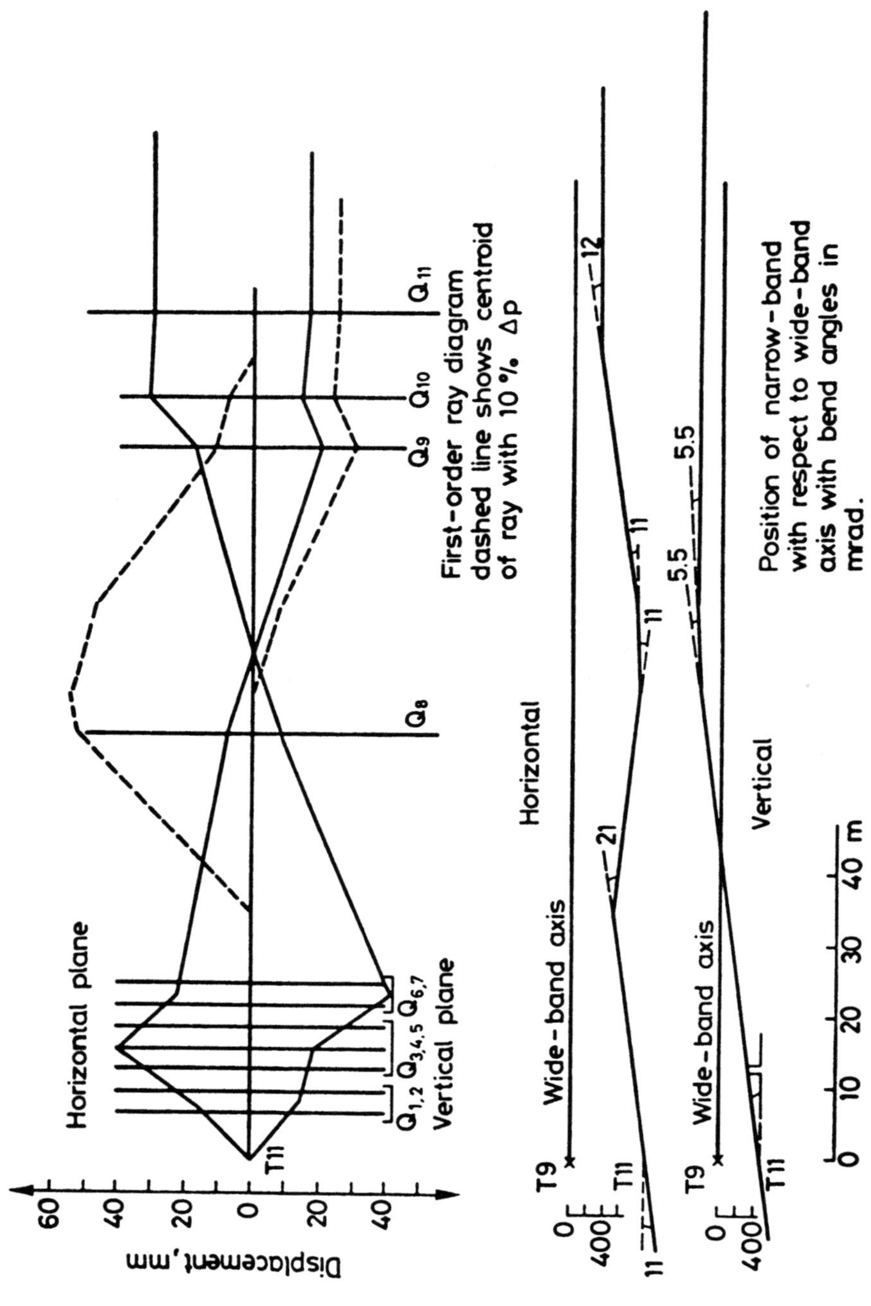

Fig. 2. Narrow-band optics at SPS

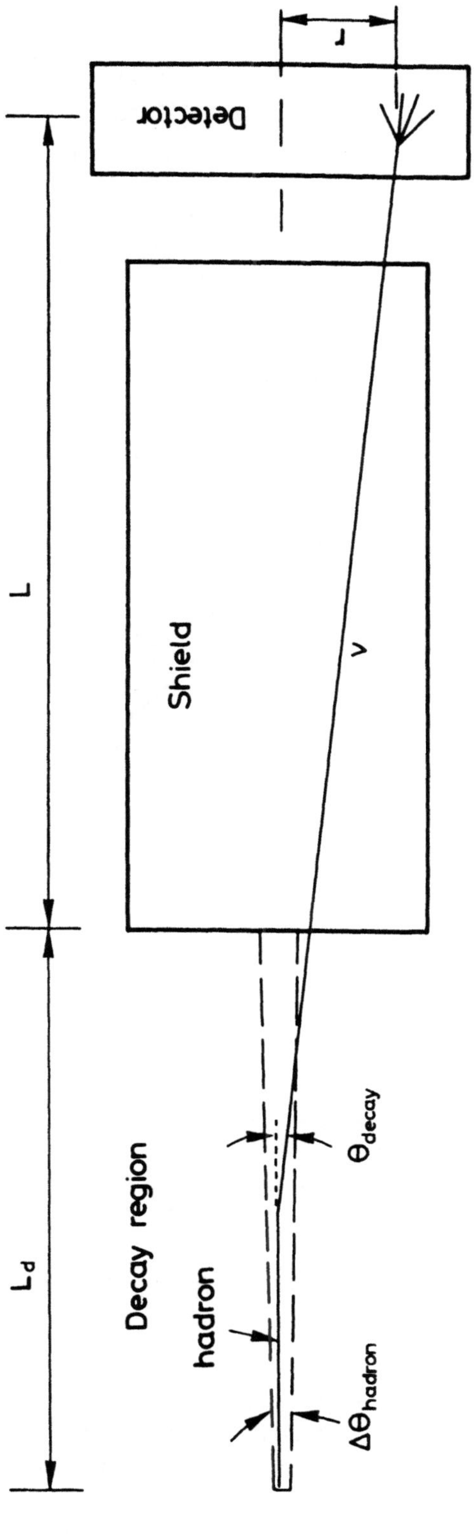

$$\Theta_{decay} \simeq r/(L + L_d/2)$$

Fig. 3. Geometry underlying neutrino energy determination

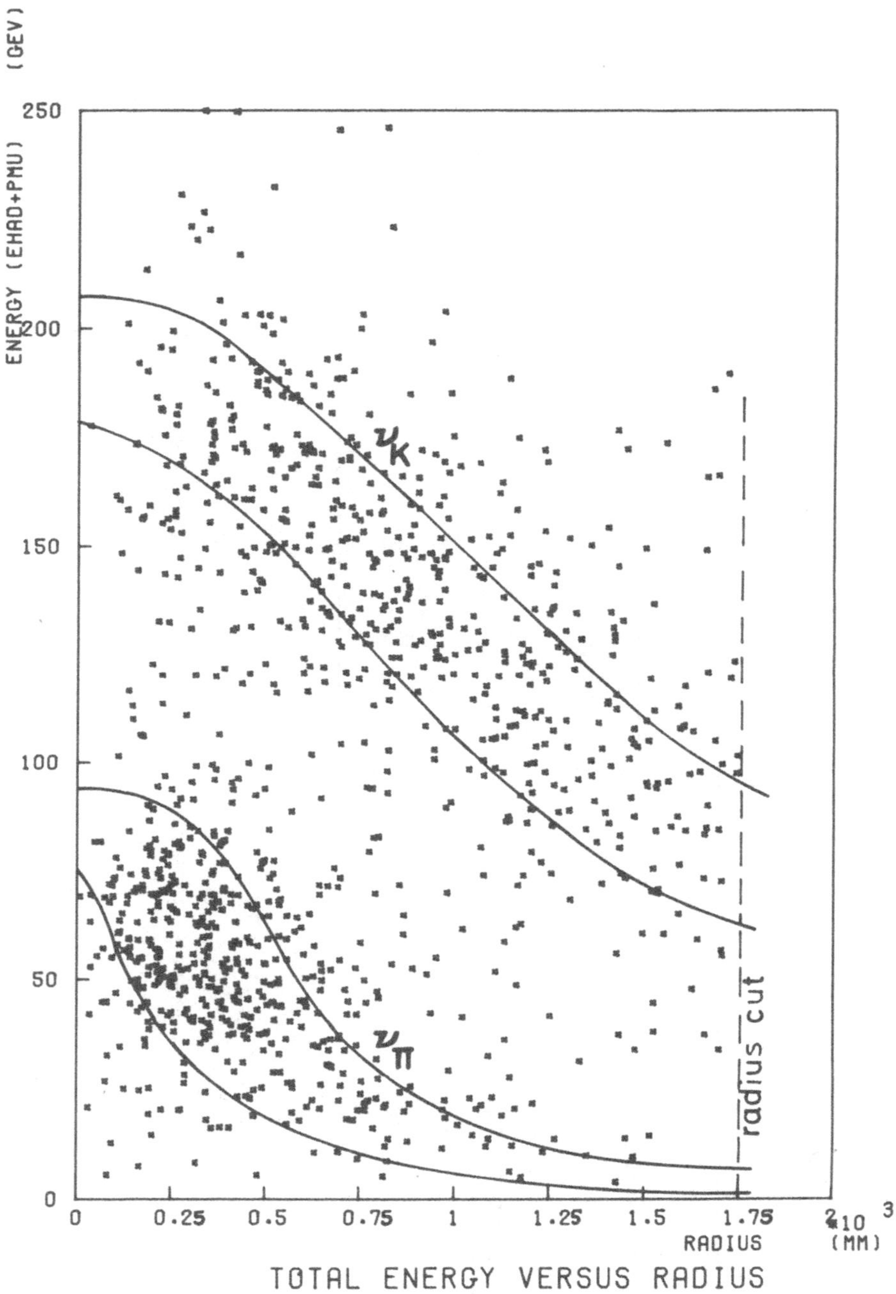

Fig. 4. Observed energy for charged current events versus the impact radius in the detector, illustrating the beam kinematics. The bands show the kinematical regions for pion and kaon neutrinos

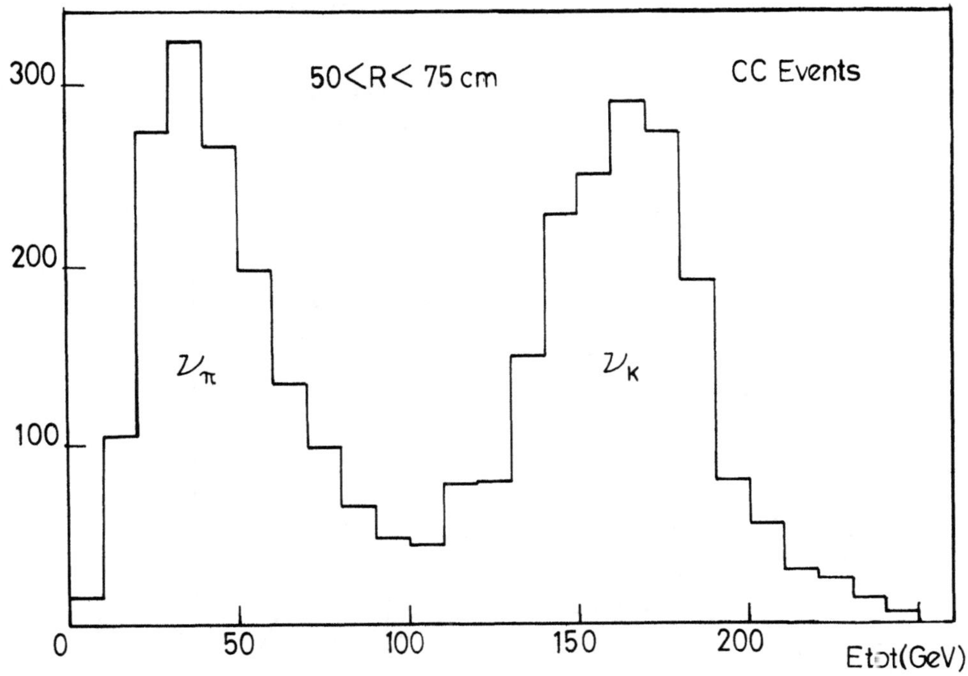

Fig. 5. Neutrino event rates as a function of the measured energy $(E_\mu + E_n)$ in a particular radial annulus, as observed in the CDHS experiment

The beam monitoring devides are the following :

a) a beam current transformer (BCT), with accurate absolute calibration, which measures the absolute hadron flux and is placed immediately following the beam-forming elements;
b) a Cerenkov counter capable of separating π's, K's and protons, placed just after the BCT and in front of the decay tunnel;
c) a monitor calorimeter just in front of the shield, in which the energy of the hadron beam is largely converted into heat so that the total hadron flux can be measured;
d) solid-state detectors placed in pits at various depths in the shield, to sample the muon flux; these detectors are intercalibrated, and an absolute calibration is obtained in a hadron beam by means of track counting in nuclear emulsions;
e) various ionization and secondary emission chambers to monitor the beam positions and focusing properties.

The three absolute monitors at present agree (disagree) to 10 %. The uncertainties of the K/ ratios are also 5-10% at present. It is hoped that these uncertainties will diminish with time.

An over-all view of the CDHS detector[5] is shown in Fig. 7. It is composed of 19 magnetized iron modules and triple plane

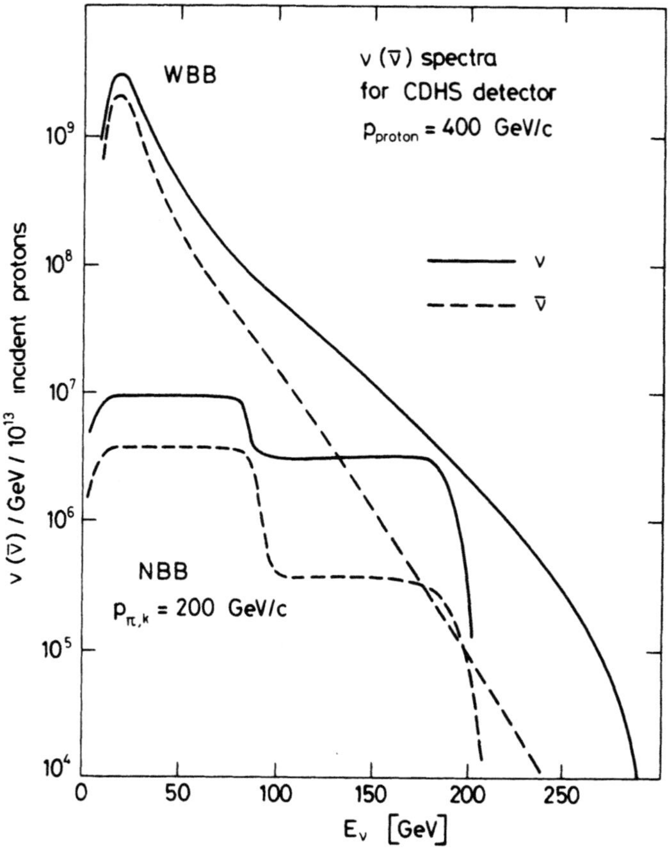

Fig.6. Wide-band and narrow-band neutrino and antineutrino spectra as seen by the CDHS detector

drift chambers. Each of these modules combines the functions of neutrino target, hadron shower detector, and muon spectrometer. Each is composed of iron plates 3.75 m in diameter, with an iron thickness of 75 cm, corresponding to a target weight of 65 tons. The iron is magnetized with ∿20,000 ampere turns to have a toroidal field, which falls from ∿ 20 kG at 15 cm radius to ∿ 16 kG at the outer edge. The first seven of these modules are made of fifteen 5 cm plates, the last twelve of five 15 cm thick plates. Between each pair of plates is a layer of eight scintillator strips viewed from both sides. The ratio of right to left pulse-height serves to measure the horizontal position of the shower, and the sum of right and left pulse-heights, after correction for attenuation using the position measurement, is used to measure the hadron shower energy calorimetrically. The accuracy of horizontal and vertical position measurement is at present ∿15-20 cm. No information on the shower direction is obtained. The energy of electrons and photons which may be produced in neutrino reactions is included in the measured hadron shower energy.

The response of these calorimeters to hadrons and electrons of several energies has been measured[6] in a smaller, magnetized iron, test calorimeter (60 cm x 80 cm sensitive area, 150 cm long), which uses the same scintillator, phototubes, and electronic circuits as the larger detector. Some of the results are shown in Figs. 8 and 9. The resolution for hadrons is reasonably described by the formula $\Delta E_h/E_h = 0.8/\sqrt{E_h}$ (GeV) for 5 cm sampling. For electron (photon) showers it is better by a factor of two. For the 15 cm sampling, the resolution is measured to be poorer by a factor of $\sqrt{3}$.

The calibration of the phototubes and pulse-height circuits is accomplished on the basis of cosmic-ray muon events which are recorded between machine bursts.

The response of the muon spectrometer has not been calibrated. It is deduced from the measured magnetization and the calculated multiple scattering and known drift chamber measurement errors. The momentum resolution is ±7 to ±15 %, depending on track length and the absolute error should not exceed 1-2 %.

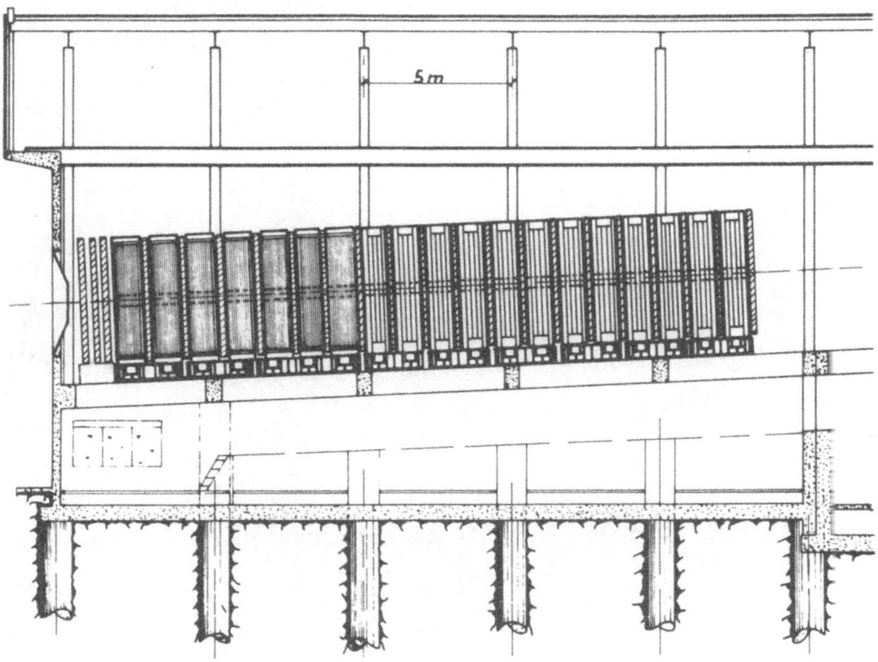

Fig.7. CDHS detector

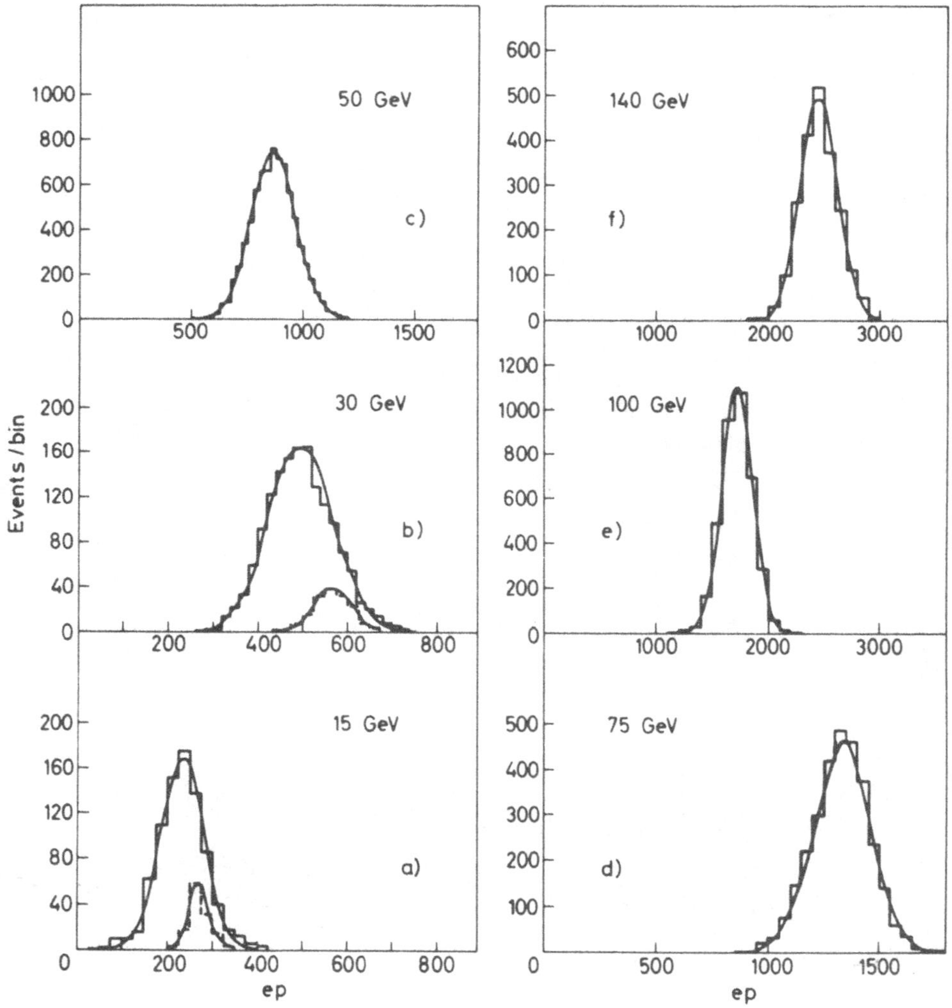

Fig. 8. Test calorimeter results on the resolution for hadrons and electrons at several energies

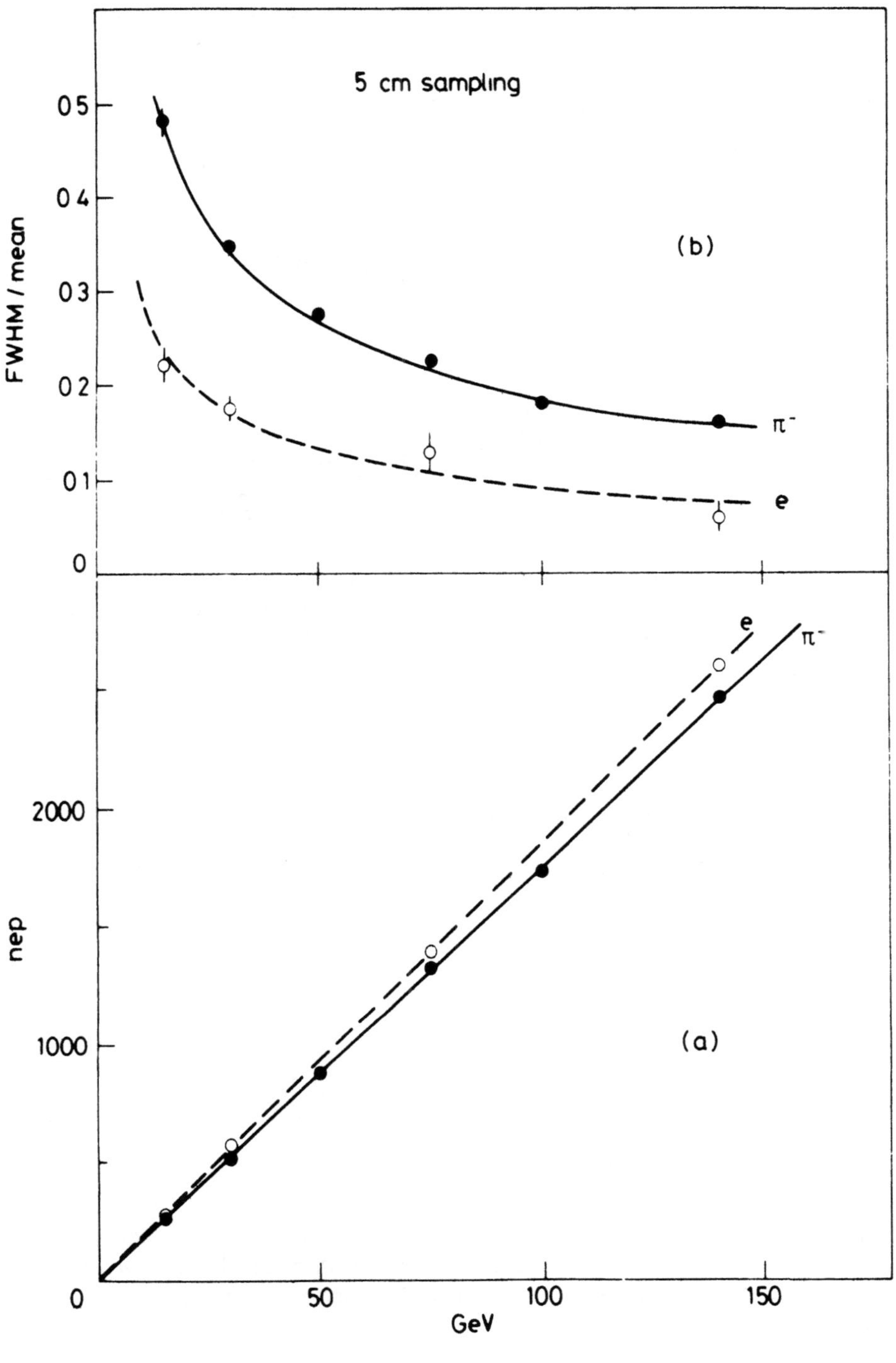

Fig. 9a. Test calorimeter results on the resolution for hadrons and electrons as a function of energy

Fig. 9b. The average pulse-height as a function of the hadron (electron) energy

3. INCLUSIVE CHARGED NEUTRINO SCATTERING AND THE "HIGH-y ANOMALY"

3.1 Kinematics and quark-parton model

We begin by recalling some of the underlying kinematical and theoretical notions :

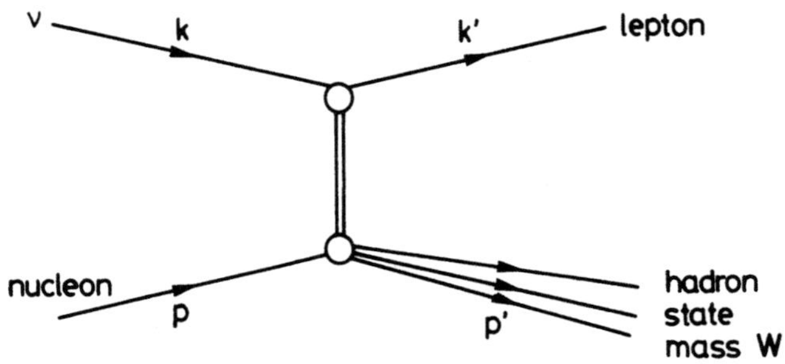

The incident neutrino of momentum k, energy E_ν, collides with a hadron of momentum p to produce a final lepton of momentum k' at laboratory angle θ , and the hadronic state p', with energy E_h, and mass W.

$$S \equiv (k + p)^2 \simeq 2E_\nu M$$

$$Q^2 \equiv -(k-k')^2 = 4E_\mu E_\nu \sin^2 \theta/2$$

$$\nu \equiv p\cdot q \simeq ME_h$$

$$W = \sqrt{2\nu - Q^2}$$

$$x \equiv Q^2/2\nu \simeq 2E_\nu E_\ell \sin^2 \frac{\theta}{2} /ME_h ; \quad 0 \leq x \leq 1$$

$$y \equiv \nu/k\cdot p \simeq E_h/E_\nu = \text{inelasticity}; \quad 0 \leq y \leq 1 .$$

the inclusive process will be described by the variables x, y and E_ν . Some contours for fixed laboratory lepton angle and W are shown in Fig. 10 in the x-y plot and, for comparison, in the Q^2-ν plot.

We continue by recalling some results of neutrino V,A theory and the quark parton model[7]. If the weak interaction is describable by V and A currents, and if polarization effects can be ignored, the inclusive differential cross-sections are described by three

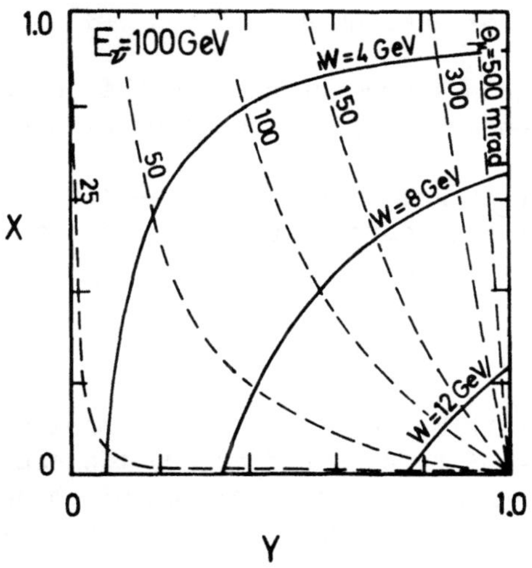

Fig. 10. Lines of constant final hadron mass W and lines of constant muon angle in the x-y plot

form factors that are functions of Q^2 and ν :

$$\frac{d^2\sigma_{\nu,\bar{\nu}}}{dQ^2 d\nu} = \frac{G^2}{2\pi M}\ \frac{E_\ell}{E_\nu}\left[\cos^2\frac{\theta}{2}\, W_2(Q^2,\nu) + 2\sin^2\frac{\theta}{2}\, W_1(Q^2,\nu) \mp \frac{(E_\nu+E_\ell)}{M}\sin^2\frac{\theta}{2}\, W_3(Q^2,\nu)\right] .$$

At high energy and momentum transfer, if there are no high mass thresholds, scaling must be expected, and the form factors can be written in terms of the unique variable x :

$$F_1(x) = W_1(Q^2,\nu)$$

$$F_2(x) = \nu W_2(Q^2,\nu)$$

$$F_3(x) = \nu W_3(Q^2,\nu) \quad ,$$

and

$$\frac{d\sigma_{\nu,\bar{\nu}}}{dx\ dy} = \frac{G^2 M E_\nu}{\pi}\left[(1-y)F_2 + xy^2 F_1 \mp (y - y^2/2)\ x\ F_3\right] .$$

The main consequences of scaling are :

i) the cross-sections are proportional to the neutrino energy; and
ii) the x- and y-dependences are independent of neutrino energy.

In the quark-parton model, the cross section is assumed to be the incoherent sum of the scattering on almost free quark constituents of the nucleon. Then

$$d^2\sigma/dxdy = \sum_j f_j(x)d\sigma_j/dy ,$$

where $f_j(x)$ is the distribution in the fractional nucleon momentum carried by quarks of type j, and $d\sigma_j/dy$ the neutrino cross-section of the j-type quark. Some examples of these elementary cross-sections in the G.I.M.[8] model are listed in Table 1.

TABLE 1
Neutrino quark cross sections

Process	$d\sigma/dy$ in units of G^2ME_ν/π
* $\nu + d \to \mu^- + u$	$2 \times \cos^2 \theta_c$
$\nu + s \to \mu^- + u$	$2 \times \sin^2 \theta_c$
$\nu + s \to \mu^- + c$	$2 \times \cos^2 \theta_c$
* $\nu + d \to \mu^- + c$	$2 \times \sin^2 \theta_c$
$\nu + \bar{u} \to \mu^- + \bar{d}$	$2 \times \cos^2 \theta_c(1-y)^2$
$\nu + \bar{u} \to \mu^- + \bar{s}$	$2 \times \sin^2 \theta_c(1-y)^2$
* $\bar{\nu} + u \to \mu^+ + d$	$2 \times \cos^2 \theta_c(1-y)^2$
* $\bar{\nu} + u \to \mu^+ + s$	$2 \times \sin^2 \theta_c(1-y)^2$
$\bar{\nu} + \bar{s} \to \mu^+ + \bar{c}$	$2 \times \cos^2 \theta_c$
$\bar{\nu} + \bar{d} \to \mu^+ + \bar{c}$	$2 \times \sin^2 \theta_c$
$\bar{\nu} + \bar{d} \to \mu^+ + \bar{u}$	$2 \times \cos^2 \theta_c$
$\bar{\nu} + \bar{s} \to \mu^+ + \bar{u}$	$2 \times \sin^2 \theta_c$

The collisions dominated by "valence" quarks are marked *. All others proceed via "sea" quarks.

It follows in this model that

$$2xF_1 = F_2 = \frac{18}{5} F^{e,d} = q + \bar{q}$$

$$-xF_3 = q - \bar{q} \; ;$$

where

$$q \equiv x\big(u(x) + d(x) + s(x) + c(x)\big) = (F_2 - xF_3)/_2,$$

$$\bar{q} \equiv x\big(\bar{u}(x) + \bar{d}(x) + \bar{s}(x) + \bar{c}(x)\big) = (F_2 + xF_3)/2.$$

$u(x)$, $d(x)$, $s(x)$ and $c(x)$ are the distributions, in the fractional momentum of the proton, of up, down, strange, and charmed quarks, respectively, and $F^{ed}(x)$ is the structure function measured in electron-deuteron collisions.

Then

$$\frac{d^2\sigma_\nu}{dxdy} = \frac{G^2 M E_\nu}{\pi} \left(q + \bar{q}(1-y)^2\right)$$

and

$$\frac{d^2\sigma_{\bar{\nu}}}{dxdy} = \frac{G^2 M E_\nu}{\pi} \left(q(1-y)^2 + \bar{q}\right)$$

These expressions separate very clearly the effect of left-handed hadronic currents, the terms in q, and right-handed hadronic currents, the terms in $\bar{q}$. The q term is composed of the "valence" contribution $q-\bar{q}$, which is supposed to dominate, and the "sea" distribution $\bar{q}$, which is supposed to be less important. The $\bar{q}$ term is pure "sea".

In terms of the fractional contribution of the valence quarks,

$$B \equiv \frac{\int\big(q(x) - \bar{q}(x)\big)\, dx}{\int\big(q(x) + \bar{q}(x)\big)\, dx} ,$$

the y-dependences of the cross sections are

$$d\sigma_\nu/dy \propto (1 + B) + (1 - B)(1 - y)^2$$

and

$$d\sigma_{\nu}/dy \propto (1 + B)(1 - y)^2 + (1 - B)$$

$$R \equiv \sigma_{\bar{\nu}}/\sigma_{\nu} = (4 - 2B)/(4 + 2B) .$$

If the sea is negligibly small,

$$B = 1 , \quad R = 1/3 , \quad \langle y_{\nu} \rangle = 1/2, \quad \text{and} \quad \langle y_{\bar{\nu}} \rangle = 1/4 .$$

3.2 Summary of the experimental situation up to 1975.

In Figs. 11-13 some results of the Gargamelle and earlier FNAL experiments, in particular x-distributions[9)], y-distributions[10)] and total cross-sections as a function of the neutrino energy[11-13)], are shown.

The data are all in good agreement with the simple quark parton model, within rather large errors.

3.3 The "high-y anomaly"

The experimental results, dominantly due to the HPWF collaboration at FNAL, and to some extent confirmed by the CIT group, are shown in Figs. 14-18. The anomalous behaviour consists of the following elements :

i) a flat y-distribution for antineutrinos at $x < 0.1$[14)] (fig. 14);
ii) an increase with energy in $\langle y \rangle$ for antineutrinos (Figs. 15 and 16), equivalent to a reduction in the B parameter with energy (Fig. 17) [15-17)]
iii) an increase with energy in the total cross-section ratio $\sigma_{\bar{\nu}}/\sigma_{\nu}$ (Fig. 18) [18)].

These results generated great interest because of the far-reaching implications. They were interpreted by the HPWF group, on the one hand to show violation of the charge independence which predicts approximate equality of neutrino and antineutrino cross-sections at small y, and on the other hand as showing the existence of a new type of quark of mass approximately 5-7 GeV, and which is produced by antineutrinos through a right-handed current. In this way the "threshold" at $\sim$ 40 GeV, the increase in $\sigma_{\bar{\nu}}/\sigma_{\nu}$ with energy and the increase in $\langle y \rangle$ with energy could be understood.

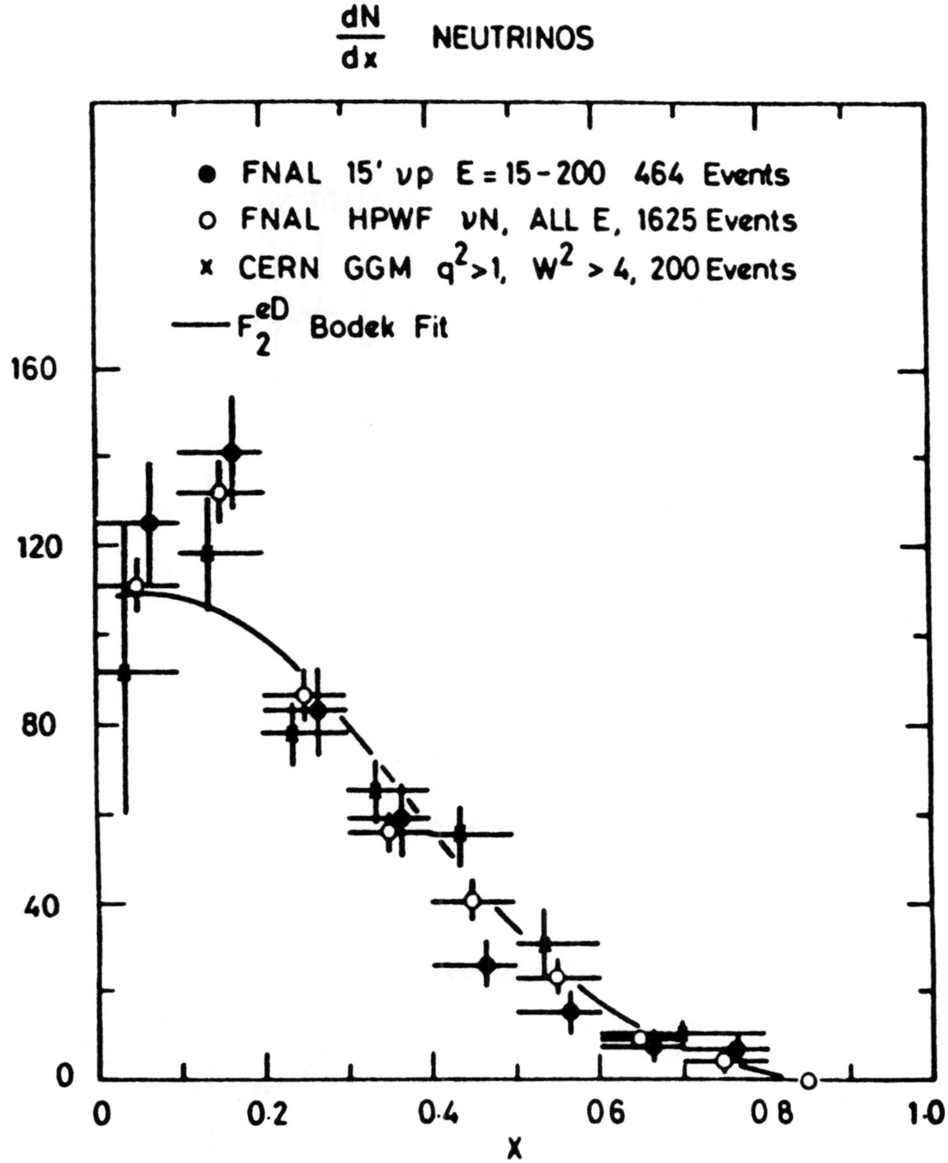

Fig. 11. Compilation of x-distributions observed in inelastic neutrino scattering [9)]

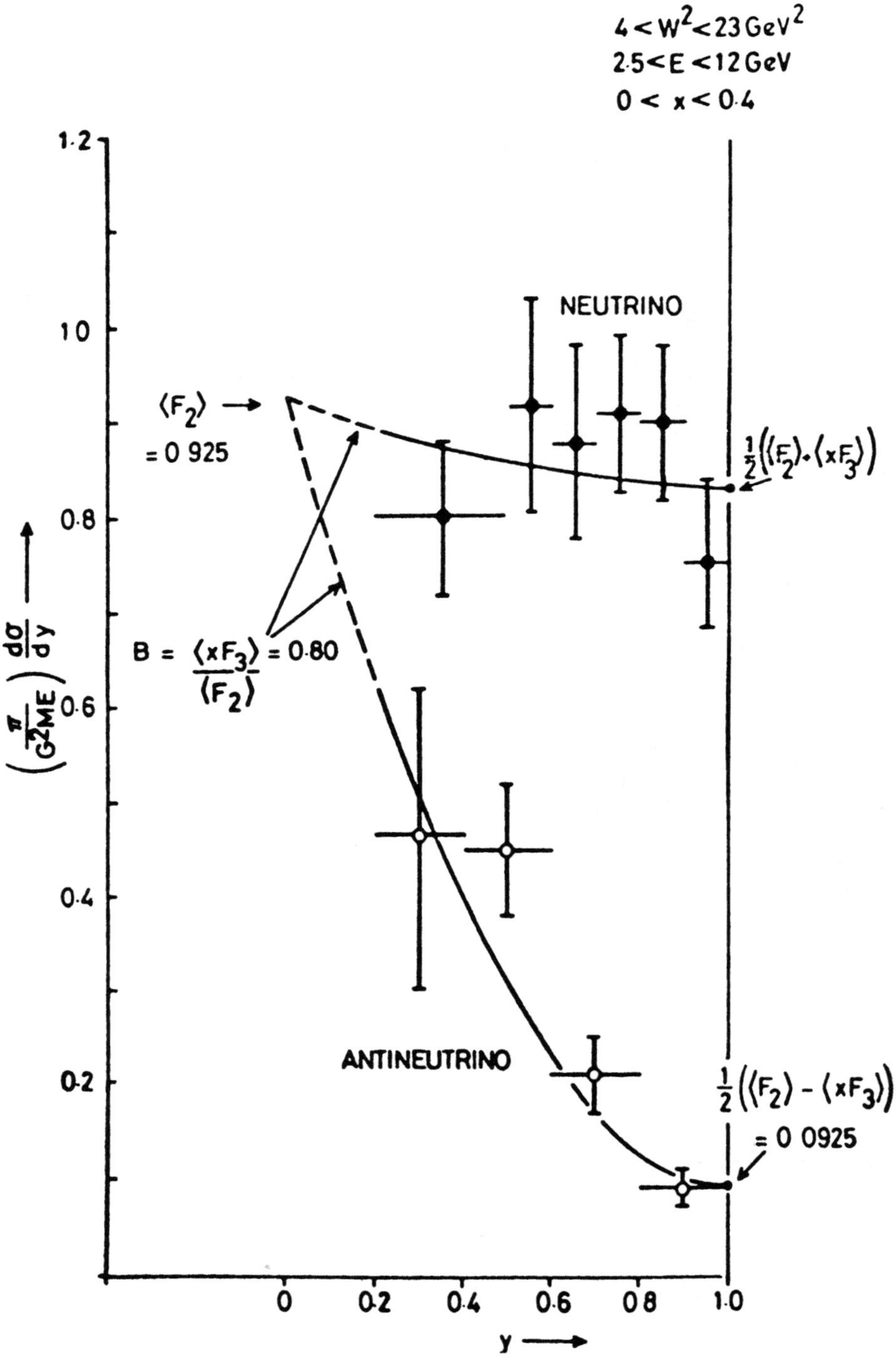

Fig. 12. y-distributions observed in Gargamelle[10)]

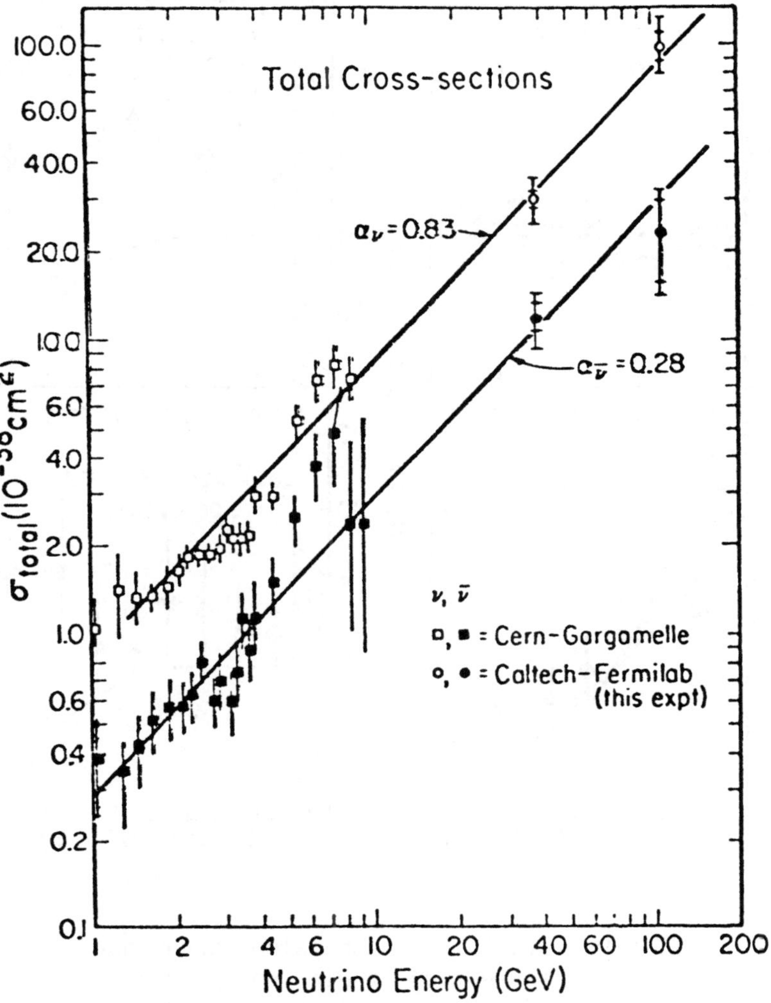

Fig. 13. Total cross-section as a function of neutrino energy. Compilation[9] of Gargamelle[11], ANL[12] and FNAL[13] results

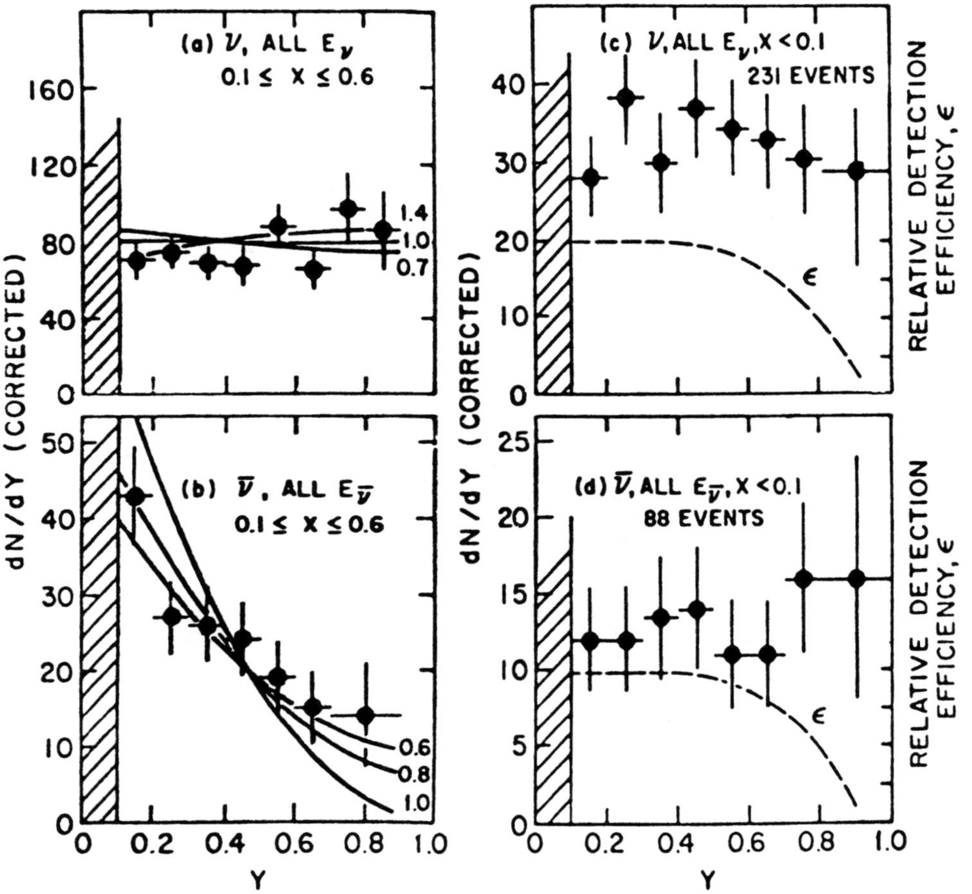

Fig.14. First HPWF results[14] on anamalous behaviour in the anti-neutrino cross-section at high energy and high y. The y-distribution for antineutrinos with $x < 0.1$ is nearly flat, much higher at large y than would be expected from scaling lower energy results.

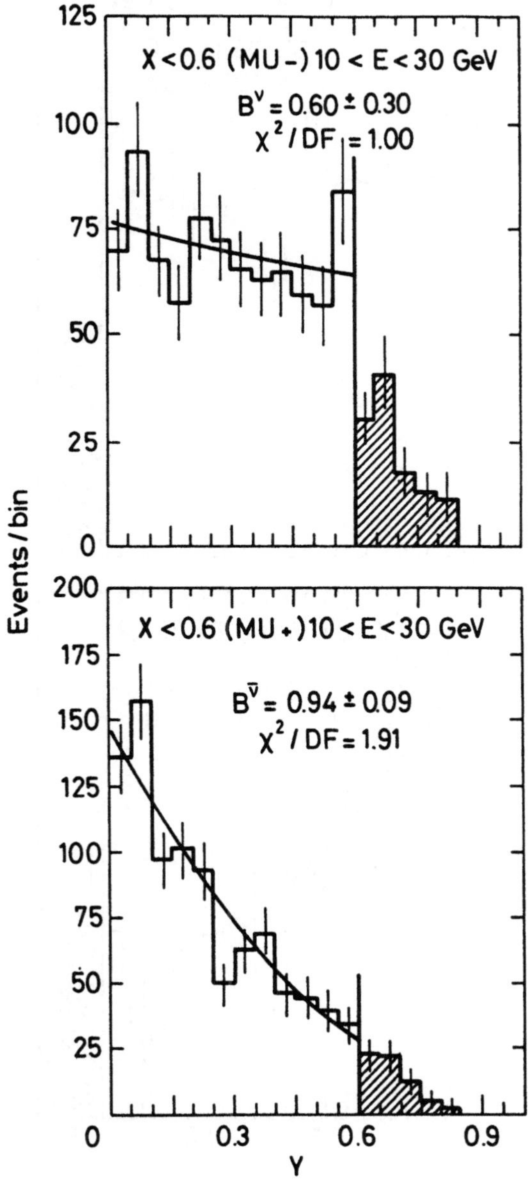

Fig. 15a : HPWF results[15)] on neutrino and antineutrino y-distributions at lower energies, $E_\nu < 30$ GeV. These are reasobably consistent with lower energy Gargamelle results.

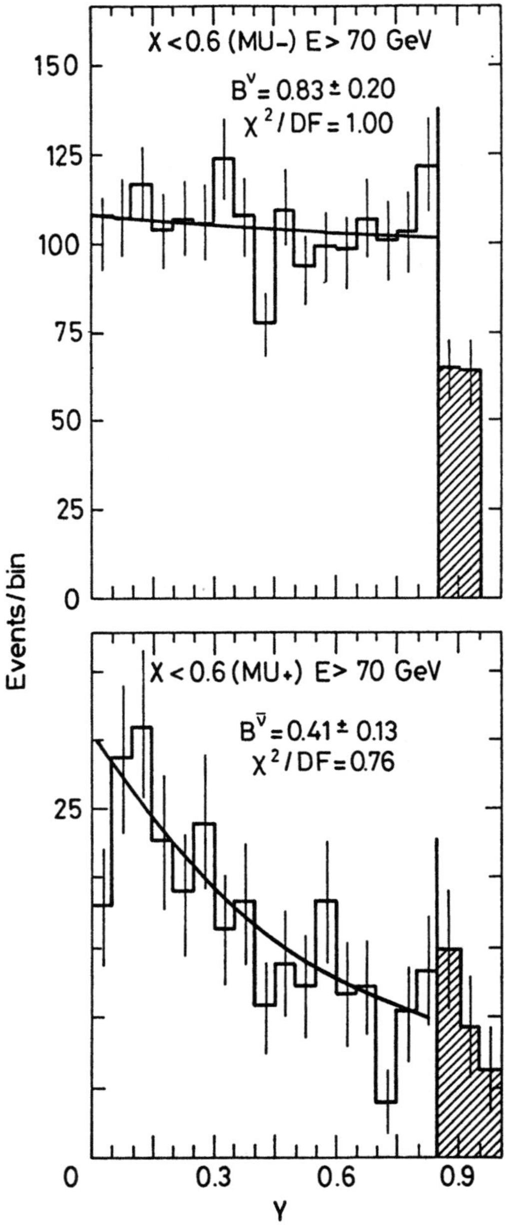

Fig. 15b : HPWF results[15] on neutrino and antineutrino y-distributions for E_ν > 70 GeV. The increase in the antineutrino cross-sections at high y is a very striking departure from scaling.

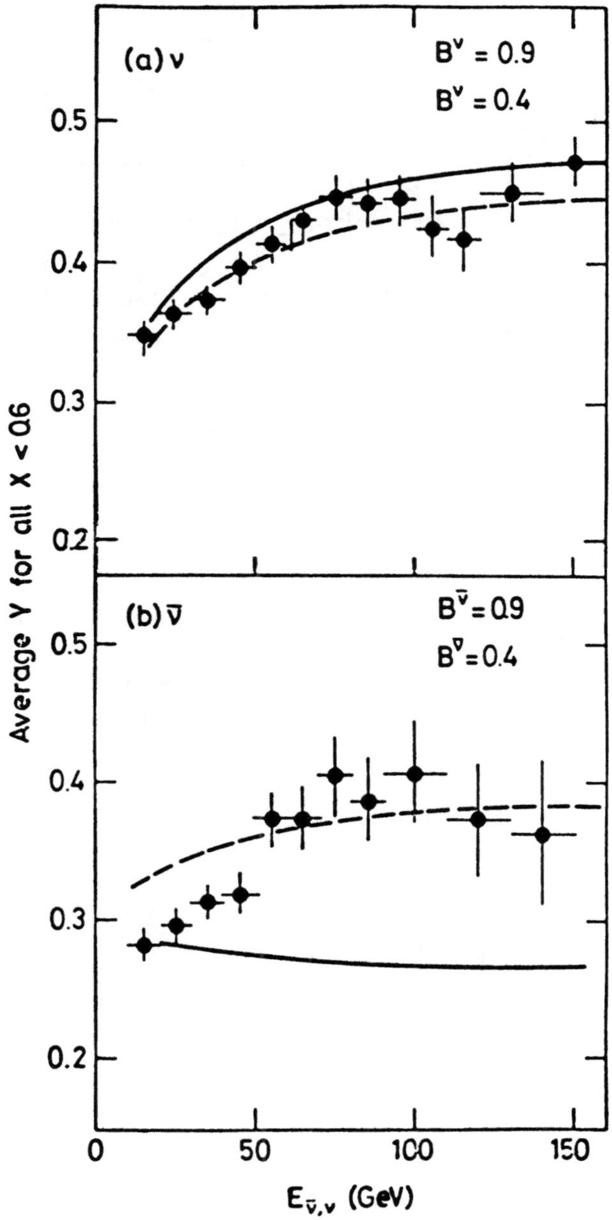

Fig. 16. HPWF result[15] for the behaviour of the average y, <y>, as a function of the energy. The rise in <y> for neutrinos is a reflection of the acceptance. The rise for antineutrinos is interpreted as a change in the B parameter from ≃ 0.9 at lower energies to ≃ 0.4 at higher energies, a dramatic violation of scaling.

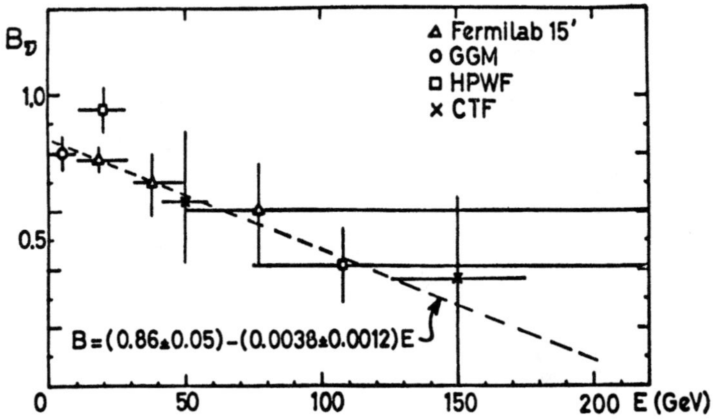

Fig. 17. Compilation [17] of $B_{\bar{\nu}}$ versus $E_{\bar{\nu}}$ of HPWF, CIT and 15' FNAL bubble chamber results

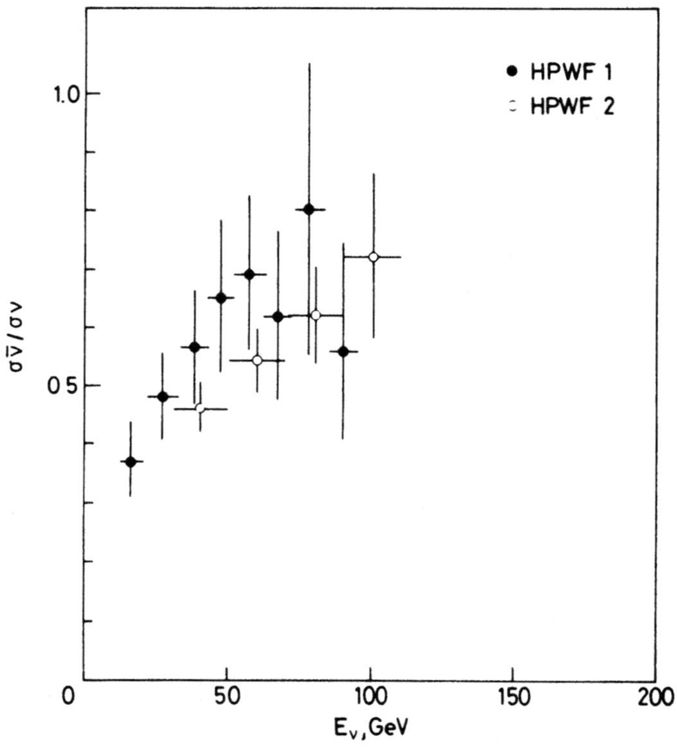

Fig. 18. HPWF result[18] on the energy variation of the total cross-section ratio $\sigma_\nu/\sigma_{\bar{\nu}}$.

3.4 Results of the CDHS collaration *) on the high-y anomaly[19]

We have analysed data obtained in 200 GeV narrow-band neutrino and anti-neutrino beams to check these results. The acceptance of this apparatus extends to higher y (see Fig. 19), and the event numbers are ∿5-10 times larger than those of the previous samples, so that both systematically as well as statistically the data are substantially superior. The energy distributions for the two samples are shown in Fig. 20. In Fig. 21 the data are shown which should be compared with Fig. 14. In particular, it should be noted that the y-distribution at small x is very far from uniform, but shows, in addition to the dominant $(1-y)^2$ distribution, about 25 % uniform contribution, not inconsistent with what can be expected from the antiquark sea at such small x. In Fig. 21, the antineutrino and neutrino distributions can be compared at small y to check charge independence. No deviation outside of the ∿10 % relative flux normalization uncertainty is observed.

The data relevant to the observed change in antineutrino y-distribution and B parameter with energy are shown in Figs.22-24. We follow HPWF in applying the cut $x < 0.6$ so that the comparison is observed. This is more clear in Fig. 23, where the experimental results for $\langle y \rangle$ are shown as a function of energy. A constant value of B, $B \simeq 0.8$, fits all antineutrino data resonably. The B-values deduced from the y-distributions are shown in Fig.24. The ratios of total cross-sections are tiven in Fig. 25 and can be seen to be independent of energy. In conclusion, these results a) do not reproduce any of the previously observed features of the "high-y anomaly" and b) show no energy variations inconsistent with scaling.

3.5 Other recent results concerning charged current reactions

In Fig. 26 the recent results[20] on charged neutrino and antineutrino cross-sections are presented, as observed in BEBC filled with a neon-hydrogen mixture, and in the same beam as for the previous experiment. The ratio of the cross-sections seems to be rising with energy, chiefly because the neutrino cross-sections seem to be falling. However, the statistics are rather limited, and the results are not inconsistent with the uniform cross-section ratio of the CDHS experiment.

---

*) The members of this group are : M.Holder, J.Knobloch, J.May, H.P.Paar, P. Palazzi, D.Schlatter,J.Steinberger,H.Suter, H.Wahl and E.G.H.Williams (CERN),F.Eisele,C.Geweniger,K.Kleinknecht, G.Spahn and H.J.Willutzki(Dortmund),W. Dorth,F.Dydak,V.Hepp, K.Tittel and J.Wotschack (Heidelberg),P.Bloch,B.Devaux,M.Grimm, J.Maillard,B.Peyaud,J.Rander, A.Savoy-Navarro and R.Turlay (CEN-Saclay),F.L.Navarria (Istituto di Fisica dell'Università, Bologna).

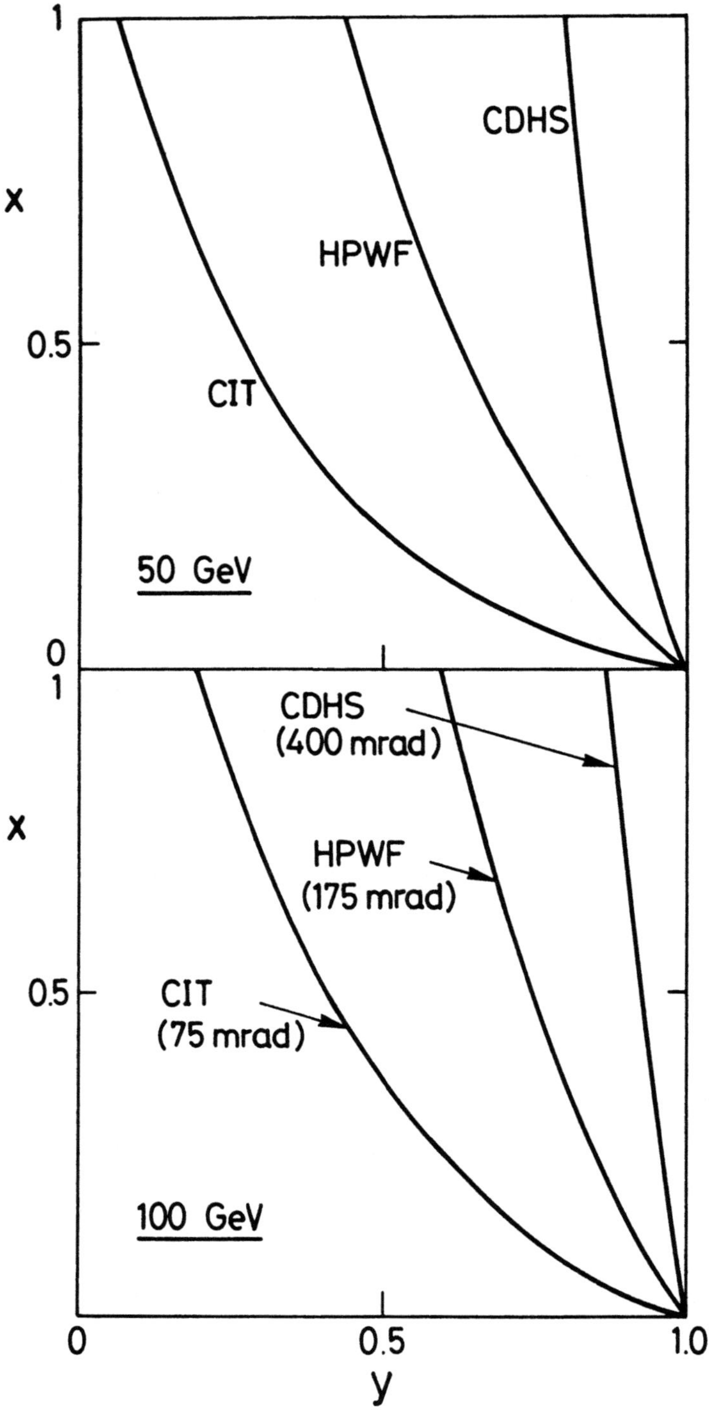

Fig. 19. Comparison of acceptances of HPWF, CIT and CDHS apparatuses at two energies.

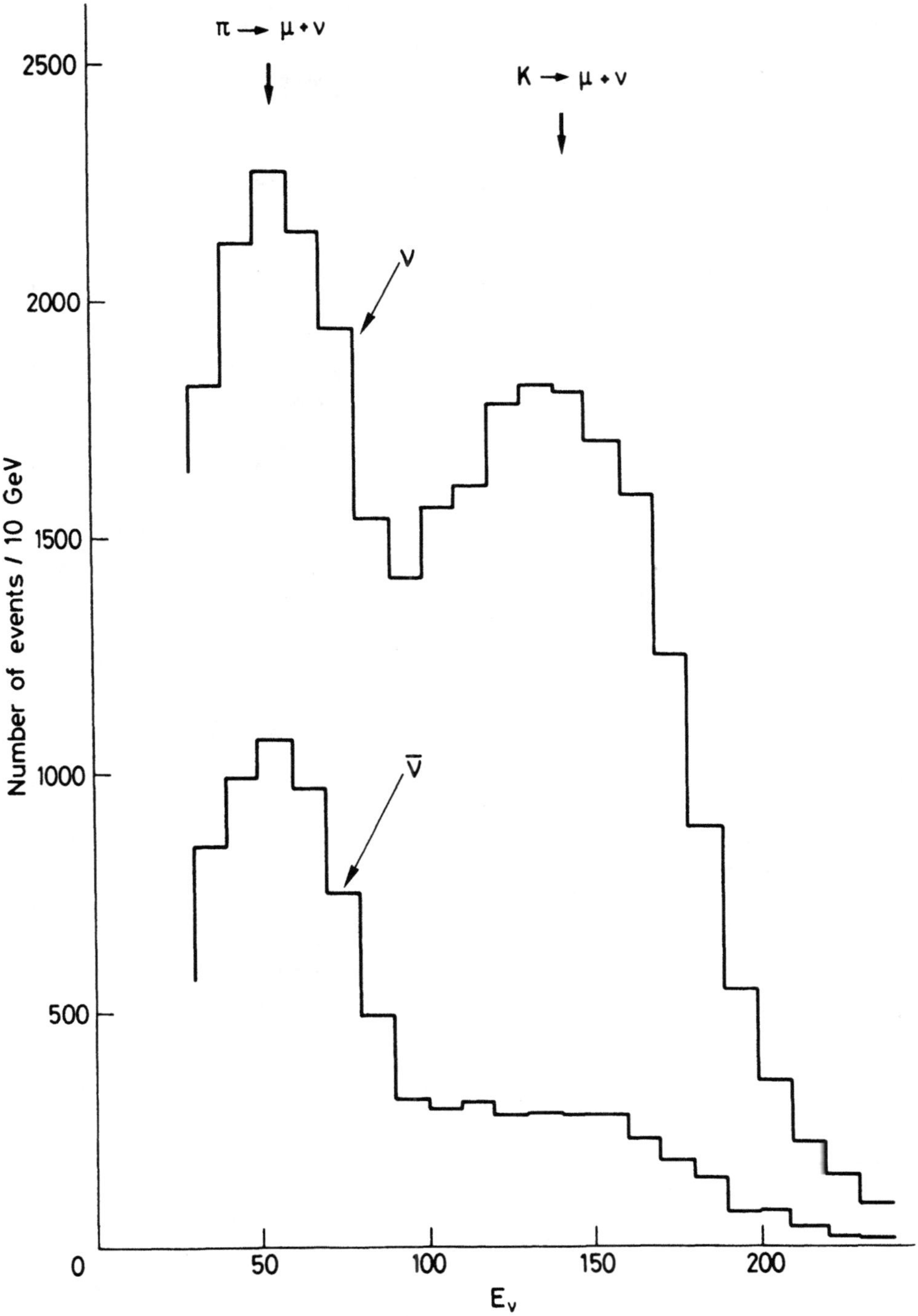

Fig. 20. Distribution in total (neutrino) energy for 36,000 neutrino and 12,000 antineutrino events.

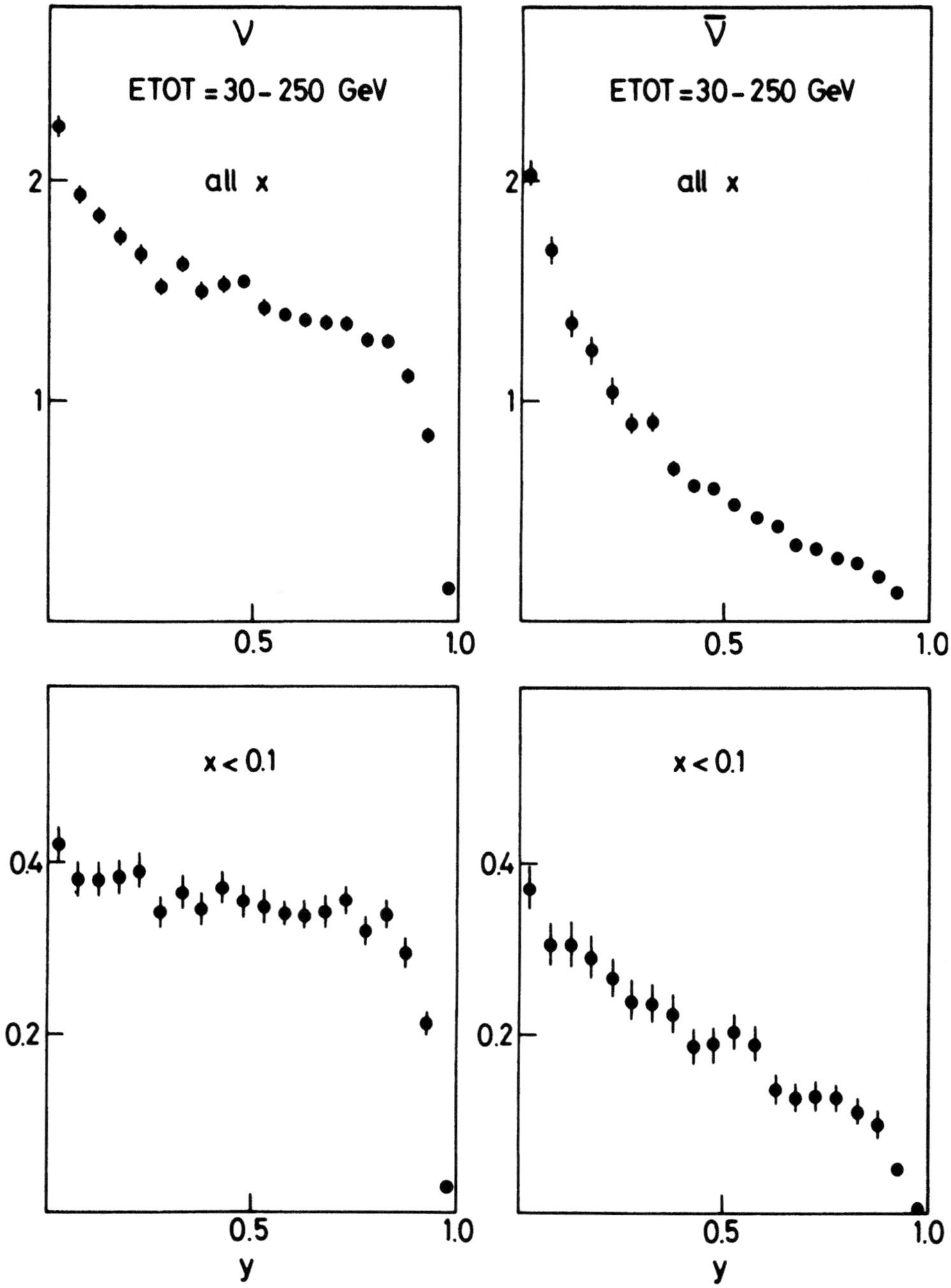

Fig. 21. Normalized y-distributions for neutrino and antineutrino data, with and without a cut x < 0.1. The distributions are normalized to the same ν and $\bar{\nu}$ fluxes subject to a relative scale error of ± 10 %.

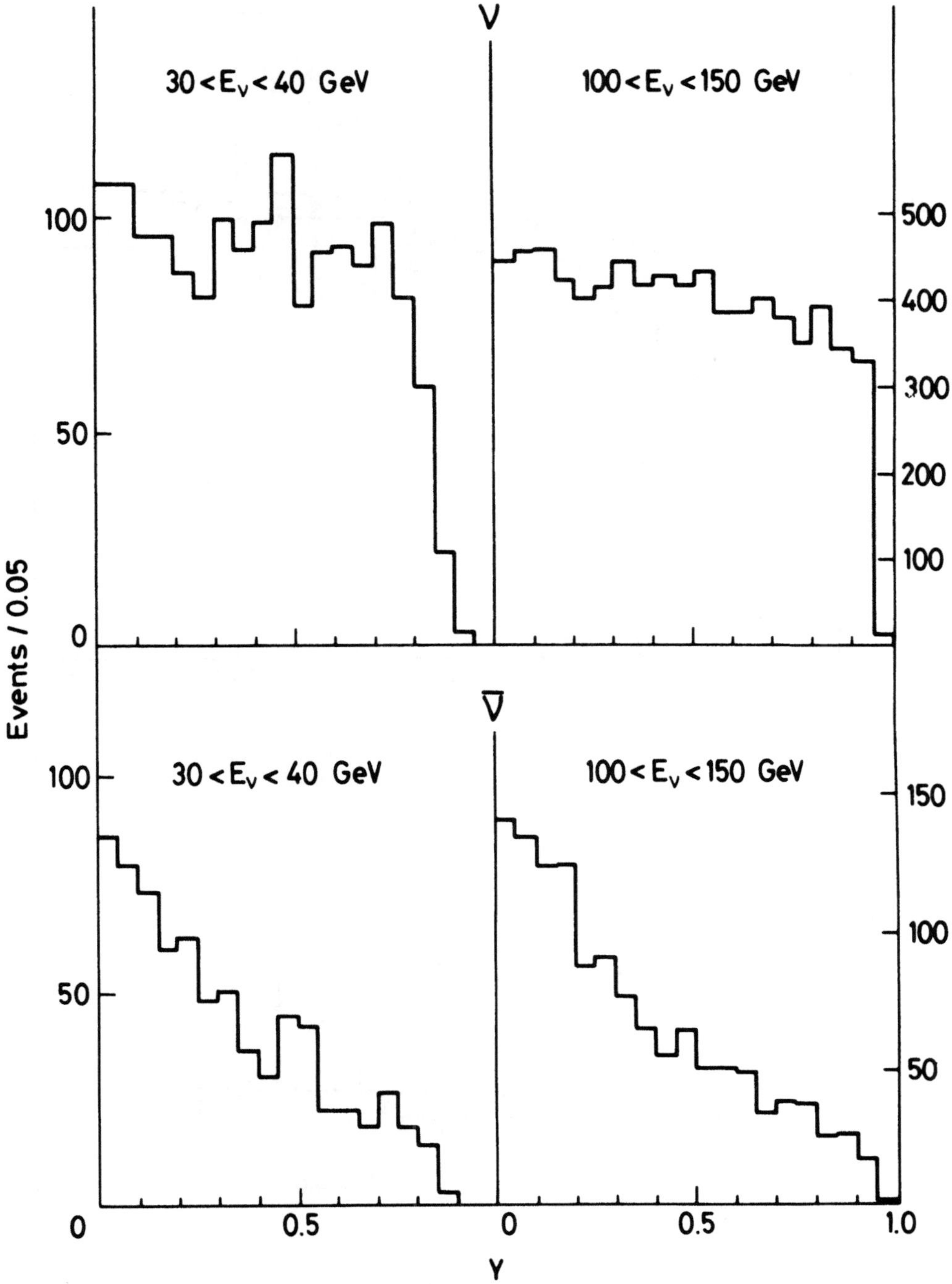

Fig.22. CDHS y-distributions in the x-interval $x < 0.6$, for neutrinos and antineutrinos in two energy regions : $30 < E_{\nu} < 40$ GeV and $100 < E_{\nu} < 150$ GeV. The rapid fall at large y, relatively more important at low neutrino energy, is due to the limited muon acceptance of the apparatus.

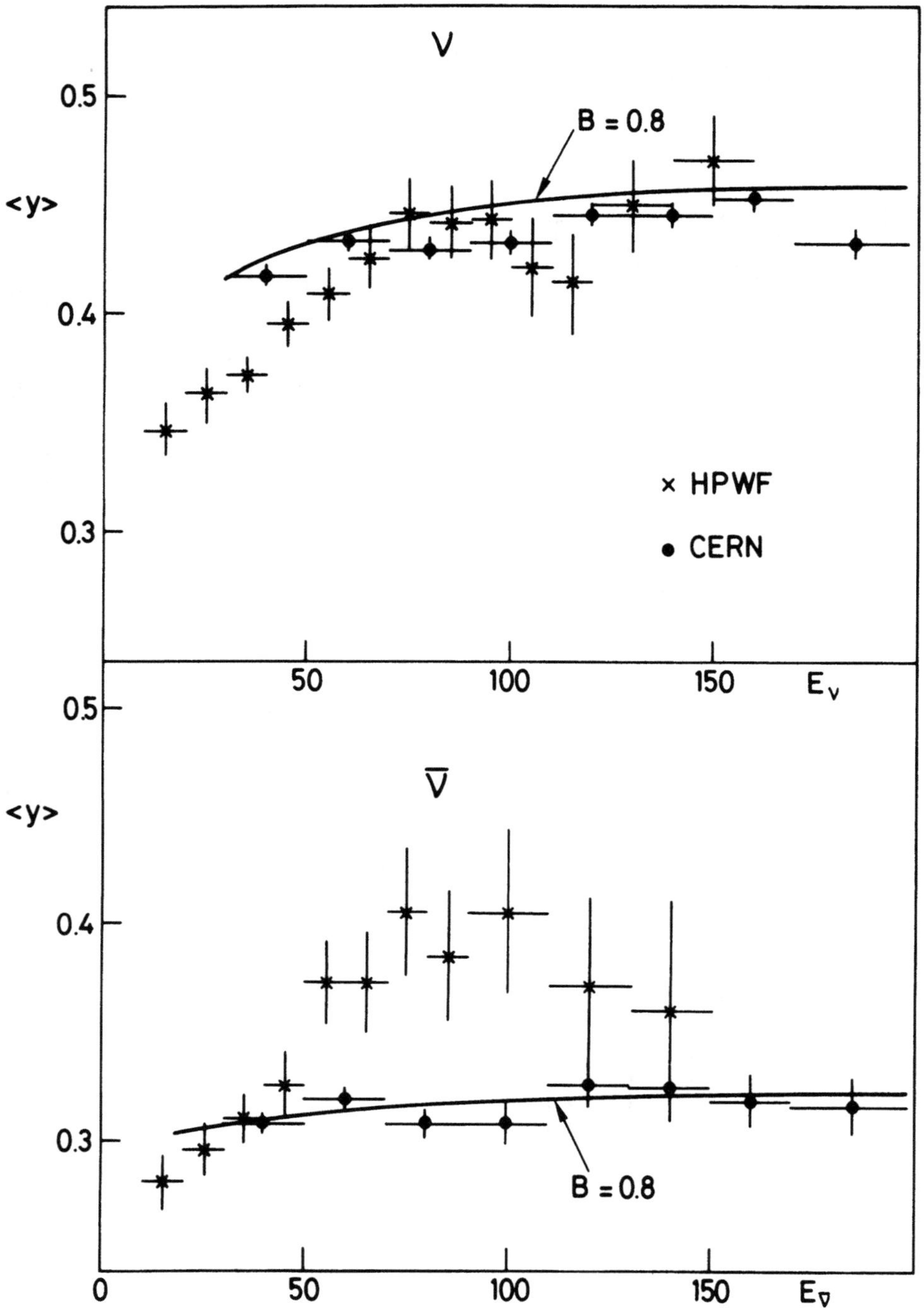

Fig.23. The average values of y as a function of neutrino energy, for neutrino and antineutrino data. The curves marked B=0.8 are calculated for the acceptance of the CDHS experiment. The HPWF data are taken from Ref. 13.

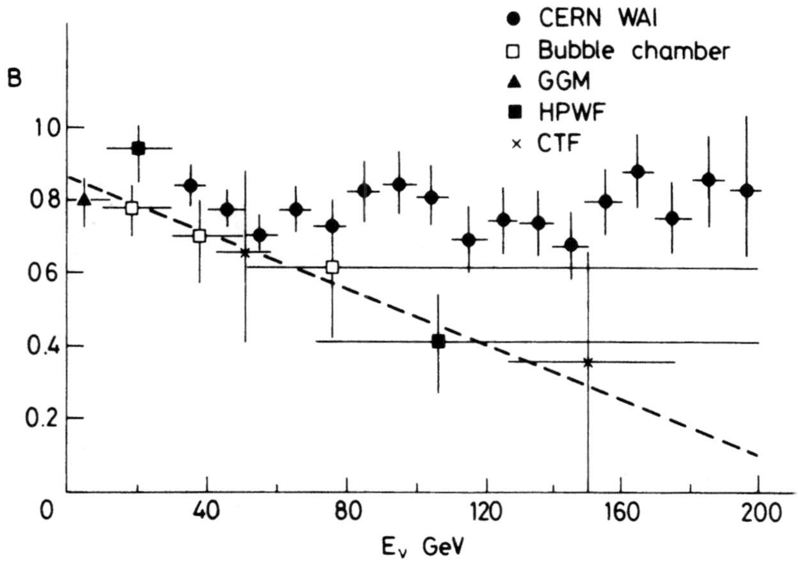

Fig.24 : The B-values as a function of antineutrino energy, from various experiments [10, 15-17,19].

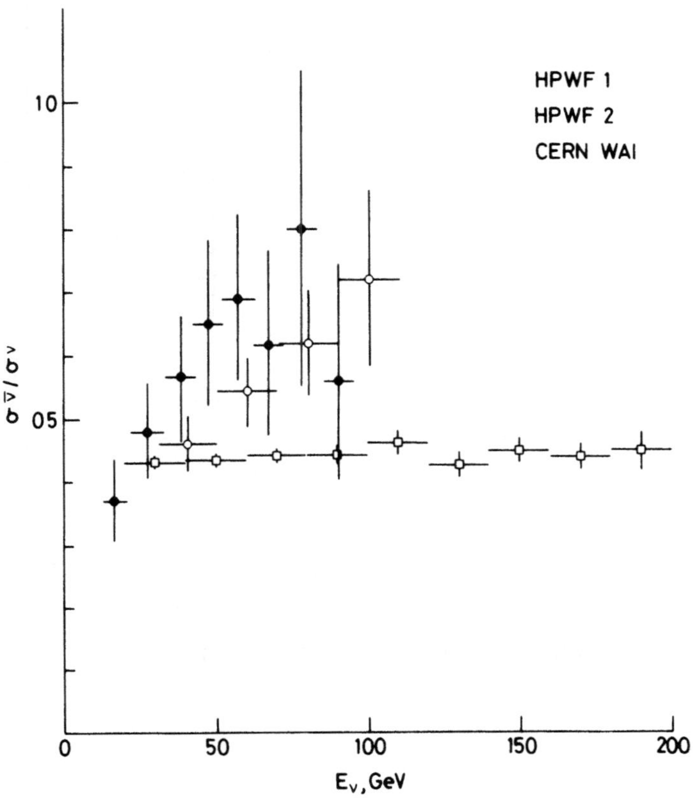

Fig.25. The ratio of neutrino and antineutrino total cross-sections as a function of neutrino energy. The data marked HPWF1 and HPWF2 are taken from ref. 16.

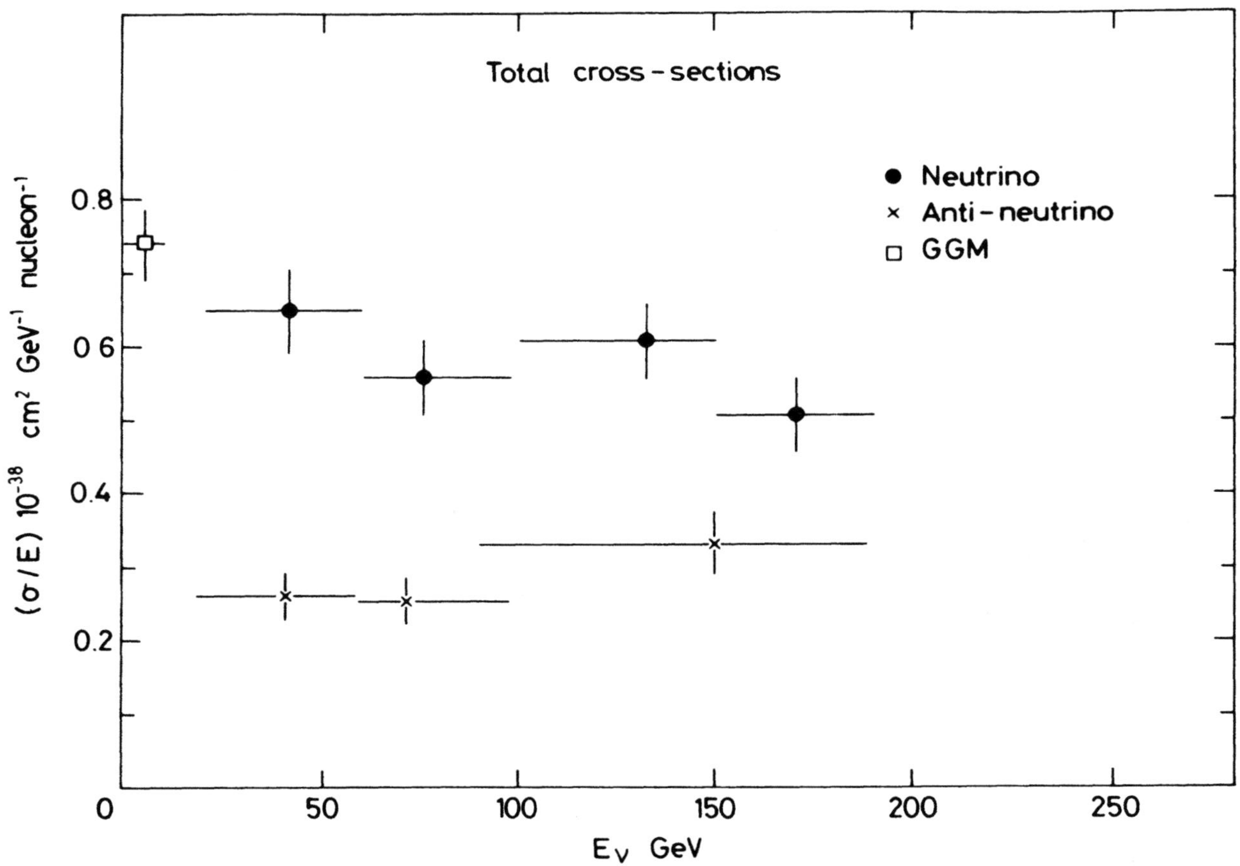

Fig. 26. Neutrino and antineutrino total cross-section results on the BEBC collaboration[20])

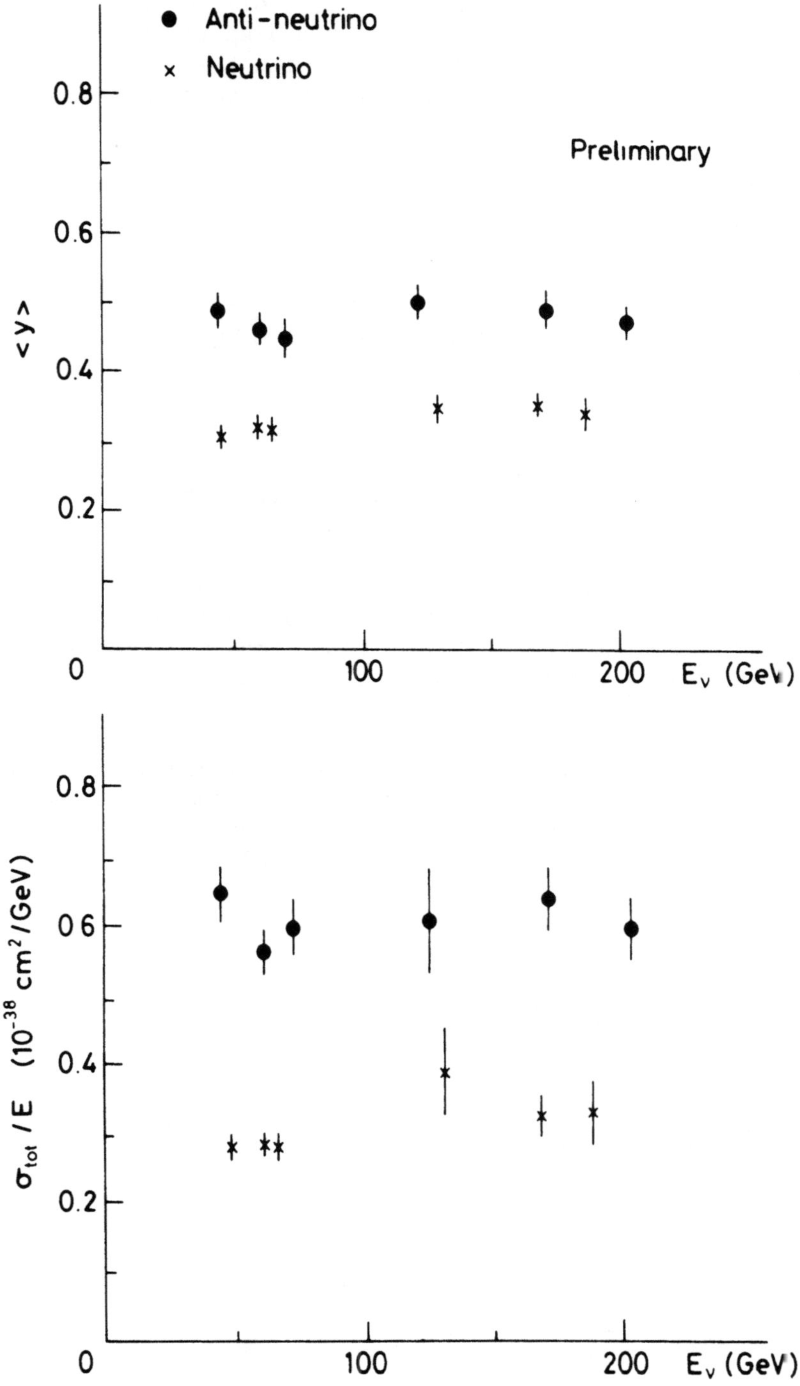

Fig. 27. Preliminary results for <y> and total cross-sections, for neutrinos and antineutrinos, from the CIT collaboration[21].

In Fig. 27, recent preliminary results of the CIT group on the dependence of <y> and the total cross-sections on energy are shown[21]. These are also consistent with the CDHS results and do not confirm the high-y anomaly. The results on total cross-sections are summarized in Table 2.

No convincing deviation from scaling, except perhaps at the lowest energy, is observed.

4. NEUTRAL CURRENTS AND THE WEINBERG-SALAM MODEL

4.1 Preliminary

Since their discovery in inclusive neutrino-nuclear reactions [2], neutral current effects have also been observed in neutrino-electron scattering[22-24] and in certain exclusive channels of neutrino-nuclear scattering[25,26]. Although both ν-e scattering and ν-nuclear exclusive scattering are potentially very informative about the structure of leptonic and hadronic neutral currents, we restrict ourselves here to new experimental results on the inclusive process.

We limit the discussion of the theoretical frame to vector and axial vector currents in the quark-parton model. However, the current is assumed to contain both left- and right-handed parts,in contrast with the charged current, which seems to be purely left-handed. Then we can write

$$\frac{d^2\sigma_\nu}{dxdy} = \frac{G^2ME_\nu}{\pi} \left(A_L + A_R(1-y)^2\right)$$

$$\frac{d^2\sigma_{\bar{\nu}}}{dxdy} = \frac{G^2ME_{\bar{\nu}}}{\pi} \left(A_L(1-y)^2 + A_R\right), \qquad (1)$$

where

$$A_L = A^- q(x) + A^+ \bar{q}(x); \qquad A^- \leftrightarrow V - A$$

$$A_R = A^+ q(x) + A^- \bar{q}(x); \qquad A^+ \leftrightarrow V + A .$$

The neutrino current is assumed to be of the form:

$$j_\mu = \bar{\nu}\gamma_\mu(1-\gamma_5)\nu,$$

Table 2

Neutrino and antineutrino total cross-sections and cross-sections ratios

	GGM $1 < E < 5$	BEBC $30 < E < 200$	CIT $30 < E < 200$	CDHS $30 < E < 200$
$\frac{\sigma_\nu}{E} \times 10^{-38} cm^2\ GeV^{-1} nucleon^{-1}$	0.74 ± 0.05	0.59 ± 0.05	0.60 ± 0.06	
$\frac{\sigma_\nu}{E} \times 10^{-38} cm^2\ GeV^{-1} nucleon^{-1}$	0.28 ± 0.03	0.27 ± 0.03	0.29 ± 0.03	
$\frac{\sigma_{\bar{\nu}}}{\sigma_\nu}$	0.38 ± 0.05	0.46 ± 0.07	0.48 ± 0.07	0.45 ± 0.05

and the current of the four quarks of the G.I.M. model :

$$j_h = C_L^u \bar{u}\gamma_\mu(1-\gamma_5)u + C_R^u \bar{u}\gamma_\mu(1+\gamma_5)u$$

$$+ C_L^d \bar{d}\gamma_\mu(1-\gamma_5)d + C_R^d \bar{d}\gamma_\mu(1+\gamma_5)d$$

$$+ C_L^s \bar{s}\gamma_\mu(1-\gamma_5)s + C_R^s \bar{s}\gamma_\mu(1+\gamma_5)s$$

$$+ C_L^c \bar{c}\gamma_\mu(1-\gamma_5)c + C_R^c \bar{c}\gamma_\mu(1+\gamma_5)c .$$

In the Weinberg-Salam model, the eight vector and axial vector coupling constants are expressed in terms of a single parameter, the Weinberg angle θ_W :

$$C_L^u = C_L^c = 1 - \frac{4}{3}\sin^2\theta_W ; \qquad C_R^u = C_R^c = -\frac{4}{3}\sin^2\theta_W$$

$$C_L^d = C_L^s = -1 + \frac{2}{3}\sin^2\theta_W ; \qquad C_R^d = C_R^g = +\frac{2}{3}\sin^2\theta_W .$$

In this model, then,

$$A^- = \frac{1}{2} - \sin^2\theta_W + \frac{5}{9}\sin^4\theta_W$$

and

$$A^+ = \frac{5}{9}\sin^4\theta_W .$$

4.2 Results from the CDHS experiment

In the foregoing discussion of the high-y anomaly, the fundamental requirement in the event selection was the existence of a reconstructed track, the muon, and the possible existence of a hadron shower was ignored. In the discussion of neutral currents the process is reversed. We require a shower (energy deposit in 1.5 m of iron) of at least 12 GeV, and the possible existence of a muon track is ignored. However, for each shower, a so-called "event-length" (EL) is established, which has as its aim the separation of the events into the two classes : neutral current (NC) and charged current (CC). In principle, this event length is the total length of the event, as evidenced by scintillator hits and drift chamber digitizings, from its beginning to its end (or to the end of the apparatus), measured in the thickness of iron. The beginning of the events accepted for the neutral current study has been restricted to the region 40 cm Fe $\leq z \leq$ 825 cm Fe. The event-length distributions are reproduced in Fig. 28. The large peak at small event lengths is due to neutral current events, the rest is due to charged current events. The step at EL=6 cm Fe is due to CC events with muons which leaves the end of the apparatus. On the basis of Fig. 25, the data are divided into "neutral current" and "charged current" events by a cut at 2.1 m of iron.

A number of corrections have to be applied to the data, as can be seen in Table 3. The cosmic-ray background can be accurately determined by looking in a time slice outside of the 23 sec beam spill. The background of CC events is determined by extrapolating the short muon events in the region 3 $\leq$ EL $\leq$ 5 m Fe to the region 0 < EL < 2.1 m Fe on the basis of the observed charged current y-distributions. The "wide-band" (WBB) background represents neutrino interactions from neutrinos with origin near the target, instead of the decay tunnel. It is evaluated on the basis of supplementary data with the hadron beam collimator closed. The K_{e3} background comes from events due to the reaction ν_e + Fe $\rightarrow$ e + ..., since electrons are indistinguishable from hadrons in our apparatus. The ν_e flux is known, and this background can be calculated precisely on the basis of e-μ universality.

The comparison with the Weinberg-Salam model is made on the basis of the expressions

$$R_\nu = \left.\frac{\sigma_{NC}}{\sigma_{CC}}\right|_\nu = \frac{1}{2} - \sin^2\theta_W + \frac{20}{27}\sin^4\theta_W ,$$

and

$$R_{\bar\nu} = \left.\frac{\sigma_{NC}}{\sigma_{CC}}\right|_{\bar\nu} = \frac{1}{2} - \sin^2\theta_W + \frac{20}{9}\sin^4\theta_W .$$

In this comparison, a rather small correction is applied to the theoretical curve by means of a Monte Carlo calculation, to account for the neglect of events with E_h < 12 GeV.

In Fig. 29, these results, as well as previous results[27-29], are compared with the predictions of the Weinberg-Salam model. From the CC- y-distributions it is known that $= (1-B)/(1+B) = \int \bar{q}(x)\,dx / \int q(x)\,dx$ is quite close to 0.1 The Weinberg angle is most accurately determined by the neutrino

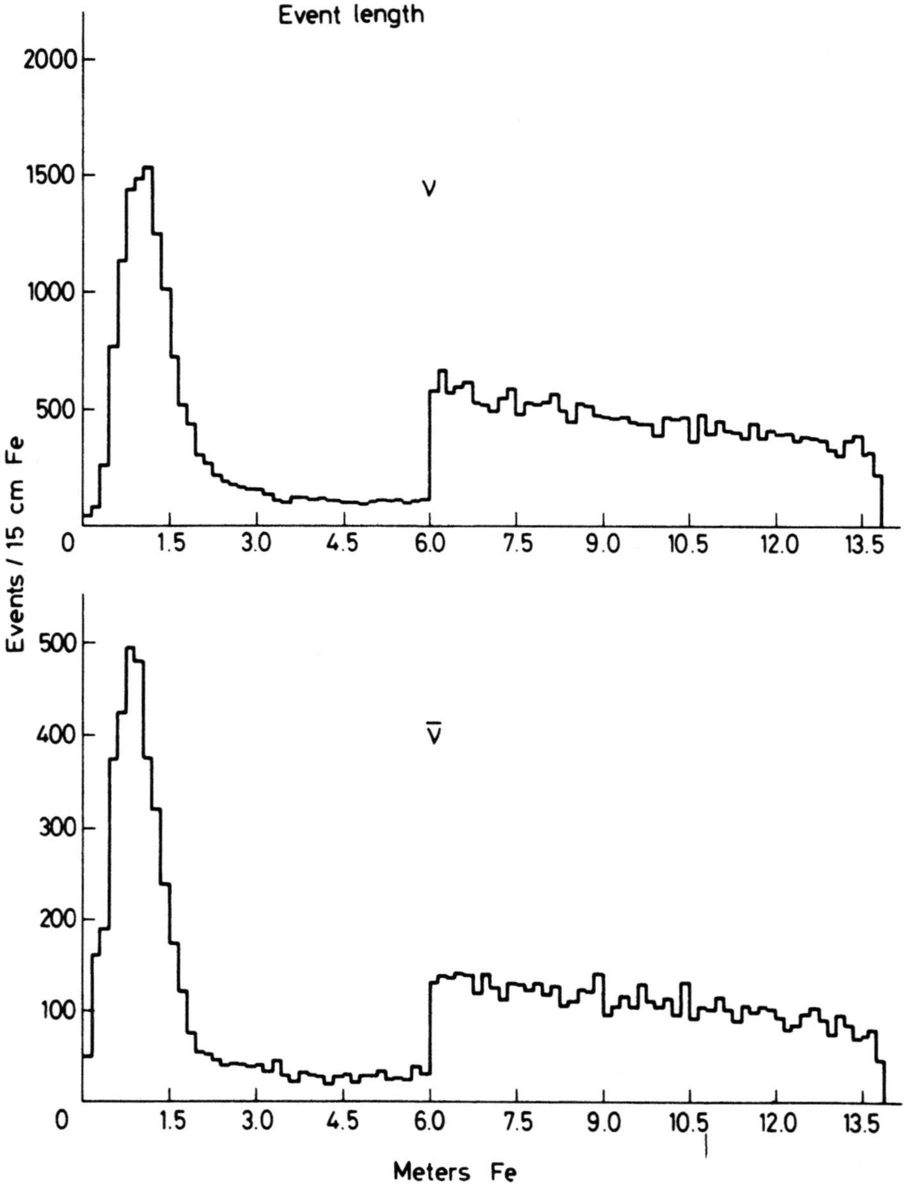

Fig. 28. "Event length" distributions for neutrinos and antineutrinos. The peak at short length is due to neutral current events, the rest due to charged current events.

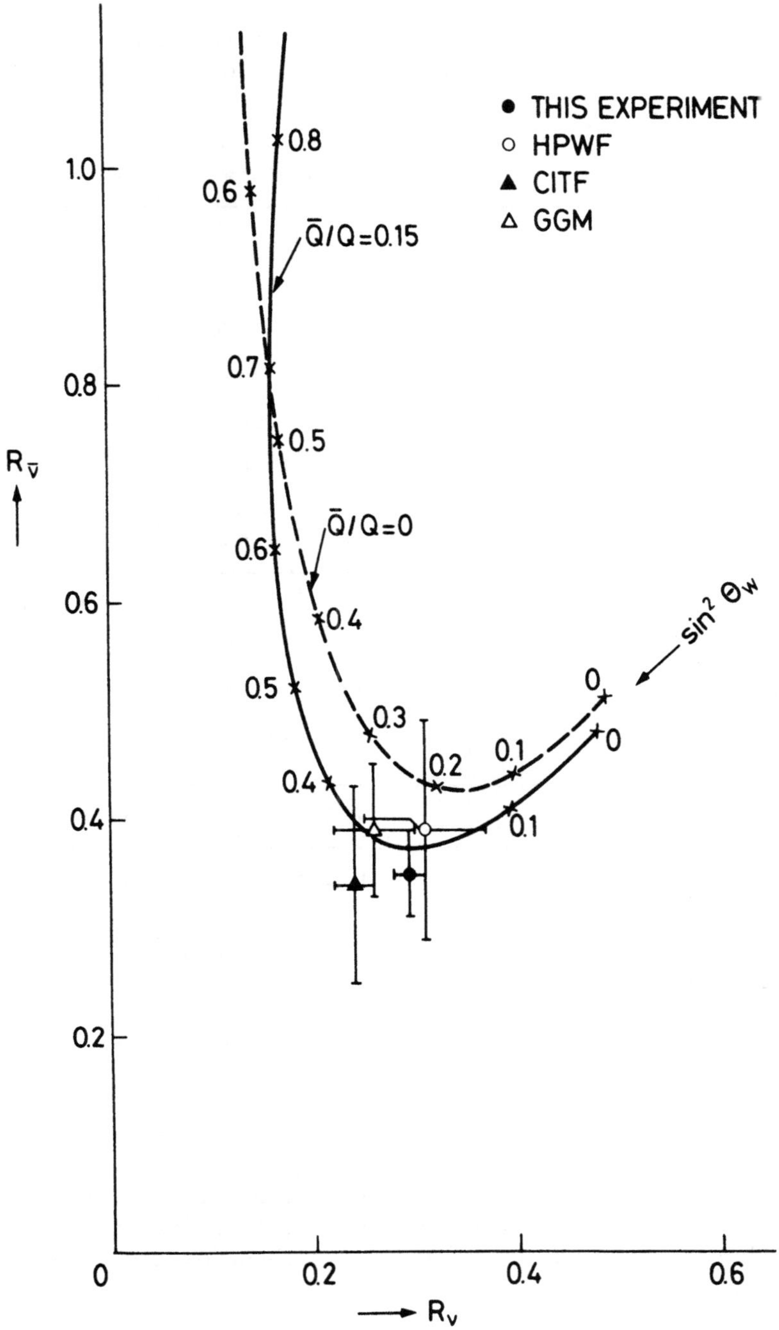

Fig. 29. The results for the ratios R_ν and $R_{\bar{\nu}}$ compared with the Weinberg-Salam model. Also shown are the results of the Gargamelle collaboration[27)], the CIT group[28)] and the HPWF collaboration[29)].

Table 3

Data reduction for the ratio of neutral current to charged current cross-sections for $E_H > 12$ GeV

	Neutrinos	Antineutrinos
NC candidates	10770 ± 104	3314 ± 58
Cosmic-ray background	-59 ± 7	-119 ± 10
WBB background	-286 ± 126	-646 ± 116
CC background	-1493 ± 64	-235 ± 49
NC with L<16 and $L>L_{cut}$	+150	+43
K_{e3} correction	-1008	-154
NC signal	8074 ± 156	2203 ± 130
CC candidates	26509 ± 163	6483 ± 81
WBB background	-239 ± 117	-323 ± 83
CC extrapolation	+1467 ± 64	+253 ± 42
NC with $L>L_{cut}$	-134	-35
CC signal	27603 ± 211	6378 ± 123
NC/CC (error only statistical-	0.293 ± 0.006	0.346 ± 0.021
NC/CC (final result, systematic error included)	0.295 ± 0.010	0.35 ± 0.03

measurement. The fact that the antineutrino result, with considerably smaller error than previous results, falls precisely on the "Weinberg nose", can be considered a check on the validity of the model.

Another step forward in our understanding of neutral currents becomes possible by making use of the hadron energy distributions at different detector radii. It will be remembered that different radii correspond to different hadron beam spectra, and the hadron energies reflect the y-distributions. The data are shown in Figs. 30 and 31.

The data can be analysed to yield information on the energy dependence of R and $R_{\bar{\nu}}$. We find

$$\frac{R_\nu(E_\nu > 100\ \text{GeV})}{R_\nu(E_\nu < 100\ \text{GeV})} = 0.99 \pm 0.1$$

and

$$\frac{R_{\bar{\nu}}(E_{\bar{\nu}} > 100\ \text{GeV})}{R_{\bar{\nu}}(E_{\bar{\nu}} < 100\ \text{GeV})} = 1.03 \pm 0.30.$$

This is in agreement with no energy variation. In order to study the y-distribution, we assume the form (1). Then

$$R_\nu(y) \propto \frac{1 + \alpha_{NC}(1-y)^2}{1 + \alpha_{CC}(1-y)^2}$$

and

$$R_{\bar{\nu}}(y) \propto \frac{(1-y)^2 + \alpha_{NC}}{(1-y)^2 + \alpha_{CC}}$$

where

$$\alpha_{CC} = \int q\, dx / \int \bar{q}\, dx$$

and

$$\alpha_{NC} = A_R/A_L .$$

If no further assumptions are made, the data of Fig. 31 yield the following result for α_{NC}, where, in the analysis, it has been assumed that $\alpha_{CC} = 0.1$:

$$\nu : \quad \alpha_{NC} = 0.21 \pm 0.14$$

$$\bar{\nu} : \quad \alpha_{NC} = 0.21 \pm 0.06 .$$

The solid lines in Fig. 31 represent this fit. The difference between α_{NC} and α_{CC} shows that the analysis of the relative hadron enery distribution of neutral and charged current events points to a non-zero V + A component of the neutral current.

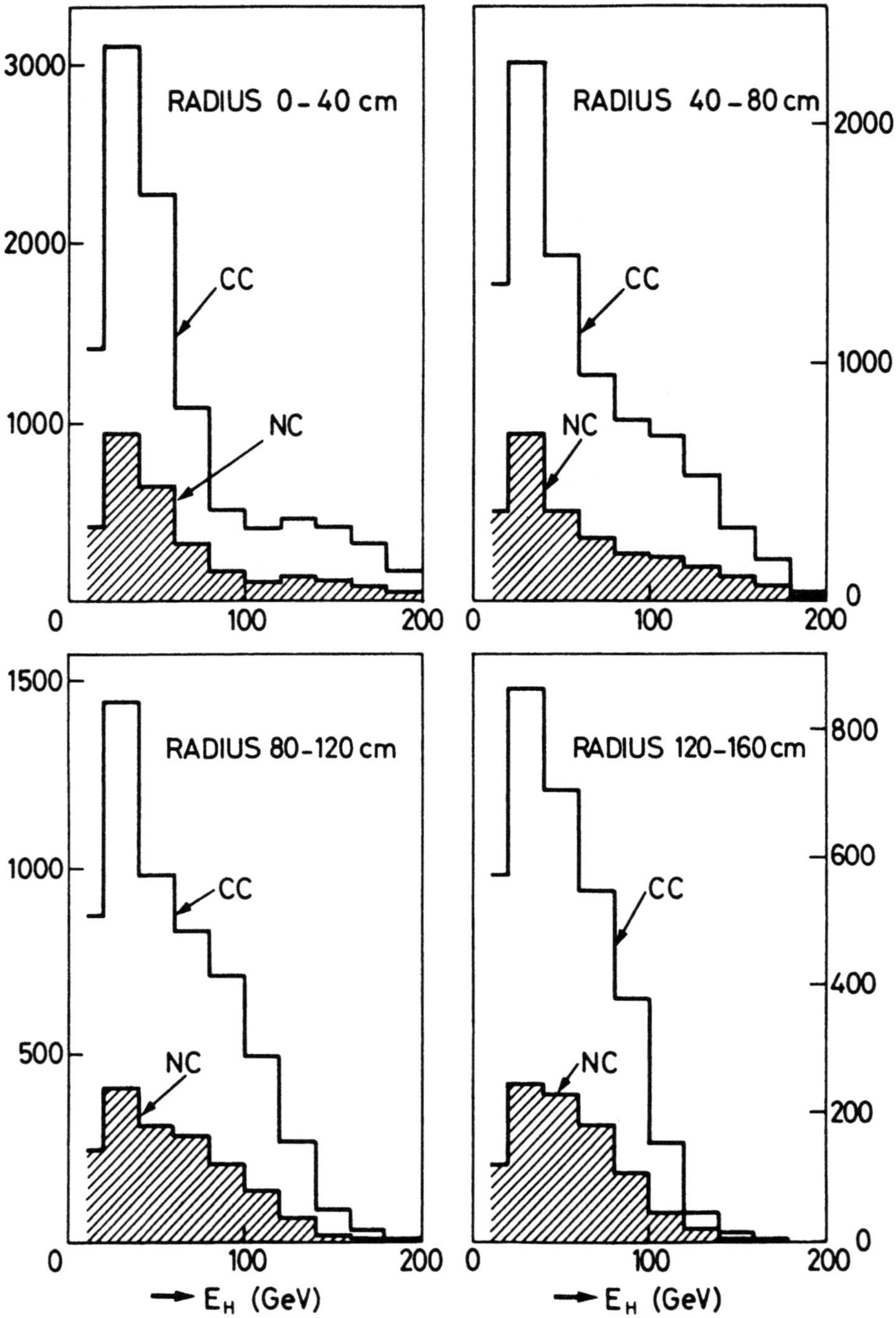

Fig. 30. Hadron energy distributions for different radial bins for CC and NC and for
a) neutrino

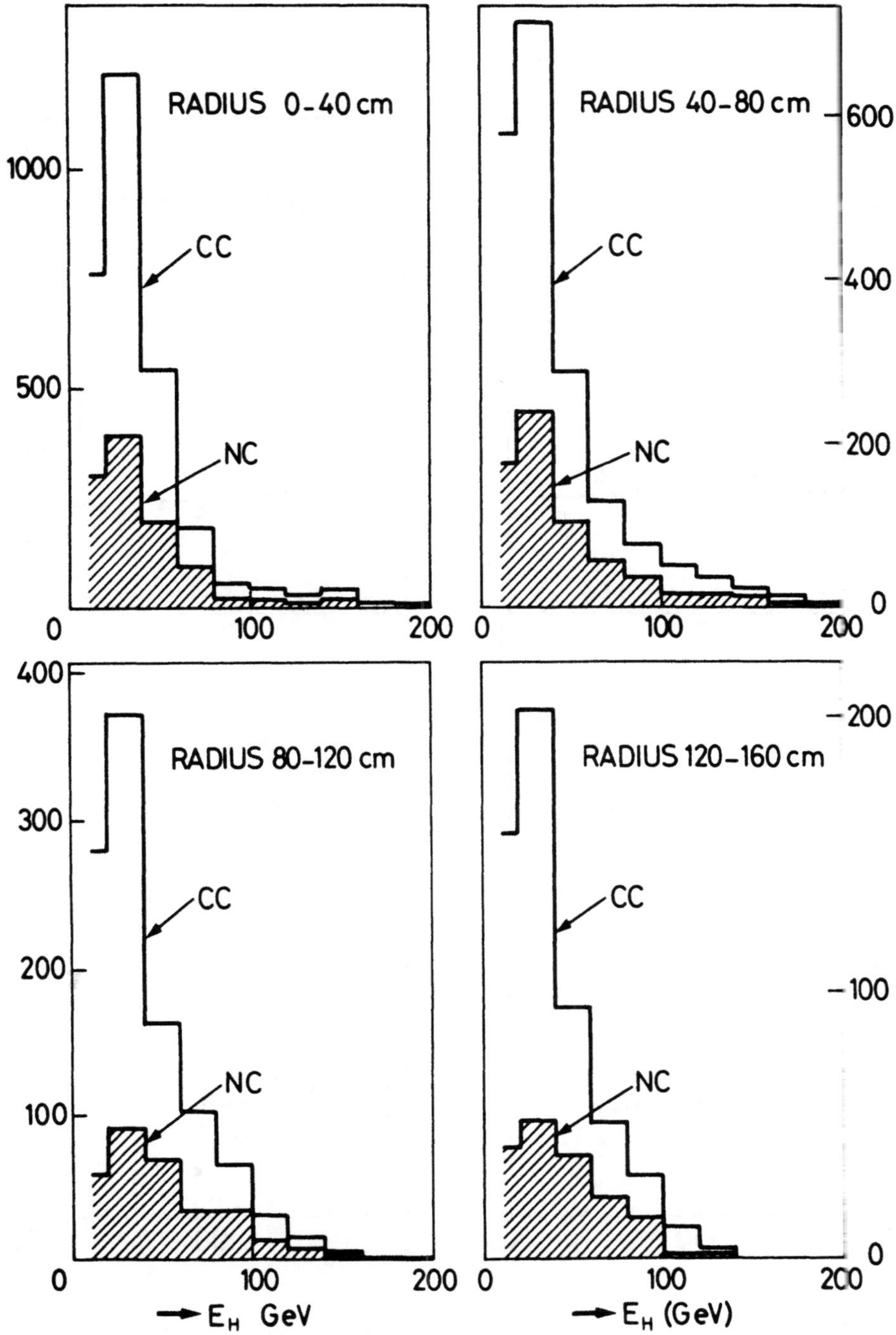

Fig. 30. Hadron energy distributions for different radial bins for CC and NC for
b) antineutrino events

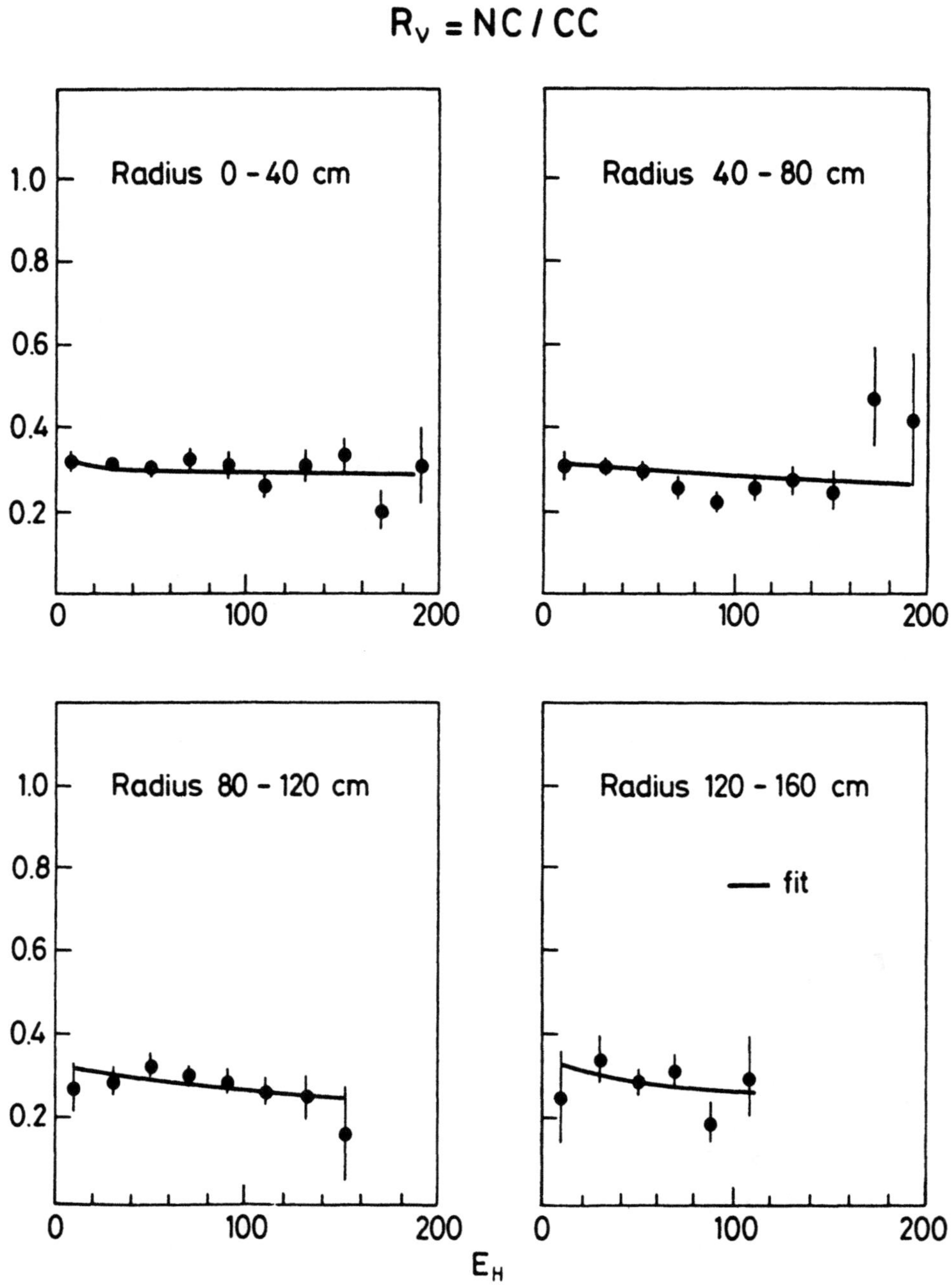

Fig. 31. Ratio of neutral to charged current events as a function of hadron energy for different radial bins, for
a) neutrinos

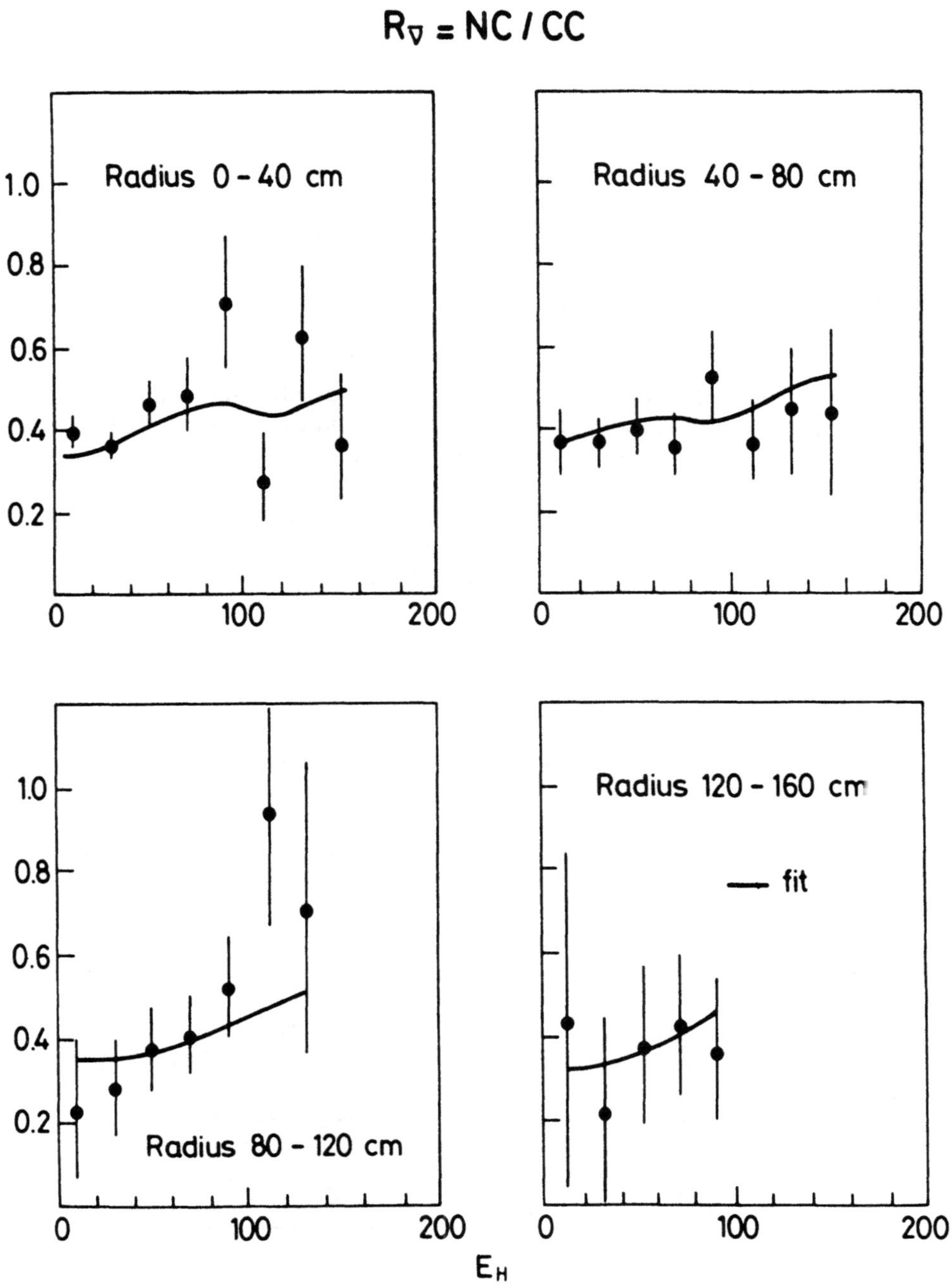

Fig. 31. Ratio of neutral to charged current events as a function of hadron energy for different radial bins, for b) antineutrinos

In the Weinberg-Salam model these values of σ_{NC} yield values of $\sin^2\theta_W$:

$$\text{from } \alpha_{NC}^{\nu} : \sin^2\theta_W = 0.22 \pm 0.13$$

and

$$\text{from } \alpha_{NC}^{\bar{\nu}} : \sin^2\theta_W = 0.25 \pm 0.05 .$$

These values, determined completely independently from the hadron energy distributions, agree well with the previously given value $\sin^2\theta_W = 0.24 \pm 0.02$, determined solely from the ratios R_ν and $R_{\bar{\nu}}$.

The data of Fig. 32 can also be used to study the question of the identity of the incident and outgoing neutrinos in neutral current reactions. If the two neutrinos are the same, then the ratio $(d\sigma_{\bar{\nu}}/dy)/(d\sigma_\nu/dy)\big|_{y=0}$ should be equal to 1 [30]. If we assume that the charged current has this property (see also the experimental verification in Section 3), then for the neutral current we find, on the basis of the data of Fig. 31 :

$$\frac{d\sigma_{\bar{\nu}}}{dy} \Big/ \frac{d\sigma_\nu}{dy}\Bigg|_{y=0} = 0.91 \pm 0.20 ,$$

in agreement with the expectation for identical neutrinos.

Finally, assuming neutrino identity, we can make a simultaneous fit to neutrino and antineutrino data, to find A_+ and A_-, the V + A and V - A compounds of the neutral current. The results, in Table 4, show that a small V+A term is required, although the uncertainty in this term is not negligible.

Table 4
Results of the neutral current study in terms of A_+ and A_-, the coefficients of V + A and V - A currents respectively

	Assumed $\alpha_{CC} = 0.1$	Assumed $\alpha_{CC} = 0.15$
A_+	0.025 ± 0.006	0.031 ± 0.006
A_-	0.283 ± 0.007	0.280 ± 0.007

In the V,A frame for neutral currents (which includes, of course, the Weinberg-Salam model), the two numbers A_+ and A_- summarize the result of this neutral current experiment.

5. "WRONG SIGN" MUONS, CHARM CHANGING NEUTRAL CURRENTS, AND THE SEA OF CHARMED PARTONS IN THE NUCLEON

If reactions with positive muons were observed among neutrino events, they could be due to the following processes :

i) dimuon events, with short-range (invisible) negative muons;
ii) neutral current reactions with subsequent $\pi^+ \to \mu^+$ or $K^+ \to \mu^+$ decay;
iii) antineutrino charged current events, due to the "wide-band" background already discussed in the last section;
iv) the neutral current, charm-changing reaction

$$\nu + u \to \nu + c, \quad c \to \mu^+ + \ldots$$

This latter would be of great interest, since it is forbidden in the G.I.M. model.

We have observed 23 such events in a sample that contained 36,000 $\bar{}$ events. The number and the energy distribution of these events are consistent with wide-band background. No event is observed above $E_{vis} = E_{\mu^+} + E_h$ greater than 100 GeV. In this region we would expect a background of 3 events from processes (i) and (ii). To find an upper limit for the ratio of the cross-section of charm-changing neutral currents to that of charm-non-changing neutral currents, we use the fact that the sample contains 1100 NC events with $E_h \gtrsim 100$ GeV, 0.15 for the muonic branching ratio, 0.5 for the muonic detection efficiency, to find

$$\frac{\sigma_{\text{charm-changing NC}}}{\sigma_{\text{normal NC}}} \lesssim \frac{2}{1100 \times 0.4 \times 0.15} = 0.04 \text{ (90 \% confidence level).}$$

6. MULTIMUONS

6.1 Preliminary

Opposite-sign dimuons produced by neutrinos were first observed in the HPWF experiment[31)], and this was quickly followed by similar results from the CIT group[32)] and by the observation of $\mu^- e^+$ events in the bubble chamber[33-35)]. The likely charmed origin of the bulk of these events was immediately clear.

Less clear were charmed origin of the bulk of these events was immediately clear. Less clear were the questions of the existence and nature of like-sign dimuons and of dimuons produced by antineutrinos. The last months have been marked by the following events :

i) the observation of trimuons[36-38]
ii) the claim[37] that the observed trimuon events, as well as certain features of dimuon events, show that new processes, involving at least two new types of heavy leptons, are occurring;
iii) substantially more extensive and better data on dimuon production by neutrinos and antineutrinos[39];
iv) more extensive bubble chamber data on $\mu^- e^+$ production by neutrinos[35].

6.2 Opposite-sign dimuon results from the CDHS experiment[39]

A dimuon event, as reconstructed off line, is shown in Fig. 32. Observed event numbers for different event classes are shown in Table 5.

Table 5
Observed event numbers for single muons, dimuon and trimuon events

	ν	$\bar{\nu}$
Normal charged current	55,000	16,000
$\mu^- + \mu^+$	257	58
$\mu^- + \mu^-$	47	0
$\mu^+ + \mu^+$	0	9
$\mu^- + \mu^- + \mu^+$	1	1/2
$\mu^+ + \mu^+ + \mu$	1	0

It should be kept in mind, in evaluating these and subsequent data, that the minimum momentum required for the detection and momentum measurement of the muons is about 4.5 GeV/c, and that this represents a serious bias for most of the measured rates and distributions. Table 6 lists averages of the more interesting parameters.

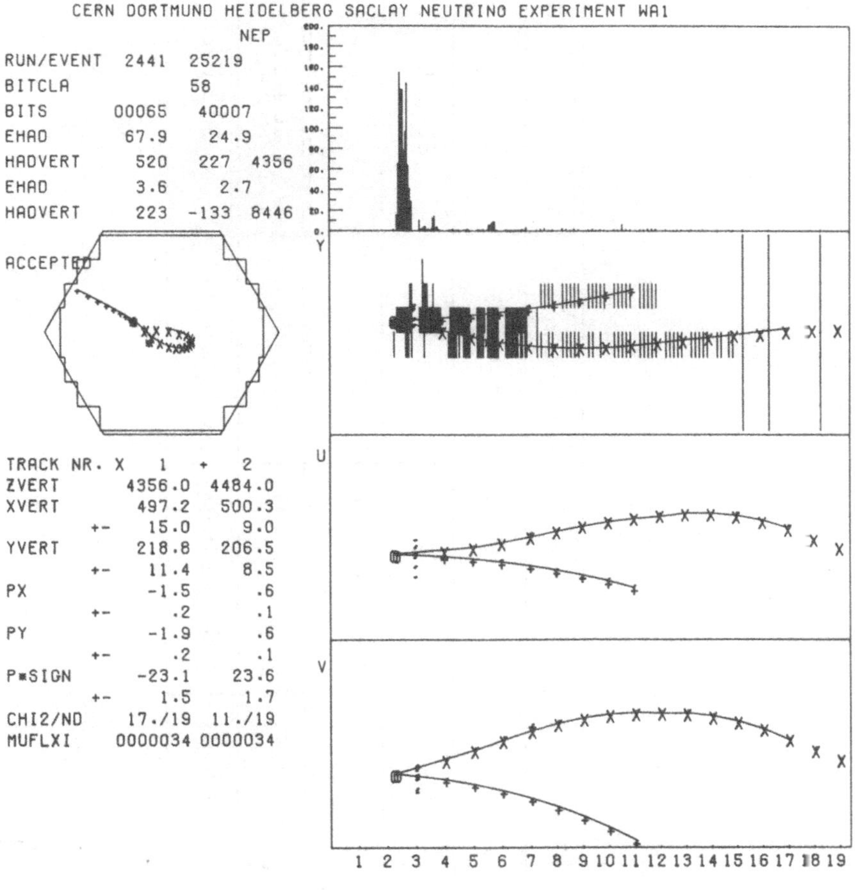

Fig. 32. A dimuon event, computer reconstructed, in the CDHS experiment [39].

Table 6

Average values of certain parameters for single muon and dimuon reactions

Reaction	$\nu \to \mu^- + X$	$\bar{\nu} \to \mu^+ + X$	$\nu \to \mu^- + \mu^+ + X$	$\bar{\nu} \to \mu^- + \mu^+ + X$	$\nu \to \mu^- + \mu^- + X$	$\bar{\nu} \to \mu^+ + \mu^+ + X$
E_{total} (GeV)	93 ± 1	70 ± 1	108 ± 3	85 ± 6	133 ± 7	97 ± 16
$E_{\mu 1}$ (GeV)	53 ± 1	50 ± 1	45 ± 2	37 ± 4	52 ± 5	53 ± 11
$E_{\mu 2}$ (GeV)	-	-	13.7 ± 1	12.4 ± 1	7.3 ± 0.6	5.7 ± 0.5
X	0.29±0.01	0.27±0.01	0.24 ± 0.01	0.15 ± 0.02	0.26 ± 0.04	0.33 ± 0.13
y	0.44±0.01	0.30±0.01	0.58 ± 0.02	0.57 ± 0.03	0.59 ± 0.03	0.48 ± 0.06

In Figs. 33-36 some results for the opposite-sign dimuons are shown. Of particular interest are the x-distributions, which are very different from those of normal CC events, and the correlations in transverse momentum of the second muon *), which show that the second muon, independent of its energy, has a close correlation with the hadron shower direction. The average transverse momentum of the second muon with respect to this direction is 0.7 GeV. All of the results are in good agreement with the expectations of the G.I.M. model, where the muon is the decay of a charmed particle, most often the D^+, in the decay mode $D \rightarrow K + \mu^+ + \nu$. We recall that the principal reactions in the G.I.M. model and their relative rates are the following :

$$\nu + d \rightarrow c + \mu^- + \ldots \propto \sin^2 \theta_c \times BR$$
$$\quad c \rightarrow \mu^+ + \ldots$$

$$\nu + s \rightarrow c + \mu^- + \ldots \quad \propto \quad \zeta_s \cos^2 \theta_c \times BR$$
$$\quad c \rightarrow \mu^+ + \ldots$$

and

$$\bar{\nu} + \bar{s} \rightarrow \bar{c} + \mu^+ \quad \propto \quad \zeta_s \cos^2 \theta_c \times BR ,$$
$$\quad \bar{c} \rightarrow \mu^-$$

where ζ_s is the momentum carried by the strange sea quarks relative to the valence d quarks : $\zeta_s = \int x\, s(x)\, dx / \int x\, d(x) dx$.

In the case of antineutrinos, the model predicts that the di-muon production is almost entirely associated with the strange sea. The corresponding x-distribution of Fig. 34 is then a measure of the x-distribution of this sea. It is strikingly different from the x-distributions normally found in experiments on neutrino, electron, and muon inelastic scattering, which are presumably dominated by valence quarks. In the case of neutrinos, the contribution from valence quarks seems to be comparable to that from the sea, so that the small factors $\sin^2\theta_c$ and ζ_s are comparable.

The excitation functions are given in Fig. 37, where also the calculated muon detection efficiencies, assuming the charm production model of Sehgal and Zerwas[40], are shown. From a comparison of

---

*) The first muon is the one whose electric charge is the same as that of normal charged current reactions.

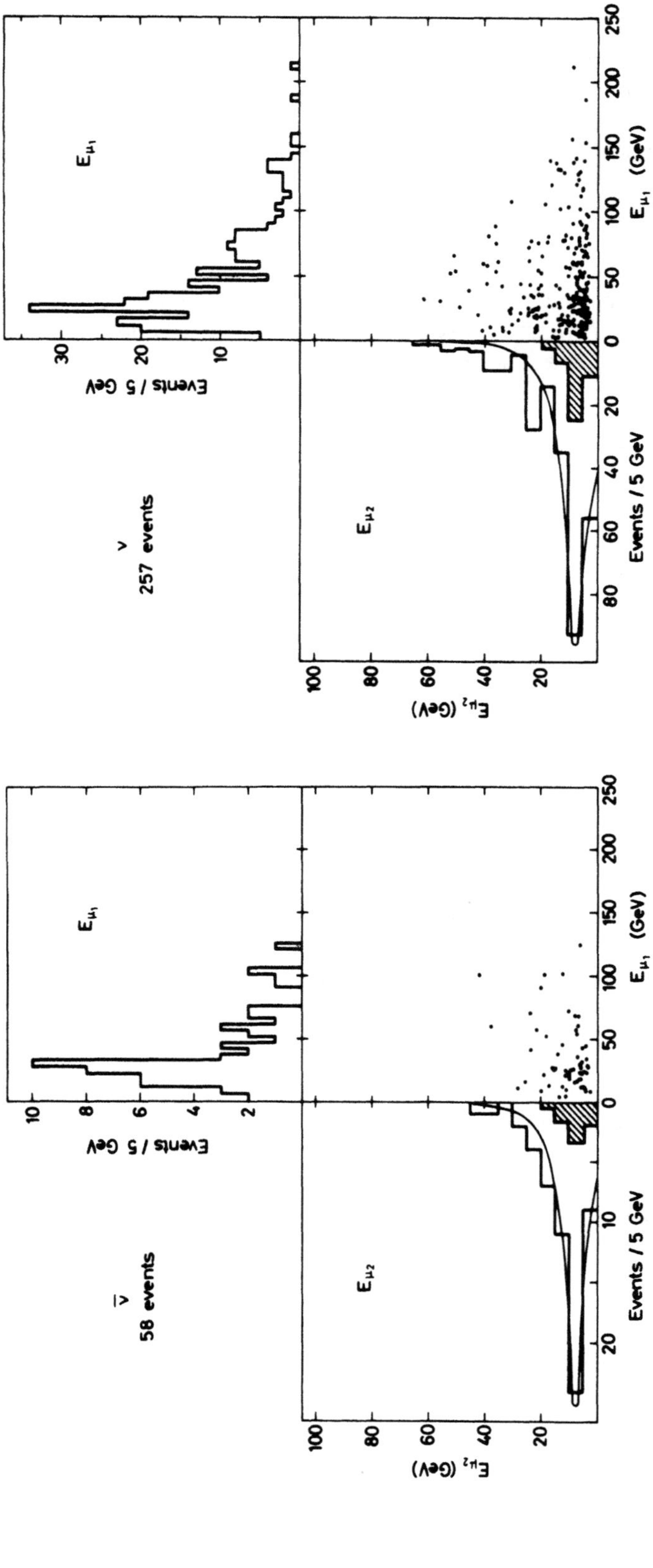

Fig.33. Correlations between the momenta of the two muons[39]. The shaded areas in the energy distributions of the second muon is the calculated $\pi \rightarrow \mu$ and $K \rightarrow \mu$ decay background. The solid curve is calculated on the basis of a charm production model.

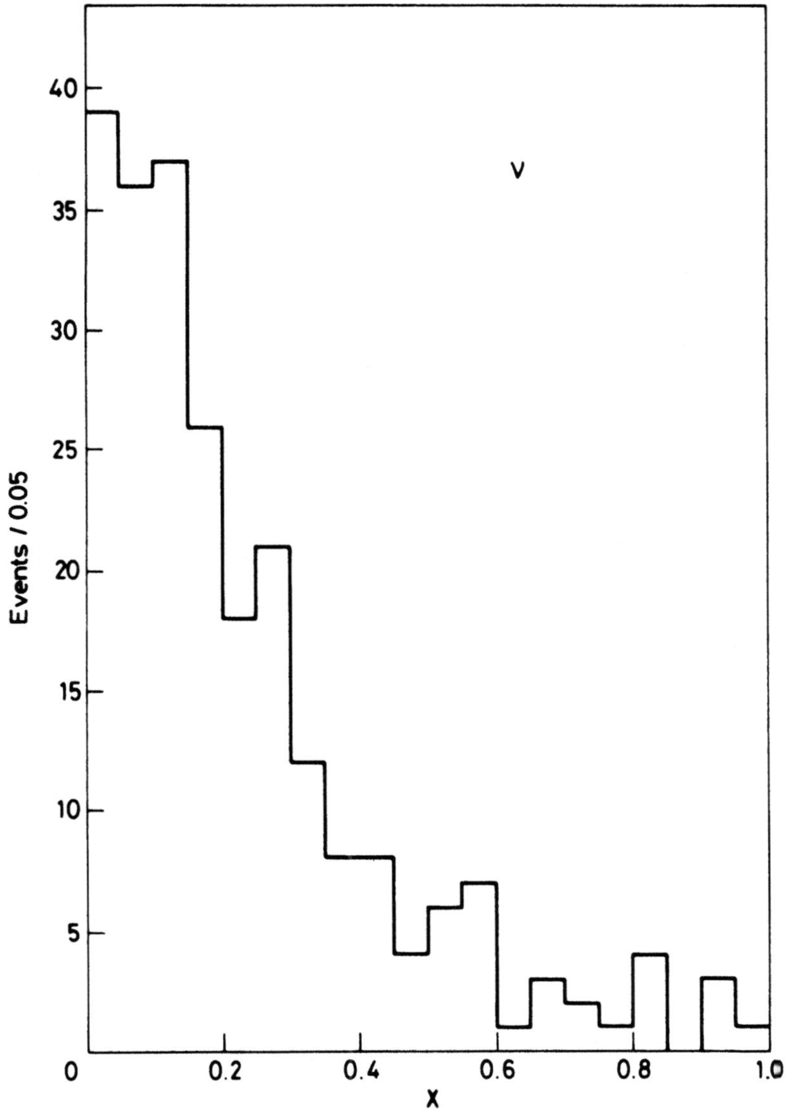

Fig. 34. x- and y-distributions for neutrino and antineutrino events. The solid curves are calculated on the basis of a charm production model. The dotted curves are the flat and $(1-y)^2$ distributions which dominate the normal charged current reactions.

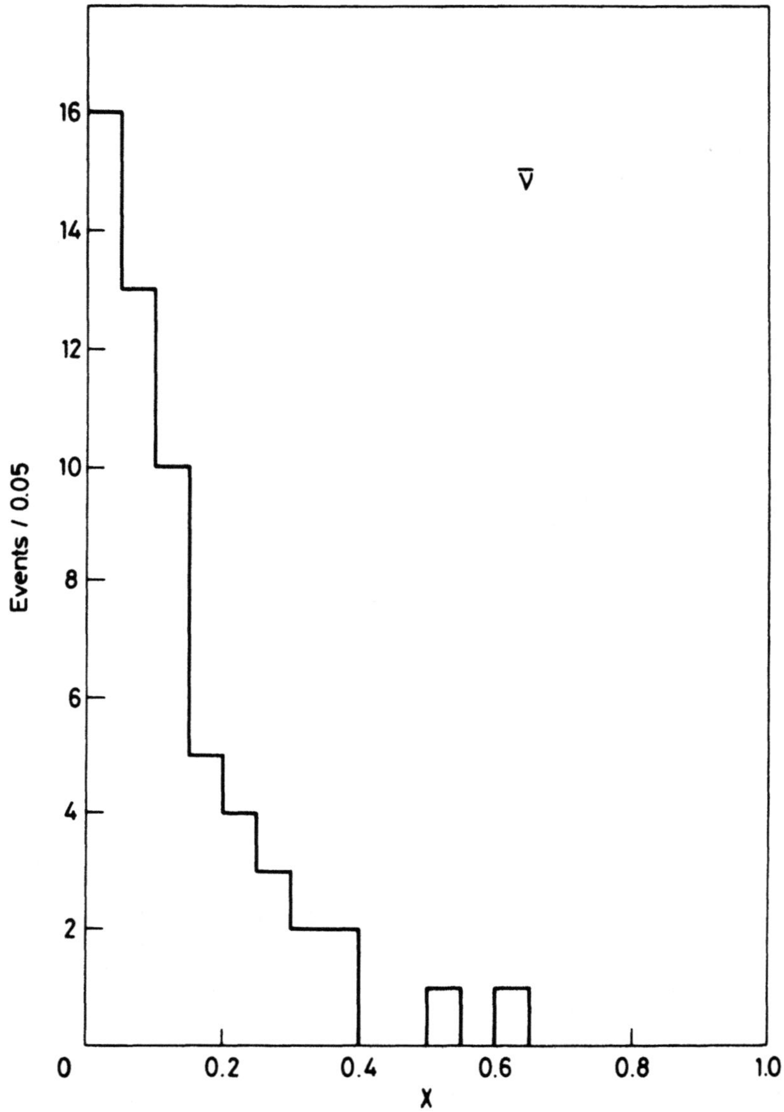

Fig. 34b

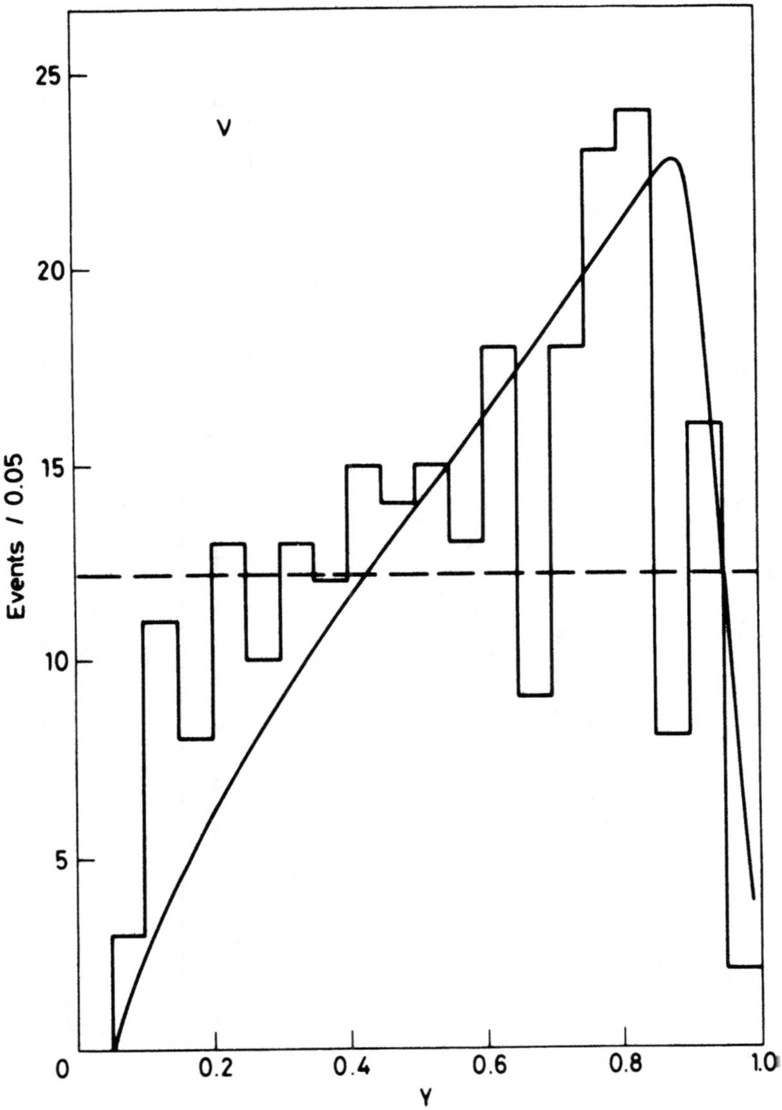

Fig. 34c

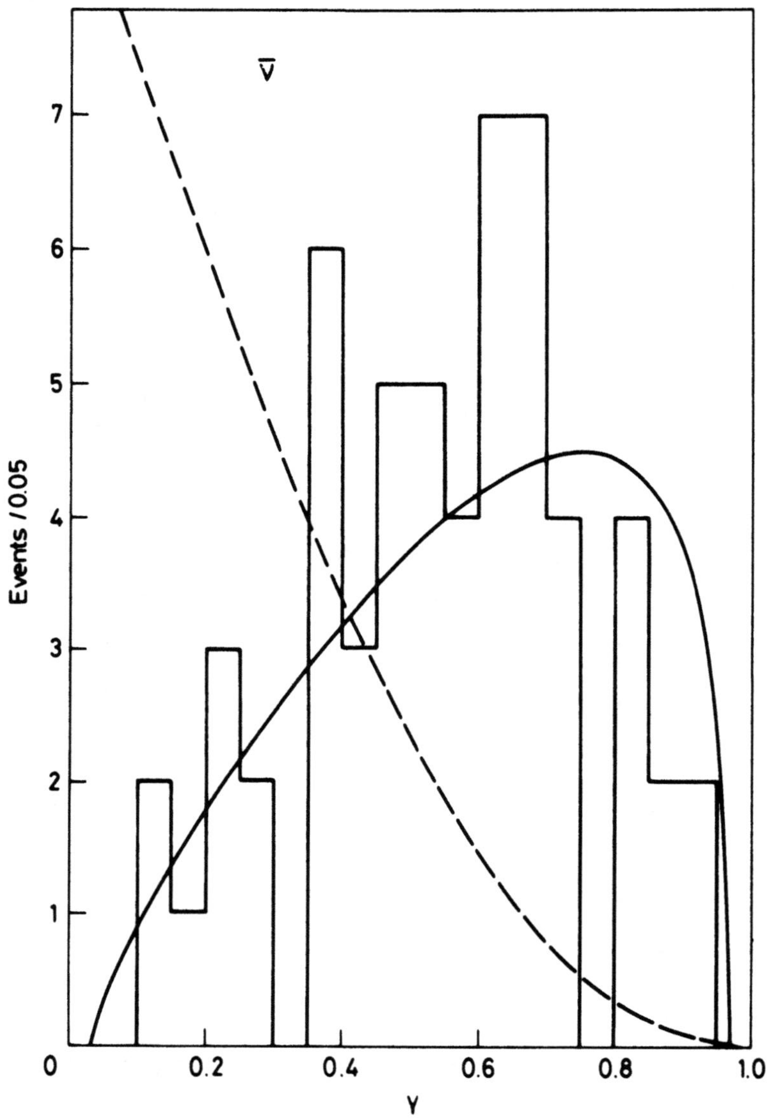

Fig. 34d

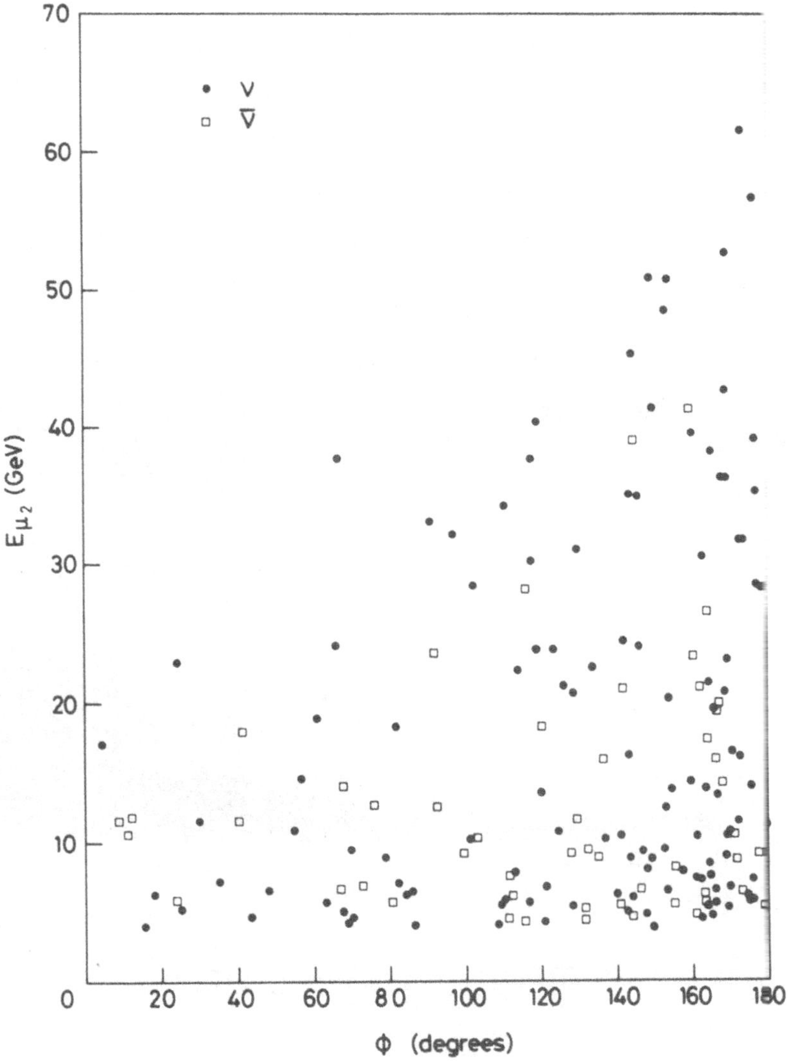

Fig. 35. Scatter plot in azimuthal angle of μ_2 versus E_{μ_2}, for neutrinos and antineutrinos. The predictions of a charm model are shown in the projections.

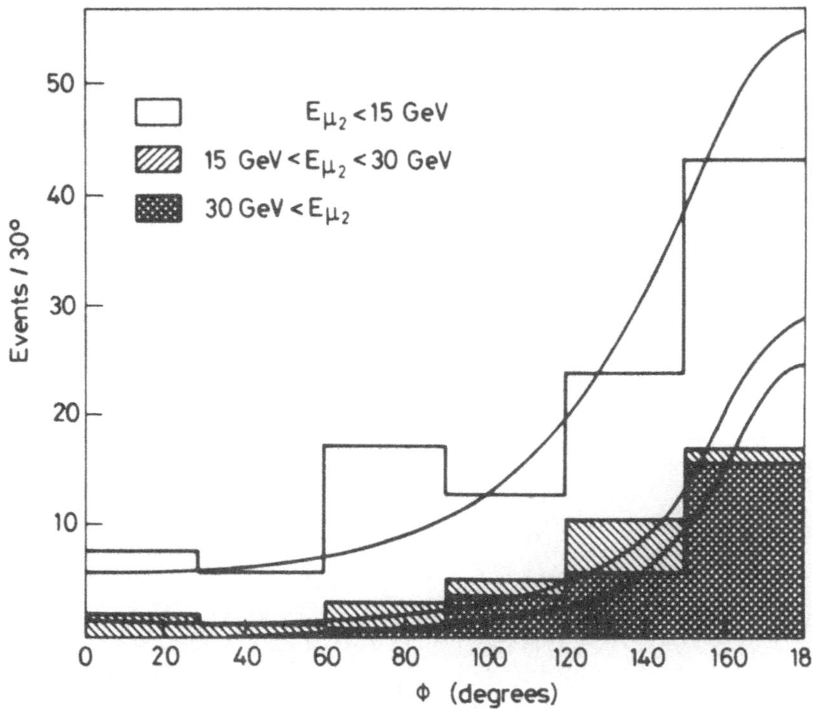
$E_{\mu_2} < 15$ GeV
15 GeV $< E_{\mu_2} < 30$ GeV
30 GeV $< E_{\mu_2}$
Events / 30°
50
40
30
20
10
0
20
40
60
80
100
120
140
160
18
Φ (degrees)

Fig. 35b

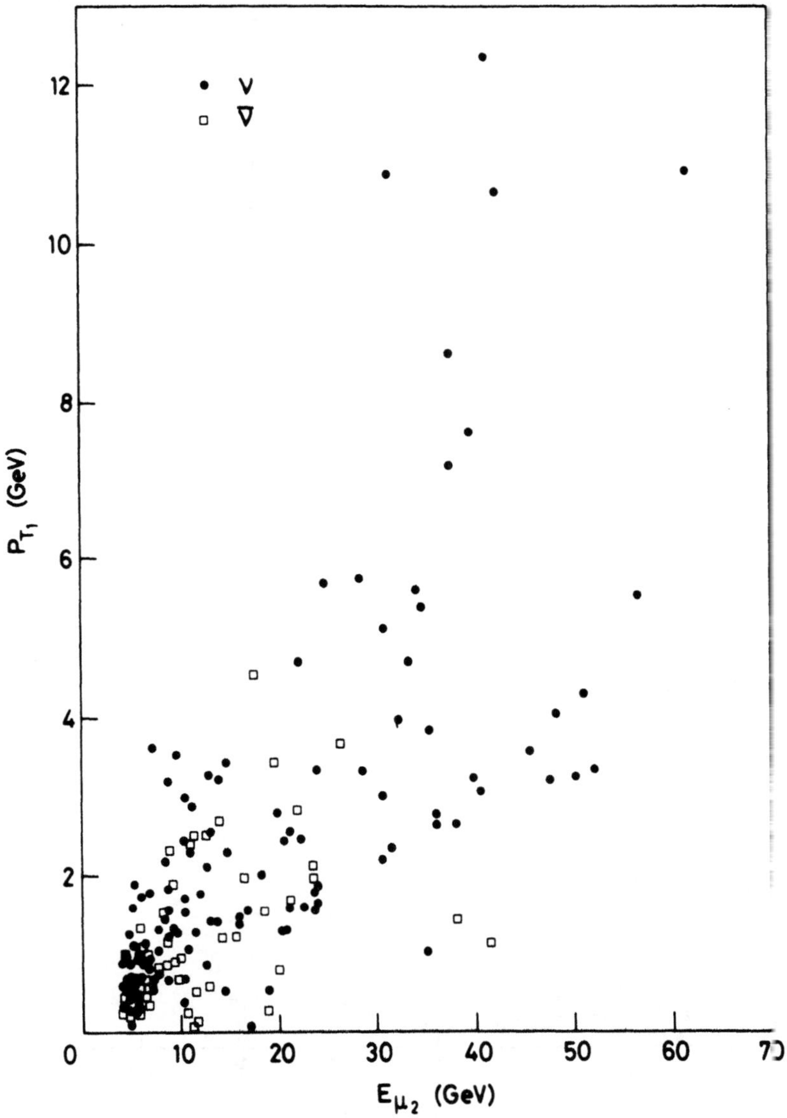

Fig. 36. Scatter plot in the transverse momentum of the second muon
a) with respect to the leading muon (p_T)

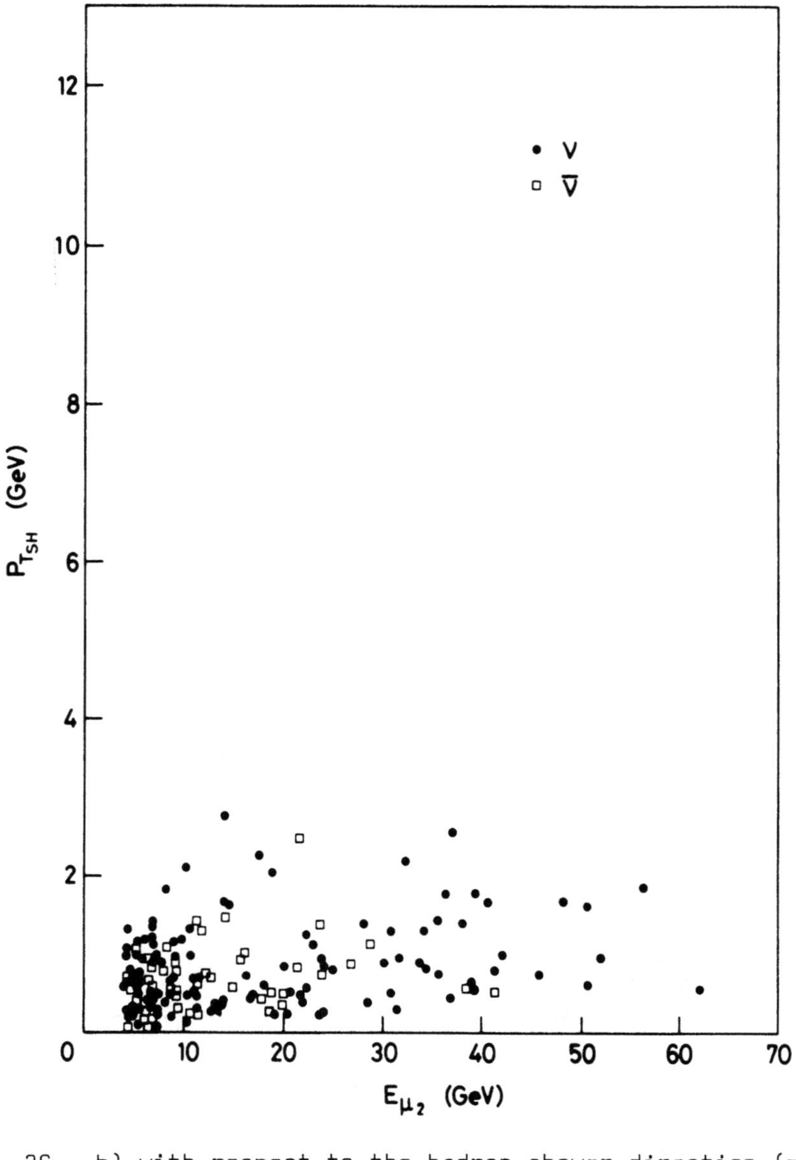

Fig. 36 b) with respect to the hadron shower direction ($p_{T_{sh}}$) versus the energy of the first muon.

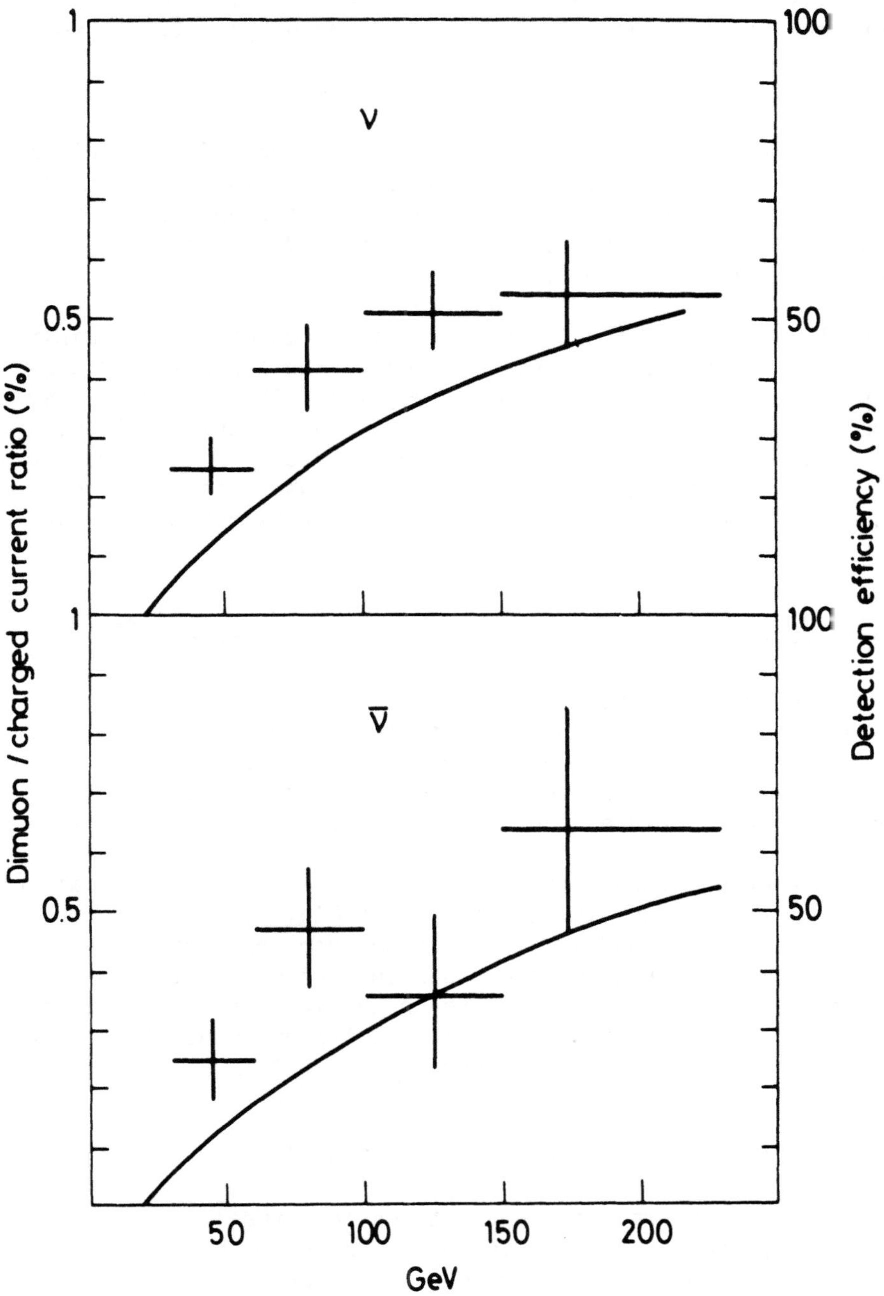

Fig. 37. The dimuon/single muon event ratios as a function of the total observed energy. The acceptance, as predicted in the charm model is also shown.

these $\bar{\nu}$ and ν production ratios corrected for the muon detection efficiencies, and the relative $\bar{\nu}/\nu$ charged cross-sections (0.43±0.02), the strange sea component in the nucleon can be directly found :

$$\zeta_s = 0.035 \pm 0.01.$$

From a comparison of the dimuon-CC ratio and the G.I.M. model production cross-sections, the charm to muon branching ratio is $\sim$0.15. Since a similar value is found in electron-positron colliding beam experiments, the argument can be turned around to furnish a quantitative check of the G.I.M. model.

Another check of the G.I.M. model is provided by the comparison of the average observed energy compared to the average neutrino energy, as deduced from the beam kinematics. This ratio is observed to be 0.92 ± 0.04, indicating that 8 ± 4% of the total energy is lost to the missing neutrino in the charm decay. This can be seen also by comparing Fig. 38 with Fig. 4, where the observed energy is plotted against detector radius for dimuon and single muon events, respectively.

6.3 Results from the FNAL 15' neon-filled bubble chamber[35)]

In a wide-band neutrino exposure, 81 events of the type $\nu + \text{Ne} \rightarrow \mu^- + e^+ + \ldots$ were identified. The main advantage of these data over those in the counter experiment are that a) even small-energy positrons can be detected, and b) the neutral strange particles can be identified.

To the extent that comparison is possible, the bubble chamber and counter results are in agreement. In Fig. 39 the e^+ energy spectrum, and the x- and y- distributions are reproduced. The observed strange-particle rate, assuming the unobserved charged particles are equal in number to the neutrals, corresponds to about one strange particle per event. All results are consistent with the expectations of the G.I.M. model.

6.4 Results from like-sign dimuons from CDHS[41)]

The raw rates are about seven times less than for opposite-sign dimuons, and the problem of background from charged current interaction with a pion or kaon decay is severe. According to our calculations $\sim$ (65 ± 20)% of the events are background of this sort. The distributions in the momentum of the lower energy muon and in y are shown for data and background calculations in Figs. 40 and 41. The calculated background distributions are quite consistent with the data. The remaining like-sign dimuon

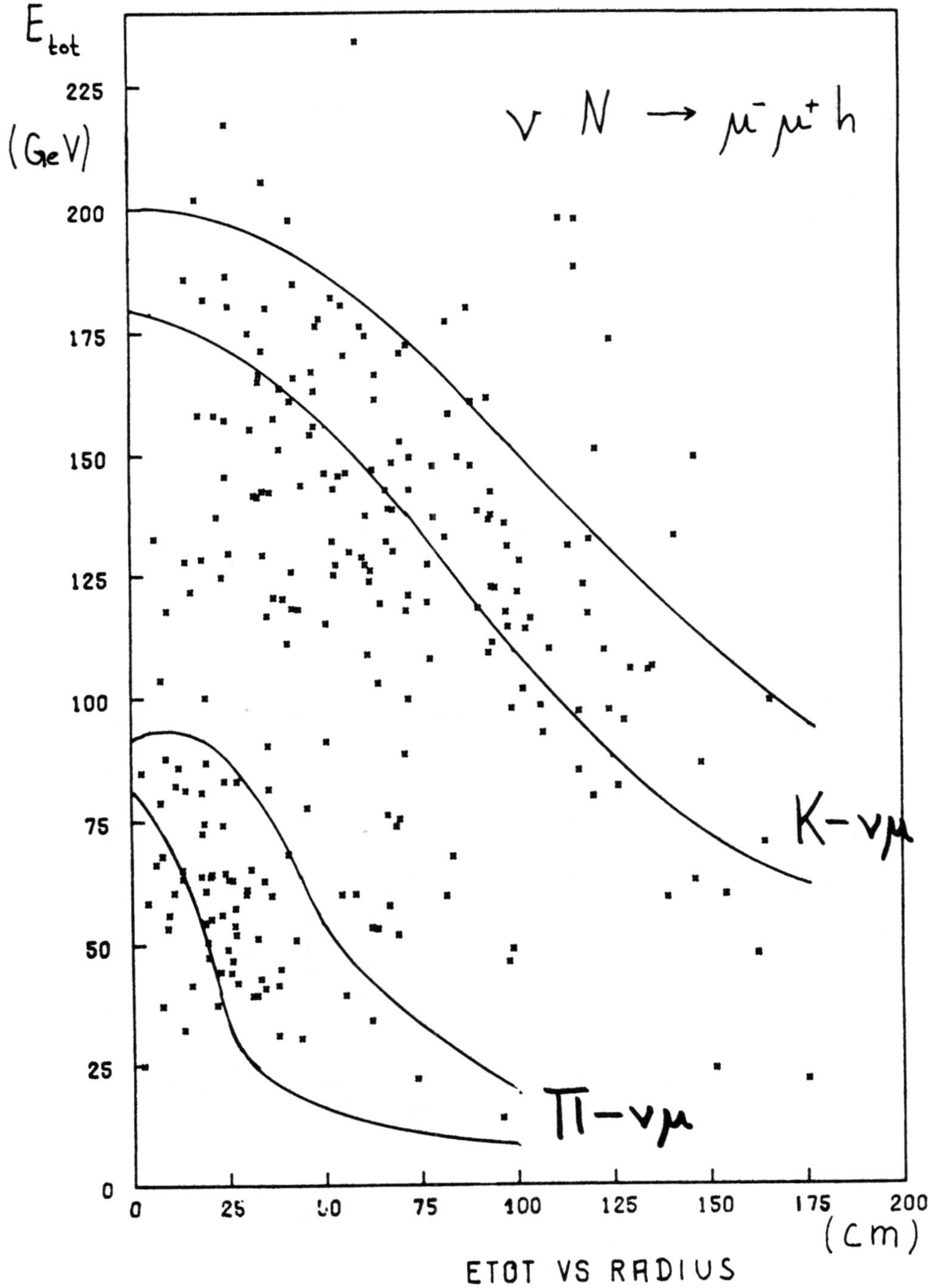

Fig. 38. Correlation of energy of opposite-sign dimuon events and detector radius. Comparison with Fig. 4 permits an estimate of the average amount of energy lost to outgoing neutrinos.

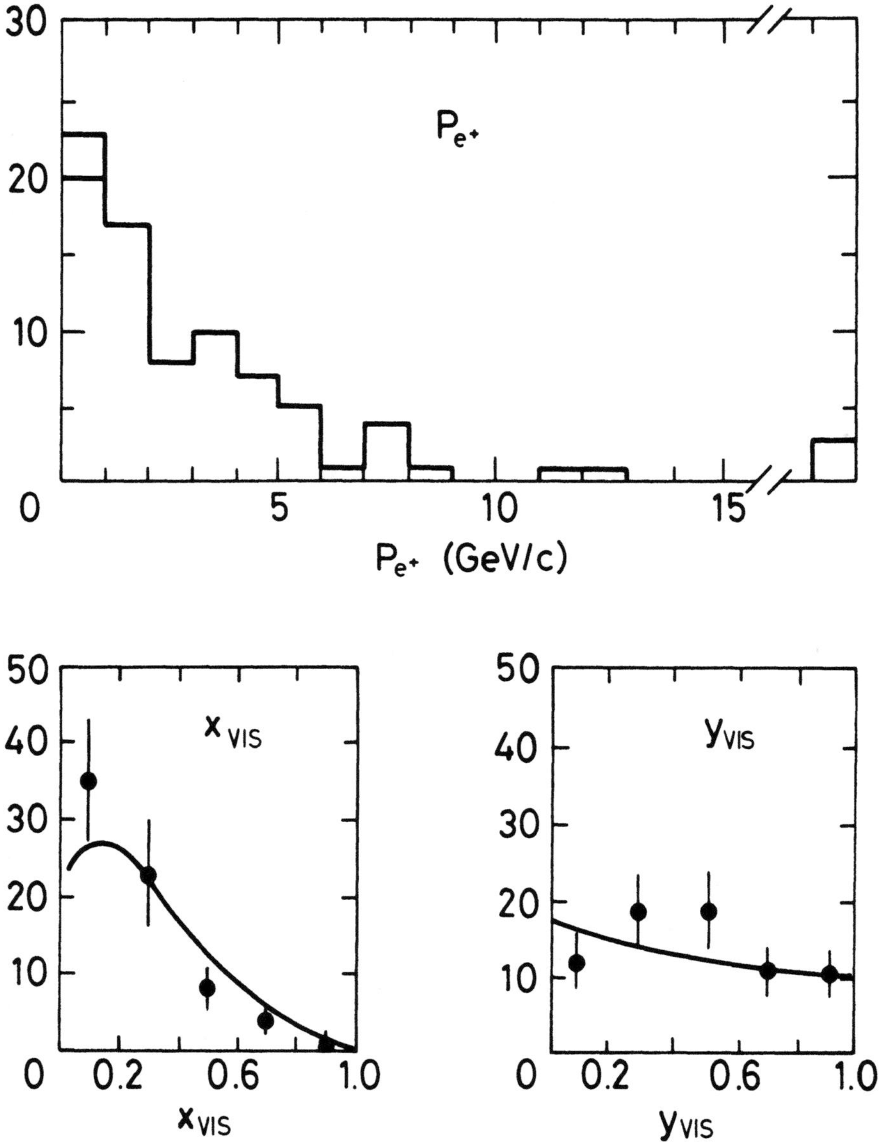

Fig. 39. Some properties of the $\mu^- e^+$ events observed in the FNAL 15' neon-filled chamber[35].
a) Positron spectrum; b) x-distribution;
c) y-distribution.

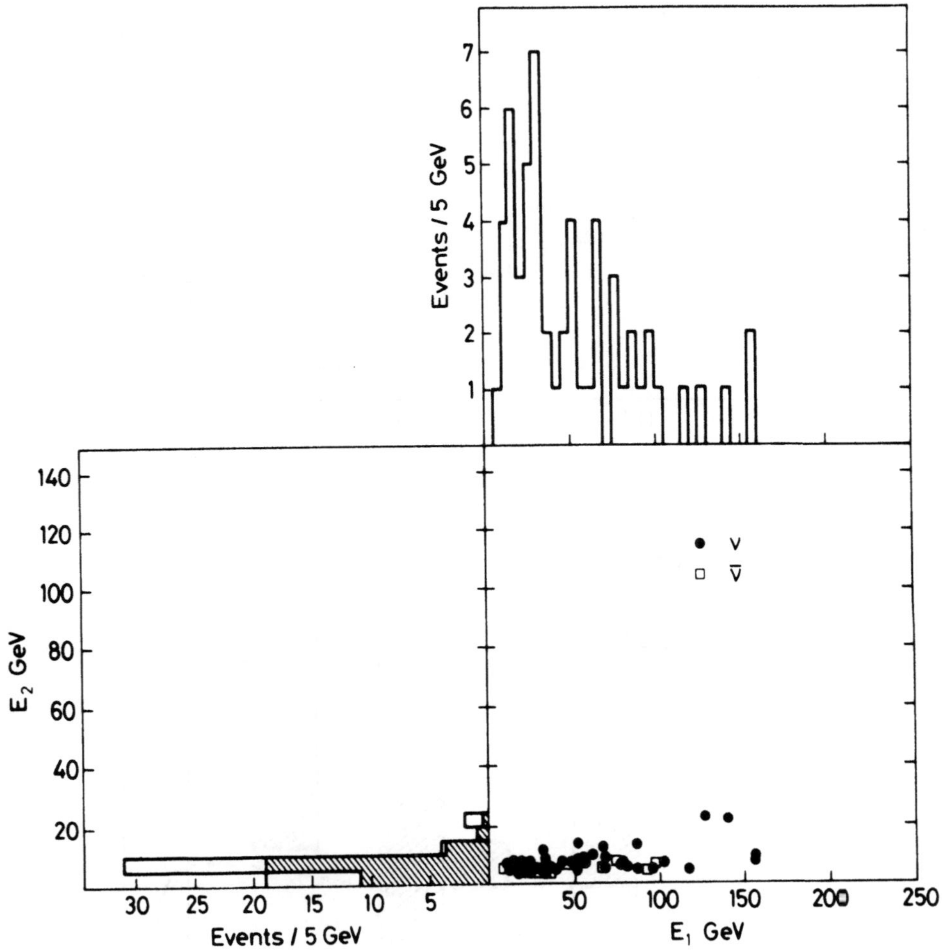

Fig. 40. Correlations in the more energetic (μ_1) and less energetic (μ_2) muons, for the CDHS like-sign dimuon events. The calculated $\pi \to \mu$ and $K \to \mu$ background is also shown.

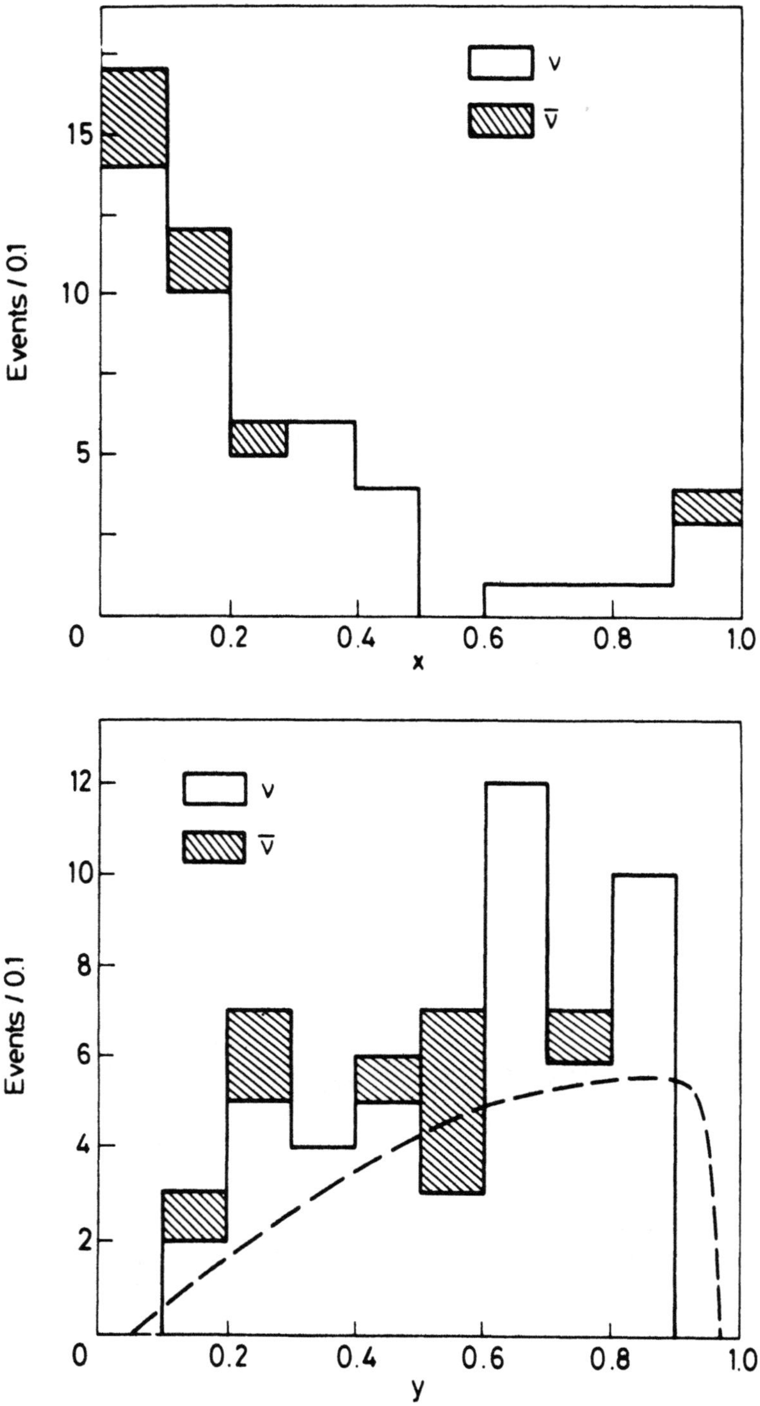

Fig.41. x- and y-distributions for the CDHS like-sign dimuon events. The solid curve in the y-distributions is the form of the expected $\pi \rightarrow \mu$ and $K \rightarrow \mu$ background.

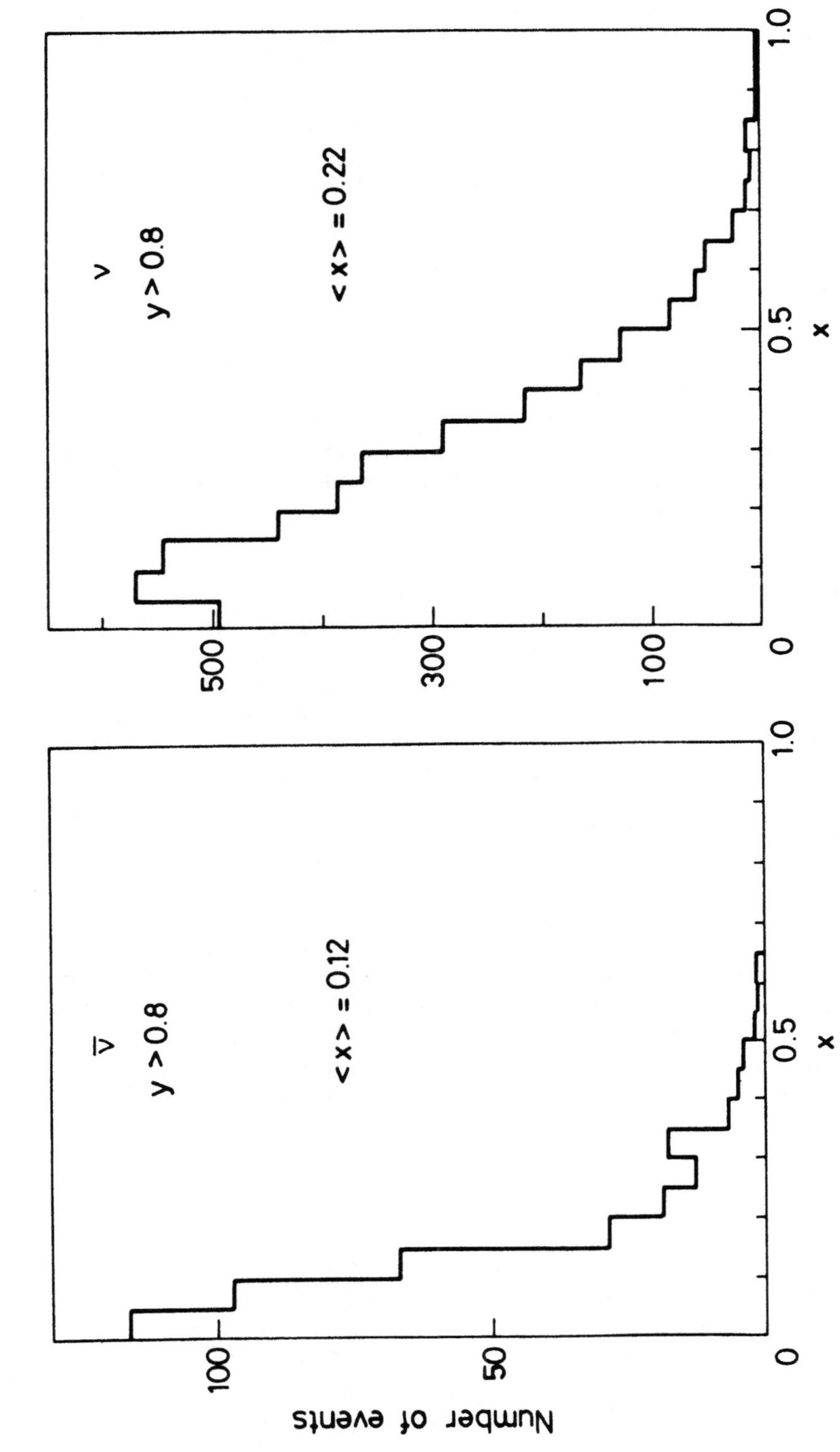

Fig. 42 : x-distributions at large y, 0.8 < y < 1.0, for neutrino and anti-neutrino CC events, with E_ν > 30 GeV.

rate is ≃ (0.05 ± 0.03) of the opposite-sign rate. We cannot be sure that it is not zero. The date are consistent with a charmed pair origin for these events, corresponding to a fraction of the order of (0.5 - 1.0) % of the charged current events. The data do not seem to be consistent with a cascading heavy lepton model of trimuon production, which predicts like-sign dimuons with both muons energetic, at rates which would have been very visible[42].

7. STRUCTURE FUNCTIONS FOR THE QUARK-ANTIQUARK "SEA" [43]

In Section 6.2 we drew attention to the fact that charm production by anti-neutrinos is expected to be purely from anti-quarks in the "sea". The corresponding x-distribution (Fig.34b) is therefore a direct measure of the x-distribution of sea quarks. It was seen to be very different from the normally observed x-distributions, which are due to valence quarks.

Another direct glimpse of the "sea" can be had by looking at normal charged current events produced by antineutrinos at large values of y. Since the quark contribution, in this case, falls as $(1-y)^2$, the sea dominates at large y. In the interval $0.8<y<1$ the valence quarks can be expected to contribute only ≃ 15 % of all events, and the sea quarks ≃ 85 %. In the neutrino case the situation is reversed, and the sea can be expected to contribute only a few per cent in this y-region. In Fig. 42 the two x-distributions are shown, for this y-interval and for all charged current events with $E_\nu > 30$ GeV. The difference is striking. The x-distribution for neutrinos at high y does not differ greatly from that at other y-values. However, the x-distribution for anti-neutrinos is remarkably shifted to low x, and indistinguishable from the x-distribution of the antineutrino dimuon events, Fig.34b. When the small contributions of valence quarks are subtracted, the distributions of Fig. 42 and Fig. 34b are well fitted with the distribution of the form $(1-x)^8$. It should be noted that, in the G.I.M. model, the "dimuon sea" is dominated by strange quarks and the "single-muon sea" by up quarks. Nevertheless the two distributions are very similar.

REFERENCES

1) G. Danby et al., Phys.Rev.Letters 9, 36 (1962).
2) F.J. Hasert et al., Phys.Letters 46B, 138 (1973).
3) A. Benvenuti et al., Phys.Rev.Letters 34, 419 (1975); B.C. Barish et al., Phys.Rev.Letters 36, 939 (1976).
4) H. Deden et al., Phys.Letters 58B, 361 (1975); J. von Krogh et al., Phys.Rev.Letters 36, 710 (1976).
5) M. Holder et al., A detector for high-energy neutrino inter-

actions, to be published in Nuclear Instruments and Methods.
6) M. Holder et al., Performance of a magnetized total absorption calorimeter between 15 GeV and 140 GeV, submitted to Nuclear Instruments and Methods.
7) See, for instance, the following reviews : C. Llewellyn Smith, Phys.Reports 3C, N°5, 261 (1972); L.M. Sehgal, ANL report HEP-PR-75-45 (1975).
8) S.L. Glashow, J. Iliopoulos and L. Maiani, Phys.Rev. D2, 1285 (1970).
9) Compilation prepared by D.H. Perkins, Proc. 1975 Internat. Symposium on Lepton and Photon Interactions at High Energies, Stanford, 1975 (Stanford Linear Accelerator Center, Stanford 1975), p. 571.
10) H. Deden et al., Nuclear Phys. B85, 269 (1975).
11) Gargamelle collaboration, Phys.Letters 46B, 274 (1973).
12) S.J. Barish et al., Proc. 1975 International Symposium on Lepton and Photon Interactions at High Energies, Stanford, 1975 (Stanford Linear Accelerator Center, Stanford, 1975) p. 511.
13) B.C. Barish et al., Phys.Rev.Letters 35, 1316 (1975).
14) B. Aubert et al., Phys.Rev.Letters 33, 984 (1974).
15) A. Benvenuti et al., Phys.Rev.Letters 36, 1478 (1976).
16) J.B. Berge et al., Scaling variable distributions for antineutrino-nucleon interactions, preprint FERMILAB/Pub-77/44-Exp (1977), submitted to Phys.Rev.Letters.
17) B.C. Barish et al., Phys.Rev.Letters 38, 314 (1977).
18) A. Benvenuti et al., Phys.Rev.Letters 37, 189 (1976).
19) M. Holder et al., Is there a high-y anomaly in antineutrino interactions ?, Phys.Rev.Letters 39, 433 (1977).
20) P.C. Bosetti et al., submitted to Phys.Letters.
21) B.C. Barish, private communication.
22) J. Blietschau et al., Nuclear Phys. B114, 169 (1976).
23) Aachen-Padua Collaboration, H. Faissner et al., Proc.Intern. Neutrino Conf., Aachen, 1976 (Vieweg, Braunschweig, 1977),p.223.
24) F. Reines, H.S. Gurr and H.W. Sobel, Proc.Intern. Neutrino Conf., Aachen, 1976 (Vieweg, Braunschweig, 1977), p. 217.
25) W. Lee et al., Phys.Rev.Letters 38, 202 (1977).
26) D. Cline et al., Phys.Rev.Letters 37, 252 and 648 (1976).
27) J. Blietschau et al., Nuclear Phys. B118, 218 (1977).
28) B.C. Barish et al., Proc.Internat.Neutrino Conf. Aachen, 1976 (Vieweg, Braunchweig, 1977), p. 289.
29) A. Benvenuti et al., Proc.Internat.Neutrino Conf., Aachen, 1976 (Vieweg, Braunschweig, 1977), p. 296.
30) L. Wolfenstein, Nuclear Phys. B91, 95 (1975).
31) A. Benvenuti et al., Phys.Rev.Letters 34, 419(1975); 35, 1199 (1975) and 35, 1203 (1975).
32) B.C. Barish et al., Proc.Internat.Symposium on La Physique du Neutrinoà Haute Energie, Paris, 1975 (CNRS, Paris, 1975), p.131.
33) H. Deden et al., Phys.Letters 58B, 361 (1975).
34) J. von Krogh et al., Phys.Rev.Letters 36, 710 (1976).

35) C. Baltay et al., Phys.Rev.Letters 39, 62 (1977).
36) B.C. Barish et al., Phys.Rev.Letters 38, 577 (1977).
37) A. Benvenuti et al., Phys.Rev.Letters 38, 1110(1977).
38) M. Holder et al., Observation of trimuon events produced in neutrino and antineutrino interactions, submitted to Phys.Letters.
39) M. Holder et al., Opposite sign dimuon events produced in narrow band neutrino and antineutrino beams, Phys.Letters 69B, 377 (1977).
40) L.M. Sehgal and P.M. Zerwas, Nuclear Phys. B108, 483 (1976).
41) M. Holder et al., Like-sign dimuon events produced in narrow-band neutrino and antineutrino beams, to be published in Phys.Letters.
42) V. Barger et al., Univ. of Wisconsin preprint COO-595 (1977).
43) M. Holder et al., in preparation.

QUARKS, HADRON SPECTROSCOPY AND BEYOND

J. Weyers

University of Louvain
Chemin du cyclotron, 2
B - 1348 LOUVAIN-LA-NEUVE

From the marvellous experimental discoveries [1] and theoretical breakthroughs [2] of the last few years has emerged a relatively coherent picture of the world of elementary particles. The "popular mythology" to which most physicists would adhere today can be summarized as follows.

1. The fundamental constituents of matter are :
   - a set of elementary fermions, hopefully finite, among which we know of 6 leptons [3] and at least 15 coloured quarks [4] with certainly more to come !
   - elementary bosons which comprise a minimum of 12 gauge bosons and presumably one or more extra (Higgs) scalars !

2. All basic interactions among the fundamental constituents are gauge interactions
   - strong interactions are described by quantum chromodynamics (QCD) which is based on the colour gauge group $SU(3)_c$ [5]. This gauge symmetry is exact hence the strong interactions are mediated by 8 massless coloured gluons. Fortunately colour confinement [6] explains why none of the fundamental constituents-gluons and quarks- which participate in the strong interactions has ever been seen !
   - weak and electromagnetic interactions are described by a gauge group which is presumably larger than $SU(2) \times U(1)$ [7]. We have thus at least four more gauge bosons (γ, Z°, $W^\pm$) and since the gauge symmetry is spontaneously broken - by the BEHHGK mechanism [8] - we are forced to introduce one - or

† These lectures are dedicated to Professor L. Bouckaert at the occasion of his seventieth birthday.

more - extra scalar particle [9].
- gravity, which will be ignored throughout these lectures, is also a gauge theory.

3. The mathematical framework to describe the fundamental constituents of matter with their basic interactions is of course relativistic quantum field theory with the key word "renormalizability" !

The ludicrous aspects of this mythology are obvious enough. It has however the merit of incorporating all the information we have gathered on "elementary processes" (after all there is some experimental evidence for each of our dogmas) and of providing us with a frame to solve the "remaining" problems.

The "deepest" of these problems is undoubtedly whether general relativity can be merged with quantum physics. The only hope along this line seems to be supersymmetry [10,11]. However no attractive phenomenological scheme or model exists as yet which incorporates the internal symmetries of the fundamental constituents in a supersymmetric theory [11,12].

Still at the philosophical level, the hopes are not negligeable that we may soon witness progress on some of the following problems :

- the proliferation of elementary fermions : leptons and quarks obviously know a lot more about each other than our mythology indicates [13]. Grand unification schemes [14] are the most naive and probably wrong approach to this problem;
- the origin and relations among the various scales present in the leptonic and hadronic sectors, e.g. the μ and τ or s, c, b, ... masses, the slopes of the Regge trajectories of new particles [15] etc...
- the "fundamental" versus "dynamical" description of spontaneous symmetry breaking. We believe, for example, that the π meson remembers its origin as a Goldstone boson [16] (fundamental π) responsible for chiral symmetry breaking while, simultaneously, behaving as a $\bar{q}q$ system (dynamical π). All models for the particle spectrum have had difficulties with the π because they neglect its double nature !

Finally at the technical and phenomenological levels many rewards and much progress is to be expected in tackling obvious problems of the mythology, a.o.,

- does QCD really confine colour ?
- will "natural symmetries" [17] finally lead to quantitative calculations of symmetry breaking effects (chiral symmetry breaking, flavour symmetry breaking, electromagnetic mass differences, etc...) ?

In the mean time it will be fun to get a more and more precise description of the hadron spectrum and its transitions (with new and old quarks), to invent the "correct" model for weak and electromagnetic interactions which incorporates the spectacular successes of the Weinberg - Salam model, to get clues on how to live with some of the old problems (the U(1) problem [18], Zweig's rule [19], CP violation [20], the $\Delta I = 1/2$ rule [21], etc...); all this while continuing to swallow more and more expected or unexpected experimental facts (no parity violation in atomic physics [22], no y anomaly, trimuons [23], etc...).

I am aware that this negligent and oversimplified description of the state of our field is at best appropriate, if at all, for the "vulgum pecus". The only reason for going through it, here, is the belief that it does not hurt to remind oneself, once in a while, of how "things look" !

The purpose of these lectures is to give a very elementary and prejudiced discussion of some of the experimental evidence and theoretical problems connected with our complicated picture of the fundamental constituents of matter. They are organized as follows: Lecture I is devoted to the constituent quark rules and lecture II to chiral symmetry and its breaking. In Lecture III the main idea and some of the spectroscopic problems related to bag models are briefly sketched and in Lecture IV we discuss one example of a grand unification theory - the SU(5) model - as an illustration of the "new spectroscopy". Some useful calculational rules for Young Tableaux are summarized in an Appendix.

I. CONSTITUENT QUARK RULES.

Hadron states at rest appear to be well described by a constituent quark model with orbital excitations [24]. Most successes of the model can be embodied in a set of "rules" :

(1) quarks appear in n flavours ($n \geq 5$) and hence the hadron spectrum will possess an approximate SU(n) symmetry. The dominant breaking mechanism of this flavour symmetry seems to be due to quark masses.

(2) quarks have spin 1/2. This enlarges the "spectrum symmetry" to an approximate SU(2n). In fact even this symmetry can be enlarged since the quark and antiquark content of hadronic systems can be described separately in terms of representations of a U(2n) algebra. If one imagines furthermore that the quarks and/or antiquarks have some relative orbital momentum one is led to a classification scheme of the hadrons based on $U(2n)_Q \otimes U(2n)_{\bar{Q}} \otimes O(3)$ which is referred to as the "rest symmetry". As is well known such a scheme works amazingly well [25].

(3) quarks come in <u>three</u> colours [26]. More precisely quarks transform as a triplet under $SU(3)_C$, the exact colour gauge symmetry which supposedly describes the basic strong interactions (quantum chromodynamics or QCD). The eight massless coloured gluons associated with this gauge symmetry are of course flavour singlets. They do not seem to play any significant role in the spectrum (i.e. labelling of states) at least for the presently known low lying hadrons.

(4) colour is confined [6]. This is the basic dogma of QCD which "explains" why single quarks or gluons are unobserved. It leads to the following allowed (i.e. colour singlet) configurations for hadrons :

- mesons can be $\bar{Q}Q$, GG, $\bar{Q}QG$, $\bar{Q}Q\bar{Q}Q$, ... etc...
- baryons can be QQQ, QQQG, $QQQ\bar{Q}Q$, ... etc...

where G stands for the gluon fields. The notation $\bar{Q}QG$ e.g. means that the $\bar{Q}Q$ pair is in a colour octet which combines with a gluon to form an overall colour singlet.

(5) quarks obey simple "transition rules" such as the Zweig-Iizuka-Okubo rule [19] (Z-I-O) and the Melosh rule [27]. The Z-I-O rule asserts basically that if hadrons are represented by pure quark graphs (i.e. no gluons), transitions which correspond to disconnected graphs are forbidden. The Melosh rule, on the other hand proclaims that hadronic transitions via photons or pions are dominantly single quark transitions.

For the familiar case of 4 flavours, quarks belong to the fundamental representation of SU(4) and have the following quantum numbers

	Q	T_3	S	C	B
u	2/3	1/2	0	0	1/3
d	-1/3	-1/2	0	0	1/3
s	-1/3	0	-1	0	1/3
c	2/3	0	0	1	1/3

The Gell-Mann Nishijima formula reads $Q = T_3 + \frac{B+S+C}{2}$.

The many successes of the constituent quark rules together with the results of lepton-hadron scattering [28] leave us with no choice but to take quarks as the basic dynamical degrees of freedom which describe hadrons. Unfortunately there is no explicit version of the quark model which succeeds in explaining the hadron spectrum and hadron transitions from quark dynamics.

A. Meson spectroscopy.

The simplest allowed configuration for mesons is $\bar{Q}Q$. We define as usual the total angular momentum, parity and charge conjugation of such states by

$$\vec{J} = \vec{L} + \vec{S} \qquad P = (-)^{L+1} \qquad C = (-)^{L+S} \tag{I.1}$$

where $\vec{S}$ is the total quark spin and $\vec{L}$ the orbital angular momentum. With, e.g., four quark flavours, we thus expect for each J^{PC} hexadecuplets of mesons

$$\underset{\sim}{4} \otimes \underset{\sim}{\bar{4}} = \underset{\sim}{1} \oplus \underset{\sim}{15} \quad \text{giving in SU(3)} \quad \underset{\sim}{1} \oplus \underset{\sim}{1} \oplus \underset{\sim}{8} \oplus \underset{\sim}{3} \oplus \underset{\sim}{\bar{3}} \tag{I.2}$$

The quark content of these representations is of course trivial to obtain; with C(charm) = +1 we have an SU(3) antitriplet $\bar{3}$: $c\bar{u}(D^\circ)$, $c\bar{d}(D^+)$, $c\bar{s}(F^+)$; with C(charm) = -1 we have the hermitian conjugate states : $\bar{c}u(\bar{D}^\circ)$, $\bar{c}d(D^-)$ and $\bar{c}s(F^-)$ which transform as an SU(3) triplet and finally C = 0 we have two SU(3) singlets and an SU(3) octet. Among these we have 4 states with T_3 = S=C=0 which are

$$\underset{\sim}{1}[SU(4)] \supset \underset{\sim}{1}[SU(3)] \; : \; \frac{1}{2}(\bar{u}u+\bar{d}d+\bar{s}s+\bar{c}c) \tag{I.3}$$

$$\underset{\sim}{15}[SU(4)] \supset \underset{\sim}{1}[SU(3)] \; : \; \frac{1}{\sqrt{12}}(\bar{u}u+\bar{d}d+\bar{s}s-3\bar{c}c) \tag{I.4}$$

$$\underset{\sim}{8}[SU(3)] \; : \; \frac{1}{\sqrt{6}}(\bar{u}u+\bar{d}d-2\bar{s}s)$$

$$8[SU(3)] \; : \; \frac{1}{\sqrt{2}}(\bar{u}u-\bar{d}d)$$

At the lowest level (L=0) the $\bar{Q}Q$ configuration for mesons with n=4 thus predicts the existence of 16 pseudoscalars $J^{PC} = 0^{-+}$ and 16 vectors $J^{PC} = 1^{--}$. Less than three years after the discovery of the J/ψ [29] these two multiplets are now essentially complete [30]. Indeed with $J^{PC} = 0^{-+}$ we have $\underset{\sim}{1} \oplus \underset{\sim}{8}$ π, K, η, η'

$\underset{\sim}{\bar{3}}$	$D^\circ(1865)$	$D^+(1876)$	$F^+(2060)$
$\underset{\sim}{3}$	$\bar{D}^\circ$	D^-	F^-
$\underset{\sim}{1}$	$\eta_c(2830)$		

while for $J^{PC} = 1^{--}$

$\underset{\sim}{1} \oplus \underset{\sim}{8}$ ρ , K*, ω, φ

$\underset{\sim}{\bar{3}}$	$D^{*\circ}(2000)$	D^{*+}	F^{*+}
$\underset{\sim}{3}$	$\bar{D}^{*\circ}$	D^{*-}	F^{*-}
$\underset{\sim}{1}$	ψ(3100)		

From the known quark structure of the P°, ω and ϕ, it follows that the ψ is a pure cc state (i.e. a mixture of $\underset{\sim}{1} \oplus \underset{\sim}{15}$ in SU(4)). For the pseudoscalars the situation is as obscure as ever : the mixing scheme is presumably different and the η_c probably contains some of the "old quarks" [31]. This pattern of mixing remains an unsolved problem.

Since the charmed quark c is so much heavier that the u, d and s quarks it is not unreasonable to expect the $\bar{c}c$ sector of the spectrum to be relatively pure (charmonium spectrum) [32,30].

At the next energy level of the $\bar{Q}Q$ configuration we expect states with L=1 and, depending on the forces between the quarks, possibly a second series of states with L=0, i.e.

L = 1	S = 0	$J^{PC} = 1^{+-}$	[B ...]	?
	S = 1	$J^{PC} = 2^{++}$	[A_2 ...]	χ(3552)
		$J^{PC} = 1^{++}$	[A_1 ...]	χ(3508)
		$J^{PC} = 0^{++}$	[δ ...]	χ(3414)
L = 0	S = 0	$J^{PC} = 0^{-+}$	[π',?, ...]	χ(3454)?
	S = 1	$J^{PC} = 1^{--}$	[ρ', ...]	ψ'(3700)

where for simplicity we have only included the I=1 state and the pure (or almost pure) $\bar{c}c$ state of each J^{PC} family. Of course none of the explicity charmed particles ($\underset{\sim}{3}$ or $\underset{\sim}{\bar{3}}$'s of SU(3)) with these quantum numbers has yet been seen.

From this labelling of states, it should be clear that the mere existence of the ψ' (3700) requires a π' and a ρ' [ρ'(1600)?].

At a still higher level of the $\bar{Q}Q$ configuration one expects states with L=2 possibly accompagnied by states with L=1 and L=0. In particular one expects at most new $J^{PC} = 1^{--}$ multiplets at this level.

The experimental situation concerning the old $\bar{q}q$ states has not changed significantly over the last few years : many states are still missing at the L=1 level (A_1, H, ...) we refer the reader to recent reviews [33] for a detailed discussion of the present status of meson spectroscopy.

It is worth pointing out that

- no low lying state exists with quantum numbers incompatible with the naive quark model expectations; there is however some evidence [34] that exotic (i.e. not $\bar{q}q$) mesons may exist higher up. All "exotic" candidates so far are of the "flavour exotic" type (e.g. I=2).

- since the lowest level (0^{-+} and 1^{--}) of mesons is essentially complete it confirms that quarks have spin 1/2;
- in e^+e^- annihilation above 3 GeV, 1^{--} resonances have been identified at 3.1(ψ), 3.7(ψ'), 3.77(ψ") and 4.4 GeV. In addition there are bumps at 3.95, 4.0 and 4.2 GeV [30]. This plethora of 1^{--} structures makes one feel a bit uncomfortable although the charmonium model has gone a long way towards explaining many of these features [35];
- the "symmetry" of the meson spectrum is in fact much larger than SU(4). With four quark flavours with spin 1/2 it is trivial to classify mesons in representations of SU(8) ⊗ O(3). With $Q \in \underset{\sim}{8}$ of SU(8) and $\bar{Q} \in \bar{\underset{\sim}{8}}$, $\bar{Q}Q$ mesons transform as ($\underset{\sim}{8} \otimes \bar{\underset{\sim}{8}}$ = $\underset{\sim}{1} \oplus \underset{\sim}{63}$; L) representations of SU(8) ⊗ O(3). The SU(4) ⊗ SU(2) content of a given SU(8) representation is easily obtained from $Q \in (\underset{\sim}{4},2)$ and $\bar{Q} \in (\bar{\underset{\sim}{4}},2)$. Hence

$$\underset{\sim}{63} \oplus \underset{\sim}{1} \supset \{ (\underset{\sim}{15},3) \oplus (\underset{\sim}{1},3) \oplus (\underset{\sim}{15},1) \oplus (\underset{\sim}{1},1)\} \qquad (I.5)$$

As was done for three quark flavours one may even wish to consider the so called rest symmetry group which now becomes $SU(8)_Q \otimes SU(8)_{\bar{Q}} \otimes O(3)$. With $Q \in (\underset{\sim}{8},1)$ and $\bar{Q} \in (1,\bar{\underset{\sim}{8}})$ of $SU(8)_Q \otimes SU(8)_{\bar{W}}$, a $\bar{Q}Q$ meson then belongs to the $(\underset{\sim}{8},\bar{\underset{\sim}{8}},L)$ representation of the rest symmetry. There is of course no difficulty in extending these considerations to the case of $n \geq 5$ quark flavours.

B. Baryon spectroscopy.

The simplest allowed quark model configuration for baryons is of course B = QQQ. For three quark flavours, the experimental evidence [36] for SU(6) ⊗ O(3) multiplets in the baryon spectrum is overwhelming. Indeed the lowest lying baryons fall neatly into a $\underset{\sim}{56}$ $L = 0^+$ of SU(6) ⊗ O(3) which contains an $\underset{\sim}{8}$ with spin 1/2 and a $\underset{\sim}{10}$ with spin 3/2. The next baryons with negative parity fall just as neatly into a

$$\underset{\sim}{70}\ L=1^- \supset \underset{\sim}{8} \text{ with } S=1/2,3/2,5/3 \text{ and } (\underset{\sim}{1},\underset{\sim}{8},\underset{\sim}{10}) \text{ with } S=1/2,3/2. \qquad (I.6)$$

Among higher positive parity baryon resonances one finds the essentially complete multiplets $\underset{\sim}{56}$ L=0 and $\underset{\sim}{56}$ L=2. There are also some indications [37] in favour of a $\underset{\sim}{70}$ L=2 and a $\underset{\sim}{70}$ L=0. Anyway the successes of the naive quark model are here very impressive indeed.

It should also be noted that the observed sequence of SU(6) ⊗ O(3) representations appears to follow the predictions of the so-called symmetric quark model [38] on $[SU(6) \otimes O(3)]_{SYM}$ namely

N = 1	$\underset{\sim}{56}$	$L = 0^+$
N = 2	$\underset{\sim}{70}$	$L = 1^-$
	$\underset{\sim}{56}$	$L = 0^+, 2^+$
N = 3	$\underset{\sim}{70}$	$L = 0^+, 2^+$
	$\underset{\sim}{20}$	$L = 1^+$

Except for this last representation, the agreement is indeed excellent ! The successes of this classification scheme give of course overwhelming evidence that the u,d and s quarks have indeed spin 1/2 and that their charges (possibly their average charges) and other internal quantum numbers correspond indeed to the standard values.

To extend the classification to four flavours does not present any difficulty. Let us consider for example the SU(4) partners of the nucleon. They belong to a $\underset{\sim}{20}$ representation corresponding to the Young tableau  Hence, with the recipes given in Appendix A, one finds

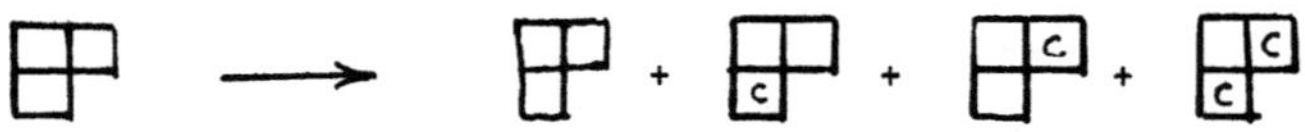

(I.7)

$\underset{\sim}{20}$ of SU(4) $\quad$ $\underset{\sim}{8} \oplus \underset{\sim}{6} \oplus \underset{\sim}{\bar{3}} \oplus \underset{\sim}{3}$ in SU(3)

The quark wave functions corresponding to these various states are easily constructed e.g. [39].

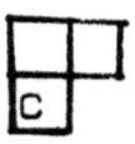

is a $\underset{\sim}{6}$ of SU(3) to which a c quark has been added. Hence, it contains - a triplet with S=0 :

$$uuc \equiv C_1^{++} \quad ; \quad \left(\frac{ud+du}{\sqrt{2}}\right)c \equiv C_1^{+} \quad ; \quad ddc \equiv C_1^{\circ} \qquad (I.8)$$

- a doublet with S =-1:

$$\left(\frac{us+su}{\sqrt{2}}\right)c \equiv S^{+} \quad ; \quad \left(\frac{ds+sd}{\sqrt{2}}\right)c \equiv S^{\circ} \qquad (I.9)$$

- a singlet with S=2:

$$ssc \equiv T^{\circ} \qquad (I.10)$$

Similarly  is a $\underset{\sim}{\bar{3}}$ of SU(3) which contains the states

$$\left(\frac{ud-du}{\sqrt{2}}\right)c \equiv C_0^{+} \quad ; \quad \left(\frac{us-su}{\sqrt{2}}\right)c \equiv A^{+} \quad ; \quad \left(\frac{ds-sd}{\sqrt{2}}\right)c \equiv A^{\circ} \qquad (I.11)$$

and 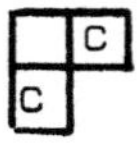 contains $ucc \equiv X_u^{++}$; $dcc \equiv X_d^{+}$; $scc \equiv X_s^{+}$ (I.12)

Repeating this decomposition for the partners of the decuplet, one obtains

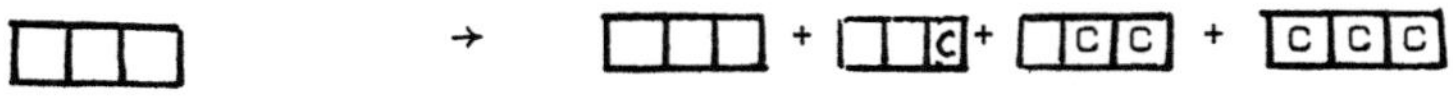 (I.13)

$\underset{\sim}{20}'$ of SU(4) $\underset{\sim}{10} \oplus \underset{\sim}{6} \oplus \underset{\sim}{3} \oplus \underset{\sim}{1}$ in SU(3)

and e.g., 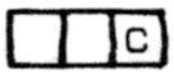 contains the states C_1^{*++}, C_1^{*+}, C_1^{*0}, S^{*+}, S^{*0} and T^{*0} whose quark wave function are easily constructed.

Combining SU(4) with $SU(2)_S$ (spin) one notes that the baryon "ground" state transforms as a $\underset{\sim}{120}$ representation of SU(8) whose Young tableau is  and which decomposes as follows into representations of SU(4) $\otimes$ $SU(2)_S$

$$\underset{\sim}{120} \supset (\underset{\sim}{20},2) \oplus (\underset{\sim}{20}',4) \qquad \text{(I.14)}$$

Since these states do not contain constituent antiquarks, this classification is easily extended to the full rest symmetry i.e.

$$(\underset{\sim}{120},1\ ;\ L=0) \text{ of } SU(8)_Q \otimes SU(8)_{\bar{Q}} \otimes O(3).$$

It will obviously take quite a while before this multiplet is experimentally completed. The situation is however far from hopeless. The lowest charmed baryon C_0^+ has presumably been found already at Brookhaven and Fermilab. [40]. In these experiments an optimist might also accept evidence for the states C_1^+ and C_1^{*+} with masses

$$m_{C_0^+} \simeq 2.25 \text{ GeV} \quad m_{C_1^+} \simeq 2.42 \text{ GeV} \quad m_{C_1^{*+}} \simeq 2.5 \text{ GeV} \qquad \text{(I.15)}$$

Are there baryon resonances which cannot be described as QQQ states ? As for the mesons there is not as yet a clear cut answer but only indications that $QQQQ\bar{Q}$ configurations might be present in the spectrum [41].

The oversimplified survey of the spectrum we have just gone through indicates that constituent quarks give us the correct counting rules for hadrons. This is certainly true for n=3 and there is no reason to expect any failure of the rules for n=4. Most of the observed hadrons do appear to behave as if they were built like $\bar{Q}Q$ or QQQ systems. It is not certain yet whether hadron resonances corresponding to $\bar{Q}Q\bar{Q}Q$ or $QQQ\bar{Q}Q$ configurations do indeed exist. In any case there is no evidence whatsoever for any other configuration. The only problem we choose to ignore while labelling states is the ground state configuration of the baryons which corresponds to a symmetric representation of SU(8) - or SU(2n) in the case of n flavours. This is of course the well known quark statistics problem which is solved by the "colour" symmetry.

C. Colour.

The first solution to the quark statistics problem was proposed by Greenberg [25]. He suggested that quarks obey "parastatistics of rank three". It can be shown that this is equivalent to assume that quarks appear in three colours (Red, Green, Blue) which transform under a coloured SU(3) and that physical states are colour singlets. From a strict spectroscopy point of view, colour does indeed solve the problem of a "symmetric" baryon ground state but the prize to pay seems heavy : one has to triple the number of basic hadronic constituents. Apart from the statistics problem, what do we gain through this proliferation ?
A couple of factors of 3 [42] :

- in the decay $\pi^\circ \to 2\gamma$
- in the value of $R = \dfrac{\sigma(e^+e^- \to h)}{\sigma(e^+e^- \to \mu^+\mu^-)}$

but these factors reflect of course the statistics of "current quarks" ! Hahn and Nambu [25] devised an ingenious scheme where colour is broken by weak and electromagnetic interactions. With $Q \in (\underset{\sim}{3}, \bar{\underset{\sim}{3}}_C)$ and $Q = T_3 + Y/2 + T_{3C} + Y_C/2$ where the c index refers to the SU(3) of colour, the quarks have integral charges as shown in the figure below :

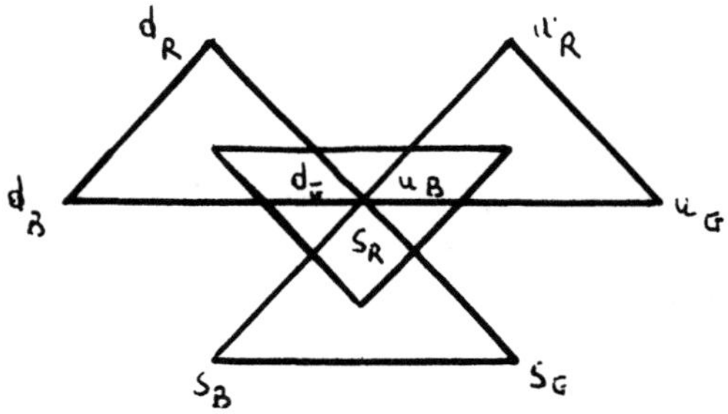

With "average" charges of 2/3, -1/3, -1/3, the Hahn-Nambu quarks are as satisfactory, if not more, as the Gell-Mann-Zweig quarks from the point of view of the hadron spectrum.

The results of deep inelastic lepton scattering[28], however, cristallized the situation and left us with very little choice indeed :
- approximate scaling which is observed experimentally can only be

reconciled with field theory in asymptotically free theories. Non abelian gauge theories are asymptotically free [5] !
- the proton's "parts" do seem to have fractional charges !
- almost half the proton's momentum is carried by neutral stuff [28,43].

The picture which emerges is that the Gell-Mann Zweig quarks are the local dynamical variables of a microscopic hadron theory and that the proton and neutrons are loosely but, as far as we know, permanently bound states of light quarks !

A gauge theory based on exact $SU(3)_c$-quantum chromodynamics- is the only field theoretic model we have to explain all these facts and the factors of three mentioned earlier. The basic entities of such a theory are thus

- quarks with an arbitrary number of "flavours" but with three colours;
- eight coloured massless gluons (the neutral "stuff").

If colour is confined in such a theory i.e. if the asymptotic states of the theory are colour singlets, neither the quarks nor the gluons would be observable ! The confinement problem is clearly the central issue in QCD and is discussed at length elsewhere in these Proceedings [6].

There is clearly no theory of the hadron spectrum based on QCD. The problem is indeed a formidable one : starting from a field theory of quarks and gluons we want to squeeze out confined hadrons and calculate their levels form first principles !

A much more naive approach which may however give us some clues as to whether QCD is indeed a viable theory of strong interactions is to avoid the confinement issue alltogether by assigning coloured fields inside hadrons which are thus treated as extended objects : this is the approach taken in "bag" models [40,45].
We will give an elementary discussion of the model in the third lecture. Let us point out however that the spectrum predicted by such models seems to be much too rich : in addition to the standard $\bar{q}q$ or qqq hadrons, there are in all bag models "unwanted" states such as $g\bar{g}$ (glueballs) [46], $\bar{q}qg$, etc...
Even if the bag is a bad phenomenological approximation to QCD, it is hard to imagine any mechanism by which the "correct" theory could avoid them. If QCD is to live, a lot of new hadrons better be found soon !

D. Masses and Couplings.

The standard assumption in breaking the flavour symmetry group is to assign all breaking effects to the quark masses. This gives for n=3

$$4m_K^2 = 3m_\eta^2 + m_\pi^2 \quad \text{(the G.M.O. formula for unmixed octets)} \qquad (I.16)$$

$$m_\rho^2 = m_\omega^2 \quad m_\phi^2 = 2m_{K^*}^2 - m_\rho^2 \quad \text{(magic mixing)} \qquad (I.17)$$

$$m_\rho^2 - m_\pi^2 = m_{K^*}^2 - m_K^2 \quad \text{(hybrid formula [47])} \qquad (I.18)$$

Similarly, for n=4, we might expect

$$m_\psi^2 = 2m_{D^*}^2 - m_\rho^2 \quad \text{or} \quad m_{D^*}^2 \cong (2.2)^2 \text{ GeV}^2 \qquad (I.19)$$

$$m_{D^*}^2 - m_D^2 = m_\rho^2 - m_\pi^2 \quad \text{or} \quad m_D^2 \cong (2.0)^2 \text{ GeV}^2 \text{ !} \qquad (I.20)$$

The self consistency is not great but should we expect a better than 10-20% agreement when the $\pi(140)$ and F(2060) mesons are put in the same SU(4) multiplet and their mass breakings are computed in "first order" ? Mass relations lead us naturally to wonder about Regge trajectories, exchange degeneracy, duality and all that. The splitting of the vector and tensor $\bar{c}c$ states is known

$$m_{\chi\,(3.55)}^2 - m_{\psi\,(3.1)}^2 = 3 \text{ GeV}^2 \qquad (I.21)$$

while for the u,d,s quark states we have e.g. $m_{f_0}^2 - m_\rho^2 \simeq 1\text{GeV}^2$. Clearly there is either no trace left of exchange degeneracy or the slope of the ψ-2^{++} trajectory is a factor of three smaller than for the old quark states. Linear and parallel exchange degenerate trajectories which were an essential ingredient of dual models do not seem to survive when going from 3 to 4 quark flavours (at least in the "low energy" region) [15].
The flavour group is presumably a reasonable vertex symmetry. For n=3, at least, we know that SU(3) works astonishingly well for resonance and Regge couplings [48]. There is no reason to expect a dramatically different situation with more flavours but at present essentially none of the flavour symmetry predictions beyond SU(3) can really be tested.

On the other hand, the total quark spin which plays such a prominent role in particle classification is of course not conserved in decay processes. Quark spin conservation forbids, e.g. decays such as $\Delta \to N\pi$ or $\rho \to \pi\pi$. Hence if a larger "approximate vertex symmetry" is to be found, the angular degrees of freedom must be described by a different group than $SU(2)_S$. Lipkin and Meskhov [49] have shown that the reflection and rotation invariance properties of collinear processes lead to invariance under W-spin or $SU(2)_W$ rather than S spin or $SU(2)_S$. As well known, for quarks $\vec{W} \equiv \vec{S} = 1/2\vec{\sigma}$ while for antiquarks $W_z = S_z$ but $W_{x,y} = - S_{x,y}$ (when moving along the z-axis). This leads trivially to the so called W-S flip relations for $q\bar{q}$ systems i.e.

$$|S=1\;;\;S_z=+1> \to |W=1\;;\;W_z=+1>$$
$$|S=1\;;\;S_z=0> \to -|W=0\;;\;W_z=0>$$
$$|S=1\;:\;S_z=-1> \to -|W=1\;;\;W_z=-1>$$
$$|S=0\;;\;S_z=0> \to -|W=1\;;\;W_z=0> \qquad \text{(I.22)}$$

Hence a π meson which is an S=0 state becomes a W=1 object while the ρ meson with helicity 0 becomes a W spin singlet. It is obvious that $\Delta \to N\pi$ and $\rho \to \pi\pi$ become allowed decay modes under $SU(2)_W$. It seems reasonable to combine $SU(2)_W$ with the flavour group to obtain the so called $SU(2n)_W$, constituents. It is a subgroup of the full rest symmetry group which might be a vertex symmetry (or a collinear symmetry). Experiment definitely tells us that this is not the case. The calssical example is $B \to \omega\pi$. The B(1235) meson is an L=1 state with S=0. Hence in helicity ± 1 it transforms as $B_{\pm1} \sim |W=1\;;\;W_Z=0\;L_Z=\pm1>$ while in helicity 0 we have $B_0 \sim |W=1\;\;W_Z=0\;\;L_Z=0>$. On the other hand $|\omega_{\pm1}\;\pi> \sim |W=1\;W_Z=\pm1> \otimes |W=1;\;W_Z=0>$ and $|\omega_0\pi> \sim |W=0> \otimes |W=1;\;W_Z=0>$. The $SU(6)_W$ predictions are thus $B_{\pm1} \not\to \omega_{\pm1}+\pi$ and $B_0 \to \omega_0+\pi$. This is precisely the opposite of what is seen experimentally : the B is observed to decay almost entirely in helicity ± 1 [50]. There are of course many more "bad" predictions which follow from $SU(6)_W$, in particular in the baryon spectrum [36]. All these disasters are completely cured by the "Melosh transformation" [27]. From this point of view pionic decays of hadron states are described by matrix elements of the divergence of the axial current (PCAC) which, the Melosh transformation tells us, transforms as follows under $SU(6)_W$ [51]; constituents.

$$|\pi_{Melosh}> \sim A'|\underset{\sim}{35};W=1,\;W_Z=0;L_Z=0> + B'|\underset{\sim}{35};W=1,W_Z=\pm1;L_Z=\mp1> \qquad \text{(I.23)}$$

Similarly for γ transitions, the Melosh transformation gives for the dipole operator $D = \int d^4x\delta(x^+)\vec{x}_\perp\, j_{e.m}(x)$ the following decomposition [52] under the constituent $SU(6)_W$:

$$D_\perp \sim A\{\underset{\sim}{35};W=0,W_Z=0;\Delta L_Z=\pm1\} + B\{\underset{\sim}{35};W=1,W_Z=\pm1;\Delta L_Z=0\}$$
$$+ C\{\underset{\sim}{35};W=1,W_Z=0;\Delta L_Z=\pm1\} + D\{\underset{\sim}{35};W=1,W_Z=\mp1;\Delta L_Z=\pm2\} \qquad \text{(I.24)}$$

From this parametrization [53] of decay amplitudes (π or γ) one derives of course a whole set of predictions relating amplitudes at different resonances, including phase relations [54]. All the predictions agree rather spectacularly with experiment [55,56]. In fact $SU(6)_W$ à la Melosh is as good a vertex symmetry as SU(3). With these Melosh rules the phenomenological picture of the lowest lying SU(6) multiplets is quite pretty and quantitatively as well as qualitatively rather well understood. The "unsolved" problems are essentially the theoretical determination of the mixing between states in the same multiplet and the explicit calculation of the various coefficients A', B', A, B, C, D, which are left free by the Melosh transformation.

Once again, there is no problem in generalizing these considerations to more than three quark flavours bur significant tests of $SU(2n)_W$ à la Melosh are not yet possible.

The most spectacular vertex rule remains however the ZOI rule [19]. The rule states that hadronic transitions described by disconnected quark diagrams are forbidden, e.g.

is allowed but is forbidden.

QCD does of course allow the forbidden diagrams in higher orders of the small (?) quark gluon coupling constant, e.g.

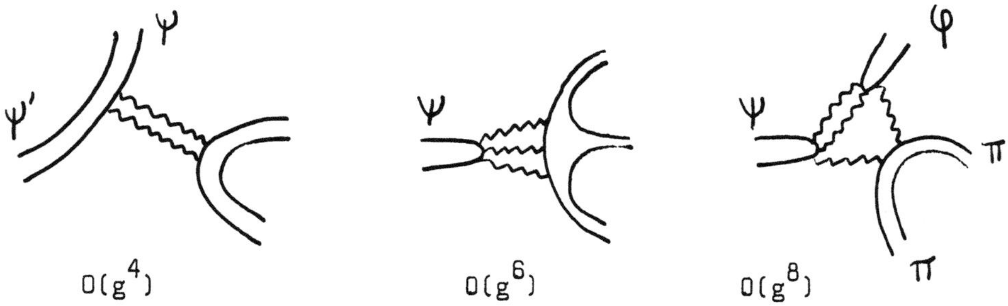

$O(g^4)$ $O(g^6)$ $O(g^8)$

If g is small (asymptotic freedom), the more gluons you need the stronger the suppression will be. This is certainly correct qualitatively as shown by the detailed comparison [57] of $\psi \to \omega\pi\pi$ where three gluons must be exchanged with $\psi \to \phi\pi\pi$ where four gluons are needed. Other "variables" which determine the suppression factor are the quark masses, the energy carried by the gluons [58] etc... Zweig's rule does work but precise quantitative tests of the rule. are still badly needed not to mention an explanation of why it works so well ! Yet another mystery.

What have we learned from all these elementary considerations ?

- The quark rules which were enumerated at the beginning of this lecture have their raison d'être, they work but despite their simplicity we do not have a viable dynamical explanation for many of them;
- QCD emerges as a possible candidate for a hadron theory (in fact the only candidate !) but some nasty spectroscopic questions will have to be answered, e.g. where are the gg or $\bar{q}qg$ states which according to the theory should exist ?

It remains to be proven of course that QCD can indeed confine colour.

Hadron spectroscopy in itself will thus remain a very challenging field but with the proliferation of quarks and leptons it is clear that the emphasis is shifting towards a "spectroscopy of fundamental fermions" : how many are they, how are they related, what determines their scales and mixings (Cabibbo angles) etc... This new spectroscopy promises to be fun !

II. CURRENT QUARKS, CHIRAL SYMMETRY AND ALL THAT.

Our purpose here again is to give an elementary discussion of the so-called "current quarks" i.e. the local relativistic quark fields as they are seen through electromagnetic and weak probes and their connection with the structure of the strong hamiltonian. This will of course bring us back to QCD.

The starting point of all these considerations is the observation that the weak and electromagnetic currents are made of several pieces which carry specific quantum numbers. For example :

$$J_\mu^{e.m} = J_\mu^{\Delta I=1; e.m} + J_\mu^{\Delta I=0; \, e.m} + \ldots \qquad \text{(II.1)}$$

$$J_\mu^{weak} = J_\mu^{\Delta I=1; \Delta S=0} + J_{\mu 5}^{\Delta I=1; \Delta S=0} + J_\mu^{\Delta I=\frac{1}{2}; \Delta S=1} + J_{\mu 5}^{\Delta I=1/2; \, \Delta S=1} + \ldots \qquad \text{(II.2)}$$

This decomposition of the currents led, twenty years ago, to the CVC assumption [59], i.e. the operators T_i defined by

$$T_1 + iT_2 = \int J_o^{\Delta I=1;\Delta S=0} d^3x \tag{II.3}$$

$$T_3 = \int J_o^{\Delta I=1;e.m} d^3x \tag{II.4}$$

are the isospin generators, hence $[T_i, T_j] = i\varepsilon_{ijk}T_k$ and $[T_i, H_{strong}] = 0$. In other words, a symmetry of the strong interactions is defined through currents which are measurable in weak and electromagnetic transitions.

It was Gell Mann's idea [60] to extend this procedure to the other pieces of the currents. In this way one derives an equal time current algebra which among other things has taught us a lot about the structure of the strong interactions. In the context of a quark model, one has for vector currents :

$$F_i(t) = \int d^3x\, q^+(x) \frac{\lambda_i}{2} q(x) \tag{II.5}$$

$$\text{and } [F_i(t), F_j(t)] = if_{ijk}F_k(t) \tag{II.6}$$

i.e. the algebra of SU(n) with n the number of quark flavours; similarly for axial currents one defines

$$F_{i5}(t) = \int d^3x\, q^+(t) \frac{\lambda_i}{2} \gamma_5 q(x)$$

which transform as the adjoint representation of SU(n)

$$[F_i(t), F_{j5}(t)] = if_{ijk}F_{k5}(t). \tag{II.7}$$

The algebra closes through

$$[F_{i5}(t), F_{j5}(t)] = if_{ijk}F_k(t) \tag{II.8}$$

i.e. the vector and axial vector charges, in a quark model generate a chiral SU(n) $\otimes$ SU(n) algebra. If one includes the flavour singlets $F_0(t) = \int d^3x\, q^+(x)q(x)$ and $F_{05}(t) = \int d^3x q^+(x)\gamma_5 q(x)$ the algebra gets naturally enlarged to U(n) $\otimes$ U(n).

One cannot insist enough on the theoretical relevance and phenomenological implications of these algebraic structures :

- the chiral algebra could be an exact property of the theory despite the fact that the flavour group is badly broken. This is to be contrasted with "constituent quark" symmetries which are, at best, approximate properties of the particle spectrum;

- with the help of PCAC [61], the SU(2) ⊗ SU(2) chiral algebra leads to the Adler Weisberger relation [62] which is, as well known, remarkably succesful;
- algebras do set scales since they are non linear. They allow, for example, to define the concept of universality of the weak interactions [60]. From the leptonic currents, one defines

$$L_1 + iL_2 = \int d^3x\, \nu_e^+ \left(\frac{1+\gamma_5}{2} \right) e + e \leftrightarrow \mu \qquad \text{(II.9)}$$

which satisfy $[L_1, L_2] = i\varepsilon_{ijk}L_k$. Requiring the same algebra for the quark sector gives a meaning to universality. It also leads naturally to a unification of weak and electromagnetic interactions in the fermion sector of the theory; $\begin{pmatrix}\nu_e\\ e^-\end{pmatrix}$ $\begin{pmatrix}\nu_\mu\\ \mu^-\end{pmatrix}$ do transform as doublets as well as $\begin{pmatrix}u\\ d_c = d\cos\theta + s\sin\theta\end{pmatrix}_L$ and $\begin{pmatrix}c\\ s_c\end{pmatrix}_L$. The full impact of this remark becomes clear only in the context of a gauge theory which, of course, goes much deeper and unifies the boson sector of the theory as well;
- partons are of course the local relativistic quark fields which make up the weak and electromagnetic currents;
- last but not least chiral (S)U(n) ⊗ (S)U(n) is an approximate symmetry of the strong interactions. Since it cannot be a "multiplet symmetry" it will be realized "à la Goldstone".

It should be clear from these remarks that current algebra has played a central role in the development of many of the basic ideas we believe in today. In this lecture we will restrict ourselves to the chiral symmetry aspect and limit our discussions to the case of 3 quark flavours. The interested reader will find no difficulty in extending these considerations to n flavours.

The SU(3) ⊗ SU(3) chiral symmetry limit is thus assumed to correspond to eight massless pseudoscalars and (ordinary) SU(3) realized as a conventional multiplet symmetry. This raises several fundamental questions :

- the double nature of the pseudoscalars : they are Goldstone bosons i.e. "elementary" fields in the usual formulation of spontaneous symmetry breaking but, of course, we do want them to be $\bar{q}q$ bound states. No coherent formulation of this "duplicity" is known;
- the so-called U(1) problem which refers to the mechanism by which invariance under the "ninth" axial charge, $F_{05}(t)$ is broken. The problem is of course present in SU(2) ⊗ SU(2) as well : in the limit of massless (Goldstone) pions how does one avoid a massless η. The problem is not at all academic : if

the masses of the four pseudoscalars are simultaneously generated, Weinberg [63] has shown that, under quite general conditions, $m_\eta \leq \sqrt{3}\, m_\pi$! A possible solution to this problem has been recently suggested by 't Hooft [64].

A. Chiral symmetry breaking.

Following Gell-Mann, Oakes and Renner[65] we assume that the strong interaction hamiltonian density can be decomposed as follows :

$$H = H_o + H' \tag{II.10}$$

with H_o chiral symmetric i.e. $[F_i, H_o] = [F_{i5}, H_o] = 0$ and H' the breaking term. This decomposition follows naturally from QCD. Indeed the quark Lagrangian may be written as

$$\mathcal{L} = \bar{q}\, \gamma^\mu \,(i\partial_\mu - g \frac{\lambda_k^c}{2} G_{k\mu})q - \bar{q}\, M\, q \tag{II.11}$$

(c refers to the colour group, $G_{k\mu}$ are the coloured gluons, M is a matrix in flavour space).

Leaving aside the question of how quark masses are generated, one may assume for M the following form :

$M = \begin{pmatrix} m & o & o \\ o & m & o \\ o & o & m_s \end{pmatrix}$ where $m = m_u = m_d$ in the isospin symmetry limit. Obviously in such a theory :

$$H' = m(\bar{u}u + \bar{d}d) + m_s\, \bar{s}s \tag{II.12}$$

and it transforms as a $(\underset{\sim}{3},\underset{\sim}{\bar{3}}) + (\underset{\sim}{\bar{3}},\underset{\sim}{3})$ representation of $SU(3) \otimes SU(3)$. Defining $s_i = \bar{q}\lambda_i q$; $p_i = \bar{q}\lambda_i\gamma_5 q$ one finds indeed :

$$[F_i, s_j] = if_{ijk}s_k \qquad [F_{i5}, s_j] = -id_{ijk}p_k$$

$$[F_i, p_j] = if_{ijk}p_k \qquad [F_{i5}, p_j] = id_{ijk}s_k \tag{II.13}$$

and

$$H' \cong s_o + cs_8$$

where $s_o \sim \bar{u}u + \bar{d}d + \bar{s}s$; $s_8 \sim \bar{u}u + \bar{d}d - 2\bar{s}s$ and

$$c = -\sqrt{2}\, \left(\frac{m_s - m}{m_s + 2m}\right) \tag{II.14}$$

This elegant theory of chiral symmetry breaking has many well known consequences, in particular :

- for pseudoscalar masses one obtains easily :

$$\frac{m_\pi^2}{m_K^2} = \left(\frac{\sqrt{2} + c}{\sqrt{2} - \frac{1}{2}c} \right) \frac{< 0|p_\pi|\pi >}{< 0|p_\mu|K >} \qquad \text{(II.15)}$$

- pionic "σ terms" have the structure

$$[F_{5\pi}, \partial_\mu J^\mu_{5\pi}] \approx (\sqrt{2} + c)(\sqrt{2}s_o + s_8) \qquad \text{(II.16)}$$

The interesting limits to consider are :

- chiral SU(3) ⊗ SU(3) which corresponds to $m = m_s = 0$ (The U(1) problem in this case is whether $m_{\eta'} \neq 0$!).
- SU(3) which clearly corresponds to $m_s = m$ or $c = 0$;
- SU(2) ⊗ SU(2) which one obtains when $m = 0$ or $c = -\sqrt{2}$ ($m_\pi = 0$).

At this level of the theory, quark masses are of course free parameters which we will try to evaluate. The spontaneous breaking of the weak and electromagnetic gauge group does play a role here. In our present formulation, quark masses are generated via couplings to the Higgs scalars. This does not however give us any indication of the scales involved. Anyway to test the Gell-Mann Oakes Renner (GMOR) theory of chiral symmetry breaking we have to make additional assumptions about the transformation properties of the matrix elements of the s's and p's. The simplest hypothesis is of course that the matrix elements of these operators have their "natural" SU(3) structure e.g. between pseudoscalar mesons $|P_i>$,

$$< P_i|s_j|P_k > \sim d_{ijk} \qquad < 0|p_j|P_k > \sim \delta_{jk} \qquad \text{(II.17)}$$

This follows uniquely if hadronic states transform irreducibly under the spacelike charges but is known to be wrong when SU(3) is broken [66] ! Anyway, given this hypothesis, GMOR conclude that :

$$c = -\sqrt{2} \left(\frac{m_K^2 - m_\pi^2}{m_K^2 + \frac{1}{2} m_\pi^2} \right) \simeq -1.25 \qquad \text{(II.18)}$$

which is very close to the chiral SU(2) ⊗ Su(2) limit.

An immediate consequence is that all pionic σ terms are predicted to be small since they are proportional to $c + \sqrt{2}$.

Even for the best known $(\pi N)\sigma$ term, phenomenological values [67] differ by factors of 2 or 3 ! There is a potential problem here which has led several authors [68] to give up Eq. (II.17) while still keeping the general framework of the GMOR theory. Before attacking the main objective of this lecture which is to "determine" the quark masses, let us discuss in a bit more detail the assumption underlying Eq (II.17).

B. Transformation properties on the light cone.

It is well known that in the case of a broken symmetry (of the flavour SU(3) type) the light like charges [69] continue to annihilate the vacuum. Hence the light plane representation of the quark model seems the most appropriate for the discussion of the problem at hand. For any four component object we define as usual

$$V^{\pm} = \frac{V^0 \pm V^3}{\sqrt{2}} , \quad \overline{V} = (V^1, V^2)$$

Introducing the projection operators $P_{\pm} = 1/2\ (1 \pm \alpha_z)$ one defines the fields $q_{\pm}(x) = P_{\pm}q(x)$. On the light plane $x^+ = 0$, $q_+(x)$ gives the only two independent components of the quark field and for these components canonical anticommutation relations hold [70]

$$\{q_+^+(x), q_+(y)\}_{x^+=y^+} = \frac{1}{\sqrt{2}} P_+ \delta(x^- - y^-)\delta^2(x_\perp - y_\perp) \qquad (II.19)$$

$$\{q_+^+(x), q_+^+(y)\}_{x^+=y^+} = \{q_+(x), q_+(y)\}_+ = 0$$

The non-independent components $q_-(x)$ are related to $q_+(x)$ by a constraint equation which depends on the dynamics of the problem. With an appropriate representation of the γ matrices such that $q_+(x) = \begin{pmatrix} q_+ \\ 0 \end{pmatrix}$ and $q_-(x) = \begin{pmatrix} 0 \\ q_- \end{pmatrix}$, the constraint equation reads for QCD :

$$q_- \sim \vec{\sigma}_\perp \cdot (\vec{\nabla}_-^{-1} + ig\nabla_-^{-1}\vec{G}_\perp)q_+ + \mathcal{M}\nabla_-^{-1} q_+ \qquad (II.20)$$

Operators which depend only on the canonical components q_+ are called "good" operators while those which depend on q_-, the dependent components, are called "bad" operators [71]. Clearly for bad operators, algebraic properties such as commutation relations will be dependent on the dynamics of the problem ! The light plane operators of chiral symmetry are given by :

$$F_i^L = \sqrt{2} \int d^4x\delta(x^+)\ q_+^+(x) \frac{\lambda_i}{2} q_+(x) \qquad (II.21)$$

and

$$F^L_{i5} = \sqrt{2} \int d^4x\delta(x^+)\; q_+^+(x)\, \frac{\lambda_i}{2}\, \sigma_z\, q_+(x) \;. \qquad \text{(II.22)}$$

They are good operators and the algebra they generate, SU(3) ⊗ SU(3), depends only on the canonical anticommutators. The scalar (s_i) and pseudoscalar (p_i) densities on the other hand are bad operators; one finds easily :

$$s_i \sim q_+^+\lambda_i\vec{\sigma}_\perp\cdot(\vec{\nabla}_\perp\, \nabla_-^{-1} + ig\nabla_-^{-1}\vec{G}_\perp)q_+ + q_+^+\, \lambda_i\;\; M\nabla_-^{-1}\, q_+ + \text{h.c.} \qquad \text{(II.23)}$$

$$p_i \sim q_+^+\lambda_i\sigma_z\vec{\sigma}_\perp\cdot(\vec{\nabla}_\perp\, \nabla_-^{-1} + ig\nabla_-^{-1}\vec{G}_\perp)q_+ + q_+^+\, \lambda_i\sigma_3\;\; M\nabla_-^{-1}q_+ + \text{h.c.}$$

The remarkable property of the F^L_i is that it annihilates the vacuum even when SU(3) is broken : it is meaningful to classify hadron states under the F^L_i whether SU(3) is exact or not !

The crucial point is of course that the algebraic properties of s_i and p_i depend on the dynamics : the presence of the quark matrix M modifies the transformation properties of the scalar and pseudoscalar densities in the presence of SU(3) breaking ! The GMOR assumption amonts thus to neglecting the effect of quark masses in the matrix elements of s_i and p_i. It should be possible to estimate in QCD whether such an assumption is reasonable or not. To the best of my knowledge this has not been done.

The chiral SU(3) ⊗ SU(3) of light charges is a subgroup of $SU(6)_{W;\,currents}$ which is generated by all good operators in a quark theory [72]. As well known [73] this algebra cannot be identified with $SU(6)_{W;\,constituents}$ which is a subgroup of the "full rest symmetry" under which hadrons have simple transformations properties. The two SU(6) algebras are related by the Melosh transformation [27]. Unfortunately this transformation is only known for free quarks. It should however be possible in a theory like QCD to define the Melosh transformation perturbatively (after all g is supposed to be small inside hadrons). A systematic analysis of all phenomenological consequences of such a transformation would be very rewarding : it would relate the parameters A', B', A, B, C, D defined in the first lecture to chiral symmetry breaking parameters, σ terms, mass formulae, etc...

C. Quark masses.

The GMOR value of c gives us a first relation among quark masses, namely $m_s/m \approx 25$.

A second relation follows from the Gell Mann Okubo mass formulae. As well known most of the "usual" hadron splittings are due to the term $m_s\bar{s}s$ in H'. First order perturbation theory then suggests, for a hadron h :

$$m_h \simeq m_o + m_s < \bar{s}s >_h \qquad \text{(II.24)}$$

At this level of approximation it is reasonable to assume that $< \bar{s}s >_h$ is proportional to the number of strange quarks present in the hadron h i.e. $< \bar{s}s >_h \alpha N_{sh}$.
With the GMOR SU(3) assumptions, the proportionality factor does not depend on the quark flavour. The mass splittings among different multiplets even suggest that this factor does not depend on the hadron h ! This common renormalization factor can then simply be reabsorbed in the definition of the quark masses (which we still denote by m_s and m) leading to :

$$m_s \simeq 150 \text{ MeV}$$

and consequently $m \simeq 6$ MeV.
These values agree with other "calculations" of the quark masses[74] !
All these calculations are of course questionable if only because one makes assumptions about the matrix elements of bad operators. Once again a Melosh type transformation for QCD would be useful in this case.

The previous analysis can readily be extended to the case where electromagnetic (and weak) effects are included in the quark masses. If we simply assume $m_u \neq m_d$ the chiral symmetry breaking hamiltonian H' becomes $H' = m_u\,\bar{u}u + m_d\,\bar{d}d + m_s\,\bar{s}s$ or equivalently :

$$H = H_{GMOR} + \chi s_3$$

with

$$s_3 = \bar{u}u - \bar{d}d \qquad \text{and} \qquad \chi = \sqrt{3}\,\frac{m_u - m_d}{m_u + m_d - 2m_s} \qquad \text{(II.25)}$$

Let us ignore for a moment the effects of virtual photons and intermediate bosons. In first order in H', we obtain in the usual way [75] :

$$\mu_0^2(\pi^+) = \mu_0^2(\pi^\circ)\ \alpha\ m_u <o|\bar{u}u|o> + m_d <o|\bar{d}d|o>$$

$$\mu_0^2(K^+)\ \alpha\ m_u <o|\bar{u}u|o> + m_s <o|\bar{s}s|o>$$

$$\mu_0^2(K^\circ)\ \alpha\ m_d <o|\bar{d}d|o + m_s <o|\bar{s}s|o> \qquad \text{(II.26)}$$

where the proportionality factor is given by current algebra. Making the simplest SU(3) assumption, once again for bad operators (!), we expect :

$$<o|\bar{u}u|o> = <o|\bar{d}d|o> = <o|\bar{s}s|o> \tag{II.30}$$

To this value of the pseudoscalar masses we still have to add the weak and electromagnetic contributions coming from virtual exchanges, i.e.

$$\mu^2(P_i) = \mu_0^2(P_i) + \mu_{JJ}^2(P_i) \tag{II.31}$$

For a virtual photon Dashen's theorem [76] in SU(3) ⊗ SU(3) relates the "JJ" contributions as follows :

$$\mu_{JJ}^2(\pi^\circ) = \mu_{JJ}^2(K^\circ) = 0 \tag{II.32}$$

$$\mu_{JJ}^2(K^+) = \mu_{JJ}^2(\pi^+) .$$

Hence we obtain

$$X = -\frac{\sqrt{3}}{2} \frac{(m_{K^+}^2 - m_{K^\circ}^2) - (m_{\pi^+}^2 - m_{\pi^\circ}^2)}{m_{K^+}^2 - m_{\pi^+}^2} \approx .02 \tag{II.33}$$

or equivalently

$$\frac{m_d}{m_u} \approx 2 \qquad \text{and} \qquad \frac{m_s}{m_d} \approx 20$$

which, when combined with the estimate of m_s give us finally :

$$m_s \simeq 150 \text{ MeV} \qquad m_u \simeq 4 \text{ MeV} \qquad m_d \simeq 8 \text{ MeV}$$

The precise values of the masses should not be taken too seriously but only as order of magnitudes ! (The splittings of the baryon octet and decuplet for example are respectively 190 MeV and 145 MeV : this gives an idea of the uncertainty in the determination of m_s).

Let me insist on the obvious fact that the masses we are talking about are (proportional to) the parameters in the QCD Lagrangian and not the "constituent quark" masses. These last quantities are rather ill defined : they receive their largest contribution from the spontaneous breakdown of the chiral algebra rather than from the mass terms in the Lagrangian. An educated guess, derived from m_B (the s_0 part of H') ≃ 1150 MeV, gives for constituent quarks $m_{u,d}^c \approx 1150/3 + m_{u,d} \approx 380$ MeV and $m_s^c \approx 530$

MeV which agree of course with naive quark model estimates.

The parameters which are of theoretical interest are of course those appearing in the Lagrangian. It is interesting to point out that χ (II.25), which is a combination of these parameters, is directly measurable in transitions such as $\psi' \to \psi\pi^0$ [77].

Several striking facts in relation to the values of the mass parameters are worth noticing :

- the scales which are remarkably similar to the lepton scales m_d, $m_u \approx O(m_e)$; $m_s \approx O(m_\mu)$. This correlation subsists at the next level $m_c \approx O(m_{\tau^-})$! We have of course no understanding of this phenomenon but it unmistakably points towards a deep connection between leptons and quarks. This will obviously be the central problem of the next level of "elementary particle" spectroscopy.
- the huge breaking of isospin (by a factor of 2!) in the mass parameters. It was pointed out by Weinberg [75] that this may lead to observable effects in πN scattering lengths.
- why is $m_d > m_u$? If we impose that $m_d - m_u$ be calculable, in the sense of Weinberg [78] and positive, this puts severe constraints on "acceptable" gauge models for weak and electromagnetic interactions. In particular it excludes the Weinberg Salam model as will be discussed in Lecture IV.

Chiral symmetry and its breaking has taught us several important aspects of quark physics :

- chiral $SU(2) \otimes SU(2)$ is almost as good a symmetry as $SU(2)$;
- reliable estimestes of matrix elements of current quark operators, in particular the bad operators are urgently needed. In QCD, at least perturbatively, this does not seem an untractable problem (Melosh transformation);
- the GMOR scheme fares remarkably well;
- the quark masses, which are beyond QCD, show striking analogies with lepton masses.

It should be obvious how to generalize these considerations to $SU(n) \otimes SU(n)$.

III. BAG MODELS.

As repeatedly mentioned in the previous lectures, QCD is our only candidate for a hadron theory. If the theory can be solved, the hadron spectrum should eventually be calculable from "first principles". The basic idea behind bag models [44] is that, if a theory like QCD does indeed confine colour, hadrons will look like extended objects in which quarks and gluons are trapped. Such an idea is quite appealing for several reasons :

- it brings us back at long last to a space-time picture of the hadrons which is intuitively quite satisfying;
- it provides us with a calculational scheme where quarks are treated at all times as relativistic particles. This is a definite improvement over all versions of the non relativistic quark model;
- it is an interesting model in its own right : the quantum mechanics of extended systems is by no means a trivial problem.

Two classes of solutions have been discussed in the context of bag models :

- "spherical bags" which seem a natural configuration for the lightest states built out of u, d and s quarks in an S wave. Keeping QCD in the back of our minds, hadrons are thus described in 0th order of the quark gluon coupling constant as a spherical cavity containing free quarks (or gluons !). The first order corrections in g will then split degenerate levels (Λ-Σ e.g.), renormalize static properties, induce deviations from scaling ... etc... Presumably they will also modify the shape of the bag but it is hoped that this effect is negligible since it cannot really be calculated !
- "elongated bags" which are considered for heavy quarks or high angular momenta. We will restrict our discussion to the spherical solutions : they illustrate the successes and problems of the model well enough for our purposes.

The presentation given here relies heavily, on a recent review of the subject given by J. Kuti [79].

A. Quarks in a cavity.

As a first step towards "bagged QCD" we discuss the level structure of hadrons corresponding to quarks confined to a spherical static cavity [80]. The hamiltonian of such a system is defined by :

$$H = \int_{\text{sphere}} d^3\vec{r} \; : q^+(\vec{r})\{-i\vec{\alpha}\cdot\vec{\nabla} + \beta M\}q(\vec{r}) : \qquad \text{(III.1)}$$

where M is the quark mass matrix in flavour space (n=3). For simplicity we shall assume that its eigenvalues are $m_u = m_d = 0 \neq m_s$. A summation over colour indices is understood in eq. (III. 1). The eigenstates of this hamiltonian obviously correspond to solutions of the Dirac equation in a "square well" potential with $V = 0$ for $r < R$ and $V = \infty$ for $r > R$ (the radius of the sphere). For single quark eigenstates the solutions are easily found by repeating step by step the procedure followed in all textbooks for the hydrogen atom. Hence the solutions are labelled by $\varepsilon J \ell M$ where ε, $J(J+1)$ and M are the eigenvalues of H, J^2 and J_2 and $\ell(\ell+1)$ is the eigenvalue of L^2 for the upper components. Of course $\ell = J \pm 1/2$ and the lower components are eigenfunctions of L^2 but with eigenvalue $\ell'(\ell'+1)$ where $\ell' = J \mp 1/2$. The radial equation yields as solutions spherical Bessel functions and the boundary condition is easily shown to be at $r = R$:

$$\vec{n} \cdot \vec{\gamma}\, \psi = i\psi \qquad \text{with} \qquad \vec{n} = \frac{\vec{r}}{r} \tag{III.2}$$

Introducing "Dirac's quantum number" $K = \ell(\ell+1) - J(J+1) - 1/4$ one obtains for a massless quark in the state $J = 1/2$, either

$$nS^{\frac{1}{2}}\ (\kappa=-1, \ell=0) \quad : \quad \psi_{\varepsilon 1/2 0 M} \begin{pmatrix} i j_0(x_{n,-1} \frac{r}{R})\, u \\ -j_1(x_{n,-1} \frac{r}{R})\, \vec{\sigma} \cdot \vec{n}\, u \end{pmatrix} \tag{III.3}$$

$$nP^{\frac{1}{2}}\ (\kappa=1, \ell=1) \quad : \quad \psi_{\varepsilon 1/2, 1, M} \begin{pmatrix} i j_1(x_{n,1} \frac{r}{R})\, \vec{\sigma} \cdot \vec{n}\, u \\ j_0(x_{n,1} \frac{r}{R})\, u \end{pmatrix} \tag{III.4}$$

where U is a two component Pauli spinor. The boundary condition yields the eigenvalues condition for the $x_{n,\kappa}$

$$j_0(x_{n,\kappa}) = - j_1(x_{n,\kappa}) \tag{III.5}$$

with the energy given by

$$\varepsilon_{n,\kappa} = \frac{x_{n,\kappa}}{R} \tag{III.6}$$

The first few positive solutions of (eq. III.5), labelled by positive n, are

$$\begin{array}{llll} \text{for } \kappa = -1 & x_{1,-1} = 2.04 & x_{2,-1} = 5.40 \ \ldots \\ \text{for } \kappa = +1 & x_{1,1} = 3.81 & x_{2,1} = 7 \ \ldots \end{array} \tag{III.7}$$

The negative roots of(eq.III.5) labelled by negative n are seen to obey $-x_{n,\kappa} = x_{-n,-\kappa}$ which gives trivially the "negative eigenvalues". These negative energy states are of course reinterpreted as antiquark solutions.
The boundary condition $\vec{n} \cdot \vec{\gamma}\psi = i\psi$ guarantees charge conservation :

$$\vec{n} \cdot \vec{j} = \vec{n}\,\bar{\psi}\,\vec{\gamma}\,\psi = -\,\bar{\psi}\psi = \bar{\psi}\psi \qquad \text{(III.8)}$$

since $\bar{\psi}\,\vec{n}\cdot\vec{\gamma} = -i\bar{\psi}$ at $r = R$

and hence no current flows through the surface.
Since $\bar{\psi}\psi$ vanishes on the surface, its derivative points along the normal $\vec{n}$

$$\frac{\partial}{\partial x^R}(\bar{\psi}\psi) = 2n_\kappa P \qquad \text{(III.9)}$$

and the momentum flow accross the surface is easily seen to be

$$n_\kappa T^{\kappa\ell} = n^\ell P \qquad \text{(III.10)}$$

P is interpreted as the quark pressure on the walls of the cavity. In the bag model it will be balanced by surface tension and "bag" pressure terms which determine dynamically the size and shape of the bag.

With the quark orbitals for J = 1/2 it is a trivial exercise in group theory to build the wave functions of the baryons. The spectrum one obtains goes as follows :

$(1S^{1/2})^3$	$N(1/2^+)$	$\Delta(3/2^+)$
$(1S^{1/2})^2(1P^{1/2})$	$2N(1/2^-)N(3/2^-)$	$\Delta(1/2^-)\Delta(3/2^-)$
$(1S^{1/2})^2(2S^{1/2})$	$2N(1/2^+)N(3/2^+)$	$\Delta(1/2^+)\Delta(3/2^+)$
$(1S^{1/2})(1P^{1/2})^2$	$2N(1/2^+)N(3/2^+)$	$\Delta(1/2^+)\Delta(3/2^+)$
$(1S^{1/2})(1P^{1/2})(2S^{1/2})$	$4N(1/2^-)2N(3/2^-)$	$2\Delta(1/2^-)\Delta(3/2^-)$
$(1P^{1/2})^3$	$N(1/2^-)$	$\Delta(3/2^-)$

.
.
.

The lowest states are of course to be identified with the nucleon and Δ. The fact that they are degenerate reflects the crudeness of our approximation. For these states one can easily compute static properties such as μ_p, μ_n and g_A. One obtains the following rather remarkable results : μ_p = 2.9 (exp. 2.79) $\mu_n/\mu_p=-2/3$, g_A = 1.09 (exp. 1.23). The higher spectrum for the non strange baryons is nevertheless obviously wrong : it shows no signs of the SU(6) multiplets which so spectacularly agree with experiment ! We did however commit several crimes in computing this spectrum :

- we did not take into account the "center of mass" motion;
- we neglected quark orbitals of the type 1 $P^{3/2}$.

The first problem is highly non trivial. As for the second one it is easily computed that 1 $P^{3/2}$ states lead to a non symmetric quark pressure term (eq.III 9-10). Hence it is highly unlikely that a spherical cavity is a good approximation for these orbitals: more precisely, in the bag model where the shape of the bag is determined dynamically in such a way that the bag pressure and, possibly, surface tension cancel the quark pressure, we do not expect a spherical solution to be acceptable. On the other hand we do know that the lowest lying baryons with negative parity fill in neatly a $\underset{\sim}{70}$ L = 1^- of SU(6) $\otimes$ O(3). One of the members of this representation is an $\underset{\sim}{8}$ with J = 5/2. This is unattainable without 1 $P^{3/2}$ quark orbitals ! Let us ignore the fact that the spherical approximation is not valid for some of the quark orbitals and compute nevertheless the possible levels of three quarks in the configurations $(1\ S\ 1/2)^2(1\ P1/2)$ and $(1\ S\ 1/2)^2(1\ P\ 3/2)$. One finds easily that they correspond to all N's and Δ's contained in two SU(6) multiplets namely :

$\underset{\sim}{56}$ L=1 i.e. $N(3/2^-)N(1/2^-)$

$\Delta(5/2^-)\Delta(3/2^-)\Delta(1/2^-)$

and

$\underset{\sim}{70}$ L=1 i.e. $2N(1/2^-)2N(3/2^-)N(5/2^-)$

$\Delta(1/2^-)\Delta(3/2^-)$

This is precisely the level structure of the non relativistic harmonic oscillator quark model where, however, the $\underset{\sim}{56}$ L = 1 is <u>rigorously</u> a genuine "center of mass" effect.

Presumably the same effect occurs in our cavity model i.e. the $\underset{\sim}{56}$ L = 1 corresponds to an L = 1 excitation of the center of mass of three quarks in their ground state $\underset{\sim}{56}$ with respect to the fixed spherical cavity ! In the "real" bag model, as we shall discuss shortly, the situation becomes infinitely more complicated due to the dynamics of the surface and the impossibility of taking the center of mass degrees of freedom rigcrously into

account [81] !
Since we will eventually consider quarks interacting with gluons in the bag, it is useful at this stage to consider gluons confined to a cavity. One finds the boundary conditions :

$$\vec{n} \cdot \vec{E}_i = \vec{n} \times \vec{B}_i = 0 \qquad \text{at} \quad r = R$$

where $\vec{E}_i$ and $\vec{B}_i$ are the colour electric and magnetic fields. These conditions are easily understood if one visualized the cavity as a "medium" with $\varepsilon = \mu = 1$ while outside the cavity one has $\varepsilon = 0$, $\mu = \infty$. The reflection coefficient ($\mathcal{R}$) on the surface between the media is given by

$$\mathcal{R} = \left(\frac{1 - \varepsilon_2}{1 + \varepsilon_2} \right)$$

when $\varepsilon_2 \to 0$, $\mathcal{R} \to 1$ i.e. there is total reflection on the surface and the gluons are confined [79].

It is then straightforward to solve ("colour") Maxwell's equations with these boundary conditions and to obtain

- transverse magnetic (TM) modes with the lowest eigenfrequency $x_0(TM) = 4.49$
- transverse electric (TE) modes with $x_0(TE) = 2.74$.

B. The bag.

In the bag model, hadrons are pictured as a small space-time domain where quarks and gluons are confined. An intuitive physical picture of how this may occur is the following [79] : assume that the physical vacuum is characterized by a two phase structure. In the "normal" phase it does not allow colour propagation but in the abnormal (or hadron) phase, quarks and gluons propagate normally. A hadron is then a vacuum bubble (bag) in abnormal phase filled with quarks and gluons. The impermeability of the bag surface against coloured fields is expressed by boundary conditions which are the same, in the static limit, as for the cavity model.

Inside the bag, quark and gluon dynamics are governed by QCD. The shape and size of the bag are determined dynamically : the pressure exerted by the quark and gluon fields on the boundary of the bag is balanced by volume and surface energies which are interpreted as (normal) vacuum pressure on the bag and surface tension. The boundary of the bag is transparent to weak and electromagnetic fields which are supposed to propagate normally in both vacuum phases. The two phase vacuum is a genuine strong interaction effect.

The relativistic action W for the bag is given by :

$$A = \int dt \int_{bag} d^3x\{-\frac{1}{4}F^i_{\mu\nu}F^{\mu\nu}_i + \bar{q}(i\partial\!\!\!/- M)q - g\bar{q}\frac{\lambda^c_i}{2}\gamma^\mu q G_{i\mu}\}$$

$$- \sigma\int dt \int_{surface} d^2s\sqrt{1-v_\perp^2} - B\int dt \int_{bag} d^3x$$

In the first term i refers to the colour index, $G_{i\mu}$ to the gluons and M is the quark matrix in flavour space. The second and third terms correspond to the bag variables and $v_\perp$ is the velocity of a bag surface element along its normal. In the original MIT-Bag model [44] the surface tension term was ignored ($\sigma = 0$).

Variations of the quark gluon fields inside the bag give the usual equations of QCD; variations of these fields variables on the surface of the bag give the covariant generalization of the boundary conditions for the static cavity, i.e.

$$n_\mu F^{\mu\nu}_i = 0$$

and

$$n_\mu\gamma^\mu q = -iq$$

with n_0 the surface velocity along the normal n to the surface. Since there is no colour electric flux through the bag's surface ($\vec{n}\cdot\vec{E}_i = 0$ on the surface), Gauss's theorem guarantees that only colour singlet states exist in the bag !

Finally, variation of the surface variables in the action give the equations of motion for the boundary which we shall not discuss here. Let us note that the theory is relativistically invariant and that momentum is of course conserved. In the limit of a static bag ($\sigma = 0$) with only quarks inside ($g = 0$) one recovers of course the well known MIT quadratic boundary condition

$$1/2(\vec{n}\cdot\vec{\partial})(\bar{\psi}\psi) = B.$$

Altough rather complicated, this bag description of the hadrons is quite appealing for several reasons :

- it is fully relativistic;
- it incorporates confinement if with ad hoc prescriptions;
- it is amenable to several physically reasonable approximations. For example for a spherical surface the last two terms of the action correspond to the Lagragian :

$$L(t) = -\sigma 4\pi\rho^2 \sqrt{1-\dot{\rho}^2} - B\frac{4\pi}{3}\rho^3$$

with $\rho(t)$ the bag radius. In zeroth order in the quark-gluon coupling constant one can solve in adiabatic approximation the bag equations for the nucleon [79] : one finds that the surface of the bag undergoes "fluctuations" but the probability distribution for finding the bag with radius ρ is sharply peaked around the classical radius. This is of course not unexpected but is nevertheless a quite remarkable achievment.
It justifies the "static bag" approximation where surface effects are neglected all-together and a sharp boundary is assumed. We recover in this way the classical MIT-Bag analysis [82] which is sufficiently well known not to be repeated here.
At this level of the analysis the N and Δ are still degenerate but for higher states of 39 baryons, the situation remains hopelessly complicated. As far as the center of mass variables are concerned, Rebbi [81] made a brave attach on the problem and showed that most of the 56 $L = 1^-$ states which we found in the cavity model are indeed translation modes of the ground state. There is however unavoidable mixing with higher excited states (also 56 $L = 1^-$ of course)

Furthermore, as mentioned in the cavity approximation, higher quark orbitals necessarily demand non spherical bags for which there is at present no known solution. This unfortunately prevents us from developing a reliable phenomenology for the states in the 70 $L = 1^-$.

An important question we have to face concerns the relation of the bag with QCD. The difficulty is of course to relate the bag variables to QCD. Whichever way this may be done we do expect the bag spectrum for the hadrons (i.e. level labelling) to show a lot of similarities with the, as yet, uncalculable spectrum predicted by QCD. And this brings us to a vital question, certainly as far as bag phenomenology is concerned : where are all the missing states predicted by the bag ? By this I mean of course the new types of hadrons such as "gluonium" mesons (GG) or (GGG) or mesons of the type $\bar{Q}QG$! In the cavity model the masses and spin parities of the lowest gluonium states are readily computed : one finds among others a 0^{++}(TE)(TE) glueball with mass 1096 Mev and a 0^{+-}(TE)(TE)(TE) glueball at 1644 Mev. This last one, in particular, is a "second kind" exotic [83] ! A similar analysis can be made for $\bar{Q}QG$, $\bar{Q}QGG$, etc... Once again, the lowest of these states should be around 1000-1500 Mev with plenty of second kind exotic flavour octets, etc... In my opinion these states are the really crucial predictions of the bag model (and perhaps of QCD as well) and they must be found ! There are of course several other spectroscopic applications or implications of the bag model which are quite interesting but will not be discussed here : "ordinary" exotic mesons, baryonium (i.e. a diquark and an antidiquark in a cigar shaped bag), linearity of Regge trajectories, charmonium, etc... [84].

From the equations of the bag model we indicated how to compute the masses of the low-lying hadrons in 0th order of the quark-gluon coupling constant g and with $m_u = m_d = 0$. There is no difficulty in including strange quarks with $m_s \neq 0$. It is slightly harder to include higher order corrections in g since in addition to one gluon exchange between two quarks one has also "self-mass" corrections corresponding to the diagram  . Clearly the intermediate quark can be in any of the eigenmodes of the bag which makes this diagram very hard to estimate (higher quark states do not correspond to a spherical bag). One hopes that the Lamb shift type slittings of hadronic levels due to this graph are small and that the main cause of level splitting is due to gluon exchange i.e. to the graph

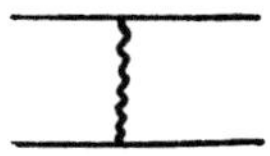

Consider for example the colour magnetic part of this diagram. Clearly $\vec{B}_i^k \propto \lambda_i^k \vec{\sigma}^k$ where $\vec{B}_i^k$ is the contribution of the kth quark to the colour magnetic field. Hence the colour magnetic energy shift has the form

$$\Delta E_M \propto \sum_i \sum_{k,\ell} (\vec{\sigma}^k \lambda_i^k)(\vec{\sigma}^\ell \lambda_i^\ell) \; f\,(m_k, m_\ell, R)$$

the function f depends on the details of the wave functions and will not be discussed here. The spin and colour structure is however quite interesting. Using the fact that for a colour singlet meson $(\lambda_i^{(1)} + \lambda_i^{(2)})|M> = 0$ and $\sum_i (\lambda_i)^2 = 16/3$ one finds

$\sum_i \lambda_i^{(1)} \lambda_i^{(2)} = -16/3$ and similarly for baryons $\sum_i \lambda_i^{(k)} \lambda_i^{(\ell)} = -8/3$ $(k \neq \ell)$. Hence the magnetic energy shift takes the form

$$\Delta E_M = \lambda \sum_{k \neq \ell} \vec{\sigma}_k \cdot \vec{\sigma}_\ell \; \mu_{k\ell}$$

where λ is a constant (λ = 1 for baryons and 2 for mesons). We have thus a spin-spin force which splits N-Δ and π-ρ. This sign is the same for mesons and baryons (which is indirect evidence for a non abelian gauge structure). As mentioned before $\mu_{k\ell}$ depends on the wave functions. It turns out in the bag model that $|\mu_{k\ell}|$ is smaller for a (u,s) or (d,s) pair than for equal mass quarks. This then leads to a secondary effect in SU(3) breaking which qualitatively accounts for the Λ-Σ splitting [85] ! It is quite remarkable that in the bag model with essentially two free parameters (g and m_s) a qualitatively correct picture of mass splittings emerges !

The same analysis goes through for the (colour) electric energy shifts. The resulting spectrum of light mesons and baryons can be found in reference [79]. The most serious "problem" with this spectrum concerns the π which comes out much too heavy ($\sim$ 300 Mev). This is of course the usual difficulty in quark models for the π, i.e. how does one take into account its double nature of near Goldstone boson and $\bar{q}q$ bound state ! There are also problems with isoscalar mesons (η,η') due to $\bar{q}q$ annihilation graphs but once again these are not too different from the standard quark model difficulties. Many other applications of the bag have of course been made (current algebra, scaling, etc...). The interested reader is referred to the literature [82].

I believe that it is fair to conclude that the bag model per se is a definite improvement over non relativistic quark models. There are clearly very difficult calculational problems once one tries to go beyond the spherical approximation but it is nevertheless quite interesting to observe that for the ground states finer effects such as mass splittings emerge naturally and are qualitatively correct. This makes it definitely worthwhile to develop reasonable approximations for higher states of the conventional spectrum and to extend the phenomenology to pionic and photonic transitions among these states [86]. It remains to be understood to which extent the bag can be viewed as an approximation to QCD and most important of all to determine experimentally whether the much richer spectrum of hadrons which emerges from the bag is indeed present in nature.

IV. OUTLOOK.

In the previous lectures we reviewed the basic rules of hadron spectroscopy and gave some of their experimental justifications and phenomenological implications. As epitomized in the bag model, hadron spectroscopy remains as fascinating as ever. However it is quite clear that we are on the doorstep of a new domain of "elementary particle" spectroscopy : with the proliferation of leptons and quarks, the similarity of their reponses to weak probes, the apparent relation between their mass scales, there is no doubt that these two classes of fundamental constituents of matter are much more intimately connected than was believed up to now. This new spectroscopy of quarks and leptons will have to account for their number and hopefully lead to an understanding of their relation. The most naive approach would be to view quarks and leptons as composite structures of yet more "fundamental" objects.

Disgusting as this possibility seems it should not be excluded a priori ! Fortunately QED comes to our rescue : the g-2 factors for the electron and muon make this hypothesis highly unlikely : at least the leptons seem to be elementary objects.

A hardly less naive "idea" is to juxtapose the fundamental fermions : leptons and quarks sit next to each other in some representation of some big gauge group which is then carefully broken so that they effectively continue to ignore each other as much as possible. This is the Grand Unification Theory [87]. Despite its naivety it has some predictive power. We will briefly discuss the "build your own" strategy in a moment. But before speculating on where we want to go, it might be wise to recall how we got where we are !

All weak and electromagnetic interactions of our elementary particles seem to be correctly described by the Weinberg-Salam model. As you all know the model is based on the gauge group SU(2) $\otimes$ U(1), all left handed leptons and quarks appear as doublets of the type :

$$\begin{pmatrix} \nu_\ell \\ \ell^- \end{pmatrix}_L \quad \text{for leptons} \quad \begin{pmatrix} q^{2/3} \\ q^{-1/3} \end{pmatrix}_L \quad \text{for each quark colour.}$$

Right handed quarks and charged leptons all correspond to singlets. To avoid the so-called anomalies the number of lepton doublets must be equal to 1/3 of the number of quark doublets (colour being taken into account).

With the benefit of hindsight, the key words which made the transition possible from the Feynmann Gell-Mann theory of β decay to the Weinberg Salam gauge theory of weak and electromagnetic interactions seem to have been calculability, facts and simplicity! They led to the predictions of new currents, new quarks and eventually to the striking analogies between leptons and quarks which hint at their deeper connection.

Where do we stand with respect to some of the facts which we might want to explain or calculate ?

- Parity Violation in Atomic Physics [22].

The experimental absence of parity violating effects in atoms with an upper limit which is now smaller by a factor of 20 than the prediction of the Weinberg-Salam model seems to be a fact. It demands extra currents beyond those of WS : V + A leptonic currents are the simplest possibility.

- No y anomaly in high energy $\bar{\nu}$ scattering [88].

This fact requires essentially that the W bosons which couple to left- and right-handed currents do not mix.

The simplest gauge groups which can take into account both these facts are either $SU(2)_L \otimes SU(2)_R$ or $SU(3) \otimes U(1)$ [89]

- CP Violation [20].

It is well known by now that the introduction of a CP violating term does not put severe restrictions on a gauge theory of weak and electromagnetic interactions : it can be done with more Higgs particles, V + A currents or, most simply, CP violation can be incorporated uniquely in the standard WS model with 6 quarks [90].

- Second class currents.

It is highly non trivial to introduce second class currents in the context of a gauge theory [91]. It is not yet clear that one should worry about the problem since experimental results are inconclusive [92].
It is interesting to note [93] that the question may be settled through some hadronic decay modes of the τ^- such as

$$\tau^- \rightarrow B^- + \nu_\tau \qquad \text{or} \qquad \tau^- \rightarrow \delta^- + \nu_\tau$$

- The $\Delta I = 1/2$ rule.

Two classes of facts in the decay modes of strange particles (with charm = 0) are understood by this rule :
a) non leptonic decays are considerably enhanced (by a factor of 20) compared to semi leptonic ones;
b) in non leptonic modes transitions with $\Delta I = 1/2$ dominate (again by a factor of 20) those with $\Delta I = 3/2$.

The present indications are that for charmed particles the first enhancement may be considerably less important than for strange particles [30] !

I do not know of any convincing explanation of these facts despite the plethora of models which rely either on a "dynamical enhancement" mechanism or on the presence of extra terms in the Hamiltonian. A typical example [21] is to add a term of the form

$$\bar{s}\gamma_\mu(1+\gamma_5)Q\bar{Q}\gamma^\mu(1-\gamma_5)d + \text{h.c.}$$

where Q is some heavy quark (c, b, t, ...). Such a term does have a parity violating component which transforms like λ_7 and hence will allow $K \rightarrow 2\pi$. It does not seem likely though that virtual $\bar{Q}Q$ pairs in the low lying strange mesons and hyperons should play the dominant role in their decay modes !

- Theoretical prejudices.

This is of course a "fact" to be discussed with some care ! Let us look back at our fundamental Lagrangians for QCD and some given gauge theory for weak and electromagnetic interactions. In the quark world, for example, the only parameters we are left with are the quark masses and their mixings in the weak currents (Cabibbo type angles). It does seem quite attractive from a theoretical point of view to make as many of these parameters as possible calculable. Let me illustrate what I mean by a simple example.

Suppose one has a theory where u and d quarks have equal bare masses. Despite the fact that the theory is renormalizable, the u-d mass difference, although finite is in general not calculable. The reason is of course the (infinite) counterterms. One is sometimes able to impose an extra symmetry on the Lagrangian which guarantees that the (infinite) counterterms are equal. Such a symmetry has been called a natural symmetry by Weinberg [78]. More generally a property of the Lagrangian will be called natural if it is independent of a particularly judicious (or sneaky) choice of the parameters of the theory. An example of a natural property is provided by the GIM mechanism : the suppression of the strangeness changing neutral current does not depend on the value of the Cabibbo angle or the parameters of the scalar potential. As will be discussed in detail by J. Ellis if one imposes ab initio that all neutral current be naturally flavour diagonal to order $O(G_F\alpha)$ we are left with very few (in fact two) acceptable gauge models !

Let me elaborate a bit on the u-d mass difference. In the W.S. model their masses are generated by the terms $f_1\bar{u}u\Phi^\circ + f_2\bar{d}d\Phi^\circ$ and in general will be unrelated : $m_u = f_1 <\phi^\circ>$, $m_d = f_2 <\phi^\circ>$. Several discrete symmetries can be imposed which guarantee that $m_u = m_d$ independently of the choice of parameters of the theory. If one calculates $m_u - m_d$ to $O(\alpha)$ it obviously will be positive [94]. One may then ask for which theories $m_u - m_d$ will be naturally negative [95]. The answer is that the theory must have more than two neutral bosons and right handed currents [96] !

These two examples illustrate the power of "naturalness" as well as the fact that it is a matter of choice to decide which properties one wishes to be "natural". Different choices lead to different theories !

Many attempts are made at present to calculate [97] the Cabibbo angle along these lines (e.g. find a theory where $\theta_c \sim m_d/m_s$ is "natural" to order $O(\alpha)$). Naturalness may give us important clues on how to relate quarks and leptons : one may e.g. require that lepton and quark masses (or mass scales) be naturally related !

Anyway it should be clear from all these facts that in the context of our present ideas, the so-called anomalies and naturalness are the only theoretical connections that we can establish between leptons and quarks.
Brilliant ideas are urgently needed !

Let me end by sketching how to construct grand unified theories and illustrate in the simplest example (SU(5)) what some of their consequences are.

In the mid thirties stiking analogies could have been observed between the known leptons and the "elementary" hadrons : indeed with the two doublets $\begin{pmatrix}\nu_e \\ e\end{pmatrix}$ $\begin{pmatrix}P \\ N\end{pmatrix}$ and the benefit of hindsight, it is remarkable to note that :

- leptons and nucleons respond similarly to the weak currents;
- they are all pointlike and have J = 1/2;
- their mass differences are obviously of electromagnetic origin;
- they know about each other and conspire to cancel the anomalies in the gauge theory of their weak and electromagnetic interactions !

Forty years later the same observations can be made with however some new features

- quarks have replaced nucleons;
- there are several generations of leptons and quarks and the generation gaps, on the quark side, are filled by flavour mixing (Cabibbo angles);
- there are additional analogies in masses and mass scales.

Since the connections between the fundamental fermions cannot be dismissed as accidental, one may try to put them together in representations of some bigger group which would unify all interactions. Such an approach has been extremely successful for the quarks but in the present case it appears as extremely naive and artificial ! Anyway it does have some general structural consequences which are worth pointing out.

The basic strategy to build your own Grand Unified Theory is quite easy to follow :

1) pick a gauge group which contains
   - SU(2) x U(1) or a suitable extension there of as gauge group of the weak and electromagnetic interactions;
   - $SU(3)_c$ or ... (?) as gauge group of the strong interactions.

2) put the fermions (quarks and leptons) in one or a few representations of the "big" group;

3) turn the spontaneous symmetry breaking machinery on to reduce effectively the Grand Unified group to a direct product of the gauge groups chosen for the basic interactions (i.e. throw in as many Higgs particles as you need !).

In all such theories which are based on "big" groups which are not direct products of the weak and strong gauge groups

- the additive quantum numbers B and L are not simultaneously conserved. In particular the stability of the proton seems "accidental" [98];
- one predicts "leptoquarks", i.e. bosons which transform leptons into quarks and vice versa. These objects will dramatically alter our expectations for lepton-hadron collisions at sufficiently large energies [99];
- all interaction couplings constants are related by Clebsch-Gordon coefficients;
- the Weinberg "angle " is predicted;

Let me illustrate these remarks in the simplest Grand Unification model. The Grand Unified group must be at least of rank four since it contains $SU(3)_c \otimes SU(2) \otimes U(1)$. If it is not a direct product, the simplest candidate is SU(5) which was considered by Georgi and Glashow [87]. One obtains easily the decomposition of the basic SU(5) representation under $SU(3)_c \otimes SU(2)$

$$\underset{\sim}{5} = (\underset{\sim}{3},1) + (\underset{\sim}{1},2)$$

and, from

$$\underset{\sim}{5} \times \underset{\sim}{5} = \underset{\sim}{15} + \underset{\sim}{10} \qquad \underset{\sim}{5} \times \underset{\sim}{\bar{5}} = \underset{\sim}{24} + \underset{\sim}{1},$$

$$\underset{\sim}{10} = (\underset{\sim}{\bar{3}},1) + (\underset{\sim}{1},1) + (\underset{\sim}{3},2)$$

$$\underset{\sim}{15} = (\underset{\sim}{6},1) + (\underset{\sim}{1},3) + (\underset{\sim}{3},2)$$

$$\underset{\sim}{24} = (\underset{\sim}{8},1) + (\underset{\sim}{1},1) + (\underset{\sim}{1},3) + (\underset{\sim}{3},1) + (\underset{\sim}{3},2)$$

this last representation, which corresponds to the gauge fields, thus contains 12 leptoquarks in addition to the 8 coloured gluons and the 4 weak and electromagnetic bosons.
Each "fermion generation" neatly corresponds to the representations

$$\underset{\sim}{10} + \underset{\sim}{\bar{5}} \quad \text{with} \quad \underset{\sim}{\bar{5}} : \bar{d}_L + (\nu_e, e^-)_L$$

$$\underset{\sim}{10} : \bar{u}_L + e^+_L + (u,g)_L$$

The unrenormalized Weinberg angle is predicted to be $\sin^2{}_W = 3/8$. With a particular choice of Higgs representations one finds mass relations of the form

$$\frac{m_e}{m_d} \approx \frac{m_\mu}{m_s} \approx \frac{m_\tau}{m_b}$$

as discussed in the lectures by J. Ellis.

I leave it to your appreciation to decide whether one has gained much insight into the fundamental fermions'world !

APPENDIX A : P_n, SU(n) and Young Tableaux.

A Young Tableau is a set of boxes legally arranged i.e.
- the tableau has no "holes"
- the number of boxes per line does not increase from top to bottom of the tableau.

Boxes in the same line or in the same column are to be symmetrized or antisymmetrized respectively.
For the permutation group of n elements (P_n) there is a one to one correspondance between Young tableaux with n boxes and irreducible representations of P_n. For example P_2 has two irreducible representations

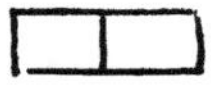 (symmetric)

 (antisymmetric)

P_3 has three irreducible representations.

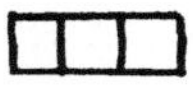 (symmetric)

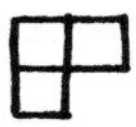 (mixed)

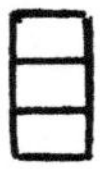 (antisymmetric)

Similarly P_4 has five irreducible representations etc...
For the special unitary groups (SU(n)), the rules are equally simple
- a single box corresponds to the fundamental representation of SU(n) (dimension n);
- with the symmetrization and antisymmetrization convention defined earlier it is clear that a Young tableau corresponding to a representation of SU(n) may never contain more than n boxes per column. Furthermore since on a vector space of dimension N, there is one and only one totally antisymmetric tensor built out of N vectors, it is obvious that a "full column" i.e. a column with exactly n boxes corresponds to the singlet representation (1) of SU(n).

Two Young tableaux T_1 and T_2 are said to be equivalent for SU(n) ($T_1 \approx T_2$) if they differ by full columns (i.e. with n elements). There is a one to one correspondance between equivalence classes of tableaux and irreducible representations of SU(n).

For example in SU(4)

$\approx$

Consider a Young Tableau T_1 which corresponds to an irreducible representation R_1 of SU(n). Draw the diagram complementary to T_1 (i.e. the diagram which makes full columns out of the columns of T_1) and rotate it by 180°:it becomes a legal young Tableau $\overline{T}_1$,and corresponds to the complex conjugate representation $\overline{R}_1$ of Su(n).In particular, since the Young Tableau which contains a single box corresponds to the fundamental representation, n, of SU(n) (dimension n), the tableau $\overline{T}_1$ which corresponds to the representation $\bar{n}$ will contain n-1 boxes in a single column.

For example in SU(3)  $\leftrightarrow$ $\underset{\sim}{3}$ and  $\leftrightarrow$ $\underset{\sim}{\bar{3}}$

while in SU(4)  $\leftrightarrow$ $\underset{\sim}{4}$ and  $\leftrightarrow$ $\underset{\sim}{\bar{4}}$

It is now a simple exercise to "prove" from the preceeding rules that to the adjoint representation of SU(n) (of dimension n^2-1) corresponds the equivalence class of the Young tableau which has two incomplete columns with n-1 boxes in the first and one box in the second. This representation is of course self conjugate.

For example, in SU(3) $\underset{\sim}{8}$ $\leftrightarrow$  , while in SU(4) 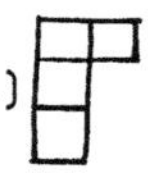 $\leftrightarrow$ $\underset{\sim}{15}$

The dimension d of the irreducible representation of SU(n) associated with a given Young tableau is given by :

$$d = \frac{N}{D}$$

where, to calculate the numerator N, one fills each box of the tableau with a number according to the following prescription : n in each box of the main diagonal, n+1 (resp. n-1) in each box of the diagonal just above (resp. below) it, n+2 (resp. n-2) in each box of the next higher (resp. lower) diagonal and so on ...

N is the product of all these numbers.

For example

n	n+1	n+2
n-1	n	n+1
n-2		

$$N = (n-2)(n-1)n^2(n+1)^2(n+2)$$

On the other hand, the denominator **D** is the product of the length l_i of the hooks centered on each box i of the diagram

$$\text{i.e.} \quad D = \prod_{\text{all boxes}} l_i$$

where l_i is computed as follows : draw from the center of box i a horizontal and vertical line which go respectively to the right end and to the bottom end of the diagram. The "length" of this hook is the number of boxes which it crosses :

e.g. for the diagram :

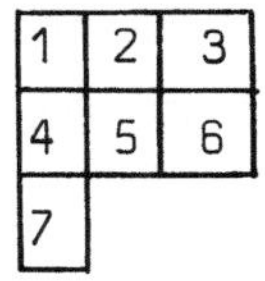

one obtains :

$l_7 = 1;\ l_6 = 1;\ l_5 = 2;\ l_4 = 4;\ l_3 = 2;\ l_2 = 3$

and $l_1 = 5$.

For P_n, the following formula holds

$$d(P_n) = \frac{n!}{D}$$

with D defined as before.
To give but two examples, one finds thus that

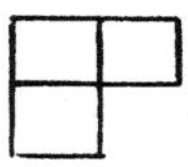

$\rightarrow$ for SU(2) : $d = \frac{1.2.3}{1.1.3} = 2$

for SU(3) : $d = \frac{2.3.4}{3} = 8$

for SU(6) : $d = \frac{5.6.7}{3} = 70$

for P_3 : $d = \frac{3!}{3} = 2$

and similarly

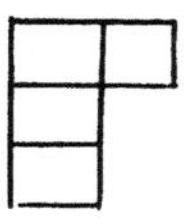

$\rightarrow$ for SU(2) : $d = \frac{0.1.2.3}{1.1.2.4} = 0$

for SU(4) : $d = \frac{2.3.4.5}{8} = 15$

for P_4 : $d = \frac{4!}{8} = 3$

The reader may find it amusing to check the following theorem : "The number of elements of P_n is equal to the sum of the squares of the dimensions of its irreducible representations" !

It is often of some interest to know which representations of SU(n-1) are contained in an erreducible representation of SU(n) associated to a Young tableau with p_1 boxes on the first line, p_2 boxes on the second etc... : they correspond to all Young tableaux with l_1 boxes on the first line, l_2 on the second line etc... provided,

$$p_n \le l_{n-1} \le \;\dots\; l_3 \le p_2 \le l_1 \le p_1$$

Finally as a last application of Young tableaux, let me recall the rules for decomposing the product of two irreducible representations of SU(n) given by their Young tableaux T_1 and T_2.

1. label by a each box on the first line of T_2, by b each box on the second line and so on;
2. add to T_1 in all possible ways all boxes of T_2. The irreducible representations which appear in the decomposition of the direct product $T_1 \times T_2$ are all those which correspond to the lexicographically distinct Young tableaux obtained by such a construction provided that

   - the resulting tableau is legal;
   - going from right to left on the first line, then from right to left on the second line ... at each moment along this path, the number of boxes labelled by a is greater or equal to the number of boxes labelled by b which it self is greater or equal to the number of boxes labelled by c ... (rule I);
   - from left to right on each line, lexicographical order must be respected; (rule II);
   - from top to bottom, all boxes must be lexicographically distinct and strict lexicographical order must be respected !
     (rule III).

As an example (in SU(3))

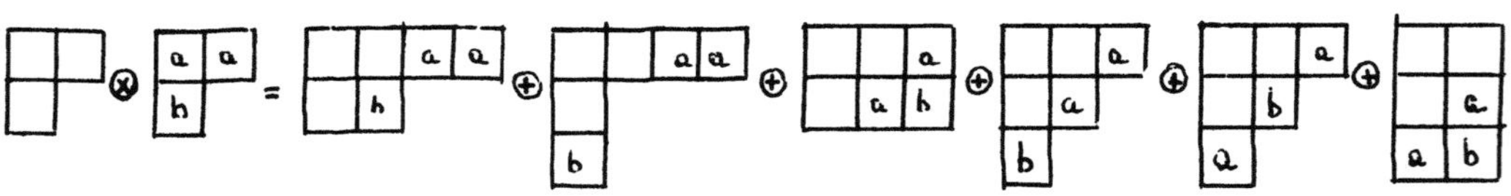

$$\underset{\sim}{8} \otimes \underset{\sim}{8} = \underset{\sim}{27} \oplus \underset{\sim}{10} \oplus \underset{\sim}{\overline{10}} \oplus \underset{\sim}{8} \oplus \underset{\sim}{8'} \oplus \underset{\sim}{1}$$

The following tableaux e.g. violate one of the three rules

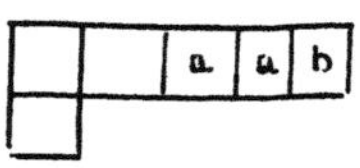

violates rule I

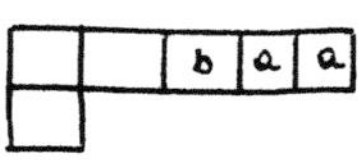

violates rule II

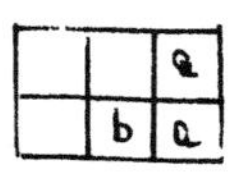

violates rule III

For proofs of most assertions made in this Appendix see, e.g. reference 100.

REFERENCES and FOOTNOTES

1) For a review see the lectures by J. Ellis, L. Lederman, C. Llewellyn-Smith, J. Steinberger and B. Wiik, these Proceedings and references quoted therein.

2) C.N. Yang and R.J. Mills, Phys. Rev. 96, 191 (1954).
G. 't Hooft, Nuclear Phys. B35, 167 (1971).
S.L. Glashow, J. Iliopoulos and L. Maiani, Phys. Rev. D2, 1285 (1970).

3) We assume that the heavy lepton τ^- is accompanied by its own neutrino. For a discussion of this point see the lectures by J. Ellis and B. Wiik in these Proceedings. The discovery of the τ^- is due to M. Perl et al., Phys. Rev. Letters 35, 1489 (1975).

4) We assume of course that the new quark which builds up the Υ(9.6) appears in three colours. For the discovery of the Υ(9.6) see the lectures by L. Lederman in these Proceedings.

5) H. Fritzsch, M. Gell-Mann, and H. Leutwyler, Phys. Lett. B47, 365 (1973).
D.J. Gross and F. Wilczek, Phys. Rev. D8, 3497 (1973).
S. Weinberg, Phys. Rev. Letters 31, 494 (1973).

6) For a review of the problem and possible solutions see, e.g., F. Englert, lectures in these Proceedings;
D. Gross, lectures in these Proceedings.

7) S. Weinberg, Phys. Rev. Letters 19, 1264 (1967).
A. Salam, Proc. 8th Nobel Symposium, Stockholm 1968 (ed. N. Svartholm).

8) R. Brout and F. Englert, Phys. Rev. Letters 13, 321 (1964).
P.W. Higgs, Phys. Rev. Letters 12, 132 (1964).
G.S. Guralnik, C.R. Hagen and T.W.B. Kibble, Phys. Rev. Letters 13, 585 (1964).

9) I ignore the so called "dynamical" symmetry breaking mechanism where the elementary scalar particle of reference 8) is replaced by a bound state. To the best of my knowledge no coherent formulation of this appealing idea exists as yet.

10) J. Wess and B. Zumino, Nucl. Phys. B70, 39 (1974); B78, 1 (1974); Phys. Letters 49B, 52 (1974).

11) For a recent review, see e.g. P. Fayet et S. Ferrara, Phys. Reports 32C, 249 (1977).

12) The problem of incorporating a U(n) internal symmetry may have been solved by M. Kaku, P.K. Townsend and P. Van Nieuwenhuizen Phys. Rev. Letters 39, 1109 (1977).

13) The only theoretical relation between leptons and quarks, at

present, is through the so-calles "anomaly" of
S.L. Adler, Phys. Rev. 177, 2426 (1969).
J.S. Bell and R. Jackiw, Nuovo Cimento 51A, 47 (1969).

14) For recent reviews see e.g.
H. Fritzsch Lectures given at the International Summer Institute for Theoretical Physics, Bielefeld, Germany (1976), to be published.
M.S. Chanowitz, J. Ellis and M.K. Gaillard, CERN preprint TH 2312 (1977) and references quoted therein.

15) See e.g. K. Igi FNAL preprint (1977).

16) J.Goldstone,Nuovo Cimento 19, 15 (1961).

17) S. Weinberg, Phys. Rev. D7, 2887 (1973).

18) For an extremely lucid discussion see R.J. Crewther, CERN preprint TH 2350 (1977).

19) G. Zweig, unpublished.
J. Iizuka, Supplement to Progress of Theoret. Phys. 37-38, 21 (1966).
S. Okubo, Phys. Lett. 4, 14 (1963).

20) See e.g. the lectures by J. Ellis, these Proceedings where references can be found.

21) See e.g. B.W. Lee in Proceedings of the International Symposium on Lepton and Photon Interactions at High Energies, W.T. Kirk Editor, Stanford (1975), and references quoted therein.

22) P. Baird et al., Nature 264, 528 (1976).
P. Sandars, Proceedings of the International Symposium on Lepton and Photon Interactions at High Energies, Hamburg 1977.
See also V. Telegdi, lectures in these Proceedings.

23) B.C. Barish et al., Phys. Rev. Letters 38, 577 (1977).
A. Benvenuti et al., Phys. Rev. Letters 38, 1110 (1977).
See also J. Steinberger, lectures in these Proceedings.

24) G. Zweig, CERN preprints TH 401 and 412 (1964) unpublished.
M. Gell-Mann, Phys. Letters 8, 214 (1964).
R.H. Dalitz, Les Houches Lectures (1965) (Gordon and Breach, N.Y. 1965).

25) For a review see e;g. J.L. Rosner, Physics Reports 11C (1974) and references quoted therein.

26) O.W. Greenberg, Phys. Rev. Letters 13, 598 (1964).
M.Y. Han and Y. Nambu, Phys. Rev. 139 B 1006 (1963).
M. Gell-Mann, Quarks CERN preprint TH 1543 (1972).

27) H.J. Melosh, Phys. Rev. D9 1095 (1974).
See also reference 25).

28) See e.g. the lectures by C. Llewellyn Smith, these Proceedings.

29) J.J. Aubert et al., Phys. Rev. Letters 33, 1404 (1974).
J.E. Augustin et al., Phys. Rev. Letters 33, 1406 (1977).

30) See e.g. the lectures by B. Wiik, these Proceedings.

31) H. Harari, Phys. Letters 60B, 172 (1976).

32) E. Eichten et al., Phys. Rev. Letters 34, 369 (1975)
T. Applequist et al., Phys. Rev. Letters 34, 365 (1975).

33) See e.g. J.L. Rosner, Physics Reports 11C (1974).
A.J.G. Hey and D. Morgan, Rutherford Report RL-77-060/A (1977).
K. Lanius in Proceedings of the XVIIIth International Conference on High Energy Physics Tbilisi (1976).

34) G.F. Chew, LBL-5391 report (1976).

35) D. Horn and J.E. Mandula, Caltech preprint (1977).
M. Bander, G.L. Shaw, P. Thomas and S. Meshkov, Phys. Rev. Letters 36, 695 (1976).

36) See e.g. J.L. Rosner, Physics Reports 11C (1974).

37) R.H. Dalitz, Oxford Conference on Baryon Resonances (1976).
D. Faiman, J. Rosner and J. Weyers, Nuclear Phys. B57, 45 (1973).

38) R.H. Dalitz, Les Houches Lectures (1965)(Gordon and Breach, N.Y. 1965).
M. Jones, R.H. Dalitz and R.R. Horgan, Nuclear Phys. B129, 45 (1977).

39) M.K. Gaillard, B.W. Lee and J.L. Rosner, Rev. of Mod. Phys. 47, 277 (1975).

40) E.G. Gazzoli et al., Phys. Rev. Letters 34, 1125 (1975).
See also the lectures by J. Ellis for further references.

41) See e.g. A.J.G. Hey and D. Morgan reference 33).

42) W.A. Bardeen, H. Fritzsch and M. Gell-Mann in "Scale and Donformal Symmetry in Hadron Physics", J. Wiley and Sons, New York,(1973).

43) It is worth pointing out that we have no information concerning the spin structure of the neutral stuff !

44) A. Chodos, R.L. Jaffe, K. Johnson, C.B. Thorn, V.F. Weisskopf Phys. Rev. D9, 3471 (1974).
A. Chodos, R.L. Jaffe, K. Johnson, C.B. Thorn, Phys. Rev. D10, 2599 (1974).
A. Chodos, R.L. Jaffe, K. Johnson, J.Kiskis, Phys. Rev. D12, 2060 (1975).

P. Gnädig, P. Hasenfratz, J. Kuti, A.S. Szalay, Phys. Lett. Letters 64B, 62 (1976).

45) G. Preparata and N. Craigie, Nuclear Physics B102, 468 (1976).
N. Craigie and G. Preparata, Nucl. Phys. B102, 497 (1976).

46) P.G.O. Freund and Y. Nambu, Phys. Rev. Letters 34, 1645 (1975).
R.L. Jaffe, K. Johnson, Phys. Letters 60B, 201 (1976).

47) For a derivation of this formula, based on duality see e.g. M. Ademollo, G. Veneziano, S. Weinberg, Phys. Rev. Letters 22, 83 (1969).
R. Brout, F. Englert and C. Truffin, Phys. Letters 29B, 590, 686 (1969).

48) See e.g. V. Barger in Proceedings of the XVIIth International Conference on High Energy Physics, London (1974).

49) H. Lipkin and S. Meshkov, Phys. Rev. Letters 14, 670 (1965).

50) See e.g. U. Karshon et al. Weizmann Institute report WIS 73/44 (1974).

51) See e.g. F. Gilman, M. Kugler and S. Meshkov, Phys. Rev. D9 715 (1974).
A.J.G. Hey, J.L. Rosner and J. Weyers, Nucl. Phys. 44B, 263 (1973).
and references quoted by J.L. Rosner 25).

52) A. Love and D.V. Nanopoulos, Phys. Lett. 45B 507 (1973),
F.J. Gilman and I. Karliner, Phys. Rev. D10, 2194 (1974),
A.J.G. Hey and J. Weyers, Phys. Letters 40B, 69 (1974).

53) H.J. Lipkin, Phys. Reports 8C (1973) pointed out that these parametrizations correspond to the most general single quark transition operator with appropriate quantum numbers.

54) See e.g. D. Faiman, J.L. Rosner, Phys. Letters 45B, 357 (1973).

55) D. Faiman and D.E. Plane, Nuclear Phys. B50, 379 (1972)

56) A.J.G. Hey, P.J. Litchfield and R.J. Cashmore, Nucl. Phys. B95, 516 (1975) and Nucl. Phys. B98, 237 (1975).

57) G. Feldman, SLAC PUB (1977).

58) H. Harari in Proceedings of the International Symposium on Lepton and Photon Interaction at High Energies, W.T. Kirk Editor, Stanford (1975).

59) R.P. Feynman and M. Gell-Mann, Phys. Rev. 109, 193 (1958).

60) M. Gell-Mann, Phys. Rev. 125, 1067 (1962).

61) M. Gell-Mann and M. Levy, Nuovo Cimento 16, 705 (1960).
Y. Nambu, Phys. Rev. Letters 4, 380 (1960).

62) S.L. Adler, Phys. Rev. Letters 14, 1051 (1965).
W.I. Weisberger, Phys. Rev. Letters 14, 1047 (1965).

63) S. Weinberg, Phys. Rev. D11, 3583 (1975).

64) G. 't Hooft, Phys. Rev. D14, 3432 (1976).
See, however, reference 18).

65) M. Gell-Mann, R. Oakes and B. Renner, Phys. Rev. 175, 2195 (1968).
R. Carlitz, D. Heckathorn, J. Kaur and W.K. Tung, Phys. Rev. D11, 1234 (1975).

66) N. Fuchs, Purdue University Preprint (1976).

67) See e.g. D.C. Moir, R.J. Jacob and G.E. Hite, Nucl. Phys. B 103, 477 (1976).
P. Gensini, SLAC PUB, 1944-1945-1967 (1977).

68) J.F. Gunion, P.C. Mc Namee and M.D. Scadron, SLAC-PUB 1847 (1976).

69) J. Jersak and J. Stern, Nucl. Phys. B7, 413 (1968),
K. Bardakei and M.B. Halpern, Phys. Rev. 176, 1686 (1968).

70) See e.g. H. Leutwyler Springer Tracts in Modern Physics 50, 29 (1969).
H. Sazdjian and J. Stern, Nucl. Phys. B94, 163 (1975).
J. Kogut and D. Soper, Phys. Rev. D1, 2901 (1970).

71) M. Gell-Mann in Strong and Weak Interactions - Present Problems Academic Press, Inc. New York (1966).
K. Bardakei and G. Segré, Phys. Rev. 153 (1967).
S. Fubini and G. Furlan, Physics, 1, 299 (1965).

72) R. Dashen and M. Gell-Mann, Phys. Letters 17, 142 (1965).

73) See e.g. J.S. Bell, Acta Phys. Austriaca, Suppl. 13, 395 (1974).
and references quoted therein.

74) H. Leutwyler, Phys. Letters 48B, 431 (1974).

75) See e.g. T. Das et al., Phys. Rev. Letters 18, 759 (1967).
S.L. Glashow and S. Weinberg, Phys. Rev. Letters 20, 224 (1968).

76) R. Dashen, Phys. Rev. 183, 1245 (1969).
See also P. Langacker and H. Pagels, Phys. Rev. D8, 4620 (1973).

77) G. Segrè and J. Weyers, Phys. Letters 62B, 91 (1976).

78) S. Weinberg, Phys. Rev. Letters : see also Ref. 17).

79) J. Kuti, Lectures given at the CERN-JINR School of Physics (1977).

80) P.N. Bogoliukov, Ann. Inst. Poincaré 8, 163 (1967).

81) C. Rebbi, Phys. Rev. D14, 2362 (1976).

82) R.L. Jaffe, Oxford Conference on Baryon Resonances (1976).
K. Johnson, Scottish Universities, Summer School (1976).
and references quoted therein.

83) By this we mean a resonance with J^{PC} unaccessible to a $\bar{Q}Q$ system.

84) See e.g. J. Kuti, reference 79 and
A.J.G. Hey, Phenomenological Aspects of the MIT Bag (Southampton 1977).

85) T. Applequist et al., Phys. Rev. Letters 34, 365 (1975).
T.A. De Grand, R.L. Jaffe, K. Johnson and J. Kiskis, Phys. Rev. D12, 2060 (1975).

86) T.A. De Grand, Ann. Phys. 101, 496 (1976).
A.J.G. Hey, B.R. Holstein and D.P. Sidhu (Southampton 1977).

87) J. Pati and A. Salam, Phys. Rev. D8, 1240 (1973).
H. Georgi and S.L. Glashow, Phys. Rev. Letters 32, 438 (1974).
H. Fritzsch and P. Minkowski, Ann. Phys. 93, 222 (1975).
F. Gürsey and P. Sikine, Phys. Letters 60B, 177 (1976) and
Phys. Rev. Letters 36, 775 (1976).
M. Abud, F. Buccella, H. Ruegg and C.A. Savoy, Phys. Letters 67B, 313 (1977).
M. Yoshimura, preprint (1977).
K. Inoue et al., preprint (1977).

88) J. Steinberger, lectures in these Proceedings.

For $SU(2)_L \otimes SU(2)_R$ see e.g.

H. Fritzsch and P. Minkowski, Nucl. Phys. B103, 61 (1976).
R.N. Mohapatra and J.C. Pati, Phys. Rev. D11, 566, 2558 (1975).
A. de Rujula, H. Georgi and S.L. Glashow, HUTP-77/A029 (1977).

89) For $SU(3) \otimes U(1)$ see e.g.

G. Segré and J. Weyers, Phys. Letters 63B, 99 (1977).
B.W. Lee and S. Weinberg, Phys. Rev. Letters 38, 1237 (1977).
P. Langacker and G. Segré, Phys. Rev. Letters 39, 259 (1977), and references quoted therein.

90) M. Kobayashi and K. Maskawa, Progress of Theoret. Phys. 49, 652 (1973).

91) See e.g. P. Langacker, Phys. Rev. D15, 2386 (1977).

92) V. Telegdi, Lectures in these Proceedings.

93) C. Leroy and J. Pestieau, Louvain Preprint (1977).

94) See e.g. A. Duncan, P. Schattner, Phys. Rev. D7, 1861 (1973).
D.Z. Freedam, W. Kummer, Phys. Rev. D7, 1829 (1973).

The reason is of course clear : the mass of the Z° simply plays the role of the cut-off in the Feynman-Speisman calculation. (Phys. Rev. 94, 500 (1954).

95) S.Y. Pi, Phys. Rev. D7, 3750 (1973).
J. Lieberman, Phys. Rev. D9, 1749 (1974).

96) M. de Crombrugghe, H. Haut, J. Weyers, Phys. Letters 71B, 400 (1977).

97) S. Weinberg, quoted by H. Fritzsch, next paper.
H. Fritzsch, Phys. Letters 70B, 436 (1977).
A. de Rujula, H. Georgi, S.L. Glashow HUTP 77/A029 (1977).

98) The usual procedure to guarantee the proton's stability is through superheavy bosons. This is probably not a self consistent procedure (see ref. 99). An interesting alternative is to make the proton absolutely stable by a global U(1) symmetry. See P. Langacker, G. Segré and A. Weldon, UPR-0082 T, (1977) and references quoted therein.

99) Mass ratios of the gauge bosons cannot be made arbitrarily large as shown by E. Gildener, Phys. Rev. D14, 1667 (1976).

100) e.g. M. Hamermesh, "Group Theory and its Applications to Physical Problems" (Addison-Wesley, 1962).

DEEP INELASTIC PHENOMENA

C.H. Llewellyn Smith

Department of Theoretical Physics

1, Keble Rd - Oxford OX1 3NP

INTRODUCTION

These lectures were supposed to serve as an introduction to the parton model and ideas associated with asymptotic freedom which could then be used freely by later speakers. To keep the attention of those to whom it was already familiar, I decided that while introducing the parton model, I would attempt simultaneously to review all its applications to processes involving leptons. It turned out that the introduction was probably too compressed for the uninitiated and the review covered too many topics too quickly to be followed easily. However, I hope that the written version of this "introductory review", which is contained in part I of these notes, will prove useful.

Part II is devoted to asymptotic freedom, which forms the basis for the parton model in my opinion. As well as the standard formal approach using the operator product expansion, I have discussed the lessons which can be learned from perturbation theory in some detail. This reveals the physics more clearly than the formal approach, in my view. Furthermore perturbation theory is an essential laboratory for studying whether asymptotic freedom can be used to make predictions for processes about which the operator product expansion yields no information. Some of these applications are discussed briefly in the last sections of part II (this subject is developing extremely rapidly and I have included some work done since the Cargèse school ended).

A. Classical Applications of the Parton Model [1)].

1. Preliminaries.

In the classical deep inelastic processes $e(\mu)N \rightarrow e(\mu)$ ... and $\nu N \rightarrow \mu$ ..., illustrated in fig. 1, the lepton vertices are well known and one particle exchange is a good approximation at energies now available. These processes therefore measure total current-nucleon cross-sections. The currents can have helicity +1, −1, 0 and we denote the corresponding cross-sections by σ_R, σ_L and σ_0; in the electromagnetic case $\sigma_R = \sigma_L = \sigma_T$ since parity is conserved. These cross-sections (which are defined precisely e.g. in refs 1 and 2, where details of the kinematics may be found) are functions of the variables ν and Q^2 defined in fig. 1. The third classical process is $e^+e^- \rightarrow$ hadrons, where the total cross-section depends only on q^2.

The dogma of the parton model is that at large ν and Q^2 currents scatter incoherently off point-like constituents of the nucleon, called partons, which behave "as if" free during the interaction. It is assumed that final state interactions can be neglected in calculating the total cross-section. This is illustrated in fig. 2.

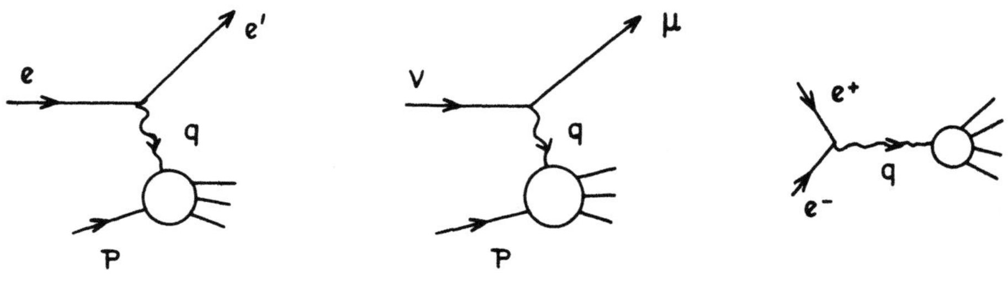

Measure : $\sigma_{T,0}^{\gamma^*}(\nu, Q^2)$ $\sigma_{L,R,0}^{W}(\nu, Q^2)$ $\sigma_{e^+e^-}(q^2)$

[Variables : $\nu = q \cdot p$, $Q^2 = -q^2 = \vec{q}^{\,2} - q_0^2$]

FIG. 1 : The classical deep inelastic processes.

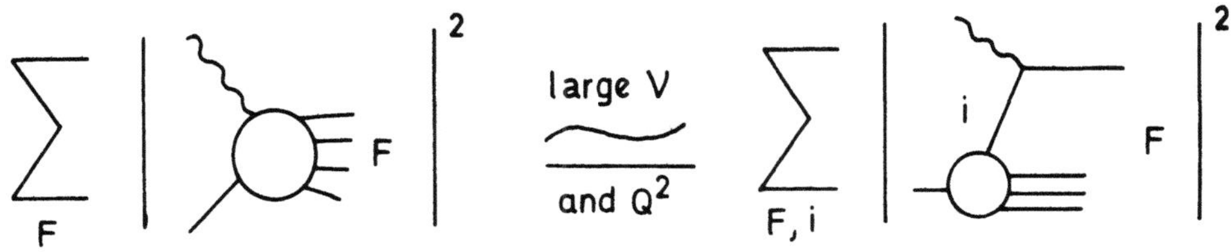

FIG. 2 : Parton model (impulse approximation) diagrams.

The physics behind this dogma depends on the fact that the time τ which controls deep inelastic scattering is of order $1/\sqrt{Q^2}$ (technically $\sigma_{\gamma^* p}$ is proportional to the imaginary part of the forward amplitude for $\gamma^* p \to \gamma^* p$; if γ^* is absorbed at x_μ and re-emitted at y_μ then deep inelastic scattering probes this amplitude in the limit $|x_\mu - y_\mu| \to 0$).
It might be argued that
a) As $Q^2 \to \infty$, the strong interactions have no time to rearrange the constituents during the scattering processes and the current "sees" a frozen state of non-interacting partons. This proposition is false in renormalizable field theories as we shall see in part II.

b) The final state interactions, which must play an essential role in the formation of the final state if partons are confined quarks, act on a time scale $\tau' \gg \tau$. They may therefore be neglected in calculating the total cross-section. This proposition might be true e.g. it is true in two dimensional QCD, a super renormalizable model with exact confinement in which the parton model works [3], and there is some truth in it in non-relativistic potential models [4]. A classical analogy may help : consider a particle attached to a point by an elastic band which is initially limp. When the particle is struck, it responds as a free particle. Later the band becomes taut and restrains its. However, this is irrelevant in calculating the total cross-section which is proportional to the probability that something happens. Once the particle moves off something has happened so the total cross-section is determined by the initial free particle response.

Undaunted by the fact that one part of the dogma fails in renormalizable field theories, let us move on to explore its consequences.

2. Calculations.

Let the momentum $\vec{p}_i$ of the ith parton be $\vec{p}_i = x\vec{p} + \vec{p}_T$, where $\vec{p}$ is the target momentum and $\vec{p} \cdot \vec{p}_T = 0$. We will use the frame

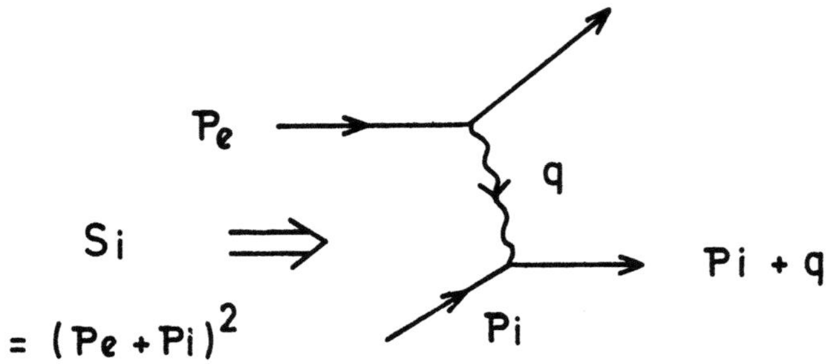

FIG. 3 : Kinematics in electron-parton scattering.

in which $q_0 = 0$, where $|\vec{p}| \simeq \nu\sqrt{Q^2} \to \infty$ in the Bjorken limit ν, $Q^2 \to \infty$ with $x = Q^2/2\nu$ fixed. According to the dogma, $|\vec{p}_T|$ will be fixed independent of ν and Q^2. Thus $p^i_\mu \simeq xp_\mu$ in the Bjorken limit. The ep cross-section is then

$$\frac{d^2\sigma}{dQ^2 d\nu} = \sum_i \int dx f_i(x) \frac{d^2\sigma_i}{dQ^2 d\nu} , \tag{1}$$

where $f_i(x)$ is the probability of finding a parton of type i with fraction x of the momentum.

The elastic e-parton cross-section $d\sigma_i$ may be written

$$\frac{d\sigma_i}{dQ^2} = \frac{4\pi\alpha^2 Q_i^2}{Q^4} g_i(Q^2/s_i), \tag{2}$$

where $Q_i|e|$ is the parton charge and we have assumed that Q^2 and $s_i = (p_e + p_i)^2 \simeq 2p_e.p \simeq xs$ are large (see fig. 3 for the kinematics), so that the parton mass can safely be neglected in Born approximation. Eq. 2 then follows from dimensional analysis. The function g_i, which depends on the spin of the parton i is easy to calculate and is given by

$$\begin{aligned} &[1 + (1 - Q^2/s_i)^2] / 2 \quad &&\text{for spin } 1/2, \\ &(1 - Q^2/s_i) &&\text{for spin } 0. \end{aligned} \tag{3}$$

The dependence on $Q^2/s_i = \dfrac{1-\cos\theta^{ei}_{c.m.}}{2}$ will be explained later (note the normalization to the Rutherford cross-section as $Q^2 \to 0$ in both cases).

For elastic scattering $(q + p_i)^2 = p_i^2$ or $Q^2 = 2q \cdot p_i \simeq 2x\nu$. Thus we may write

$$\frac{d^2\sigma_i}{dQ^2 d\nu} = \frac{d\sigma_i}{dQ^2} \; \delta(\nu - Q^2/2x) \tag{4}$$

which may be substitued into eqn 1 to give

$$\frac{d^2\sigma}{dQ^2 d\nu} = \sum_i f_i(x) \frac{2x^2}{Q^2} \frac{d\sigma_i}{dQ^2} \tag{5}$$

with $x = Q^2/2\nu$.

Comparison with the general form

$$\frac{d^2\sigma}{dQ^2 d\nu} = \frac{(\nu - Q^2/2)}{4\pi^2\alpha} \frac{4\pi\alpha^2}{Q^2\nu^2} \left[(1-y)\sigma_s + \frac{[1+(1-y)^2]}{2} \sigma_r\right]$$

$$= 4\pi\alpha^2/Q^4\nu[(1-y)F_2 + xy^2F_1] \tag{6}$$

(valid for $\nu^2 >> M^2Q^2$; $y \equiv 2\nu/s\text{-}M^2 \simeq Q^2/s_i$, at high energy) yields the parton model formulae for σ_Q and σ_T, or equivalently the dimensionless structure functions F_1 and F_2, in terms of Q_i^2 and g_i.

3. Results for ep → e + ...

a) Since there are no parameters with dimensions in the parton model calculations, they necessarily yield

$$\nu_{Q,T}(\nu, Q^2) = \frac{1}{Q^2} \; \Sigma_{Q.T}(x)$$

or equivalently, that F_1 and F_2 depend only on x.
Theoretically this Bjorken scaling is far from obvious. In general it is certainly not true that masses etc... can be neglected at high energy and that cross-sections depend only on the large dynamical variables with dimensions. This would imply (e.g.) the wrong result that $\sigma_{pp}^{tot} \sim 1/s$.
In this case the radius of the proton is another quantity with dimensions which is obviously relevant and could be used in the construction of σ_{pp}^{tot}. However, the distances which are important in $Im(\gamma^* p \to \gamma^* p) \propto \sigma_{\gamma^* p}$ decrease like $1/\sqrt{Q^2}$ and it might therefore be thought that $\sigma_{\gamma^* p}$ could exhibit Bjorken scaling. But in fact, as everyone who has calculated non-trivial Feynman diagrams knows,

logarithmic divergences occur in the limit $m \to 0$. This implies logarithmic violations of Bjorken scaling in renormalizable theories in perturbation theory. We will return to this in part II.

Experimentally it was thought at one time that scaling might turn out to be exact. Some of the SLAC-MIT data which led to this conclusion are shown in fig. 4.

Subsequently improved data from the SLAC-MIT group [6] and data at higher Q^2 from FNAL [7] have shown that $F_2(x,Q^2)$ decreases with Q^2 at large x and increases with Q^2 at small x, as shown in figs. 5. Some - perhaps all - of the decrease at large x can be attributed to the wrong choice of scaling variable. For example, suppose that

$$F_2(x,Q^2) \quad = \quad f(x' = \frac{x}{1 + M^2x/Q^2})$$

then F_2 would decrease with Q^2 at fixed x since f decreases with increasing x'. As we shall see in sec. II.4, there are theoretical reasons to prefer some variable like x' and at least some of the decrease observed at large x is probably due to this effect. The rise at small x is also difficult to interpret as an unambiguous indication of scaling violation since some of it must be due to the associated production of charmed particles (see section II.4). However, if it were all due to charmed particle production this would probably have shown up in trimuon production by now. In fact I believe that the data show genuine scaling violation at both large and small x, although data at higher Q^2 are needed to establish this completely.

Nevertheless I will continue to discuss the naive scaling parton model since many of the predictions survive the introduction of scaling violations.

b) It follows from comparing eqns 3, 5 and 6 that [9]

$$\frac{\sigma_0}{\sigma_T} = \infty \text{ for spin 0 partons}$$
$$= 0 \text{ for spin 1.2 partons.}$$

This can be understood by helicity arguments since the parton and virtual photon meet head on in the frame we employ, if we neglect p_T, as illustrated in fig. 6.

Taking the z axis along $\vec{p}$, a spin zero parton has $J_z = 0$ initially and finally and can therefore not absorb a virtual photon of helicity +1. Hence $\sigma_T = 0$ in this case. For a spin 1/2 parton, however, the bare electromagnetic vertex $u\gamma_\mu u$ conserves helicity at high energy and $\Delta J_z^{parton} = \pm 1$, which can only be supplied by a transverse photon. Hence $\sigma_0 = 0$ in this case. The data, some of which are shown in fig. 7, suggest that $\sigma_0/\sigma_T << 1$ and that

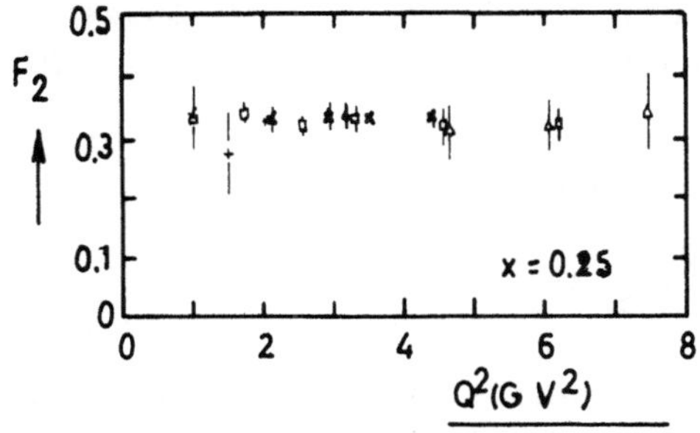

FIG.4 : $F_2(x,Q^2)$ for ep scattering for x=0.25 as a function of Q^2 for W > GeV [5]

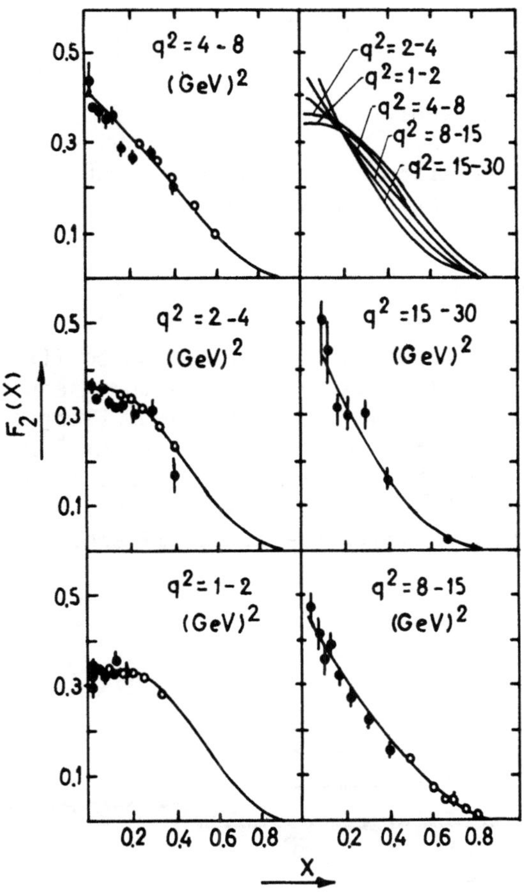

FIG. 5 : $F_2(x,Q^2)$ for ep scattering vs. x for various Q^2 ranges[7]. ● FNAL data. O SLAC-MIT data. The lines, which are fits are collected in the corner to show the Q^2 dependence.

this ratio could vanish as $Q^2 \to \infty$ 10). We therefore assume that all electrically charged partons have spin 1/2 - presumably they are quarks.

c) The formula above yield

$$F_2(x) = \sum_i Q_i^2 x f_i(x).$$

Since the data (fig. 5) suggest that $F_2(x)$ is not zero in the limit $x \to 0$, it follows that $f_i(x) \sim 1/x$ at small x. i.e. partons have a Bremsstrahlung spectrum. Of course, the kinematic assumptions we have made are only justified for $xP >> M_{parton}$ or $x >> M^2{}_{parton} / s_{\gamma^* p}$. Integrating from this point we see that the parton multiplicity

$$< n_i > = \int f_i(x)dx \sim \int \frac{dx}{x} \sim \ln s_{\gamma^* p}$$

grows with increasing energy.

d) Consider a configuration of N partons. If the protons momentum were shared equally among them we would have

$$< x >_N = \frac{1}{N}$$

In any case, we would expect the contribution to f(x) to be confined to small x for very large N and only to extend to large x for small N i.e. large (small) x probes small (large) N. This is borne out by comparing F_2^{ep} and F_2^{en} (fig. 8). At large x, the fact that some of the partons must be different in p and n is significant and $F_2^{ep} \neq F_2^{en}$. At small x, however, the parton multiplicity is high and the difference is significant.

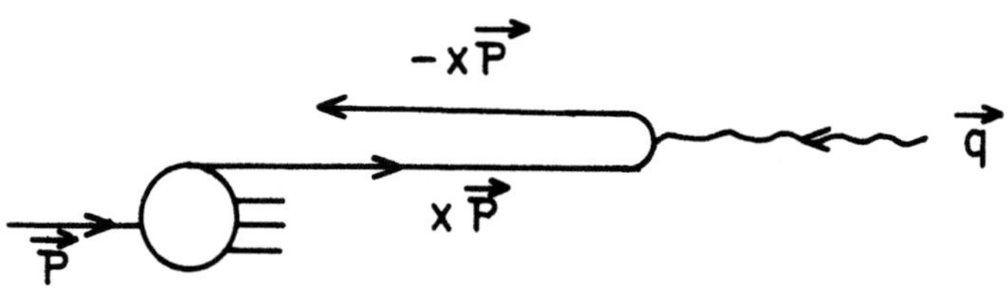

FIG. 6 : γ^* - N scattering in the γ^* parton Breit frame accordo,g tp the naive parton model.

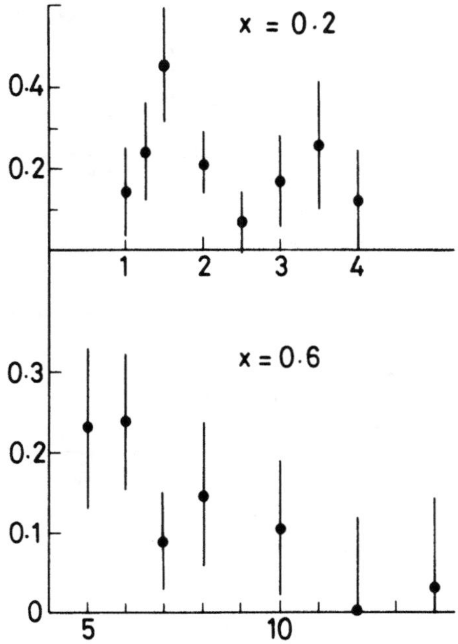

FIG. 7 : σ_0/σ_T vs$\cdot Q^2$ (in GeV^2) at two of the eleven values of x for which data are given in ref.8. The errors are statistical only (systematic errors decrease from about = 0.13 at Q^2=1 to ± 0.09 at Q^2=4 for x = 0.2 and from ± 0.06 at Q^2=5 to ± 0.03 at Q^2 for x=0.6)

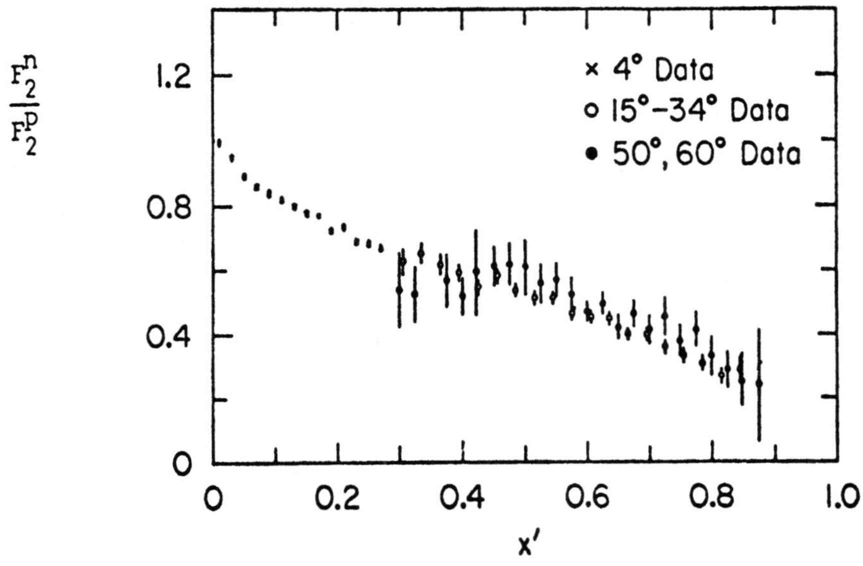

FIG. 8 : F_2^{en}/F_2^{ep} vs.x'. Data from SLAC-MIT experiments [1]) (the Q^2 range of the data is roughly 2-20 GeV^2).

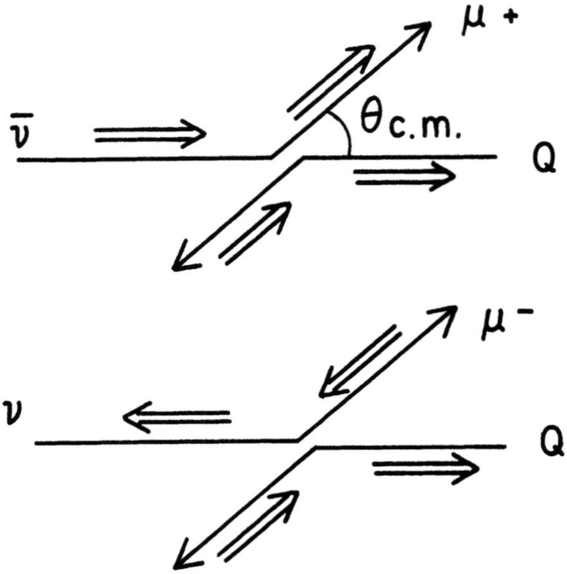

FIG. 9 : Helicities in $\nu q \to \mu q'$ and $\bar{\nu} q \to \bar{\mu} q'$.

4. Results for $\nu N \to \mu + \ldots$

We now assume that partons are quarks and, for simplicity, assume just three flavours u, d and s and put $\theta_C = 0$. The basic processes are

$$\nu d \to \mu^- u \,, \qquad \nu \bar{u} \to \mu^- \bar{d}$$

$$\bar{\nu} u \to \mu^+ d \,, \qquad \bar{\nu} \bar{d} \to \mu^+ \bar{u}$$

and the elementary cross-sections to be fed into eqn 5 are

$$\frac{d\sigma}{dQ^2}^{\nu Q,\ \bar{\nu}\bar{Q}} = \frac{G^2}{\pi}$$

$$\frac{d\sigma}{dQ^2}^{\nu \bar{Q},\ \bar{\nu} Q} = \frac{G^2}{\pi}(1-y)^2. \qquad (7)$$

Remarks :

1. Since Fermi's constant $G = 10^{-5} / M_p^2$ these formulae have the right dimensions and in fact it is necessary that $d\sigma/dQ^2 =$

$G^2 g(y = Q^2/s_i)$.

2. The y dependence is easy to understand from helicity conservation. Consider $\bar{\nu}Q$ scattering in the $\bar{\nu}Q$ centre of mass frame where $2y = 1 - \cos\theta_{c.m.}$ (fig. 9). Initially $J_z = +1$ along the $\bar{\nu}$ direction; finally $J_z = +1$ along the μ^+ direction (assuming a V - A quark current and neglecting the quark and muon masses).

This process is therefore forbidden when $\theta_{c.m.} = \pi$ or $y = 1$ (explicit calculation is needed to find the power of $1 - y$). νQ scattering, however, is not forbidden at any angle and is actually isotropic.

3. With right handed couplings to new quarks the helicity factors for ν and $\bar{\nu}$ must obviously be exchanged and

$$\frac{d\sigma}{dy}[\nu Q, \bar{\nu}\bar{Q}]_{RH} \sim (1-y)^2$$

$$\frac{d\sigma}{dy}[\bar{\nu}Q, \nu\bar{Q}]_{RH} \sim 1$$

We denote the distribution of up, down, ... quarks in the proton by $u(x)$, $d(x)$, ... The distributions $u^n(x)$, $d^n(x)$ in the neutron are related by charge symmetry of the strong interactions

$$u^n(x) = d(x)$$

$$d^n(x) = u(x) \quad \text{etc...}$$

On isoscalar targets, neutrinos therefore measure

$$d(x) + d^n(x) = d(x) + u(x) \equiv Q(x)$$

and

$$\bar{u}(x) + \bar{u}^n(x) = \bar{u}(x) + \bar{d}(x) \equiv \bar{Q}(x)$$

and antineutrinos measure the same quantities. Combining ecns 5 and 7 we therefore find

$$\frac{d^2\sigma}{dxdy} = \frac{G^2 s}{2\pi}\left[xQ(x) + x\bar{Q}(x)(1-y)^2\right]$$

$$\frac{d^2\sigma^{\bar{\nu}}}{dxdy} = \frac{G^2 s}{2\pi}\left[x\bar{Q}(x) + xQ(x)(1-y)^2\right] \qquad (8)$$

A simple first guess would be to assume that the nucleon consists of just three quarks ($\bar{Q} = 0$). This would imply

1) $\frac{d\sigma^{\nu}}{dy} = c, \quad \frac{d\sigma^{\bar{\nu}}}{dy} = c(1-y)^2$

2) $\frac{\sigma^{\bar{\nu}}}{\sigma^{\nu}} = \frac{1}{3}$ (equivalent to 1, assuming scaling and charge symmetry of the weak currrent).

3) $\frac{\sigma^{\nu n}}{\sigma^{\nu p}} = 2$ since $\int udx = 2\int ddx$ if $\bar{Q} = 0$

4) $\sigma^{\nu} = \frac{G^2 s}{2\pi} \approx 1.5\ E\nu \times 10^{-38}\ cm^2\ Gev^{-1}$

since the momentum conservation sum rule

$$\sum_i \int xf_i(x)dx = 1$$

implies

$$\int xQ(x)dx = 1$$

if there are just three quarks in the nucleon.

The first three results are in approximate agreement with the data [12] although some $\bar{Q}$ contribution is required by the y distributions at small x and the related fact that $\sigma^{\bar{\nu}}/\sigma^{\nu}$ is slightly bigger than 1/3 (the data give $\int x\bar{Q}/\int xQ \simeq 0.05$-$0.15$). However, the assumption that $\int x(Q)dx = 1$ overestimates σ^{ν} by a factor of about 2. This suggests that avout one half of the proton's momentum is carried by gluons which do not interact weakly, which are needed to mediate the strong interquark force in any renormalizable theory (they were first introduced in the parton model to explain the fact that $F_2^{\gamma^*}$ is implausibly small even if the partons are non-integrally charged quarks [13]).

There are several possible sources of discrepancies from eqn 8 :

1. A non-zero Cabibbo angle and the effects of $\Delta C = 1$ currents should be included (we leave this as an exercise).

2. Scaling violations will produce changes - see section II.4.

3. Couplings to new quarks can produce radical changes;e.g. the most extreme case (not now required up to 200 GeV)is the existence of a b quark with charge - 1/3 with a right handed coupling (V+A) to the u quark, which would change $d\sigma^{\bar{\nu}}$ dramatically above threshold.

5. General tests of the quark parton model.

We now turn to more general and incisive tests of the quark parton model, which have the virtue that they survive the introduction of the scaling violations predicted by asymptotically free gauge theories (see section II.3). First we give the general form of the differential neutrino cross-section in terms of σ_L, σ_R and σ_0, or equivalently, the three standard structure functions F_1, F_2 and F_3 :

$$\frac{d^2\sigma^{\nu,\bar{\nu}}}{dxdy} = \frac{(\nu - Q^2/2)}{\pi} \times \frac{G^2 s}{2\pi} \left[2(1-y)\sigma_0 + \sigma_{L,R} + \sigma_{R,L}(1-y)^2\right]$$

$$= \frac{G^2 s}{2\pi} \left[(1-y)F_2 + y^2 \times F_1 \mp y\left(1 - \frac{1}{2} y\right) \times F_3\right] \qquad (9)$$

Comparison with the formulae above then gives several important results.

1. $\sigma_0/\sigma_{L,R} = 0$ for spin 1/2 partons, as already discussed for electro-production. For neutrino production, assuming charge symmetry for the $\Delta S = \Delta C = 0$ weak current and that $\Delta S = \Delta C = 0$ processes dominate the total cross-section, we have $\sigma_i^{\nu n} = \sigma_i^{\bar{\nu} p}$, $\sigma_i^{\nu n} = \sigma_i^{\nu p}$ or equivalently $F_i^{\nu \overset{p}{n}} = F_i^{\bar{\nu} \overset{n}{p}}$. The fact that $\sigma^{\bar{\nu}}/\sigma^{\nu}$ is close to one third for large nuclei then requires $\sigma_0/\sigma_{L,R} \ll 1$ assuming approximate scaling.

2. The parton distributions must be consistent with the correct B, Y, Q, I_3, S, C, ... of the target. For example, there must be three more quarks than antiquarks in the proton i.e.

$$3 = \int dx[u(x) + d(x) + s(x) + c(x) + \ldots - \bar{u}(x) - \bar{d}(x) - \bar{s}(x) - \bar{c}(x) \ldots]$$

$$= \int dx \, [u(x) + d(x) - \bar{u}(x) - \bar{d}(x)]$$

using the zero strangeness, charm, etc... of the nucleon to eliminate the integrated contributions of $s - \bar{s}$, $c - \bar{c}$, ...
The last line can easily be expressed in terms of measured structure functions as [14]

$$-\frac{1}{2} \int F_3(x)^{\nu(n+p)}_{\Delta S=\Delta C=0} \, dx = 6$$

(where we have taken a factor of $\cos^2\theta_C$ out of $F_{\Delta S=\Delta C=0}$ by definition). Relevant data are shown in fig. 10. Unfortunately the

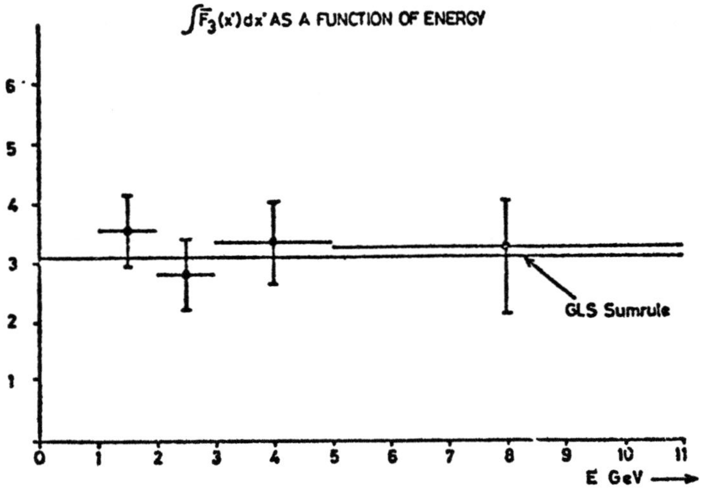

FIG. 10 : $-1/2\int F_3^{\nu n+\nu p}$ x' from the Gargamelle experiments[15] as a function of the neutrino energy ($\langle Q^2 \rangle \sim 0{,}2E$).

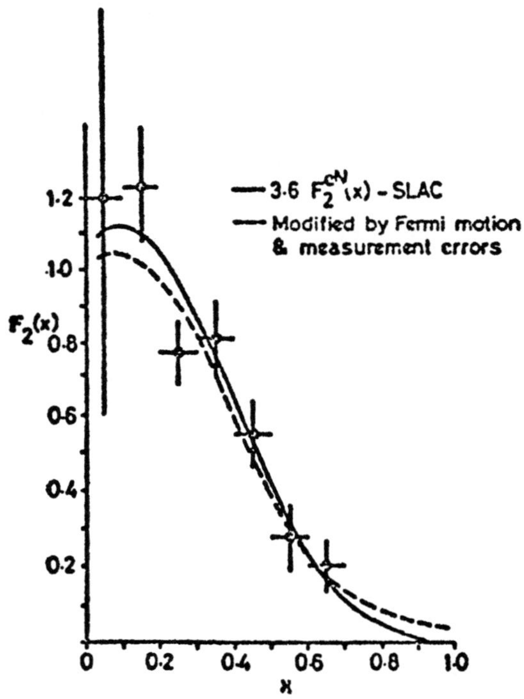

FIG. 11 : $F_2^{\nu N}$ vs. x as measured in the Gargamelle experiment[15] for Q^2> compared to SLAC-MIT data for 18/5 F_2^{ed} in the same Q^2 range.

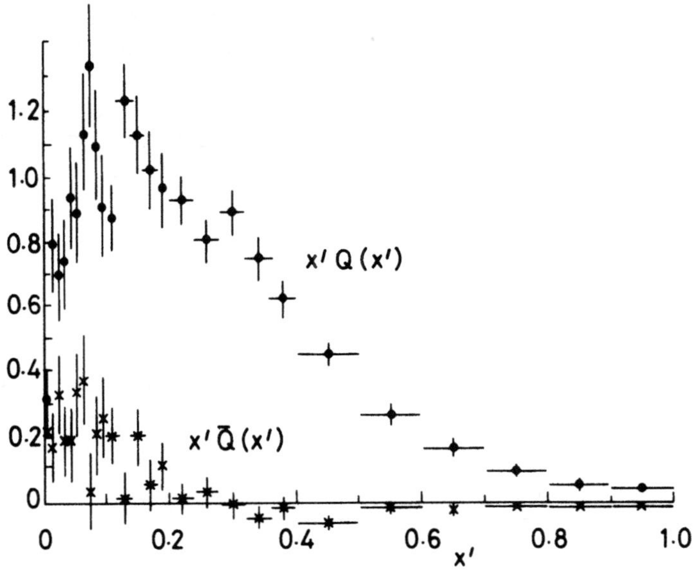

FIG. 12 : Quark and antiquark momentum distribution as a function of x' as measured in the Gargamelle experiment [15].

small x contribution (which is important) corresponds to small Q^2 so the agreement should not be taken too seriously but tests at much higher Q^2 will shortly become possible. The other sum rules which can be derived in this way (which are catalogued in ref. 2) are much harder to test since they involve the difference of structure functions for neutrons and protons.

3. The quark parton model requires [13]

$$F_2(x)^{\nu(n+p)}_{\Delta S=\Delta C=0} < \frac{18}{5} F_2(x)^{e(p+n)} \qquad (10)$$

The left hand side is given by

$$F_2 = 2x[u(x) + d(x) + \bar{u}(x) + \bar{d}(x)]$$

whereas the right hand side is given by

$$\frac{18}{5} F_2^{ep+en} = 2x[u + d + \bar{u} + \bar{d}] + \frac{4x}{5}[s+\bar{s}] + \frac{16x}{5}[c+\bar{c}\ ]+ \ldots$$

Eqn 10 becomes an equality if $s = \bar{s} = c = \bar{c} = \ldots = 0$.
The data (fig. 11) respect the inequality and show that c, $\bar{c}$, s and $\bar{s}$ are small - except perhaps near $x = 0$, where we expect discrepancies. These same data have been used to construct $xQ(x)$ and $x\bar{Q}(x)$ with the result shown in fig. 12, which supports the idea that there are few constituents at large x but many (with $Q \simeq \bar{Q}$) at small x.

The fact that F_2^{ν}/F_2^{e} is given correctly by the quark model at large x strongly supports the picture of incoherent scattering off (effectively) non-integrally charged particles. Clearly if the correct ratio is obtained with fractional quark charges we would expect to obtain the wrong result in models with integral charges. Averaging over protons and neutrons the Q_i^2 factors in the parton formula for $F_2^{\gamma *}$ give

$$\left|\sum_i f_i Q_i^2\right|_p + \left|\sum_i f_i Q_i^2\right|_n = 2\sum_i f_i \left(I_3^2 + \frac{Y^2}{4}\right)_i$$

The first term (if known separately) could be used to construct F_2^{ν} by using CVC and making an isorotation. Alternatively we may write

$$\frac{<Y^2>_{eff}}{4<I_3^2>_{eff}} = \frac{4F_2^{ep+en} - F_2^{\nu p+\nu n}}{F_2^{\nu p+\nu n}}$$

The data clearly show that this ratio is small. It can be argued that this is exactly what we would expect, given continuity from small to large Q^2, since the fairly successful ρ, ω, ϕ dominance model of photo-production gives :

$$\frac{\sigma(\gamma N \to I = 0)}{\sigma(\gamma N \to I = 1)} \simeq \frac{1}{9}$$

However, I do not think that this belittles the quark-parton picture; rather it shows that photo-production also suggested small average hypercharge for the constituents.

There are other interesting parton model inequalities such as

$$4 > \frac{4(u+\bar{u})+d+\bar{d}+s+\bar{s}+{}'(c+\bar{c})}{4(d+\bar{d})+u+\bar{u}+s+\bar{s}+4(c+\bar{c})} = \frac{F_2^{ep}}{F_2^{en}} \geqslant \frac{1}{4}$$

which we will not discuss further here.

6. $\nu N \to \nu + \ldots\ldots$

We will now assume that the parton model works and show how it can be used to determine neutral current couplings (the consistency with other determinations then provides an indirect check of the model). We can define

$$\mathcal{L}_{eff}^{NC} = \frac{G}{\sqrt{2}}\, \nu\gamma_\lambda(1-\gamma_5)\nu \sum_i \bar{\psi}_i [\varepsilon_i^L(1-\gamma_5)+\varepsilon_i^R(1+\gamma_5)]\gamma_\lambda \psi_i$$

where i runs over all fermions.

For an I = 0 nucleus the parton model calculations above immediately give

$$\frac{d^2\sigma^{\nu\to\nu}}{dxdy} = \frac{G^2 s}{2\pi}\left[xQ|\varepsilon_L|^2+x\bar{Q}|\varepsilon_R|^2+(1-y)^2\{x\bar{Q}|\varepsilon_L|^2+xQ|\varepsilon_R|^2\}\right]$$

$$\frac{d^2\sigma^{\bar{\nu}\to\bar{\nu}}}{dxdy} = (\ \ldots\ \varepsilon_L \leftrightarrow \varepsilon_R\ \ldots)$$

where

$$|\varepsilon_{L,R}|^2 \equiv |\varepsilon_{L,R}(u)|^2 + |\varepsilon_{L,R}(d)|^2$$

and we have neglected all but u, d, $\bar{u}$ and $\bar{d}$ quarks. Since $\bar{Q}$ and Q are determined by charged current experiments, neutral current cross sections determine the couplings $|\varepsilon_L|^2$ and $|\varepsilon_R|^2$ if we accept the parton model, as discussed in Steinberger's lectures at this school. However, scaling violation will introduce errors into this determination [16].

7. $\vec{e}\vec{N} \to e + \ldots.$

We now consider inclusive scattering of polarized electrons or muons on a polarized target. This involves no new theoretical ideas but leads to interesting results which test the model. Some data are available for $Q^2 < 3$ GeV^2 and data at much higher Q^2 should be forthcoming from a CERN experiment in a few years time.

We begin by discussing the kinematics using the fact that $\sigma_{\gamma^* N}$ is the imaginary part of the forward $\gamma^* N \to \gamma^* N$ amplitude (optical theorem), as illustrated in fig. 13. Conservation of J_z is equivalent to helicity conservation for the forward amplitude. The table can therefore be used to deduce that in general there are ten amplitudes or structure functions. However, parity and time reversal relate them in an obvious way so that

$$10 \overset{P}{\to} 5 \overset{T}{\to} 4$$

i.e. there are two new structure functions, usually called G_1 and G_2, in addition to W_1 and W_2 which survive spin averaging (explicit formulae are given e.g. in ref. 17). They can be separated e.g. by studying $d\sigma^{\uparrow\uparrow} - d\sigma^{\uparrow\downarrow}$ as a function of E_e at fixed ν and Q^2; (where the first (second) arrow denotes the electron (nucleon) spin along the electron beam direction). The parton formula eqn 1 obviously generalizes to

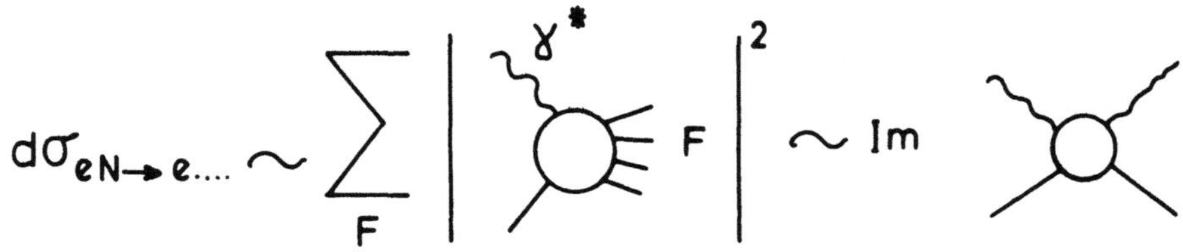

FIG. 13 : Illustration of the optical theorem.

$$\frac{d^2\sigma^{\downarrow\uparrow}}{dQ^2 d\nu} = \frac{Q^2}{\nu} \sum_i \left[f_i^{\uparrow}(x) \frac{d^2\sigma_i^{\downarrow\uparrow}}{dQ^2 d\nu} + f_i^{\downarrow}(x) \frac{d^2\sigma^{\downarrow\downarrow}}{dQ^2 d\nu} \right]$$

where $f_i^{\uparrow}(f_i^{\downarrow})$ is the probability of finding parton i with momentum $x\vec{p}$ polarized parallel (anti-parallel)to the nucleon. The $d\sigma_i$ are given by

$$\frac{d\sigma_i^{\downarrow\uparrow}}{dQ^2} = Q_i^2 \frac{4\pi\alpha^2}{Q^4}$$

$$\frac{d\sigma_i^{\downarrow\downarrow}}{dQ^2} = Q_i^2 \frac{4\pi\alpha^2}{Q^4} (1-y)^2$$

for spin 1/2 partons. The origin of the y dependence should be obvious to the reader by now and the normalization is of course fixed by the necessity of reducing to Rutherford scattering as $Q^2 \to 0$(spin averaging of $d\sigma_i$ yields eqns 2 and 3). Clearly

$$d\sigma^{\downarrow\uparrow} - d\sigma^{\uparrow\uparrow} \sim \frac{1}{2} \sum_i Q_i^2 [f_i^{\uparrow}(x) - f_i^{\downarrow}(x)]$$

in this model. This combination is actually $g_1(x)$ (the scaling function corresponding to G_1)which satisfies the sum rule

$$\int g_1(x)dx = \frac{1}{2} \sum_i (n_i^{\uparrow} - n_i^{\downarrow})Q_i^2 \ .$$

Using the relation $Q_i^2 = 2/3B_i + 1/3I_3^i + 1/6Y_i$, valid for non-charmed quarks, we can see that parts of the right hand side are measured. For a helicity +1 proton in the $\vec{P} \to \infty$ frame we have

$$\langle P,\lambda=1|\psi(o)\gamma_\lambda(1\pm\gamma_5)I_3\psi(o)|P,\lambda=1\rangle = 2P_\lambda \sum_i I_3^i n_i^{\uparrow,\downarrow}$$

or

$$\langle P|\bar{\psi}(o)\gamma_\lambda\gamma_5\lambda^\alpha\psi(o)|P\rangle = 2P_\lambda(-\frac{g_A}{2g_V}) = 2P_\lambda\sum_i I_3^i(n_i^\uparrow - n_i^\downarrow)$$

More generally for a diagonal matrix λ^α :

$$\sum_i (n_i^\uparrow - n_i^\downarrow)\ \lambda_{ii}^\alpha = Z^\alpha$$

where

$$\langle P|\psi\gamma_\lambda\gamma_5\lambda^\alpha\psi|P\rangle = 2P_\lambda Z^\alpha.$$

This leads to Bjorken's sum rule [18]

$$\int (g_1^p - g_1^n)\ dx = \frac{1}{6}\ \left|\frac{g_A}{g_V}\right|.$$

From Cabibbo theory we also know that $Z^y \simeq 0.24$ but

$$Z^B = \frac{1}{3}\sum_i (n_i^\uparrow - n_i^\downarrow)$$

is not measured. The common sense guess $Z^B = 1/3$ leads to

$$\int g_1^p dx \simeq 0.24, \quad \int g_1^n dx \simeq 0.02.$$

The suggestion that $g_1^{ep} > 0$ is quite non-trivial since the corresponding quantity is negative in photo-production [19]. However, it is borne out by the available data [20], which are unfortunately still rather meagre but are compatible with an SU(6) type spin distribution for the valence quarks.

8. $e^+e^- \to \ldots.$

According to the parton dogma the cross section is given by fig. 14 which yields

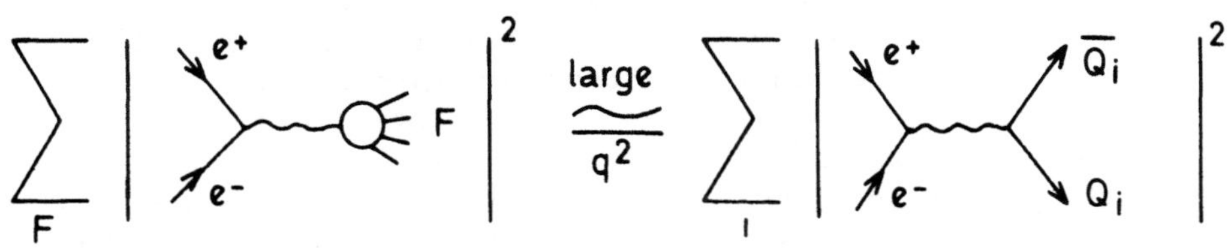

FIG. 14 : Parton diagrams for $e^+e^- \to$ anything.

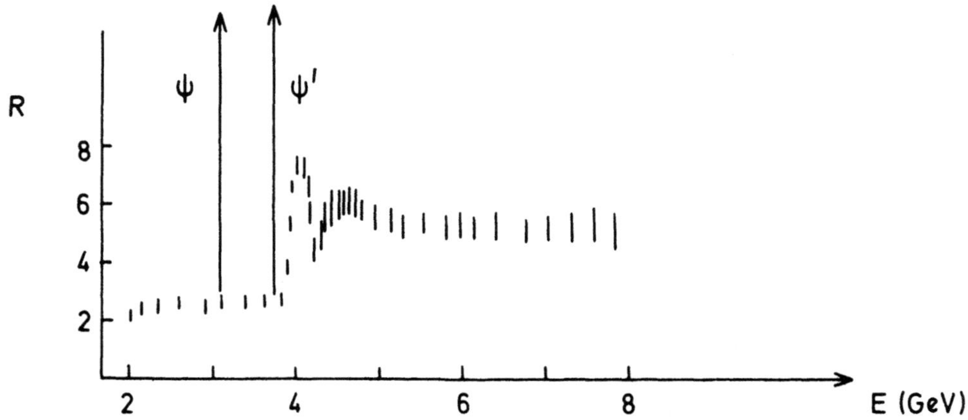

FIG. 15 : Sketch of $R = \sigma(e^+e^- \to \text{hadrons}) / \sigma(e^+e^- \to \mu^+\mu^-)$ vs.E_{cm} (symbolic - not real data).

$$R_{\bar{e}e} = \frac{\sigma(\bar{e}e \to \text{hadrons})}{\sigma(\bar{e}e \to \bar{\mu}\,\mu)} = \sum_i Q_i^2$$

for spin 1/2 partons, where the sum runs over colour as well as flavour indices. Of course it cannot be correct to sum incoherently over colour indices if only colour singlets are produced. However, it produces the same result as summing coherently and then projecting onto a colour singlet. The data are sketched symbolically in fig. 15. They are compatible with the 3 quark value (R_{ee} = 2) below charm threshold and could converge on the 4 quark value (R_{ee} = 3 1/3) at high energy if we allow for one unit of R^{expt} due to heavy lepton production.

We now consider the angular distribution relative to the beam axis in $\bar{e}e \to H \ldots$ Assuming one photon exchange

$$d\sigma_{\bar{e}e \to H \ldots} \sim \bar{\sigma}_0 \sin^2\theta + \bar{\sigma}_T\,(1 + \cos^2\theta)$$

where the $\bar{\sigma}$'s depend on Q^2 and on $E_H / \sqrt{Q^2}$ in the centre of mass frame. If the e^- and e^+ have antiparallel polarization of P each

$$d\sigma \sim (\bar{\sigma}_T + \sigma_0) + (\sigma_T - \sigma_0)(\cos^2\theta + P^2 \sin^2\theta \cos\phi)$$

where ϕ is the azimuthal angle relative to the axis of polarization. The ϕ dependence provides no new information in principle but is very useful in practice since 2π acceptance in ϕ is possible.

Much stronger assumptions than those used up to now are needed to apply the parton model to the final states in $\bar{e}e$ or lepto-production. However, it is reasonable to expect that if the hadrons exhibit any pronounced θ/ϕ dependence it should reflect that of the quarks in fig. 14. In fact it is known that the final state consists of 2 collimated jets of particles and their angular dependence does seem to be that of spin 1/2 particles, which is

$$1 + \cos^2\theta$$

(This can be inferred from helicity arguments, as illustrated in fig. 16 where we consider a left handed incident e^-, which automatically requires the e^+ to be right handed). This suggests that fast particles, which follow the jet (quark) axis most closely, should have $\overline{\sigma}_0/\overline{\sigma}_T = 0$. This is in agreement with the data which are shown in fig. 17.

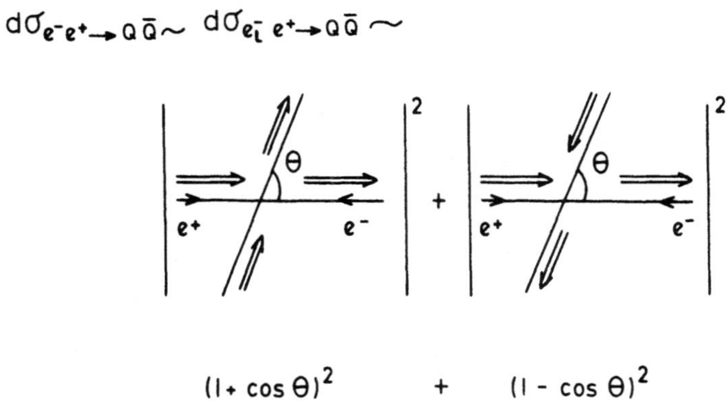

FIG. 16 : Helicities in the two amplitudes which contribute incoherently to $e^-_L e^+ \to Q\bar{Q}$ and the implied angular dependence.

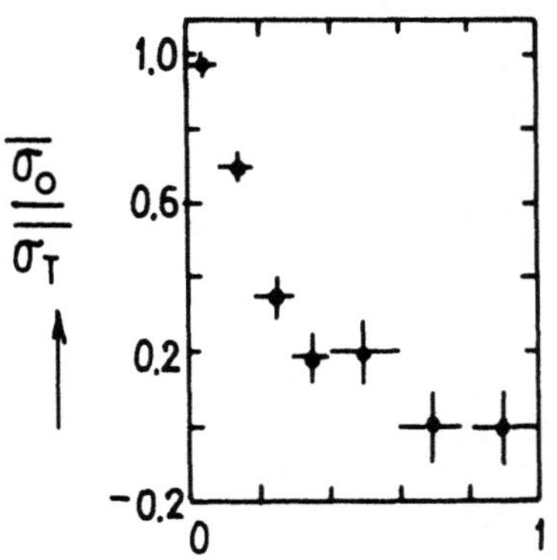

FIG. 17 : σ_0 / σ_T vs. E_H/E_{beam} in $e^+e^- \to H + \ldots$ at $2E_{beam} = 7.4$ GeV [21].

B. More Speculative Applications of the Parton Model.

1. $pp \to \mu\bar{\mu}$, W, Z .....

According to Drell and Yan [22], these processes are dominated by the diagrams in fig. 18. This configuration is supposed to be favoured because in $q\bar{q} \to \mu\bar{\mu}$ the quark and antiquark can be near mass shell, whereas in Bremsstrahlung processes a quark must become very virtual, so that its propagator suppresses the amplitude. Accepting this dogma, the cross section is given by

$$\frac{d\sigma}{dQ^2} = \sum_i \iint dx dx' f_i(x) \bar{f}_i(x') \frac{d\sigma_i}{dQ^2} \tag{11}$$

when

$$\sigma_i = Q_i^2 \frac{4\pi\alpha^2}{3Q^2}$$

so that

$$\frac{d\sigma_i}{dQ^2} = Q_i^2 \frac{4\pi\alpha^2}{3Q^2} \delta(xx's - Q^2).$$

Note that

1) If the quarks are coloured, eqn 11 must be divided by 3 [23].

2) In the naive model, $Q_T^2 = 0$ and $Q_3 = (x-x')|\vec{P}|$, so eqn 11 can be used to calculate $d^2\sigma / dQ^2 dQ_3$.

3) The naive model implies the scaling law

$$\frac{d\sigma}{dQ^2} = \frac{1}{Q^4} f(Q^2/s)$$

since no scale enters. It is very important to know whether this scaling law is valid experimentally since we could then use it to estimate W and Z production rates for proposed new strorage rings. Certainly the naive formula (eqn 11) cannot be right if scaling is violated in deep inelastic scattering; we shall return to this point in section II5.

4) The data agree qualitatively with the model but uncertainties in the input quark distributions and systematic errors make more precise statements impossible at present (see the lectures by Lederman at this school).

5) According to this model πN and $\bar{N}N$ collisions should be a much more copious source of μ-pairs than NN collisions since the nucleon contains so few antiquarks. The model also predicts that as $Q^2/s \to 1$, and hence $xx' \to 1$, the ratio of the π^+ N

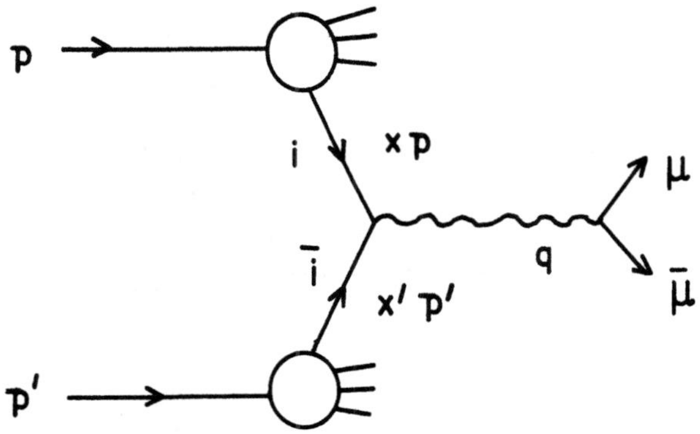

FIG. 18 : Diagrams which are supposed to dominate the Drell-Yan process $pp \to \mu\mu$ ... .

to the $\pi^- N$ cross section should tend to 1/4, since the former is dominated by $\bar{d}d$ annihilation ($Q^2 = 1/9$) and the latter by $\bar{u}u$ annihilation ($Q^2 = 4/9$). The trend of the little data that now exists supports these predictions.

2. <u>A three photon process</u>.

Consider the diagrams for the process $ep \to e\gamma + \ldots$ which are shown in fig. 19. The relative sign of the amplitudes due to the first diagram and the Bethe Heitler terms obviously changes depending on whether the incident particle is an e^- or an e^+ Therefore $d\sigma(e^-p \to e^-\gamma\ldots) - d\sigma(e^+p \to e^+\gamma\ldots)$ measures the interference term, illustrated in fig. 20. It has been argued [24] that when Q^2, $\tilde{Q}^2$, $\nu = q\cdot p$, $\tilde{\nu} = \tilde{q}\cdot p$ and $|Q^2 - \tilde{Q}^2|$ become very large, the

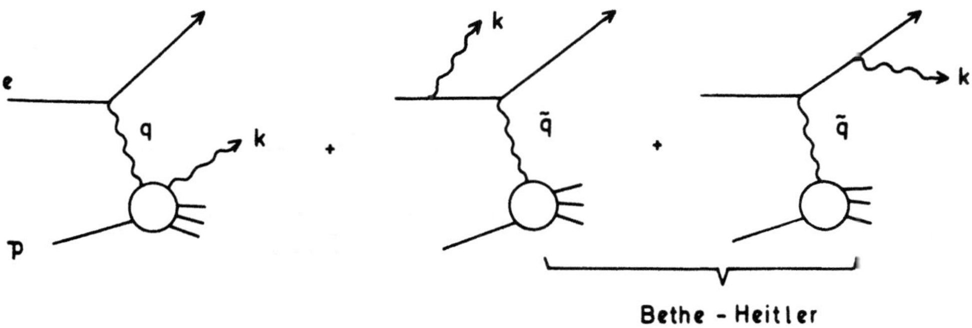

FIG. 19 : Diagrams which contribute to $ep \to e\gamma$ ... .

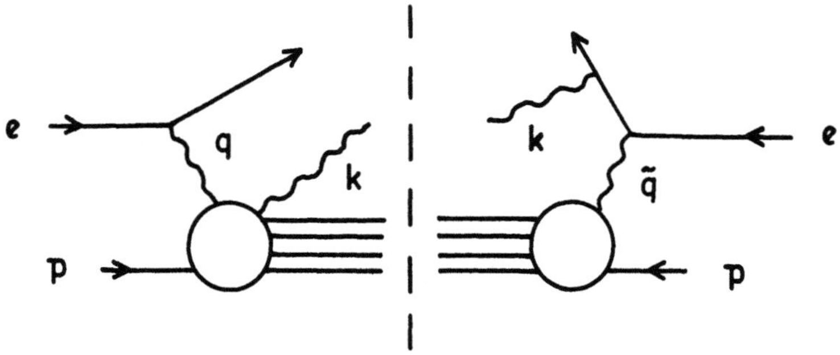

FIG. 20 : The interference term in $d\sigma(ep\to e\gamma)$ which determines $d\sigma(e^-p\to e^-\gamma)-d\sigma(e^+p\to e^+\gamma)$.

interference term can be calculated using the parton model (fig. 21). The argument is that in this case there exist frames in which the components of $\vec{q}$, $\vec{\tilde{q}}$ and $\vec{k}$ transverse to $\vec{P}$ all become infinite as Q^2 etc... are scaled to infinity. The transverse momenta at the hadronic vertices can be small in the parton diagrams, since $\vec{\tilde{q}}_T = \vec{q}_T + \vec{k}_T$, but in all other diagrams the hadronic vertices are forced to sustain large transverse momenta. The basic parton model assumption of a rapid damping in transverse momenta then ensures that the parton diagrams dominate the process.
In the naive parton model, this will obviously lead to some scaling law for $d\sigma_{e^-} - d\sigma_{e^+}$ which can be expressed in terms of a scaling function [24]

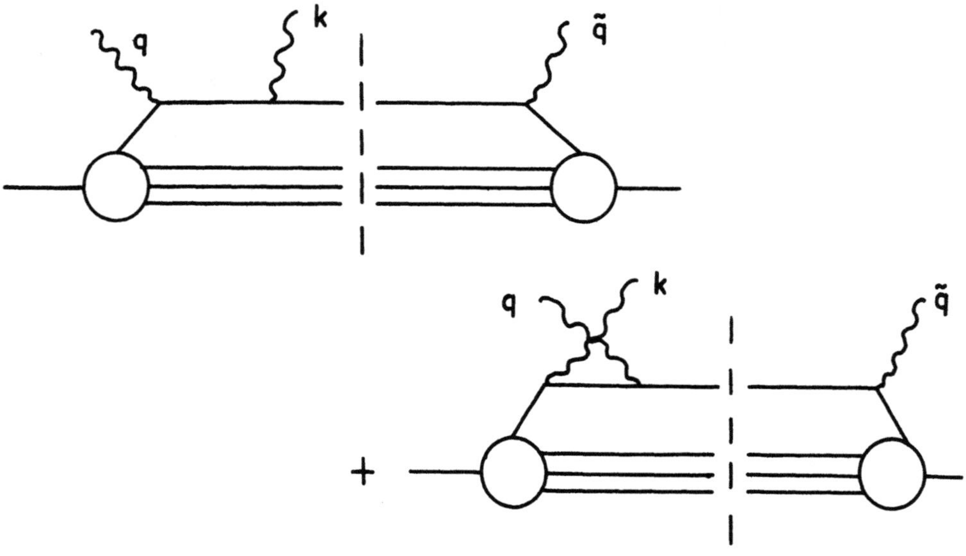

FI . 21 : Parton diagrams supposed to dominate the interference term in fig. 20.

$$V(x) = \sum_i Q_i^3 f_i(x) = \frac{8}{27}(u-\bar{u}) - \frac{1}{27}(d-\bar{d}) + \ldots$$

The fact that the cross-section depends on Q_i^3 is very interesting for two reasons. First, it is particularly sensitive to the quark charges. Second, V is odd under charge conjugation and therefore depends on the quantities $q_i - \bar{q}_i$ which are all constrained by conservation laws (e.g. $\int(u-\bar{u})dx = 2$). Hence V satisfies the sum rule

$$\int V^{ep}(x)\, dx = \frac{5}{9}.$$

Data have been taken with $Q^2 = 1.5 - 3.3$ GeV^2, $\tilde{Q}^2 = 0.75 - 2.0$ $|Q^2-\tilde{Q}^2| > 0.75$ and $x = 0.12 - 1.0$. Using a model to extrapolate to zero, they give [24]

$$\int V^{ep}(x)dx = 0.89 \pm 0.34.$$

The error is such that at present this experiment cannot distinguish integer and non-integer charges, nor has it tested the scaling law, but the interference term does have the expected order of magnitude and sign. However, in contrast to the processes fiscussed in part A, and possibly also to the Drell-Yan process, it seems likely that the parton picture and the sum rule do not survive the introduction of the scaling violations anticipated in asymptotically free theories [25] (see section II.5); verification of the predictions of the naive model would therefore be particularly interesting.

3. Semi-inclusive lepto-production and e^+e^- annihilation.

The parton model has been used to make predictions for the final states in lepton induced processes by assuming that the cross-section factorises into pieces describing production and fragmentation of quarks, as illustrated in fig. 22. These diagrams cannot be taken too literally since they imply quark production. Nevertheless factorisation of the cross-section may be a good approximation if the back flow of quark quantum numbers is a soft isoscalar process. We assume that the function D_i^h which describes the fragmentation of $q_i(k) \to h(p') + \ldots$ is rapidly damped in $(\vec{p}_T^{\,\prime})^2$, where $\vec{p}_T^{\,\prime} \cdot \vec{k} = 0$, and depends only on $(\vec{p}_T^{\,\prime})^2$ and $Z = |\vec{p}^{\,\prime}|/|\vec{k}|$. The assumptions of scaling and of a transverse momentum cut off in $q_i \to h \ldots$ are natural since the same assumptions have already been made about $h \to q_i + \ldots$ in the naive parton model.

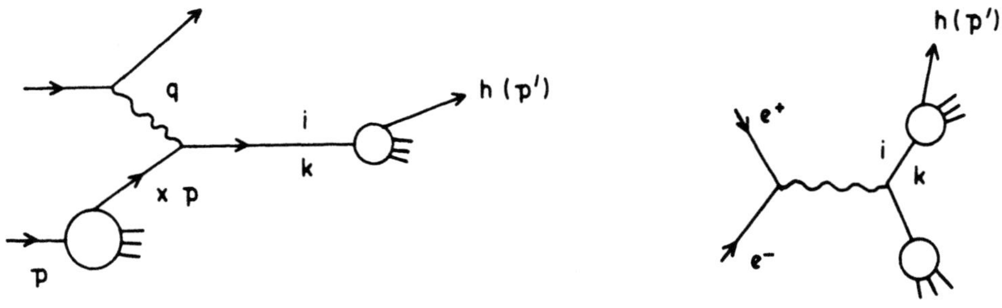

FIG. 22 : Parton diagrams for $e(\nu)N \rightarrow e(\mu)h$ ... and $e^+e^- \rightarrow h$ ....

The model predicts that the final states consist of two jets of particles, along $x\vec{p} + \vec{q}$ and $\vec{p}$, in eN and νN interactions and of two back to back jets of particles, with a $1 + \cos^2\theta$ distribution relative to the beams, in e^+e^- annihilation. This expectation is borne out by the data [26]. The properties of the jets are predicted to be independent of the production process. Some data which test this prediction are shown in fig. 23. Furthermore the model makes very definite predictions about properties such as charge ratios since isospin conservation and charge conjugation invariance of the fragmentation process lead to relations such as

$$D_u^{\pi^+} = D_d^{\pi^-} = D_{\bar{d}}^{\pi^+} = D_{\bar{u}}^{\pi^-}$$

$$D_u^{\pi^+} + D_u^{\pi^-} = 2\, D_u^{\pi^0}$$

$$D_u^{\pi^-} < 3\, D_u^{\pi^+}$$

First we consider e^+e^- annihilation. The model predicts

$$\frac{1}{\sigma(Q^2)} \frac{d\sigma_{e^+e^- \rightarrow h + \ldots}}{dz} = \sum_i \frac{Q_i^2 (D_i^h + D_{\bar{i}}^h)}{\sum Q_i^2}$$

where $z = E_h/E$ beam. As well as scaling, the model immediately predicts [28]

$$d\sigma_{\pi^0} = d\sigma_{\pi^\pm} \cdot$$

This is quite non-trivial; isospin conservation alone only gives [29]

$$\frac{1}{4} \langle N_{\pi^+} + N_{\pi^-} \rangle - \frac{1}{2} \leq \langle N_{\pi^0} \rangle \leq \frac{3}{2} \langle N_{\pi^+} + N_{\pi^-} \rangle - 1,$$

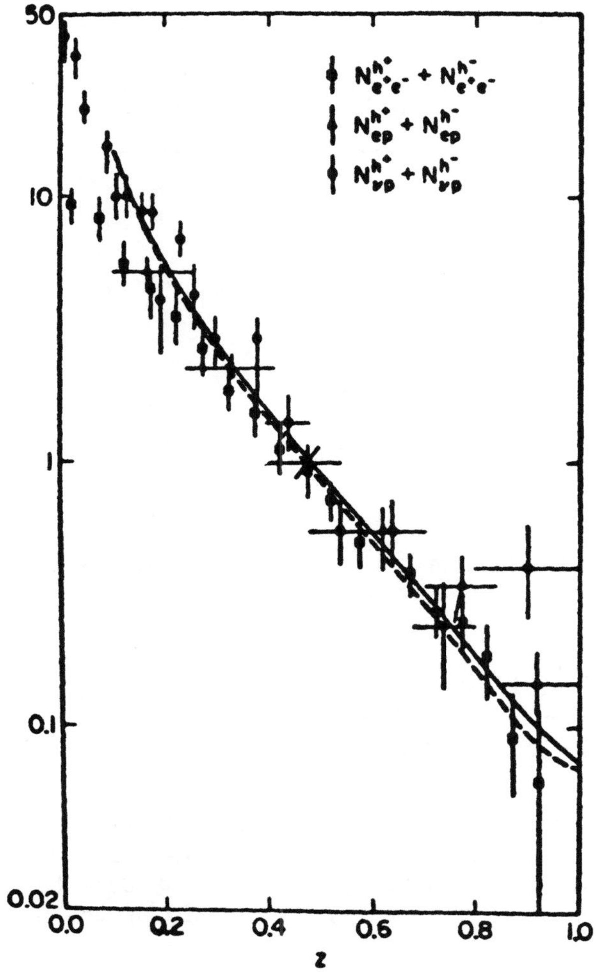

FIG. 23 : Fragmentation functions into the sum of positive and negative particles determined i e^+e^- annihilation at $E_{c.m.} = 3$ GeV, ep scattering for $Q^2 = 1\text{-}2$ GeV^2, $12 < W^2 < 30$ GeV^2 and νp interactions for $30 < E < 200$ GeV. According to the quark parton model, the data should all lie on a single curve, if we neglect the small s, $\bar{s}$ and c, $\bar{c}$ contributions (absent in e^+e^- at E=3) and Δs, $\Delta c = 1$ contributions to p [27].

for $e^+e^- \to N\pi$'s (N even), and leads to even weaker results for a general final state.

Next we consider $\nu N \to \mu h + \ldots$. Putting $\theta_c = 0$ and neglecting charm, the model gives

$$\frac{1}{\frac{d\sigma^{\nu}}{dx}} \frac{d^2\sigma^{\nu p\to\mu\pi^{\pm}}}{dxdz} = \frac{d(x)D_u^{\pi^{\pm}}(z) + \frac{1}{3}\bar{u}(x)D_{\bar{d}}^{\pi^{\pm}}(z)}{d(x) + \frac{1}{3}\bar{u}(x)}$$

$$= D_u^{\pi^{\pm}}(z)$$

$$\frac{1}{\frac{d\sigma^{\bar{\nu}}}{dx}} \frac{d^2\sigma^{\bar{\nu}p\to\bar{\mu}\pi^{\pm}}}{dxdz} = D_d^{\pi^{\pm}}(z) = D_u^{\pi^{\mp}}(z) .$$

The model therefore predicts $d\sigma^{\nu\to\pi^{\pm}} = d\sigma^{\bar{\nu}\to\pi^{\mp}}$ (this is automatic for an $I = 0$ target but not in general) and that

$$\eta(z) = \frac{D_u^{\pi^+}(z)}{D_u^{\pi^-}(z)} = \frac{1/\sigma \; d\sigma^{\nu\to\pi^+}}{1/\sigma \; d\sigma^{\nu\to\pi^-}}$$

should be independent of x. The data (fig. 24) support this prediction and show that $\eta(z)$ increases with z (as we would expect) with $\langle\eta\rangle \approx 2$ to 3. There is not yet sufficient data to answer the interesting question of whether this ratio or the D functions themselves scale.

In $eN \to eh + \ldots$ the model gives

$$\frac{1}{\sigma}\frac{d\sigma^{\to h}}{dz} = \frac{\sum_i Q_i^2 f_i(x) D_i^h(z)}{\sum_i Q_i^2 f_i(x)}$$

and the discussion is more complicated since all the quarks participate. However, the sum rules for $\int f_i(x)dz$ can be used to derive simple and precise predictions such as [31]

$$\frac{\int dx(d\sigma_n^{\pi^+} - d\sigma_n^{\pi^-})}{\int dx(d\sigma_p^{\pi^+} - d\sigma_p^{\pi^-})} = \frac{2}{7}$$

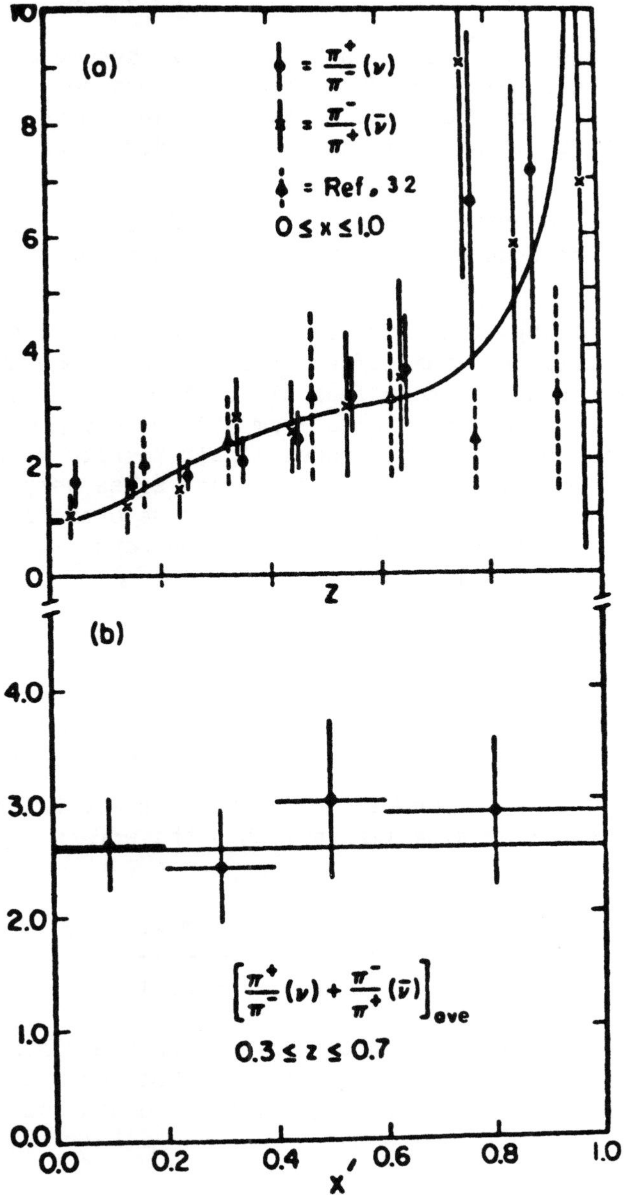

FIG. 24 :

a) $D_u^{\pi+} / D_u^{\pi-}$ determined in ν and $\bar{\nu}$ induced reactions[30] as a function of z, averaged over x. The curve is a fit due to Sehgal[32] and the triangles are his determination of the same quantity from electronproduction data.

b) $D_u^{\pi+} / D_u^{\pi-}$ as a function of x' averaged over z.

independent of z, but such predictions are quite hard to test.

A tremendous amount of data has recently become available and the most profitable procedure is probably to see if reasonable $f_i(x)$ (constrained to satisfy the quark sum rules) and D_i^h can be found which fit it all simultaneously. Here we only remark that

a) Such fits are possible - see e.g. ref. 32 or the recent review by Sehgal [33] who uses the latest data presented at the Hamburg conference.

b) As well as making simultaneous fits to the data, the $f_i(x)$ and $D_i^h(z)$ can be extracted from different experiments and compared directly. This has been done recently by Martin and Osborne [34] who used data on π^+ and π^- yields from SLAC electroproduction experiments on protons and neutrons. Their results must be interpreted cautiously, since the data do not agree very well with that obtained in other experiments [35], but nevertheless they are very interesting. Assuming $\overline{u}(x) = \overline{d}(x)$, the data can be used to obtain $\eta = D_u^{\pi^+} / D_u^{\pi^-}$ from

$$\frac{(\pi_p^+ - \pi_p^-) - (\pi_n^+ - \pi_n^-)}{(\pi_p^+ + \pi_p^-) - (\pi_n^+ - \pi_n^-)} = \frac{5(\eta-1)}{3(\eta+1)}$$

where $\pi_p^+ = d^2\sigma_{\gamma^* p \to \pi^+ \ldots} / dxdz \ldots$ etc). The value of η determined in this way is independent of x and consistent with the value obtained from neutrino experiments within the errors (± 25%). Similarly

$$\frac{(\pi_p^+ - \pi_p^-) - (\pi_n^+ - \pi_n^-)}{(\pi_p^+ - \pi_p^-) + (\pi_n^+ - \pi_n^-)} = \frac{5(\ u(x) - d(x)\ )}{3(\ u(x) + d(x)\ )}$$

gives a value for $u(x) / d(x)$ consistent with that obtained in inclusive experiments. Finally we remark that these data can be used to test the model by assuming arbitrary quark charges and constructing

$$r = \sqrt{\frac{1+y}{1-y}}$$

with

$$y = \frac{(\pi_p^+ + \pi_p^-) - (\pi_n^+ + \pi_n^-)}{\pi_p^+ + \pi_p^- + \pi_n^+ + \pi_n^-} \quad \frac{\pi_p^+ - \pi_p^- + \pi_n^+ - \pi_n^-}{(\pi_p^+ - \pi_p^-) - (\pi_n^+ - \pi_n^-)}$$

In the valence quark approximation, which is presumably valid as $x \to 1$, $r = |Q_u|/|Q_d|$. The data (fig. 25) are consistent with the expected value of 2.

TABLE I

Helicities amplitudes in the scattering of virtual photons on polarized nucleons.

Initial Helicities γ	Initial Helicities N	Possible transitions	Final Helicities γ	Final Helicities N
→	→	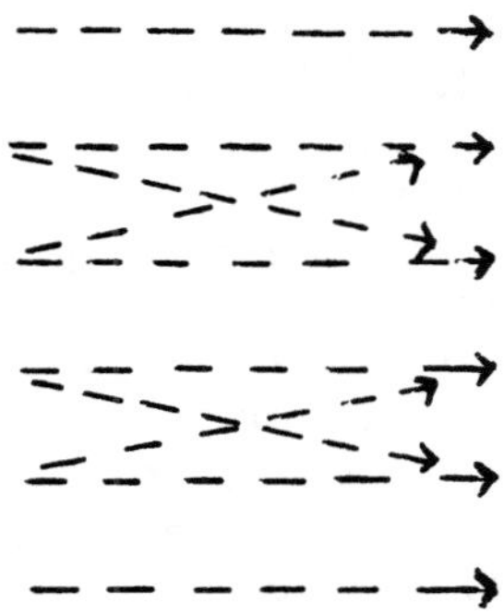	→	→
→	←		→	←
0	→		0	→
0	←		0	←
←	→		←	→
←	←		←	←

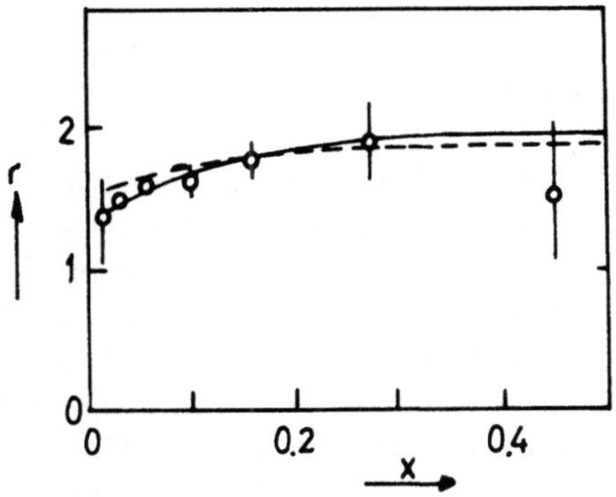

FIG. 25 : The combination of $\pi_p^{\pm}$ and $\pi_n^{\pm}$ discussed in the text which is equal to $|Q_u|/|Q_d|$ in the valence quark approximation [34].

c) Having determined the $f_i(x)$ and the $D_i^h(z)$, they can be used in turn to determine the neutral current couplings $\varepsilon_{L,R}^i$ (defined in section I.A6) from data on $\nu N \to \nu + \pi^{\pm}$. The inclusive cross-section only determines $|\varepsilon_{L,R}(u)|^2 + |\varepsilon_{L,R}(d)|^2$ on isoscalar targets but $\pi^{\pm}$ production measures $D_d^{\pi^\pm}|\varepsilon_{L,R}(u)|^2 + D_d^{\pi^\pm}|\varepsilon_{L,R}(d)|^2$. This allows an easy separation of $\varepsilon(u)$ and $\varepsilon(d)$ since $D_u^{\pi^+} / D_u^{\pi^-}$ is about 3. The results are in remarkable agreement with the Weinberg Salam model [33,36].

4. Hadronic Processes at large P_T.

It is now clear that in pp collisions particles are produced in jets at large P_T, the transverse momentum being compensated by another "opposite side" jet [37]. An obvious question is whether these are quark fragmentation jets. It has long been clear that if large P_T production is due to hard quark quark scattering, $d\sigma_{qq}$ does not satisfy the simple scaling law expected from naive dimensional analysis. This has led to the development of other models. However, there is now sufficient data to ask directly whether the jets have the same properties as those observed in lepton induced processes. Some months ago new data from a Caltech experiment [38] and a comprehensive analysis of all existing data by Feynman, Field and Fox [39] suggested the answer yes. However, recent data from the British French Scandinavian experiment [40] at the ISR cast doubt on this. In particular they find that the properties of the away side jet depend on the quantum numbers of the near side trigger particle, which is impossible if flavour independent qq scattering, followed by independent quark fragmentation, is responsible. On the other hand, since we know that 50% of the nucleon's momentum is carried by glue, qg and gg scattering should also be considered. If qg scattering is important (as expected by some authors [41]) it would produce the effects observed e.g. the ratio of positive to negative particles observed at large P_T opposite a large $P_T K^-$ trigger is much larger than for a $\pi^{\pm}$ trigger [42]. However this would also be a feature of models in which the basic process is quark meson scattering.

In conclusion, the situation is at present very complicated and a proper review would require many lectures. However, we can expect that in the next two or three years experiment will show whether hard scattering of elementary constituents is responsible for large P_T hadronic processes. Meanwhile theorists should decide whether they expect them to be described by the parton model (suitably modified to accomodate scaling violations). If so, should it work in the P_T range already explored (in which case, does it fit the data ?) or only in unexplored regions ?

II. Asymptotic freedom etc.

1. Lessons from Perturbation Theory.

Before becoming embroiled in the formal complexities of the renormalization group, it is instructive to consider deep inelastic scattering in perturbation theory. First we consider a theory with scalar fields with a ϕ^3 interaction in which the parton picture is valid and Bjorken scaling occurs. We then show that in theories with spin 1/2 particles scaling is broken by powers of log Q^2 in perturbation theory; in the next section we shall see that, when we go beyond perturbation theory, these logs sum up to powers of log Q^2 in asymptotically free theories and to powers of Q^2 in other models.

First, then, we consider a theory with scalar fiels ϕ^+, ϕ^- and ϕ^o with interaction $\lambda m\phi^+\phi^-\phi^o$. To zeroth order in λ, the structure function F_2 for scattering on a ϕ^+ is of course $\delta(1-x)$. To first order in λ there are two diagrams (fig. 26) which give

$$F_2 = \frac{\lambda^2 m^2 x^2}{16\pi^2} \int_{-\frac{m^2 Q^4}{s(s+Q^2)}}^{-(s+Q^2)} dt \left[\frac{1}{(t-m^2)^2} + 0\left(\frac{1}{st}\right) + 0\left(\frac{1}{s^2}\right)\right]$$

$$= \frac{\lambda^2 x^2 (1-x)}{16\pi^2 (1-x+x^2)} + 0\left(\frac{\lambda^2 m^2 \ln Q^2}{Q^2}\right)$$

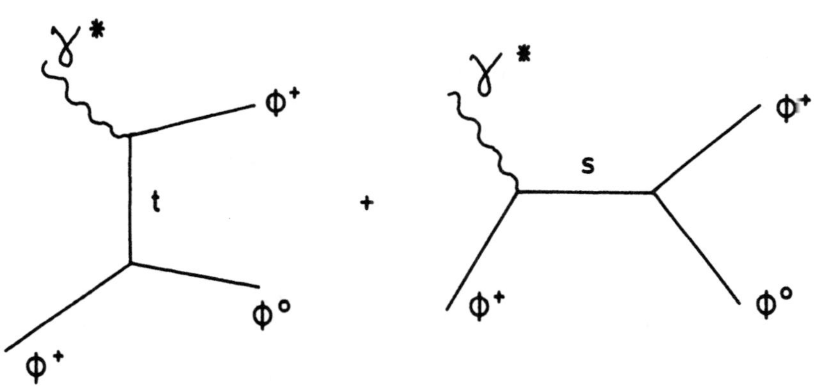

FIG. 26 : Diagrams contributing to $\gamma\phi \to \phi\phi$ to order λ.

$$F_T = \frac{\lambda^2 m^2 x(1-x)}{16\pi^2 Q^2} \int dt \left[\frac{1}{t-m^2} + O(\frac{1}{s}) \right]$$

where $F_T = F_1$. Note that

1) F_2 scales in this super-renormalisable theory when $\sqrt{Q^2} >> \lambda m$ - the scale set by the interaction.

2) The contribution which scales comes from the region $t \approx 0$ in the first diagram - i.e. it corresponds to scattering from an almost real particle with small transverse momentum and four momentum approximately equal to xp_μ, so the parton model works. Comparing the calculated value of F_2 with the naive parton model (p.m.) expression

$$F_2^{p.m.} = \int f(x,P_T^2)\ \delta(x - Q^2/2\nu)\ d\,x\ dP_T$$

we find a parton probability distribution

$$f(x,\ P_T^2)\ dP_T = \frac{\lambda^2 m^2 x^2 dt}{16\pi^2 (t-m^2)^2} + \ldots.$$

3) The second diagram gives a contribution of order $\lambda^2 m^2/s-m^2 = \lambda^2 m^2/2\nu(1-x)$, which is negligible in the scaling limit except as $x \to 1$ - as expected in the parton model.

4) F_T vanishes in the scaling limit - as expected for a spin zero parton.

5) For real or virtual scalar particles of momentum $\tilde{p}_\mu$ which absorb a photon of momentum q_μ, the current is $(2\hat{p}+q)_\mu$ and the contribution to F_T is proportional to $(\varepsilon^T.(2\hat{p}+q))^2 \sim (\vec{p}_T)^2$ in frames in which $\vec{p}$ and $\vec{q}$ are parallel. This parton-like result actually works quantitatively for the leading contribution to F_T. For a free particle of momentum $xp + p_T$ ($p_T^0 = 0$, $\vec{p}_T \cdot \vec{p} = 0$), it is easy to show that for $E >> |\vec{p}_T|$

$$F_T^{free} = \frac{x\vec{p}_T^{\,2}}{Q^2}\ \delta(x - Q^2/2\nu)$$

Folding this with $f(x,p_T^2)$, the naive parton model then gives

$$F_T^{p.m.} = \frac{1}{x} \int f(x,\ p_T^2) dx dp_T\ F_T^{free}$$

(the factor $1/x$ occurs because the definition of F_T [F_T^{free}] assumes covariant normalization to $2E[2E^{free} \simeq 2xE]$particles per unit volume, whereas f assumes normalization to one particle per unit volume). Substituting $f(x,p_T^2)$ and F_T^{free} into this expression and using $\vec{p}_T^2 = (1-x)t + O(t^2/Q^2) + O(m^2)$, we obtain the correct result for the leading contributions to F_T (which are of order $\lambda^2 m^2 \ln Q^2/Q^2$) although the model fails for the order $\lambda^2 m^2/Q^2$ terms.

To summarize : in ϕ^3 theory the naive parton model works for F_2 and F_T, with corrections which vanish like m^2/Q^2 and $1/\ln Q^2$ respectively (this is also true for $\gamma^* \phi^0 \to \phi^+ \phi^-$). Although the parton diagram dominates F_2 and $\langle p_T^2 \rangle$, it would not dominate $\langle p_T^4 \rangle$, for which the model would fail completely since $p_T^4 \sim t^2$ (setting $m = 0$) which kills the t^{-2} singularity which is the root of the success of the parton model in this case.

In super-renormalizable ϕ^3 theory scaling occurs when $\sqrt{Q^2} \gg \lambda m$ - the scale set by the interaction. In four dimensions, however, there are no super-renormalizable theories involving spin 1/2 particles. The best behaved theories are renormalizable and have dimensionless coupling constants. There is then no scale set by the interaction and no Bjorken scaling. In fact, as is plausible on dimensional grounds, the expression for F_2 involves

$$g^2 \int \frac{dt}{t - m^2}$$

which gives rise to $\ln Q^2/m^2$ scaling violations.

We now consider this explicitly in the Abelian gluon model, in which a massless neutral vector meson (gluon) V_λ is coupled to a charged fermion with interaction $g V_\lambda \bar{\psi} \gamma_\lambda \psi$. The order g diagrams for $\gamma^* \psi$ scattering are shown in fig. 27. Recalling the well known result that there are only ln m singularities in QED as $m \to 0$, it

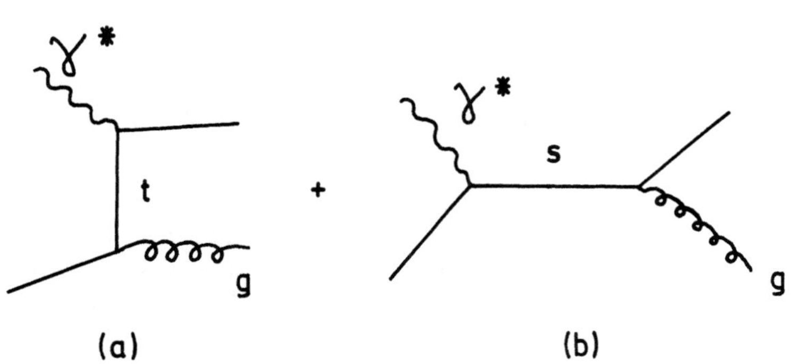

FIG. 27 : Diagrams contributing to $\gamma\psi \to \psi g$ to order g.

is clear that the $(t-m^2)^{-2}$ singularity from the propagator must be softened by factors in the numerator. In fact, explicit calculation reveals that :

$$F_2 = \frac{g^2 x(1+x^2)}{8\pi^2(1-x)} \ln(Q^2/m^2) + g(x) + O(m^2/Q^2)$$

Note that :

1) The result does not scale.

2) Both (diagram a)^2 and the a - b interference term could give $(t-m^2)^{-1}$ singularities and might therefore contribute to the leading term. In fact this is the case with some choices of gauge for the vector gluon. However, in the Coulomb gauge ($\varepsilon \cdot k = 0$) in which no contributions from unphysical states are included, the interference term is not singular and the "parton diagram" dominates the (non-scaling) result by log Q^2.

3) It turns out that F_L (which is zero in the naive parton model) scales with

$$F_L = \frac{g^2 x^2}{4\pi^2} + O\left(\frac{m^2}{Q^2}\right)$$

By using the fact that $F_L \sim p^\mu p^\nu W_{\mu\nu}$, it is easy to show that F_L comes entirely from diagram a, in which $|\vec{p}_T| \neq 0$ for the particle which absorbs the virtual photon. However, in contrast to F_T in ϕ^3 theory, the simple-minded parton connection between F_L and $\vec{p}_T^{\,2}$ does not work quantitatively. The parton distribution must be identified with the piece with the $(t-m^2)^{-1} \sim (\vec{p}_T^{\,2})^{-1}$ singularity, which dominates F_2. However, this singularity is killed when we calculate $\langle\vec{p}_T^{\,2}\rangle$ and the other non-singular contributions are equally important (nor can we connect F_L with $\langle\vec{p}_T^{\,2}\rangle$ of the particles in the final state, which gets contributions from both diagrams).

We conclude that :

a) Scaling is violated in perturbation theory in theories with spin 1/2 particles.

b) As $Q^2 \to \infty$ the contribution which dominates F_2 comes from a region $|t| < \varepsilon Q^2$, where ε is any small fixed number. The "struck parton" in diagram a, which dominates in Coulomb gauge, therefore has an invariant mass which is small relative to Q^2 (although in absolute terms it becomes infinite as $Q^2 \to \infty$) and we can treat it as a particle with momentum xp_μ in first approximation. Thus the parton interpretation survives in leading log approximation to this order in Coulomb gauge.

c) The parton picture gets more dubious the more detailed the questions we ask. In particular, although the naive $F_T \leftrightarrow \langle \bar{p}_T^2 \rangle$ relation works in ϕ^3 theory (by a log), the analogous $F_L \leftrightarrow \langle \vec{p}_T^2 \rangle$ relation fails for spin 1/2 particles.

We now turn to formal methods which enable us to go beyond perturbation theory. (recently I have succeeded in examining the leading contributions to all orders [43]. The results of this section generalize in asymptotically free theories and the dominant diagrams, which reproduce the results of the next sections, have a parton like structure in Coulomb gauge. The advantage of this approach is that it makes it possible to investigate the nature of the final states in deep inelastic processes. This is discussed briefly in Section II.6).

2. The Renormalization Group and The Operator Product Expansion[44].

Preliminaries

In perturbation theory

$$\int x^{n-2} F_2(x,Q^2)\,dx \xrightarrow{Q^2\to\infty} \sum_p (g^2)^p \sum_{i=o}^{p} a_i^{n,p} (\ln Q^2)^i$$

$$+ \frac{1}{Q^2} \sum_p (g^2)^p \sum_{i=o}^{p} b_i^{n,p} (\ln Q^2)^i + \ldots$$

in renormalizable theories (this is proved below). In this section we shall discuss techniques which enable us to analyse the asymptotic behaviour of the sum of all the terms in the first line (quantitatively in the case of asymptotically free gauge theories; qualitatively for fixed point theories) and to determine the coefficients $a_p^{n,p}$ explicitly.

There are two ingredients :

1) The renormalization group equations which apply to Green functions with external momenta $p_1, p_2 \ldots p_n$ in the "deep Euclidean limit"

$$p_i = (p_i^o, \vec{p}_i) \to \lambda(ip_i^o, \vec{p}_i)$$

$$\lambda \to \infty$$

2) The operator product expansion which connects this information to the behaviour of deep inelastic processes in the physical region.

The Operator Product Expansion

The operator product expansion expresses the product of local operators A(y) and B(x) in terms of operators $O_i(z)$ in the limit $(x - y) \to 0$:

$$A(y)B(x) \xrightarrow{(x-y) \to 0} \sum_i C_i(y-x)O_i(o)$$

where the C's are (generally singular) C number functions. It is essentially a statement of the completeness of the operators O_i (it applies for simple products, time ordered products, commutators etc...) [45]. It is important to realize that it is a genuine operator statement i.e. taking matrix elements

$$< \alpha|AB|\beta > \to \sum_i C_i < \alpha|O_i|\beta >$$

the C's are independent of α and β. This is analogous to the multipole expansion of

$$V(\mathbf{r}) = \int \frac{\rho(r')d^3r'}{|\vec{r}-\vec{r}'|} = \frac{Q}{r} + \frac{\vec{r}\cdot\vec{d}}{\mathbf{r}^3} + \ldots$$

in terms of standard potentials with ordered singularities, where only the Q, $\vec{d}$ ... depend on the particular problem considered. Similarly it is useful to order the singularities of the C's. If we define the dimensions of the operators A, B and O_i through

$$C_i(x-y) \xrightarrow{(x-y) \to 0} \left(\frac{1}{x-y}\right)^{d_A + d_B - d_{O_i}}$$

then the operator O_i with smallest dimension d_{O_i} will dominate the operator product in this limit.

In more detail we can write

$$A(y)B(0) \xrightarrow{y \to 0} \sum_i \sum_n C_n^i(y^2)y_{\mu_1}y_{\mu_2} \ldots y_{\mu_n} O^i_{\mu_i \ldots \mu_n}$$

where

$$C_n^i(y^2) \xrightarrow{y \to 0} \left(\frac{1}{y^2}\right)^{\frac{d_A+d_B-n-d_{O_n^i}}{2}}$$

In perturbation theory, the C_n^i have this form with the d's given by the naive (mass) dimensions of the operators (d = 3/2 for $\bar{\psi}$ ψ, d=2 for ϕϕ etc.) if we neglect polynomials in $\ln y^2$. If we supposed that the $\ln y^2$ factors were absent (which is only true in free field theory) or were independent of n (which is not true in perturbation theory), we could write

$$\sum_{i,n} C^i_n(y^2) y_{\mu_1} \ldots y_{\mu_n} O^i_{\mu_1 \ldots \mu_n} = \sum_i C^i(y^z) D(y)$$

This is the "light cone expansion" which enjoyed a considerable vogue at one time. However, it is now believed that such an expansion is true only in free field theory; it is therefore of no great interest.

Strategy of Applications to Deep Inelastic Scattering.

The deep inelastic structure functions W_1 and W_2 are given by

$$W_{\mu\nu} = (-g_{\mu\nu} + \frac{q_\mu q_\nu}{q^2}) W_1 + (p_\mu - \frac{q_\mu \nu}{q^2})(p_\nu - \frac{q_\nu \nu}{q^2}) \frac{W_2}{M^2}$$

$W_{\mu\nu}$ is the imaginary part of the forward virtual Compton amplitude

$$T_{\mu\nu} = i\int d^4y \, e^{iq\cdot y} \theta(y) < p|[J_\mu(y), J_\nu(0)]|p >$$

(this is the optical theorem). The strategy followed in using the operator product expansion/renormalization group is

1) Show that in the deep inelastic limit $W_{1,2}$ are controlled by $[J_\mu(y), J_\nu(0)]$ as $y \to 0$. Their behaviour is therefore determined by the behaviour of the most singular coefficients $C^i_n(y)$ in the operator product expansion.

2) Note that the same $C^i_n(y)$ describe deep inelastic scattering off a real nucleon and off a virtual quark since the C's are independent of the states used.

3) Consider the highly non-physical problem of deep inelastic scattering off a virtual quark in the deep Euclidean limit, in which $(p_{quark})^2 \to -\infty$ as well as $q^2 \to -\infty$. The renormalization group equations apply to this process and determine the behaviour of the dominant $C^i_n(y)$.

For simplicity we consider scalar currents $J(y)$. In the deep Euclidean limit $q_0 \to i\infty$ so it is obvious that the region $y_0 \to 0$ dominates the virtual Compton amplitude

$$T(\nu,Q^2) = i\int d^4y \, e^{iq\cdot y} \theta(y_0) < p|[J(y), J(0)]|p >$$

However

$$[J(y), J(0)] = 0$$

unless $y_o^2 > \vec{y}^2$, so $y_o \to 0$ forces $y \to 0$. We can therefore use the operator product expansion, which we write in the form

$$\theta(y_o)[J(y), J(0)] \to \sum_i \sum_n C_n^i(y^2) y_{\mu_1} y_{\mu_2} \cdots y_{\mu_n} O^i_{\mu_1 \cdots \mu_n}$$

we define

$$\int e^{iq\cdot y} C_n^i(y^2) y_{\mu_1} \cdots y_{\mu_n} d^4y \equiv \frac{\tilde{C}_n^i(Q^2) q_{\mu_1} \cdots q_{\mu_n}}{(Q^2/2)^n}$$

and

$$< p|O^i_{\mu_1 \cdots \mu_n}|p > = p_{\mu_1} \cdots p_{\mu_n} M_n^i + g_{\mu_1\mu_2} p_{\mu_3} \cdots \text{etc.}$$

Then in the deep Euclidean limit we have

$$T(\nu, Q^2) \to \sum_{i,n} \tilde{C}_n^i(Q^2) \left(x^{-n} M_n^i + O\left(\frac{x^{-n+2}}{Q^2} \right) \right)$$

where $x = Q^2/2\nu$ is the usual Bjorken variable. In this limit $x \to \infty$ and we are far from the physical region where $|x| < 1$ (and this series is divergent). However, we can connect the behaviour of the structure function

$$W = \frac{1}{2\pi} \text{Im } T$$

in the physical region to the deep Euclidean behaviour of T by using the fact that T is an analytic function of x with a cut from -1 to +1. (T is an analytic function of ν with a cut from $2\nu = Q^2$ to ∞ [neglecting M_π] and a cut for the crossed process from $-\infty$ to $2\nu = -Q^2$. It is therefore analytic in x with a cut from - 1 to +1. It is convenient to consider functions which are either even [e.g. F_2^e, $F_2^\nu + F_2^{\bar{\nu}}$]or odd [$F_2^\nu - F_2^{\bar{\nu}}$] when $x \to -x$). Integrating around the contour in Fig. 28 with the contour either shrunk around the cut, where the discontinuity is W, or at infinity we have

$$2\int_{-1}^{+1} x^{p-1} W(x,Q^2) dx = \frac{1}{2\pi i} \int_C x^{p-1} T(x,Q^2) dx$$

$$\xrightarrow{Q^2 \to \infty} \sum_i \tilde{C}_p^i(Q^2) M_p^i$$

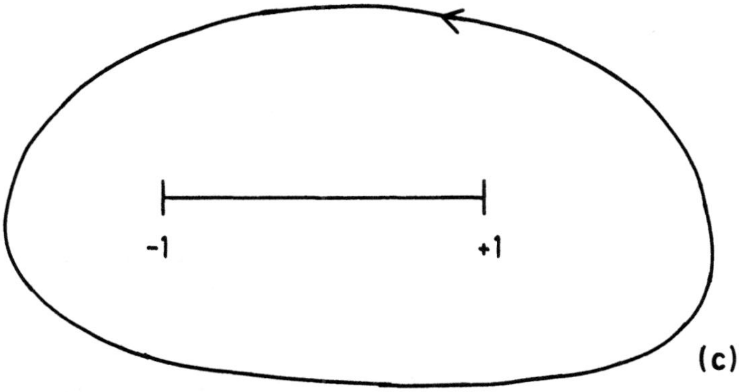

FIG. 28 : Contour in the x plane used to connect moments of structure functions with the deep Euclidean behaviour of the forward virtual Compton amplitude.

Thus as $Q^2 \to \infty$ the moments of the structure function are controlled by the coefficients of the operators of lowest dimension in the operator product expansion.

The Renormalization Group.

We now introduce the renormalization group equations which we shall use to study the $\tilde{C}_n^i(Q^2)$. We start from the observation that in the deep Euclidean limit masses are irrelevant. Consider some Feynman diagram with all external momenta space-like. If we Wick rotate all the loop momenta over which we must integrate $(q \to (iq_0, \vec{q}))$, all the internal momenta k_i will be space like. We then have $k_i^2 = (k_i^0)^2 - \vec{k}^2 \leqslant 0$ and can set all masses m_i equal to zero without introducing singularities (this is more or less obvious for simple diagrams but the general proof is hard). It might seem that with $m_i = 0$ there are no dimensional parameters and amplitudes would scale in terms of ratios of invariants in a way determined by dimensional analysis. This is not the case. The Feynman integrals still have ultraviolet divergences which must be regulated in some way (e.g. by introducing a cut off at $k_i^2 = -\lambda^2$) which necessarily introduces a parameter with dimensions [46].

We will regulate by "subtraction" at some Euclidean point characterized by a mass μ. For example, in QED (with $m_e \neq 0$) if the photon and electron propagators are given by

$$D_{\mu\nu}(q) = -\frac{d(q^2)}{q^2} g_{\mu\nu} + \text{gauge terms}$$

$$S(p) = \frac{1}{A(p^2)\not{p} + B(p^2)}$$

we define $d(-\mu^2) = A(-\mu^2) = 1$ (this is a choice of normalization for the fields - any other finite value would do) and $B(-\mu^2) = b$. We also define the value of the coefficient of γ_μ in the $\gamma\bar{e}e$ vertex to be g when the squares of all the external momenta are equal to $-\mu^2$. We can then subtract the integral expressions for the two point and three point functions at the points where they are defined. This renders them finite. In fact, with these definitions, all amplitudes are finite functions of the two parameters b and g - the theory is renormalized (defining d, A, B and the $\gamma\bar{e}e$ vertex amounts to wave function, mass and coupling constant renormalization). The two parameters b and g can be adjusted to give the measured values of the two parameters usually used in QED (m_e and e) - or any other two quantities which are accurately measured; all other quantities are then predicted in terms of these parameters.

The renormalization group equations express the fact that μ is an arbitrary parameter which is introduced purely to define the theory but is without physical significance. If we change the value of μ, all Green functions are left unchanged if we suitably adjust d, A, b and g (or corresponding quantities in other theories). This invariance of the physics under changes of μ is particularly simple to express in the deep Euclidean region, where we can drop the m_i and the n point Green function is given by

$$\Gamma^{(n)} = \Gamma^{(n)}(p_i, \mu, g, \phi),$$

where ϕ is a generic field and the dependence on ϕ indicates that $\Gamma^{(n)}$ depends on the normalization of ϕ. Changing μ to $\mu + \delta\mu$

$$\Gamma^{(n)}(p_i,\mu,g,\phi) = \Gamma^{(n)}\left(p_i,\mu + \delta\mu, g + \beta(g)\frac{\delta\mu}{\mu}, \left(1 + \gamma(g)\frac{\delta\mu}{\mu}\right)\phi\right)$$

where, because this relation is independent of p_i, β and γ are dimensionless functions which can only depend on g(γ is called the anomalous dimension of ϕ). Making a Taylor expansion, we obtain the renormalization group equation (RGE) of Peterman and Stueckelburg and Gell-Mann and Low.

$$\left[\mu\frac{\partial}{\partial\mu} + \beta(g)\frac{\partial}{\partial g} + n\gamma(g)\right]\Gamma^{(n)} = 0$$

(with several fields we obtain matrix equations; another complication occurs in gauge theories where there will in general be an extra term expressing the fact that we must also change the gauge when we change μ if we want to leave unchanged the Γ^n, which are in general gauge dependent). The industrious theoretical student is urged to study also another derivation of the RGE which starts from the scheme in which Green functions are calculated in terms of bare masses and coupling constants, the divergences being absorbed in multiplicative functions of the cut off [47] (this leads first to the Callan-Symanzik equation and then to the RGE).

Before showing that the Wilson coefficients $\tilde{C}_n^i(Q^2)$ which control the moments of the deep inelastic structure functions satisfy an equation of the RGE type, we indicate how $\beta(g)$ and $\gamma(g)$ may be calculated. If we calculated the two and three point functions in QED with $m_i = 0$, ordinary dimensional analysis shows that we would find

$$\Gamma^{(2)}_{\psi\psi} = \frac{1}{\not p}[1 + g^2 f(p^2/\mu^2) + \ldots$$

$$\Gamma^{(2)}_{A_\mu A_\nu} = -\frac{g_{\mu\nu}}{p^2}[1 + g^2 k(p^2/\mu^2) + \ldots$$

$$\Gamma_{A_\mu\psi\psi} = \gamma_\mu(g + g^3\psi(p_i/\mu) + \ldots$$

Substituting these expressions into the RGE we obtain

$$g^2\mu\frac{\partial f}{\partial\mu} + 2g\beta f + 2\gamma_\psi + \ldots = 0$$

$$g^2\mu\frac{\partial k}{\partial\mu} + 2g\beta k + 2\gamma_A + \ldots = 0$$

$$g^3\mu\frac{\partial\psi}{\partial\mu} + \beta + (2\gamma_\psi + \gamma_A)g + \ldots = 0$$

From these equations we deduce

a) that

$$\gamma_i = A_i g^2 + \ldots$$

$$\beta = Bg^3 + \ldots$$

A knowledge of f, k and ψ would give us the A_i and B explicitly (e.g. $A_\psi = (1/2)\mu\,\partial f/\partial\mu$).

b) f, k and ψ must have the form constant $\times\ln(p^2/\mu^2)$ (we could guess that, to any order, the lowest order results are only modified by logarithms from the success of the renormalization programme; if power modifications were introduced, diagrams would become more divergent upon adding propagator and vertex corrections).

RGE for the Wilson Coefficients.

We now wish to show that the local operator matrix elements M_n^i, the Compton amplitude T and the Wilson coefficients $\tilde{C}_n^i$ satisfy the RGE. This is not obvious since we have only shown that the RGE apply to Green functions. We must first show how to calculate current and operator matrix elements in terms of Green functions. For operators $O_n^i(x)$ we use the standard trick of changing the Lagrangian

$$L \to L + \sum_{n,i} J_n^i(x) O_n^i(x)$$

where $J_n^i(x)$ are C number functions. We then calculate Green functions in the presence of the extra term, functionally differentiate with respect to J_i^n and set $J_n^i = 0$ to obtain

$$< \alpha|O_n^i|\beta > \; = \; \frac{\delta}{\delta J_n^i} \; \Gamma^J_{\alpha,\beta}\Big|_{J=0}$$

$\Gamma^J_{\alpha,\beta}$ is invariant when we change $\mu \to \mu + \delta\mu$ provided we also change $g \to g + \beta\, \delta\mu/\mu$, $\phi_i \to (1+\gamma_i\delta\mu/\mu)\phi_i$ and $J_i^n \to (\delta_{ij} + \gamma_{ij}^n\,\delta\mu/\mu)$ J_n^i i.e. the matrix elements of O_i^n satisfy the RGE

$$([D + \gamma_\alpha + \gamma_\beta]\delta_{ij} + \gamma_{ij}^n) < \alpha|O_{ij}^n|\beta > \; = 0$$

where

$$D \equiv \mu \frac{\partial}{\partial\mu} + \beta \frac{\partial}{\partial g} \, .$$

Note the possibility of "mixing" between different operators with the same Lorentz structure (same n) and the same internal quantum numbers. If two operators O_n^i and O_n^j are allowed by charge conjugation and SU(m) invariance (where m is the number of flavours) to couple to the same C number source J_n^i, then Green functions will only remain invariant when $\mu \to \mu + \delta\mu$ if we turn on an order $\delta\mu$ coupling between J_n^i and O_n^j i.e. $\gamma_{ij}^n \neq 0$. This should become clearer below when we indicate how the matrix elements of O_n^i are renormalized.

In the case of interest to us, there are two types of operators which can mix : SU(m) flavour singlet operators which are bilinear in the quark fields, of the type $\bar{\psi}\gamma_\mu D_\nu \psi$, $\bar{\psi}\gamma_\mu D_\nu D_\eta \psi$ etc., where D_μ is the covariant derivative, and bilinears in the gluon fields, of the type $F^{\alpha\mu} F^{\nu\alpha}$, $F^{\alpha\mu} D^\nu F^{\eta\alpha}$ etc. (we are only concerned with colour singlet operators). We denote these by $O_{n,s}^i$.

where s stands for singlet and i = 1 for quark operators and i = 2 for gluon operators. We consider the matrix elements of these operators between one quark (α = 1) or one gluon (α = 2) states $|\alpha,p>$ of momentum p. When calculated perturbatively, these matrix elements turn out to be logarithmically divergent. We adopt the definition

$$< \alpha,p|O^i_{n,s}|\alpha,p >|_{p^2 = -\mu^2} = \delta_{i\alpha}$$

to render them finite (where Lorentz indices are suppressed). We then find

$$< \alpha,p|O^i_{n,s}|\alpha,p > = \delta_{i\alpha} + g^2 b^n_{i\alpha} \ln(-p^2/\mu^2) + \ldots$$

which gives the γ^n_{ij} when substituted into the RGE. We now reconsider operator mixing. Except in lowest order, quark operators will in general have non-zero gluon matrix elements and vice-versa (i.e. $b^n_{i\alpha} \neq 0$ for $i \neq \alpha$) but we can define them to be zero at $p^2 = -\mu^2$. If we then change $\mu \to \mu + \delta\mu$, we can check using the equation above that the redefinition/mixing $O^i_{n,s} \to O^{i'}_{n,s} = (\delta_{ij} - 2\delta\mu/\mu \; g^2 b^n_{ij}) O^j_{n,s} + O(g^4)$ is necessary if the matrix elements are not to change.

The Compton amplitude T_α on a target α satisfies the RGE

$$(D + 2\gamma_J + 2\gamma_\alpha)T_\alpha = 0$$

in the deep Euclidean region, where γ_J is the anomalous dimension of the current. In fact γ_J is zero for vector and axial vector currents, which are conserved or would be conserved if there were no mass terms in the Lagrangian, because their matrix elements are not renormalized [48]. We apply the RGE to the moments of T_α, using the asymptotic expansion

$$x^{n-1} T_\alpha \to \sum_i \tilde{C}^i_n(Q^2) M^{i,\alpha}_n$$

Using the RGE

$$([D + 2\gamma_\alpha]\delta_{ij} + \gamma^n_{ij}) M^{i,\alpha}_n = 0$$

derived above, and then setting $p^2 = -\mu^2$ so that $M^{i,\alpha}_n = \delta_{i\alpha}$, we obtain

$$[D\delta_{ij} - \gamma^n_{ij}] \tilde{C}^j_n = 0$$

Finally introducing the variable $t = 1/2 \ln Q^2/\mu^2$ we get

$$\left[\left(-\frac{\partial}{\partial t} + \beta(g) \frac{\partial}{\partial g} \right) \delta_{ij} - \gamma^n_{ij} \right] \tilde{C}^j_n(t) = 0$$

which is the form in which we shall actually use the RGE.

Two Applications.

We already know that in the Abelian gluon model for a "quark" of unit charge

$$F_2(x,Q^2) = \delta(1-x) + g^2 f(x) t + \ldots$$

We now wish to use the fact that the moments of F_2 satisfy the RGE to verify the statements in the first paragraph of this section and calculate an anomalous dimension. Since $f(x) \sim 1/1-x$ as $x \to 1$, its moments are actually singular. However, this infrared divergence is cancelled by another in the $O(g^2)$ corrections to the elastic from factor which modify the $\delta(1-x)$ contribution to F_2.

In fact the net result is that

$$\int dx x^{n-2} F_2 = 1 + \frac{g^2 t}{4\pi^2} \left[\frac{2}{n(n+1)} - 1 - 4 \sum_{j=2}^{n} \frac{1}{j} \right]$$

Note that the $O(g^2 t)$ term vanishes for $n = 1$. This has a simple parton-model interpretation : to this order there is no pair creation and the number of quarks in a quark ($\int q(x)dx = \int x^{-1} F_2 (x,Q^2)dx$) is equal to one. In fact we could easily have used this requirement to construct the moments uniquely from $f(x)$.

With two "quarks" p and n with $Q_p = 1$, $Q_n = 0$, $F_2^{en} = 0$ in this order so we know (to $O(g^2)$) the moments of the non-singlet structure function $F_2^{NS} = F_2^{ep} - F_2^{en}$ which satisfy the unmixed RGE

$$\left[-\frac{\partial}{\partial t} + \beta \frac{\partial}{\partial g} - \gamma^n \right] \int x^{n-2} F_2^{NS} dx = 0$$

without mixing. Substituting the expression above we obtain

$$\gamma^n = \frac{g^2}{4\pi^2} \left[\frac{-2}{n(n+1)} + 1 + 4 \sum_{j=2}^{n} \frac{1}{j} \right]$$

(the j^{-1} term is characteristic of all gauge theories since it is a direct consequence of the $(1-x)^{-1}$ soft gluon emission singularity). It is now straightforward to show that the moments of F_2 have the form stated at the beginning of the section. Furthermore, given

$$\beta = bg^3 + \ldots$$

we can easily find all the coefficients $a_p^{n,p}$ or equivalently solve the RGE in leading log approximation to obtain [49]

$$\int x^{n-2} F_2^{NS} dx \to (1-2bg^2 t)^{\gamma_n / 2bg^2}$$

As a second example we shall find exact solutions of the RGE for $t \to \infty$ (in the non-singlet case for simplicity). First we solve the equation

$$\left(-\frac{\partial}{\partial t} + \beta \frac{\partial}{\partial g} \right) f = 0.$$

The solution is

$$f = f(\bar{g}\ (g,t))$$

where $\bar{g}$ satisfies the ordinary differential equation

$$\frac{d\bar{g}}{dt} = \beta(\bar{g})$$

subject to the boundary condition $\bar{g}(t = 0) = g$. The fact that this is a solution can be checked by noting that solutions with different g's are related by a translation of the origin in t :

$$\bar{g}(g,t + \delta t) = \bar{g}(g + \delta g,t)$$

where $\delta g = \beta(g)\delta t$ (this is illustrated in fig. 29) so that

$$\beta(\bar{g}) = \frac{\partial \bar{g}(g,t)}{\partial t} = \beta(g) \frac{\partial \bar{g}}{\partial g}$$

The solution of the complete equation is then obviously given by

$$C_n(t) = f(\bar{g}(g,t))\exp\left[-\int_0^t \gamma^n(\bar{g}(g,t'))dt' \right]$$

$$= f(\bar{g}(g,t))\exp\left[-\int_g^{\bar{g}(g,t)} \frac{\gamma^n(g')}{\beta^n(g')} dg' \right]$$

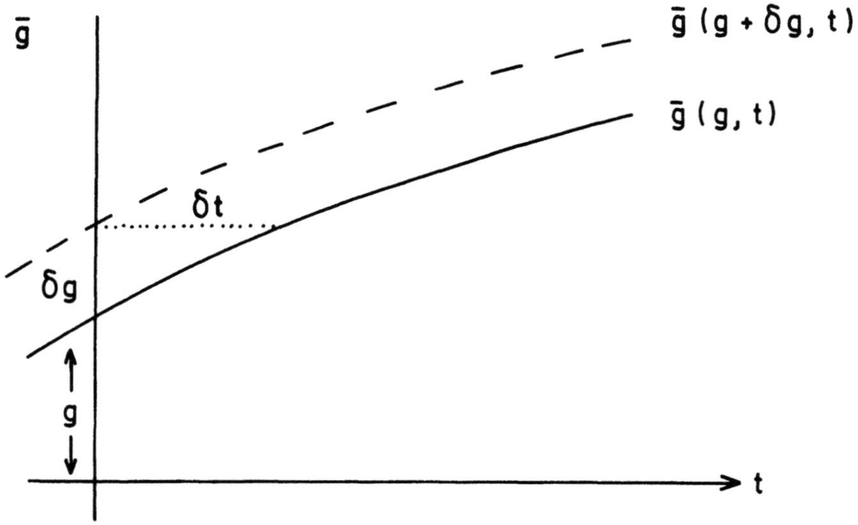

FIG. 29 : Plot showing that $\bar{g}(g+\delta g,t) = \bar{g}(g,t+\delta t)$ with $\delta g = \beta(g)\delta t$.

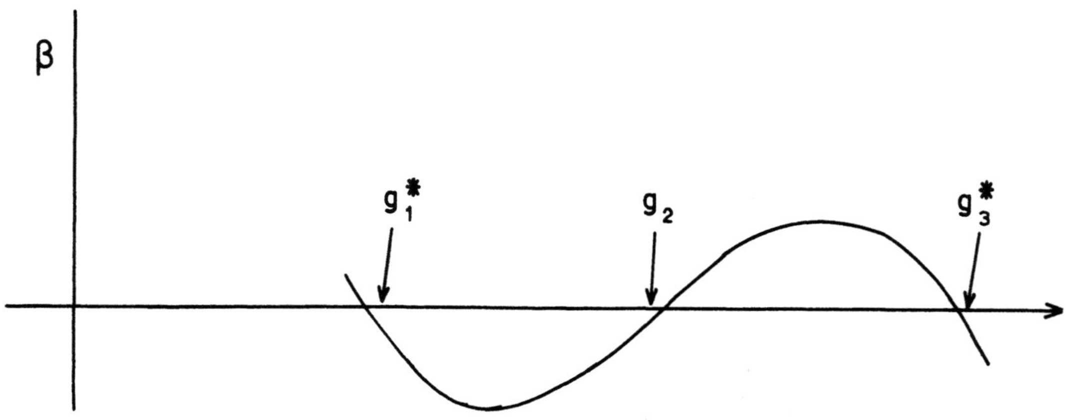

FIG. 30 : Possible behaviour of the β function in a fixed point theory.

Let us try to understand qualitatively what is going on. Since C depends only on Q^2/μ^2, the effect of changing $Q^2 \to Q^2\,(1 + 2\,\delta\mu/\mu)$ at fixed μ is the same as that of changing $\mu \to \mu - \delta\mu$ at fixed Q^2. However, this is equivalent to changing $g \to g + \beta\,\delta\mu/\mu$, $\psi \to (1+\gamma_\psi\,\delta\mu/\mu)\psi$ etc at fixed μ. Hence the effect of increasing Q^2 by δQ^2 is the same as that of changing $g \to g + \beta\delta Q^2/2Q^2$ and rescaling fields at fixed Q^2.

Thus the limit $Q^2 \to \infty$ can be studied by considering a theory with an effective coupling constant $\bar{g}$ which either increases or decreases with Q^2 depending on the sign of β.

Suppose $\beta(g)$ has the shape illustrated in fig. 30. If $g_1^* < g < g_2$ initially, β is everywhere negative and $\bar{g}$ decreases as Q^2 increases until asymptotically it reaches the "fixed point" $\bar{g} = g_1^*$ where $\beta = 0$. If $g_2 < g < g_3^*$ initially, $\bar{g}$ increases with Q^2 tending to the fixed point at g_3^* asymptotically. If the zero at $g = g^*$ is linear and we start from reasonably close by we can put

$$\beta(g) \simeq b(g^* - g)$$

$$\gamma^n \simeq \gamma^n(g^*).$$

The RGE is then solved with

$$\bar{g}(t) = g^* + e^{-bt}(g-g^*)$$

and

$$C_n(t) \to f_n(g^*)e^{-\gamma^n(g^*)t} \sim (Q^2)^{\gamma^n(g^*)/2}\ .$$

Thus in theories with fixed points at $g \neq 0$, the moments of the structure functions behave like powers of Q^2 asymptotically.

Now consider the case illustrated in fig. 31 and suppose that initially g is to the left of any zeros in β. In this case $\bar{g} \to 0$ as $t \to \infty$. This is asymptotic freedom. It only occurs in non-Abelian gauge theories (in all other cases β is positive for small g). For $t > t_0$ with t_0 large enough that $\bar{g}(t_0) \ll 1$ we can put

$$\beta \simeq -bg^3$$

$$\gamma^n \simeq c^n g^2$$

where b and c^n can be calculated perturbatively as discussed above. Then

$$\bar{g}^2 = \frac{g^2}{1 + 2bg^2 t}$$

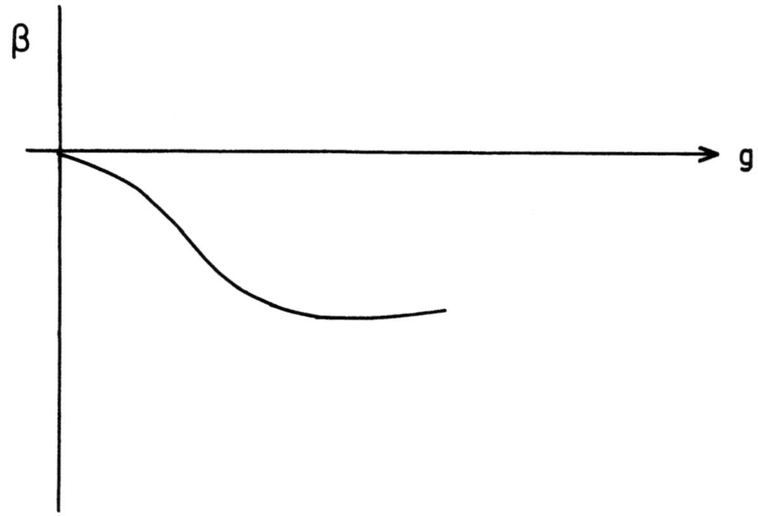

FIG. 31 : Behaviour of β in an asymptotically free theory.

and

$$C_n(t) \rightarrow f_n(0)(t)^{C^n/2b}$$

i.e. the moments of the structure functions behave like powers of $\ln Q^2$ in asymptotically free theories.

In the non-singlet case the solution of the RGE is a little harder [44,45]. However, asymptotically we can effectively diagonalize the γ^n_{ij} and decouple the RGE to obtain

$$\tilde{C}^{\prime i}_{n,s} = K_{ij}\,\tilde{C}^j_{n,s}$$

$$\rightarrow f^i_n\,(\bar{g}(g,t))\,\left[\frac{\alpha_s(t)}{\alpha_s(t_o)}\right]^{C^i_n}$$

where $\alpha_s(t) = \bar{g}^2(t)\,/\,4\pi$ and K_{ij} is a (non-unitary) matrix which can easily be constructed explicitly. Calculated perturbatively in $\bar{g}$, $\tilde{C}^{glue}_{n,s} = O(\bar{g}^2)$ which expresses the fact that to zeroth order deep inelastic scattering measures the quark distribution. Normalizing $\tilde{C}^{quark}_{n,s}(t_o) = 1$, we have $f^i_n(\bar{g}) = K_{i\,\text{quark}} + O(\bar{g}^2)$ so that

$$\int x^{n-2}F^s_2(x)\,dx \rightarrow K^{-1}_{ij}\,K_{j\,\text{quark}}\left[\frac{\alpha_s(t)}{\alpha_s(t_o)}\right]^{C^j_n} < p|O^i_n|p > + O(\bar{g}^2)$$

We see that, by definition, the moments measure the matrix elements of the quark operator or, equivalently, the quark distribution at $t = t_0$.
However, for $t \neq t_0$ they are sensitive to the $t = t_0$ quark and gluon distributions. We can say that deep inelastic scattering at any t measures only the quarks (since t_0 is arbitrary) but the distribution of quarks at one t depends on the distribution of glue at other t's and vice-versa; this interpretation is discussed in detail in the next section.

Turning to $\bar{e}e$ annihilation, we can apply the RGE directly to the photon propagator as $q^2 \to -\infty$. Assuming that the analytic continuation of the dominant contribution as $q^2 \to -\infty$ gives the dominant contribution as $q^2 \to +\infty$, we can take the discontinuity of the photon propagator to obtain the famous result that

$$\frac{\sigma(e^+e^- \to \text{hadrons})}{\sigma(e^+e^- \to \mu^+\mu^-)} = \sum_i Q_i^2 + O(\bar{g}^2)$$

in asymptotically free theories, where i runs over all quarks and the $O(\bar{g}^2)$ term is calculable. The continuation should be valid provided there are no oscillations asymptotically (it could be spoiled by terms like $e^{-\sqrt{-q^2}}$). If there are oscillations, the prediction may still apply for the cross-section averaged over an energy range large compared to the oscillations [50].

3. The Scale Invariant Parton Model.

We have already assembled all the machinery we need to make predictions for the behaviour of the structure functions and confront them with experiment. First, however, we shall discuss an intuitive picture of deep inelastic scattering developed by Kogut and Susskind [51] (following work of Wilson, Kadanoff and Polyakov) which gives considerable insight into the formal results.

When virtual photons of momentum Q_1 are scattered from a target they are able to resolve constituents of size proportional to $1/\sqrt{Q_1^2}$ (for fixed $x \neq 0$) but are not sensitive to substructure on a much smaller scale. When the momentum is increased, however, substructure may be revealed. Further increases may reveal further substructure, as illustrated in fig. 32. This process of successively resolving clusters made of smaller clusters which are made of smaller clusters etc. is easiest to visualize if we suppose that there is a relatively sharp transition between length scales, as occurs in the successive resolution of molecules, atoms, nuclei, nucleons, quarks ... (although quark pairs are of course unveiled continuously in a quark field theory as quarks are probed at shorter and shorter distances).

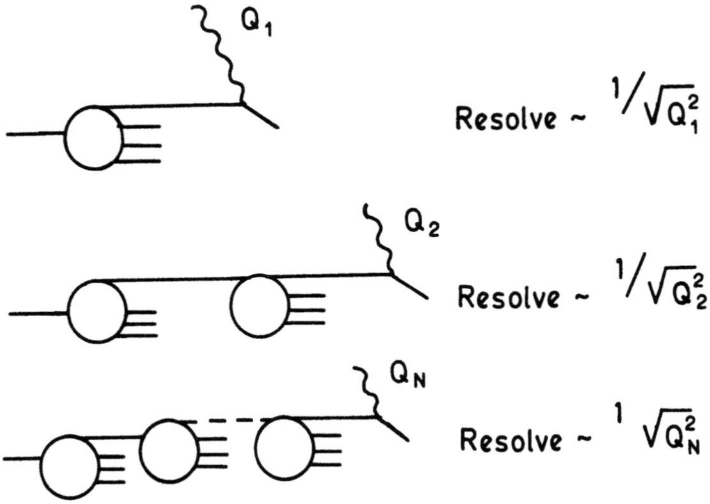

FIG. 32 : Resolution of successive layers of substructure by virtual photons of increasing momentum.

In order to discuss scattering at momentum Q_N we need an N-dependent effective Hamiltonian H_N which describes the $(N-1)^{th}$ layer in terms of the Nth layer constituents, e.g. at the stage at which quasi-elastic scattering from single nucleons dominates we need a Hamiltonian which describes the nucleus in terms of nucleons but we can ignore the existence of nucleon substructure. At the quark level, the nature of the constituents involved does not vary with Q^2 but the Hamiltonian $H(Q^2)$ will depend on an effective coupling constant $g(Q^2)$ which changes with Q^2 as different sizes of cluster are resolved. We consider three possibilities as $Q^2 \to \infty$:

a) $g^2 \to 1/(Q^2)^P$. In this case the constituents would behave like free particles at large Q^2 and the naive parton model would be valid. We have already seen that this can only occur in superrenormalizable theories in which there is a scale in the interaction. However, there are no superrenormalizable theories involving spin 1/2 particles. In renormalizable theories like QED or QCD the coupling constant is dimensionless which means that there is no distance scale below which interactions can be neglected; a virtual photon never has enough resolving power to reveal totally naked partons. However, by imposing a transverse momentum cut off on a renormalizable theory (which spoils the locality and Lorentz covariance), it can be made to behave like a superrenormalizable theory and the parton picture will be valid for $Q^2 << (\vec{p}_T{}^2)_{max}$; this was the basis of an interesting investigation of parton models due to Drell, Levy and Yan [52].

b) $g^2 \to g_*^2 \neq 0$. This "fixed point" behaviour could occur in any field theory. If it occurs, the relation of one layer to the next becomes universal in a sense we shall discuss below.

c) $g^2 \to 1/\ln Q^2$. This is asymptotic freedom. It can only occur in non-Abelian gauge theories.

Before exploring the consequences of the Q^2 dependent parton picture corresponding to cases b) and c) we consider the notion of an effective coupling constant in a little more detail. A classical analogue is given by the interaction between two charges in di-electric. The potential is

$$V = \frac{Q_1 Q_2}{4\pi\varepsilon r}$$

where ε differs from ε_0 due to polarization of the intervening molecules which shields the charges (so $\varepsilon > \varepsilon_0$). For r very much less than the molecular spacing d, however, there will be no shielding and $\varepsilon \to \varepsilon_0$. We can define an effective charge

$$Q_{eff}(r) = Q\sqrt{\frac{\varepsilon_0}{\varepsilon}}$$

which is equal to Q for $r \ll d$ and decreases for $r \sim d$ to a constant value for $r \gg d$. In terms of Q_{eff}, the potential is given by the same expression as in a vacuum

$$V = \frac{Q^1_{eff}\, Q^2_{eff}}{4\pi\varepsilon_0 r} .$$

This is very similar to the situation in field theory. We can either use a fixed coupling constant, defined, e.g., for on shell particles and zero momentum transfer $q^2 = 0$, which will be "shielded" for $q^2 \neq 0$ by vertex corrections etc. which we must calculate (this is analogous to using the first expression for V above), or we can calculate an effective coupling constant in terms of which the vertices are simple. For example, in the diagram in fig. 33 we can replace everything inside the dotted line by the bare vertex with coupling $g(p^2)$ if all lines have invariant mass p_i^2 or order p^2 which is large and space-like (by a dotted line cutting a vertex correction, I mean that if, e.g., the complete propagator is $A(p^2)/\not{p}$ then $\sqrt{A}$ is absorbed in $g(p^2)$; space like momenta are needed to avoid poles and cuts).

In QED vacuum polarization diagrams (e.g. fig. 34) give rise to shielding and the effective charge decreases with separation, tending to its experimental value as $r \to \infty$ (vertex corrections cancel corrections to the fermion propagator because of the Ward identity); in the language of the last section β is positive for

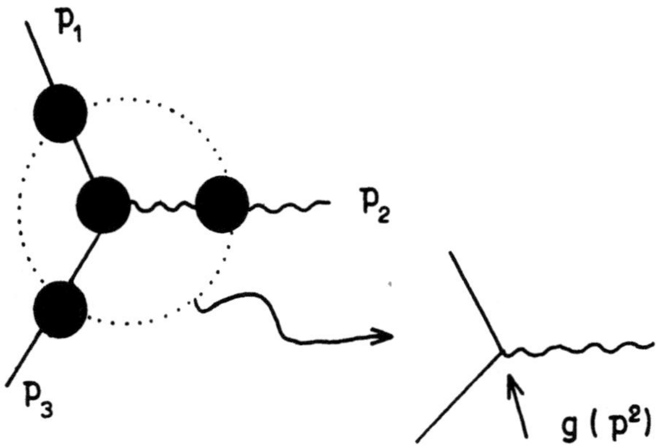

FIG. 33 : Absorption of vertex and propagator corrections in the effective coupling constant $g(p^2)$ for large spacelike momenta.

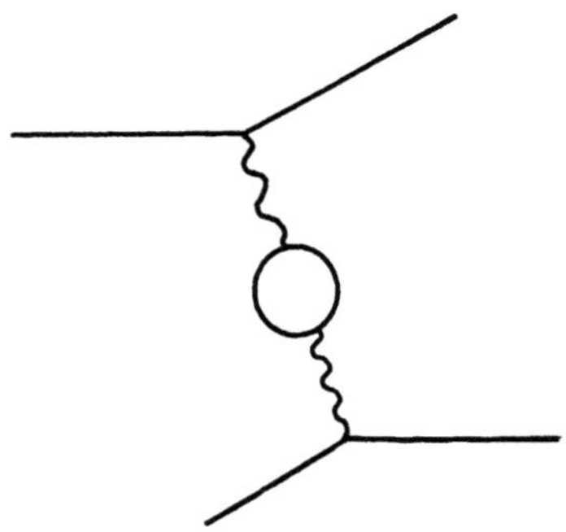

FIG. 34 : Vacuum polarization diagram which shields charges in QED.

small g. In non-Abelian gauge theories, however, there are additional diagrams, such as that in fig. 35. Detailed calculation shows that there is antishielding; in our previous language $\beta < 0$ for small g. In fact the antishielding is infinite so we need $Q_{eff} \xrightarrow{r \to 0} 0$ if it is to be finite for $r \neq 0$. No-one has yet succeeded in "explaining" this result physically (we cannot even say which diagrams are responsible, since their relative contributions are gauge dependent, but we might imagine that the diagram in fig. 35 "spreads out" the charge density making it vanish as $r \to 0$).

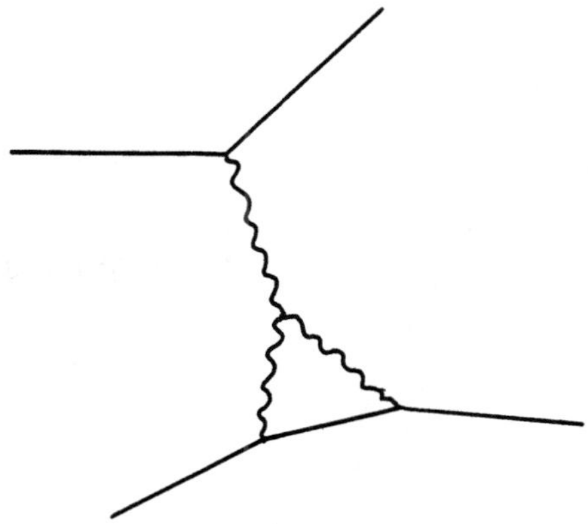

FIG. 35 : Example of an additional diagram in QCD.

Keeping a discrete index N, we can define probability distributions for partons of type i at the Nth level ($F_N^i(x)$) and the probability $f_N^{ij}(y)$ of resolving parton type j with momentum $y\vec{K}$ at the Nth level inside parton type i with momentum $\vec{K}$ at the $(N-1)^{th}$ level. This is illustrated in fig. 36. In terms of these distributions, this picture has the following implications :

a) Momentum conservation reads

$$\sum_i \int x F_N^i(x)\,dx = 1$$

However, since the number of partons increases indefinitely with N we expect

$$< x >_N = \sum_i \int x^2 F_N^i(x)\,dx \xrightarrow[N \to \infty]{} 0$$

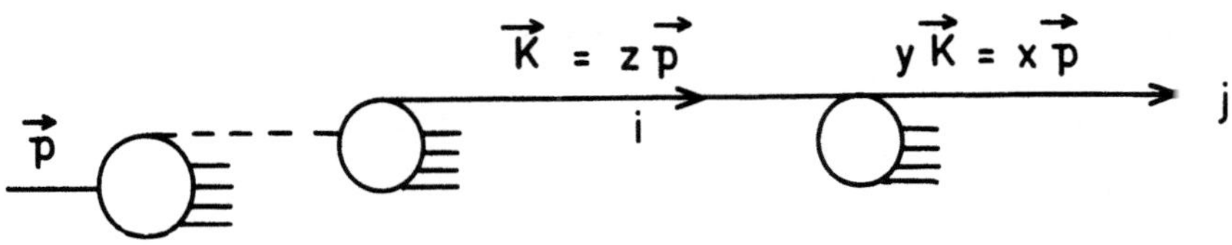

FIG. 36 : Probabilities in the "scale invariant" parton model.

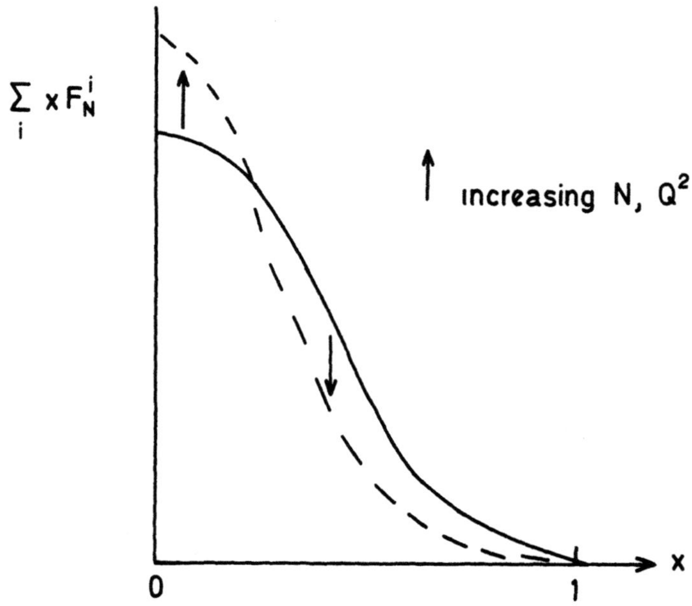

FIG. 37 : Expected Q^2 dependence of $F_2 \sim \Sigma\ x\ F_N^i$.

Thus $\sum_i xF_N^i$ will behave in the way illustrated in fig. 37. Since $F_2^e = \sum_i Q_i^2 xF_N^i$, the picture predicts that F_2^e also increases at small x and decreases at large x with increasing Q^2 - a trend seen in the data.

b) Since the distributions must yield the correct conserved quantum numbers B, I_3, Q ... for all N, the parton sum rules and inequalities will be satisfied.

c) The probabilities F_{N-1}, F_N and f_N are obviously related by

$$F_N(x) = \int_0^1 F_{N-1}(x/y) f_N(y) \frac{dy}{y}$$

(we temporarily ignore the matrix nature of f_N^{ij} for simplicity). Taking moments of this equation we obtain

$$\frac{\int x^p F_N(x)\,dx}{\int x^p F_{N-1}(x)\,dx} = \int y^p f_N(y)\,dy$$

Since specific models will tell us something about f, we see why it is the behaviour of the moments of the structure function which will be predicted directly.

In the case of a fixed point theory ($g_N^2 \rightarrow g_*^2$)

$$\int y^p f_N(y)\,dy = q_p$$

is independent of N (note that $q_1 = 1$, $q_p < 1$ for $p > 1$, if there is a single species of constituent). The ratio of the size of successive clusters is assumed to be N-independent

$$\frac{R_N}{R_{N+1}} = \Lambda$$

and we have

$$\frac{R_o}{R_N} = \sqrt{\frac{Q^2}{Q_o^2}} = \Lambda^N$$

N is therefore related to Q^2 by

$$N = \frac{1}{2}\,\frac{\ln Q^2/Q_o^2}{\ln\Lambda}$$

and

$$\frac{\int x^p F_N(x)\,dx}{\int x^p F_o(x)\,dx} = (q_p)^N = \left(\frac{Q^2}{Q_o^2}\right)^{\frac{\ln q_p}{2\ln\Lambda}}$$

- we have recovered the result that the moments decrease as powers of Q^2 in a fixed point theory.

In the case of non-Abelian gauge theories which are asymptotically free

$$g^2(t) = c/t$$

where $t = 1/2\ \ln Q^2/\mu^2$ and c is calculable. Since g is small for large Q^2, the function f(y,t) which relates the distribution F at t and t + δt can be calculated perturbatively from graphs such as the one in fig. 38 [53]. A prescription for finding f(y,t) corresponding to this diagram is given in ref. 53. Another is given by noting that we want the distribution probed by virtual photons. But the relevant diagrams (fig. 39) are just those from which we calculated the dominant contribution to F_2 (in Coulomb gauge) in

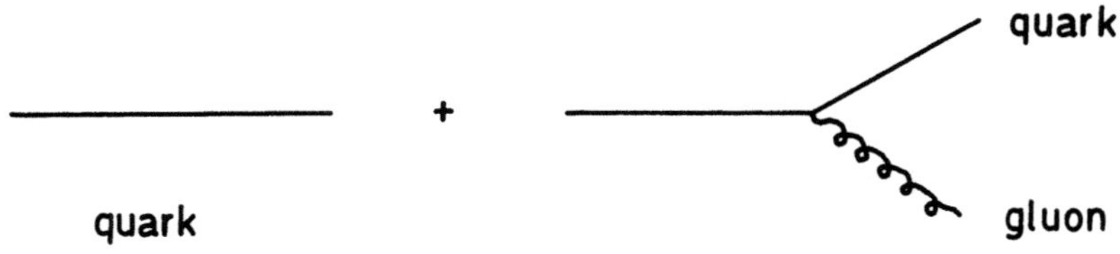

FIG. 38 : $q \to q + (qg)$ transition as $t \to t + \delta t$.

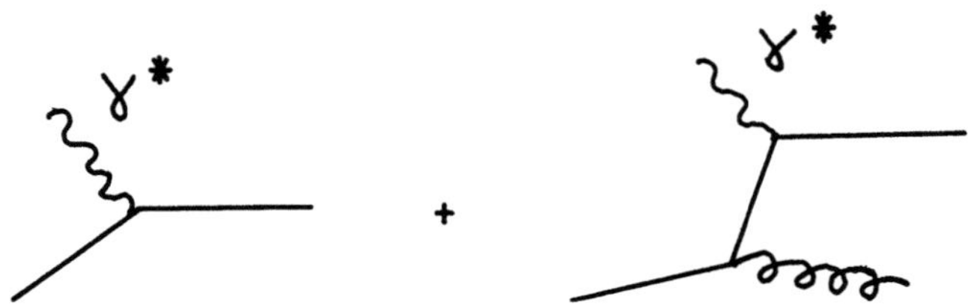

FIG. 39 : $\gamma^* q \to q + qg$ transition.

ordinary perturbation theory in section B1. We can interpret our previous result, using $g_{eff} = g(t)$, to obtain the quark distribution at t which evolves from a δ function at t_o (i.e. $f(y,t_o) = \delta(1-y)$

$$f(y,t_o + \delta t) = \delta(1-y) + g^2(t)k(y)\delta t$$

where $\delta t = 1/2 \ln(Q^2/Q_o^2)$ and

$$k(y) = \frac{1}{4\pi^2}\,\frac{1+y^2}{1-y}$$

The F's therefore satisfy

$$F(x,t+\delta t) = F(x,t) + g^2(t)\int F(x/y,t)k(y)\frac{dy}{y}\,\delta t$$

or

$$\frac{dF(x,t)}{dt} = g^2(t)\int F(x/y,t)k(y)\frac{dy}{y} \quad .$$

Taking moments

$$\frac{d}{dt}\int x^p F(x,t)\,dx = \frac{A_p}{t}\int x^p F(x,t)\,dx$$

where $A_p = c\int y^p k(y)\,dy$ is known. The solution of this differential equation is

$$\frac{\int x^p F(x,t)\,dx}{\int x^p F(x,t_o)\,dx} = \left(\frac{t}{t_o}\right)^{A_p}$$

and we have recovered the result that the moments of the structure functions decrease as calculable powers of log Q^2 in asymptotically free gauge theories (as discussed in the previous section, the moments of k(y) are actually singular but this infrared singularity is cancelled by another in the $O(g^2)$ corrections to the $\delta(1-y)$ term which we have ignored).

These intuitive arguments really give an interpretation rather than a derivation of the renormalization group equations (from which we can work backwards to the interpretation [52]). However, this picture can in fact be derived by showing that as $t \to \infty$ the dominant diagrams are ladders evaluated using g_{eff} [43].

Up to now, we have ignored a complication due to the fact that the function f is actually a matrix f_{ij} for the probability of finding parton type j in parton type i (the f already calculated is f_{qq} and $f_{qg}(x) = f_{qq}(1-x)$; f_{gg} and f_{gq} can be calculated in a similar manner). This is the mixing problem discussed in the last section which we now interpret in parton language. The most annoying consequence of the coupling between different species is that the neutral gluons, whose effects are not felt at t, give rise to effects at $t + \delta t$ since they can dissociate into $Q\bar{Q}$ pairs, as shown in fig. 40. The effects of gluons cancel out in the case of the so-called non singlet structure functions such as $F_2^{ep} - F_2^{en}$ (since the glue is isoscalar) and F_3 (which is proportional to $Q(x,t) - \bar{Q}(x,t)$). In general, however, it means that data are needed at two different values of t as boundary conditions to solve the differential equations. The solution of this mixing problem reveals not only that truly asymptotically all

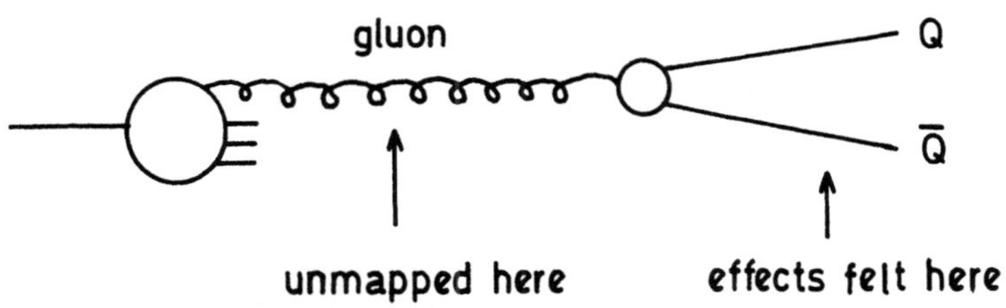

FIG. 40 : The effect of the neutral glue in creating charged $Q\bar{Q}$ pairs.

species of quark and antiquark are equally probable (as expected infinitely far down the quark fragmentation chain) but that the momentum is shared between quarks/antiquarks and gluons in the ratio 3m/16 in a model with m flavours and three colours. This gives rise to sum rules [54] for the limits of $\int F_2^{e,\nu} dx$, as $t \to \infty$.

It turns out that these limits are not expected to be reached until astronomical values of Q^2, but it is important to note that the asymptotic limits are below the measured values of these moments which are therefore expected to decrease with Q^2.

4. Scale Breaking in Leptoproduction.

In principle the best way to test the predictions of asymptotic freedom is to measure the moments of the structure functions and compare them with the theory. This is now becoming possible but it requires good data at all x (which is hard to obtain both at very small x for finite Q^2, since $Q^2 = 2x\nu$, and also for x close to 1 where the rate is very small). An alternative procedure is [55].

1) Make a fit to some simple function of x at $Q^2 = Q_0^2$ and construct $M_n(Q_0^2)$ as an analytic function of n where

$$M_n(Q^2) \equiv \int x^{n-2} F(x,Q^2)\,dx$$

2) In the non-singlet case, $M_n(Q^2)$ is then given by

$$M_n(Q^2) = M_n(Q_0^2) \left(\frac{\ln Q_0^2/\lambda^2}{\ln Q^2/\lambda^2} \right)^{d_n}$$

where d_n are known. In the singlet case the construction of $M_n(Q^2)$ is a little more complicated but can be done using input at two values of Q^2 or by guessing a shape for the glue distribution (the normalization is known from the energy momentum sum rule).

3) Invert the Mellin transform numerically to obtain

$$F(x,Q^2) = \frac{1}{2\pi i} \int_{c-i\infty}^{c+i\infty} \frac{dn}{x^{n-1}} M_n(Q^2)$$

There are two difficulties with this procedure. The first is that the moments $M_n(Q^2)$ are expected to approach the asymptotic behaviour given by QCD non-uniformly in n as Q^2 increases.

This is due to terms of order $\alpha_s \ln n (\alpha_s \equiv g^2_{eff}/4\pi)$ (which can be calculated [56]) and also contributions from subdominant terms in the operator product expansion which probably behave like nM_o^2/Q^2 (terms such as these sum up to give the contributions of resonances [57]). This means that the predictions for M_n get worse as n increases at fixed Q^2. Consequently the prediction of $F(x,Q^2)$ will be unreliable in the region where x is close to 1, which is controlled by large n.

The second difficulty, already mentioned in section I.A.3., concerns the choice of scaling variable which has a strong influence on the behaviour of $F(x,Q^2)$ for $x \approx 1$ and on the large n moments. It has been argued [58] that we can make some progress in choosing a scaling variable by following Nachtmann [59] and using operators of definite spin in the Wilson expansion. For example, for the leading second rank tensors, we could use operators $O_{\mu\nu}$ with nucleon matrix elements proportional to $p_\mu p_\nu - 1/4\, g_{\mu\nu} M^2$. Contracting with $q_\mu q_\nu / (Q^2/2)^2$ we would get $(\hat{\xi})^{-2}$ in the asymptotic expansion of the Compton amplitude where we had x^{-2} previously, where

$$\hat{\xi} = \frac{x}{1 + \frac{x^2 M^2}{Q^2}} \xrightarrow{Q^2 \to \infty} x$$

In fact a complicated analysis is now needed which shows [59] that the matrix elements of the operators of definite spin are related to the moments

$$\int \xi^{n-2} F_2(x,\xi)\, d\xi$$

of the structure functions where

$$\xi = \frac{2x}{1 + \sqrt{1 + \frac{4x^2M^2}{Q^2}}} \simeq \hat{\xi}$$

The behaviour of the ξ moments is expected to be [57]

$$A_n(Q^2) + \sum_{k=1}^{\infty} B_{n,k}(Q^2) \left(\frac{nM_o^2}{Q^2} \right)^k$$

where the explicit n dependence in the second term is a conjec-

ture. A_n and B_n depend logarithmically on Q^2 and A_n can in principle be calculated to any desired order in α_s. The difference of x and ξ moments (or moments with any other variables which tend to x as $Q^2 \to \infty$) is of the same order -$(\text{mass})^2/Q^2$ - as the contribution of thP B terms which we cannot calculate. It therefore seems that nothing has been gained.
However, it can be argued [57,58] that ratios of nucleon matrix elements of operators of different dimensions but fixed spin can be estimated using dimensional analysis with the scale set by $(\langle r^2\rangle_{proton})^{1/2} \sim (250\text{MeV})^{-1}$. This amounts to saying that if $M_o \sim 250\text{MeV}$, $B_{n,k}/A_n$ would be of order one. Certainly for operators of indefinite spin M would enter and upset these estimates since $M^2 >> (250\text{MeV})^2$. Thus ξ moments might have a bitter chance of approaching their asymptotic behaviour more rapidly than x moments since the corrections could be of order $(250\text{MeV})^2/Q^2$ rather than M^2/Q^2.

In any case if we ignore these problems and invert the moment relations we get predictions such as those in fig. 41 for the

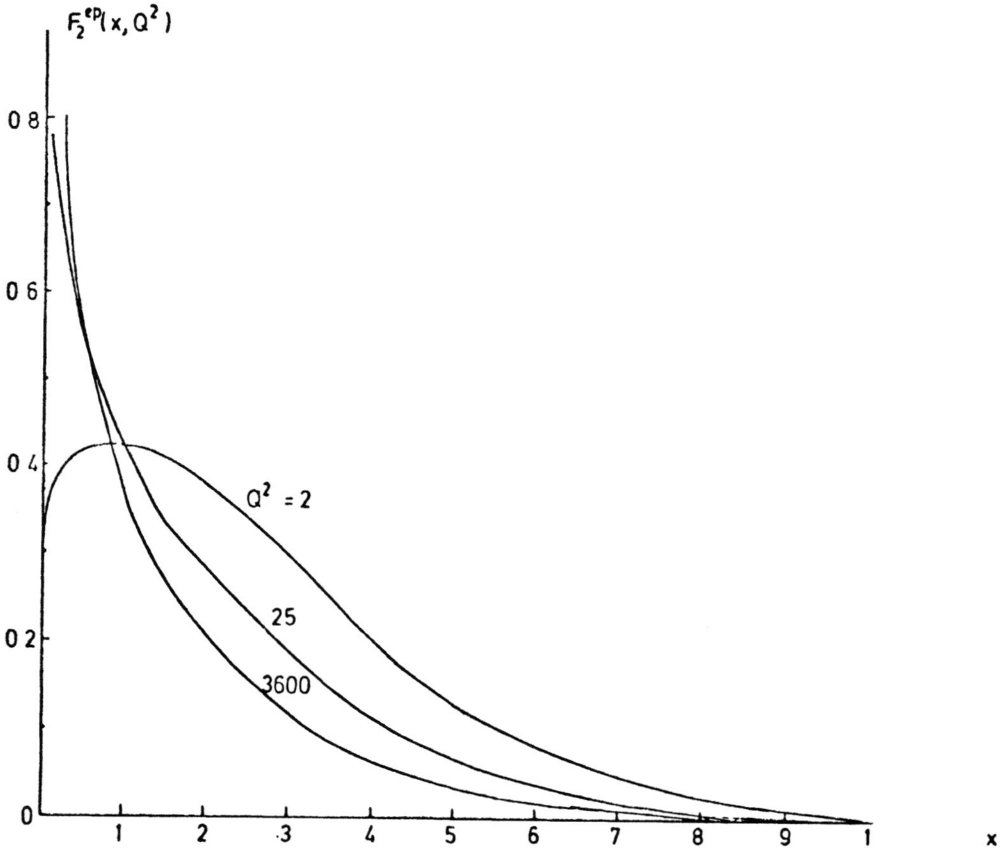

FIG. 41 : Evolution of $F_2^{ep}(x,Q^2)$ as a function of Q^2 according to one QCD calculation [60]

evolution of $F_2^{ep}(x,Q^2)$ as a function of Q^2. Other results of such detailed calculations are [55]

1) $F_2(x,Q^2)$ varies quite rapidly for modest Q^2 but the variation slows down and becomes almost imperceptible at very large Q^2 (as expected since the moments behave as powers of log Q^2).

2) The results are insensitive to "reasonable" variations in the shape of the glue distribution.

3) The density of charmed quarks increases rapidly with Q^2 at small x and is probably responsible for a substantial part of the observed increase of F_2^e, in which case associated production of charmed particles should be seen (a signature being di and trilepton events). However, in detail, this prediction is very sensitive to unknown mass corrections of order M_c^2/Q^2.

4) Although theory should confront experiment for the structure functions (or preferably their moments), it is interesting to note the predictions for quantities like σ^ν. In fact QCD predicts that σ^ν/E should decrease with increasing E (quite rapidly for $E \lesssim 100$ GeV and then very slowly), despite the increasing charm contribution.

It would be inappropriate to review all the predictions of QCD here. We only note that they agree quite well with almost all leptoproduction data with a value of about 0.2GeV^2 for the parameter λ^2 which sets the scale in the variable $\ln Q^2/\lambda^2$ [55,62,63]. The only exception is that the predictions for F_L are systematically below the data. However, the theory predicts that

$$\int x^{n-2}F_L dx = \frac{4\alpha_s(Q^2)}{3\pi(n+1)}\int x^{n-2}F_2 dx + O(\alpha_s^2) + O\left(\frac{M_o^2}{Q^2}\right)$$

(this is for non-singlets - the expression is more complicated for singlets), so we need $M_o^2 / \alpha_s Q^2 << 1$ before we can neglect the contribution of non-leading operators whereas in the case of the predictions for F_2 we only require $M_o^2 / Q^2 << 1$. [64] Furthermore, due to the $(n+1)^{-1}$ factor, corrections to the predicted moments of F_L become increasingly important as $n \to \infty$ $(x \to 1)$. In fact, because of this factor the leading term gives

$$R = \frac{F_L}{F_T} \sim \alpha_s(1 - x)$$

while parton ideas suggest that there are non-leading contributions to R which are finite as $x \to 1$ equal to

$$\frac{4\langle p_T^2\rangle}{Q^2 + 2\langle p_T^2\rangle}$$

where $< p_T^2 >$ is for the bound constituents (and is not to be confused with the p_T^2 of particles produced in large Q^2 collisions [43]). While these corrections allow the theory to survive comparison with experiment for the present (see, e.g., ref. 62), the large values of R recently observed [10] must turn out to decrease like $1/Q^2$ to the values predicted asymptotically if the theory is to remain viable.

5) Scale Breaking in $pp \to \mu\bar{\mu}$, Z, W, ...

We cannot use the operator expansion to analyse $pp \to \mu\bar{\mu} + \ldots$ in the limit $M_{\mu\mu}^2 = Q^2 \to \infty$. The reason is that the matrix elements of the local operators O_i in the expression

$$C_i(Q^2) < p,p|O_i|p,p >$$

depend on s, so we cannot order their contributions according to how rapidly the C_i vanish as $Q^2 \to \infty$ since $s > Q^2$ in the physical region and $s \to \infty$ also in this limit.

The only tool at present available to analyse this process is perturbation theory. It turns out that in leading log approximation the cross-section does seem to have the parton model form (eqn 11) to all orders in the Abelian gluon model [65,66]. This form is also found in QCD in calculations which have been carried out up to second order [67,68]. (Furthermore, eqn 11 is valid in two dimensional QCD [69], in which scaling is exact and quarks are confined) [70]. It therefore seems reasonable to explore the consequences of the conjecture [71,66] that a parton like expression holds in the real world.

Eqn 11 must of course be modified in some way if scaling is violated. Perturbation theory suggests that the quark/anti-quark distributions on the right hand side should be those measured experimentally (or calculated using QCD) at space-like $q^2 = -Q^2$. In this case $Q^4\, d\sigma/dQ^2$, which is a function of $\tau = Q^2/s$ only in the naive model, would be expected to rise with Q^2 for $\tau \lesssim 0.01$ and fall for $\tau \gtrsim 0.01$ since $Q^2 = xx's$ and the quark/anti-quark distributions rise for $x \lesssim 0.1$ and fall for $x \gtrsim 0.1$. The predictions for μ pair, W and Z production are analysed in detail in refs 66, 72.

It is very important to test the prediction of scaling violations in μ-pair production because

1) Any scaling violation might seriously affect predictions of W/Z production at future machines.

2) The prediction tests the very interesting hypothesis that QCD predictions can be abstracted from perturbation theory.

6. Jet Production.

The idea of abstracting results from perturbation theory can be applied to other processes. For example, we can see whether the parton prediction for the three photon process discussed in section IB2 might be valid in QCD when suitably modified to include scaling violations. In this case, the answer seems to be no [25].

We can also use this idea to examine the final states in lepto-production and $\bar{e}e$ annihilation. In perturbation theory the final states consist of quarks and gluons. Nevertheless, perturbation theory may be a useful guide to what happens at very short times and distances. Consider $\bar{e}e$ annihilation, for example. QCD predicts that (with corrections of $O(\alpha_s)$) the final state consists of two jets of quarks and gluons. If at very short distances (10^{-15} cm say) all the energy starts out in two jets, we might expect these jets to survive the confining forces, which are supposed to convert the quarks and gluons into hadrons at larger distances. Two jets are observed experimentally so it is not inconsistent to assume that the jets predicted by QCD survive confinement. In fact QCD predicts a dominant two jet structure in lepto-production and $pp \to \mu\bar{\mu}$ as well as $\bar{e}e$ annihilation and makes very definite predictions for 3 jet contributions of order α_s in all these processes. The application of QCD to jet production is a very new subject which the reader can follow in the literature [73]; we also refer the reader to the future literature for an answer to the important question of whether this approach can be used to describe the production of large p_T jets in hadronic collisions.

Table

Helicities amplitudes in the scattering of virtual photons on polarized nucleons.

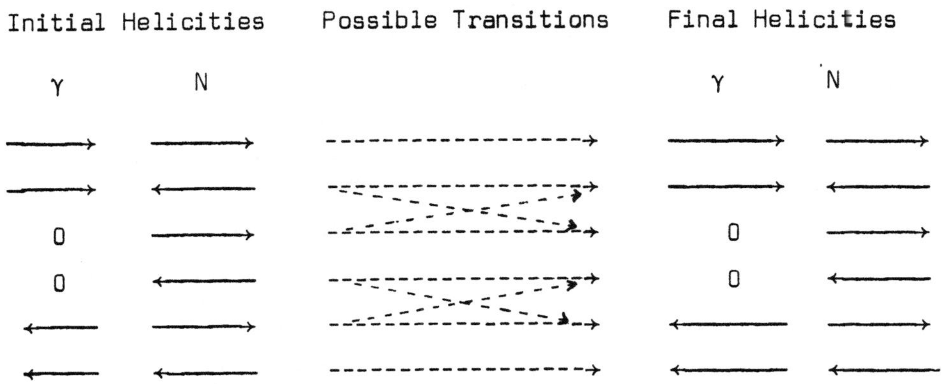

Acknowledgements

I am very grateful to Nathan Isgur and Ash Carter for comments on the manuscript.

REFERENCES

1) Previous reviews of various applications of the parton model include
R.P. Feynman "Photon Hadron Interactions"(Benjamin (1972)
J. Ellis in "Weak and Electromagnetic Interactions"ed. R.Balian CH. Llewellyn Smith, North Holland, 1977.

2) C.H. Llewellyn Smith Physics Reports 3C, n°5, 1972.

3) M. Einhorn, Phys. Rev. D14, 3451, 1976.

4) J.S. Bell, CERN TH-2291, 1977.

5) The older data are reviewed by J.I. Friedman and H.W. Kendall. Ann. Rev. Nucl. Science 22, 203, 1972.

6) More recent data reviewed by R.E. Taylor. Proc. 1975 Stanford Conference.

7) H.L. Anderson et al. Phys. Rev. Lett. 38, 1450, 1977. See the forthcoming Proc. 1977, Hamburg Lepton Symposium for the latest results.

8) E.M. Riordan et al. SLAC. Pub 1634, 1973.

9) C.G. Callan and D.J. Gross, Phys. Rev. Lett. 22, 156, 1969.

10) Data reported at the Hamburg Conference (August 1977) apparently suggest that σ_0/σ_T may be bigger than previously believed. The implications are discussed in section B4.

11) W.B. Atwood, SLAC report 185, 1975.

12) For reviews of neutrino data see D.H. Perkins. Reports on Prog. in Physics 40, 409, 1977.
P. Musset. Physics Reports (in press).
B.C. Barish in "Weak and Electromagnetic Interactions", loc. cit. and Physics Reports (in press).
For the latest SPS date see : M. Holder et al., Phys. Rev. Lett. 39, 433, 1977 and P.C. Bosetti et al., Phys. Lett 70B, 273, 1977 and Proc. 1977 Hamburg Lepton Symposium (in press).

13) C.H. Llewellyn Smith, Nucl. Phys. B17, 277, 1970.

14) D.J. Gross & C.H. Llewellyn Smith, Nucl. Phys. B14, 337,1969.

15) H. Deden et al., Nucl. Phys. B85, 269, 1975.

16) A.J. Buras and K.J.P. Gaemers Physics Letters, 71B, 106, 1977.
I. Hinchliffe. Oxford preprint 60/77.

17) The kinematics are analysed (e.g.) in R.L. Heimann Nucl. Phys. B64, 429, 1973. Where references to earlier work may be found.

18) J.D. Bjorken, Phys. Rev. 148, 1467, 1966 and D1, 1376, 1969.

19) F.E. Close, F. Gilman and I. Karliner, Phys. Rev. D6, 2533, 1973.
F.E. Close, Nucl. Phys. B80, 269, 1974.

20) M.J. Alguard et al., Phys. Rev. Lett. 37, 1261, 1976.

21) R.F. Schwitters et al., Phys. Rev. Lett. 35, 1320, 1975.

22) S.D. Drell and T.M. Yan, Phys. Rev. Lett. 25, 316, 1970 (see also ref. 52).

23) R.L. Kingsley and C. Nash, Nuovo Cimento Letters 4, 802, 1972.

24) S.J. Brodsky, J.F. Gunion and R.L. Jaffe, Phys. Rev. D6, 2487, 1972. Results of an experiment on this process are reported in D.L. Faucher et al., Phys. Rev. Lett. 38, 800,1977.

25) K.H. Craig, Oxford University thesis (in preparation).

26) See papers in Proc. 1977 Hamburg Lepton Symposium and Proc. 1975, Stanford Lepton Symposium.

27) This curve is taken for ref 32 where references to the experiments are given (the curves are fits made in ref 32).

28) J. Ellis and C.H. Llewellyn Smith - footnote 36 of C.H.Llewellyn Smith in Proc. 4th Int. Conf. on High Energy Collisions, Oxford 1972 (RHEL 72.001).

29) C.H. Llewellyn Smith and A. Pais, Phys. Rev. D6, 2625, 1972.

30) M. Haguenauer in Proc. XVII Int. Conf. on High Energy Physics, London, 1974, e.d. J.R. Smith (RHEL Publication, 1974).

31) M. Gronau, F. Ravndal and Y. Zarmi, Nucl. Phys. B51, 611,1973.

32) R.D. Field and R.P. Feynman, Phys. Rev. D15, 2590, 1976. (in Fig. 24, the points with Δ are from L. Sehgal, Nucl. Phys. B90, 471, 1975).

33) L.M. Sehgal, Aachen preprint PITHA nr 81, 1977 to be published in Proc. 1977 Hamburg Lepton Symposium.

34) J.F. Martin and L. Osborne, Phys. Rev. Lett, 38, 1193, 1977.

35) W.A. Loomis et al., Phys. Rev. D (in press).

36) Pham Quang Huang and J.J. Sakurai, UCLA preprint 77/TEP/17.

37) See, e.g. P. Darriulat in Proc. XVIII Int. Conf. on High Energy Physics, Tblisi, 1976 and Proc. VIII Int. Conf. on Multiparticle Dynamics, Kaiserburg, France 1977.

38) C. Bromberg et al., Phys. Rev. Lett. 38, 447, 1977 and Caltech preprint CALT 68-63-1977.

39) R.P. Feynman, R.D. Field and G.C. Fox, Nucl. Phys. B128, 1, 1977 and ref. 32.

40) R. Moller in Proc. 1977 Rencontre de Moriond (to be published).

41) R. Cutler and D. Sivers. Argonne report ANL-HEP-PR-77-40.

42) C.H. Llewellyn Smith (unpublished).

43) C.H. Llewellyn Smith, to be published.

44) For an excellent review see H.D. Politzer Physics Reports 14C 129, 1974. Refs 54 and 50 are to some of the original literature.

45) W. Zimmerman in "lectures on Elementary Particles and Quantum Field Theory" ed. S. Deser et al. (MIT press 1971).

46) For excellent reviews of renormalization see C.G. Callan in "Methods in Field Theory" ed. R. Balian and J. Zinn-Justin, North Holland 1976. S. Coleman in "Properties of the Fundamental Interactions" part C, ed. A. Zichichi, Editrice Compositori - Bologna, 1973.

47) S. Coleman in "Properties of the Fundamental Interactions" part A, ed. A. Zichichi, Editrice Compositori - Bologna, 1973.

48) K. Symanzik in "Fundamental Interactions at High Energies" ed. A. Perlmutter et al., (Gordon and Breach, 1970). See S. Coleman (ref. 46) for a pedagogical exposition of this theorem.

49) N. Christ, B. Hasslacher and A.H. Mueller, Phys. Rev. D6, 3543, 1972. This result was first derived by studying diagrams explicitly in all orders by V.N. Gribov and L.N. Lipatov Sov. J. Nucl. Phys. 15, 438, 1972.

50) T. Appelquist and H. Georgi, Phys. Rev. D8, 4000, 1973. A. Zee, Phys. Rev. D8, 4038, 1978. For a discussion of the problem of continuing to time like q^2 see R.G. Moorehouse, M.R. Pennington and G.G. Ross, Nucl. Phys. B124, 285, 1977 and references therein.

51) J. Kogut and L. Susskind, Phys. Rev. D9, 697 and 3391, 1974.

52) S.D. Drell and T.M. Yan Annals of Physics 66, 578, 1971 and references therein.

53) G. Altarelli and G. Parisi, Nucl. Phys. B126, 298, 1977. See also K.J. Kim and K. Schilcher Mainz preprint 172-TH 7712.

54) D. Gross and F. Wilczek, Phys. Rev. D8, 3633, 1973 and D9, 980, 1974.
H. Georgi and H.D. Politzer, Phys. Rev. D9, 416, 1974.

55) This is essentially what is done in
M. Glück and E. Reya, Phys. Rev. D14, 3035, 1976 and Mainz preprint MZ-TH 7619.
A. Buras, Nucl. Phys. B125, 125, 1977 and A. Buras and K.J. Gaemers, CERN TH-2322 and 2368.
I. Hinchliffe and C.H. Llewellyn Smith, Nucl. Phys. B128, 93, 1977.
C.G. Fox, Nucl. Phys. B131, 107, 1977.

56) E.G. Floratos, D.A. Ross and C.T. Sachrajda, Nucl. Phys. B129, 66, 1977, and A.J. Buras, E.G. Floratos, D.A. Ross and C.T. Sachrajda, CERN TH-2340, 1977.

57) A. De Rujula, G. Georgi and H.D. Politzer, Annals of Physics 103, 315, 1977.

58) A. De Rujula, H. Georgi and H.D. Politzer, Phys. Rev. D15, 2495, 1977 and references therein.

59) O. Nachtmann, Nucl. Phys. B63, 237, 1973.

60) I. Hinchliffe and C.H. Llewellyn Smith, ref. 55.

61) The effect of QCD scaling violations on charged current neutrino cross-sections is discussed in
I. Hinchliffe and C.H. Llewellyn Smith, Phys. Letters 70B, 247, 1977.
A. Buras, CERN TH-2285, 1977 and ref. 55.
G.C. Fox, Caltech preprint CALT 68-619.
For neutral current cross-sections see ref. 16.

62) For a review/survey see O. Nachtmann, Heidelberg preprint HD-THEP-77-13, to be published in Proc. 1977, Hamburg Lepton Symposium.

63) For an analysis of moments of the neutrino structure functions see a forthcoming paper based on results from the BEBC neutrino experiment by the ABCLOS (Aachen etc) collaboration. For an analysis of moments of F_2^e, see H. L. Anderson et al., Chicago preprint submitted to Phys. Rev. Letters.

64) An investigation of the choice of scaling variable, the correct combinations of structure functions to which the predictions apply and the higher orders in α_s is obviously especially important in the case of F_L. A careful discussion of these points may be found in A.J. Buras et al., ref 56.

65) This was pointed out explicitly in ref. 66 where the leading log calculations of structure functions due to Gribov and Lipatov (ref 49) and of μ pair production due to Halliday (Nucl. Phys. B103, 343, 1976), were confronted.

66) I. Hinchliffe and C.H. Llewellyn Smith, Phys. Lett. 66B, 281, 1977.

67) C.T. Sachrajda, CERN TH-2416, 1977.

68) H.D. Politzer, Nucl. Phys. B129, 301, 1977.

69) J. Kripfganz and M.G. Schmidt, CERN TH-2266, 1977.

70) For a discussion of deep inelastic processes in this model see ref. 3. For a review of 2D QCD see
J. Ellis, in Proc. 1977 Zakopane School, to be published in Acta Physica Polonica.

71) J. Kogut, Phys. Lett. 65B, 377, 1976.

72) I. Hinchliffe and C.H. Llewellyn Smith, ref 55.
J. Kogut and J. Shigemitsu, Nucl. Phys. B129, 461, 1977.

73) Work on jets may be found in J. Ellis, M.K. Gaillard and G.G. Ross, Nucl. Phys. B111, 253, 1976.
E.G. Floratos, CERN TH-2261, 1976 - to be published in Nuovo Cimento.
T.A. De Grand, Y.J. Ng and S.H.H. Tye, SLAC PUB 1950, 1977.
H.D. Politzer, ref 68.
G. Sternman and S. Weinberg, Phys. Rev. Lett., 39, 1436, 1977.
H. Georgi and M. Machacek, Phys. Rev. Lett. 39, 1237, 1977.
E. Farhi, Phys. Rev. Lett., 39, 1587, 1977.
H.D. Politzer, Phys. Letters (in press).
K.H. Craig and C.H. Llewellyn Smith Phys. Letters (in press)
H. Georgi and H.D. Politzer, Phys. Rev. Lett., 40, 3, 1978.
C.H. Llewellyn Smith - ref 43.

CHARM, APRES-CHARM, AND BEYOND

J. Ellis

CERN - GENEVA

PREFACE

The purpose of these lectures is to review current phenomenology of the weak and electromagnetic interactions from a gauge theoretical prospective. Particular attention will be paid to the recent results presented by experimental lecturers at this school [1-3]. Free use will be made of the quark-parton and gauge technology developed by other theoretical lecturers [4,5]. It will not often enter explicitly , but where a choice is necessary we will assume that strong interactions are also described by a gauge theory, namely quantum chromodynamics (QCD)[6].

It should be emphasized that, because so much is being discovered about the weak interactions so quickly, their phenomenology is necessarily subjective. While an attempt has been made to reflect divergent currents of opinion, the presentation is necessarily tinted by the author's prejudices. For alternative points of view, see, for example, the lecture notes in Ref. 7. Many of the points developed in these lectures arise from discussions or work done with Mary K. Gaillard, to whom I am very grateful.

1. CHARM

1.1. Introduction.

The objectives of this first lecture are twofold.

- To remind you of the motivations for charm[8]. While the physics is familiar, the language used will be geared towards a unified discussion of the world beyond charm.
- To convince you that experiments have now confirmed essentially every aspect of the charm-changing weak current discussed by Glashow, Iliopoulos and Maiani[9]. Charm is now ready for the textbooks.

So it now seems reasonable to pigeon-hole charm as the latest piece of "old" physics -- it is fairly well confirmed, and the motivations for it were derived[9] from the weak interactions of the old-hat u, d and s quarks, to which we now turn.

1.2. The disease.

It has been known for decades that there are weak interactions of strength $O(G_F) \sim 10^{-5}/m^2(\text{proton})$, which proceed via charged currents. It has long been conjectured that these current-current interactions are mediated by at least one charged vector boson $W^\pm$. In the past few years we have uncovered[10] neutral weak interactions which probably also proceed via vector and axial currents, and are also of strength $O(G_F)$ [2]. One presumes neutral currents are also mediated by at least one neutral vector boson Z^0. A common prejudice is that these weak interactions should be unified with electromagnetic interactions, with an intrinsic coupling strength $O(\alpha)$ [11]. The weak interactions are supposed to be weak at present energies, because their carrier bosons are very massive :

$$O(G_F) = \frac{g^2_{W,Z}}{m^2_{W,Z}} , \quad \frac{g^2_{W,Z}}{4\pi} = O(\alpha) \qquad (1.1)$$

so that

$$m_{W,Z} = O(50 \text{ to } 100) \text{ GeV} \qquad (1.2)$$

However, there seems to be no direct experimental evidence that the prejudices (1.1), (1.2) are indeed correct. Upper limits on corrections to quark-lepton weak universality from higher order contributions to β and μ decay only indicate that $m_{W^\pm} \lesssim O(400)$ GeV [12]. Demanding that effective neutral interactions due to double $W^\pm$ exchanges not be larger than the observed neutral weak cross-sections also gives $m_{W^\pm} \lesssim O(400)$ GeV. These bounds would still allow $g_W^2 / 4\pi >> \alpha$, and I know of no other bounds which enforce unification of weak and electromagnetic interactions. Nevertheless, we will assume henceforth that such a unified theory does exist.

We will assume that unification is achieved by a gauge theory with spontaneous symmetry breaking, of the type shown by 't Hooft[13] to be renormalizable. Such a theory must be based on a group G with at least four generators corresponding to the $W^{\pm}$, Z^0 and photon to which we are now committed. The smallest suitable group is therefore SU(2) x U(1), which was first studied by Glashow[14], though he did not put in the spontaneous symmetry breaking introduced into the game by Weinberg, Salam and Ward[15].

There were a collection of embarrassments to the gauge framework summarized by the numbers

$$\frac{\Gamma(K_L^o \to \mu^+\mu^-)}{\Gamma(K_L^o \to \text{all})} \sim 10^{-8}, \quad \frac{\Gamma(K^{\pm} \to \pi^{\pm}e^+e^-)}{\Gamma(K^{\pm} \to \text{all})} = (2.6 \pm 0.5)\times 10^{-7}$$

$$\frac{\Gamma(K^{\pm} \to \pi^{\pm}\nu\bar{\nu})}{\Gamma(K^{\pm} \to \text{all})} < 0.6\times 10^{-6}, \quad \left|\frac{m_{K_L^o} - m_{K_S^o}}{m_K}\right| = 0.7 \times 10^{-14} \qquad (1.3)$$

which indicate that strangeness-changing $|\Delta S|=1$ neutral currents and $|\Delta S|=2$ transitions are suppressed far below the $O(G_F)$ of normal charged and neutral weak interactions. The reason why the results (1.3) were embarrassing can be seen by considering a charged boson $W^{\pm}$ corresponding to a generator $F^{\pm}$ of the gauge group G, which has the associated quark current

$$J^{\mu}_{\pm} \equiv \bar{q}\gamma^{\mu}\frac{1}{2}(1-\gamma_5)T^L_{\pm}q + \text{(possible right-handed } 1/2\,(1+\gamma_5) \text{ pieces)} \qquad (1.4)$$

In a gauge theory there would also be a neutral boson W^0 coupled to the generator $F^0 \equiv [F^{\pm}, F^-]$ which has the associated quark current

$$J^{\mu}_{o} \equiv \bar{q}\gamma^{\mu}\frac{1}{2}(1-\gamma_5)T^L_{o}q + \text{(possibly } L \leftrightarrow R) \qquad (1.5)$$

where

$$T^L_o \equiv [T^L_+, T^L_-] \qquad (1.6)$$

The Cabibbo charged current corresponding to (1.4) has a piece coupling

$$\begin{pmatrix} u \\ d\cos\theta_c + s\sin\theta_c \end{pmatrix}_L \qquad (1.7)$$

in a doublet, and the commutator (1.6) then yields a contribution

$$J^\mu_o \ni \bar{s}\gamma\mu\,\frac{1}{2}\,(1-\gamma_5)d\,\sin\theta_c\cos\theta_c \qquad (1.8)$$

which would give rise to $|\Delta S| = 1$ neutral interactions and $|\Delta S|=2$ transitions in $O(G_F)$. These would be much larger than the values and limits (1.3), which indicate that such effects are actually $O(G_F^2)$. This suppression is even far below that expected for fourth-order weak processes, which are expected, in general, to be of the same order as radiative corrections to μ decay, namely $O(G_F\alpha)$. Therefore a cunning way of removing both $O(G_F)$ and $O(G_F\alpha)$ $|\Delta S| = 1$ and $|\Delta S| = 2$ effects had to be found. Every schoolboy now knows that the panacea[9] is to introduce a new charge 2/3 quark with a weak charged current coupling

$$\begin{pmatrix} c \\ s\cos\theta_c - d\sin\theta_c \end{pmatrix}_L \qquad (1.9)$$

This would certainly introduce a term in the commutator (1.6) which precisely cancels (1.8) and removes $O(G_F)$ $|\Delta S| = 1$ neutral currents. Also, the (-) sign in (1.9) suggests that the fourth-order loop diagrams will meet with cancellations. Let us now look in more detail how these tricks are worked, and whether there are any alternatives to charm.

1.3. The cure.

Let us pose the problem more generally[16] than is needed for the specific case of strangeness-changing neutral currents in the SU(2) x U(1) model. Consider a theory based on a group G with neutral bosons Z_i^0 coupled to generators F_i^0, with coupling matrices $Y_i^{L,R}$

$$J^\mu_{z_i} \equiv \bar{q}\gamma^\mu\,\frac{1}{2}\,(1-\gamma_5)Y_i^L q + \text{(possible right-handed piece)} \qquad (1.10)$$

In order to avoid flavour-changing neutral currents in the spontaneously broken theory, it is first necessary that there be no flavour-changing neutral currents in the basic, symmetric, theory [e.g. no "kaonic" currents $\bar{s}\gamma^\mu 1/2\ (1-\gamma_5)d$]. This condition is clearly obeyed in the simple SU(2) x U(1) model. In the broken theory the quarks acquire a non-zero mass matrix M

$$\mathcal{L}_{mass} = -\bar{q}\,\frac{1}{2}(1-\ _5)Mq - \bar{q}\,\frac{1}{2}(1+\ )M^{+}q \qquad (1.11)$$

This will not, in general, be diagonal, and the two different helicities of quarks must be unitarily transformed :

$$\tilde{q} = \frac{1}{2}(1-\gamma_5)U^{L}q + \frac{1}{2}(1+\gamma_5)U^{R}q \qquad (1.12)$$

These changes in the definitions of the quarks reflect a mismatch between the weakly-interacting "Cabibbo" quarks and the strongly-interacting quarks which are eigenstates of the world's mass-matris. The neutral current coupling matrices Y_i, defined in (1.10), get transformed by the redefinitions (1.12)

$$\tilde{Y}_i^{L,R} = U^{L,R}Y^{L,R}\,[U^{L,R}]^{-1} \qquad (1.13)$$

In order to avoid flavour neutral currents in second order (fig. 1), we need $\tilde{Y}_i^L$ to be diagonal. Of course, we could impose this requirement by fiat, choosing the quark masses and mixing angles sneakly so that the $\tilde{Y}_i^L$ just happen accidentally to be diagonal. This seems somewhat artificial, and in the absence of any theory of masses and mixing angles, it seems reasonable to demand that $\tilde{Y}_i^{L,R}$ be diagonal for all a priori equally likely quark mass matrices M. This means that U^L and U^R should also be allowed to be arbitrary unitary matrices, subject of course to the condition that they commute with the electric charge

$$[U^{L,R}, Q] = 0 \qquad (1.14)$$

But if the $\tilde{Y}_i^{L,R}$ are diagonal for all possible $U^{L,R}$ obeying the conditions(1.14), the $\tilde{Y}_i^{L,R}$ must themselves be functions of the quark charge Q

$$Y_i^{L,R} = Y_i^{L,R}(Q) \qquad (1.15)$$

As an example, consider the relevant case of $SU(2)_L \times U(1)$, where the only Y_i apart from Q itself is just T_3^L. The condition (1.15) then requires that

$$T_3^L = T_3^L(Q) \qquad (1.16)$$

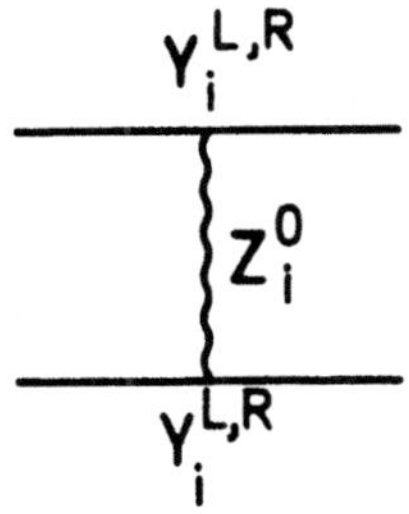

Fig. 1 : A second-order neutral weak interaction mediated by the vector boson Z_i.

If this condition is not respected, there will be flavour-changing neutral interactions in $O(G_F)$. Consider the old Cabibbo theory, which had a weak doublet (1.7) and a singlet $(s\cos\theta_C - d\sin\theta_C)_L$. In this case

$$\left.\begin{array}{l} d\cos\theta_C + s\sin\theta_C \\ s\cos\theta_C = d\sin\theta_C \end{array}\right\} \text{ has } \; T_3^L \left\{\begin{array}{l} -\frac{1}{2} \\ 0 \end{array}\right. \tag{1.17}$$

and we get the first-order flavour-changing neutral current (1.8).

Let us go on to the fourth-order effects [16] : the various relevant diagrams are illustrated in Fig. 2. They have the effect of renormalizing the coupling matrices Y_i of the neutral weak bosons Z_i(Fig. 2a), and of generating new effective weak interactions of $O(G_F\alpha)$ by double $W^{\pm}$ exchange (Fig. 2b), and we should ensure that all these new neutral interactions are also diagonal. It turns out that the condition (1.15) is sufficient to ensure that the interactions due to Figs. 2a) and b are diagonal for any $U^{L,R}$. But the double W exchange gives an effective neutral coupling proportional to

$$(\Delta Y^{L,R})_{\alpha,\alpha} \propto 3\left[(T_\alpha^{L,R})^2 - (T_{3_\alpha}^{L,R})^2 \right] \pm 5T_{3_\alpha}^{L,R} \tag{1.18}$$

where we have introduced an isospin $\underline{T}_\alpha$, generated by the bosons $W_\alpha^{\pm}$ and closing on $T_{3\alpha}$. If we demand once more that (1.18) be

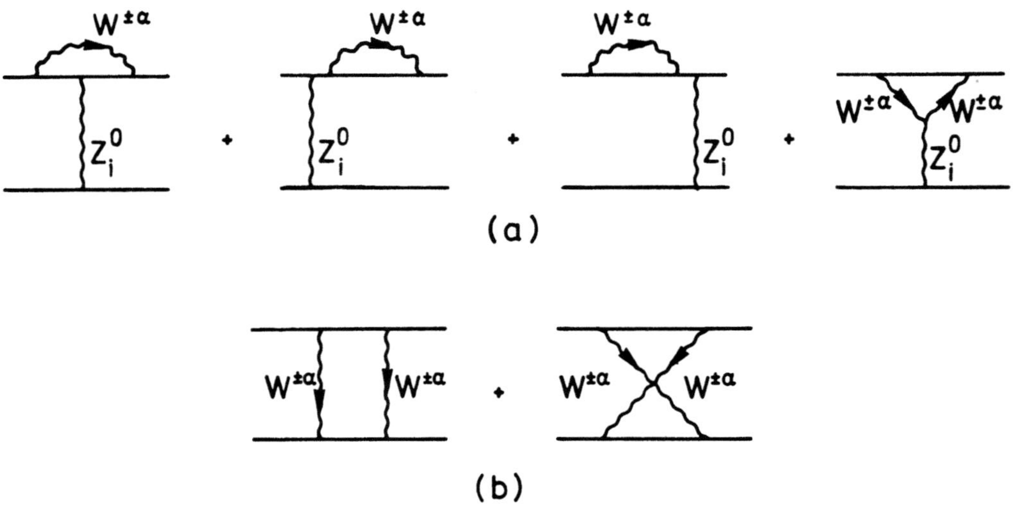

Fig. 2 : Fourth-order neutral weak interactions corresponding to (a) renormalization of the Z_i coupling; (b) double W exchange.

diagonal, whatever the choice of quark mass matrix, then we clearly find by analogy with (1.15) that

$$(T_\alpha^{L,R})^2 = f_\alpha^{L,R}(Q) \tag{1.19}$$

So all quarks of the same charge and helicity must have the same value of this funny isospin $\underline{T}_\alpha$ that we introduced.

In the case of $SU(2)_L \times U(1)$, this means that the rotated quark $(s \cos\theta_C - d \sin\theta_C)_L$ that Cabibbo left out of his current (1.7) should not only have $T_3^L = -1/2$, but should actually be a member of an SU(2) isodoublet, rather than an isoquartet or bigger representation. Hence there should be a quark to occupy the spot X in

$$\begin{pmatrix} s \cos\theta_C^{\;X} - d \sin\theta_C \end{pmatrix} \tag{1.20}$$

This is the quark called "charm" [8], which was first shown by Glashow, Iliopoulos and Maiani [9] to have all these nice flavour concerning properties.

This concludes the motivation for charm, but it does not rule out the possible existence of other quarks, or of a bigger gauge group than $SU(2)_L \times U(1)$, just as long as the conditions (1.16) and (1.19) are respected. We will return to this point in some detail in Lecture 3, but let us briefly consider what happens if there are more than four quarks.

1.4. The charged coupling matrix.

Let us consider models with just Q = 2/3 (call them P) and Q = -1/2 (call them N) quarks. Let us suppose they come in doublets of $SU(2)_L \times U(1)$ and just have left-handed couplings, which ensures that all flavour-changing neutral currents are $O(G_F^2)$. The charged current (1.4) can then be written in the form

$$J_-^\mu = \bar{P}\,\gamma^\mu 1/2(1-\gamma_5)V^L N \tag{1.21}$$

where the matrix V^L can be expressed as

$$V^L = U_P^L\, U_N^{L^{-1}} \tag{1.22}$$

in terms of the restrictions to P and N quarks of the unitary matrix U^L introduced in (1.12) to diagonalize the quark mass matrix. We notice that the individual reshuffles of P and N quarks are not observable, just the difference $U_P^L U_N^{L-1}$ between them. The matrix V^L is an $n_D \times n_D$ unitary matrix, where n_D is the number of doublets, and so has n_D^2 parameters. But one of these is an unobservable over-all phase, and (n_D-1) parameters can be absorbed by redefining the relative phases between different P quarks, and another (n_D-1) can be absorbed via the N quarks. The charged weak coupling matrix therefore has $(n_D-1)^2$ observable parameters [17].

Let us consider a few cases : if $n_D=1$, the above counting gives zero free parameters, and that corresponds to the parameter-free coupling of one $SU(2)_L$ doublet

$$\begin{pmatrix} P \\ N \end{pmatrix}_L \tag{1.23}$$

If $n_D=2$, the above counting gives one free parameter, and this can readily be identified with the Cabibbo angle of the standard GIM[9] charm model

$$J_- = (\bar{u}, \bar{c})\frac{1}{2}(1-\gamma_5)\begin{pmatrix} \cos\theta_c & \sin\theta_c \\ -\sin\theta_c & \cos\theta_c \end{pmatrix}\begin{pmatrix} d \\ s \end{pmatrix} \tag{1.24}$$

Notice that redefining the quark phases enables us to reduce the unitary matrix V to be orthogonal. Such is not the case when $n_D \geq 3$: for $n_D = 3$ there are four parameters in the matrix, three Euler angles and a complex phase. A convenient parametrization of the coupling matrix, due to Kobayashi and Maskawa[17], is

$$J_-^\mu = (\bar{u},\bar{c},\bar{t})\frac{1}{2}(1-\gamma_5)\begin{pmatrix} C_1 & -S_1C_3 & -S_1S_3 \\ S_1C_2 & C_1C_2C_3-S_2S_3e^{i\delta} & C_1C_2S_3+S_2C_3e^{i\delta} \\ S_1S_2 & C_1S_2C_3+C_2S_3e^{i\delta} & C_1S_2S_3-C_2C_3e^{i\delta} \end{pmatrix}\begin{pmatrix} d \\ s \\ b \end{pmatrix} \tag{1.25}$$

The phase δ introduces complex weak couplings which have the potential of generating CP violation[17]. The above argument indicates that in a model with six or more quarks CP violation is as natural and expectable as the Cabibbo angle. Since it is overwhelmingly probable (see Lectures 2 and 4) that the world has at least six quarks, CP violation due to (1.25) is almost inevitable. Of course CP violation may also arise in other ways -- perhaps because of the strong interactions [18], or right-handed currents, or through

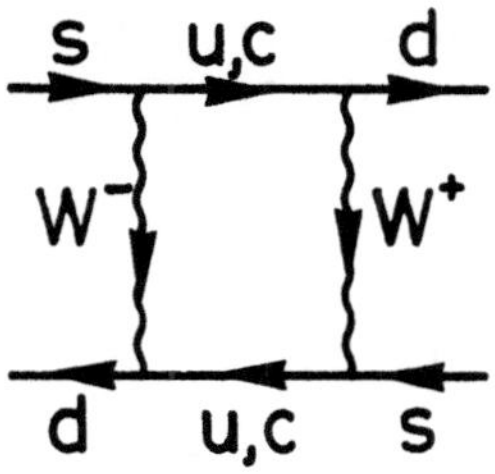

Fig. 3 : Diagram used to estimate m_c from the K_1-K_2 mass difference.

the Higgs system[19]. In Lecture 4 we will return to analyse the coupling matrix (1.25) in more detail.

1.5. The mass of the charmed quark.

Let us, for the moment, concentrate on a world with just four quarks -- we will see later how extra quarks need not affect the analysis -- and estimate the mass of the charmed quark from data on strangeness-changing neutral effects. We saw from Section 1.2 that the GIM model has $|\Delta S| = 1$ neutral currents and $|\Delta S| = 2$ transitions $O(G_F^2)$, basically because of cancellations between u and c quarks. Indeed these effects would vanish if $m_u = m_c$, and are $\propto G_F^2(m_c^2 - m_u^2)$ for $m_{c,u} >> m_W$. So we might be able to estimate m_c^2 from the experimental value of some flavour-changing neutral process. It turns out the best estimate comes from the $K_1 - K_2$ mass difference [20]. The relevant box diagram of Fig. 3 is easily calculated to be

$$\approx \frac{g^4}{8} \sin^2\theta_c \cos^2\theta_c \int \frac{d^4k}{(2\pi)^4} \left(\frac{1}{k^2-m_c^2} - \frac{1}{k^2-m_u^2}\right) \left(\frac{1}{k^2-m_W^2}\right) \quad (\gamma \text{ factors}) \tag{1.26}$$

The integral is easily evaluated in the limit $m_u^2 << m_c^2 << m_W^2$ and gives an "effective Lagrangian"

$$L_{eff} = - \frac{G_F}{\sqrt{2}} \frac{\alpha}{4\pi} \left(\frac{m_c}{38 GeV}\right)^2 \cos^2\theta_c \sin^2\theta_c \, [\bar{s}\gamma_\mu \tfrac{1}{2}(1-\gamma_5)d] [\bar{s}\gamma_\mu \tfrac{1}{2}(1-\gamma_5)d] \tag{1.27}$$

Sandwiching this expression between K_0 and $\overline{K}_0$ states, and inserting the vacuum in the middle gives an order of magnitude estimate

$$\Delta m_K \approx \frac{1}{m_K} < \overline{K}_0 | - \mathcal{L}_{eff} | K_0 > =$$

$$\frac{G_F}{\sqrt{2}} f_K^2 m_K \frac{\alpha}{4\pi} \left(\frac{m_c}{38\text{GeV}} \right)^2 \cos^2\theta_c \sin^2\theta_c \qquad (1.28)$$

Using the experimental value $\Delta m_K = (0.7 \times 10^{-14})\ m_K$ gives the estimate

$$m_c \sim (1.5 \text{ to } 2)\ \text{GeV} \qquad (1.29)$$

Gaillard and Lee [20] showed that this value is consistent with all the other results and limits (1.3) on $|\Delta S| = 1$ neutral currents. You may wonder [21] about the effects of asymptotic freedom on the calculation of $\mathcal{L}_{eff}$ (1.27), or question the estimate (1.28) of the matrix element. But the estimate (1.29) turned out [22] to be correct.

1.6. Is charm proved ?

Everybody feels that recent experimental data strongly support the simple $SU(2)_L \times U(1)$ GIM charm model : to what extent can we regard the model as proved ? Let us systematically check phenomenologically the salient features of the charm-changing charged current

$J_\mu^{\Delta C=1} =$	g	$\bar{c}$	$\gamma_\mu \frac{1}{2}(1-\gamma_5)$
	normal weak coupling strength	new quantum number	left-handed V-A current

$$(s \cos\theta_c \quad - \quad d \sin\theta_c) \qquad (1.30)$$

$\Delta c = \Delta s = 1$	$\Delta c = 1, \Delta s = 0$
dominant	present, ∿5%

Each of the five features listed above is vital for the GIM[9] cancellation to work.

- *New quantum number (quark)* ? It should be conserved by the strong and electromagnetic interactions, but can be produced singly or decay via weak interactions. This has been amply demonstrated both in production tests [$e^+e^- \to D\bar{D} + X$ charmed particle pairs[23,2)], $\nu + N \to \mu^- + C + X$ gives charmed particles singly [24,3)]. That the lowest-lying states decay weakly has also been confirmed by the observation of parity violation [25,3)] and of semileptonic decays [26,3)] with substantial branching ratios. The existence of a new quantum number or quark is completely established.

- *Normal weak coupling strength ?* The most direct way of estimating the charm weak coupling strength is from decays. The most straightforward way [27)] of estimating the charm lifetime is to assume that the charmed quark decays as if it were free, and scale up the rate for $\mu \to e\nu\bar{\nu}$

$$\Gamma(C \to X) \simeq \frac{G_F^2 m_c^5}{192\pi^3} A \qquad (1.31)$$

where A is a counting factor introduced to allow for all the different final states in Fig. 4. We may write A as

$$A = \underset{(e\bar{\nu}_e)}{1} + \underset{(\mu\bar{\nu}_\mu)}{1} + \underset{(\text{colour})}{3} \times \underset{(u\bar{d})}{\bar{A}_C} \qquad (1.32)$$

The factor $\bar{A}_C$ is a non-leptonic enhancement factor, due to the renormalization of the basic diagram by the strong interactions [28)]. The analogous quantity for strange particle decays is very large experimentally. Its origin is unclear, but for several reasons $\bar{A}_C$ is believed to be much smaller [29)]. A reasonable bracket seems to be $1 \leq \bar{A}_C \leq 3$ which, when inserted into (1.31) and (1.32), yields a lifetime estimate [27)] of

$$\tau \text{ (charm)} \sim \text{few} \times 10^{-13} \text{ seconds} \qquad (1.33)$$

Since non-leptonic decays are not much enhanced relative to semileptonic decays, the latter should have substantial branching ratios[29)]

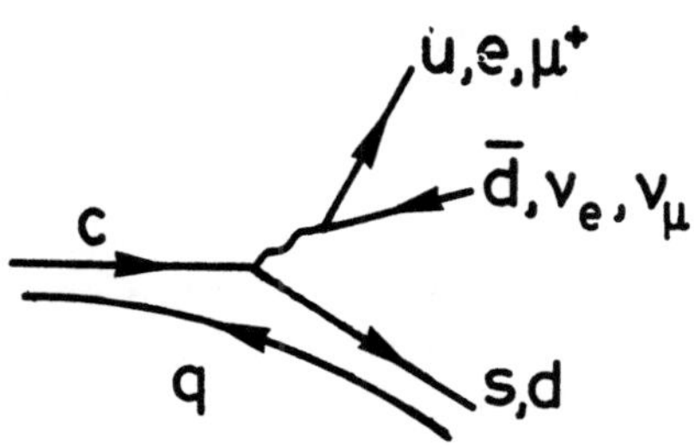

Fig. 4 : Free quark diagram model for charmed particle decay.

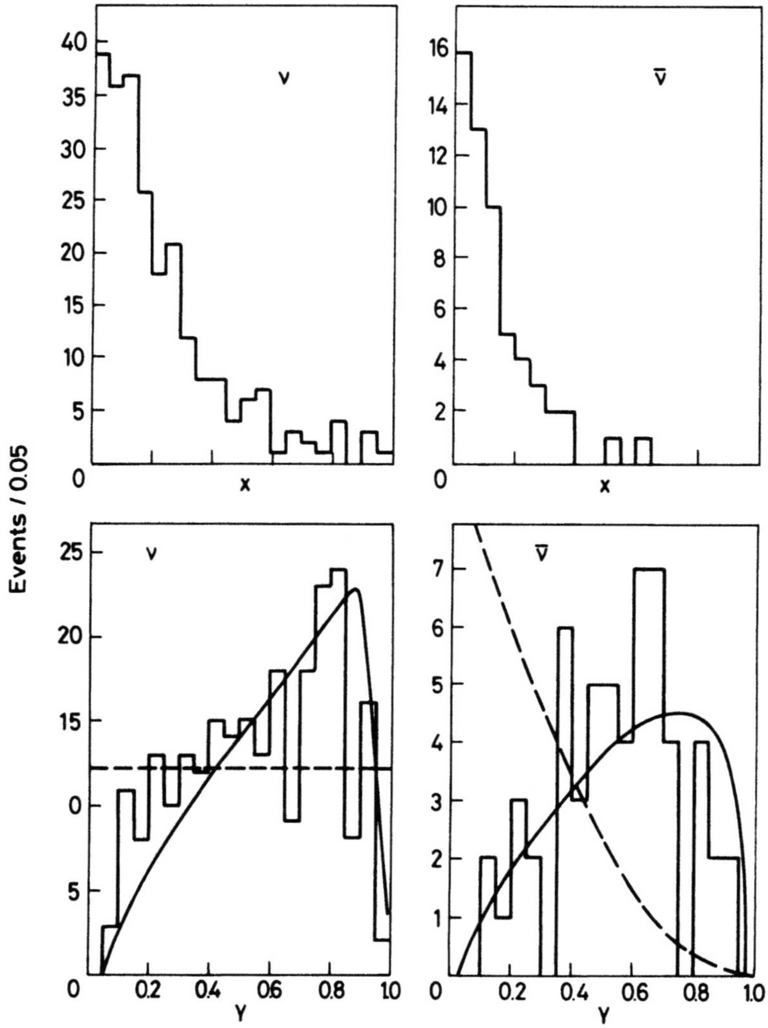

Fig. 5 : X and Y distributions for $\nu \to \mu^-\mu^+$ and $\nu \to \mu^+\mu^-$ events from the CDHS experiment 31). The solid curves are Monte Carlo calculations modifying a flat Y distribution for the experimental acceptance.

$$\frac{\Gamma(C \to e^+ + \nu_e + X)}{\Gamma(C \to X)} \approx \frac{\Gamma(C \to \mu^+ + \nu_\mu + X)}{\Gamma(C \to X)} = \frac{1}{2+3\bar{A}_c} \approx (10 \text{ to } 20)\% \qquad (1.34)$$

For charmed particles with momenta of a few GeV, which are typical of high energy experiments, the lifetime (1.33) would yield a decay path of O(100) microns. There is a general absence in high energy experiments of "short" tracks of ≥ 1mm length, suggesting that τ(charm) is not more than an order of magnitude longer than the estimate (1.33), and indicating that the charm weak coupling strength is not exceedingly small. There is even one event [30] from a FNAL neutrino experiment using an emulsion target which has a short track of length $\sim$ 182 microns, corresponding to a lifetime of the order of (1.33). There is also evidence from e^+e^- and neutrino collisions that the semileptonic branching ratio of charmed particles [2,3,26,31] is comparable with (1.34). For example, the DASP group [26] estimate from looking at prompt electrons produced above charm threshold,

$$\frac{\Gamma(C \to e^+ + \nu_e + X)}{\Gamma(C \to X)} = (11 \pm 3)\%$$

As far as charm production is concerned, the simple fact that ν and $\bar{\nu}$ cross-sections at FNAL and SPS energies are not [32] much larger than a linear extrapolation

$$\sigma^{\nu,\bar{\nu}} \approx (\text{constant}) \times E_{\nu,\bar{\nu}} \qquad (1.35)$$

of the lower energy Gargamelle data indicates that a very large charm weak coupling strength is ruled out. To go further we have to use the quark-parton model and details of the charm current (1.30). The fundamental weak processes expected to contribute to charm production are tabulated in Table 1, together with their expected X ($\equiv Q^2/2m\nu$) and Y ($\equiv E_h/E_\nu$) distributions, and cross-sections relative to $\sigma(\nu + N \to \mu^- + X)$. We see that charm production is expected to be O(10%) of both ν and $\bar{\nu}$ cross-sections, leading -- on the basis of the semileptonic branching ratio estimate (1.34) -- to dilepton cross-sections of (1 to 2)%. These predictions are completely consistent with the data [31].

- *Left-handed current ?* It is in principle possible to distinguish V ± A currents in charm decays, but this requires more precise data than are available at present. However, the V - A nature of (1.30) can be checked from the dilepton events in ν and $\bar{\nu}$ collisions, which are expected to come predominantly from charm decays, and hence reflect the flat Y distributions expected in Table 1 which are characteristic of a (V - A) current. Present data on $\nu \to \mu^- e^+$ [33], $\mu^-\mu^+$ and $\bar{\nu} \to \mu^+\mu^-$ [31] (Fig. 5) all favour flat Y distributions from charm, and we conclude that the charm-changing current is indeed predominantly left-handed.

- *Large $\Delta C = \Delta S = 1$ coupling ?* There is ample evidence from many experiments that charmed particles like to decay into strange particles. There are $\nu + N \to \mu^- + e^+ + (K, \Lambda)$ events [24], $D \to K n\pi$ decays [23], a jump in $e^+e^- \to K + X$ [34] at 4 GeV centre-of-mass energy corresponding to about two strange particles per charm pair production event, and $e^+e^- \to e + K + X$ coincidences [34]. Particularly significant is the observation of the decays $D^\pm \to K^\mp\pi^\pm\pi^\pm$. These exotic final states are just what you would expect from the fundamental decay chain

$$(c\bar{d}) \to (s u\bar{d})\bar{d} \;\approx\; (s\bar{u}(u\bar{d})(u\bar{d})$$

and indicate that c is coupled to s and not $\bar{s}$, so that $\Delta C = +\Delta S$, not $-\Delta S$.

- *Small $\Delta C = 1$, $\Delta S = 0$ coupling ?* This is clearly more difficult to dig out, but there is some indication of its presence. The CDHS neutrino experiment finds [31] (see Fig. 5) that

$$< X >_{\nu\to\mu^-} \; > \; < X >_{\nu\to\mu^-\mu^+} \; > \; < X >_{\bar{\nu}\to\mu^+\mu^-} \tag{1.36}$$

which is consistent with the estimates of Table 1, where about half of charm production by ν proceeds off valence quarks, which have large <X> and dominate $\nu \to \mu^-$ events, whereas all $\nu \to \mu^+\mu^-$ events proceed off sea quarks with small <X>. Also, by comparing the relative strengths of $\bar{\nu} \to \mu^+\mu^-/\bar{\nu} \to \mu^+$ and $\nu \to \mu^-\mu^+/\nu \to \mu^-$, they estimate [2,31] comparable cross-sections for d valence and s sea production of $\nu \to \mu^+\mu^-$ events. Clearly a much more direct test of the small $\Delta S = 0$ component of (1.30) would be observation of a Cabibbo suppressed decay mode of the D's, such as $D^0 \to \pi^+\pi^-$. One would expect

$$\frac{\Gamma(D^0\to\pi^+\pi^-)}{\Gamma(D^0\to K^-\pi^+)} = 0\,(\tan^2\theta_c) \approx 5\% \tag{1.37}$$

Experimentally, there is an upper limit of 8% [3], and further developments in this direction are eagerly awaited.

A summary of different tests of the charm-changing current (1.30) is presented in Table 2. We see that every significant ingredient has successfully passed at least one test, and many have been proved over and over again. It would be nice to measure more precisely the charm lifetime, verify the V - A structure in charmed particle decays as well as production, and check for the small $\Delta C=1$ $\Delta S = 0$ bit in D decays. However, we would conclude that GIM charm [9] is so well established that it is almost ripe to be put in the textbooks. It can be regarded as a well-understood part of quark physics, which can now be used in probes of dynamical models -- asymptotic freedom, charmonium, etc...
Yesterday's sensation is tomorrow's tool.

2. APRES-CHARM

2.1. Introduction.

Despite the great experimental success of the GIM charm model, there are ample indications that there are more discoveries to be made. These may be categorized as :

- *More leptons ?* There are very strong indications [35] that a heavy lepton -- the τ -- with mass $\sim$ 1.9 GeV has been discovered in e^+e^- annihilation experiments. There is also the possibility [36] that other, heavier leptons may be showing up in neutrino collisions.

- *More quarks ?* The most solid experimental evidence is, of course, the Υ (9.5) resonance [1,37]. But there were theoretical reasons before this discovery to anticipate at least two more quarks. These included the need [38] to cancel "anomalous" triangle diagrams in order to restore renormalizability of the weak-electromagnetic gauge theory, the now-defunct high Y anomaly [39,40,2] in neutrino scattering, CP violation [17], and attempts to construct more unified gauge theories [41].

- *More bosons ?* The standard $SU(2)_L \times U(1)$ Weinberg-Salam-Ward gauge group has various theoretical problems [it is not simple it treats right and left helicities asymmetrically, it was invented by someone else], and experimental problems [the absence [42] of parity violation in atomic physics]. So many people have proposed bigger gauge groups [$SU(2) \times U(1) \times U(1)$ [43], $SU(2)_L \times SU(2)_R \times U(1)$ [43,44], SU(3) [45], $SU(3) \times U(1)$ [46], $SU(3) \times SU(3)$ [47], SU(4) [48] ...] each of which introduces more gauge bosons.

In this lecture, we will go through the various apres-charm possibilities, consider some of their experimental implications, and try to assess the present evidence for their existence.

2.2. Possible new leptons.

There are a considerable number of different types of heavy leptons that may exist, and we first categorize some of them, so that we have some theoretical orientation when we look at the phenomenology.

2.2.1. Sequential leptons.

As far as we know, the muon has just the same weak and electromagnetic interactions as the electron, but is simply heavier. To quote Paul Tsai [49] : "Since muons exist in nature for no apparent reason, it is possible that other heavy leptons may also exist in nature." There may exist one or more extra lepton doublets to be added sequantially to those we know, love, and do not understand

$$\begin{pmatrix} \nu_e \\ e^- \end{pmatrix}_L , \begin{pmatrix} \nu_\mu \\ \mu^- \end{pmatrix}_L ; \begin{pmatrix} \nu_\tau \\ \tau^- \end{pmatrix}_L \quad ?, \ldots \tag{2.1}$$

Such a lepton would have its own lepton number, which would be conserved (at least approximately, exactly if all ν's have identical masses).

2.2.2. "Gauge" leptons.

By this term we mean leptons which participate essentially in the cancellations of bad high-energy behaviour necessary for the renormalizability of gauge theories. Consider, for example, the high-energy behaviour of the reaction $\nu_\mu \bar{\nu}_\mu \to W^+W^-$. The t-channel μ^- exchange graph yields a cross-section growing like s, whereas we need $\sigma \sim 1/s$ to become renormalizable [4,13,50]. We can either add in an s-channel Z^0 exchange, or a u-channel M^+ exchange, or another t-channel M-exchange, or some combination (see fig. 6). Weinberg and Salam [14,15] chose the neutral current

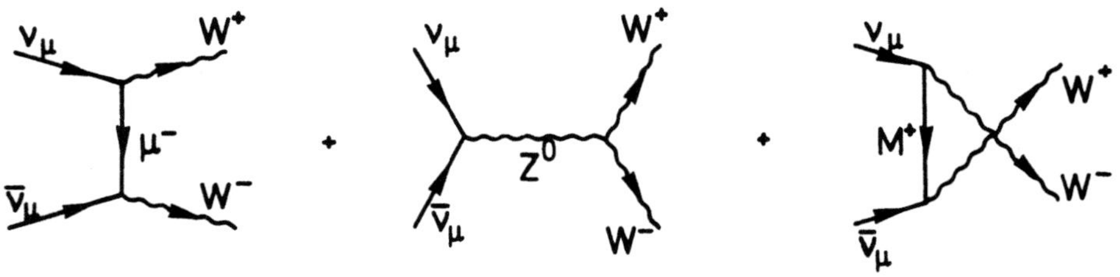

Fig. 6 : Different diagrams which may conspire in gauge theories to give suitable high-energy behaviour for $\nu_\mu \bar{\nu}_\mu \to W^+W^-$

option; Georgi and Glashow [51] chose the heavy lepton option. Neutral currents exist [10], so the Weinberg-Salam [14,15] model is favoured, but it is quite possible that either or both of the other possibilities are also realized [52]. We will use different names [53] for the M^+ and M^- options.

2.2.3. Para-leptons.

The positively charged lepton has electron or muon number +1. Examples are [51,54]

$$\begin{pmatrix} M^+ \\ \nu_\mu \\ \mu^- \end{pmatrix}_L \; ; \; \begin{pmatrix} M^+ \\ \dfrac{\nu_\mu + \nu'_\mu}{\sqrt{2}} \end{pmatrix}_L \; , \; \begin{pmatrix} \dfrac{\nu_\mu - \nu'_\mu}{\sqrt{2}} \\ \mu^- \end{pmatrix}_L \qquad (2.2)$$

2.2.4. Ortho-leptons.

The negatively charged lepton has electron or muon number -1. An example is an SU(3) triplet [55]

$$\begin{pmatrix} \mu^- & & \nu_\mu \\ & M^- & \end{pmatrix}_L \qquad (2.3)$$

2.2.5. Mono-leptons.

Although we usually put leptons in non-trivial multiplets, one can certainly imagine [56] introducing a new singlet charged lepton field. It could then mix with the e^- and μ^- and so decay

$$\begin{pmatrix} \nu_e \\ e^- \end{pmatrix}_L \; , \; \begin{pmatrix} \nu_\mu \\ \mu^- \end{pmatrix}_L \; , \; \left(l^- \right)_L \qquad (2.4)$$

mixing

2.2.6. Neutral leptons.

So far we have only discussed charged leptons, but massive neutral leptons may also occur. Just as an M^+ might be introduced to make a gauge cancellation in $\nu_e\bar{\nu}_e \to W^+W^-$, so an E^0 might be introduced as a t-channel exchange into $e^+e^- \to W^+W^-$. One can imagine possibilities like [57]

$$\begin{pmatrix} E^0 \\ e^- \end{pmatrix}_R \; , \; \begin{pmatrix} E^+ \\ \nu_e \cos\alpha + E^0 \sin\alpha \\ e^- \end{pmatrix}_L \qquad (2.5)$$

Finally, we should remember that if we discover a new lepton, we should be on guard for novel intermediate bosons mediating its decays. For example, in the SU(3) x U(1) model (2.3) [55] the $\nu_\mu \leftrightarrow M^-$ and $\mu^- \leftrightarrow M^-$ transitions are mediated by new $W_K^\pm$ and W_K^0 bosons, respectively. In particular, we should be on guard for charged currents which are not V - A, and neutral currents which change flavour ($\mu^- \leftrightarrow M^-$, for example).

2.3. Decays of heavy leptons.

Let us now quickly catalogue [49,58] the likely decay modes of heavy leptons for confrontation with experiment. We will use a fundamental lepton current

$$J_\mu = \bar{f}_2\gamma_\mu[g_L \frac{1}{2}(1-\gamma_5) + g_R \frac{1}{2}(1+\gamma_5)] f_1 \tag{2.6}$$

2.3.1. Leptonic decays.

These can be simply scaled up from the μ decay rate

$$\Gamma(\mu^- \to \nu_\mu e^- \bar{\nu}_e) \simeq \frac{G_F^2 m_\mu^5}{192\pi^3} \tag{2.7}$$

and have rates [58] (for $m_{l^0} < m_{l^-}$)

$$\frac{\Gamma(l \to l^0 e^- \bar{\nu}_e)}{\Gamma(\mu^- \to \nu_\mu e^- \bar{\nu}_e)} = \frac{G_F'^2}{G_F^2} \left(\frac{m_{l^-}}{m_{\mu^-}} \right)^5 \tag{2.8}$$

$$\left\{ \frac{|g_R|^2 + |g_L|^2}{|g_L^\mu|^2} f_1\left(\frac{m_{l^0}}{m_{l^-}} \right) + \frac{2\mathrm{Re}\, g_R^* g_L}{|g_L^\mu|^2} f_2\left(\frac{m_{l^0}}{m_l} \right) \right\}$$

where

$$f_1(z) = (1-z^4)(z^4-8z^2+1)+24z^4\ln(1/z) \quad \to|1 \quad z\to 0 \tag{2.9}$$

and

$$f_2(z) = 4z(1-z^2)^3 - 6z(1+z^2)[1-z^4-4z^2\ln(1/z)]$$

$$\to 0 \quad \text{as} \quad z\to 0 \tag{2.10}$$

We have allowed in (2.8) for the possibility that a new W' boson is responsible for l^- decays. We see from (2.9) and (2.10) that the $l^- \to l^0$ transition has to look either left- or right-handed when $m_{l^0}/m_{l^+} \to 0$. From the upper limit of $\approx$ 500 MeV on the mass of any l^0 produced in τ^- decay [3] we have $f_1(m_{l^0}/m_{\tau^-}) < 1/4$ and we will henceforth neglect m_{l^0}/m_{l^-} and such interference terms. The decay rates for left-handed and right-handed leptons are, of course, the same. On the other hand, the decay spectra are different, with a right-handed lepton producing a softer final state e or μ on the average. This can be understood from Fig. 7, where we have indicated the e^-, $\bar{\nu}_e$, l^0 momenta corresponding to maximal e^- energy, and shown how the lepton helicities are aligned. The maximum e^- energy configuration for right-handed l^- decay has helicity 3/2 and so is forbidden. [This sort of argument can be extended to charm or other quark decays, and may eventually yield tests [59] of the V ± A nature of quark currents]. For anything except a para-electron or para-muon

$$\frac{\Gamma(l^- \to l^0 e^- \bar{\nu}_e)}{\Gamma(l^- \to l^0 \mu^- \bar{\nu}_\mu)} \approx 1 \quad \text{because} \quad \frac{m_{l^-}}{m_{\mu^-}} >> 1 \tag{2.11}$$

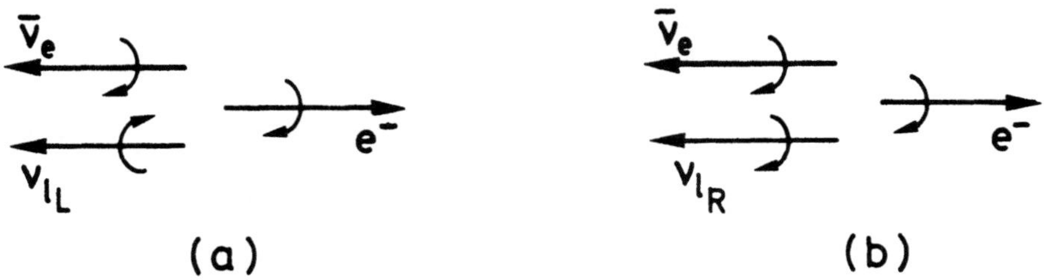

Fig. 7 : Configurations giving maximal e^- energy in $l^- \to \nu_l e^- \bar{\nu}_e$ decay (a) for a left-handed l^-; (b) for a right-handed l^-.

However, for a para-muon or para-electron decay there are two diagrams contributing and two identical neutrinos in the final state [58]. Therefore

$$\Gamma \frac{(M^+ \to \nu_\mu \mu^+ \nu_\mu)}{(M^+ \to \nu_\mu e^+ \nu_e)} \approx 2 \quad , \quad \frac{\Gamma(E^+ \to \nu_e e^+ \nu_\mu)}{\Gamma(E^+ \to \nu_e e^+ \nu_e)} \simeq \frac{1}{2} \tag{2.12}$$

2.3.2. Inclusive semihadronic decays.

The simplest way to estimate these is to use the estimate (2.8) with $(e^- \bar{\nu}_e)$ replaced by free quarks $(\bar{u}d)$, etc... with the difference that the rate should be multiplied by three to take account of the three quark colours. For a very heavy lepton which was anything except a para-lepton one would then have (assuming W' is V ± A)

$$\Gamma(l^- \to l^0 e^- \bar{\nu}_e) : \Gamma(l^- \to l^0 \mu^- \bar{\nu}_\mu) : \Gamma(l^- \to l^0 \bar{u}d) \approx 1:1:3 \tag{2.13}$$

For a lepton with mass ∿ 1.9 GeV, semihadronic decays are expected to be (60 to 70)%, including a continuum accounting for ∿25%.

2.3.3. Exclusive semihadronic decays.

These decay rates are given by the spectral functions $\rho_1{}^{V,A}$ and $\rho_2{}^{V,A}$ of the weak current

$$\begin{aligned}\sum_X <0|J_\mu^{V^+,A^+}(o)|X><X|J_\mu^{V^-,A^-}(0)|0>(2\pi)^3\delta^4(q-p_X) \\ \equiv \rho_1^{V,A}(q^2)(q_\mu q_\nu - q^2 g_{\mu\nu}) + \rho_2^{V,A}(q^2) q_\mu q_\nu\end{aligned} \tag{2.14}$$

The contribution of the $\rho^\pm$ can be estimated directly by CVC

$$\rho_1^\rho \simeq \frac{m_\rho^2}{2\gamma_\rho^2}\,\delta(q^2 - m_\rho^2) \quad , \quad \rho_2^\rho = 0 \tag{2.15}$$

where $\gamma_\rho^2/4\pi \simeq 0.6$ experimentally. Similarly, the $\pi^\pm$ contribution can be taken directly from π decay

$$\rho_1^\pi = 0 \quad , \quad \rho_2^\pi = f_\pi^2 \delta(q^2 - m_\pi^2) \tag{2.16}$$

where $f_\pi \simeq 0.9\ m_\pi$. In general, we have in the limit $m_{W'} >> m_{l^-} >> m_{l^0}$

$$\frac{d\Gamma}{dq^2}(l^- \to l^0 + \text{Hadrons}) \underset{\sim}{\sim} \frac{G_F'^2 m_{l^-}^2}{16\pi}\left(1 - \frac{q^2}{m_{l^-}^2}\right)^2 \left(\frac{|g_R|^2 + |g_L|^2}{|g_L^\mu|^2}\right) g(q^2) \tag{2.17}$$

where

$$g(q^2) \simeq [C_V^2 \rho_1^V(q^2) + C_A^2 \rho_1^A(q^2)][1 + \frac{q^2}{m_{l^-}^2} - \frac{2q^4}{m_{l^-}^4}]$$

$$+ [C_V^2 \rho_2^V(q^2) + C_A^2 \rho_2^A(q^2)][1 - \frac{q^2}{m_{l^-}^2}] \tag{2.18}$$

We have allowed for the possibility that the W' boson responsible for l^- decays does not necessarily couple to a V ± A quark current

$$J_\mu^{W'} = \bar{u}\gamma_\mu [C_V + C_A\gamma_5]d + \ldots \tag{2.19}$$

The relative frequency of π^- and ρ^- decays is, of course, a sensitive measure of the (V, A) content of the current responsible for l^- decay. For a conventional V ± A current, a lepton with mass $\sim$ 1.9 GeV would have relative branching ratios [49,53,58]

$$\Gamma(l^- \to l^0 e^- \bar{\nu}_e) : (l^- \to l^0 \pi^-) : (l^- \to l^0 \rho^-) \underset{\sim}{\sim} 1 : \frac{1}{2} : \frac{3}{2} \tag{2.20}$$

If the l^- decays via the usual Cabibbo current, it will have decays to strange particles (e.g. $l^0 K^-$) suppressed by the usual factor

$$\frac{\Gamma(l^- \to l^0 K^-)}{\Gamma(l^- \to l^0 \pi^-)} \underset{\sim}{\sim} \frac{\sin^2\theta_c}{\cos^2\theta_c} \frac{f_K^2}{f_\pi^2} \text{(phase space)} \underset{\sim}{\sim} \frac{1}{20} \tag{2.21}$$

2.3.4. Radiative decays.

Decays of the type $l^- \to e^-\gamma$ or $\mu^-\gamma$ are forbidden for a sequential lepton in a left-handed doublet with a neutrino with mass $\simeq m_{\nu_e}, m_{\nu_\mu}$. But many other leptons have radiative decays occuring at some level. Consider, for example [60], the radiative decay of an ortho-electron E^- mediated by a right-handed coupling to a massive neutral ortho-electron E^0

$$\frac{\Gamma(E^- \to e^-\gamma)}{\Gamma(E^- \to \nu_e l^- \bar{\nu}_e)} \gtrsim \frac{6\alpha}{\pi}\left(\frac{m_{E^\circ}}{m_{E^-}}\right)^2 \tag{2.22}$$

Observation of $\tau^- \to e^-\gamma$ or $\mu^-\gamma$ would be an exciting indicator of hidden lepton structure.

2.3.5. Neutrinoless decays.

A conventional sequential lepton always decays via a charged current, and always leaves a neutrino in the final state. On the other hand, in some models there are off-diagonal neutral lepton currents, which could give rise to heavy lepton decays. For example [56], the mono-lepton (2.4) would violate the leptonic analogues of the flavour-conservation conditions (1.16), (1.19). Therefore, decays of the type

$$\tau^- \to \left.\begin{matrix} e^- \\ \mu^- \end{matrix}\right\} e^+e^-,\ \left.\begin{matrix} e^- \\ \mu^- \end{matrix}\right\} \mu^+\mu^-,\ \left.\begin{matrix} e^- \\ \mu^- \end{matrix}\right\} \pi^+\pi^-,\ \ldots$$

could occur. From the validity of (μe) universality, etc ..., the mixings of such a mono-lepton are constrained [56] so that

$$\frac{\Gamma(\tau^- \to 3 \text{ charged particles})}{\Gamma(\tau^- \to \text{all})} \gtrsim 10\% \tag{2.23}$$

If we combine all the above estimates [49,53,58], we arrive at the following expectations for the branching ratios of a sequential charged lepton with mass $\sim$ 1.9 GeV in a isodoublet with its own massless neutrino

$$\left.\begin{matrix} e^-\bar{\nu}_e\nu_\tau \sim 17\%, & \mu^-\bar{\nu}_\mu\nu_\tau \sim 17\%, & \pi^- + \nu_\tau \sim 9\% \\ \rho^-\nu_\tau \sim 25\%, & A_1^- + \nu_\tau \sim 8\%?, & \text{continuum}+\nu_\tau \sim 25\%\ ? \\ K^-\nu_\tau \sim \frac{1}{2}\%, & e^- + \gamma \sim 0\%, & \mu^- + \gamma \sim 0\% \end{matrix}\right\}$$

$$3 \text{ charged particles} \sim 0\% \tag{2.24}$$

Since A_1 and continuum states have relatively small branching ratios, the great majority of τ final states are expected to contain just one charged particle and ≤ 2 γ's. In the next section we will see how the predictions (2.24) compare with the experimental data.

2.4. What is the τ particle ?

Evidence for the pair production $e^+e^- \to \tau^+\tau^-$ of a new particle first came from the (μe) events [35] of the SLAC-LBL group. Much more data has now accumulated [61], and a tentative identification can be attempted [3].

The cross-sections for (μe) and related prompt lepton + one-prong events show a threshold around 4 GeV. They are therefore presumably nothing to do with old pre-charm physics. On the other hand, they do not seem to be charm. There are no peaks in the cross-sections at 4.03 or 4.41 GeV, and the DASP group argue convincingly that their $e^\pm$ + one-prong and $e^\pm$ + multiprong events have different origins. The smooth behaviour of the cross-section betokens a particle without strong interactions. The relatively rapid rise above threshold militates against a spin 0 or spin 1 assignment for the τ, and spin 1/2 is perfectly consistent with the $e^\pm$, $\mu^\pm$ spectra which seem to come from three-body decays. The relative rates of e^+e^- : $\mu^\pm e^\mp$: $\mu^+\mu^-$ events are consistent with the 1 : 2 : 1 expected if (2.12) holds : a para-electron or para-muon is therefore disfavoured. As discussed in the next section, ν and $\bar{\nu}$ scattering experiments exclude an ortho- or para-muon assignment : an ortho-electron is still possible. The decay rate into three charged leptons is < 0.6 %, eliminating the mono-lepton hypothesis. The $e^-\bar{\nu}_e\nu$, $\mu^-\bar{\nu}_\mu\nu$ branching ratios seem to be O(15)%, in accord with (2.24). The decay e^-, μ^- spectra tentatively favour a (V - A) charged current for the τ^-. Most semihadronic decays of the τ seem to have a charged multiplicity of one; the $\rho\nu$ and $A_1\nu$ decay modes seem to have been identified. The DASP group find $(\tau^- \to K^- \ldots)/(\tau^- \to \pi^- \ldots) << 1$, as needed by the Cabibbo suppression (2.21). On the other hand, the DASP group fail to find the $\pi\nu$ decay at the expected level. There are upper limits of 2.6% (1.3%) on the decays $\tau^- \to e^-\gamma(\mu^-\gamma)$.

We conclude it is overwhelmingly probable that the τ is a heavy lepton, very likely a sequential (V - A) lepton. The one remaining thing to check out is the decay $\tau^- \to \pi^-\nu_\tau$, whose decay rate is calculable [49,53,58] with a distinctive pion momentum spectrum. Theorists will be very embarrassed if it does not exist: for the rest of these lectures we will assume the τ to be a sequential (V-A) heavy lepton.

2.5. Heavy lepton production by neutrinos ?

Either ortho- or para-muons could, in principle, be produced by ν_μ and $\bar{\nu}_\mu$ beams. After an initial threshold effect, the cross-sections at high energies would be comparable to the conventional $\nu_\mu \to \mu^-$, $\bar{\nu}_\mu \to \mu^+$ cross-sections. An ortho- or para-muon with mass $\sim$ 2 GeV would have caused FNAL-CERN SPS neutrino cross-sections to be almost twice as large as those measured in Gargamelle. Such seems not to be the case, so that the τ could not be an ortho- or para-muon. But still heavier leptons may exist, and the observation of 3μ events [36] in high-energy neutrino collisions has suggested to some people the existence of a new lepton or leptons with mass(es) 0 (5 to 10) GeV [62].

If, for example, a heavy charged ortho-muon M^- were produced [63], which subsequently decayed into a neutral lepton M^0, one might observe the chain (see Fig. 8).

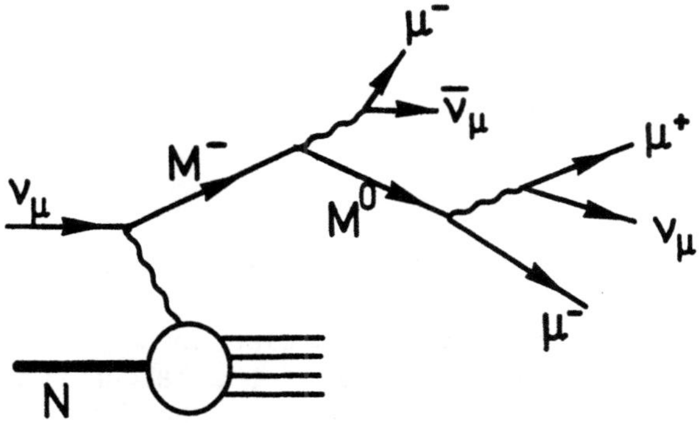

Fig. 8 : Heavy lepton cascade production of 3μ events in neutrino collisions.

$$\nu \to M^- \quad \llcorner\!\!\to M^\circ + \mu^- + \bar{\nu}_\mu \quad \llcorner\!\!\to \mu^- + \mu^+ + \nu_\mu \tag{2.25}$$

The production of M^- relative to μ^- is suppressed by the phase-space factor ϕ [58] shown in Fig. 9. To see a multimuon event one must also pay the price of a leptonic branching ratio factor B of the order of 0.1 to 0.2.

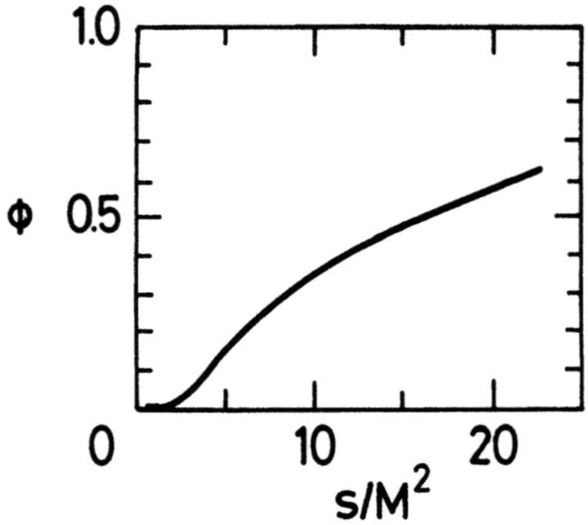

Fig. 9 : Threshold suppression factor for left-handed heavy lepton production by neutrinos in a simple parton model, taken from Ref. 57.

There should be $\mu^-\mu^-$ [64] and $\mu^-\mu^+$ events as well as $\mu^-\mu^+\mu^-$ events, which may also have been seen. The final-state leptons would tend to come out at azimuthal angles around the beam direction which were opposite to those of the hadron shower. This, in principle, allows a distinction to be drawn between trilepton events, due to heavy leptons, or more mundane hadronic explanations, such as Drell-Yan, charm-pair production, bottom- or top-quark production. The invariant masses of μ pairs in 3μ events, and other momentum distributions, suggest [36,63] a parent M^- mass O (7 to 10) GeV and an M^0 mass O (3 to 5) GeV. The cross-section claimed for 3μ events, and the phase-space suppression factor ϕ of Fig. 9 then suggest that the $\nu \rightarrow M^-$ has a normal weak coupling strength. To obtain this in an SU(2) x U(1) model requires an *ad hoc* modification of weak universality [65]. Alternatively, one may [66] postulate a bigger gauge group.

It may be excessive to proliferate new leptons and bosons on the strength of a handful of events, particularly since two other experiments [62] have not yet confirmed the characteristics of the FHPRW 3μ events. On the other hand, some of the FHPRW events are extremely dramatic and unimpeachable. In a few months the world should have a large stock of 3μ events, and the existence of M^- and/or M^0 should be resolved.

2.6. Do more leptons entail more quarks ?

It has been commonplace to use the existence of the τ to motivate the imminence of more quarks [67]. This motivation occurs at two levels. Philosophically, one may believe there should be equal numbers of quarks and leptons -- this was indeed one of the reasons advanced [8] for charm

$$\begin{array}{ccc} \begin{pmatrix} \nu_e \\ e^- \end{pmatrix}_L & \begin{pmatrix} \nu_\mu \\ \mu^- \end{pmatrix}_L & \begin{pmatrix} \nu_\tau \\ \tau^- \end{pmatrix}_L \\ \updownarrow & \updownarrow & \updownarrow\ ? \\ \begin{pmatrix} u \\ d_C \end{pmatrix}_L & \begin{pmatrix} c \\ s_C \end{pmatrix}_L & \begin{pmatrix} t \\ b_C \end{pmatrix}_L \end{array} \qquad (2.26)$$

On a more formal level, just adding a new lepton multiplet to the old $(\nu_e, e^-, \nu_\mu, \mu^-, u, d, c, s)_L$ model would destroy renormalizability [13], because it would generate "anomalies" [38].

The formal proofs of renormalizability of gauge theories depend on very precise relations and cancellations between different couplings which are expressed by Ward-Takahashi identities. For example, there is an identity which says

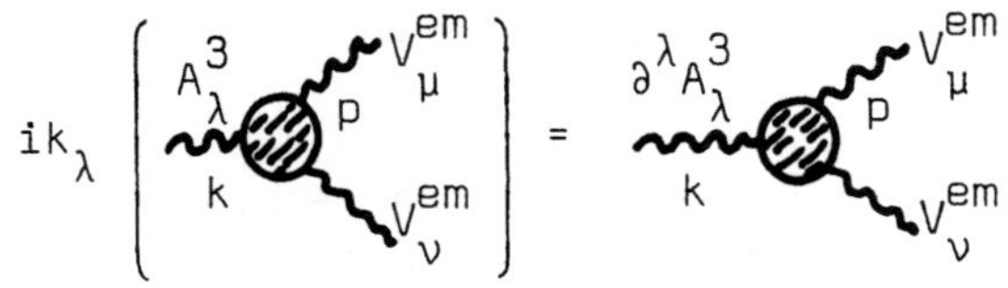

(2.27)

This relation is not necessarily true in a field theory with fermions. In general, there is [68] an extra "amomalous" piece

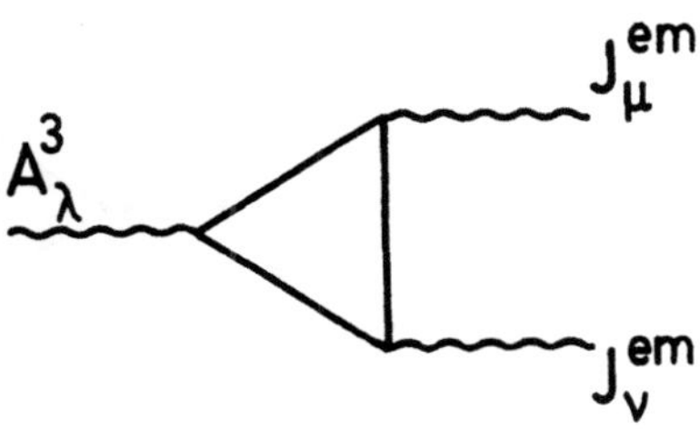

Fig. 10 : Basic Adler-Bell-Jackiw [68] "anomalous" triangle diagram.

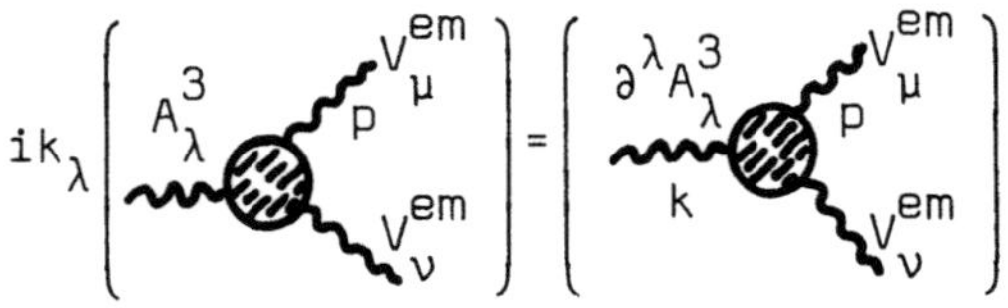

$$+ \frac{S}{2\pi^2} p^\xi k^\tau \varepsilon_{\xi\tau\mu\nu} \qquad (2.28)$$

where S arises from the triangle diagrams of Fig. 10, and has the value

$$S = \sum_{\text{fermions } f} g_A^f Q_f^2 \qquad (2.29)$$

Another way of stating the problem is to recall that renormalizability requires gauge symmetry, which in turn requires conserved currents. The triangle diagrams create an "anomalous" divergence term in a current coupled to a gauge boson

$$\partial^\lambda A_\lambda^3 \ni \frac{S}{8\pi^2} \varepsilon_{\mu\nu\xi\tau} F_{em}^{\mu\nu} F_{em}^{\xi\tau} \qquad (2.30)$$

which would be present even if all the spontaneous symmetry breaking were switched off.

A specific example of a troublesome result of the "anomaly" (2.28) or (2.30) is given by Fig. 11. The "anomalous" triangle piece in the three-boson vertex destroys a delicate Higgs-ghost cancellation. When squared, as in Fig. 11b, the "anomaly" generates a mass for the direct channel intermediate boson which it would not otherwise have possessed [39]. Since gauge bosons are only supposed to derive their masses from the Higgs mechanism, this is bad karma.

We should therefore seek to cancel the "anomalies" [69] (2.29) due to different fermions in the theory. For conventional lepton and quark doublets, we see that

$$\left.\begin{aligned} \begin{pmatrix} \nu_e \\ e^- \end{pmatrix}_L &\Longrightarrow \quad \Delta S = \left(-\frac{1}{2}\right)\left(0-(-1)^2\right) = \frac{1}{2} \\ \begin{pmatrix} u \\ d \end{pmatrix}_L &\Longrightarrow \quad \Delta S = \left(-\frac{1}{2}\right)\left(\left(\frac{2}{3}\right)^2 - \left(-\frac{1}{3}\right)^2\right) = -\frac{1}{6} \end{aligned}\right\} \qquad (2.31)$$

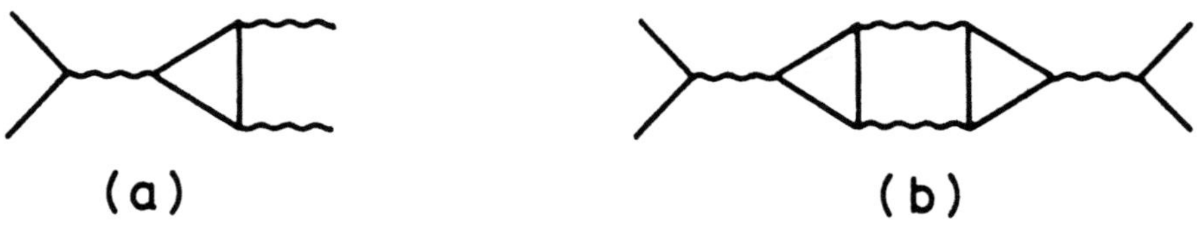

Fig. 11 : (a) Diagram with an "anomalous" triangle giving bad high-energy behaviour, which gives (b) an unwanted boson mass. See the second paper in Ref. 38.

The quark anomaly should be multiplied by three to take account of colour, and then $(\nu_e, e^-, u, d)_L$ cancel [38]. This cancellation would be maintained in the $SU(2)_L \times U(1)$ model by the addition of $\begin{pmatrix}\nu_\mu \\ \mu^-\end{pmatrix}_L$ and $\begin{pmatrix}c \\ s\end{pmatrix}_L$. If confirmed, the existence of a doublet $\begin{pmatrix}\nu_\tau \\ \tau^-\end{pmatrix}_L$ would upset the "anomalous" applecart, and need to be cancelled by something else. The simplest possibility would be a $\begin{pmatrix}t \\ b\end{pmatrix}_L$ doublet, but there are others. For example, there might be a right-handed lepton doublet somewhere : $\begin{pmatrix}\nu_\xi \\ \xi^-\end{pmatrix}_R$. And of course the τ might not be part of a left-handed doublet, and might not even give a net "anomaly", e.g.

$$\begin{pmatrix}T^+ \\ \nu_\tau \\ \tau^-\end{pmatrix}_L \Longrightarrow \Delta S = 0 \qquad (2.32)$$

There are other models in which the anomaly cancellation does not require a quark-lepton conspiracy. For example, the now-discredited vector-like $SU(2) \times U(1)$ models [70] had quark and lepton "anomalies" cancelling separately

$$\left.\begin{array}{l} \begin{pmatrix}u \\ d\end{pmatrix}_L \begin{pmatrix}c \\ s\end{pmatrix}_L \begin{pmatrix}t \\ b\end{pmatrix}_L \Longrightarrow \Delta S = 3/2 \\ \begin{pmatrix}u \\ b\end{pmatrix}_R \begin{pmatrix}c \\ s\end{pmatrix}_R \begin{pmatrix}t \\ d\end{pmatrix}_L \Longrightarrow \Delta S = -3/2 \end{array}\right\} \qquad (2.33)$$

(These models also had $\geq$ six quarks !).

In summary, it seems very reasonable to demand that a respectable gauge theory be "anomaly"-free [71], and quite plausible that the τ generates a net non-zero "anomaly". The most natural reaction would be to cancel it by another pair of quarks, but this is by no means a logical necessity.

2.7. Other evidence for après-charm.

Until the $\Upsilon(9.5)$ [1,37] there were three more phenomenological reasons for something beyond charm, two of which seem to have evaporated for the time being, but one remains as a very serious conflict with the standard $SU(2)_L \times U(1)$ model.

2.7.1. High Y anomaly ?

Until a short while ago, experiments [39] suggested a flattening of the Y distribution in $\bar{\nu} + N \to \mu^+ + X$ at high energies and a rise in the ratio of $\sigma(\bar{\nu} \to \mu^+)/\sigma(\nu \to \mu^-)$. In the naive quark parton model with left-handed currents, $\bar{\nu} + N \to \mu^+ + X$ would proceed predominantly by

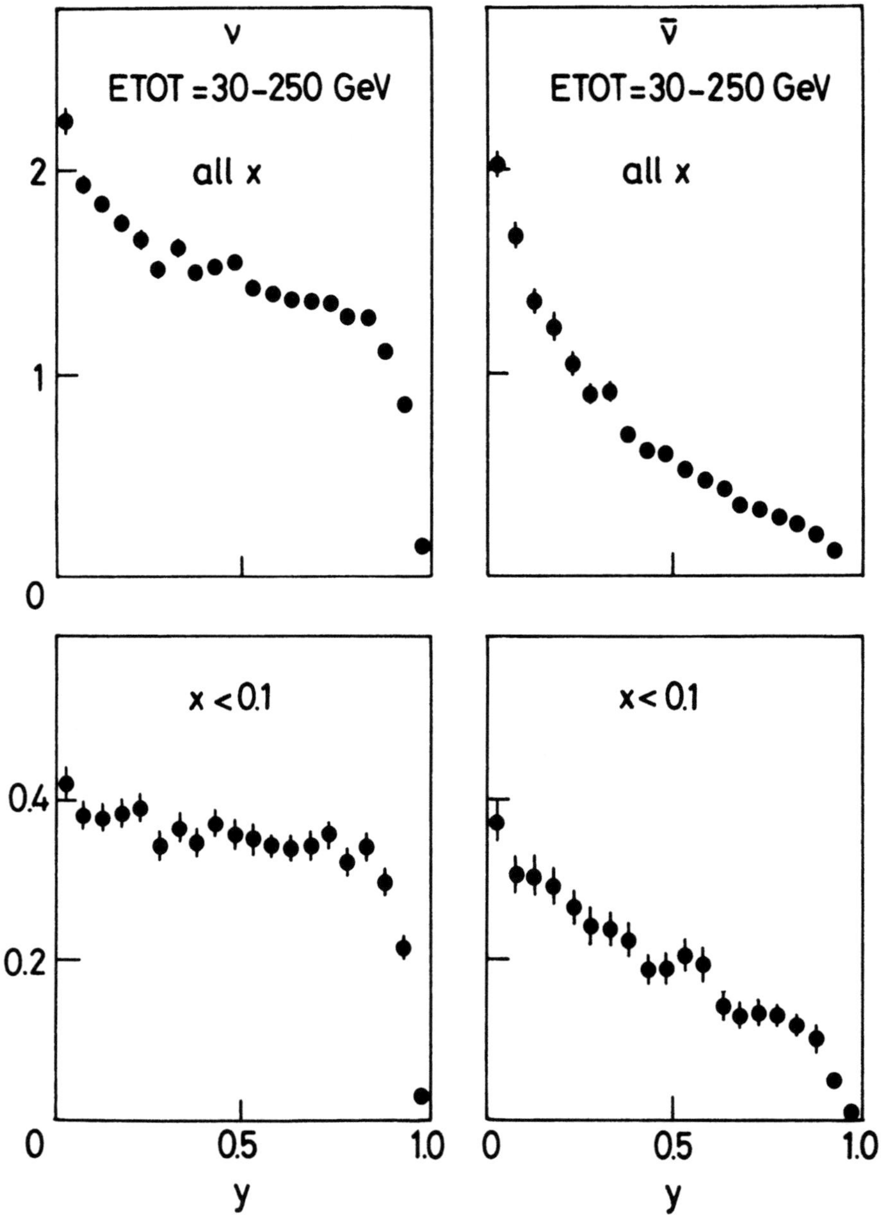

Fig. 12 : Y distributions from antineutrinos in the CDHS experiment 40).

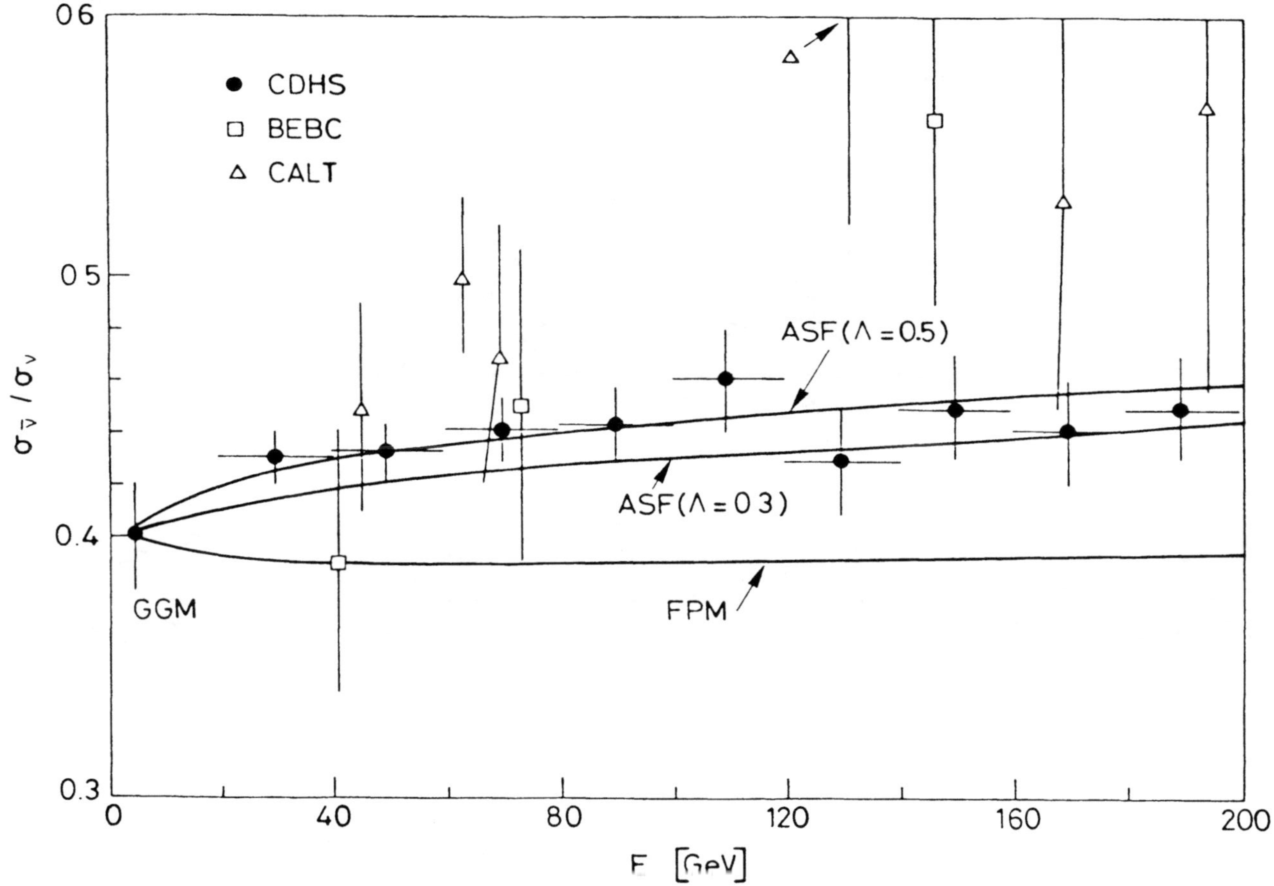

Fig. 13 : Comparison of recent data 40,32) on $\sigma(\nu\to\mu^+)/\sigma(\nu\to\mu^-)$ compared with a calculation of Buras and Gaemers 72).

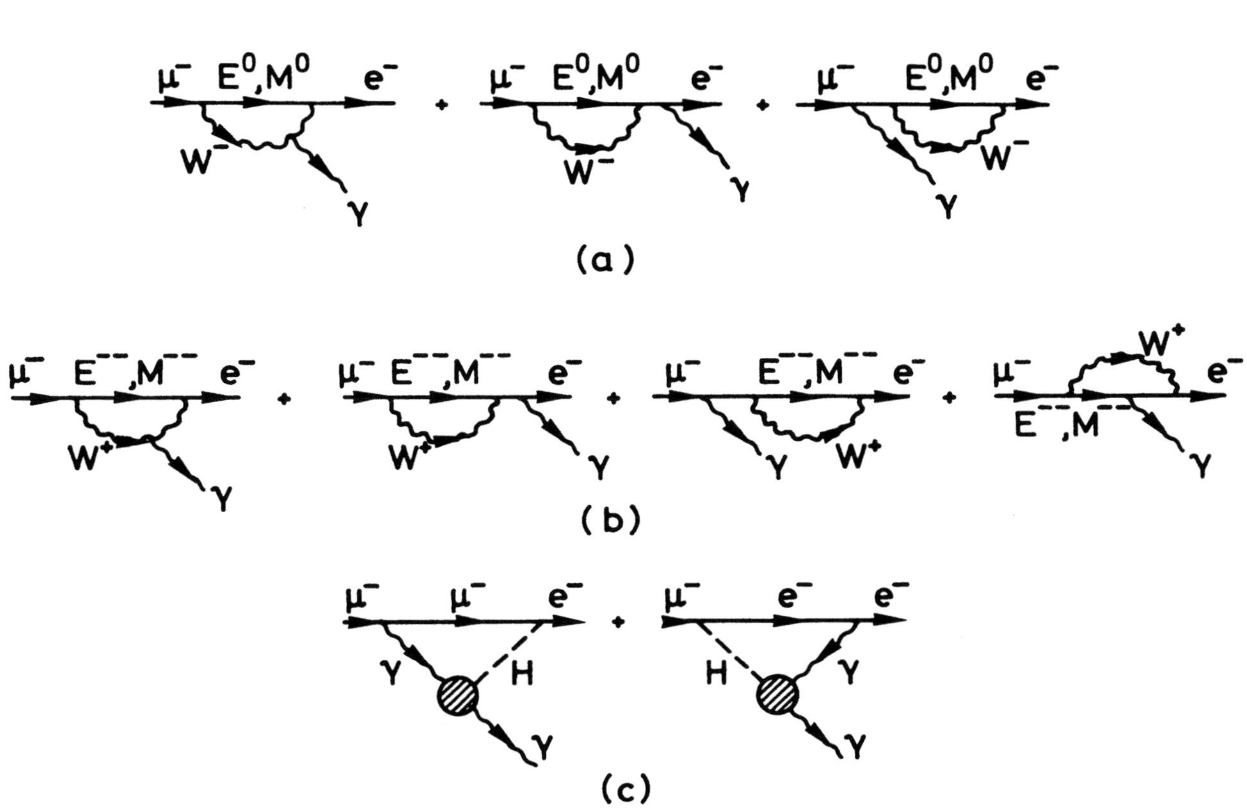

Fig. 14 : Models for $\mu \to e\gamma$ decay which involve (a) heavy right-handed neutral leptons [74]; (b) left-handed doubly-charged leptons [75]; (c) Higgs bosons [76].

$$\bar{\nu} + u \to \mu^+ + d \implies \frac{d\sigma^{\bar{\nu}}}{dY} \propto (1-Y)^2 \qquad (2.34)$$

giving

$$\frac{\sigma(\bar{\nu} \to \mu^+)}{\sigma(\nu \to \mu^-)} \approx 1/3 \qquad (2.35)$$

The expectations (2.34), (2.35) were more or less confirmed by Gargamelle. To explain the high-energy flattening of $d\sigma^{\bar{\nu}}/dY$, some theorists proposed a doublet $\begin{pmatrix} u \\ b \end{pmatrix}_R$ which would give

$$\bar{\nu} + \mu \to \mu^+ + b \implies \frac{d\sigma^{\bar{\nu}}}{dY} \propto \text{flat} \qquad (2.36)$$

and

$$\frac{\sigma(\bar{\nu} \to \mu^+)}{\sigma(\nu \to \mu^-)} \simeq \frac{4}{3} \qquad (2.37)$$

once the b quark threshold was well past. The HPWF data[39)] were fitted with $m_b \sim 5$ GeV.

The CDHS[2,40)] (Fig. 12) and other [32)] experiments have conclusively excluded any new quark with large right-handed couplings to the u quark and a mass < O(8) GeV. Figure 13 compares the latest data on $\sigma(\bar{\nu} \to \mu^+)/\sigma(\nu \to \mu^-)$ with a calculation of Buras and Gaemers [72)] using GIM charm and asymptotic freedom. The absence of a b quark with right-handed couplings to u quarks does not exclude the existence of a b quark with just left-handed couplings, or with right-handed couplings to other quarks. We will return to this point in Lecture 4.

2.7.2. $\mu \to e\gamma$?

This decay, which is forbidden in the standard $SU(2)_L \times U(1)$ model with neutrinos of identical mass, was the subject of evanescent rumours during the past year. Although it has disappeared for now, this decay may well (probably will ?) return some day. Almost any model except the standard model would expect it. Possible models include :

- $\begin{pmatrix} \nu_\tau \\ \tau^- \end{pmatrix}_L$ with m_{ν_τ} very large [73)];
- $\begin{pmatrix} E^0 \\ e^- \end{pmatrix}_R$ and $\begin{pmatrix} M^0 \\ \mu^- \end{pmatrix}_R$ doublets which mix [74)] (see Fig. 14a);
- a $\begin{pmatrix} \nu_e \\ e^- \\ E^{--} \end{pmatrix}_L$ triplet [75)] (see Fig. 14b);

- off-diagonal Higgs couplings [76] (see Fig. 14c).

All these models could generate branching ratios

$$\frac{\Gamma(\mu^- \to e^- + \gamma)}{\Gamma(\mu^- \to e^- + \bar{\nu}_e + \nu_\mu)} = O(10^{-9}) \qquad (2.38)$$

compared with the latest limits of 3.6×10^{-9} and 1.6×10^{-9} from TRIUMF and SIN experiments, respectively. Regardless of the present, it will be a crucial probe of gauge theories to push to the limit experimental upper bounds on muon number violating processes such as $\mu^- \to e^-\gamma$, $\mu^- \to e^-e^+e^-$, and anomalous μ capture : $\mu^- + Z \to e^- + Z'$.

2.7.3. The absence of parity violation in atomic physics [42].

The electronic levels of atoms should, in general, exhibit parity violation because of Z^0 exchanges through the coupling combinations $g_A^e g_V^{nucleus}$ or $g_V^e g_A^{nucleus}$.
Boson exchange is a very weak, short-range interaction so that its effects are difficult to see [77]. Present experiments [42] use heavy nuclei, in which the coherent combination $g_A^e g_V^{nucleus}$ is dominant. The parity violation is then proportional to a parity violating charge

$$Q_{PV} = g_A^e \, [(2Z+N)g_V^u + (2N+Z)g_V^d] \qquad (2.39)$$

In the standard $SU(2)_L \times U(1)$ model

$$g_A^e = 1, \; g_V^u = 1 - \frac{8}{3}\sin^2\theta_W, \; g_V^d = -1 + \frac{4}{3}\sin^2\theta_W \qquad (2.40)$$

The experiments which have announced results so far have used bismuth (Z = 83, N = 126), for which the CDHS value [2] of $\sin^2\theta_W = 0.24 \pm 0.02$ implies.

$$Q_{PV} \approx -123 \qquad (2.41)$$

The recent atomic physics calculations of Sandars [78] and the "CDHS value" (2.41) of Q_{PV} would then suggest an optical rotation in the Oxford experiment of about -1.3×10^{-7}, whereas the Oxford experiment finds $(+2.7 \pm 4.7) \times 10^{-8}$.
The Seattle group finds $(-0.7 \pm 3.2) \times 10^{-8}$ for a transition where the effect should be about -1.0×10^{-7}.

This contradiction may be resolved in any combination of at least three different ways : the Weinberg-Salam $SU(2)_L \times U(1)$

model is wrong, and/or the atomic theory is shaky, and/or there is something not understood affecting the experiments.

Possible changes in the weak interaction theory would include adding is an $\begin{pmatrix} E^0 \\ e^- \end{pmatrix}_R$ doublet [74], or embedding the e^- in a left-handed triplet $\begin{pmatrix} \nu_e \\ e^- \\ E^{--} \end{pmatrix}_L$ [75], either of which would set

$$g_A^e = 2[(T_3^L)^e + (T_3^R)^e] = 0 \tag{2.42}$$

Another possibility is to invoke two or more Z^0's [43,44]. The inequality [2] of neutral current cross-sections $\sigma(\nu \to \nu) \neq \sigma(\bar{\nu} \to \bar{\nu})$ does not imply parity violation in the neutral current, but it does imply contributions from voth vector and axial currents. These could couple to different vector bosons Z_V^0 and Z_A^0, whose exchanges would interfere in ν and $\bar{\nu}$ scattering [43]. The atomic physics effect is due to single Z exchange, and either of Z_V or Z_A would conserve parity. Introducing a second Z^0 boson requires increasing the rank of the weak and electromagnetic gauge group, and going to a group $\geq SU(2) \times U(1) \times U(1)$. We turn to some possibilies in the next section.

My own inclination is to take the null atomic physics results very seriously, but to await the results of experimens on simpler atoms such as caesium or hydrogen before concluding that the standard $SU(2)_L \times U(1)$ model is wrong.

2.8. More bosons ?

The standard $SU(2)_L \times U(1)$ gauge model is theoretically unsatisfactory because, amongst other things :

- It is not based on a simple group, so that the fact that the weak coupling $g^2/4\pi = O(\alpha)$ is accidental at our present level of understanding.
- It treats right- and left-handed fermions asymmetrically.
- It offers no understanding of quark and lepton masses and mixing angles.

For these and for the phenomenological reasons discussed in previous sections, many people have proposed other groups, necessarily bigger than $SU(2)_L \times U(1)$ [43,48]. The minimal modification is to go to $SU(2) \times U(1) \times U(1)$ [43], but this does not resolve the theoretical problems listed above, and should probably only be regarded as a stepping-stone to a bigger and better group. Attractive models can be built using $SU(2)_L \times SU(2)_R \times U(1)$ [44]. They can easily accommodate small atomic physics parity violation, and can be tuned to have a small high Y anomaly compatible with recent experiments. The group $SU(2)_L \times SU(2)_R \times U(1)$ is not simple, but it does afford the opportunity to estimate weak mixing angles [79]. A possible group [46], which has attracted some interest recently is $SU(3) \times U(1)$. An example of a model based on this group is that due to Lee and Weinberg [66], which has triplets

$$\left.\begin{array}{l}
\begin{pmatrix} d & u' \\ b' & \end{pmatrix}_L \begin{pmatrix} s & c' \\ h' & \end{pmatrix}_L \begin{pmatrix} b' & t \\ & d \end{pmatrix}_R \begin{pmatrix} h' & g' \\ & s \end{pmatrix}_R \\
\begin{pmatrix} e^- & \nu_e \\ E^{-\prime} & \end{pmatrix}_L \quad (e \leftrightarrow \mu,\tau)_L \quad \begin{pmatrix} E^- E^{\circ\prime} \\ e^- \end{pmatrix}_R \quad (e \leftrightarrow \mu,\tau)_R
\end{array}\right\} \tag{2.43}$$

A similar model was proposed previously by Segré and Weyers [46] : the Lee-Weinberg version was motivated primarily by the neutrino 3μ events [66].

The examples of this and previous sections show how commonplace it is to "fix" the standard $SU(2)_L \times U(1)$ model by adding in new fermions, new representations, or more group. It would be nice to have some simple or elegant criterion for restricting the freedom of choice provided us by the power of the gauge theory framework. We turn to such a possible criterion [16] in Lecture 3.

3. NATURAL FLAVOUR CONSERVATION.

3.1. Introduction.

The recipe for constructing a gauge theory of the weak and electromagnetic interactions is [11]

- choose a group G;
- put fermions into your favourite representations R_i of G;
- add some Higgs multiplets and break the gauge symmetry down to a U(1) for electromagnetism.

Provided you have no net "anomalies" [38], you have [13] a renormalizable gauge theory. The trouble with this procedure is its power : any G and R_i will do. It would be nice to have some theoretical constraints on gauge phenomenology, in the same way that duality constrained Regge phenomenology. One possibility, proposed by Paschos, Glashow and Weinberg [16], is to generalize the GIM charm cancellation to all flavour-changing neutral currents. The idea is that the suppression of $|\Delta S| = 1$ neutral currents and $|\Delta S| = 2$ transitions to $O(G_F^2)$ is so striking that it should be "natural", and not just the result of a sneaky choice of quark masses or mixing angles. We know now that charm-changing neutral currents are small [3]

$$\frac{\Gamma(D \to e^+ + e^- + X)}{\Gamma(D \to e^+ + \nu_e + X)} \ll 1, \quad \frac{\sigma(\nu+N \to \nu + C\,(\hookrightarrow \mu^+) + X)}{\sigma(\nu+N \to \nu+X)} < 10^{-3} \tag{3.1}$$

Also, since $D^0 \leftrightarrow \bar{D}^0$ mixing is $\lesssim$ 20% experimentally, it must be that the $|\Delta C|$ = 2 transition

$$< D^0 \,|- \mathcal{L}_{eff}|\, \bar{D}^0 > = O(G_F^2) \tag{3.2}$$

indicating that a GIM-style cancellation works for $\Delta C \neq 0$ neutral currents. The "naturalness" requirement will be that properties analogous to (3.2) hold for all flavours, and for all quark masses, mixing angles, Higgs couplings and potentials, etc...

We will see that this requirement is very restrictive -- in the absence of $|\Delta Q| \geq 2$ charged currents, new exactly conserved quantum numbers or very exotic quark charges, it forces [80] us into the gauge groups $SU(2)_L \times [U(1)]^{1 \text{ or } 2}$, with quarks in identical [16] doublet representations $\begin{pmatrix} u \\ d \end{pmatrix}_L$, $\begin{pmatrix} c \\ s \end{pmatrix}_L$, $\begin{pmatrix} t \\ b \end{pmatrix}_L$, ... The paucity of "natural" weak and electromagnetic gauge groups encourages one [80] to study their embeddings in grand unified theories including the strong interactions : $SU(2)_L \times U(1) \subset SU(5)$ [81] and $SU(2)_L \times U(1) \times U(1) \subset SU(10)$ [82,80], respectively. We present calculations of $\sin^2\theta_W$, α_{strong}, the proton lifetime, m_s/m_μ, and m_b/m_τ in these models. A "natural" version of SU(5) predicts [80]

$$m_b \approx (5.0 \text{ to } 5.8) \text{ GeV} \tag{3.3}$$

-- not a disastrous prediction if the $\Upsilon(9.5)$ turns out to be bottomonium.

3.2. Naturalness conditions [16]

Many of the conditions necessary for natural flavour conservation were derived in Section 1.3, when we were introducing charm. They included :

a) *The absence of flavour-changing neutral bosons in the symmetric theory before symmetry breakdown.*

This is necessary for $O(G_F)$ $\Delta F \neq 0$ neutral currents to be absent in the spontaneously broken theory. In general it requires the Z_i^0 boson coupling matrices $Y_i^{L,R}$ to be diagonal. Condition (*a*) is trivially satisfied in $SU(2)_L \times U(1)$ models, and trivially disobeyed in many other models -- e.g. the "K^0" - like boson in an $SU(3) \times U(1)$ model with a quark triplet

$$\begin{pmatrix} d & & u \\ & s & \end{pmatrix} \tag{3.4}$$

would be disallowed.

b) Quark mixing should not generate $\Delta F \neq 0$ neutral currents in second-order $O(G_F)$.

We saw in Section 1.3 that this condition required that the weak coupling matrices be functions of the quark charge

$$\gamma_i^{L,R} = \gamma_i^{L,R}(Q) \tag{1.15}$$

An example of a theory disobeying (1.15) was the old Cabibbo theory

$$\begin{pmatrix} u \\ d\cos\theta_c + s\sin\theta_c \end{pmatrix}_L \qquad (s\cos\theta_c - d\sin\theta_c)_L \tag{3.5}$$

which had the $|\Delta S| = 1$ neutral current

$$\bar{s}\gamma_\mu \frac{1}{2}(1-\gamma_5)\cos\theta_c \sin\theta_c \tag{1.8}$$

An example of a theory obeying (1.15) is the vector-like theory [70]

$$\left.\begin{array}{ll} \begin{pmatrix} u \\ d_c \end{pmatrix}_L \begin{pmatrix} c \\ s_c \end{pmatrix}_L \begin{pmatrix} t \\ b_c \end{pmatrix}_L & T_3^{L,R}(Q = 2/3) = +1/2 \\ \begin{pmatrix} u \\ b' \end{pmatrix}_R \begin{pmatrix} c \\ s' \end{pmatrix}_R \begin{pmatrix} t \\ d' \end{pmatrix}_R & T_3^{L,R}(Q = -1/3) = -1/2 \end{array}\right\} \tag{3.6}$$

In general, conditions *(a)* and *(b)* imply that all neutral generators belong to the mutually-commuting, simultaneously diagonalizable Cartan subalgebra of the weak and electromagnetic gauge group, and all quarks of the same charge and helicity must have the same group-theoretical weight.

In deriving (1.15) it was implicitly assumed that there was no absolutely conserved quantum number besides Q to commute with the quark mass matrix. Models exist where this assumption is relaxed [83,66], but we will not pursue them here.

c) Quark mixing should not generate $\Delta F \neq 0$ neutral currents in $O(G_F\alpha)$.

We saw in Section 1.3 that this condition on the fourth-order graphs of Fig. 2 implied that

$$(\underline{T}_\alpha^{L,R})^2 = f_\alpha^{L,R}(Q) \tag{1.19}$$

for any α isospin subgroup of the weak and electromagnetic gauge group. It is easy to convince oneself that the condition (1.19) implies that representations containing quarks of the same charge must overlap completely. Thus we have

$$\text{GOOD} \quad \begin{pmatrix} u \\ d \end{pmatrix}_L , \begin{pmatrix} c \\ s \end{pmatrix}_L \quad \text{or} \quad \begin{pmatrix} -1/3 & 2/3 \\ & -4/3 \end{pmatrix}_R , \begin{pmatrix} -1/3' & 2/3' \\ & -4/3' \end{pmatrix}_R \tag{3.7}$$

$$\text{BAD} \quad \begin{pmatrix} u \\ d \end{pmatrix}_L , \begin{pmatrix} h \\ c \\ s \\ g \end{pmatrix}_L$$

Here and in Section 1.3 we have so far only considered the effects of gauge boson exchanges, but we should perhaps also worry about Higgs exchanges, and demand

d) Neutral Higgs boson exchanges have $\Delta F = 0$.

If the quarks of any given charge get their masses from more than one Higgs, those Higgses will, in general, have $\Delta F \neq 0$ couplings. An amplitude (Fig. 15) for Higgs exchange is

$$A_{Higgs} = O\left(\frac{g^2_{H\bar{q}q}}{m^2_H} \right) \tag{3.8}$$

and the typical magnitude of Higgs-$\bar{q}q$ couplings is

$$O\left(g \frac{m_q}{m_W} \right) \tag{3.9}$$

where g is the usual semiweak coupling constant

$$\frac{g^2}{8m^2_W} = \frac{G_F}{\sqrt{2}} \tag{3.10}$$

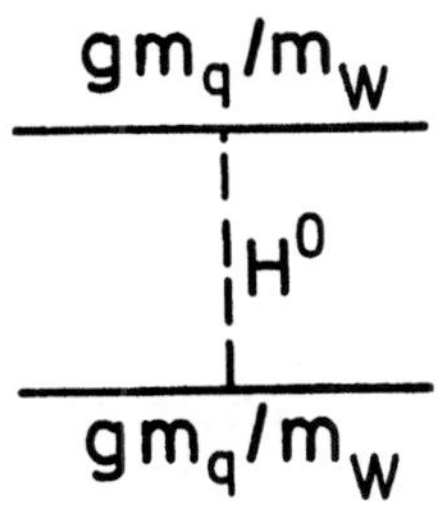

Fig. 15 : Neutral Higgs exchange amplitude.

Combining (3.8) to (3.10) we see that

$$A_{Higgs} = O\left(G_F \frac{m_q^2}{m_H^2} \right) \tag{3.11}$$

Depending on the Higgs mass, the Higgs amplitude (3.11) could well be larger than the double W exchange

$$A_{WW} = O\left(G_F^2 m_q^2\right) \tag{3.12}$$

The Higgs exchange will only be $O(G_F^2)$ if

$$m_H \gtrsim O(1/G_F) \simeq 300 \text{ GeV} \tag{3.13}$$

The Higgs bosons may well be heavier than (3.13), but we have no idea [84].

The only experimental bounds on the Higgs mass come from its non-appearance in phenomena involving nuclei. The absence of $0^{++} \to H + 0^{++\prime}$ nuclear transitions, the canonical behaviour of neutron-nucleus scattering and of muonic X-rays all imply that $m_H \gtrsim 20$ MeV [84]. There is no phenomenological restriction from high-energy physics. There are theoretical arguments [85], based on calculating radiative corrections to the Higgs potential, that in the Weinberg-Salam $SU(2)_L \times U(1)$ model with just one Higgs

$$m_H \gtrsim \left| \frac{3\alpha^2}{16\sqrt{2}\, G_F} \frac{2+\sec 4\theta_W}{\sin^4\theta_W} \right|^{\frac{1}{2}} \gtrsim 5.7 \text{ GeV} \tag{3.14}$$

$$\text{for } \sin^2\theta_W \lesssim 0.27$$

At the other extreme, the Higgs system would become strongly interacting and uncalculable for $m_H \gtrsim 1000$ GeV [86]. Probably m_H is within a factor of 10 of the $SU(2)_L \times U(1)$ values of

$$\left.\begin{aligned} m_{W^\pm} &= (72 \text{ to } 82) \text{ GeV} \\ m_{Z^0} &= (84 \text{ to } 92) \text{ GeV} \end{aligned}\right\} \text{ for } \sin^2\theta_W = 0.24 \pm 0.03 \tag{3.15}$$

This leaves the relevance of condition (*d*) an open question. People often construct models in which the Higgs structures are very complicated [e.g. the vector-like lidels [70]], as opposed to the pristine simplicity of the $SU(2)_L \times U(1)$ model (only one physical H^0 is needed). Since the Higgs particles will have to emerge from a grander synthesis some day, perhaps we should not let them get out of hand now.

3.3. Restrictions on the weak and electromagnetic group [80].

Let us start with an arbitrary group $G = \Pi_i G_i$, where the G_i are simple subgroups or U(1) factors, and analyse systematically the implications of conditions (*a*) to (*d*) of the previous section.

3.3.1. Rank $G_i \geq 2$?

The smallest simple rank 2 group is SU(3). But we have already seen in (3.4) that in the usual Gell-Mann/Zweig charge assignment $\begin{pmatrix} -1/3 & 2/3 \\ & -1/3 \end{pmatrix}$ there is an off-diagonal neutral current, which wou
current, which would be excluded by condition (*a*). If we want $\Delta F = 0$ neutral currents, we must choose a charge assignment like $\begin{pmatrix} -1/3 & 2/3 \\ & -4/3 \end{pmatrix}$ for the fundamental triplet. We then have a $|\Delta Q| \geq 2$ charged current. This problem is obviously more acute if we go to higher rank groups [SU(4) $\Longrightarrow$ $|\Delta Q| \geq 3$, etc ...]. Thus we see that a combination of naturalness and $|\Delta 1| \leq 1$ implies rank $G_i < 2$.

3.3.2. G simple ?

Since neutral currents exist, rank G $\geq$ 2, and we run into the problems of the previous paragraph. The simplicity requirement further implies that

$$\sum_{q \,\epsilon\, \text{representation}} Q_q = 0 \tag{3.16}$$

The fractional charges of quarks imply that the dimension of the representation must be divisible by three. The condition (3.16) was obeyed by the "unnatural" Gell-Mann/Zweig $\begin{pmatrix} -1/3 & 2/3 \\ & -1/3 \end{pmatrix}$ charge assignment for the quark triplet.
It cannot be satisfied with "natural" <u>3</u>, <u>6</u> or <u>15</u>-plets of SU(3). Conceivably a higher representation of SU(3), a <u>30</u> of SO(5), or a <u>27</u> of G_2 might work, but any such theory would be outlandish, and we conclude that naturalness and simplicity are incompatible.

We are now reduced to $G = [SU(2)]^N \times [U(1)]^M$ and the next question is whether

3.3.3. G contains SU(2) x SU(2) ?

If either of the SU(2) groups is ambidextrous, i.e. acts on quarks of both helicities, then naturalness must be sacrified, in the Higgs sector at least. Therefore we only examine $SU(2)_L \times SU(2)_L$ and $SU(2)_L \times SU(2)_R$.

Suppose we placed some quarks in an (n, m) (n > 1, m > 1) matrix representation of $SU(2)_L \times SU(2)_L$. Then by condition (*b*) of the previous section all quarks in the matrix must have different charges. Hence there must be at least n x m distinct quark charges, and if the vertical W's have $|\Delta Q| = 1$, the horizontal W's must have $|\Delta Q| \geq 2$. Even the simplest (2.2) case

$$\begin{pmatrix} 2/3 & 8/3 \\ -1/3 & 5/3 \end{pmatrix}$$

is unattractive. Even if we took quarks in (n,1) and (1,m) representations of $SU(2)_L \times SU(2)_L$, there would be at least (n+m) distinct quark charges. We exclude $SU(2)_L \times SU(2)_L$.

A more interesting subgroup is $SU(2)_L \times SU(2)_R$ [44]. It can be made "natural" for gauge boson exchanges by putting all quarks into identical left- and right-handed representation -- the usual

$$\begin{pmatrix} u \\ d \end{pmatrix}_L , \begin{pmatrix} c \\ s \end{pmatrix}_L , \begin{pmatrix} t \\ b \end{pmatrix}_L , \begin{pmatrix} u \\ b \end{pmatrix}_R , \begin{pmatrix} c \\ s \end{pmatrix}_R , \begin{pmatrix} t \\ d \end{pmatrix}_R \tag{3.17}$$

Unfortunately, to get realistic generalized Cabibbo mixing angles, the Higgs exchanges necessarily have $\Delta F \neq 0$. If we try a single real (2,2) representation for the model (3.17), the $Hq\bar{q}$ interaction is (where r, s label different quark doublets)

$$h_{rs}\, \bar{q}\,{}^r_L H q^s_R \; + \; h^*_{rs}\, \bar{q}\,{}^s_R H^+ q^r_L \tag{3.18}$$

Putting in vacuum expectation values

$$H_o = \begin{pmatrix} q_1 & 0 \\ 0 & q_2 \end{pmatrix} , \quad H^+_o = \begin{pmatrix} q^*_1 & 0 \\ 0 & q^*_2 \end{pmatrix} \tag{3.19}$$

the quark mass-matrix becomes

$$\bar{q}\,{}^r_L \begin{pmatrix} h_{rs}q_1 & 0 \\ 0 & h_{rs}q_2 \end{pmatrix} q^s_R + \bar{q}^{\bar{s}}_R \begin{pmatrix} h_{rs}q_1 & 0 \\ 0 & h_{rs}q_2 \end{pmatrix} q^r_L \tag{3.20}$$

We can diagonalize h_{rs} and hence the matrix (3.20) by appropriate unitary transformations on the left- and right-handed quark fields. But then the generalized Cabibbo angles vanish. We are forced into more than one Higgs multiplet, in which case it is easy to see that the Cabibbo angles are only non-zero to the extent that the coupling matrices $h_{rs}^{(1)}$ and $h_{rs}^{(2)}$ are not proportional, in which case the physical Higgs couplings necessarily have $\Delta F \neq 0$. We conclude that an $SU(2)_L \times SU(2)_R$ model [44] of the type (3.17) generally wanted is *Higgs unnatural* [80]. Of course, as discussed in the previous section, you may not care about Higgs exchanges. If you accept the previous line of reasoning, we have now reduced the possibly "natural" groups G to $SU(2)_L \times [U(1)]^M$, and the next question is :

3.3.4. $G = SU(2)_L \times [U(1)]^{M > 1}$?

The naturalness condition (*b*) of the previous section implies that all the M generalized hypercharges Y_i must be identical for fermions of identical charge and helicity. How many different Y_i can we represent on the known fermions ?

$$\begin{array}{ll} \text{quarks} : & Q = 2/3,\ -1/3,\ L \text{ and } R \\ \text{leptons} : & \left\{ \begin{array}{l} Q = 0,\ L \text{ and possibly } R \\ Q = -1,\ L \text{ and } R \end{array} \right. \end{array} \qquad (3.21)$$

To analyse this question we use the zeo "anomaly" constraint introduced in section 2.6. We saw there that, in order to cancel the $SU(2)_L \times U(1)$ anomalies, we needed equal numbers of quark and lepton doublets [38]. The one remaining freedom is in the number of right-handed neutrinos, with plausible options being : 0(all m_ν=0), or the same as the number of left-handed quark and lepton doublets. The various different "anomalous" triangles to be cancelled are shown in Fig. 16. The resulting set of mathematical conditions corresponding to (2.29) is

$$\begin{array}{lll} \Sigma (T_3^L)^3 = 0\ , & \Sigma (T_3^L)^2 Y_i^L = 0\ , & \Sigma T_3^L Y_i^L Y_j^L = 0 \\ \Sigma (T_3^L)^2 Q = 0\ , & \Sigma T_3^L Q Y_i^L = 0\ , & \Sigma Q (Y_i^L Y_j^L - Y_i^R Y_j^R) = 0 \\ \Sigma T_3^L Q^2 = 0\ , & \Sigma Q^2 (Y_i^L - Y_i^R) = 0\ , & \Sigma (Y_i^L Y_j^L Y_k^L - Y_i^R Y_j^R Y_k^R) = 0 \end{array} \qquad (3.22)$$

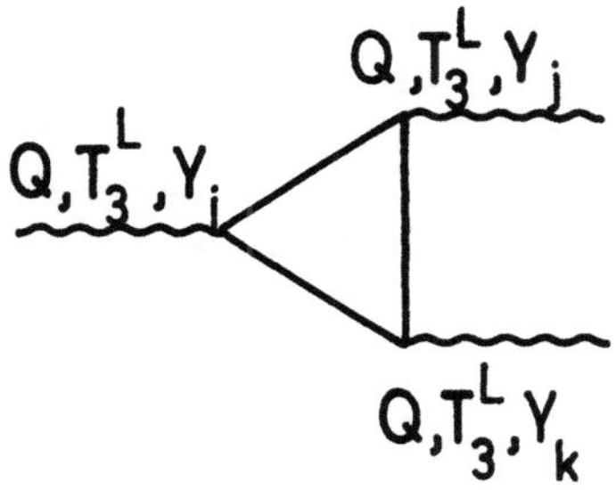

Fig. 16 : "Anomalous" triangle diagrams to be cancelled in an $SU(2)_L \times [U(1)]^M$ theory.

The most general linearly independent solutions for the fermions

$$(P_L, N_{Li};\ l^\circ_L,\ l^-_L;\ P_R;\ N_R;\ l^\circ_R,\ l^-_R) \qquad (3.23)$$

are

$$\left.\begin{array}{l} Y = (\ 1/3;\ -1\ ;\ 4/3\ ;\ -2/3\ ;\ 0\ ,\ -2\ ) \\ Y' = (\ 0\ ;\ 0\ ;\ 1\ ;\ -1\ ;\ 1\ ,\ -1\ ) \\ Y'' = (\ 0\ ;\ 0\ ;\ -7\ ;\ 5\ ;\ 1\ ,-(35)^{1/3}) \end{array}\right\} \qquad (3.24)$$

The first of these is just the usual hypercharge of $Su(2)_L \times U(1)$, the third one is blatantly absurd, and both cannot be used simultaneously. The second one, Y', looks very plausible and is indeed $\propto T^R_3$. In principle, one could have more than one gauge boson coupling to the same hypercharge Y_i, but the reasonable deduction from the "anomaly"-free solutions (3.24) is that the known fermions only allow

$$G = SU(2)_L \times [U(1)]^{1 \text{ or } 2} \qquad (3.25)$$

The $[U(1)]^2$ solution could be part of a $SU(2)_L \times SU(2)_R \times U(1)$ group, which would not be Higgs natural, but as we said in section 3.2, that may not be anything to worry about.

In reaching the conclusion (3.25) we have tried to emphasize the subjective aesthetic judgements that went into deriving it. Nevertheless, you may find it amusing that the naturalness requirement forces you into such a theoretical straight-jacket. Everyone is of course free to ask why naturalness should be imposed in the first place, and there are many holes in the derivation of (3.25) which are big enough to push an interesting new model through.

It is open season on naturalness.

3.4. Grand unified theories.

Let us take some time out from our weak phenomenology to discuss theories [81,82,80] which unify the strong, weak, and electromagnetic interactions. Many approaches to this problem are possible, of course [87]. Here we will study a resolutely simple-minded approach in which the strong, weak, and electromagnetic gauge theories are unified via a gauge group, which is either simple or else admits a discrete symmetry, so that it has a unique coupling constant. How can such a naive approach work since the strong (QCD) and weak/electromagnetic coupling constants are very different

$$\left.\begin{array}{l} \alpha_{strong}(Q^2 = 10) \approx 0.2 \text{ to } 0.3 \\ \alpha \sim 1/137 \end{array}\right\} \qquad (3.26)$$

and they should differ only by a Clebsch-Gorden coefficient of the

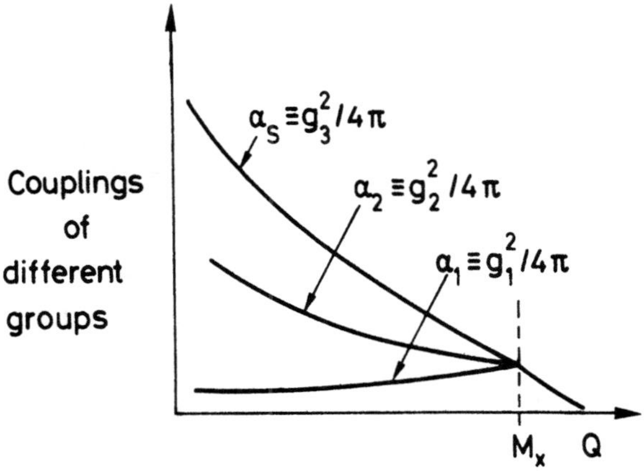

Fig. 17 : Sketch of how strong, weak, and electromagnetic coupling constants may approach each other and be unified.

order of one in a unified theory ? The trick is that the renormalization group assures us that coupling constants vary with the Q^2 at which they are measured, and that asymptotic freedom suggests that the strong interaction coupling decreases as Q^2 increases, whereas a U(1) coupling increases slightly with Q^2 [88]. So perhaps the disparate values (3.26) can be brought together at some sufficiently high Q^2, as shown in Fig. 17, which would correspond to the M_X^2 of heavy gauge bosons X which unify the strong, weak, and electromagnetic interactions. These bosons would mediate interactions such as $q+q \to \bar{q}+$lepton, which violate baryon number conservation. In most cases M_X, the grand unification mass GUM, is $\geq 10^{15}$ GeV [89], and these effects are very small. In fact you may wonder [6] why one bothers at all with grand unified theories at the present time. Several justifications are possible :

- You may find constraints on the possible unified weak and electromagnetic theories, or on QCD.
- You may get relations between the different interactions.
- You have to try it sometime, and maybe now is the time, since the other interactions may be getting solved in short order.

The reasons for discussing grand unified theories at this point in the lectures are first that in (3.25) we have just a couple of preferred weak/electromagnetic gauge theories, and secondly that if the simplest SU(5) grand unified theory is made "natural" it enables one to predict [80] the mass of the next charge -1/3 quark. The prediction is not inconsistent with the subject of Lecture 4 !

So we will base grand unified theories on groups which contain

$$SU(3)_{colour} \times SU(2)_L \times U(1)$$
$$\text{or } SU(3)_{colour} \times SU(2)_L \times U(1) \times U(1) \qquad (3.27)$$

Such groups must therefore have rank ≥ 4 or 5, respectively, and we examine these two limiting cases.

3.4.1. Rank 4 [81].

There are nine rank 4 groups which can have just one coupling constant

$$[SU(2)]^4,\ [SO(5)]^2,\ [SU(3)]^2, [G_2]^2, SO(8), SO(9), Sp(8), F_4, SU(5) \qquad (3.28)$$

The first two are ruled out because they do not contain an SU(3) subgroup. The $SU(3)_{colour} \times SU(2)$ representation content of the (u, d, e^-, ν_e) fermions is

$$(3,2) + 2(3^*,1) + (1,2) + (1,1) \qquad (3.29)$$

if we admit no ν_{eR}. The (c, s, ν_μ, μ^-) and other fermion generations have the same content. The representation (3.29) is complex and only $[SU(3)]^2$ and SU(5) among the groups (3.28) admit complex representations. The choice $SU(3)_{colour} \times SU(3)$ is impossible because leptons (colour singlets) and quarks must be in different representations, the electromagnetic charge is a generator of the weak SU(3) which does not allow fractional charges, and the sum of quark charges would have to be zero. Therefore the only possible rank 4 group is SU(5) [81].

The reducible representations $\underline{\bar{5}} + \underline{10}$ has exactly the $SU(3)_{colour} \times SU(2)$ content (3.29).

$$(\bar{d}^R, \bar{d}^Y, \bar{d}^B; \bar{e}, \nu_e)_L; \qquad \frac{1}{\sqrt{2}}\begin{pmatrix} 0 & \bar{u}^B & -\bar{u}^Y & -u^R & -d^R \\ -\bar{u}^B & 0 & \bar{u}^R & -u^Y & -d^Y \\ \bar{u}^Y & -\bar{u}^R & 0 & -u^B & -d^B \\ u^R & u^Y & u^B & 0 & -e^+ \\ d^R & d^Y & d^B & e^+ & 0 \end{pmatrix} \qquad (3.30)$$

The "anomalies" of the irreducible representations (3.30) cancel. Identical representations could be used for the (c, s, μ^-, ν_μ) and (t, b, τ^-, ν_τ) generations, and these could mix in the usual way. If we introduce the quantity D by

$$Q = T_3^L + DT_0 \qquad (3.31)$$

where T_0 is a normalized generator of SU(5), then the representation content (3.30) implies that $D^2 = 5/3$. The neutral weak interaction mixing angle θ_W, defined so that the neutral Z^0 boson couples to $T_3^L - \sin^2\theta_W Q$, then takes the value [80]

$$\sin^2\theta_W = \frac{1}{1+D^2} = 3/8 \qquad (3.32)$$

at the grand unification mass GUM. The value (3.32) is not a phenomenological disaster, since it gets renormalized at present energies [89], as we will see in the next section. SU(5) is not yet an unacceptable grand unified theory.

3.4.2. Rank 5 [82,80].

The only rank 5 groups which can admit a unique coupling constant are

$$[SU(2)]^5,\ SO(10),\ SO(11),\ SP(10) \text{ and } SU(6) \qquad (3.33)$$

Of these, $[SU(2)]^5$ has no SU(3) subgroup, while SO(11) and SP(10) have no 16-dimensional representation suitable for the fundamental fermions ($u_{L,R}$, $d_{L,R}$, $e^-_{L,R}$, $\nu_{eL,R}$). [Remember that ν_R is needed in order to represent the second hypercharge Y' in (3.24)]. SU(6) has no $\underline{15}$ or $\underline{16}$ representation suitable to contain the $SU(3)_{colour} \times SU(2)$ representations (3.29) which are needed. We are left with SO(10)[82,80] which can be broken down to $SU(4) \times SU(2)_L \times SU(2)_R$, and has a 16-dimensional representation with the acceptable $SU(3)_{colour} \times SU(2)_L \times SU(2)_R$ decomposition

$$\underline{16} = (3,2,1) + (3^*,1,2) + (1,2,1) + (1,1,2) \qquad (3.34)$$

If the neutral weak mixing angle θ_W is defined by the photon field being

$$A^\mu = \sin\theta_W (W_3^L + W_3^R)_\mu + \sqrt{\cos 2\theta_W}\; X_\mu \qquad (3.35)$$

then at the grand unification mass we again have $\sin^2\theta_W = 3/8$. Generally, SO(10) stands to $SU(2)_L \times SU(2)_R \times U(1)$ just as SU(5) stands to $SU(2)_L \times U(1)$. We will not analyse SO(10) further here : see Ref. 80 for details.

3.5. Phenomenological consequences.

To see what implications such a grand unified theory can have for present-day phenomenology, let us look at the SU(5) model. To get from the grand unification mass GUM (M_X) to present energies μ, we need to know the variation [89] in the SU(3), SU(2) and U(1) coupling constants (we assume $|g_i| \ll 1$)

$$\frac{1}{g_{GUM}^2} = \frac{1}{g_i^2(\mu)} + 2b_i \ln(\mu/M_X) \quad (i = 1,2,3) \tag{3.36}$$

where the b_i are defined by

$$\mu \frac{\partial}{\partial\mu} g_i(\mu) \simeq b_i g_i^3(\mu) + O(g_i^5) \tag{3.37}$$

We will neglect Higgs contributions and finite-mass effects, so that

$$b_1 = \frac{f}{24\pi^2}, \quad b_2 = -\frac{1}{16\pi^2}\left[\frac{22}{3} - \frac{2}{3}f\right], \quad b_3 = -\frac{1}{16\pi^2}\left[11 - \frac{2}{3}f\right] \tag{3.38}$$

where f is the number of quark (lepton) flavours. It is then easy to show that the grand unification mass M_X obeys

$$\ln(M_X/\mu) = \frac{12\pi}{22(1+3D^2)\alpha}\left(1 - \frac{\alpha}{\alpha_s(\mu)}(1+D^2)\right) \tag{3.39}$$

where $\alpha_s(\mu) \equiv g_3^2(\mu)/4\pi$ and $D^2 = 5/3$. Equation (3.39) comes from knitting the strong and semiweak coupling constants together. Looking at the relative variation of SU(2) and U(1) coupling constants we get a renormalized version of (3.32)

$$\sin^2\theta_W(\mu) = \frac{1 + 2D^2\alpha/\alpha_s(\mu)}{1 + 3D^2} \tag{3.40}$$

Not only can we estimate coupling constant renormalization, but also mass renormalization using the anomalous dimension of the fermion mass operator

$$\gamma_m \simeq -\frac{g_3^2}{2\pi^2} - \frac{9}{32\pi^2} g_2^2 - \frac{3}{8\pi^2} g_1^2 \tag{3.41}$$

We find [90]

$$\ln\left(\frac{m_q(\mu)}{m_l(\mu)}\right) = \ln\left(\frac{m_q(M_X)}{m_l(M_X)}\right) + \frac{4}{11-\frac{2}{3}f}\ln\left(\frac{\alpha_s(\mu)}{\alpha_{GUM}}\right) + \frac{3}{2f}\ln\left(\frac{\alpha_1(\mu)}{\alpha_{GUM}}\right) \tag{3.42}$$

To calculate m_q/m_l at present energies, we need to know m_q/m_l at the grand unification mass M_X. A hypothesis which fixes this is the "naturalness" requirement [80] that the fermions only get their masses from one Higgs multiplet. To give the fermions (3.30) we need either a $\underline{5}$ or $\underline{45}$ of Higgs fields : the obvious thing is to take a $\underline{5}$. Then [81]

$$\frac{m_d(M_X)}{m_e(M_X)} = \frac{m_s(M_X)}{m_\mu(M_X)} = \frac{m_b(M_X)}{m_\tau(M_X)} = 1 \tag{3.43}$$

If we choose $\alpha_s(\mu=\sqrt{10}$ GeV) to lie in a phenomenologically favoured range, we find that if we use just the first-order expressions (3.36) and (3.41), neglect Higgs meson contributions, and use finite-mass corrections to the b_i (3.38) and γ_m(3.41), then

$$\alpha_s(Q^2=10)=(0.19 \text{ to } 0.32) \Longrightarrow \begin{cases} M_X = (1 \text{ to } 4) \times 10^{16}\text{GeV} \\ \sin^2\theta_W \gtrsim 0.20 \end{cases} \tag{3.44}$$

We estimate that $M_X \sim 2 \times 10^{16}$ GeV implies a proton lifetime $\gtrsim 10^{38}$ years [91], to be compared with the present lower limit of 2×10^{30} years [92]. If we just use six flavours and assume m_τ = 1.9 GeV, then for the same range of α_s

$$m_b = (5.0 \text{ to } 5.8) \text{ GeV} \tag{3.45}$$

while going to eight (ten) flavours would increase [90] the estimate (3.45) by 20% (40%).

None of the estimates (3.44), (3.45) is disastrous. We are encouraged that the CDHS experiment reports [2] $\sin^2\theta_W=0.24\pm0.02$, somewhat lower than previous estimates, and hope the Υ(9.5)[1,37] turns out to be bottomonium. It is perhaps premature to conclude that there are only six quarks !

4. BOTTOMOLOGY AND TOPOLOGY.

4.1. Introduction.

The discovery of Υ(9.5) [1,37] very probably signals the presence of one or more new quarks. In this lecture we will discuss in some detail the properties of the state (or states) Υ(9.5), and then go on to discuss the properties of $t\bar{q}$ and $b\bar{q}$ naked top or bottom states. The specific questions we address [93] are :

- Can Υ(9.5) be something other than a new -onium (bound states of a new heavy quark-antiquark pair) ?
- If it is -onium, why is it so broad ? Are there two or more unresolved peaks, as strongly indicated [1] by the latest available data ?

- Is there any preference for it to be bottomonium ($b\bar{b}$: Q_b=-1/3) rather than toponium ($t\bar{t}$: Q_t = 2/3) ?

Analysis of these questions is independant of the weak interactions of the new quark or quarks. Data of the CDHS [40] and other [32] recent neutrino experiments show no evidence of right-handed currents. Accordingly we study the properties of the six-quark SU(2) x U(1) model [17] with just left-handed currents [94,95]. We start with the implications for the old physics, in particular, CP violation in the K_0-$\bar{K}_0$ system and the electric dipole moment of the neutron [95]. Finally we discuss [93] the properties of naked top and bottom particles, including decay modes, lifetimes, B_0-$\bar{B}_0$ and T_0-$\bar{T}_0$ mixing, CP violation, and production by neutrinos and antineutrinos.

4.2. Alternatives to the -onium hypothesis.

When the J/ψ was discovered there was considerable discussion whether it was indeed the bound state of a new heavy quark. It may indicate progress in our understanding that no such debate rages over the Υ(9.5), which in any case now seems to be resolved into ≥ 2 peaks. Nevertheless, for the sake of completeness, we should rehearse some of the alternatives.

4.2.1. A Z^0 boson ?

The CDHS and other experiments find [2] a near constancy of the neutral current to charged current ratio $\sigma(\nu+N \to \nu+X)/\sigma(\nu+N \to \mu^-+X)$ as a function of energy. Since the charged current cross-section is known to arise almost linearly with E_ν, we conclude there are no Z^0 propagator effects corresponding to $m_Z \sim 10$ GeV. Indeed,a bpund m_{Z^0}>(20 to 30) GeV can probably be established.

4.2.2. A Higgs boson ?

It is common to the Higgs bosons occuring in many models that $\sigma(H)BR(H \to \mu^+\mu^-)$ is very small, and that the decays violate (μe) universality, since Higgs couplings are correlated with fermion masses. Specifically, in the Weinberg-Salam $SU(2)_L \times U(1)$ with a single Higgs doublet [84]

$$\frac{\sigma(H)BR(H \to \mu^+\mu^-)}{\sigma(pp \to \mu^+\mu^- + X)} \gtrsim (10^{-1\pm1}) \left(\frac{M_H}{\Delta M}\right) BR\,(H \to \mu^+\mu^-) \qquad (4.1)$$

where M_H is the Higgs mass. ΔM is the mass resolution in a given experiment, and $pp \to \mu^+\mu^- + X$ is the virtual photon continuum background to a Higgs search. For M_H > 4GeV, we estimate [84]

$$BR(H \to \mu^+\mu^-) \approx \frac{m_\mu^2}{3m_C^2 + m_\tau^2} \approx \frac{1}{800} \qquad (4.2)$$

and

$$\frac{BR(H \to e^+e^-)}{BR(H \to \mu^+\mu^-)} \approx \frac{m_e^2}{m_\mu^2} \approx \frac{1}{4 \times 10^4} \qquad (4.3)$$

The magnitude of the T(9.5) signal and the consistency with (μe) universality in its decays suggest that it is not a Higgs boson.

4.2.3. A single broad resonance ?

The width of the T(9.5) bump is $\sim$ 1 1/2 GeV, compared with an experimental resolution of $\sim$ 1/2 GeV. The line shape is not fitted well by a single Breit-Wigner resonance smeared with a Gaussian resolution, but can the hypothesis of a single wide state be discarded ? The CFS experiment [37] finds that

$$\frac{\sigma(T)BR(T \to \mu^+\mu^-)}{\sigma(J/\psi)BR(J/\psi \to \mu^+\mu^-)} \approx \frac{1}{3 \times 10^4} \qquad (4.4)$$

If $\Gamma(T \to \mu^+\mu^-)$ has a value characteristic of other vector mesons $\lesssim$ 10 keV, then the production cross-section for a state of width 1 1/2 GeV would have to be

$$\sigma(T) \geq \frac{1}{10} \sigma(J/\psi) \qquad (4.5)$$

which seems absurdly high for such a heavy state. If we allow $\Gamma(T \to \mu^+\mu^-)$ to approach the limit imposed by the success of QED muon magnetic moment calculations [96]

$$\Delta\left(\frac{g-2}{2}\right)_\mu = \frac{1}{3\pi}\left(\frac{g_V^2 + g_A^2}{4\pi}\right)_{T \to \mu^+\mu^-} \left(\frac{m_\mu}{m_T}\right)^2 < 10^{-8} \qquad (4.6)$$

then we find

$$\Gamma(T \to \mu^+\mu^-) < 3\text{MeV} \Longrightarrow B(T \to \mu^+\mu^-) < 3 \times 10^{-3} \qquad (.47)$$

This would in turn imply that

$$\sigma(T) \gtrsim 10^{-3} \sigma(J/\psi) \qquad (4.8)$$

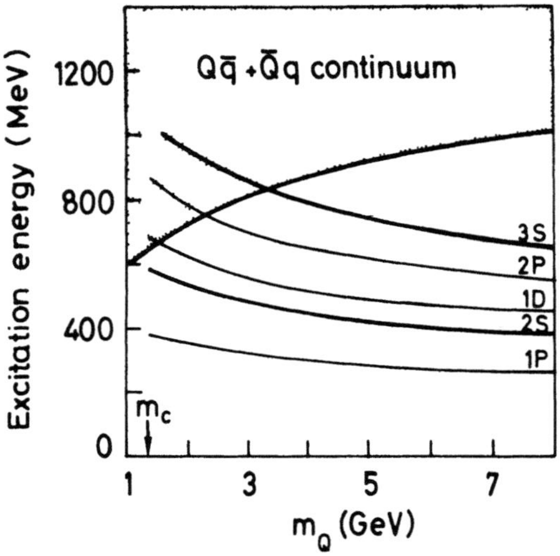

Fig. 18 : Relative locations of different -onium levels and continuum for different heavy quark masses, taken from Eichten and Gottfried [99].

While not so absurd as (4.5), this cross-section seems implausibly hight to people [97] modelling the production of very massive states in hadron-hadron collisions. While not rigorously excluded, the hypothesis of a single broad Υ state seems rather unlikely. We therefore move on to the next section.

4.3. Properties of new-onia.

The properties of the bound states of new heavy quarks can be studied using the theoretical machinery developed for the charmonium system [98,99]. The idea is to treat the lowest-lying $b\bar{b}$ or $t\bar{t}$ states non-relativistically, using a Schrödinger equation with a potential of the generic form

$$V(r) = -\frac{4}{3}\alpha_s(r)\frac{1}{r} + \frac{r}{a^2} \tag{4.9}$$

The first part of (4.9) is given by asymptotic freedom at short distances, while the form of the second is motivated by quark-confining models, with the parameter a to be fixed by calculations for the $c\bar{c}$ system [99]. Such a model enables one to predict the masses and separations of $b\bar{b}$ or $t\bar{t}$ states, as well as the magnitudes of radiative transitions. This has been done *in extenso* by Eichten and Gottfried [99]. They expect that for a new quark of mass ∿ 5 GeV :

- there should be three narrow $^3S_1$ states below threshold:
- $m(2^3S_1) - m(1^3S_1) \simeq 420$ MeV,

$$m(3^3S_1) - m(2^3S_1) \simeq 330 \text{ MeV} \qquad (4.10)$$

- P and D states should lie between the S states just as they do for charmonium.

Their spectra are shown in Fig. 18.

The results of their calculations for the dominant decay modes of the first three $^3S_1$ $b\bar{b}$ (J_B, J_B', J_B'') and $t\bar{t}$ (J_T, J_T', J_T'') states are listed in Table 3. Their calculations have been supplemented by estimates [93] of the cascade decays $J_B' \rightarrow J_B\pi\pi$, etc...

We know from the ($\pi\pi$) mass spectrum in $\psi' \rightarrow J/\psi\pi\pi$, and believe, on the basis of chiral symmetry and the Adler zero, that the matrix element for $V' \rightarrow V\pi\pi$ is suppressed close to the $\pi\pi$ threshold. If we take a matrix element

$$\tilde{g}_{V'V}\varepsilon'_\mu\varepsilon^\mu[(p_{\pi 1}+p_{\pi 2})^2 - 4m_\pi^2] \qquad (4.11)$$

which accords with these prejudices, then

$$\frac{\Gamma(V' \rightarrow V\pi\pi)}{\Gamma(\psi' \rightarrow J/\psi\pi\pi)} \simeq \left(\frac{\tilde{g}_{V'V}}{\tilde{g}_{\psi' J/\psi}}\right)^2 \times (3 \times 10^{-3}) \qquad (4.12)$$

For $m_{V'} - m_V \simeq 400$ MeV, and $m_V \sim 10$ GeV. If we assume that $\tilde{g}_{V'V} \lesssim \tilde{g}_{\psi' J/\psi}$, then Eq. (4.12) suggests that

$$\Gamma(J_B' \rightarrow J_B\pi\pi),\ \Gamma(J_B'' \rightarrow J_B'\pi\pi),\ \Gamma(J_T' \rightarrow J_T\pi\pi),\ \Gamma(J_T'' \rightarrow J_T'\pi\pi) < \tfrac{1}{3}\text{keV} \qquad (4.13)$$

We suspect that the decays $J_B'' \rightarrow J_B\pi\pi$, $J_T'' \rightarrow J_T\pi\pi$ would also be small, suppressed by orthogonal wave functions or something. We are therefore able to complete the decay mode table of Eichten and Gottfried [99], and estimate $B(J_B, J_B', J_B'', J_T, J_T', J_T'' \rightarrow \mu^+\mu^-)$. We see that all three J_B states have comparable ($\mu^+\mu^-$) branching ratios. The same is almost true of the J_T states, but in any case

$$\frac{BR(J_T' \rightarrow \mu^+\mu^-)}{BR(J_T \rightarrow \mu^+\mu^-)} > \frac{B(\psi' \rightarrow \mu^+\mu^-)}{B(J/\psi \rightarrow \mu^+\mu^-)} \gtrsim \frac{1}{7} \qquad (4.14)$$

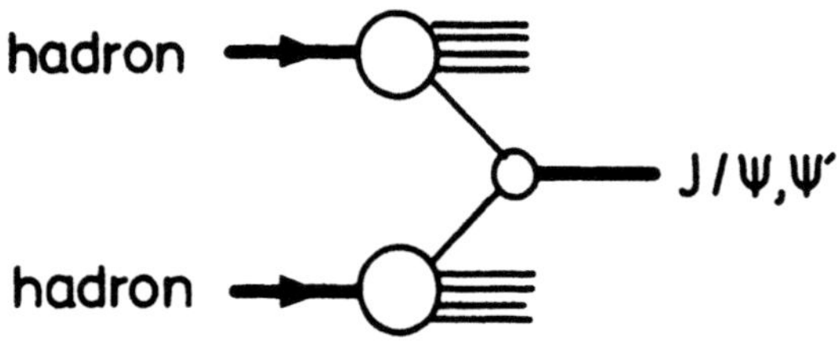

Fig. 19 : Drell-Yan model for J/ψ and ψ' production.

This means that the 2^3S_1 and 3^3S_1 states of a new-onium may be less invisible in hadronic collisions than was the ψ' relative to the J/ψ.

The numerical estimates listed above and in Table 3 are, of course, sensitive to the precise form of the quark-antiquark potential and level spacings. The latest data [1)] suggest that $m(2^3S_1) - m(1^3S_1) \gtrsim 600$ MeV, so that the potential (4.9) has to be modified. However, the relevant qualitative features of Table 3 -- narrow total widths and large $\mu^+\mu^-$ branching ratios for J_B, J_B' and J_B'' -- may be less model-dependent. The decay rate $\Gamma(V' \to V\pi\pi)$ would be much larger, but it may still avoid drowning out at the competing decay modes. We therefore turn to the

4.4. Hadronic production of new-onia.

Before calculating bottomonium and toponium production, we first warm up on charmonium. It looks as if some sort of Drell-Yan type constituent fusion model may be responsible for hadronic production of the J/ψ. The cross-section has a linear A-dependence, redolent of the impulse approximation, and the p_T and x_F distributions are not grossly dissimilar from the surrounding lepton-pair continuum. But it is not clear what hadronic constituents collide to form the J/ψ. Presumably they are not $c\bar{c}$ pairs, because there are not very many third μ's produced in association with the J/ψ, as would be expected from the decays of charmed particles emanating from the c or $\bar{c}$ spectator quarks [100)]. The fact [101)] that $\sigma(\bar{p}p \to J/\psi+X) \neq \sigma(pp \to J/\psi+X)$ excludes models where the colliding constituents are some sort of gluons [102)]. On the other hand, $\sigma(p\bar{p} \to J/\psi+X)/\sigma(pp \to J/\psi+X) <<$ the order of 50, which would be expected [103)] from the simple fusion of ordinary $q\bar{q}$ pairs, and the p_T of the J/ψ is slightly different [1)] from that of the Drell-Yan continuum. The most reasonable possibility may be that the J/ψ is produced by a combination of "gluon-gluon" and $q\bar{q}$ collisions, with "G-G" dominating in pp collisions, and $q\bar{q}$ dominating in $p\bar{p}$ collisions.

Regardless of the details of the production mechanism, it is possible to calculate [93] the ratio of J/ψ and ψ' production by assuming :

- J/ψ and ψ' are made by collisions of the same set of constituents : $A\bar{A} \to J/\psi, \psi'$, as shown in Fig. 19;
- The decays $J/\psi \to A\bar{A}$ and $\psi' \to A\bar{A}$ yield fractions F, F', respectively, of the total decay of J/ψ and ψ'.

The first assumption implies that $\sigma(J/\psi)$ and $\sigma(\psi')$ are scaled by their (mass)$^{-3}$, and proportional to a common luminosity function for $A\bar{A}$ collisions $\mathcal{L}(\tau)$: $\tau = m_{J/\psi} / \sqrt{s}$ or $m_{\psi'} / \sqrt{s}$. The second assumption then yields

$$\frac{\sigma(pp\to\psi'+X)BR(\psi'\to\mu^+\mu^-)}{\sigma(pp\ J/\psi+X)BR(J/\psi\to\mu^+\mu^-)} \approx \left(\frac{m_{J/\psi}}{m_{\psi'}}\right)^3 \frac{\mathcal{L}(m_{\psi'}/\sqrt{s})}{\mathcal{L}(m_{J/\psi}/\sqrt{s})} \frac{F'}{F} \frac{\Gamma(\psi'\to\mu^+\mu^-)}{\Gamma(J/\psi\to\mu^+\mu^-)} \tag{4.15}$$

If we now identify the decays $J/\psi, \psi' \to A\bar{A}$ with decays into the usual u-d-s hadrons, then the data on J/ψ and ψ' decays [3] imply that

$$F \sim 0.86 , \quad F' \sim 0.1 \text{ to } 0.2 \tag{4.16}$$

Eyeballing the J/ψ excitation function as a function of $\sqrt{s}$ in order to fix the shape of the luminosity function $\mathcal{L}$, we find

$$\mathcal{L}(m_{\psi'} = 3.7/\sqrt{s} = 27) \approx \frac{3}{4} \mathcal{L}(m_{J/\psi} = 3.1/\sqrt{s} = 27) \tag{4.17}$$

Putting (4.16), (4.17), and the known decay rates $\Gamma(J/\psi, \psi' \to \mu^+\mu^-)$ into Eq. (4.15), we find

$$\frac{\sigma(pp \to \psi'+X)BR(\psi' \to \mu^+\mu^-)}{\sigma(pp \to J/\psi+X)BR(J/\psi \to \mu^+\mu^-)} \approx \left(2\tfrac{1}{2} \text{ to } 5\right)\% \tag{4.18}$$

This range is not grossly incompatible with experimental data which indicate [1,104] a ratio of about 2 1/2 %. We therefore conclude that we do not have a dysfunctional model for -onium production, and proceed to apply it to bottomonium and toponium.

Let us assume that J_B, J_B' and J_B'' are produced by Drell-Yan collisions of constituents $A_B\bar{A}_B$. Defining F_B and F_B' analogously to the F and F' defined earlier, Table 3 implies that

$$F_B \sim 0.89, \quad F_B' \sim 0.53, \quad \frac{\Gamma(J_B' \to \mu^+\mu^-)}{\Gamma(J_B \to \mu^+\mu^-)} \sim 0.6 \tag{4.19}$$

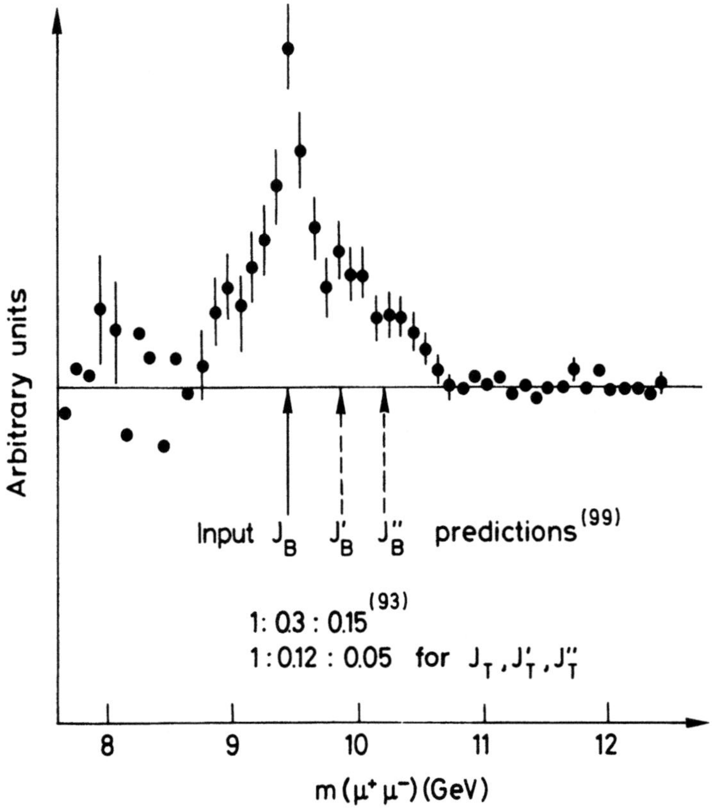

Fig. 20 : Lineshape of the $\Upsilon(9.5)$ [1,37] compared with the calculations of Refs. 99 and 93. This graph was shown at the Summer School : a more recent plot was not available at the time of writing.

For the bottomonium analogue of Eq. (4.15) we need to make an assumption about the luminosity function $L_B(\tau)$ for $A_B\bar{A}_B$ collisions. We suppose it has a similar threshold to $L(\tau)$

$$L_B(\tau) \propto L(\tau) \tag{4.20}$$

but not necessarily the same normalization. Eyeballing the J/ψ excitation curve then yields the ratios of luminosities for J_B, J_B', and J_B'' production. Putting all the factors into the analogues of (4.15), we find

$$\sigma(J_B)BR(J_B\to\mu^+\mu^-) : \sigma(J_B')BR(J_B'\to\mu^+\mu^-) : \sigma(J_B'')BR(J_B''\to\mu^+\mu^-) \tag{4.21}$$

$$\approx 1 : 0.3 : 0.15$$

$$\sigma(J_T)BR(J_T\to\mu^+\mu^-) : \sigma(J_T')BR(J_T'\to\mu^+\mu^-) : \sigma(J_T'')BR(J_T''\to\mu^+\mu^-)$$

$$\approx 1 : 0.12 : 0.05$$

Thus there can be substantial production of the higher J_B' and J_B'' (or J_T' and J_T'') states relative to J_B (or J_T) : higher than the ratio of ψ' to J/ψ. The precise numbers in (4.20) are rather sensitive to details of the -onium calculations, in particular to the strengths of the radiative decay modes, but the qualitative conclusion seems relatively model-independent. For instance, if one assumed that all J_B, J_B' and J_B'' production was indirect, via intermediate P_B/χ_B production and decay, similar predictions about the relative production of J_B, J_B' and possibly J_B'' could probably be drawn. It will be interesting to see the T(9.5) structure analysed with better mass resolution !

It is grossly premature to speculate whether T is bottomonium, or toponium, or a superposition of the two. Nevertheless, the low signal-to-background ratio and over-all magnitude of the cross-section might favour the buttomonium hypothesis. The data, shown in Fig. 20, seem to like two peaks with relative cross-sections of 1:0.4, but the uncertainties in estimating the ratios (4.20) make a charge assignment foolhardy. If anyone believed the SU(5) estimate of the bottom quark mass given in the previous lecture, then he or she would have a preference for the assignment of the T !

4.5. Six left-handed quarks ?

So there seems to be at least one new quark with a mass $\sim$ 5 GeV. On the other hand, the CDHS experiment [2,40] saw no evidence for a quark of this mass with any substantial right-handed coupling to the usual u or d quarks. It could well be that the new quark or quarks have right-handed couplings to u or d at some low level, or that there are substantial right-handed couplings between quarks as yet unseen. But the most economical hypothesis is that there are six quarks, t and b, as well as the old u, d, s, c quarks, all with only left-handed couplings. This is the Kobayashi-Maskawa [17] extension of the $SU(2)_L \times U(1)$ [14,15] GIM [9] model which was discussed in Section 1.4 : the $\begin{pmatrix} t \\ b \end{pmatrix}_L$ doublet cancels the "anomaly" due to the presumed $\begin{pmatrix} \nu_\tau \\ \tau \end{pmatrix}_L$ lepton doublet (see Section 2.6) [38]. This model is *prima facie* conflict with the atomic physics parity violation experiments [42] discussed in Section 2.7. For the moment let us take a "wait and see" attitude : essential ingredients in the conflict are atomic physics theory and experiment, of which I for one understand nothing. It may be legitimate to wait for the results of the theoretically simpler caesium or hydrogen experiments before finally giving up on the $SU(2)_L \times U(1)$ model.

As discussed in Section 1.4, the charged weak coupling matrix for six left-handed quarks is characterized by four parameters [17]: three Euler angles θ_i and a complex phase :

$$J^{\mu}_{-} = (\bar{u},\bar{c},\bar{t})\gamma^{\mu}\frac{1}{2}(1-\gamma_5) \begin{pmatrix} C_1 & -S_1C_3 & -S_1S_3 \\ S_1C_2 & C_1C_2C_3-S_2S_3e^{i\delta} & C_1C_2S_3+S_2C_3e^{i\delta} \\ S_1S_2 & C_1S_2C_3+C_2S_3e^{i\delta} & C_1S_2S_3-C_2S_3e^{i\delta} \end{pmatrix} \begin{pmatrix} d \\ s \\ b \end{pmatrix} \tag{4.22}$$

where we have written $C_i(S_i) = \cos\theta_i$ $(\sin\theta_i)$, $i = 1,2,3$. The angles must obey certain conditions in order that the KM model does not conflict with the Cabibbo-GIM phenomenology of the familiar u, d, s, c quarks. It is clear that

$$\theta_1 \approx \theta_c \implies S_1^2 \approx 0.05, \quad C_1^2 \approx 0.95 \tag{4.23}$$

From the observed success of Cabibbo weak universality between the μ decay coupling and the sum of $(u \leftrightarrow d)$ and $(u \leftrightarrow s)$ couplings, we know [in the language of (4.22)], that

$$C_1^2 + S_1^2C_3^2 = 1.001 \pm 0.004 \tag{4.24}$$

Therefore, at the one-standard-deviation level we have [94,95]

$$S_1^2S_3^2 < 0.003 \implies S_3^2 < 0.06 \tag{4.25}$$

but S_3^2 could, of course, be much less than this limit. There is also a constraint on S_2, which comes from the success of the Gaillard and Lee [20] calculation of the charmed quark mass. The diagrams they calculated must now be supplemented by ones involving the t quark as shown in Fig. 21. The factor $m_c^2 \sin^2\theta_c \cos^2\theta_c$ in their calculation (1.28) of Δm_K gets modified:

$$\sin^2\theta_c\cos^2\theta_c m_c^2 \rightarrow$$
$$S_1^2C_1^2C_3^2[C_2^4M_c^2+S_2^4m_t^2 + \frac{2S_2^2C_2^2m_t^2m_c^2}{m_t^2-m_c^2}\ln\left(\frac{m_t^2}{m_c^2}\right)] \tag{4.26}$$

If we demand that the second and third terms in the square bracket of (4.26) not be larger than the first Gaillard-Lee term, so as to retain their estimate of m_c, then we find [93]

$$S_2^2 < \eta\ln\eta + \sqrt{(\eta\ln\eta)^2+\eta} \quad : \quad \eta \equiv m_c^2/m_t^2 \tag{4.27}$$

Since we now know that $m_t \geq 5$ GeV, S_2^2 is contrained to be small

$$\left.\begin{array}{lll} m_t = 5\text{ GeV} & \Longrightarrow & S_2^2 < 0.2 \\ m_t = m_W & \Longrightarrow & S_2^2 < 0.03 \end{array}\right\} \qquad (4.28)$$

We therefore know phenomenologically that all the Euler angles in the KM matrix (4.22) must be not much larger than the Cabibbo angle, but we have no theoretical understanding of this remarkable fact. Note that our present knowledge does not constrain the phase δ, which could, in principle, be quite large.

4.6. CP violation for (u,d, s, c) states).

As noted in section 1.4, the phase δ introduces CP violation and this was the original motivation of Kobayashi and Maskawa for going to six quarks. It may well be that CP is violated in other ways as well -- vacuum tunnelling [18], right-handed currents [105] Higgs'[19], or whatever -- but as soon as you have six quarks a contribution from the KM phase seems inevitable.

4.6.1. The K_0 - $\overline{K}_0$ system [106].

CP violation can arise in two ways -- from a contribution of c and t exchange to the imaginary part Im M_{12}^K of the (K_0, $\overline{K}_0$) mass matrix, and from phases in individual decay amplitudes which involve heavy quarks (e.g. $s\overline{d}\rightarrow c\overline{c}, t\overline{t}$). The latter source is systematically suppressed by the $\Delta I=1/2$ rule (as in any milli-weak theory) and by a sort of "Zweig rule" which suppresses heavy quark couplings to light meson states [93]. The KM model therefore reproduces superweak predictions (where the only CP violation is in the mass matrix) to a high degree of accuracy. The classical CP violating parameter [106] ε_K is then found from Fig. 21 to be

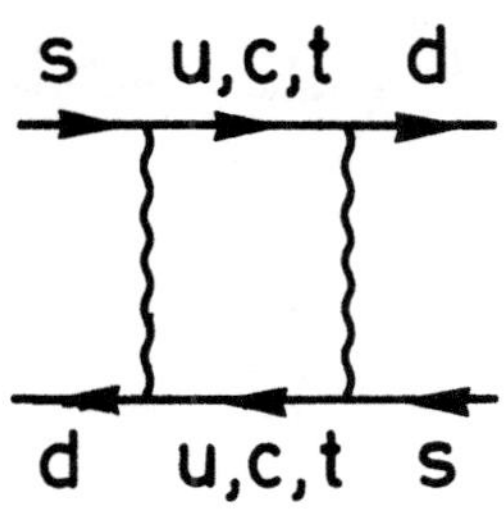

Fig. 21 : Diagrams giving CP violation in the K_1-K_2 mass matrix.

$$|\varepsilon_K| \simeq \left| \frac{\mathrm{Im} M_{12}^K}{\Delta m_k} \right| \simeq 2S_2S_3\sin\delta[-\ln\eta - 1 + \frac{S_2^2}{\eta}] \qquad (4.29)$$

Experimentally, $|\varepsilon_K| \simeq 2\times10^{-3}$: theoretically (4.29) shows it to be the product of three angle factors, two of which [(4.25), (4.28)] are already known for other reasons to be small. The phase parameter δ is still unfixed. On the basis of (4.25),(4.28) and (4.29), one can only conclude

$$m_t = \left\{ \begin{matrix} 5\mathrm{GeV} \\ m_W \end{matrix} \right. \Longrightarrow |\sin\delta| > \left\{ \begin{matrix} 3\times10^{-3} \\ 4\times10^{-4} \end{matrix} \right. \qquad (4.30)$$

More direct ways of getting at δ will be found in the $B_0 - \bar{B}_0$ and $T_0 - \bar{T}_0$ naked top and bottom analogues of the $K_0 - \bar{K}_0$ system.

4.6.2. The $D_0 - \bar{D}_0$ system.

In the $SU(2)_L \times U(1)$ generalized GIM model the mixing of D_0 ($\equiv c\bar{u}$) and $\bar{D}_0$ ($\equiv \bar{c}u$) mesons is expected to be very small [93, 107]. This is because they can only communicate via non-strange final states, which are suppressed by the Cabibbo angle

$$\left| \frac{\Delta m_D}{\Gamma_D} \right| \sim \left| \frac{\Delta\Gamma_D}{\Gamma_D} \right| \lesssim \sin^2\theta_c \left(\frac{m_s^2 - m_d^2}{m_c^2} \right) \sim 10^{-2} \qquad (4.31)$$

One of the clearest manifestations of CP violation would be from asymmetries in the "wrong-sign" leptons produced by decays of mesons which were produced as D_0 or $\bar{D}_0$, but then mixed. Unfortunately, Eq. (4.31) implies that "wrong-sign" leptons appear at a rate $\lesssim 10^{-4}$! The CP violating parameter ε_D is no larger than for the $K_0 - \bar{K}_0$ system

$$|\varepsilon_D| \approx \left| \frac{\mathrm{Im} M_{12}^D}{\Delta m_D} \right| \approx 2S_2S_3\sin\delta[\ln(m_b^2/m_s^2) - 1] \approx 2\times10^{-3} \qquad (4.32)$$

Thus the KM model foresees no drama in the $D_0 - \bar{D}_0$ system.

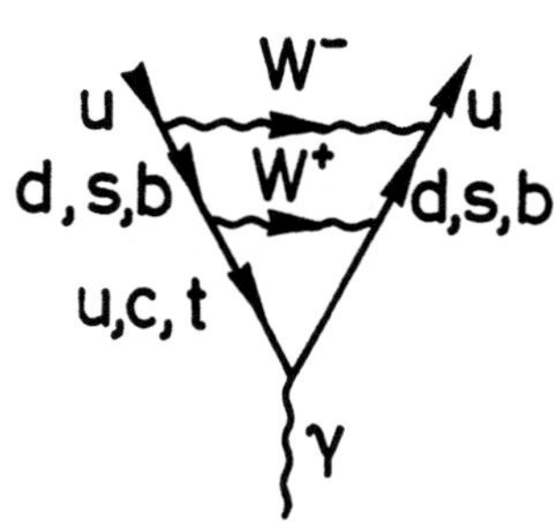

Fig. 22 : A diagram contributing to the electric dipole moment of the neutron.

4.6.3. The neutron electric dipole moment.

Estimating this quantity involves double boson exchange diagrams like that shown in Fig. 22. The effect vanishes if $m_t = m_c$ or $m_b = m_s$, or if $m_{u,d} = 0$. One may estimate 95)

$$\left(\frac{D}{e}\right)_n \sim m_{u,d} \frac{G_F \alpha}{\pi^3} \frac{(m_t^2 - m_c^2)(m_b^2 - m_s^2)}{m_W^4} \ln^2 \left(\frac{m_W^2}{m_t^2 \text{ or } m_b^2} \right) s_1^2 s_2 s_3 \sin\delta \qquad (4.33)$$

Using the value of $s_2 s_3 \sin\delta$ required for CP violation in the $K_0 - \overline{K}_0$ system (4.33) yields

$$\left(\frac{D}{e}\right)_n \sim (10^{-26} \text{ to } 10^{-27}) \frac{m_u}{m_n} \frac{m_t^2 m_b^2}{m_W^4} \ln^2 \left(\frac{m_W^2}{m_t^2 \text{ or } m_b^2} \right) \text{cm} \qquad (4.34)$$

$$\lesssim (10^{-28} \text{ to } 10^{-29}) \text{cm} \quad \text{if one of } m_t, m_b \sim 5\text{GeV} \qquad (4.35)$$

By comparison, the present experimental limit on (D/e) is $O(10^{-24})$cm 108), and presently planned experiments have an expected sensitivity of $O(10^{-26})$cm. If these experiments find something, KM CP violation is not the whole story.

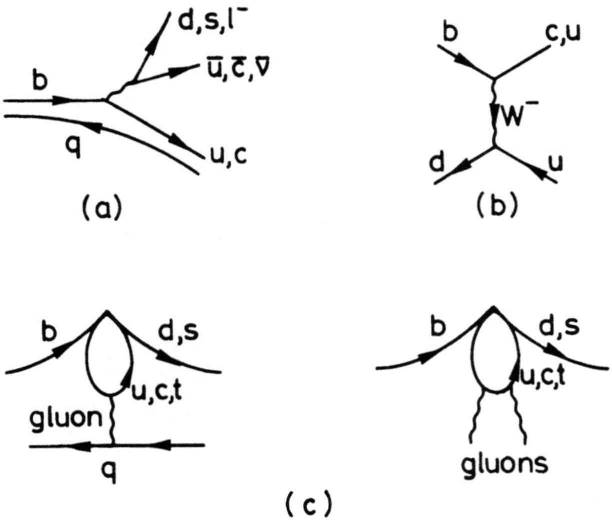

Fig. 23 : Diagrams contributing to b decay : (a) free quark decay; (b) annihilation; (c) "penguin" diagrams.

4.7. The decays of naked top and bottom states.

To estimate these we may calculate the diagrams indicated in Fig. 23 : the non-leptonic enhancement factors are probably even smaller than they seem to be in the case of charm, and a free quark approximation should be reliable. The annihilation diagrams (Fig. 23b) and the "penguin" diagrams (Fig. 23c) are expected to be smaller than the diagrams with four-fermion final states. In the case of bottom decays, the muon decay formula analogue (1.31) can be extended to give

$$\Gamma_{(a)}(B\to C+X) \approx (S_2^2+S_3^2+2S_2S_3\cos\delta)F\left(\frac{m_c^2}{m_b^2}\right)5\,\frac{G_F^2 m_b^5}{192\pi^3} \qquad (4.36)$$

where

$$F(X) = 1-8X+8X^3-X^4-12X^2\ln X \approx \frac{1}{3} \text{ for } X \equiv \frac{m_c^2}{m_b^2} \text{ if } m_b \sim 5\text{GeV}$$

The ratio of decays with and without charmed particles is then

$$\frac{\Gamma_{(a)}(B\to C+X)}{\Gamma_{(a)}(B\to \text{no } C+X)} \approx \frac{F(m_c^2/m_b^2)}{1}\left(\frac{S_2^2+S_3^2+2S_2S_3\cos\delta}{S_1^2S_3^2}\right) \approx O(6) \qquad (4.37)$$

if S_3^2 and $(S_2^2 + S_3^2 + 2S_2S_3\cos\delta)$ are the same order of magnitude, as expected. We conclude that most bottom decays should contain charmed particles if $m_b < m_t$ [109]. Similar arguments lead to strange particles dominating top decays if $m_t < m_b$ [109], and top decays would then look very much like charm decays, except for the greater energy release. If $m_b < m_t$, we can estimate the bottom lifetime from the formula (4.36), finding

$$\tau(\text{bottom}) \approx 10^{-14}/(S_2^2+S_3^2+2S_2S_3\cos\delta) \gtrsim 10^{-13} \text{ seconds} \qquad (4.38)$$

where we used the bounds (4.25), (4.28). The top lifetime would be similarly long if $m_t < m_b$. Thus it seems that, like charm, top and bottom particles should live long enough to leave visible tracks in emulsions -- if only one could discover some way of producing them !

The charm technology of Section 1.5 can also be recycled to estimate the mass differences, mixing, and CP violation parameters for $B_0(\equiv b\bar{d})$, $\bar{B}_0(\equiv \bar{b}d)$, $T_0(\equiv t\bar{u})$ and $\bar{T}_0(\equiv \bar{t}u)$ mesons. Using the same tricks as for the K_0-$\bar{K}_0$ system, one finds

$$\Delta m_B \approx \frac{1}{m_B}\langle B_0|-\mathcal{L}_{eff}|\bar{B}_0\rangle \approx \frac{G_F}{\sqrt{2}} f_B^2 m_B \frac{\alpha}{4\pi}\left(\frac{m_t}{38\text{GeV}}\right)^2 S_1^2S_2^2\cos 2\delta \qquad (4.39)$$

where f_B is the meson decay constant analogous to f_π, f_K. Various estimates [93] suggest $f_B \sim 500$ MeV, in which case comparing (4.36) and (4.39) yields

$$\frac{\Delta m_B}{\Gamma_B} \approx \frac{m_t^2}{700 GeV^2} \tag{4.40}$$

and B_0-$\bar{B}_0$ mixing could be substantial if $m_t \gtrsim 7$ GeV, as one would expect if the $\Upsilon(9.5)$ turns out to be bottomonium. Equation (4.40) may eventually enable us to estimate m_t from bottom phenomenology. Recycling the CP violation technology of the previous section of this Lecture provides the dramatic result that the B_0-$\bar{B}_0$ CP violating parameter

$$|\varepsilon_B| \approx \left| \frac{I mM_{12}^B}{\Delta m_B} \right| \approx \tan 2\delta \tag{4.41}$$

As emphasized before, we do not know how big δ may be : all we can say, on the basis of (4.25), (4.28) and (4.29), is that CP violation is probably much larger for B_0-$\bar{B}_0$ mesons than for the K_0-$\bar{K}_0$ system. Indeed, if we compare the rates of "wrong-sign" leptons coming from states prepared initially as B_0 or $\bar{B}_0$, we could perhaps see dramatic effects if

$$\frac{\Gamma(''B_0''\to l^+\nu x)}{\Gamma(''B_0''\to l^-\bar{\nu} x)} \Big/ \frac{\Gamma(''\bar{B}_0''\to l^-\bar{\nu} x)}{\Gamma(''\bar{B}_0''\to l^+\nu x)} - 1 = O(\frac{1}{10} \text{ to } 1) \tag{4.42}$$

if δ were as large as present limits allow. The top and bottom meson systems will certainly provide crucial tests of the six left-handed quark model of CP violation.

4.8. Production of t and b by ν and $\bar{\nu}$.

Using the usual naive quark-parton model, it is easy to estimate the characteristics of left-handed t- and b-production by ν and $\bar{\nu}$. Unfortunately, there are only discouraging upper limits on the production cross-section -- to be blamed on the successes of current GIM phenomenology and the small angles (4.25), (4.28) -- and derisory lower bounds $O(10^{-5})$ of the total charged current cross-sections. The results are listed in Table 4.

A few points are worth noticing. Bottom production by $\bar{\nu}(\nu)$ can yield charmed (anticharmed) particles in the decays, whereas the normal GIM [9] model only yields anticharmed (charmed) particles in $\bar{\nu}(\nu)$ collisions. Secondly, because of the plethora of charmed particles in bottom decays, and also the possibility of bottom production off a $c\bar{c}$ pair in the nucleon sea, there is the prospect of three or four muon events in $\bar{\nu}$ and ν collisions, resulting from multiple semi-leptonic decays such as

$$\left.\begin{array}{lll} \bar{\nu} + c \to \mu^+ + b & & \\ \quad + \bar{c} & \hookrightarrow c+\mu^- +\bar{\nu}+X & \\ \quad \hookrightarrow \mu^- +\bar{\nu}+X & \quad \hookrightarrow s+\mu^+ +\nu+X & \end{array}\right\} \to \mu^+\mu^-\mu^+\mu^- +X \qquad (4.43)$$

Upper limits on the rates for such processes are also listed in the table, assuming semileptonic branching ratios O(15%). Although the rates are small, the theoretical upper limits are comparable with the present experimental sensitivities [36,62], and the CDHS experiment will soon have many million wide-band neutrino events to play with. Questing for left-handed tops and bottoms in neutrino collisions may not be completely hopeless. On the other hand, the unravelling of the top and bottom mysteries must presumably await the generation of e^+e^- colliding ring machines, CESR, PETRA and PEP.

4.9. Messages.

There are no conclusions possible in a state of such rapid flux, but there are a few messages which may be distilled from these lectures.

- The classical GIM charm seems to be established in essentially all its details.
- A heavy charged lepton with mass ∿ 1.9 GeV almost certainly exists. It may be a sequential lepton with a V-A coupling, or it may be an orthoelectron.
- The existence of this heavy lepton almost certainly implies the existence of more flavours beyond charm.
- The most serious experimental indicators for going beyond the $SU(2)_L \times U(1)$ Weinberg-Salam model are the atomic physics parity violation experiments, and the observation of trimuon events in neutrino collisions.
- Natural flavour conservation to $O(G_F^2)$ greatly restricts your freedom of model-building, forcing you into $SU(2)_L \times [U(1)]^{1 \text{ or } 2}$, unless you want something like $|\Delta Q| \geq 2$ charged currents as a new exactly conserved quantum number.
- The most naive SU(5) grand unified model of the strong, weak, and electromagnetic interactions is not yet grossly incompatible with experiment.
- The Υ(9.5) bump is probably two or three unresolved states, the -onia of one or more new quarks.

TABLE 1 : Charm production by ν, $\bar{\nu}$.

Fundamental subprocess	X distribution	Y distribution	Coupling strength	Fraction of total cross-section	Dimuon signals
$\nu + d \to \mu^- + c$	all X	flat	$\sin^2 \theta_c$	$\sim$ 5 %	$-+$ (1/2 to 1)%
$\nu + s \to \mu^- + c$	small X	flat	$\cos^2 \theta_c$	$\sim$ 5 %	$-+$ (1/2 to 1)%
$\nu + \bar{c} \to \mu^- + \bar{d}, \bar{s}$ (c spectator)	← Negligible →				
$\bar{\nu} + \bar{s} \to \mu^+ + \bar{c}$	small X	flat	$\cos^2 \theta_c$	$\sim$ 10%	$+-$ (1 to 2)%
$\nu + \{^{\bar{d}}_{c} \to \mu^+ + \{^{\bar{c}}_{s,d}$	← Negligible →				

TABLE 2 : Test of charm

Property to be tested	Test in production	Test in decay	Established?
New quantum number (quark)	(23,24)	(25,16)	
Normal weak coupling strength	(32)	? (30)	
(V-A) current	(31,33)	?	
Large $\Delta C = \Delta S = 1$	?	(23,24,34)	
Small $\Delta C=1$, $\Delta S=0$	? (31)	? (3)	?

TABLE 3 : Decay modes of bottomonia and toponia.

Mode \ State	J_B	J'_B	J''_B	J_T	J'_T	J''_T	
e^+e^-	0.7	0.4+	0.4-	2.7	1.7	1.5	Decay widths in keV
$\mu^+\mu^-$	0.7	0.4+	0.4-	2.7	1.7	1.5	
$\tau^+\tau^-$	0.7	0.4+	0.4-	2.7	1.7	1.5	
$\gamma^* \rightarrow$ hadrons	2.8	1.7	1.5	10.8	6.8	6.0	
direct hadrons	13.7	8.8	7.8	13.7	8.8	7.8	
$\gamma\chi_{B,T}$	-	8	12.5	-	30	50	
$J_{B,T}\pi\pi$	-	$\lesssim$ 0.3	small?	-	$\lesssim$ 0.3	small?	
$J'_{B,T}\pi\pi$	-	-	$\lesssim$ 0.1	-	-	$\lesssim$ 0.1	
Total	∿19	∿20	∿23	∿33	∿51	∿68	
BR $\rightarrow \mu^+\mu^-$	3.5%	2.1%	1.6%	8.2%	3.3%	2.2%	

Only take the last two lines seriously ! -- see text.

Table 4

Bottom and top production by ν, $\bar{\nu}$ [a]

Bottoms

Fundamental subprocess	X distribution	Y distribution	Coupling strength	Fraction of total cross-section	Multimuon signals		
					2μ	3μ	4μ
$\bar{\nu} + u \to \mu^+ + b$, $b \to c$	all X	$(1 - Y)^2$	$s_1^2 s_3^2$	$< 3 \times 10^{-3}$ (combined with next row: $< ½\%$)	$+-$ $< 1 \times 10^{-3}$	$+-+$ $< 1.5 \times 10^{-4}$	$+-+-$ $< 2 \times 10^{-5}$
$\bar{\nu} + c \to \mu^+ + b$ ($\bar{c}$ spectator), $b \to c$	very small X	$(1 - Y)^2$	$(s_2^2 + s_3^2 + 2s_2s_3 \cos\delta)$	$< 2 \times 10^{-3}$	$++$ $< 7 \times 10^{-4}$	$+--$ $< 5 \times 10^{-5}$	
$\nu + \bar{u} \to \mu^- + \bar{b}$, $\bar{b} \to \bar{c}$	small X	$(1 - Y)^2$	$s_1^2 s_3^2$	$< 3 \times 10^{-5}$ (combined with next row: $< 7 \times 10^{-4}$)	$-+$ $< 2 \times 10^{-4}$	$-+-$ $< 3 \times 10^{-5}$	$-+-+$ $< 3 \times 10^{-6}$
$\nu + \bar{c} \to \mu^- + \bar{b}$ (c spectator), $\bar{b} \to \bar{c}$	very small X	$(1 - Y)^2$	$(s_2^2 + s_3^2 + 2s_2s_3 \cos\delta)$	$< 7 \times 10^{-4}$	$--$ $< 1 \times 10^{-4}$	$-++$ $< 1.5 \times 10^{-5}$	

Tops

Fundamental subprocess	X distribution	Y distribution	Coupling strength	Fraction of total cross-section	Multimuon signals		
					2μ	3μ	4μ
$\bar{\nu} + \bar{d} \to \mu^+ + \bar{t}$	all X	flat	$s_1^2 s_2^2$	$< 1.5 \times 10^{-4}$ (combined with next row: $< 5 \times 10^{-3}$)	$+-$ $< 7 \times 10^{-4}$		
$\bar{\nu} + \bar{s} \to \mu^+ + \bar{t}$	small X	flat	$(s_2^2 + s_3^2 + 2s_2s_3 \cos\delta)$	$< 5 \times 10^{-3}$	$++$ $< 2 \times 10^{-4}$		
$\nu + d \to \mu^- + t$	all X	flat	$s_1^2 s_2^2$	$< 3 \times 10^{-3}$ (combined with next row: $< 8 \times 10^{-3}$)	$-+$ $< 10^{-3}$		
$\nu + s \to \mu^- + t$	small X	flat	$(s_2^2 + s_3^2 + 2s_2s_3 \cos\delta)$	$< 5 \times 10^{-3}$	$< 3 \times 10^{-4}$		

a) In the upper (lower) half of the above table we assume the b (t) quark to be the lighter one. The like sign dimuon events due to t production require very substantial $T^0 - \bar{T}^0$ mixing.

- A model with six left-handed quarks gives an elegant mechanism for CP violation and distinctive predictions for the fifth and sixth quarks.
- Needless to say, it will be some time before the mysteries of tops and bottoms, heavy leptons, and new gauge bosons will be resolved !

REFERENCES

1) L.M. Lederman, Lectures presented at this School and talk presented at the 1977 International Symposium on Lepton and Photon Interactions at High Energies, Hamburg, 1977.

2) J. Steinberger, Lectures presented at this School and talks by P. Bloch, K. Kleinknect and K. Tittel, presented at the Hamburg Symposium, *loc. cit.*

3) B.H. Wiik, Lectures presented at this school. See also B.H. Wiik and G. Wolf, Electron-positron interactions, Lectures presented at the 1976 Les Houches Summer School, DESY preprint 77/01(1977) and many talks presented at the Hamburg Symposium, *loc. cit.*

4) M. Lévy, Lectures presented at this School.
J. Weyers, Lectures presented at this School.

5) C.H. Llewellyn Smith, Lectures presented at this School. See also J. Ellis, Deep hadronic structure, Lectures presented at the 1976 Les Houches Summer School, CERN preprint (1977).

6) H.D. Politzer, Phys. Reports 14C, 129 (1974).
W. Marciano and H. Pagels, Quantum chromodynamics - a review, Rockefeller preprint COO-2232B-130 (1977).

7) A. de Rújula, The current weak current, Lectures presented at the 1976 Les Houches Summer School, Harvard preprint HUTP-77/A006 (1977).
H. Fritzsch, The world of flavour and colour, Lectures presented at the 1977 Varenna Summer School, CERN preprint TH 2359 (1977).

8) Y. Hara, Phys. Rev. 134, B701 (1964)
Z. Maki and Y.Ohnuki, Progr. Theor. Phys. 32, 144 (1964).
D. Amati, H. Bacry, J. Nuyts and J. Prentki, Phys. Letters 11, 190 (1964) and Nuovo Cimento 34, 1732 (1964).
B.J. Bjørken and S.L. Glashow, Phys. Letters 11, 255 (1964).

9) S.L. Glashow, J. Iliopoulos and L. Maiani, Phys. Rev. D2, 1285 (1970).

10) F. Hasert et la., Phys. Letters, 46B, 121, 138 (1973).

11) E.S. Abers and B.W. Lee, Phys. Reports 9C, 1 (1973).
J. Iliopoulos, An Introduction to gauge theories, CERN 76-11 (1976).

12) M. Veltman, Lectures presented at the 17th Cracow School of Theoretical Physics, Zakopane, Poland, 1977.

13) G. 't Hooft, Nuclear Phys. B35, 167 (1971).

14) S.L. Glashow, Nuclear Phys. 22, 579 (1971).

15) S. Weinberg, Phys. Rev. Letters, 19, 1264 (1967).
A. Salam, Proc. 8th Nobel Symposium, Stockholm 1968 (ed. N. Svartholm) (Almqvist and Wiksells, Stockholm, 1968)p. 367.

16) E.A. Paschos, Phys. Rev. D 15, 1966 (1977).
S.L. Glashow and S. Weinberg, Phys. Rev. D 15, 1958 (1977).

17) M. Kobayashi and K. Maskawa, Progr. Theor. Phys. 49, 652 (1973).
See also L. Maiani, Phys. Letters 62B, 183 (1976).

18) C.G. Callan, R.F. Dashen and D.J. Gross, Phys. Letters 63B, 334 (1976).
R.D. Peccei and H.R. Quinn, Stanford University Preprint ITP-572 (1977).

19) T.D. Lee, Phys. Reports 9C, 148 (1974).
S. Weinberg, Phys. Rev. Letters 37, 657 (1976).

20) A.I. Vainshtein and I.B. Khriplovich, JETP Letters 18, 141 (1973).
M.K. Gaillard and B.W. Lee, Phys. Rev. D 10, 897 (1974).

21) See, for example, M.K. Gaillard, B.W. Lee and R.E. Shrock, Phys. Rev. D 13, 2674 (1976).

22) J.J. Aubert et al., Phys. Rev. Letters 33, 1404 (1974).
J.E. Augustin et al., Phys. Rev. Letters 33, 1406 (1974).

23) G. Goldhaber et al., Phys. Rev. Letters 37, 255 (1976).
I. Peruzzi et al., Phys. Rev. Letters 37, 569 (1976).

24) E.G. Cazzoli et al., Phys. Rev. Letters 34, 1125 (1975).
G. Blietschau et al., Phys. Letters 60B, 207 (1975).
J. Von Krogh et al., Phys. Rev. Letters 36, 710 (1976).

25) J.E. Wiss et al., Phys. Rev. Letters 37, 1531 (1976).

26) W. Braunschweig et al., Phys. Letters 63B, 471 (1976).
J. Burmester et al., Phys. Letters 64B, 369 (1976).

27) M.K. Gaillard, B.W. Lee and J.L. Rosner, Rev. Mod. Phys. 47, 277 (1975).

28) K. Wilson, Phys. Rev. 179, 1499 (1969).
M.K. Gaillard and B.W. Lee, Phys. Rev. Letters 33, 108 (1974).
G. Altarelli and L. Maiani, Phys. Letters 52B, 351 (1974).

29) J. Ellis, M.K. Gaillard and D.V. Nanopoulos, Nuclear Phys. B100, 313 (1975).

30) E.H.S. Burhop et al., Phys. Letters 65B, 299 (1976).

31) M. Holder et al., Phys. Letters 69B, 377 (1977).
Dimuon events in neutrino collisions were discovered by
A. Benvenuti et al., Phys. Rev. Letters 34, 419 (1975).

32) B.C. Barish et al., Caltech preprint CALT 68-606 (1976).
D.C. Cundy, Talk presented at the Neutrino 1977 Conference, Elbrus, USSR, 1977.

33) C. Baltay et al., Phys. Rev. Letters 39, 62 (1977).

34) J. Burmester et al., Ref. 26. See also
R. Brandelik et al., Phys. Letters 67B, 363 (1977).
J. Burmester et al., Phys. Letters 67B, 367 (1977).

35) M.L. Perl et al., Phys. Rev. Letters 35, 1489 (1975).

36) B.C. Barish et al., Phys. Rev. Letters 38, 577 (1977).
A. Benvenuti et al., Phys. Rev. Letters 38, 1110 (1977).

37) S. Herb et al., Phys. Rev. Letters 39, 252 (1977).

38) C. Bouchiat, J. Iliopoulos and Ph. Meyer, Phys. Letters 38B, 519 (1972).
D.C. Gross and R. Jackiw, Phys. Rev. D 6, 477 (1972).

39) B. Aubert et al., Phys. Rev. Letters 33, 984 (1974).
A. Benvenuti et al., Phys. Rev. Letters 36, 1478 (1976).
A. Benvenuti et al., Phys. Rev. Letters 37, 189 (1976).
B.C. Barish et al., Phys. Rev. Letters 38, 314 (1977).

40) M. Holder et al., Phys. Rev. Letters 39, 433 (1977).

41) See, for example, P. Ramond, Nuclear Phys. B110, 214 (1976).
F. Gürsey and P. Sikivie, Phys. Rev. Letters 36, 775 (1976).

42) P. Baird et al., Nature 264, 528 (1976).
P. Sandars, Talk presented at the 1977 International Symposium on Lepton and Photon Interactions at High Energies, Hamburg, 1977.

43) See, for example, H. Fritzsch and P. Minkowski, Ann. Phys. (NY) 93, 193 (1974) and Nuclear Phys. B103, 61 (1976).
Also G.G. Ross and T. Weiler, private communication.

44) R.N. Mohapatra and J.C. Pati, Phys. Rev. D 11, 566, 2558 (1975).
G. Senjavonic and R.N. Mohapatra, Phys. Rev. D 12, 1502 (1975).
A de Rujula, H. Georgi and S.L. Glashow, Harvard preprint HUTP-77/A002 (1977).

45) See, for example, H. Fritzsch and P. Minkowski, Phys. Letters 63B, 99 (1976).

46) G. Segré and J. Weyers, Phys. Letters 65B, 243 (1976) and references therein.

47) J.D. Bjorken and K. Lane (unpublished).

48) T.C. Yang, DESY preprint 77/39 (1977).

49) Y.-S. Tsai, Phys. Rev. D 4, 2821 (1971).

50) C.H. Llewellun Smith, Phys. Letters 46B, 233 (1973).
J.S. Bell, Nuclear Phys. B60, 427 (1973).
J.M. Cornwall, D.N. Levin and G. Tiktopoulos, Phys. Rev. Letters 30, 1268 (1973) and Phys. Rev. D 10, 1145 (1974).

51) See, for example, H. Georgi and S.L. Glashow, Phys. Rev. Letters 28, 1494 (1972).

52) See for example, H. Fritzsch, M. Gell-Mann and P. Minkowski, Phys. Letters 59B, 256 (1976).

53) C.H. Llewellyn Smith, Oxford preprint OUTP 76-33 (1976) is a nice review.

54) J. Prentki and B. Zumino, Nuclear Phys. B47, 99 (1972).

55) See, for example, J. Schechter and Y. Ueda, Phys. Rev. D 8, 484 (1973).

56) D. Horn and G.G. Ross, Phys. Letters 67B, 460 (1977).
G. Altarelli, N. Cabibbo, L. Maiani and R. Petronzio, Phys. Letters 67B, 463 (1977).

57) For a sampler of models, see J.D. Bjorken and C.H. Llewellyn Smith, Phys. Rev. D 7, 887 (1973).

58) H.B. Thacker and J.J. Sakurai, Phys. Letters 36B, 103 (1971); see also the calculations in Ref. 57.

59) A. Ali and T.C. Yang, Phys. Letters 65B, 275 (1977).
A. Buras and J. Ellis, Nuclear Phys. B111, 431 (1976).
G. Köpp, L.M. Sehgal and P.M. Zerwas, Nuclear Phys. B123, 61, 77 (1977).

60) F.A.Wilczek and A. Zee, Nuclear Phys. B106, 461 (1976).

61) M. Cavalli-Sforza et al., Phys. Rev. Letters 36, 558 (1976).
M.L. Perl et al., Phys. Letters 63B, 466 (1976).
G.J. Feldman et al., Phys. Rev. Letters 38, 117 (1977).
J. Burmester et al., Phys. Letters 68B, 297 (1977).
J. Burmester et al., Phys. Letters 68B, 301 (1977).
R. Brandelik et al., DESY preprint 77/36 (1977).
A. Barbaro-Galtieri et al., Berkeley preprint LBL-458 (1977).
M.L. Perl et al., preprint SLAC-PUB-1997/LBL-6731 (1977).

62) However, the events of B.C. Barish et al., Ref. 36, and of the CDHS collaboration, M. Holder et al., Observation of trimuon events produced in neutrino and antineutrino interactions, submitted to Phys. Letters B, do not cry out for a heavy lepton explanation, and are consistent with a more conventional hadronic source.

63) A. Benvenuti et al., Phys. Rev. Letters 38, 1183 (1977).
V. Barger et al., Phys. Rev. Letters 38, 1190 (1977).

64) A. Benvenuti et al., Phys. Rev. Letters 35, 1199 (1975).
M. Holder et al., Like-sign dimuon events produced in narrow-band neutrino and antineutrino beams, submitted to Phys. Letters B.

65) See, for example, V. Barger, D.V. Nanopoulos and R.J.N. Phillips, Wisconsin preprint COO-597 (1977).

66) B.W. Lee and S. Weinberg, Phys. Rev. Letters 38, 1237 (1977).
P. Langacker and G. Segré, Phys. Rev. Letters 39, 259 (1977).

67) H. Harari, Proc. 1975 Internat. Symposium on Lepton and Photon Interactions at High Energies, Stanford 1975 (ed. W.T. Kirk) (SLAC, Stanford, 1975), p. 317.

68) S.L. Adler, Phys. Rev. 177, 2426 (1969).
J.S. Bell and R. Jackiw, Nuovo Cimento 51A, 47 (1969).
S.L. Adler, in Lectures on elementary particles and quantum field theory, 1970 Brandeis Summer Institute (eds. S. Deser, M. Grisaru and H. Pendleton) (MIT Press, Cambridge, 1971).
R. Jackiw, Field theoretic investigations, in Lectures in current algebra and its applications, by S.B. Treiman, R. Jackiw and D.J. Gross (Princeton University Press, N.J.,1972) p. 97.

69) The word "anomaly" has been put in quotation marks because it has belatedly been realized that such extra terms in the divergence equations reflect fundamental classical features of differential operators on manifolds, which are expressed by the Atiyah-Singer and related theorems. See, for example,

S.W. Hawking, Phys. Letters 60A, 81 (1977).
N.K. Nielsen and B. Schroer, CERN preprint TH 2317 (1977);
R. Jackiw and C. Rebbi, MIT preprint CTP-619 (1977);
Since the "anomalies" are now known to be rather canonical, perhaps we should find a new word for them.

70) A. de Rujula, H. Georgi and S.L. Glashow, Phys. Rev. Letters 35, 69 (1975).
F.A. Wilczek, A. Zee, R.L. Kingsley and S.B. Treiman, Phys. Rev. D 12, 2768 (1975).
H. Fritzsch, M. Gell-Mann and P. Minkowski, Ref. 52.
S. Pakvasa, W.A. Simmons and S.F. Tuan, Phys. Rev. Letters 35, 702 (1975).

71) This point of view is not universal. In particular, J.D. Bjorken stresses that the observable effects of "anomalies" at present energies are negligible. The issue is rather one of principle.

72) A.J. Buras and K.F. Gaemers, CERN preprint TH. 2368 (1977). See also A.J. Buras, Nuclear Phys. B125, 125 (1977) and references therein.

73) B.W. Lee, S. Pakvasa, R.E. Shrock and H. Sugawara, Phys. Rev. Letters 38, 977, 1230 (E) (1977).
H. Fritzsch, Phys. Letters 67B, 451 (1977).

74) T.P. Cheng and L.F. Li, Phys. Rev. Letters 38, 381 (1977).

75) F.A. Wilczek and A. Zee, Phys. Rev. Letters 38, 531 (1977).

76) J.D. Bjorken and S. Weinberg, Phys. Rev. Letters 38, 622 (1977).

77) M.A. Bouchiat and C.C. Bouchiat, Phys. Letters 48B, 111 (1974) and references therein.

78) This work was reported in Ref. 42.
I.B. Khriplovich, Proc. 18th Internat. Conf. on High-Energy Physics, Tbilisi, USSR, 1976 (JINR, Dubna, 1977) p. B180 discusses previous calculations.

79) For a recent flurry of activity, see
A. De Rujula, H. Georgi and S.L. Glashow, Harvard preprint HUTP-77/A028 (1977);
F.A. Wilczek and A. Zee, Discrete flavour symmetries and a formula for the Cabibbo angle, Princeton preprint (1977).
H. Fritzsch, CERN preprint TH. 2358 (1977).

80) M.S. Chanowitz, J. Ellis and M.K. Gaillard, CERN preprint TH 2312 (1977).

81) H. Georgi and S.L. Glashow, Phys. Rev. Letters 32, 438(1974).

82) H. Fritzsch and P. Minkowski, Ref. 43.
H. Georgi, Particles and fields 1974, APS/DPF Williamsburg meeting (ed. C.E. Carlson) (AIP, New York, 1975), p. 575.
S.L. Glashow, Harvard preprint HUTP-77/A005 (1977).

83) E. Ma, Phys. Letters 65B, 468 (1976).
See also B.W. Lee and S. Weinberg, Ref. 66.

84) J. Ellis, M.K. Gaillard and D.V. Nanopoulos, Nuclear Phys. B106, 292 (1976).

85) A.D. Linde, JETP Letters 23,64 (1976).
S. Weinberg, Phys. Rev. Letters 36, 294 (1976).

86) M. Veltman, Second threshold in weak interactions, Utrecht preprint (1977).
B.W. Lee, C. Quigg and H.B. Thacker, Phys. Rev. Letters 38, 883 (1977).
C.E. Vayonakis, New threshold of weak interactions, Athens preprint (1977).

87) Alternatives to the approach mentioned here include, for example, the possibility that hadrons are weak gauge solitons -- see, for example,
E. Corrigan, D.I. Olive, D.B. Fairlie and J. Nuyts, Nuclear Phys. B106, 475 (1976); and the use of supersymmetry and/or supergravity -- see, for example
B. Zumino, CERN preprint TH. 2356 (1977).

88) Strictly speaking, it is not necessary that the strong interactions remain asymptotically free at very high Q^2. The differences between the gauge boson contributions to the β functions control the way that different group couplings approach one another, with fermions cancelling because they form a grand multiplet -- see Eq. (3.38).

89) H. Georgi, H.R. Quinn and S. Weinberg, Phys. Rev. Letters 33 451 (1974).

90) A.J. Buras, J. Ellis, M.K. Gaillard and D.V. Nanopoulos are studying these questions in greater detail.

91) Reference 90 finds that the proton lifetime is $O(10^4)$ longer than the dimensional estimate $\tau(\text{proton}) \sim M^5/m_p^4$ which was made in Ref. 89 and used uncritically in Ref. 80.

92) F. Reines and M.F. Crouch, Phys. Rev. Letters 32, 493 (1974). Their method was only sensitive to proton decays which produce a muon. In the model discussed here, a large fraction of final states probably do contain a muon -- see Ref. 90.

93) J. Ellis, M.K. Gaillard, D.V. Nanopoulos and S. Rudaz, CERN preprint TH. 2346 (1977).

94) H. Harari, Phys. Letters 57B, 265 (1975) and Ann. Phys. (NY) 94, 391 (1975).
S. Pakvasa and H. Sugawara, Phys. Rev. D 14, 305 (1976).

95) J. Ellis, M.K. Gaillard and D.V. Nanopoulos, Nuclear Phys. B109, 213 (1976).

96) J. Bailey et al., Phys. Letters 68B, 191 (1977).

97) F. Halzen and S. Matsuda, CERN preprint TH. 2354 (1977).

98) T. Appelquist and H.D. Politzer, Phys. Rev. Letters 34, 43 (1975) and Phys. Rev. D 12, 1404 (1975).

99) E. Eichten and K. Gottfried, Phys. Letters 66B, 286 (1977).

100) J.G. Branson et al., Phys. Rev. Letters 38, 580, 791(E) (1977).

101) M.J. Corden et al., Phys. Letters 68B, 96 (1977).

102) C.E. Carlson and R. Suaya, Phys. Rev. D 15, 1416 (1977).
S.D. Ellis, M.B. Einhorn and C. Quigg, Phys. Rev. Letters 36, 1263 (1976).
Gluon-gluon collisions cannot give J/ψ or ψ' directly because of charge conjugation, but the hard reaction could be accompanied by the exchange of a soft gluon which fixes up the quantum numbers so that GG → J/ψ need not proceed through P_c/χ intermediate states.

103) H. Fritzsch, Phys. Letters 67B, 217 (1977).

104) H.D. Snyder et al., Phys. Rev. Letters 36, 1415 (1976).

105) B.W. Lee, Phys. Rev. D 15, 3394 (1977) gives references.
See also refs. 43, 44, 70.

106) K. Kleinknecht,Proc. 17th Internat. Conf. on High-Energy Physics, London 1974 (ed. J.R. Smith) (Rutherford Laboratory Chilton, Didcot, 1974), p. III-23.

107) A. Pais and J.B. Treiman, Phys. Rev. D 12, 2744 (1975).

108) N.F. Ramsey, private communication to B.W. Lee, quoted in Ref. 105. The latest published number is 10^{-23} cm.
W.B. Dress, P.D. Miller and N.F. Ramsey, Phys. Rev. D 7, 3147 (1973).

109) See also, H. Harari, Ref. 94, in a model where the t and b quarks had much lower masses than are discussed nowadays.

AN INTRODUCTION TO GAUGE THEORIES

Maurice LEVY

Laboratoire de Physique Théorique et Hautes Energies

Université Pierre et Marie Curie, Paris

I. INTRODUCTION.

All lectures on gauge theories start with a discussion of the reasons why these theories have been the subject of so much interest in the last seven or eight years. On the one hand, these justifications are less and less necessary as time goes on; on the other, the reasons are changing. To respect to this tradition, I shall list briefly a few of the recent achievements of gauge theories :

- The success of quantum electrodynamics (QED), the first gauge theory (or the second, if one wants to consider Einstein's theory of gravitation as a particular gauge theory).
- The unification of weak and electromagnetic interactions, using Yang-Mills fields (1), and the various predictions which came out of it, particularly the existence of neutral currents.
- The proof of the renormalizability of massive gauge theories (abelian and non abelian).
- The predictions of massless gauge theory of strong interactions (chromodynamics), particularly asymptotic freedom and, perhaps, confinement.
- The possibility to unify strong interactions with the others (weak and electromagnetic).
- The development of supersymetric theories, and recent progresses in the theory of super-gravity.

There is a great deal of arbitrariness in the choice of topics for an introduction to gauge theories. Excellent such introductions exist already (2); furthermore, similar lectures were given in Cargèse in 1975 by R. Gastmans (3). I have tried to cover slightly different topics than in those lectures, which are available. In particular, I will not discuss weak interactions at all. The detailed treatment of renormalization which will be given here is rather formal, and entirely based on the Lagrangian formalism. A few models will be discussed as examples, mostly for pedagogical purposes. But they will be rather different from those which are found in existing reviews or lecture notes.

These lectures are organized as follows : an introductory treatment of abelian gauge theories is given in sec. II, including the gauge field mass generation mechanism (Higgs fields). Sec. III contains a similar treatment of non-abelian theories, using a formalism which parallels very closely that of abelian theories. Sections IV and V deal with the renormalization problem for both kinds of theories. Section VI is devoted to a unified model of strong and electromagnetic interactions. Finally, sec. VII deals with chiral gauge theories, and their relation with other chiral models (like the σ-model).

Although it is clearly possible to couple many different "matter" fields to the gauge vector fields, I shall limit myself, throughout these lectures, to a spinor field, which can represent successively the electron field (QED), the nucleon field (strong interaction SU(2) theory), a triplet of quark fields (SU(3)) or even a set of several quark multiplets ("coloured" SU(3) theory). This does not preclude, of course, the introduction of the scalar Higgs field needed for mass generation.

II. ABELIAN GAUGE THEORY.

1) Massless gauge field (QED).

We start with an abelian vector field $a_\mu(x)$, with its corresponding field strengths $f_{\mu\nu}(x)$:

$$f_{\mu\nu}(x) = \partial_\mu a_\nu(x) - \partial_\nu a_\mu(x) \ . \qquad \text{(II.1)}$$

(In the following, small letters indicate unrenormalized fields, whereas capital letters will later represent renormalized ones). The field a_μ is coupled to a simple Dirac spinor field ψ through the covariant derivative :

$$D_\mu \psi = \partial_\mu \psi + ie_o a_\mu \qquad \text{(II.2)}$$

where e_o is the unrenormalized charge. The Lagrangian is written as follows

$$\mathcal{L}_1 = -\frac{1}{4} f_{\mu\nu} f^{\mu\nu} + \bar{\psi}(i\partial^\mu D_\mu - m_o)\psi \qquad \text{(II.3)}$$

It is invariant in the following infinitesimal gauge transformations :

$$a_\mu(x) \to a_\mu(x) + \partial_\mu\lambda(x),$$
$$\psi(x) \to [1 - ie_o\lambda(x)]\psi(x), \qquad (II.4)$$
$$\bar\psi(x) \to \bar\psi(x)[1 + ie_o\lambda(x)],$$

where $\lambda(x)$ is an infinitesimal function of the coordinates. (In all these lectures, we shall consider mostly infinitesimal transformations, for simplicity). Under the transformations (II.4), the covariant derivative $D_\mu\psi$ changes as follows :

$$D_\mu\psi \to [1 - ie_o\lambda(x)]D_\mu\psi \qquad (II.5)$$

so that $f_{\mu\nu}$, $\bar\psi D_\mu\psi$ and $\bar\psi\psi$ are separately invariant.

For any tansformation of the type :

$$\Phi_i \to \Phi_i - i\varepsilon\lambda_i{}^j\Phi_j, \qquad (II.6)$$

where the Φ_i's are any of the fields coupled in the Lagrangian, a Noether's current (4) can be defined through the relation :

$$j_\mu = -i\,\frac{\delta L}{\delta(\partial_\mu\Phi_i)}\,\lambda_i{}^j\Phi_j \qquad (II.7)$$

and is automatically conserved :

$$\partial^\mu j_\mu = 0. \qquad (II.8)$$

Here, the Noether's current is simply :

$$j_\mu = e_o\bar\psi\gamma_\mu\psi \qquad (II.9)$$

2) Massgeneration of the vector field (5).

We introduce a complex scalar field ϕ, and define its covariant derivative exactly as for ψ :

$$D_\mu\phi = (\partial_\mu + ie_oa_\mu)\phi\ . \qquad (II.10)$$

In a transformation :

$$\phi \to (1 - ie_o\lambda)\phi,$$
$$\phi^+ \to \phi^+(1 + ie_o\lambda), \qquad (II.11)$$

it transforms as :

$$D_\mu\phi \to (1 - ie_o\lambda)D_\mu\phi.$$

We can then add the following invariant Lagrangian :

$$L_2 = D_\mu\phi^+ D^\mu\phi - V(\phi^+\phi), \tag{II.12}$$

where V acts as a "potential" and is only a function of the product $\phi^+\phi$. It can happen that V has a minimum at a non-vanishing positive value. For example, limiting ourselves (for reasons of renormalizability) to fourth powers of ϕ, we can write :

$$V = -a\phi^+\phi + b(\phi^+\phi)^2, \tag{II.13}$$

with $a > 0$ and $b > 0$. This has a minimum for

$$\phi^+\phi = v^2 = \frac{a}{2b} . \tag{II.14}$$

Shifting the value of ϕ :

$$\begin{aligned} \phi &\to \phi + v, \\ \phi^+ &\to \phi^+ + v, \end{aligned} \tag{II.15}$$

(v is real and positive : $v = (a/2b)^{1/2}$), we get :

$$V = b[(\phi^+\phi)^2 + 2(\phi+\phi^+)\phi^+\phi + (\phi^++\phi)^2] \tag{II.16}$$

If we now write :

$$\begin{aligned} \phi &= \frac{\phi_1 + i\phi_2}{\sqrt{2}}, \\ \phi^+ &= \frac{\phi_1 - i\phi_2}{\sqrt{2}}, \end{aligned} \tag{II.17}$$

V is finally equal to

$$V = 2b\phi_1^2 + \sqrt{2}b\phi_1(\phi_1^2+\phi_2^2) + \frac{b}{4}(\phi_1^2+\phi_2^2)^2. \tag{II.18}$$

We see that ϕ_1 is now a massive field with mass 4b and that ϕ_2 is massless. In L_2, the kinetic term has become :

$$D_\mu\phi^+ D^\mu\phi \to D_\mu\phi^+ D^\mu\phi + ie_0 v[D_\mu\phi^+ a^\mu - a_\mu D^\mu\phi] + e_0^2 v^2 a_\mu^2, \tag{II.19}$$

so that the a_μ field has also acquired a mass $2e_0^2v^2$. (In practice, the masses of ϕ_1 and a_μ can be fixed by choosing appropriately the parameters a and b).

The field ϕ_2 can be eliminated by writing, instead of (II.17):

$$\phi = \chi e^{i\theta}, \tag{II.20}$$

where χ is invariant and θ transforms as :

$$\theta \rightarrow \theta - e_o\lambda.$$

Then $\phi^+\phi = \chi^2$, and if we shift $\chi \rightarrow \chi + v$, where v^2 is the minimum of $V(\chi^2)$, χ becomes massive and θ can be eliminated from the Lagrangian by an appropriate gauge transformation $(\lambda=-e_o^{-1}\theta)$, which however fixes the gauge.

We are thus left with :
- a massive vector field a_μ;
- a massive real scalar field χ;
- massless "Goldstone" boson (5) which can be eliminated in a particular gauge.

The S-matrix elements are unchanged in the transformation $\phi \rightarrow \phi + v$. Only the definition of perturbation theory is different. For a discussion of the independence of S-matrix elements on fields transformations, the reader is referred to the lectures of 't Hooft and Veltman (6) in the proceedings of the 1973 Louvain Summer Institute.

III. NON-ABELIAN GAUGE THEORIES.

We shall assume here that there is an internal symmetry group G, of dimension N and that the spinor ψ, representing the matter fields, is in a representation (n) of this group. The corresponding N structure (n x n) matrices are t_i, (i=1, ...N). The gauge fields a^i_μ will be assumed to belong to a representation (N) of the group.
For example : in SU(2), n=2, N=3, the t_i matrices are the Pauli matrices τ_1, τ_2, τ_3;
- for SU(3), n=3, N=8, $t_i=\lambda_i$,
- for U(3), n=3, N=9, $t_i=\lambda_i$ (i = 1,2, ... 8) and $t_o=\lambda_o \sqrt{2/3} \times 1$,
- etc ...

We use the following rules for the matrices t_i :

Normalization : $\mathrm{Tr}(t_i t_j) = 2\delta_{ij}$, (III.1)

Structure relation : $[t_i, t_j] = 2if_{ijk}t_k$. (III.2)

For any field $\{\Phi_i\}$ in the N representation, we construct the matrix

$$\Phi = \frac{1}{\sqrt{2}} t^i\Phi_i, \tag{III.3}$$

such that

$$\mathrm{Tr}\Phi^2 = \sum_i \Phi^i\Phi_i. \tag{III.4}$$

(For example, we shall write gauge field as a matrix :

$$a_\mu = \frac{1}{\sqrt{2}}\, a^i_\mu t_i\ , \qquad \text{(III.5)}$$

and also the corresponding field strengths $f_{\mu\nu}$).

The construction of a gauge invariant Lagrangian involves two covariant derivatives :

In the n representation : $D_\mu\psi = (\partial_\mu + \frac{i}{\sqrt{2}} g_o a_\mu)\psi,$ (III.6)

In the N representation : $\mathcal{D}_\mu\Phi = \partial_\mu\Phi + \frac{i}{\sqrt{2}}\ g_o[a_\mu,\Phi]$ (III.7)

In particular :

$$\mathcal{D}_\mu a_\nu = \partial_\mu a_\nu + \frac{i}{\sqrt{2}} g_o[a_\mu,\ a_\nu].$$

Our initial Lagrangian will be

$$L_1 = -\frac{1}{4}\,\mathrm{Tr} f_{\mu\nu}f^{\mu\nu} + \bar\psi(i\gamma^\mu D_\mu - m_o)\psi, \qquad \text{(III.8)}$$

where

$$f_{\mu\nu} = \partial_\mu a_\nu - \partial_\nu a_\mu + \frac{i}{\sqrt{2}} g_o[a_\mu,\ a_\nu] \qquad \text{(III.9)}$$

In these equations, g_o is an unrenormalized "charge" corresponding to the field a_μ.
L_1 is invariant under the following gauge transformations :

$$\psi \to (1 - \frac{i}{\sqrt{2}} g_o\lambda)\psi \qquad \text{(III.10)}$$

$$\bar\psi \to \bar\psi\ (1 + \frac{i}{\sqrt{2}} g_o\lambda),$$

where λ is a matrix representing N independent infinitesimal transformations λ_i :

$$\lambda = \frac{1}{\sqrt{2}}\, t_i\lambda^i, \qquad \text{(III.11)}$$

$$\begin{aligned} a_\mu &\to a_\mu + \mathcal{D}_\mu\lambda \\ &= a_\mu + \partial_\mu\lambda + \frac{i}{\sqrt{2}} g_o[a_\mu,\lambda] \end{aligned} \qquad \text{(III.12)}$$

It follows that $D_\mu\psi$ transforms like ψ :

$$D_\mu\psi \to (1 - \frac{i}{\sqrt{2}} g_o\lambda)D_\mu\psi \qquad \text{(III.13)}$$

and that the field strengths transform as

$$f_{\mu\nu} \rightarrow f_{\mu\nu} - \frac{i}{\sqrt{2}} g_o[\lambda, f_{\mu\nu}]. \qquad \text{(III.14)}$$

In order to generate masses for the fields a_μ, we shall introduce a set of complex scalar fields ϕ. This can be done in many different ways depending on the physical requirements for the masses of the gauge fields a_μ^i. Here, we shall choose, as an example, a mechanism such that all a_μ^i's will, in first approximation, acquire the same mass. This mechanism would be, of course, unsuited to models which tend to unify weak and electromagnetic interactions, because in these models, one field at least must remain massless, to represent the photon. What we shall describe could be, for example, used to represent the vector octet containing the ρ, K^* mesons considered as a set of gauge fields in a semi-phenomenological Lagrangian.

This "symmetric Higgs model" involves n complex multiplets, each of dimension n : $\phi^\alpha \equiv \{\phi_i^\alpha\}$, i=1, ... n and α=1, ...n. Each of these multiplets will transform like ψ (or $\bar{\psi}$).

$$\phi^\alpha \rightarrow (1 - \frac{i}{\sqrt{2}} g_o\lambda)\phi^\alpha ,$$
$$\phi^{\alpha+} \rightarrow \phi^{\alpha+}(1 + \frac{i}{\sqrt{2}} g_o\lambda) \qquad \text{(III.15)}$$

Then

$$D_\mu\phi^\alpha = (\partial_\mu + \frac{i}{\sqrt{2}} g_o a_\mu)\phi^\alpha$$

transforms as :

$$D_\mu\phi^\alpha \rightarrow (1 - \frac{i}{\sqrt{2}} g_o\lambda)D_\mu\phi^\alpha, \text{ etc ...} \qquad \text{(III.16)}$$

The Higgs-Lagrangian can then be written :

$$L_2 = \sum_\alpha D_\mu\phi^{\alpha+}D^\mu\phi^\alpha - V(\phi^{\alpha+}\phi^\alpha). \qquad \text{(III.17)}$$

Actually, the n multiplets ϕ^α can be arranged as a n x n matrix by putting them side by side :

$$\phi = (\phi^1\phi^2 \ldots \phi^n) = \begin{pmatrix} \phi_1^1 & \phi_1^2 & \cdots & \phi_1^n \\ \phi_2^1 & \phi_2^2 & \cdots & \phi_2^n \\ \vdots & \vdots & \ddots & \vdots \\ \phi_n^1 & \phi_n^2 & \cdots & \phi_n^n \end{pmatrix} \qquad \text{(III.18)}$$

Although it is, like a_μ, a n x n matrix, ϕ does <u>not</u> transform like a_μ :

$$\phi \to (1 - \frac{i}{\sqrt{2}} g_o \lambda)\phi. \tag{III.19}$$

Then L_2 can be rewritten :

$$L_2 = \mathrm{Tr}\ \{\ D_\mu \phi^+ D^\mu \phi - V(\phi^+ \phi)\ \}, \tag{III.20}$$

where

$$D_\mu \phi = (\partial_\mu + \frac{i}{\sqrt{2}} g_o a_\mu)\phi \tag{III.21}$$

If v^2 is the minimum of $V(\phi^+\phi)$, v is a n x n matrix which can be choosen diagonal. We can then make the shift :

$$\phi \to \phi + v\ , \quad \phi^+ \to \phi^+ + v\ , \tag{III.22}$$

and obtain, depending on the structure of v, different mass schemes for a_μ. If we take v proportional to the unit matrix, L_2 will contain a mass term :

$$\frac{1}{2} g_o^2 v^2 \mathrm{Tr} a_\mu^2\ ,$$

so that all the a_μ^i's acquire the same unrenormalized mass $g_o^2 v^2$.

Reduction in the number of the fields.

The above sheme clearly involves too many fields ϕ_i^α. An economic sheme would need only N Goldstone bosons, plus one real scalar Higgs field. Instead of N+1 fields, we have introduced actually $2n^2$ fields (each of the ϕ_i^α is complex). For example, in SU(3), we would need only 9 fields and we have introduced 18. However, nothing prevents us to establish a relation among the n multiplets ϕ^α, provided we maintain the transformation law (III. 19).

We can have the following realization :

$$\phi = \chi\, e^{-\frac{i}{\sqrt{2}} g_o \theta}\ , \tag{III.23}$$

where χ is proportional to the unit matrix and where θ is a N-representation matrix :

$$\theta = \frac{\theta_i \lambda^i}{\sqrt{2}}\ , \quad (i = 1, \ldots N). \tag{III.24}$$

Altogether, we have therefore N + 1 real scalar fields. We choose invariants and we determine the transformation law of θ in order to have (III.19). This can be obtained by a power series in g_o :

$$\theta \to \theta + \delta\theta_0 + \frac{g_0}{\sqrt{2}}\delta\theta_1 + \frac{g_0^2}{2}\delta\theta_2 + \ldots \qquad \text{(III.25)}$$

It is then easy to find :

$$\delta\theta_0 = \lambda,$$

$$\delta\theta_1 = -\frac{i}{2}[\lambda, \theta], \qquad \text{(III.26)}$$

$$\delta\theta_2 = -\frac{1}{12}(\theta^2\lambda + \lambda\theta^2 - 2\theta\lambda\theta), \text{ etc } \ldots$$

If λ is traceless, like in SU(3), we verify that Eqs (III.25, 26) are compatible with a choice of θ traceless. If λ contains also a term proportional to the unit matrix, then this will only appear in $\delta\theta_0$ (which will correspond to the abelian part of the transformation). The N-Goldstone bosons θ_i can be eliminated by a special gauge transformation. We are then left, in that particular gauge, only with χ and the shift $\chi \to \chi + v$, where v^2 is the minimum of $V(\chi)$ generates the vector field mass.

In the case of SU(3), we can also choose $v = 1/\sqrt{2}\,\lambda_0 v_0 + 1/\sqrt{2}\,\lambda_8 v_8$, where v_8 would be a small breaking parameter. This would lead directly to the Gell-Mann - Okubo mass relation for the vector fields a_μ^i.

IV. RENORMALIZATION OF ABELIAN GAUGE THEORIES.

1. Introduction.

As was emphasized in Sec. I, part of the interest in gauge theories comes from the fact that they are renormalizable both in the case where the gauge particles are massless or when they acquire a mass through the Higgs mechanism.

The initial proofs (7, 8) used functional methods and an analysis of Feynman graphs. A good summary is given in the review article of B.W. Lee and Abers (2) and in the lectures of 't Hooft and Veltman in the proceedings of the Louvain Summer Institute of 1973 (6).

It would be too long to go over the functional techniques in the present lectures. However, another approach has since been followed by Becchi, Rouet and Stora (9),based, up to a point, on Lagrangian methods, and using the Ward-Takahasi (10) and the Slavnov-Taylor identities (11). But even this method, which is completely rigorous, is too complicated to be reproduced in a short set of lectures. We shall therefore follow the so called "formal method", used for example in the recent paper by R. Brandt (12), which will enable us to continue in a straightforward manner the analysis of the previous chapters.

This method is not completely rigourous, in as much as it manipulates freely the renormalization constants Z_i and δm, which are normally infinite. However, it is quite acceptable if one uses first a regularization technique preserving entirely the form of the Ward-Takahashi (WT) and the Slavnov-Taylor (ST) identities). Such a technique has been proposed by 't Hooft and Veltman (13) : it is the dimensional regularization method, which has been well discussed, including applications, in Gastman's lectures at Cargèse in 1975 (3). The dimensional regularization orginates from the fact that Feynman integrals are, in general, finite for continuous dimensions ν such that $1 < \nu < 4$. They can therefore be calculated for continuous dimensions, using standard formulae which can be found in Gastman's lectures. The quantities of interest have a pole for $\nu = 4$, coming for example from coefficients like $\Gamma(2 - \nu/2)$, etc... Their elimination by standard renormalization techniques makes then all physical quantities finite. There are some complications when the Feynmann integrals contain Dirac matrices, especially γ_5 matrices, which prevent the extension of the Dirac algebra to ν dimensions. However, methods exist to solve this problem.

In the following, we shall assume that all infinite quantities which appear are regularizable, and that this does not affect the form of the W.T. identities.

2) Quantization in the Lorentz gauge.

We start with the Abelian case, which is well-known from the renormalization of QED, because the comparison wit the Yang-Mills problem will be very instructive. We continue to study the case of a gauge field a_μ coupled to a spinor ψ, as in section II. The initial Lagrangian is given by Eq (II.3). The corresponding classical field equations are as follows :

$$\partial^\mu f_{\mu\nu} = j_\nu = e_o \bar{\psi}\gamma_\nu\psi ,$$
$$(i\gamma^\mu\partial_\mu - m_o)\psi = e_o\gamma^\mu a_\mu\psi. \tag{IV.1}$$

Here, j_ν is the Noether's current defined in Eq. (II.17). It is generated by the one parameter Lie group

$$a_\mu \to a_\mu ,$$
$$\psi \to (1 - ie_o\lambda)\psi, \tag{IV.2}$$

where λ is a constant. We can generalize this to the infinite dimensional group (gauge group) defined by Eqs (II.4). The corresponding current is, up to an infinitesimal constant (we can change, for example, $\lambda(x)$ into $\varepsilon\lambda(x)$, where ε is an infinitesimal parameter and $\lambda(x)$ is finite) :

$$s_\mu(\lambda) = f_{\mu\nu}\partial^\nu\lambda - j_\mu\lambda, \tag{IV.3}$$

which is also conserved.

$$\partial^\mu s_\mu(\lambda) = 0 . \tag{IV.4}$$

Actually, the equations of motion give $s_\mu = \partial^\nu(f_{\mu\nu}\lambda)$, so that Eq (IV.4) follows simply from the antisymmetry of $f_{\mu\nu}$. The canonical momentum π_μ, conjugate to a_μ, is given by

$$\pi_\mu = - f_{o\mu}, \tag{IV.5}$$

so that π_o vanishes. In order to quantize the theory, we have to add a "gauge fixing" term, which formally breaks gauge invariance, but which does not affect the measurement of physical quantities. Here, we shall use a generalized class of Lorentz gauges by adding to L_1 :

$$L_2 = - \frac{1}{2\alpha_o} (\partial^\nu a_\nu)^2, \tag{IV.6}$$

($\alpha_o = 1$, would correspond to the usual Lorentz gauges leading to the equation of motion $\Box a_\nu = j_\nu$). The parameter α_o will have to be renormalized. Later in these lectures, we shall show how it is possible to work also in the Landau gauge corresponding to $\alpha_o \to 0$.

The Lagrangian $L = L_1 + L_2$, is still invariant on the one-dimensional Lie group (IV.2), with the same conserved Noether's current j_μ. The equations of motion are now :

$$\begin{aligned} &\partial^\mu f_{\mu\nu} + \frac{1}{\alpha_o}\partial_\nu(\partial^\mu a_\mu) = j_\nu, \\ &(i\gamma^\mu\partial_\mu - m_o)\psi = e_o\partial^\mu a_\mu\psi \end{aligned} \tag{IV.7}$$

The first equation lead (by multiplying with ∂^ν) to :

$$\Box(\partial^\mu a_\mu) = 0 \tag{IV.8}$$

so that $(\partial^\mu a_\mu)$ behaves like a free field. The gauge group (II.4) now works only for those functions $\lambda(x)$ satisfying the condition :

$$\Box\lambda = 0 . \tag{IV.9}$$

The corresponding current $t_\mu(\lambda)$ can be broken into two parts :

$$t_\mu = s_\mu + \tilde{t}_\mu, \tag{IV.10}$$

where $s_\mu = \partial_\mu(f_{\mu\nu}\lambda)$ as before and

$$\tilde{t}_\mu = \frac{1}{\alpha_0}\,[(\partial a)\partial_\mu\lambda - \partial_\mu(\partial a)\lambda] \equiv \frac{1}{\alpha_0}[(\partial a)\,\overleftrightarrow{\partial}_\mu\lambda], \tag{IV.11}$$

which involves only the "free" quantities (∂a) and λ. All the three "currents" t_μ, s_μ, $\tilde{t}_\mu$ are separatly conserved.

The canonical conjugate momentum is now :

$$\pi_\mu = -f_{0\mu} - \frac{1}{\alpha_0}\; g_{\mu 0}(\partial a), \tag{IV.12}$$

so that $\pi_0 = - 1/\alpha_0(\partial a)$ and the theory can be quantized in the standard manner.

The "charge" T associated with the current t_μ

$$T = \int d\vec{x}\,t_0(x) = \tilde{T} = \int d\vec{x}\,\tilde{t}_0(x), \tag{IV.13}$$

is the generator of the gauge group (II.4) :

$$\begin{aligned} \psi(x) &\to \psi(x) - i[T,\ \psi(x)] \\ a_\mu(x) &\to a_\mu(x) - i[T,\ a_\mu(x)]; \end{aligned} \tag{IV.14}$$

(obviously, the "charge" associated with s_μ vanishes).

From the gauge group (II.4), one can deduce easily the "unrenormalized" W.T. identities. For example, defining the propagators :

$$s(p) = -\,i\int d^4x\; e^{ipx} < 0|T\psi(x)\overline{\psi}(0)|0 > \tag{IV.15a}$$

$$d_{\mu\nu}(k) = -\,i\int d^4x\; e^{ikx} < 0|Ta_\mu(x)a_\nu(0)|0 > \tag{IV.15b}$$

and the vertex function :

$$\begin{aligned} f\,(p,k) &\equiv e_0 s(p+k)\gamma^\nu(p,k)s(p)d_{\mu\nu}(k) \\ &= \int d^4x d^4y\; e^{ipx+iky} < 0|T\psi(x)\overline{\psi}(0)a_\mu(y)|0 >, \end{aligned} \tag{IV.16}$$

one gets, using the equal time commutators and Eq. (IV.8), the two identities :

$$k^{\mu}d_{\mu\nu}(k) = \frac{\alpha_o k_\nu}{k^2 - i\varepsilon}, \qquad (IV.17)$$

$$k_\nu \gamma^\nu = s^{-1}(p) - s^{-1}(p+k) \qquad (IV.18)$$

3) Alternative approach to gauge invariance.

We shall now discuss an alternative approach to gauge invariance which is relatively trivial in abelian theories, but which will become essential in the Yang-Mills case.

One can restore the complete invariance, partially broken by L_2, by adding to the Lagrangian a third term

$$L_3 = \frac{1}{2}\partial^\mu c \partial_\mu c, \qquad (IV.19)$$

when c is a free massless scalar field. At this stage, it is unnecessary to specify its statistics, because it never interacts. We assume, for the time being, that it obeys the usual Bose-Einstein statistics. Then, the complete Lagrangian

$$L = L_1 + L_2 + L_3 \qquad (IV.20)$$

is invariant under a one-parameter Lie group :

$$\begin{aligned} a_\mu &\to a_\mu + \varepsilon_o \partial_\mu c, \\ \psi &\to (1 - i\varepsilon_o e_o c)\psi, \\ c &\to c - \frac{\varepsilon_o}{\alpha_o}(\partial a); \end{aligned} \qquad (IV.21)$$

(actually, L changes by a total divergence, which does not affect the equations of motion).

The Noether's current associated with (IV.21) is :

$$t_\mu = [f_{\mu\nu} + \frac{1}{\alpha_o} g_{\mu\nu}(\partial a)]\partial^\nu c - j_\mu c = s_\mu + \tilde{t}_\mu, \qquad (IV.22)$$

where $s_\mu = \partial^\nu (f_{\mu\nu} c)$ and

$$\tilde{t}_\mu = \frac{1}{\alpha_o}[(\partial a)\overleftrightarrow{\partial}_\mu c]. \qquad IV.23)$$

The one-parameter group (IV.21) can be used, in a very simple meaner, to generate the W.T. identities. This can be done by simply stating that some specified Green's functions are invariant under the group.

If we write, for example,

$$\delta < 0|Ta_\mu(x)c(y)|0 > = 0, \tag{IV.24}$$

we obtain, through Eqs (IV.21) :

$$0 = \varepsilon_0 < 0|T\partial_\mu c(x)c(y)|0 > - \frac{\varepsilon_0}{\alpha_0} < 0|Ta_\mu(x)\partial a(y)|0 >.$$

This clearly relates $\partial^\nu d_{\mu\nu}(k)$ to the free propagator of the scalar field c, which leads to the identity (IV.17).

Similarly, writing :

$$\delta < 0|T\psi(x)\overline{\psi}(y)c(z)|0 > = 0, \tag{IV.25}$$

gives :

$$\begin{aligned} 0 = -i\varepsilon_0 e_0 &< 0|Tc(x)\overline{\psi}(y)\psi(x)c(z)|0 > \\ &+ i\varepsilon_0 e_0 < 0|T\psi(x)\overline{\psi}(y)c(y)c(z)|0 > \\ &- \frac{\varepsilon_0}{\alpha_0} < 0|T\psi(x)\overline{\psi}(y)\partial a(z)|0 > . \end{aligned}$$

Using the definition of the free propagator of c :

$$\frac{1}{k^2-i\varepsilon} = - i \int d^4xe^{ikx} < 0|Tc(x)c(0)|0 >,$$

one gets, with a little manipulation of the Green's functions apprearing in (IV.26), the W.T. identity for the vertex obtained in (IV.18).

Consequently, gauge invariance can be transformed into an ordinary Lie-group invariance, using a slightly enlarged Lagrangian, involving a free scalar field, which can be quantized in the standard manner. This observation will become crucial for non-abelian theories.

4) Renormalization.

By power counting, it can be seen that all primitive divergences can be absorbed in five renormalization constants δ_m and $Z_1 \ldots Z_4$:

$$\Psi = Z_2^{-1/2}\psi,\ A_\mu = Z_3^{-1/2}a_\mu,\ e = Z_1^{-1}Z_2Z_3^{1/2}e_o,\ \alpha = Z_4^{-1}\alpha_o\ ,$$

$$m = m_o + \delta m. \tag{IV.27}$$

Then, Eq. (IV.17) gives :

$$k^\mu D_{\mu\nu}(k) = \frac{Z_4}{Z_3}\ \alpha\ \frac{k_\nu}{k^2 - i\varepsilon} \tag{IV.28}$$

so that :

$$Z_3 = Z_4\ . \tag{IV.29}$$

The second W.T. identity (IV.18) yields similarly :

$$Z_1 = Z_2. \tag{IV.30}$$

Consequently, Eqs (IV.27) can now be reexpressed in terms of 3 constants only : Z_1, Z_3 and δ_m :

$$\Psi = Z_1^{-1/2}\psi,\ A_\mu = Z_3^{-1/2}a_\mu,\ e = Z_3^{1/2}e_o,\ \alpha = Z_3^{-1}\alpha_o,$$

$$m = m_o + \delta m. \tag{IV.31}$$

The renormalized equations of motion can be written :

$$(i\partial - m)\Psi = eA\Psi - \delta m\Psi\ ,$$

$$\partial^\mu F_{\mu\nu} + \frac{1}{\alpha}\,\partial_\nu(\partial.A) = \frac{Z_1}{Z_3}\,e\bar{\Psi}\gamma_\nu\Psi + \frac{1}{\alpha}\,(1 - \frac{1}{Z_3})\,\partial_\nu(\partial.A) \equiv J_\nu,$$

$$F_{\mu\nu} = \partial_\mu A_\nu - \partial_\nu A_\mu\ . \tag{IV.32}$$

Here, J_ν is the renormalized current, which is finite due to cancellation of divergences.

The renormalized commutators of the A field are no longer covariants in our generalized Lorentz-gauge :

$$[\dot{A}_o(x),\ A_o(y)]_{x_o=y_o} = i\alpha\delta(\vec{x} - \vec{y}),$$

$$[\dot{A}_i(x),\ A_j(y)]_{x_o=y_o} = -\ i\delta_{ij}Z_3^{-1}\delta(\vec{x} - \vec{y}), \tag{IV.33}$$

$$[\dot{A}_o(x),\ A_i(y)]_{x_o=y_o} = i(\alpha - \frac{1}{Z_3})\partial_i\delta(\vec{x} - \vec{y}).$$

However, the renormalized Green functions are all finite :

$$S(p) = Z_1^{-1}s(p) = -i\int d^4x e^{ipx} < 0|T\Psi(x)\overline{\Psi}(o)|0 >, \qquad \text{(IV.34a)}$$

$$D_{\mu\nu}(k) = Z_3^{-1}d_{\mu\nu}(k) = -i\int d^4x e^{ikx} < 0|TA_\mu(x)A_\nu(o)|0 > \qquad \text{(IV.34b)}$$

$$\Gamma_\mu(p,k) = Z_1\gamma_\mu(p,k) . \qquad \text{(IV.34c)}$$

The renormalized vertex function corresponding to (IV.16) is :

$$\begin{aligned} F_\mu(p,k) &= eS(p+k)\Gamma^\nu(p,k)D_{\mu\nu}(k) = Z_1^{-1}Z_3^{-1/2}f_\mu(p,k) \\ &= \int d^4x d^4y e^{ipx+iky} < 0|T\Psi(x)\overline{\Psi}(o)A_\mu(y)|0 >. \end{aligned} \qquad \text{(IV.35)}$$

If we <u>define</u>

$$\Lambda(x) = Z_3^{-1/2}\,\lambda(x), \qquad \text{(IV.36)}$$

the gauge transformation laws become :

$$\begin{aligned} &A_\mu \to A_\mu + \partial_\mu\Lambda, \\ &\Psi \to (1 - ie\Lambda)\Psi, \\ &\Box\Lambda = 0. \end{aligned} \qquad \text{(IV.37)}$$

It is only because $Z_1 = Z_2$ that the structure of these transformations remains the same, leading to identical renormalized W.T. identities. The corresponding Noether's current (conserved) :

$$T_\mu = (Z_3F_{\mu\nu} + \frac{1}{\alpha}\,g_{\mu\nu}(\partial\cdot A))\partial^\nu\Lambda - Z_1e\overline{\Psi}\gamma_\mu\Psi\Lambda \qquad \text{(IV.38)}$$

can be decomposed into two separately conserved currents :

$$S_\mu = Z_3\partial^\nu(F_{\mu\nu}\Lambda), \qquad \text{(IV.39)}$$

and

$$\tilde{T}_\mu = \frac{1}{\alpha}\,(\partial\cdot A)\overleftrightarrow{\partial}_\mu\Lambda \qquad \text{(IV.40)}$$

$\tilde{T}_\mu$ is finite, but not T_μ. However, since S_μ has zero charge the charge T associated with T_μ :

$$T = \int d\vec{x}\; T_o(x), \qquad \text{(IV.41)}$$

is finite.

In the alternative approach, the non-interacting scalar field c remains unrenormalized. We can introduce a renormalized group parameter :

$$\varepsilon = Z_3^{-1/2} \varepsilon_o, \qquad \text{(IV.42)}$$

so that, the transformation laws (IV.21) become :

$$A_\mu \to A_\mu + \varepsilon \partial_\mu c,$$
$$c \to c - \frac{\varepsilon}{\alpha} (\partial \cdot A), \qquad \text{(IV.43)}$$
$$\Psi \to (1 - i\varepsilon e A)\Psi.$$

Again, the Ward as Slavnov identities (IV.29, 30) guarantee that these transformation laws retain the same form, yielding the renormalized W.T. identities, etc...

5) Quantization in the Landau gauge.

We have see that, in the generalized Lorentz gauge, the photon propagator is given by :

$$d_{\mu\nu}(k) = (g_{\mu\nu} - \frac{k_\mu k_\nu}{k^2}) d(k) + \alpha_o \frac{k_\mu k_\nu}{k^4} \qquad \text{(IV.44)}$$

Going to the Landau gauge corresponds therefore to choosing $\alpha_o = 0$. This cannot be done conveniently if we use the gauge fixing Lagrangian L_2, defined by Eq. (IV.6). To overcome this difficulty, we can follow the approach of Brandt, Ng Wing Chiu and Young (14), which consists first in replacing the complete Lagrangian $L = L_1 + L_2 + L_3$, defined by Eqs (III.3), (IV.6) and (IV.19) with a completely equivalent first order Lagrangian :

$$L = -\frac{1}{2} f_{\mu\nu}(\partial^\mu a^\nu - \partial^\nu a^\mu) + \frac{1}{4} f_{\mu\nu} f^{\mu\nu} - b(\partial^\nu a_\nu)$$
$$+ \frac{\alpha_o}{2} b^2 + L_3 + \overline{\psi}(i\gamma^\mu D_\mu - m_o)\psi, \qquad \text{(IV.45)}$$

where $f_{\mu\nu}$, a_μ and b must be considered as independent dynamical variables (b is another auxiliary scalar field). The Eqs. of motion are :

$$f_{\mu\nu} = \partial_\mu a_\nu - \partial_\nu a_\mu, \qquad \text{(IV.46a)}$$

$$\partial^\mu (f_{\mu\nu} + \partial_\mu b) = e_o \overline{\psi}\gamma_\nu \psi, \qquad \text{(IV.46b)}$$

$$\alpha_o b = (\partial \cdot a). \qquad \text{(IV.46c)}$$

We can then take formally the limit $\alpha_o \to 0$, so that L becomes simply :

$$\mathcal{L} = -\frac{1}{2} f_{\mu\nu}(\partial^\mu a^\nu - \partial^\nu a^\mu) + \frac{1}{4} f_{\mu\nu} f^{\mu\nu} - b(\partial\cdot a) + \mathcal{L}_3 + \bar{\psi}(i\partial^\mu D_\mu - m_o)\psi. \quad \text{(IV.47)}$$

The last Eq. of motion (IV.46c) becomes simply :

$$\partial^\nu a_\nu = 0. \quad \text{(IV.48)}$$

The conjugate momentum to a_μ is now :

$$\pi_\mu = \frac{\delta \mathcal{L}}{\delta(\partial_o a_\mu)} = -f_{o\mu} - g_{\mu o} b, \quad \text{(IV.49)}$$

so that

$$\pi_i = -f_{oi}, \qquad \pi_o = -b, \quad \text{(IV.50)}$$

and the theory can be quantized without difficulty.

V. RENORMALIZATION OF NON-ABELIAN GAUGE THEORIES.

1) Quantization in the generalized Lorentz gauge.

We start with the Lagrangian $\mathcal{L}_1$ defined by Eq. (III.8). As in the abelian case, we add, in order to be able to quantize the theory, the gauge fixing Lagrangian :

$$\mathcal{L}_2 = -\frac{1}{2\alpha_o} \mathrm{Tr}(\partial^\mu a_\mu)^2. \quad \text{(V.1)}$$

The canonical conjugate momentum to a^i_μ becomes :

$$\pi^i_\mu = -f^i_{o\mu} - \frac{1}{\alpha_o} g_{\mu o}(\partial^\nu a^i_\nu), \quad \text{(V.2)}$$

leading to the usual equal time commutators :

$$[\pi^i_\mu(x), a^j_\nu(y)]_{x_o=y_o} = -ig_{\mu\nu}\delta^{ij}\delta(\vec{x}-\vec{y}),$$

$$\{\psi_i(x), \psi^+_j(y)\}_{x_o=y_o} = \delta_{ij}\delta(\vec{x}-\vec{y}). \quad \text{(V.3)}$$

However, unlike the abelian case, where the introduction of $\mathcal{L}_2$ preserved a limited form of gauge invariance (for functions $\lambda(x)$ satisfying $\Box\lambda = 0$). The invariance of the non-abelian Lagrangian

$L = L_1 + L_2$ is now completely destroyed. The equivalent condition to $\Box\lambda = 0$ can now be written :

$$\partial^\mu D_\mu \lambda(x) = 0 \quad \text{or :} \tag{V.4}$$

$$\Box\lambda + \frac{i}{\sqrt{2}} g_o \partial^\mu [a_\mu, \lambda] = 0$$

Since this equation involves the quantized field a_μ, there is no non-trivial c-number solution. Introducing operator solutions for λ immediately leads us to the alternative approach, described in §3 of the previous chapter. We have to introduce a set of N fields $c_i(x)$, satisfying the condition :

$$\partial^\mu D_\mu c_i(x) = 0. \tag{V.5}$$

Introducing a set of N Lagrange multiplyers $\overline{c}^i$, we can implement these conditions by introducing :

$$L_3' = - \sum_i \overline{c}^i \partial^\mu D_\mu \; c_i(x). \tag{V.6}$$

We can actually construct as before, matrices $c(x)$ and $\overline{c}(x)$ (belonging to the N-representation of the group like a_μ) :

$$C(x) = \frac{t^i c_i}{\sqrt{2}}, \quad \overline{C}(x) = \frac{\overline{c}^i t_i}{\sqrt{2}}, \tag{V.7}$$

and replace L_3' with a Lagrangian which is equivalent, up to a total divergence :

$$L_3 = \mathrm{Tr}\,\partial_\mu \overline{C} D_\mu C. \tag{V.8}$$

The Lagrangian $L = L_1 + L_2 + L_3$ is invariant under an N-parameter Lie group corresponding to N constants λ_i :

$$\begin{aligned} a_\mu &\to a_\mu + \frac{i}{\sqrt{2}} g_o [a_\mu, \lambda], \\ \psi &\to (1 - \frac{i}{\sqrt{2}} g_o \lambda)\psi, \\ c &\to c + \frac{i}{\sqrt{2}} g_o [c, \lambda], \\ \overline{c} &\to \overline{c} + \frac{i}{\sqrt{2}} g_o [\overline{c}, \lambda] \end{aligned} \tag{V.9}$$

which leaves invariant separetly L_1, L_2 and L_3. Indeed, for constant λ

$$\delta(\partial^\nu a_\nu) = -\frac{i}{\sqrt{2}} g_o[\lambda, \partial^\nu a_\nu] \tag{V.10}$$

and

$$\delta D_\mu c = -\frac{i}{\sqrt{2}} g_o[\lambda, D_\mu c]$$

$$\delta\partial_\mu \bar{c} = -\frac{i}{\sqrt{2}} g_o[\lambda, \partial_\mu \bar{c}]. \tag{V.11}$$

However, it was observed by Becchi, Rouet and Stora (9) that L is also invariant under a one-parameter Lie-group, which they called the Slavnov group :

$$\begin{aligned} a_\mu &\to a_\mu + \bar{\varepsilon}_o D_\mu c, \\ \psi &\to (1 - \frac{i}{\sqrt{2}} g_o \bar{\varepsilon}_o c)\psi, \\ \bar{c} &\to \bar{c} - \frac{1}{\alpha_o} \bar{\varepsilon}_o(\partial\cdot a), \\ c &\to c - \frac{i}{2\sqrt{2}} g_o\bar{\varepsilon}_o[c, c], \end{aligned} \tag{V.12}$$

<u>provided one imposes to c and $\bar{c}$ to be anticommuting operators.</u> $\bar{\varepsilon}_o$ is a gauge "parameter", anticommuting with c and $\bar{c}$. The last of Eqs (V.12) can be written explicitly :

$$c_i \to c_i + \frac{1}{2} f_{ijk} g_o \bar{\varepsilon}_o c^j c^k, \tag{V.13}$$

which is compatible with $\{c_i, c_j\}_+ = \delta_{ij}$. L_1 and $L_2 + L_3$ are separately invariant (actually $L_2 + L_3$ is only invariant up to a total divergence) :

$$\begin{aligned} \delta(D_\mu c) &= 0, \\ \delta(\partial_\mu \bar{c}) &= -\frac{1}{\alpha_o} \bar{\varepsilon}_o \partial_\mu(\partial\cdot a), \end{aligned} \tag{V.14}$$

so that :

$$\delta L_2 = -\frac{1}{\alpha_o} \bar{\varepsilon}_o \mathrm{Tr}(\partial a)\partial^\mu D_\mu c,$$

$$\delta L_3 = -\frac{1}{\alpha_o} \bar{\varepsilon}_o \mathrm{Tr}[\partial^\mu(\partial a)\cdot D_\mu c], \tag{V.15}$$

and

$$\delta L_2 + \delta L_3 = -\frac{\bar{\varepsilon}_o}{\alpha_o}\partial^\mu \mathrm{Tr}[(\partial a)\cdot D_\mu c]. \tag{V.16}$$

If we had not imposed the anticommutativity of the c-components, c would have been invariant and δL_3 would have contained an additional piece :

$$\frac{i}{\sqrt{2}} g_o \mathrm{Tr}\partial_\mu \bar{c}\cdot[D_\mu c, c] \neq 0.$$

It can be proved that, if the gauge symmetry is unbroken or broken spontaneously (Higgs mechanism), unitarity of the S-matrix is not violated in the sector of the physical particles (ψ, a_μ, ϕ).

This property was found empirically through the detailed examination of loops appearing in Feynman graphs. It is understood very well with the analysis of the Slavnov-transformations. The c field, because it is a scalar field obeying Fermi statistics is carried a "ghost" field.

The Lagrangian L_3 looks unsymmetrical in c and $\bar{c}$. However, by a little manipulation, one can rewrite L_3, up to a total divergence in a symmetric form :

$$L_3 = \frac{1}{2}\mathrm{Tr}[\partial_\mu \bar{c} D_\mu c + D_\mu \bar{c}\partial_\mu c] + \frac{i}{2\sqrt{2}} g_o \mathrm{Tr}[\bar{c}\cdot c]\cdot\partial a. \tag{V.17}$$

It must be noted, nevertheless, that the transformation laws of c and $\bar{c}$ under the Slavnov group (V.12) are not symmetric. There exists, of course, another group where c and $\bar{c}$ are inverted, with a group parameter ε_o instead of $\bar{\varepsilon}_o$. The quantization of the c and $\bar{c}$ fields is done through the usual equal-time anti-commutators :

$$\{\dot{c}_i(x), c_j(y)\}_{x_o=y_o} = \{\dot{\bar{c}}_i(x), \bar{c}_j(y)\}_{x_o=y_o} = 0, \tag{V.18}$$

$$\{\dot{\bar{c}}_i(x), c_j(y)\}_{x_o=y_o} = i\delta_{ij}\delta(\vec{x}-\vec{y}). \tag{V.18}$$

There is a conserved Noether current t_μ :

$$t_\mu = [f_{\mu\nu} + \frac{1}{\alpha_o} g_{\mu\nu}(\partial\cdot a)]D^\nu c \tag{V.19}$$
$$+ \frac{i}{2\sqrt{2}}\partial_\mu \bar{c}[c, c] + \frac{g_o}{2}\bar{\psi}\gamma_\mu c\psi,$$

which is such that $t = \int d\vec{x} t_o(x)$ can serve as a generator of the Slavnov group.

One can get easily the W.T. identities simply by noting, as in the abelian case, that the appropriate Green functions are invariant. For example, by writing that :

$$\delta < 0|T\partial a^i(x), \bar{c}^j(o)|0 > = 0. \tag{V.20}$$

One obtains :

$$\bar{\varepsilon}_o < 0|T\partial^\nu D_\mu c^i(x)\bar{c}^j(o)|0 > - \frac{1}{\alpha_o}\bar{\varepsilon}_o < 0|T\partial a^i(x), \partial a^j(o)|0 > = 0$$

which relates the propagator $d^{ij}_{\mu\nu}$ to the propagator of the c-field:

$$k^\mu k^\nu d^{ij}_{\mu\nu} = \alpha_o \delta_{ij}, \tag{V.21}$$

where

$$d^{ij}_{\mu\nu}(k) = (g_{\mu\nu} - \frac{k_\mu k_\nu}{k^2}) d^{ij} + \alpha_o \frac{k_\mu k_\nu}{k^2} \frac{\delta^{ij}}{k^2 - i\varepsilon} \tag{V.22}$$

We can therefore write $d^{ij}_{\mu\nu}$ in terms of a transverse propagator $d^{ij}(k)$. The general definition of $d^{ij}_{\mu\nu}$ is :

$$d^{ij}_{\mu\nu} = - i \int d^4x e^{ikx} < 0|Ta^i_\mu(x)a^j_\nu(o)|0 >. \tag{V.23}$$

Similarly we can obtain a W.T. identity relating the vector-vector-vector vertex to the vector-ghost-ghost vertex by writing that :

$$\delta < 0|Ta^i_\mu(x)\bar{c}^j(y)\partial \cdot a^\ell(o)|0 > = 0 \tag{V.24}$$

To write the corresponding identity, which was derived independently by J.C. Taylor (11) and A.A. Slavnov (11), we have to define the ghost-propagator $g^{ij}(p)$:

$$g^{ij}(p) = - i \int d^4x e^{ipx} < 0|Tc^i(x)\bar{c}^j(o)|0 >, \tag{V.24}$$

the vector-ghost-ghost vertex function :

$$F^{ijk}(p,q) = -\int d^4x d^4y e^{ipx+iqy} < 0|Tc^i(x)\bar{c}^j(o)a^k_\mu(y)|0 >$$
$$= - g^{i\ell}(p)(\gamma^\nu)^{\ell mn}(p,q)g^{jm}(p+q)d^{kn}_{\mu\nu}(q), \tag{V.26}$$

and the three vector-vertex function :

$$G^{\ell mn}_{\alpha\beta\gamma}(p,q) = \gamma^{ijk}_{\mu\nu\rho}(p,q)d^{il}_{\alpha\mu}(p)d^{jm}_{\beta\nu}(q)d^{kn}_{\gamma\rho}(p+q)$$
$$= \int d^4x d^4y \; e^{ipx + iqy} < 0|Ta^\ell_\alpha(x)a^m_\beta(y)a^n_\gamma(o)|0 > . \tag{V.27}$$

Related to these functions, we also have to define :

$$\sigma^{ij}_{\mu\nu}(p)g^{jk}(p) = g_o\int d^4xe^{ipx} < 0|T[a_\mu(x), c(x)]^i\bar{c}^k(o)|C > \tag{V.28}$$

and

$$[\gamma^{ijk}_{\mu\nu}(p,q) - \sigma^{i\ell}_{\mu}(p)g^{\ell m}(p)\gamma^{mjk}_{\nu}(p,q)]g^{jn}(p)d^{kr}_{\rho\nu}(p+q) \tag{V.29}$$

$$= ig_o\int d^4xd^4ye^{ipx+iqy} < 0|T[a_\mu(x),c(x)]^i\bar{c}^n(y)a^r_\rho(o)|C >.$$

We then obtain the Slavnov-Taylor relation :

$$(g_{\mu\rho} - \frac{p_\mu p_\rho}{p^2})\gamma^{ijk}_{\rho\nu}(p,q)g^{j\ell}(q)q^\nu$$

$$= - (g_{\mu\sigma} - \frac{p_\mu p_\sigma}{p^2})\gamma^{i\ell k}_{\sigma\nu\rho}(p,q)d(p)\frac{q^\nu q^\rho}{q^2}. \tag{V.30}$$

2. Renormalization.

The renormalized Lagrangian will involve 6 Z functions and a mass renormalization term :

$$A_\mu = Z_3^{-1/2}a_\mu, \quad \Psi = Z_2^{-1/2}\psi, \quad C = \tilde{Z}_3^{-1/2}c,$$

$$g = Z_3^{3/2}Z_1^{-1}g_o, \quad \tilde{g} = \tilde{Z}_3Z_3^{1/2}\tilde{Z}_1^{-1}g_o, \tag{V.31}$$

$$\alpha = Z_4^{-1}\alpha_o, \qquad m = m_o + \delta m,$$

since there is no apriori reason why the coupling constants of the a_μ and of the c fields will remain the same when radiative corrections are taken into account. However, the two W.T. identities which we have written previously give us two relations between the Z_s' : Eq. (V.21) tells us that :

$$Z_3 = Z_4, \tag{V.32a}$$

and Eq. (V.30) yields :

$$\frac{Z_1}{Z_3} = \frac{\tilde{Z}_1}{\tilde{Z}_3}. \tag{V.32b}$$

These are the well-known Slavnov-Taylor relations. The relations for α and $\tilde{g}$ can be rewritten :

$$\alpha = Z_3^{-1}\alpha_o, \quad \tilde{g} = g, \tag{V.33}$$

the others remaining unchanged. However, since we do not have $Z_1 = Z_2$, as in QED, the gauge transformations(III.10) and (III.12) do not remain finite after renormalization. If we choose, for example,

$$\Lambda = Z_3^{-1/2}\lambda,$$

we get :

$$\Psi \to [1 - i \frac{Z_1}{Z_3}\frac{g}{\sqrt{2}} \Lambda]\Psi, \tag{V.34a}$$

$$A_\mu \to A_\mu + \partial_\mu\Lambda + \frac{ig}{\sqrt{2}} \frac{Z_1}{Z_3}[A_\mu, \Lambda] \tag{V.34b}$$

Had we chosen $\Lambda = Z_1 Z_3^{-3/2}\lambda$, the right hand side of (V.34a) and the last term of (V.34b) would have become finite, but the second term of (V.34b) would then become infinite. This shows that the gauge transformations can be made finite only for λ = cte, which is the only case, actually, where we have an invariance group for the Lagrangian $L_1 + L_2 + L_3$ (see Eqs. (V.9)).

On the other hand, it can be rectified, by writing

$$\bar{E}_o = \tilde{Z}_3^{-1/2} Z_3^{1/2} \bar{\varepsilon}_o, \tag{V.35}$$

that the Slavnov group transformations retain the same form after renormalization.

We see now that the non Abelian gauge theory can be renormalized, but with the addition of the pieces L_2 and L_3, which correspond finally to a theory quite different from the original Yang-Mills formalism (1). Thus, the initial motivation of non-Abelian gauge theories has been lost in the renormalization process. Perhaps, a way out of this difficulty could be found in non-local formalisms, where manifestly gauge-invariant fields can be constructed from the outset (15).

It is possible to go to the Landau gauge, as in the abelian case, by introducing a set of auxilliary scalar fields

$$b = \frac{1}{\sqrt{2}} \lambda^i b_i , \tag{V.36}$$

in the N-representation, with the Lagrangian :

$$L = -\frac{1}{2}\,\mathrm{Tr} f_{\mu\nu}\,[\,\partial^\mu a^\nu - \partial^\nu a_\mu + \frac{ig_0}{\sqrt{2}}[a^\mu, a^\nu]\,]$$

$$+ \frac{1}{4}\,\mathrm{Tr} f_{\mu\nu} f^{\mu\nu} - \mathrm{Tr} b(\partial a) + L_3 + \overline{\psi}(i\gamma^\mu D_\mu - m_0)\psi, \tag{V.37}$$

where again $f_{\mu\nu}$, a_μ and b must be thought of as independant dynamical (matrix) variables. In the Slavnov transformation for $\overline{c}$, one must replace $1/\alpha_0$ (∂a) with b. The rest is unchanged.

VI. EXAMPLE OF MODEL BUILDING : UNIFICATION OF STRONG AND ELECTROMAGNETIC INTERACTIONS.

After having discussed the basis of gauge theories, it is of interest to analyze a few models, in order to exhibit some of the features which come out of the application of these theories to specific physical situations. It is customary, in lectures like these, to discuss in detail the Salam-Weinberg model, which was the first successful unification of weak and electromagnetic interactions. For this, the reader is referred to the previous Gastman's lectures in Cargèse (3).

Here, we shall study instead another unified model of interest : that which describes together strong and electromagnetic interactions. Such a model would be essential for any serious discussion of the electromagnetic properties of hadrons, since it is necessary, to obtain finite results, to introduce the electromagnetic field as part of the gauge fields responsible for strong interactions. If the electromagnetic interaction is added in the usual "ad hoc" manner, to the strong interaction Lagrangian, it is well known that unrenormalizable infinities appear.

1) Unification through the SU(2) x U(1) group.

This model has already been discussed by 't Hooft. We show it here in order to exhibit the marked differences which arise when we go to the SU(3) x U(1) model. We first start with the usual SU(2) non abelian gauge model, coupling a gauge triplet b^i_μ(i = 1,2,3) to a spinor doublet ψ. We introduce as usual the field matrix

$$b_\mu = \frac{1}{\sqrt{2}}\,(\vec{\tau}\cdot\vec{b}), \tag{VI.1}$$

where the τ_i's are the usual Pauli matrices and the corresponding field strength matrix :

$$g_{\mu\nu} = \partial_\mu b_\nu - \partial_\nu b_\mu + \frac{i}{\sqrt{2}} g_o[b_\mu, b_\nu] \quad \text{(VI.2)}$$

Defining the covariant derivative of ψ as :

$$D_\mu\psi = (\partial_\mu + \frac{i}{\sqrt{2}} g_o b)\psi, \quad \text{(VI.3)}$$

we can write the invariant unrenormalized Lagrangian :

$$\mathcal{L}_1 = -\frac{1}{4} Tg_{\mu\nu}g^{\mu\nu} + \bar{\psi}(i\not{D} - m_o)\psi. \quad \text{(VI.4)}$$

If we introduce now an infinitesimal gauge matrix

$$\varepsilon = \frac{1}{\sqrt{2}} (\vec{\tau}\cdot\vec{\varepsilon}), \quad \text{(VI.5)}$$

the gauge transformations are :

$$\begin{aligned} b_\mu &\rightarrow b_\mu + \mathcal{D}_\mu\varepsilon, \\ g_{\mu\nu} &\rightarrow g_{\mu\nu} - \frac{i}{\sqrt{2}} g_o[\varepsilon, g_{\mu\nu}] \\ \psi &\rightarrow (1 - \frac{i}{\sqrt{2}} g_o\varepsilon)\psi, \\ D_\mu\psi &\rightarrow (1 - \frac{i}{\sqrt{2}} g_o\varepsilon)\, D_\mu\psi \,; \end{aligned} \quad \text{(VI.6)}$$

where we have defined the <u>other</u> covariant derivative

$$\mathcal{D}_\mu\varepsilon = \partial_\mu\varepsilon + \frac{i}{\sqrt{2}} g_o[b_\mu, \varepsilon]. \quad \text{(VI.7)}$$

The same mass of the three fields b^i_μ can be generated by adding one complex Higgs doublet

$$\phi = \begin{pmatrix} \phi_1 \\ \phi_2 \end{pmatrix} \quad \text{(VI.8)}$$

which transform as

$$\phi \rightarrow (1 - \frac{i}{\sqrt{2}} g_o\varepsilon)\phi. \quad \text{(VI.9)}$$

We add the Higgs Lagrangian :

$$\mathcal{L}_2 = D_\mu\phi^+ D_\mu\phi - V(\phi^+\phi), \quad \text{(VI.10)}$$

and make the shift $\phi \rightarrow \phi + v$, where

$$v = \begin{pmatrix} 0 \\ v_2 \end{pmatrix} \qquad \text{(VI.11)}$$

is such that V_2^2 minimizes the function V(Z). This generates in L_2 a mass term $1/2g^2v_2^2b_\mu^2$. We would have obtained exactly the same result if we had followed the general theory of Section III, which would have required two complex scalar doublets. This is due to the special properties of the Pauli matrices.

We now go over to the SU(2) x U(1) group by adding a singlet vector field a_μ coupled to ψ through the modified covariant derivative (instead of (VI.3)) :

$$D_\mu\psi = \partial_\mu\psi + \frac{i}{\sqrt{2}} g_0 b_\mu \psi + \frac{i}{\sqrt{2}} g'_0 a_\mu \psi, \qquad \text{(VI.12)}$$

the corresponding field strengths being defined as :

$$f_{\mu\nu} = \partial_\mu a_\nu - \partial_\nu a_\mu, \qquad \text{(VI.13)}$$

and the Lagrangian (VI.4) being replaced with :

$$L_1 = -\frac{1}{4} \mathrm{Tr} g_{\mu\nu} g^{\mu\nu} - \frac{1}{4} f_{\mu\nu} f^{\mu\nu} + \bar{\psi}(i\not{\partial} - m_0)\psi; \qquad \text{(VI.14)}$$

the gauge transformations are now :

$$\begin{aligned} b_\mu &\to b_\mu + D_\mu \varepsilon, \\ a_\mu &\to a_\mu + \partial_\mu \eta, \\ \psi &\to \psi - \frac{i}{2} g_0 \varepsilon\psi - \frac{i}{2} g'_0 \eta\psi, \end{aligned} \qquad \text{(VI.15)}$$

where η is a singlet infinitesimal gauge function. The change in $D_\mu\psi$ becomes now :

$$\delta D_\mu\psi = -\frac{i}{\sqrt{2}} g_0\, D_\mu\psi - \frac{i}{2} g'_0 \eta D_\mu\psi . \qquad \text{(VI.16)}$$

Assuming that ψ can be decomposed into a charged and a neutral part :

$$\psi = \begin{pmatrix} \psi_p \\ \psi_n \end{pmatrix}, \qquad \text{(VI.17)}$$

one can redefine the fields b_μ^i and a_μ :

$$b_\mu^1 = B_\mu^1 \quad ; \quad b_\mu^2 = B_\mu^2 \quad ; \quad b_\mu^3 = B_\mu^3 + \frac{e_0}{g_0} A_\mu, \qquad \text{(VI.18)}$$

where

$$a_\mu = cA_\mu \quad \text{and} \quad g'_o = \frac{e_o}{c} \,. \tag{VI.19}$$

The constant c can be chosen as :

$$c = 1 - \frac{e_o^2}{g_o^2} \,. \tag{VI.20}$$

Then

$$D_\mu\psi = \partial_\mu\psi + \frac{i}{\sqrt{2}} g_o B_\mu \psi + \frac{i}{2} e_o(1+\tau_3)A_\mu\psi, \tag{VI.21}$$

which shows that A_μ is only coupled to the ψ_p part of the spinor field.

An alternative method would be to diagonalize $\mathrm{Tr}(g_{\mu\nu}g^{\mu\nu}) + f_{\mu\nu}f^{\mu\nu}$ by a rotation of b^3_μ and a_μ, with an angle θ.

The Higgs field ϕ introduced before is now coupled to B_μ and A_μ as follows :

$$D_\mu\phi = \partial_\mu\phi + \frac{i}{\sqrt{2}} g_o B_\mu \phi + \frac{i}{2} e_o(1+\tau_3)A_\mu\phi. \tag{VI.22}$$

With the choice of v which we have already made, it is clear that A_μ will remain massless.

2) Generalization to the group SU(3) x U(1).

The previous example, already well known, was given for pedagogical purposes. Going to the full group SU(3) x U(1), we shall have to introduce the full apparatus developed in section III.

We introduce an octet $b_\mu = 1/\sqrt{2}\ \lambda_i b^i_\mu$ and a singlet a_μ. Let us start first with just one triplet spinor ψ (a quark field, for example). The covariant derivative, which defines the coupling to b_μ and a_μ is simply :

$$D_\mu\psi = \partial_\mu\psi + \frac{i}{\sqrt{2}} g_o b_\mu \psi + \frac{i}{\sqrt{3}} g'_o a_\mu \psi \ ; \tag{VI.23}$$

the coupling of a_μ is defined in such a way as to make use of the λ_o matrix ($\lambda_o = \sqrt{2/3}\ I$). It can be seen easily that there is no possibility to have a_μ coupled to each component of ψ with the correct charge value, except by writing $g' = 0$ and :

$$b^i_\mu = B^i_\mu \quad (i = 1,2,4,5,6,7),$$
$$b^3_\mu = B^3_\mu + \frac{e_o}{g_o} A_\mu \quad ; \quad b^8_\mu = B^8_\mu + e_o/g_o\sqrt{3}\ A_\mu. \tag{VI.24}$$

The singlet field a_μ would play no role and it would be impossible to generate a massless field A_μ.

To get a clue to the solution of this problem, we consider the general coupling to the Higgs fields defined in section III, with three triplets ϕ^α. We can write :

$$D_\mu\phi^\alpha = \partial_\mu\phi^\alpha + \frac{i}{\sqrt{2}} g_o b_\mu \phi^\alpha + \frac{i}{\sqrt{3}} g_o' a_\mu d_\alpha \phi^\alpha, \qquad \text{(VI.25)}$$

where the coupling constant of a_μ, $g_o' d_\alpha$ is different for each of the triplets ϕ^α. This does not break the SU(3) symmetry, since each of the ϕ^α's transforms independently under this group.

On the other hand, we break the other group, the "colour group" which transforms the ϕ^α's into each other. The 3 x 3 matrix

$$[\phi] = (\phi^1, \phi^2, \phi^3) \qquad \text{(VI.26)}$$

will be shifted $[\phi] \rightarrow [\phi] + v$, where v is proportional to the unit matrix, in order to give the same mass to the B_μ octet (this is the formalism used in section III). We want to choose the d_α's in such a way that the diagonal fields ϕ_i^i have a zero charge, so that A_μ will remain massless. Writing as before :

$$a_\mu = cA_\mu, \quad g_o' = \frac{e_o}{c} , \qquad \text{(IV.19)}$$

we see that we need :

$$d_1 = - \frac{2}{\sqrt{3}} , \quad d_2 = d_3 = \frac{1}{\sqrt{3}} \qquad \text{(VI.27)}$$

Putting :

$$Q = \frac{1}{2} (\lambda_3 + \frac{1}{\sqrt{3}} \lambda_8), \qquad \text{(VI.28)}$$

we can rewrite

$$D_\mu\phi = \partial_\mu\phi + \frac{i}{\sqrt{2}} g_o B_\mu \phi + ie_o A_\mu [Q, \phi], \qquad \text{(VI.29)}$$

because, with our choice of the d_α constants,

$$\frac{1}{\sqrt{2}} d_\alpha \phi^\alpha = - \phi Q. \qquad \text{(VI.30)}$$

The fact that A_μ will remain massless if v is a diagonal matrix

is obvious. We can therefore even break the masses of the B_μ octet, by choosing v diagonal but not proportional to the unit matrix, without changing this property of A_μ.
The charge matrix of ϕ is given by :

$$\begin{pmatrix} 0 & 1 & 1 \\ -1 & 0 & 0 \\ -1 & 0 & 0 \end{pmatrix}. \qquad \text{(VI.31)}$$

The group U(1), generated by a_μ, actually acts on the "colour" index α of the fields ϕ_i^α. Going back to the spinor field, it is then tempting to define $D_\mu\psi$ exactly in the same way. One is therefore led naturally to postulate the existence of three quarks triplets, with exactly the same charges as above. These charges are those which were introduced by Salam et al. (16).

The analogous theory for SU(2) x U(1), different from the one we discussed in the previous paragraph, would make use of a matrix $[\phi]$ composed of two doublets (ϕ^1, ϕ^2), coupled to the vector fields as follows :

$$D_\mu\phi = \partial_\mu\phi + \frac{i}{\sqrt{2}} g_o B_\mu\phi + \frac{i}{2} e_o A_\mu[\tau_3, \phi], \qquad \text{(VI.32)}$$

the corresponding charge matrix being given by

$$\begin{pmatrix} 0 & 1 \\ -1 & 0 \end{pmatrix} \qquad \text{(VI.33)}$$

The shift $\phi \to \phi + v$ would be the usual one.

VII. ANOTHER MODEL : CHIRAL GAUGE THEORY.

In order to give another example of "unusual" models, we shall discuss here a chiral gauge theory. This will enable us to explain the relation between gauge theories and other types of chiral models, such as the σ model (17).

For simplicity, we shall only discuss here the abelian case. The extension to non-abelian models, using the general formalism of section III, is rather straightforward and left to the reader as an exercise.

We introduce two gauge fields : a vector field v_μ and an axial vector field a_μ. The corresponding free Lagrangian L_1 is given by :

$$L_1 = -\frac{1}{4} f_{\mu\nu}f^{\mu\nu} - \frac{1}{4} g_{\mu\nu}g^{\mu\nu} \qquad \text{(VII.1)}$$

with

$$f_{\mu\nu} = \partial_\mu v_\nu - \partial_\nu v_\mu,$$
$$g_{\mu\nu} = \partial_\mu a_\nu - \partial_\nu a_\mu. \qquad \text{(VII.2)}$$

1) Higgs coupling.

Before studying the coupling of these gauge fields to baryons, we shall first look at the mass generation mechanism. We introduce two Higgs complex fields, one scalar field called s and one pseudoscalar field p.
The corresponding covariant derivatives are simply :

$$D_\mu s = \partial_\mu s - \frac{i}{2} g v_\mu s - \frac{i}{2} g' a_\mu p,$$
$$D_\mu p = \partial_\mu p - \frac{i}{2} g v_\mu p - \frac{i}{2} g' a_\mu s. \qquad \text{(VII.3)}$$

In principle, the model can be developed by keeping $g \neq g'$. However, in non-abelian theories, the invariance of L_1 imposes the condition $g = g'$. For simplicity, and because the results can then be extended to the non-abelian case, we shall impose this condition here, although it is not strictly necessary in the framework of the present model. The gauge transformation of s and p are then (with $g = g'$) :

$$s \to s + \frac{i}{2} \varepsilon s + \frac{i}{2} \varepsilon' p,$$
$$p \to p + \frac{i}{2} \varepsilon p + \frac{i}{2} \varepsilon' s, \qquad \text{(VII.4)}$$

$$v_\mu \to v_\mu + \frac{1}{g} \partial_\mu \varepsilon,$$
$$a_\mu \to a_\mu + \frac{1}{g} \partial_\mu \varepsilon', \qquad \text{(VII.5)}$$

where ε is a scalar infinitesimal gauge function and ε' a pseudo-scalar one. The Higgs Lagrangian L_2 can be written as

$$L_2 = D_\mu s^+ D_\mu s + D_\mu p^+ D_\mu p + U(s,p) \qquad \text{(VII.6)}$$

We can build U(s, p) out of two independent invariants : a scalar one

$$I_1 = s^+ s + p^+ p, \qquad \text{(VII.7)}$$

and a pseudoscalar one :

$$I_2 = s^+ p + p^+ s, \qquad \text{(VII.8)}$$

by writting :

$$U = aI_1 - bI_1^2 - cI_2^2 \ . \qquad \text{(VII.9)}$$

Writing, as in section II,

$$s = \frac{s_1 + is_2}{\sqrt{2}} \ , \qquad p = \frac{p_1 + ip_2}{\sqrt{2}} \ , \qquad \text{(VII.10)}$$

We shift only the s_1 field :

$$s_1 \rightarrow s_1 + \lambda, \qquad \text{with} \qquad \lambda^2 = \frac{a}{b} \qquad \text{(VII.11)}$$

We generate, in such a way, two masses

$$\mu_1^2 = 2b\lambda^2, \qquad \mu_2^2 = 2c\lambda^2, \qquad \text{(VII.12)}$$

the "potential" U becoming :

$$U = -\frac{1}{2}\mu_1^2 s_1^2 - \frac{1}{2}\mu_2^2 p_1^2 - \frac{\mu_1^2}{8\lambda^2}\left[(s_1^2+s_2^2+p_1^2+p_2^2)^2 + 4\lambda s_1\,(s_1^2+s_2^2+p_1^2+p_2^2)\right] - \frac{\mu_2^2}{8\lambda^2}\left[(s_1p_1+s_2p_2)^2 + 2\lambda p_1(s_1p_1+s_2p_2)\right]. \qquad \text{(VII.13)}$$

One can see that s_1 and p_1 acquire different masses (μ_1 and μ_2 respectively), whereas s_2 and p_2 remain massless.

The choice $g = g'$ leads us, for the time being, to the same mass for the gauge fields v_μ and a_μ :

$$m_v^2 = m_a^2 = \frac{1}{4}\,g^2\lambda^2. \qquad \text{(VII.14)}$$

2) Chiral gauge transformations.

It is useful to separate the gauge transformations in such a ways as to introduce two completely different gauge groups [U(1) x U(1)] :

$$\phi_{\pm} = \frac{s \pm p}{\sqrt{2}} ,$$

$$w^{\mu}_{\pm} = \frac{v^{\mu} \pm a^{\mu}}{\sqrt{2}} , \qquad \text{(VII.15)}$$

$$E_{\pm} = \frac{\varepsilon \pm \varepsilon'}{\sqrt{2}}$$

then the gauge transformations of the four fields $\phi_{\pm}$ and $w^{\mu}_{\pm}$ are simply :

$$\phi_{\pm} \rightarrow (1 + \frac{i}{\sqrt{2}} E_{\pm})\phi_{\pm} ,$$

$$w^{\mu}_{\pm} \rightarrow w^{\mu}_{\pm} + \frac{1}{g} \partial^{\mu}E_{\pm}. \qquad \text{(VII.16)}$$

We can then eliminate the massless fields, much in the same way as before, by writting :

$$\phi_{\pm} = \rho_{\pm}\exp[\frac{i}{\lambda} (\theta \pm \theta')]. \qquad \text{(VII.17)}$$

The choice of a special gauge transformation :

$$w^{\mu}_{\pm} \rightarrow w^{\mu}_{\pm} + \frac{\sqrt{2}}{g\lambda} \partial^{\mu}(\theta \pm \theta'), \qquad \text{(VII.18)}$$

which implies that :

$$D_{\mu}\phi_{\pm} = e^{\frac{i}{\lambda} (\theta\pm\theta)} [\partial_{\mu}\rho_{\pm} - \frac{ig}{\sqrt{2}} w_{\pm\mu}\rho_{\pm}] \qquad \text{(VII.19)}$$

then leads to the complete elimination of the θ and θ' fields :

$$U = a(\rho_{+}^{2}+\rho_{-}^{2}) - b(\rho_{+}^{2}+\rho_{-}^{2})^{2} - c(\rho_{+}^{2}-\rho_{-}^{2})^{2}. \qquad \text{(VII.20)}$$

Rewriting $\rho_{\pm}$ as :

$$\rho_{\pm} = \frac{1}{\sqrt{2}} (s_0 \pm p_0 + \lambda), \qquad \text{(VII.21)}$$

the s_0 and p_0 fields acquire the masses μ_1^2 and μ_2^2 defined in (VI. 12).

3) Coupling to baryons.

We introduce one spinor field ψ, coupled to v_{μ} and a_{μ} through the covariant derivative :

$$D_\mu\psi = \partial_\mu\psi - \frac{i}{2} g v_\mu \psi - \frac{i}{2} g \gamma_5 a_\mu \psi \qquad \text{(VII.22)}$$

We can, as usual, separate ψ into two parts :

$$\psi_R = \frac{1}{2}(1 + \gamma_5)\psi, \qquad \psi_L = \frac{1}{2}(1 - \gamma_5)\psi \qquad \text{(VII.23)}$$

The corresponding covariant derivatives are :

$$\begin{aligned} D_\mu\psi_R &= \partial_\mu\psi_R - \frac{i}{\sqrt{2}} g w_{+\mu}\psi_R, \\ D_\mu\psi_L &= \partial_\mu\psi_L - \frac{i}{\sqrt{2}} g w_{-\mu}\psi_L . \end{aligned} \qquad \text{(VII.24)}$$

The gauge transformations of the ψ_R and ψ_L are then given by :

$$\begin{aligned} \psi_R &\to (1 + \frac{i}{\sqrt{2}} E_+)\psi_R, \\ \psi_L &\to (1 + \frac{i}{\sqrt{2}} E_-)\psi_L . \end{aligned} \qquad \text{(VII.25)}$$

The separation of the two groups is now complete. (ψ_R, w_+^μ, ϕ_+) and (ψ_L, w_-^μ, ϕ_-) form two differents sets of fields, coupled to each other, and transforming independently. The special gauge transformation (VII.17, 18) which eliminates the θ and θ' fields must be extended to ψ_R and ψ_L :

$$\begin{aligned} \psi_R &= e^{\frac{i}{\lambda}(\theta+\theta')}\tilde{\psi}_R, \\ \psi_L &= e^{\frac{i}{\lambda}(\theta+\theta')}\tilde{\psi}_L . \end{aligned} \qquad \text{(VII.26)}$$

The Lagrangian L_3 coupling the baryons, is then simply :

$$\begin{aligned} L_3 &= i\bar{\psi}_R \not{D}\psi_R + i\bar{\psi}_L \not{D}\psi_L \\ &\to i\bar{\tilde{\psi}}_R \not{D}\tilde{\psi}_R + i\bar{\tilde{\psi}}_L \not{D}\tilde{\psi}_L . \end{aligned} \qquad \text{(VII.27)}$$

<u>No mass term for the baryons is possible at this stage.</u>

4) Currents and symmetry breaking.

The above gauge transformations generate two separatly conserved currents, a vector current

$$j_\mu = \frac{1}{2}\bar{\psi}\gamma_\mu\psi + \frac{i}{2}[s^+(D_\mu s) - (D_\mu s^+)s] + \frac{i}{2}[p^+(D_\mu p) - (D_\mu p^+)p], \tag{VII.28}$$

and an axial vector current :

$$j^5_\mu = \frac{1}{2}\bar{\psi}\gamma_\mu\gamma_5\psi + \frac{i}{2}[p^+(D_\mu s) - (D_\mu s^+)p] + \frac{i}{2}[s^+(D_\mu p) - (D_\mu p^+)s] \tag{VII.29}$$

We have :

$$\partial^\mu j_\mu = \partial^\mu j^5_\mu = 0 \tag{VII.30}$$

In the same spirit as in the σ-model, we can break the chiral gauge group by adding to U(s,p) a term :

$$\Delta v = -\frac{d\lambda}{\sqrt{2}}(s + s^+) \tag{VII.31}$$

This creates masses for s_2 and p_2 (or θ and θ') proportional to d. We have then PCAC.
However, the baryon field ψ, which is not coupled to s and p in this model, remains massless. This is a basic difference with the σ-model.

5) Baryon mass generation.

How can we introduce here, without breaking the gauge symmetry, a mass for the baryons ? One can think of several methods, which correspond then to seperate models. Two of these methods imply the existence of new fields. The third one involves a non linear coupling, the renormalizability properties of which have not been studied.

a) Salam-Weinberg type of model.

We add to the Lagrangian a new invariant field spinor ψ_1 :

$$\delta\psi_1 = 0 , \tag{VII.32}$$

coupled to s and p as follows :

$$\mathcal{L}^a_4 = i\,\bar{\psi}_1 \partial\!\!\!/ \psi_1 - m_1\bar{\psi}_1\psi_1 + \frac{i}{\sqrt{2}} g_1\,\bar{\psi}_1(s^+ + \gamma_s p^+)\psi + \text{h.c.} \tag{VII.33}$$

Introducing

$$\psi_{1R} = \frac{1}{2}(1+\gamma_s)\psi_1 \quad , \qquad \psi_{1L} = \frac{1}{2}(1-\gamma_s)\psi_1, \qquad \text{(VII.34}$$

the last term of L_4 can be rewritten as :

$$g_1[\ \bar{\psi}_{1L}\phi_+^+\psi_R + \bar{\psi}_L\phi_+\psi_{1R} + \bar{\psi}_{1R}\phi_-^+\psi_L + \bar{\psi}_R\phi_-\psi_{1L}]. \qquad \text{(VII.35)}$$

The invariance is obvious, because ϕ_+ and ψ_R on one hand, ϕ_- and ψ_L on the other, transform in the same way.

The shift of s leads then to a "mass term" :

$$\frac{g_1\lambda}{2}\ (\ \bar{\psi}_1\psi + \bar{\psi}\psi_1), \qquad \text{(VII.36)}$$

which must be diagonalized with L_3 and the rest of L_4. The field ψ_1, then, acquires a negative mass : it can either be understood as a ghost field, or as a spinor field of opposite parity.

b) Addition of a σ-model Lagrangian.

We add to the Lagrangian $L_1+L_2+L_3$ a σ-model Lagrangian :

$$L_4^b = \frac{1}{2}\ [(D_\mu\pi)^2 + (D_\mu\sigma)^2] + A(\pi^2+\sigma^2) - B(\pi^2+\sigma^2)^2$$
$$- G\ \bar{\psi}\ (\sigma+i\gamma_s\pi)\psi. \qquad \text{(VII.38)}$$

The fields π and σ are new pseudoscalar and scalar fields, their covariant derivatives being given by :

$$D_\mu\pi = \partial_\mu\pi + ga_\mu\sigma,$$
$$D_\mu\sigma = \partial_\mu\sigma - ga_\mu\pi\ . \qquad \text{(VII.39)}$$

These derivatives depend only on a_μ. (In the non-abelian case, they would also depend on v_μ through commutators which vanish in the abelian case).
The gauge transformations of π and σ involve only the ε' gauge function :

$$\sigma \to \sigma + \varepsilon'\pi$$
$$\pi \to \pi - \varepsilon'\sigma, \qquad \text{(VII.40)}$$

so that $\pi^2 + \sigma^2$ is an invariant. (Again in the non abelian case, these transformations would also involve ε).
The shift $\sigma \to \sigma + v$, with $v^2 = A/2B$, creates a mass for $\sigma(\mu_\sigma^2=8v^2B)$, but π remains massless. It also creates a mass term for the baryons : $- Gv\bar{\psi}\psi$, much in the same way as in the traditional

σ-model. An interesting feature of this model is that we can eliminate, through an appropriate gauge transformation, not only s_2 and p_2 but also π (in other words, all three massless fields). This can be done as follows :
Writing

$$K_{\pm} = \frac{\sigma \pm i\pi}{\sqrt{2}} \quad , \tag{VII.40}$$

we can transform :

$$\begin{aligned} \phi_{\pm} &\to \frac{\rho\pm}{\sqrt{2}} \exp \left| \frac{i}{\lambda} (\theta \pm \theta') \pm i \frac{\alpha}{2v} \right| \\ K_{\pm} &\to \frac{\Sigma}{\sqrt{2}} \exp \left| \mp \frac{2i\theta'}{\lambda} \mp \frac{i\alpha}{v} \right| \end{aligned} \tag{VII.41}$$

and

$$\begin{aligned} \psi_R &\to \exp \left| \frac{i}{\lambda} (\theta + \theta') + \frac{i\alpha}{2v} \right| \tilde{\psi}_R \\ \psi_L &\to \exp \left| \frac{i}{\lambda} (\theta - \theta') - \frac{i\alpha}{2v} \right| \tilde{\psi}_L . \end{aligned} \tag{VII.42}$$

This eliminates completely the fields θ, θ' and α from the Lagrangian. We are left only with ρ_+, ρ_- and Σ. The conclusion is that the pion is not "necessary" in a chiral gauge theory, because a_μ and $\partial_\mu \pi$ play essentially the same role. However, a breaking term of the type introduced in the σ-model generates a small mass for the pion, which becomes then a physical particle. This gives a clue to the role of the pion in our mechanism.

c) Non linear chiral model.

We can dispense entirely with the introduction of new fields (either ψ, or σ and π) by building out of s and p, the Higgs fields, two combinations which transform exactly like σ and π

$$\sigma = \frac{ss^+ - pp^+}{2} \qquad \pi = - i \left(\frac{sp^+ - ps^+}{2} \right). \tag{VII.43}$$

(There are similar combinations in the non abelian case, where σ and π become matrices).

We then simply have to add to $L = L_1 + L_2 + L_3$ the interaction Lagrangian :

$$L_4^c = - G \bar{\psi} (\sigma + i\gamma_5 \pi)\psi, \tag{VII.44}$$

where σ and π are defined, in terms of s and p, by Eqs. (VII.43).

The shift $s \to s + \lambda/\sqrt{2}$, which we have already introduced in §1, then generates a mass for the ψ :

$$M_\psi = + \frac{G\lambda^2}{4} . \tag{VII.45}$$

The difficulty of this model, as we have said, is that L_4^c contains a quadratic coupling, which may not be renormalizable. But it is clear that this model is very economic.

Note

In the standard σ-model, discussed in paragraph (5,b) above, the axial vector field a_μ acquires an additional mass 1/2 g^2v^2, so that the masses of the vector and axial vector fields become different.

REFERENCES

1. C.N. Yang and R.C. Mills, Phys. Rev. 96, 191 (1954).
2. See, for example, S. Abers and B.W. Lee, Physics Reports 9C, n°1 (1973).
3. R. Gastmans, in Weak and Electromagnetic Interactions at High Energies (Cargèse 1975), Part A, pp. 109-171. Plenum Press (1976).
4. See, for example, J.D. Bjorken and S.D. Drell, Relativistic Quantum Fields (McGraw Hill, 1965) pp. 19-20.
5. P.W. Higgs, Phys. Letters 12, 132 (1964). Several other authors could be quoted here : see Ref. 3.
6. G. 't Hooft and M. Veltman, Particle interactions at very high energies (part B) pp. 177-322, Plenum Press (1974).
7. G. 't Hooft, Nucl. Phys. B33, 173 (1971); B35, 167 (1971); G. 't Hooft and M. Veltman, Nucl. Phys. B50, 318 (1972).
8. B.W. Lee, Phys. Rev. D5 823 (1972); B.W. Lee and J. Zinn-Justin, Phys. Rev. D5, 3121 (1972); D5, 3137 (1972); D5 3155 (1972); D7, 1049 (1973).
9. C. Becchi, A. Rouet and R. Stora, Commun. Math. Phys. 42, 127 (1975).
10. J.C. Ward, Phys. Rev. 78, 1824 (1950); Y. Takahashi, Nuovo Cimento 6, 370 (1957).
11. A.A. Slavnov, Theor. Math. Phys. 10, 99 (1972); J.C. Taylor, Nucl. Phys. B33, 436 (1971).
12. R. Brandt, Nucl. Phys. B116, 413 (1976).
13. G. 't Hooft and M. Veltman, Nucl. Phys. B44, 189 (1972).
14. R.A. Brandt, Ng Wing-Chiu and K. Young, Phys. Rev. D, to be published.
15. M. Lévy, Nucl. Phys. 57, 152 (1964).
16. J.C. Pati and A. Salam, Phys. Rev. 10D, 275 (1974).
17. J. Schwinger, Ann. Phys. (N.Y.) 2, 407 (1958); M. Gell-Mann and M. Lévy, Nuovo Cimento 16, 705 (1960).

ELECTRIC AND MAGNETIC CONFINEMENT SCHEMES

F. ENGLERT*

Faculté des Sciences, Université Libre de

Bruxelles, Bruxelles

1. INTRODUCTION

The theory of confinement is addressed to a paradoxical situation.

On one hand, the existence of quarks would explain in simple terms why hadrons fall into "flavour" group multiplets. Moreover, experiments in which a high momentum is transferred to the hadron probe its internal structure which from the parton model point of view appears indeed to be a composite one; the components may be identified with point-like quarks, losely bound together.

On the other hand, such entities, which otherwise should be similar to leptons, are not produced in high energy collisions. This fact could be imputable either to their high mass or to some particular properties of the forces that bind them together into hadrons. The first possibility is difficult to reconcile with the existence of a different mass scale for hadrons and with the parton interpretation of high momentum transfer experiments. The second alternative may overcome these difficulties if the binding agent confines quarks in hadrons structures; this means that the removal of a quark out of a hadron could require a large, or perhaps even infinite, amount of energy. That energy confinement may indeed follow in some cases from gauge field theory is suggested from the properties of type II-superconductors (see below) : there, magnetic energy is confined within flux tubes of constant energy per

* Permanent address : Université Libre de Bruxelles, Campus Plain, C.P. 225, Bd. du Triomphe, 1050 Bruxelles.

unit length. If quarks would have similar type of flux tubes attached to them, the presence of a quark would cost an infinite energy except if the flux tube terminates, say, on an antiquark. In such a way one would get a confined linear structure for mesons and in some instances a similar type of confinement for baryons.

There is another important reason for prospecting the possibility of such a "linear" confinement scheme. Namely, it would provide a simple explanation for the existence of linearly rising Regge trajectories. To understand this point, let us picture a meson by a thin tube of length L whose rest energy is βL . If we assume that the ends of the flux tube move with the velocity of light (a not unreasonable assumption in the limit of zero quark mass), we get for rigid rotation

$$m = E = 2 \int_0^{L/2} \frac{ds}{\sqrt{1-V_\perp^2}} \qquad \text{I.1)}$$

$$V_\perp = \frac{s}{L/2} \qquad \text{I.2)}$$

Here the transverse section of the tube has been neglected. From I.1) and I.2) it follows

$$m = \frac{\beta\pi L}{2} \qquad \text{I.3)}$$

The angular momentum of the meson is

$$\delta = 2 \int_0^{L/2} \frac{\beta\ s\ v_\perp\ ds}{\sqrt{1-V_\perp^2}} = \frac{\beta\pi L^2}{8} \qquad \text{I.4)}$$

Hence

$$\delta = \alpha' m^2 \ ; \qquad \alpha' = \frac{1}{2\pi\beta} \qquad \text{I.5)}$$

and in this very crude description one may already find the germ of the Regge phenomenon. Moreover, it is straightforward to construct the action from I.1) follows by displacement in time; this is

$$S(t,t_0) = -\beta \int_{t_0}^{t} dt \int_{-L/2}^{+L/2} \sqrt{1-v_\perp^2}\, ds \qquad \text{I.6)}$$

or

$$S = -\beta \int d\,\Sigma \qquad \text{I.7)}$$

where $d\Sigma$ is the element of the area swept by the meson "string" in Minkowsky space-time. I.6) or I.7) offers a possible covariant generalisation of I.1) in which not only rigid rotations but all possible vibrational modes are now permitted. The physics underlying such generalisation is that in a situation for which $v_{\perp} = 0$ in the rest frame, one finds always $E = \beta L$ where L is now the length of the distored tube. It is well known that I.7) is the action of the "dual string" model[1]. It does not only allow rigid rotations which engender the leading Regge trajectory but it also gives rise to daughter trajectories when vibrational modes are taken into account. The quantization of the "dual string" action I.7) leads to inconsistencies except in some critical space-time dimension which for I.7) turns out to be 26. Even then however a state of negative mass squared (tachyon) is present. Though the critical dimension may be reduced and in some cases the tachyon removed by introducing internal degrees of freedom in the "primitive" string I.7) no physically satisfactory generalization has yet been found along these lines. It might be hoped that some of these difficulties are due to an oversimplified model which after all is but a phenomenological expression of an underlying more fundamental theory. Such may be the confinement schemes based on gauge field theories. We sahll come back to this point in Chapter VI but we consider the positive feature of the string model, that is its Regge behaviour, as an encouragement for investigating quark confinement models which tend, at least at high energies, to linear structures.

These lectures are presented as follows :

In Chapter II we review the properties of a superconductor, with particular emphasis on flux tube formation. The interest of such a study follows from the above discussion. In particular we shall exhibit the peculiar gauge properties which arise in a (relativistic) superconductor when flux tubes are present. This is not only of interest by itself but it paves the way for further study of topological features of gauge field theories.

In Chapter III these topological properties are investigated. One is then led in a natural way to understand how monopoles and instantons arise in non abelian gauge theories as regular solutions of the classical field equations.

Chapter IV describes how monopoles and flux tubes may be put together to yield confinement schemes both for mesons and baryons. Since the flux tubes are analogous to those encountered in superconductors, such constructions will be referred to as magnetic confinement. Such schemes rely deeply on the topological properties of gauge fields. They are however not fully justifiable theoretically, at least within our present knowledge, and they may suffer from serious defects when confronted with direct experimental evidence.

We therefore analyze in Chapter V an alternate mechanism of confinement which, if realized in nature, would be free of such ambiguities. This we call electric confinement as it may provide an essential element in the construction of quantum chromodynamics. Electric and magnetic confinement are quite similar in their physical aspects with electric and magnetic field interchanged and in particular may be often described in semi-classical terms; this analogy is however concealed in the mathematical framework of gauge field theory. In specific models the difficulty in interpreting electric confinement is traced back to the fact that the quantum action of a gauge field may not be applicable in the classical limit because of topological effects. Once this is cured, the similarity between electric and magnetic confinement appears at the semi-classical level.

Finally in Chapter VI, we emphasize the possible implications of confinement schemes for further development of particle physics.

The following presentation of the confinement problem is the result of a continued collaboration with P. Windey; I am greatly indebted to him for his help in preparing these notes. I am also gratefull to R. Brout and C. Truffin for discussions and criticisms.

II. SUPERCONDUCTIVITY

I. The nature of the superconducting state

Superconductivity may arise in metals at sufficiently low temperature as a consequence of electron-phonon interaction[2]. The attractive interaction, mediated by phonon exchange between pairs of electrons moving near the Fermi surface with opposite spin and momenta, leads to a binding of such pairs. These "Cooper pairs" experience a Bose-Einstein condensation, and as consequence there is a macroscopic number of Cooper pairs in the ground state of a superconductor. Thus the creation and the annihilation operators for a Cooper pair acquire classical significance and hence non vanishing expectation values. These operators may be defined as

$$b^{+}_{\vec{k}} = a^{+}_{\vec{k},\uparrow} \; a^{+}_{-\vec{k},\downarrow}$$

$$b_{\vec{k}} = a_{-\vec{k},\downarrow} \; a_{\vec{k},\uparrow} \qquad \text{II.1)}$$

where $a^+_{\vec{k},\uparrow}$ creates an electron of momentum $\vec{k}$ and spin $\uparrow$; we have

$$\langle b^+_{\vec{k}} \rangle = \rho_{\vec{k}}\, e^{i\theta} \qquad \text{II.2)}$$

The occurrence of the arbitrary phase θ in II.2) implies a degenerate vacuum with respect to θ ; this degeneracy of the vaccuum characterizes a dynamical spontaneous symmetry breaking of the quantum mechanical phase group U_1.

II.2) implies that an electron propagating above the Fermi surface can be converted into a hole-type particle. The negative energy of the hole with respect to the Fermi level implies a reajustment for the energy of the single particle excitations. By solving the Schwinger-Dyson equation for the electron self-energy one easily finds in Hartree Fock approximation that the bare energy $|\varepsilon_{\vec{k}}|$ of an electron (or hole), referred to the Fermi level, becomes $E_{\vec{k}} = (\varepsilon^2_{\vec{k}} + \Delta^2)^{1/2}$ where the "mass" Δ is the superconducting gap.

The qualitative picture that emerges is thus a Bose-Einstein condensate of Cooper pairs separated by a gap Δ from the individual excitations. The coherent motions of the Cooper pairs from the supercurrents and energy momentum transfer from the supercurrents to individual excitations is inhibited by the existence of the gap.

The above discussion does not make reference to electric charge but the fact that Cooper pairs do carry electricity has far reaching consequences because we may identify θ to a constant electromagnetic phase. The degeneracy of the vacuum with respect to a phase group, U_1, would normally imply the occurrence of a normal mode with vanishing frequency as its wave-vector $q \to 0$; this would be the "massless" Nambu-Goldstone boson[3] of the problem and it would describe long range fluctuations in space and time of θ . Note that such a mode would be in complete analogy with spin-wave modes in an isotropic ferromagnet which describes long range fluctations of the magnetization $\vec{R}$; $\vec{R}$ is indeed the order parameter characterizing the vacua degenerate under the rotational group SO_3. In the present situation however the Nambu-Goldstone mode will not show up in a direct way; this is because electromagnetism is built upon gauge invariance of the second kind and that local phase fluctuation may be absorbed in the electromagnetic field by a gauge transformation. Therefore the zero-mass bound state is transferred to the electromagnetic field which then acquires a new longitudinal degree of freedom.

The phenomenon just described is a typical example of gauge invariant mass generation for Yang-Mills fields in presence of dynamical symmetry breaking[4]. The dramatic effect resulting from the photon mass m_V in a superconductor is that magnetic disturbances die off at a critical length $\ell \simeq m_V^{-1}$ so that magnetic fields are expelled out of a superconductor except for a penetration depth ℓ, this is the well known Meissner effect. We now discuss this effect more quantitatively, with special emphasis on flux tube formation and flux tube quantization.

2. Flux tubes in superconductors

a. Ginzburg-Landau formulation[5]. One can describe the ground state of a superconductor by a complex order parameter $\phi(\vec{r})$

$$\phi = \int d^3k < a^+_{\vec{k},\uparrow} \; a^+_{\vec{k},\downarrow} > \qquad \text{II.3)}$$

We restrict ourselves to zero temperature and interpret $\phi(r)$ as a phenomenological scalar field describing the behavior of Cooper pairs. In presence of a static external magnetic field one may write the following effective Lagrangian

$$L = -\frac{1}{4} F_{k\ell}F_{k\ell} - |(\partial_k - ieA_k)\phi| \; |(\partial_k + ieA_k)\phi| - V(|\phi|) \;;$$

$$F_{k\ell} = \partial_k A_\ell - \partial_\ell A_k \;;$$

$$V = -\mu^2|\varphi|^2 + \lambda|\phi|^4 \;; \qquad \lambda > 0 \qquad \text{II.4)}$$

Here the potential $V(|\phi|)$ is chosen in such a way that its minimum occurs for $|\phi| = |\phi|_c \neq 0$ thereby expressing the spontaneous breaking of the U_1 - symmetry and the related vacuum degeneracy. The electromagnetic coupling to ϕ is determined by gauge invariance of the second kind; e must than be interpreted as the charge of a Cooper pair which is thus equal to twice the electronic charge.

It is well known that when $|\phi|_c \neq 0$ the photon field A_ℓ acquires a gauge invariant mass m_V which in the semi-classical (tree) approximation is [4) 6)]

$$m_V^2 = 2e^2|\phi|_c^2 \qquad \text{II.5)}$$

This result confirms that the effective phenomenological Lagrangian II.4) correctly describes, in a semi-classical approximation, the relevant features of the Cooper pairs condensate. Of course, in substituting II.4) to the correct fundamental Lagrangian we have lost all information about the particle-like excitations and in particular about their dynamical "mass" Δ . This fact should be kept in mind when Lagrangians involving scalar fields and generalizing II.4) are used to explore analogies in elementary particle physics.

b. Flux-tube solutions. Let us look for classical solutions of II.4) when $F_{k\ell}$ varies in space; In vector notation ($\vec{A} = A_x, A_y, A_z$; $\vec{\nabla} \times \vec{A} = \vec{B}$) we have

$$\Delta\vec{A} - \vec{\nabla}(\vec{\nabla}.\vec{A}) = -ie \left|\phi^* \overleftrightarrow{\nabla} \phi - (\vec{\nabla}\phi^*)\phi\right. + 2 e^2 |\phi|^2 A \qquad \text{II.6)}$$

$$(\vec{\nabla} + ie\,\vec{A}).(\vec{\nabla} + ie\vec{A})\phi = \frac{1}{2}\frac{\partial V}{\partial|\phi|}\,\phi \qquad \text{II.7)}$$

We shall show that II.6) and II.7) admit solutions of confined tubes of magnetic flux which either close on themselves or are of infinite length. These tubes have a core in which the modulus of the scalar field rises from zero along an axis s to approximately its bulk value $|\phi|_c$; the radius of this core is given by the inverse "mass" of the scalar field m_s^{-1} ($m_s = \left(\partial^2 V/\partial\phi.\partial\phi^*\right)_{\phi=\phi_c}$). Thus at the center of this core superconductivity is destroyed and m_s^{-1} measures the coherence length of the condensate. If $m_V^{-1} >> m_s^{-1}$ one has a typical flux tube as shown in Fig. 1. This is the case which characterizes type II superconductors: an external magnetic field exceeding some critical value will penetrate the superconductor through flux tubes of radius m_V^{-1} instead of being expelled along the surface by the Meissner effect. This phenomenon happens because the loss of condensation energy in the core of the flux tube will be overcompensated by the decrease in the kinetic energy required for expelling the magnetic field.

Consider first a tube of infinite length (Fig. 1). At large distance d from the axis of the tube the energy density of the electromagnetic disturbance must go to zero. Thus both the potential and the kinetic energy of the ϕ-field vanish. Asymptotically, the relevant solutions of II.7) must tend to

$$\frac{\partial V}{\partial|\phi|} = 0 \qquad \text{or} \qquad |\phi| = |\phi|_c \qquad \text{II.8)}$$

$$(\vec{\nabla} + ie\vec{A})\,\phi = 0 \qquad \text{II.9)}$$

From II.8) we deduce $\phi = |\phi|_c \exp\left(-i\chi(\vec{r})\right)$; inserting this relation into II.9) we obtain, up to terms smaller than 1/d

$$\vec{\nabla}\chi = e\,\vec{A} \qquad \text{II.10)}$$

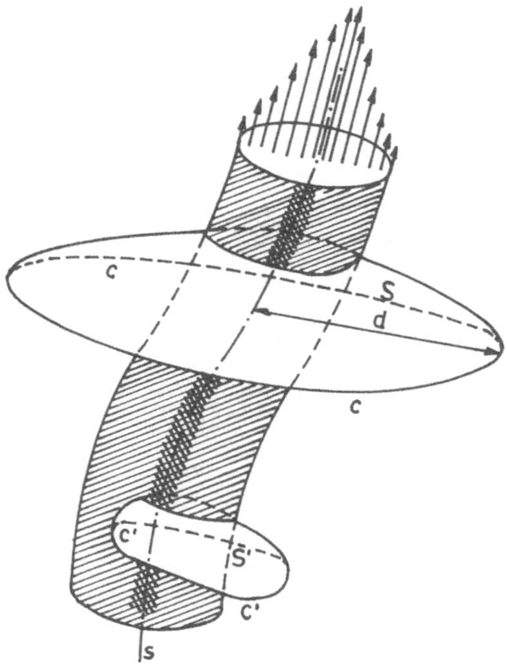

Figure 1. Flux tube ($m_S^{-1} << m_V^{-1}$)
.—.—.: s⁻ axis of the flux tube ($|\phi| = 0$)
: core where $|\phi|$ varies appreciably
→ : $\vec{B}$ - field
: region where $\vec{B} = \vec{\nabla} \times \vec{A}$ is confined

Integrating the circulation of $\vec{A}$ along a large curve enclosing the tube (Fig. 1) we get from Stokes theorem

$$\Phi = \int_S \vec{B}.\vec{dS} = \frac{1}{e} \oint_C \vec{\nabla}\chi.\vec{ds} = \frac{2\pi n}{e} \quad ; \quad n \text{ integer} \qquad \text{II.11)}$$

where we have used the single valuedness of the ϕ field. Thus the total flux Φ of the tube is quantized in units $2\pi/e$

The existence of an axis where $|\phi| = 0$ may now be understood. The phase $\chi(\vec{F})$ defines a mapping of any close curve in the superconductor onto a close curve in the group space of U_1. This group space is a circle and thus admits a discrete infinity Z of topologically distinct closed curves labelled by a winding number n . The curve C can be characterized by such an integer (for instance, for cylindrical flux tube, one may take $\chi(\vec{r}) = n\varphi$ where φ is the azimuthal angle); therefore if $n \neq 0$, C cannot be deformed into a point <u>inside</u> the superconductor ($|\phi| \neq 0$). The axis s is thus a locus of points where $|\phi| = 0$.

We how derive explicitely the flux tube solutions in the London limit $m_s/m_V \to \infty$; the effect of finite m_s will be briefly discussed afterwards. In this limit $|\phi| = |\phi|_c$ everywhere except along the axis s where $|\phi| = 0$. Taking into account II.5), the equation II.6) reduces to

$$\Delta\vec{A} - \vec{\nabla}(\vec{\nabla}.\vec{A}) = m_V^2 (\vec{A} - \frac{1}{e}\vec{\nabla}\chi) \qquad \text{II.12)}$$

and the conserved, gauge invariant supercurrent density is thus given by

$$\vec{S}_s = - m_V^2 (\vec{A} - \frac{1}{e}\vec{\nabla}\chi) \qquad \text{II.13)}$$

Strictly speaking, II.12) and II.13) are valid only outside the axis. However $\vec{\nabla}\chi$ has a 1/d singularity along the axis as seen from the previous discussion and this singularity drives the solutions of equation II.12). Therefore we shall keep II.12) and II.13) everywhere but we must then treat the derivatives of $\vec{\nabla}\chi$ along the axis in the distribution sense. Indeed we have from II.13)

$$\vec{\nabla} \times \vec{S}_s = -m_V^2\vec{B} + \frac{m_V^2}{e} \vec{\nabla}\times\vec{\nabla}\chi \qquad \text{II.14)}$$

and we cannot put the last term of II.14) equal to zero as this would violate Stokes theorem for the circulation of $\vec{\nabla}\chi$ along an arbitrary curve C' circling the axis[7] (fig. 1). In fact, we must have $(1/e) \oint_{C'} \vec{\nabla}\chi . \vec{ds} = (1/e) \oint_C \vec{\nabla}\chi . \vec{ds} = \Phi$ and, from Stokes theorem,

$$\int_{S'} (\vec{\nabla} \times \frac{\vec{\nabla}\chi}{e}) . \overrightarrow{dS'} = \Phi \quad . \qquad \text{II.15)}$$

Thus we pose[7]

$$\vec{\nabla} \times \frac{\vec{\nabla}\chi}{e} = -\vec{B}_s \qquad \text{II.16)}$$

where $\vec{B}_s$ represents the fictitious magnetic induction which would build a linear flux $\Phi = 2\pi n/e$ running along the axis of the tube in the opposite direction to the real flux. It is thus expressible as a two-dimensional δ-function in a plane perpendicular to the axis.

Explicitely

$$\vec{B}_s(\vec{r}) = \Phi \int_s d\vec{s}'\, \delta^3\, (\vec{r} - \vec{r}') \qquad \text{II.17)}$$

where the integral is taken along the axis of the tube (Fig. 2). In terms of $\vec{B}_s$ we may now write II.14) as

$$\boxed{\vec{\nabla} \times \vec{S}_s = - m_V^2\, (\vec{B} + \vec{B}_s)} \qquad \text{II.18)}$$

and thus

$$\boxed{\begin{array}{l} \Delta\, \vec{S}_s - m_V^2 S_s = m_V^2\, \vec{\nabla} \times \vec{B}_s \\ \vec{\nabla} \cdot \vec{S}_s = 0 \end{array}} \qquad \text{II.19)}$$

The "current" $\vec{\nabla} \times \vec{B}_s$ which drives the solutions of II.19) is the solenoïdal current which builds a fictitious linear solenoïd running along the s-axis and carrying the induction $\vec{B}_s$; from II.16) we see that $-\nabla\chi/e$ provides a vector potential for such a solenoïd. We shall from now on call such a linear fictitious solenoïd a <u>string</u>; we shall encounter such strings in different physical contexts in the sequel. The solution to II.19) are easily obtained once the axis s and the flux $\Phi = 2\pi n/e$ are given; these are

$$\vec{S}_s(\vec{r}) = \frac{m_V^2}{4\pi}\, \Phi \int_s d\vec{s}' \times \vec{\nabla} \left[\frac{-\exp(-m_V|\vec{r}-\vec{r}'|)}{|\vec{r}-\vec{r}'|} \right] \qquad \text{II.20)}$$

Indeed using $\vec{\nabla} \cdot \left[d\vec{s} \times \vec{\nabla}\right] = - d\vec{s}' \cdot (\vec{\nabla} \times \vec{\nabla})$ one varifies $\vec{\nabla} \cdot \vec{S}_s = 0$ and it follows from

$$\Delta\, \frac{\exp(-m_V r)}{r} = m_V^2\, \frac{\exp(-m_V r)}{r} - 4\pi\delta^3(\vec{r}) \qquad \text{II.21)}$$

that II.19) is verified with $\vec{B}_s$ given by II.17). To get $\vec{B}$ one takes the curl of II.20) and use

$$\vec{\nabla} \times d\vec{s} \times \vec{\nabla} = d\vec{s}'\Delta - (d\vec{s}' \cdot \vec{\nabla})\vec{\nabla} \qquad \text{II.22)}$$

The second term in II.22) may be written as $+(d\vec{s}' \cdot \vec{\nabla}')\vec{\nabla}$ and yields a term from the limits of integration which vanishes because the string is infinite, the first term yield, using II.21), II.18) and II.17).

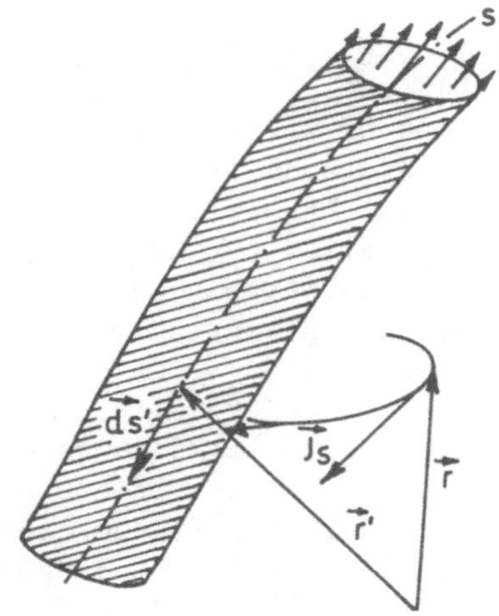

Figure 2. Supercurrent of a flux tube in the London limit.

$$\vec{B} = -\frac{m_V^2}{4\pi}\,\Phi \int_S d\vec{s}\,'\,\frac{\exp(-m_V|\vec{r}-\vec{r}\,'|)}{|\vec{r}-\vec{r}\,'|} \qquad \text{II.23)}$$

The solutions II.20) and II.23) are valid for superposition of flux tubes obtained by superposing strings. In particular they are valid for closed flux tubes with closed strings. The logarithmic divergence in II.23) is due to the limit $m_S \to \infty$. Finite m_S makes the scalar field $|\phi|$ smoothly increase from zero to $|\phi|_c$ and this provides logarithmic cut-off $\ln m_S/m_V$ which makes $\vec{B}$ finite along the axis. Hence flux tubes in a superconductor represent confined configurations with finite energy density.

3. The relativistic superconductor.

A relativistic model which shares the magnetic properties of a superconductor can easily be constructed by taking the Lorentz covariant generalization of Ginzburg-Landau Lagrangian density II.4). Thus we take

$$L = -\frac{1}{4}F_{\mu\nu}F^{\mu\nu} + [(\partial_\mu + ieA_\mu)\phi^*]\cdot [(\partial^\mu - ieA^\mu)\phi\,] - V(|\phi|) \qquad \text{II.24)}$$

$$F_{\mu\nu} = \partial_\mu A_\nu - \partial_\nu A_\mu \qquad \text{II.25)}$$

We use the notations $g^{00} = +1$, $g^{kk} = -1$; $A^1 = A_z$, ...; $F^{12} = -B_z$, ...; $F^{01} = -E_x$, ...

The Lagrangian II.24) again leads to spontaneous symmetry breaking of a U_1-symmetry and the photon acquires the mass m_V given by II.5) in the semi-classical approximation. Clearly all the flux tube solutions of II.4) studied in the preceeding section are also static solutions of the relativistic version II.24). This fact was pointed out by Nielsen and Olesen [8] who also suggested a possible application to hadron physics : the action of the "dual string" I.7) was interpreted as the phenomenological version of the corresponding section of a flux tube with constant energy β per unit length.

Of course the Lagrangian II.24) does not describe any properties of the individual excitations of a superconductor and this fact was already pointed out for its non relativistic version II.4). Nevertheless, to simplify the language we shall call a system described by II.24), II.25) a relativistic superconductor.

The main difference between a non relativistic and a relativistic superconductor is the fact that in the former the mean charge density ρ of the condensate is non vanishing while it remains in general zero in the latter. This is because ρ is proportional to $|\phi|^2$ in the non relativistic case but for a relativistic superconductor both charges exist and the charge operator is now $\rho = -ie[\phi^*\partial_0\phi - (\partial_0\phi^*)\phi]$ in absence of scalar potential. This difference clearly does not alter the static solutions but, as pointed out by Nambu [9], it deeply affects the motion of the flux tubes. In particular any solution obtained by boosting the static solutions is now permitted.

We shall close the chapter on superconductivity by discussing some particular features of gauge transformations in superconductors. As these features appear both for relativistic and non relativistic superconductors, we shall phrase the problem in its most general form, namely for the relativistic case.

4. Singular gauges and gauge strings.

In the preceeding section, flux tube quantization and the existence of an axis $|\phi| = 0$ were derived from the multivaluedness of the phase χ of the scalar field. In absence of flux tubes one can eliminate a time independent phase χ by a gauge transformation to a new potential $\vec{W}$

$$\vec{W} = \vec{A} - \frac{1}{e}\vec{\nabla}\chi \qquad \text{II.26)}$$

The gauge transformation II.26) is just the static Higgs-Kibble transformation to the unitary gauge [6]. It is therefore of interest to inquire whether a unitary gauge is available in presence of flux tubes.

The difficulty lays in the fact that when χ is multivalued, $\vec{\nabla}\chi$ is singular along the axis of the flux tube and hence $\vec{W}$ is singular on the axis. This singularity is the same as the string singularity we encountered in the study of the London limit. However, it has here a different significance because $\vec{W}$ as defined from II.26) in singular even if m_s is finite and hence does not imply a singularity in the gauge invariant supercurrent. To keep this distinction in mind we shall refer to string singularities in II.26) as to gauge strings. We now formulate superconductivity in the unitary gauge which exhibits gauge strings when flux tubes are present; this formulation will also be applicable to any singular gauge obtainable from the unitary gauge by a regular gauge transformation.

Let us first consider static solutions. From II.26) and II.15) we have

$$\vec{B} = \vec{\nabla} \times \vec{W} - \vec{B}_s \qquad \text{II.27)}$$

Hence, the magnetic part of the free electromagnetic density $\mathcal{L}_{em}$ in II.24), II.25) or equivalently in II.4) is not invariant under singular gauge transformations and takes the form

$$\mathcal{L}_{em} = -1/2\, [\vec{\nabla} \times \vec{W} - \vec{B}_s] \cdot [\vec{\nabla} \times \vec{W} - \vec{B}_s] \qquad \text{II.28)}$$

However we shall show that the Lagrangian modified according to II.28) as well as its covariant generalization does provide a correct description of the superconductor in the singular gauge when gauge string parameters are arbitrarily varied.

We introduce gauge string singularities in the relativistic Lagrangian II.24) through closed strings or strings of infinite length which are the covariant generalization of the B_s-strings appearing in II.28). To this aim we define the string tensor

$$M_{\mu\nu}(x) = -\sum_i g_i \iint d\sigma_i d\tau_i [z_\mu, z_\nu]_i\, \delta^4(x-z) \qquad \text{II.29)}$$

$$[z_\mu, z_\nu]_i = \frac{\partial z_\mu}{\partial \tau_i}\frac{\partial z_\nu}{\partial \sigma_i} - \frac{\partial z_\mu}{\partial \sigma_i}\frac{\partial z_\nu}{\partial \tau_i} \qquad \text{II.30)}$$

The sum is over all strings carrying quantized fluxes $\Phi_i = g_i$; z^μ are the Minkowskian coordinates of the world-sheet of the strings parametrized by σ_i and τ_i. In the static cases $d\tau = dz^0$ and the contribution to $M_{\mu\nu}$ of a single string ($\Phi = g$) is

$$M^{ok} = -\Phi \int dz^k\, \delta^3(x-z) \qquad \text{II.31)}$$

Comparing II.31) to II.17) we see that $\vec{B}_s = -\vec{M}$, $(\vec{M} = M^{01}, M^{02}, M^{03})$. Introducing the dual

$$M^{+\mu\nu} = 1/2\, \varepsilon^{\mu\nu\sigma\lambda} M_{\sigma\lambda} \quad ; \quad \varepsilon^{0123} = +1 \qquad \text{II.32)}$$

we express the electromagnetic field tensor $F_{\mu\nu}$ in the singular gauges in the form :

$$F_{\mu\nu} = A_{\mu\nu} + M^{+}_{\mu\nu} \qquad \text{II.33)}$$

$$A_{\mu\nu} = \partial_\mu A_\nu - \partial_\nu A_\mu \qquad \text{II.34)}$$

If we now identify A^1, A^2, A^3 to W_x, W_y, W_z we see that II.33) and II. 34) reduce to II.27) for static configurations. We therefore generalize the free action II.28) to include singular gauges in the form

$$S_{em} = -\frac{1}{4}\int (A_{\mu\nu} + M^{+}_{\mu\nu})(A^{\mu\nu} + M^{+\mu\nu})d^4x \qquad \text{II.35)}$$

This again reduces to II.28) for static configurations.

We now verify that the generalization of the free electromagnetic action according to II.35) does yield the correct equation of motions for the relativistic superconductor.

The variation of A^ν in II.35) together with any interaction Lagrangian with minimal electromagnetic coupling yields the first Maxwell equation

$$\partial_\mu F^{\mu\nu} = j^\nu \qquad \text{II.36)}$$

where j^ν is the electric current. This result may be interpreted in terms of A_μ by rewriting it as

$$\partial_\mu A^{\mu\nu} = j^\nu - \partial_\mu M^{+\mu\nu} \qquad \text{II.37)}$$

In the static case $-\partial_\mu M^{+\mu\nu}$ reduces to $\vec{\nabla} \times \vec{B}_s$ so that $\partial_\mu M^{+\mu\nu}$ is the covariant solenoïdal current required to build the string and II.37) is thus the sum of the vector potentials due to the real currents j^ν and to the solenoïdal current. However the solenoïd is fictitious and therefore its field must be substracted from $A^{\mu\nu}$ to get the correct field tensor $F^{\mu\nu}$. This is achieved by the term $M^{+\mu\nu}$ in II.33) which cancels the field of the fictitious solenoïd : in the static case $M^{+\mu\nu}$ indeed reduces to $-\vec{B}_s$.

We now verify the second Maxwell equation. From II.33) we obtain by taking the dual $F^{+}_{\mu\nu}$ and using the relation

$$(O^+)^+ = -O \qquad \text{II.38)}$$

valid for any antisymmetric tensor O.

$$\partial_\mu F^{+\mu\nu} = \partial_\mu A^{+\mu\nu} - \partial_\mu M^{\mu\nu} = -\partial_\mu M^{\mu\nu} \qquad \text{II.39)}$$

Then from II.29) and II.30

$$\partial_\mu M^{\mu\nu}(x) = + \sum_i g_i \iint d\sigma_i d\tau_i [z^\mu, z^\nu]_i \frac{\partial}{\partial z^\mu} \delta^4 (x-z)$$

$$= + \sum_i g_i \iint d\sigma_i d\tau_i [\delta^4(x-z), z^\nu]_i$$

$$= - \sum_i g_i \oint \delta^4(x-z) dz^\nu_{(i)} \qquad \text{II.40)}$$

where by Stokes theorem the $dz^\nu_{(i)}$ integral is carried over the rim of the ith sheet. As this rim vanishes for closed strings and is rejected at infinity for other ones we indeed recover the second Maxwell equation

$$\partial_\mu F^{+\mu\nu} = 0 \qquad \text{II.41)}$$

We must show that the gauge string does not introduce new degrees of freedom. Variation of the string coordinate z^μ yields

$$- \frac{1}{2} \int F_{\mu\nu} \delta M^{+\mu\nu} d^4x = - \frac{1}{2} \int F_{\mu\nu} \delta M^{\mu\nu} d^4x$$

$$= + \frac{1}{2} \sum_i g_i \iiint F^+_{\mu\nu}(x) \delta z^\lambda \frac{\partial}{\partial z^\lambda} \{\delta^4(x-z)[z^\mu, z^\nu]_i\} \, d\sigma_i d\tau_i d^4x$$

$$= + \frac{1}{2} \sum_i g_i \iint d\sigma_i d\tau_i \; \{\delta z^\lambda \frac{\partial F^+_{\mu\nu}}{\partial z^\lambda} [z^\mu, z^\nu]_i + 2 F^+_{\mu\nu} [\delta z^\mu, z^\nu]_i\}$$

$$= + \frac{1}{2} \sum_i g_i \iint d\sigma_i d\tau_i \; \{\delta z^\lambda \frac{\partial F^+_{\mu\nu}}{\partial z^\lambda} [z^\mu, z^\nu]_i - 2\delta z^\mu [F^+_{\mu\nu}, z^\nu]_i\}$$

$$+ \frac{1}{2} \sum_i g_i \iint d\sigma_i d\tau_i [2F^+_{\mu\nu} \delta z^\mu, z^\nu]_i \qquad \text{II.42)}$$

Applying Stokes theorem, the last term vanishes and one is left with

$$- \frac{1}{2} \int F^{+}_{\mu\nu} \delta M^{\mu\nu} d^4x$$

$$= \frac{1}{2} \sum_i g_i \iint d\sigma_i d\tau_i \; \{\delta z^\lambda \frac{\partial F^{+}_{\mu\nu}}{\partial z^\lambda} [z^\mu, z^\nu]_i$$

$$- 2\delta z^\mu \frac{\partial F^{+}_{\mu\nu}}{\partial z^\rho} [z^\rho, z^\nu]_i\}$$

$$= \frac{1}{2} \sum_i g_i \iint d\sigma_i d\tau_i \delta z^\lambda \left[\frac{\partial F^{+}_{\mu\nu}}{\partial z^\lambda} + \frac{\partial F^{+}_{\nu\lambda}}{\partial z^\mu} + \frac{\partial F^{+}_{\lambda\mu}}{\partial z^\nu} \right] .$$

$$[z^\mu, z^\nu]_i \qquad \text{II.43)}$$

The string equations of motion are thus

$$\left[\frac{\partial F^{+}_{\mu\nu}}{\partial z^\lambda} + \frac{\partial F^{+}_{\nu\lambda}}{\partial z^\mu} + \frac{\partial F^{+}_{\lambda\mu}}{\partial z^\nu} \right] \cdot [z^\mu, z^\nu]_i = 0$$

or equivalently

$$\frac{\partial F^{\mu\nu}}{\partial z^\mu} [z_\nu, z^\rho]^{+}_i = j^\nu [z_\nu, z^\rho]^{+}_i = 0 \qquad \text{II.44)}$$

The bracket $[z^\mu, z^\nu]_i$ represents the oriented projection of the surface element of the world-sheet on the μ-ν plane; therefore II.44) expresses that the current must be either zero or orthogonal to the dual surface; equivalently j^ν must be zero or tangential to the world-sheet of the string. In other words no electric charge may cross the string. Such a constraint is always verified in a superconductor if the gauge string is along the axis of flux tube and the removal of the phase χ in the unitary gauge precisely implies these gauge strings. We therefore conclude that II.35) is, at the classical level, the correct characterization of flux tubes in the singular gauges.

The validity of II.35) in quantum mechanics follows from the quantization condition of flux tubes II.11). The circling of a charge e of the condensate around the gauge string results in a phase increase $e\oint A_\mu dx^\mu$ and the contribution of the string potential to this phase is, from II.16) and II.11), equal to $e\phi = e(2\pi n/e) = 2\pi n$; hence the gauge string is unobservable quantum mechanically as well as classically.

The possibility of using singular gauges to describe flux tubes will be usefull in particular in the context of monopoles. We emphasize again that the gauge strings introduced here do not rely on the London limit and on the strings used in that context. What happens when $m_S \to \infty$ is that the gauge string in the unitary gauge then coïncides with the "physical" gauge invariant string which serves as a driving term to the supercurrent in II.19). To make this clear let us consider a static flux tube in the London limit, described in the unitary gauge. Now II.13) reads

$$\vec{S}_s = - m_V^2 \vec{W} \quad ; \quad \vec{\nabla} \cdot \vec{W} = 0 \qquad \text{II.45)}$$

where $\vec{W}$ is the vector potential in the unitary gauge. The equation II.12) would now lead to the incorrect result

$$\Delta\vec{W} - m_V^2 \vec{W} = 0 \qquad \text{II.46)}$$

which is inconsistent with II.19) and II.45). It is easily seen that the driving term missing in II.46) is provided in singular gauge by the string terms which yield the solenoïdal current $- \partial_\mu M^{+\mu\nu}$ in II.37). The correct equation for $\vec{W}$ is then

$$\Delta\vec{W} - m_V^2\vec{W} = - \vec{\nabla} \times \vec{B}_s \qquad \text{II.47)}$$

This now agrees with II.19).

III. MONOPOLES AND INSTANTONS

1. Introduction.

Flux tubes are localised classical solutions with finite energy density constructed from gauge field theory coupled to matter fields. These solutions however have either infinite energy (if the flux tube is infinite) or are unstable against decay into the superconductor vacuum (if the flux tube is closed). In the present section we shall examine solutions of gauge field equations or of their analytic continuation to euclidian space which have the property of being topologically stable. In fact we shall restrict ourselves to these solutions of non abelian field theories which are the natural generalization of the Dirac monopole construct in the abelian case. We thus first examine how monopoles can be introduced in electromagnetism.

2. Dirac monopoles [10)]

Monopoles may be introduced in Maxwell equation by treating symmetrically $F_{\mu\nu}$ and $F^{+}_{\mu\nu}$; namely one keeps the equation II.36) $\partial_\mu F^{\mu\nu} = j^\nu$ and one postulates

$$\partial_\mu F^{+\mu\nu} = k^\nu \qquad \text{III.1)}$$

where k^ν represents a magnetic current. It is of interest to inquire if III.1) is consistent with the description of electrodynamics in terms of a vector potential A_μ; A_μ plays indeed an essential role in ensuring gauge invariance of the second kind through minimal coupling to the matter fields.

An answer to this question is suggested by our previous study of superconductivity. Indeed we saw that a gauge string $M_{\mu\nu}$ generated by a solenoïdal current $\partial_\mu M^{+\mu\nu}$ adds a pure gauge term to $F_{\mu\nu}$; therefore an open string should contribute to $F_{\mu\nu}$ only a term depending on the extremities of the string. This is clear in the static case from Fig. 3 where two closed strings having a common part are depicted; the linearity of Maxwell's equations imply that the field at a point P outside the solenoïds is the same for the two open strings II and II' and therefore depends only on their end points A and A'. By symmetry the field issued from each extremity must then be radial and hence A and A' carry respectively magnetic charges +g and -g. These considerations lead to the following tentative expression for $F_{\mu\nu}$ in presence of monopoles.

$$F_{\mu\nu} = A_{\mu\nu} + M^{+}_{\mu\nu} \qquad \text{III.2)}$$

$$A_{\mu\nu} = \partial_\mu A_\nu - \partial_\nu A_\mu \qquad \text{III.3)}$$

$$M_{\mu\nu} = - \sum_i g_i \int_{-\infty}^{+\infty} d\tau_i \int_{-\infty}^{\sigma[z^\mu_{(i)}]} d\sigma_i [z_\mu , z_\nu] \, \delta^4(x-z) \qquad \text{III.4)}$$

where σ_i, τ_i describe the motion of an arbitrary string extending from space-like infinity to the position $z_\mu(i)$ of the monopole;

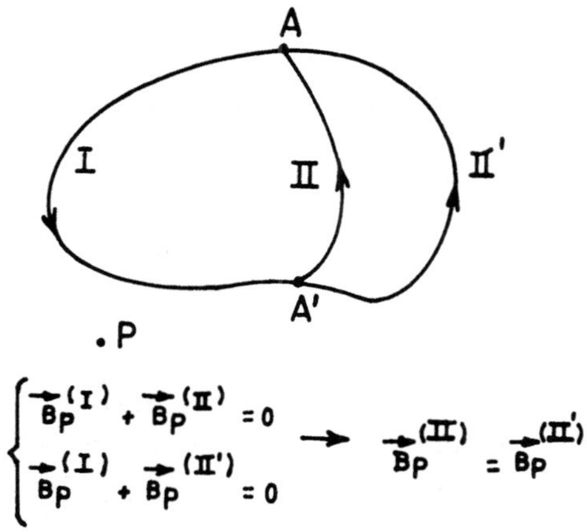

Figure 3. Independence of the $\vec{B}$-field from the shape ot an open string.

g_i are the monopole charges. The strings III.4) were introduced by Dirac and we shall refer to them as to Dirac strings.

Repeating the argument following II.37) we see that the first Maxwell equation implies that A_μ correctly describes the strings potentials and that the fictitious flux carried along the strings is cancelled in $F_{\mu\nu}$ by the $M^+_{\mu\nu}$ term. We now verify that the end points of the string yield the modified second Maxwell equation III.2). We have, following the steps leading to II.40)

$$\partial_\mu F^{+\mu\nu} = -\sum_i g_i \oint \delta^4(x-z_{(i)})dz^\nu_{(i)} \qquad \text{III.5)}$$

The contour integral is however no more zero and if the $dz^\nu_{(i)}$ element is taken along the positive time direction one gets indeed

$$\partial_\mu F^{+\mu\nu} = k^\nu \qquad \text{III.6)}$$

with

$$k^\nu(x) = \sum_i g_i \oint \delta^4(x-z_{(i)})dz^\nu_{(i)} \qquad \text{III.7)}$$

Thus III.2) and III.3) are consistant with a classical description of monopoles. As in the case of gauge strings the extension to quantum mechanics relies upon the unobservability of the string and the same unobservability condition now implies the Dirac quantization condition

$$eg_i = 2\pi n \quad ; \quad n \text{ integer} \qquad \text{III.8)}$$

where e is the smallest charge present in the system. Relation III.8) follows from the requirement that the contribution to $e\oint A_\mu dx^\mu$ due to the fictitious flux $M^+_{\mu\nu}$ should be unobservable; this phase increase eg_i can be isolated by taking an infinitesimal contour circling around the string.

The validity of the field equations II.36 and III.1) is however not yet sufficient to justify the use of Dirac strings in quantum theory. This justification would follow from a Lagrangian formulation if the action would not be affected by variation of the string coordinates, except for end point effects. However it is clear that in such a formulation we would recover equation II.44) and thus the strings are constrained to avoid all electric charges. Though this restriction is not serious classically it is not clear that such a constraint is quantum mechanically consistent. One may attempt to overcome the difficulty by a description in which the Dirac string is interpreted in terms of topological properties of gauge fields [11]; it has indeed been argued from such a topological interpretation that the string coordinates should not be varied in a Lagrangian formalism [12]. We shall not enter the debate as in any case one has not yet succeeded in reconciling Dirac monopoles with second quantization in the context of an action principle. Therefore the theoretical validity of the Dirac monopole concept is at best questionable.

However we shall see that the Dirac construction appears very usefull for the study of consistant non abelian generalizations. We shall also notice in next chapters that there are cases in which Dirac monopoles are free of constraint conditions.

To close this section we illustrate the Dirac construction by evaluating the potential due to an isolated monopole of charge g located at the origin and carrying a string along the negative z axis. In the gauge $\vec{\nabla}\cdot\vec{A} = 0$, one has

$$\Delta\vec{A} = -\vec{\nabla}\times\vec{B}_s \qquad \text{III.9)}$$

$$\vec{B}_s = g\theta(-z)\delta(x)\delta(y)\,\vec{I}_z \qquad \text{III.10)}$$

$$\vec{B} = \vec{\nabla}\times\vec{A} - \vec{B}_s \qquad \text{III.11)}$$

The solution can be derived from the method used in chapter II. It follows however straightforwardly from the symmetry around the z axis (Fig. 4) that

$$2\pi\sqrt{x^2+y^2}\;A(r,\theta) = Br^2 2\pi(1-\cos\theta)$$
$$= \frac{g}{2}(1-\cos\theta),$$

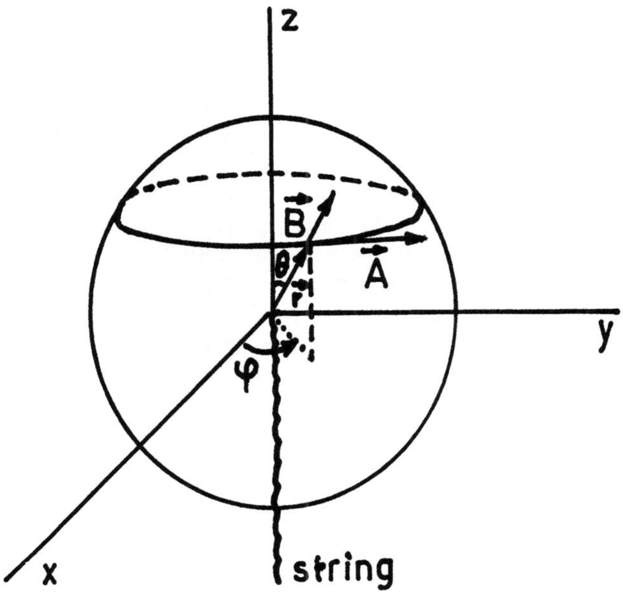

Figure 4. Potential and field from a monopole.

or

$$\vec{A} = \frac{g}{4\pi}(1 - \cos\theta)\vec{\nabla}\phi \qquad \text{III.12)}$$

$$\vec{B} = \frac{g}{4\pi}\frac{\vec{r}}{r^3} \qquad \text{III.13)}$$

One easily verifies that III.12) and III.13) do satisfy the equations III.9), III.10) and III.11). In particular the singularity along the negative z axis arises from the fact that along this axis $\vec{A} = \vec{A}_S$ with $\vec{A}_S = (g/2\pi)\vec{\nabla}\phi$ and that as in II.15) one has $\vec{\nabla} \times \vec{A}_S = (g/2\pi)\vec{\nabla} \times \vec{\nabla}\phi = g\delta(x)\delta(y)$.

3. Singular monopoles in non-abelian theories.

When the Dirac construction is carried over for non-abelian gauge fields coupled to matter fields, new features emerge which drastically change the whole monopole picture. These are :

a) For some values of the quantized magnetic charges, which depend solely on the Lie algebra of the group, the Dirac strings can be gauged out by using the enlarged set of gauge transformations available in non-abelian gauge theories. Thus for these values of the magnetic charges, the Dirac string is a fiction and no constraint condition is left. We shall call such monopoles singular 't Hooft-Polyakov monopoles for reasons which will be clear in the next section.

b) Within the classes of groups which admit the same Lie algebra but differ globally (such as SU_2 and SO_3 or SU_3 and SU_3/Z_3), one may in some cases also introduce genuine Dirac monopoles whose strings cannot be gauged out; we shall still refer to these as to Dirac monopoles. In fact Dirac monopoles can be introduced only in those groups in which there exist closed curves in group space which cannot be continuously shrunk to a point, as we shall see this happens in SO_3 and SU_3/Z_3 but not in SU_2 and in SU_3. Clearly the distinction between two groups having the same Lie algebra but differing globally is imposed by the representation of the matter fields which do couple to the gauge fields.

Consider Yang-Mills fields A^a_μ coupled minimally to matter fields belonging to unitary representations of a simple compact Lie group G. Let T^a be generators of a faithfull representation of G; we select a maximal abelian subalgebra H^i $(i=1,2,\ldots l)$ which generates a maximal abelian subgroup t of G and choose a basis where all the H^i are diagonal. From the matrix $A_\mu = A^a_\mu T^a$ one forms the field tensor matrix

$$A_{\mu\nu} = \partial_\mu A_\nu - \partial_\nu A_\mu - ie[A_\mu A_\nu] \qquad \text{III.14)}$$

where e defines the coupling constant to matter. The field equations are

$$\partial_\mu A^{\mu\nu} - ie[A_\mu A^{\mu\nu}] = j^\nu \qquad \text{III.15)}$$

where $j^\nu = j^{a\nu}T^a$ and $j^{a\nu}$ is the matter current.
Under a gauge transformation $\Omega(x) = \exp(-i\varepsilon^a(x)T^a)$, A_μ transforms to A'_μ according to

$$A'_\mu = \Omega A_\mu \Omega^{-1} + \frac{i}{e}\,\Omega\partial_\mu\Omega^{-1} \qquad \text{III.16)}$$

while

$$A'_{\mu\nu} = \Omega A_{\mu\nu}\Omega^{-1} \qquad \text{III.17)}$$

We now construct a solution to III.15) which in the limit $j^\nu=0$ is valid everywhere except at the origin; in a particular gauge called the abelian gauge [13] this solution will represent a Dirac monopole.

The abelian gauge will be defined by the requirement that $A'_{\mu\nu}$ be diagonal everywhere so that $A'_{\mu\nu} = A'^{i}_{\mu\nu} H^i$; this is possible for any solution of III.15) by a suitable choice of $\Omega(x)$. However such an $\Omega(x)$ may lead to singular A'_μ because of the second term in III.16) and we shall take advantage of this fact. Selecting a solution for which A'_μ is also diagonal and using the further gauge condition $\partial_\mu A'^\mu = 0$ we would get from III.15) in the limit $j_\nu = 0$ and is absence of gauge sigularities

$$\Box A'_\mu = 0 \qquad \text{III.18)}$$

To obtain a non trivial solution which vanishes at infinity we have to admit gauge singularities and we tentatively drive III.18) by a Dirac string term. To this aim we introduce the string matrix $M_{\mu\nu}$ which in the abelian gauge reduces to $M^i_{\mu\nu}$ which in the abelian gauge reduces to $M^i_{\mu\nu}H^i$. Here $M^i_{\mu\nu}$ is the Dirac string associated to a monopole g^i. In terms of the monopole matrix

$$g = g^i H^i \qquad \text{III.19)}$$

the solutions of III.18) with a string at rest along the negative z axis are then given by III.12); namely

$$\vec{A}' = \frac{g}{4\pi}(1-\cos\theta)\vec{\nabla}\phi \qquad \text{III.20)}$$

where $\vec{A}'$ and g are diagonal in the abelian gauge.

The validity of III.20) hinges upon the possibility of gauging away the string so that our identification of the singular terms in III.19) with a Dirac string term was indeed correct and the monopole is really a solution of the non linear equation III. 16) in the limit $j^\nu = 0$. This will indeed be the case, except at the origin, for some well defined values of g^i. Before turning to a general characterization of these values of g^i we shall work out a particular example in the group SO_3 generated by "isospin" matrices T^1, T^2, T^3. We define the abelian gauge by the choice $T^3 = H$; III.20) then reduces to

$$\vec{A}'^3 = \frac{g}{4\pi}(1 - \cos\theta)\vec{\nabla}\phi$$
$$\vec{A}'^1 = \vec{A}'^2 = 0 \qquad \text{III.21)}$$

where $g = g^3$ is now a number. We now transform $\vec{A}'^1$, $\vec{A}'^2$, $\vec{A}'^3$ to a new gauge $\vec{A}^1$, $\vec{A}^2$, $\vec{A}^3$ by choosing the gauge matrix

$$\Omega = e^{-i\phi T^3}\, e^{-i\theta T^2}\, e^{i\phi T^3} \qquad \text{III.22)}$$

From III.16)

$$\vec{A}^a T^a = \Omega^{-1}\vec{A}'^a T^a \Omega + \frac{i}{e}\Omega^{-1}\vec{\nabla}\Omega \qquad \text{III.23)}$$

The first term in III.23) rotates in isospace the isovector $\vec{A}'^a$ at each space time point by an angle θ around T^2 and subsequently by an angle ϕ around T^3 where θ and ϕ are the polar angle and the asimuthal angle in space. The nature of such an isospace rotation can be visualized by taking a constant vector in isospace directed along the 3 axis with unit length in space. If one superposes two cartesian coordinate systems (x,y,z) and (1,2,3) in space and isospace then Ω takes the isovector $\phi' = (0,0,1)$ to a new unit vector ϕ with components $\phi^a = r^a/r$ in the coordinate system just described; thus $r^1 = x$, $r^2 = y$ and $r^3 = z$ (Fig. 5). Such a situation where space and internal space are mixed in a classical solution of field theory is not new : in the cylindrical flux tube with n=1 of chapter II, one also can identify in a regular gauge the asimuthal angle in space with the phase angle in the group space of U_1. The present case is a three-dimensional generalization and one may therefore call the gauge generated from the abelian gauge by Ω a spherical gauge. One sees clearly from the picture that $\partial_\mu\Omega$ must be singular along the negative z axis. In fact the second term in III.23) introduces a term proportional to $\vec{\nabla}\phi$ and the coefficient of $\vec{\nabla}\phi$ varies from 0 to - 2/e when θ varies from 0 to π. Hence this term cancels the string singularity if

$$eg = 4\pi \qquad \text{III.24)}$$

that is for twice the minimal Dirac value. Explicitely one finds in the spherical gauge [14]

$$A^{a\mu} = \frac{g}{4\pi}\,\varepsilon^{\mu ab}\,\frac{r^b}{r^2} \qquad \text{III.25)}$$

where μ stands for a space index x,y or z and ε is the usual completely antisymmetric Levy-Civita tensor. Clearly the only singularity in III.25) is at the origin and therefore III.25) describes correctly a solution to III.15) with $j^\nu=0$, everywhere except at the origin; this may also be checked by direct substitution. We may therefore call the spherical gauge a regular gauge and the abelian gauge a singular gauge in analogy with the superconducting case when flux tubes are present. We now turn to a general characterization of the values of g in compact simple Lie groups for which a regular gauge exists.

We first have to show that the string in III.20) is really an unobservable gauge string. Circling again the string along an infinitesimal closed curve C' by a charge e or by a charge contai-

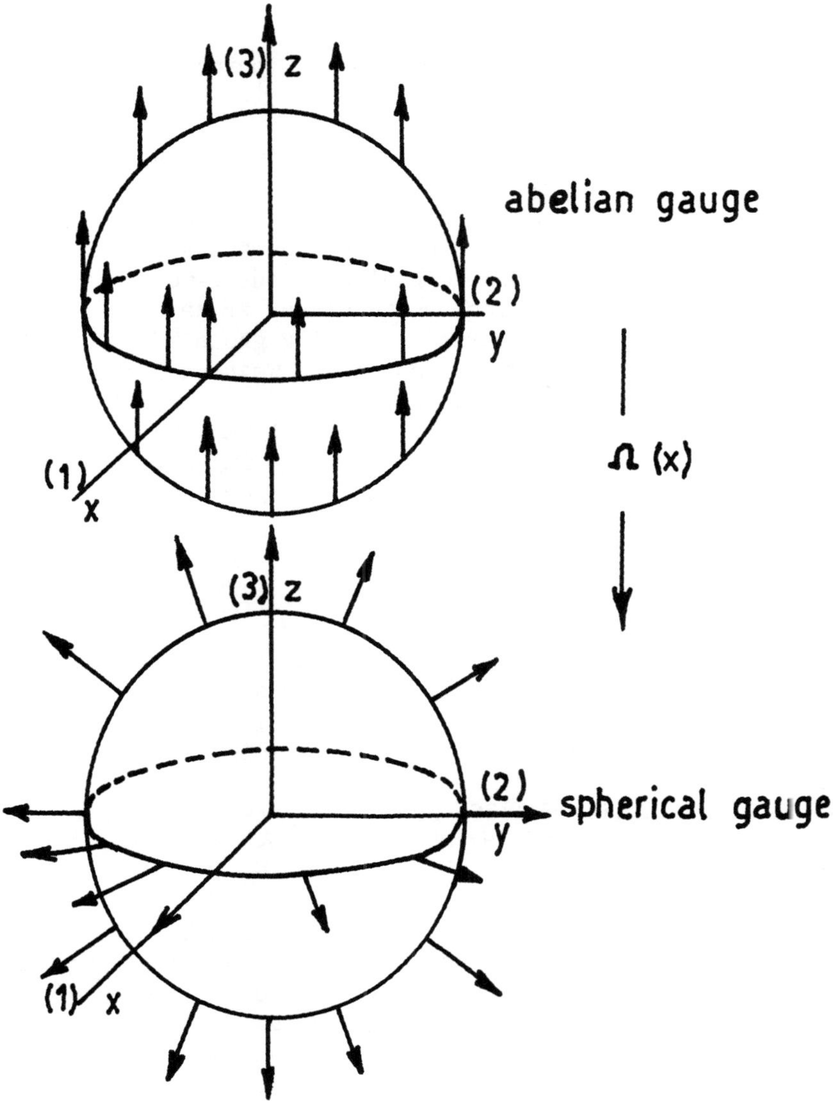

Figure 5. Gauge transformation III.22 for a constant unit sovector directed along the third isospin axis. →: ϕ' and ϕ.

ned in the matter Lagrangian we obtain the conditions which generalize III.8) namely.

$$[eg^i H^i] = 2\pi n \qquad \text{(n integer)} \qquad \text{III.26)}$$

where the symbol [ ] means that the equality holds for all eigenvalues of the matrix.

The change of phase along the infinitesimal curve defines, as in the superconductor a mapping of the curve C' onto a curve C" in group space of G. If C" can be continuously shrunk to zero, there exists a gauge transformation in G which suppresses the phase dependance of the matter fields along C' and thus removes the string. Now any closed curve in G which can be shrunk to zero corresponds to a closed curve in the universal convering group G^c of G and vice versa because G^c is simply connected. We recall that G and G^c have the same Lie algebra but that $G = G^c/E$ where E is a subgroup of the discrete center of G^c. For instance the group G^c corresponding to SO_3 is SO_2 and one has $SO_3 = SU_2/Z_2$. Thus if $(T^c)^a$ generates a faithfull representation of G^c with $(H^c)^i$ corresponding to H^i, the condition

$$[eg^i (H^c)^i] = 2\pi n \qquad \text{(n integer)} \qquad \text{III.27)}$$

implies that the string can be gauged out by a transformation in G [15]. Note that the condition III.27) expresses that the transformation generated in G^c by the parameters eg^i is the identity and hence this condition does not depend on which faithfull representation $(T^c)^a$ is used.
Moreover III.27) implies III.26) because any representation of G is also a representation of G^c (while the converse is not true if E is non trivial).

Thus III.27) expresses that III.15) does admit monopole solutions when $j^r=0$; however these solutions are generally singular at the origin; III.27) implies that the string can be gauged out but not its extremity ! Thus we refer to these solutions as to singular 't Hooft-Polyakov monopoles to distinguish them from the regular solutions we shall describe in next section. As III.27) depends only on G^c for all groups G having the same Lie algebra, the charges of the singular monopoles depend only on the Lie algebra of the group G. This establishes statement a). For instance, both for SU_2 and SO_3 we may take for $(H^c)^i$ the matrix $t_z = \begin{vmatrix} 1/2 & 0 \\ 0 & -1/2 \end{vmatrix}$ and one finds

$$g = \frac{4\pi n}{e} \qquad \text{(n integer)} \qquad \text{III.28)}$$

that is twice the Dirac value. The above example III.24) illustrates the case n = 1 in III.28).

Solution of III.26) which are not solution of III.27) may only lead to monopoles if one drives III.15) with genuine Dirac strings terms. If such strings are introduced by the matrix $M_{\mu r}$ these solutions define non-abelian Dirac monopoles. Thus genuine Dirac monopoles can be introduced only if $G = G^c/E$ when E is non trivial. E is isomorphic to the first homotopy group $\pi_1(G)$, that is the group of topological multiplication of closed curves [16]; its non trivial elements are in one to one correspondance with classes of topologically distinct closed curves which cannot be deformed one into another and which cannot shrunk into a point. This explains statement b). For instance in SU_2 no Dirac monopoles can be introduced; in $SO_3 = SU_2/Z_2$ one may in addition to III.28) introduce Dirac monopoles. Choosing $t_z = \begin{vmatrix} 1 & 0 & 0 \\ 0 & 0 & 0 \\ 0 & 0 & -1 \end{vmatrix}$ we see that Dirac monopoles in SO_3 are given by

$$g = \frac{2\pi}{e}(2n+1) \qquad \text{(n integer)} \qquad \text{III.29)}$$

they fill the "holes" between the singular 't Hooft-Polyakov monopoles exhibited in III.28).

The results obtained in this section depend on the choice of a particular base H^i and hence of a particular abelian subgroup t. However, there is an infinity of subgroups t' conjugated to t in G and we could apparently have chosen any of them. Such an arbitrariness would imply that the monopole solution constructed from t could be continuously deformed to other solutions so that this arbitrariness is deeply related to the problem of topological stability. However we shall postpone a discussion on this point until next section when monopole solutions with finite energy will be presented. The monopoles introduced in this section have indeed infinite classical energy due to the singularity at the origin; any stability consideration at this level is thus somewhat academic. Moreover as we shall see, finite j^ν may result in spontaneous breaking of symmetry and hence in restricting the choice of the subgroup t.

4. Regular monopoles and instantons.

Singular 't Hooft-Polyakov monopoles are solutions of the field equations III.15) with $j^\nu=0$ everywhere except at the origin. It is therefore of interest to inquire whether the existence of a finite matter current could remove the singularity at the classical level and therefore produce finite energy solution of the classical field equations. That this indeed happens was shown by 't Hooft and Polyakov [17].

We discuss the problem by considering first the particular case of $G \equiv SO_3$ when the matter field is an isovector; the Lagrangian density is taken to be

$$L = \mathrm{tr}\,[-\frac{1}{4} A_{\mu\nu}A^{\mu\nu} + \frac{1}{2} D_\mu\phi D^\mu\phi] - V(\phi) \qquad \text{III.30)}$$

with

$$D_\mu\phi = \partial_\mu\phi - ie[A_\mu\phi]$$

$$(\phi = \phi^a T^a) \qquad \text{III.31)}$$

and

$$V(\phi) = -\frac{\mu^2}{2}\,\mathrm{tr}\,\phi^2 + \frac{\lambda}{8}[\mathrm{tr}\phi^2]^2 \;;\; \lambda > 0 \qquad \text{III.32)}$$

As in the cas of the superconductor $V(\phi)$ leads to spontaneous breakdown of the symmetry group of the Lagragian, in the present case SO_3. The unbroken group is a group U_1 and we select a vacuum by singling out a particular subgroup generated by T^3=H. The vector mesons A^1_μ and A^2_μ acquire a degenerate mass $m^2_W = e^2 \mathrm{tr}\phi^2$ while A^3_μ remains massless.

A finite energy static solution of III.30) must again satisfy asymptotically the equations

$$\frac{\partial V}{\partial(\mathrm{tr}\phi^2)} = 0 \qquad \text{III.33)}$$

$$D_\mu\phi = 0 \qquad \text{III.34)}$$

which correspond to II.8) and II.9) in the flux tube problem. From III.34) it follows that $\phi^3(x)$ tends to a constant F at infinity where $\vec{A}^1(x)$ and $\vec{A}^2(x)$ tend to zero and III.33) implies that F must be chosen at the minimum of V. We therefore search for a solution which matches the monopole solutions III.21) for distance larger than m_W^{-1}. For the case eg = 4π the problem is most conveniently solved in the spherical gauge by introducing form factors f(r) and a(r) defined by

$$A^{a\mu} = \frac{g}{4\pi}\,\varepsilon^{\mu ab}\,\frac{r^b}{r^2}\,a(r) \qquad \text{III.35)}$$

$$\phi^a = \frac{r^a}{r}\,F\,f(r) \qquad \text{III.36)}$$

One then impose $a(\infty) = f(\infty) = 1$ and the finiteness of the total energy imposes $f(0) = a(0) = 0$. Explicit solutions for a(r) and f(r) have been found in the limit $m_S \to 0$; these are [18]

$$a(r) = 1 - \frac{m_W r}{\sinh m_W r}$$

$$f(r) = \coth m_W r - \frac{1}{m_W r} \qquad \text{III.37)}$$

In the abelian gauge, the form factors III.37) introduce transverse components $\vec{A}^1$, $\vec{A}^2$ which fall off exponentially for distance $r \gg m_W^{-1}$; there is still a string which is determined only by the gauge transformation from the spherical to the abelian gauge. Thus one may call in general the solutions III.35) and III.36) a magnetic monopole of charge $g = 4\pi/e$, but now the energy M of the static monopole is finite and explicit evaluation yields for the monopole mass

$$M = \frac{m_W}{e^2} \zeta \qquad \text{III.38)}$$

where ζ is a number of order 1 whose exact value depends as the ratio m_S/m_W. Thus at least in this particular case ($G \equiv SO_3$ and $g = 4\pi/e$), one can explicitely show that a finite matter current j^a may smear out the singularity of the singular 't Hooft-Polyakov monopole and thus leads to what one may call a <u>regular monopole</u>. In general we shall call a finite energy solution of the field equation III.15) a regular monopole if there exists an abelian gauge in which the monopole exhibits a string. Of course it would be helpfull to have at our disposal other explicit solutions for regular monopoles than the one corresponding to $G \equiv SO_3$, n=1. In fact for SO_3, n > 1 it has been proven that no spherically symmetric regular monopole can be constructed [19]; this result suggests that regular monopoles with n > 1, if they exist, are essentially formed out of interacting "basic" n=1 (or n=-1) spherically symmetric monopoles. Candidates for such basic monopoles have been classified for all compact Lie groups and there is reason to believe that regular solutions for these basic cases do exist [20].

We now discuss the topological stability of regular monopoles. In the SO_3 case with eg = 4π the symmetry was broken spontaneously by the matter current from SO_3 to U_1. It is this symmetry breaking which is responsable for the stability of the monopole. If we select a vacuum by choosing a given U_1 subgroup of SO_3, we know that the massless Yang Mills field belongs to this U_1; thus to construct the monopole we are forced to identify the abelian gauge subgroup t with U_1. As the subgroup t is now locked in U_1 the possible regular monopoles are characterized by the set of integers III.28). Therefore the magnetic charge $g = 4\pi/ne$ cannot be changed by continuous motion of matter and is absolutely conserved. This ensures in particular that the regular monopole III.35) cannot disappear into vacuum. One may repeat this argument whenever the unbroken subgroup of G is a maximal abelian subgroup : all magnetic charges defined by III.27) are strictly conserved. Such a conservation law which is not a priori a consequence of Noether's theorem is called a topological conservation law and the related charge a topological charge. Note that the same result would hold for Dirac monopoles if the latter could be treated in a consistent way.

When the unbroken subgroup is not a maximal abelian subgroup, the analysis is more complex and the magnetic charge needs not coincide with the topological charge [15]. In particular if the symmetry is not broken there is no topological charge available for regular "monopoles", even if they could be constructed, because the subgroup t is entirely "unlocked" and because there is no non-trivial topological invariant corresponding to these would-be monopoles [21]. It is worth mentioning here that Dirac monopoles in the unbroken case would fall into topologically distinc classes [11]. These correspond to the classes of non trivial topologically distinct curves in $G = G^c/E$ and these classes are in one to one correspondance with the non trivial elements of E. For instance in SU_n/Z_n these are n-1 classes of topologically distinct Dirac monopoles, the unit element of Z_n corresponding to regular (or singular) 't Hooft-Polyakov monopoles. The situation for Dirac monopoles when the symmetry is not broken is then very different for simple compact Lie group and for the group U_1 where Dirac monopoles with different n are always topologically distinct. This difference is imputable to the fact that $U_1 = T/Z$ where T is the non-compact translation group while the covering of a simple compact Lie group is always compact and has therefore always a discrete center of finite order. Whether such a drastic difference between the behaviour of Dirac monopoles in abelian and non abelian gauge group has any relevance for stability consideration depends in the firs place on the validity of the Dirac construction. At present this is not established in the abelian case, as we already mentioned; the situation in the non-abelian case is even worse [22].

To summarize, topologically stable regular monopoles may exist depending on the nature of the unbroken subgroup of a group G. These objects are classically well defined and a consistant procedure for studying the quantum corrections to these classical solution has been developed (see 21 and references there in). The stability properties of the monopoles leads to a topological conservation law for the magnetic charges. Dirac monopoles in non-abelian gauge group have interesting topological properties, especially when the group is not broken, however there is at present no sound theoretical foundations for such a construction.

The above study of regular monopoles has been carried in the framework of Minkowski space, that is in 3 + 1 dimensions. However the static monopoles may also be viewed as object defined in an Euclidian 3-dimensional space. If this space is considered as an analytic continuation from a Minkowskian 2 + 1 space which serves as underlying space for the gauge fields, regular monopoles acquire a different physical interpretation. They now represent solution of finite (analytically continued) action S instead of solution of finite energy [23]. Such solutions are called instantons and play a crucial role in statistical mechanical analysis for restoring symmetries [24] and also in quantum mechanical problems where they describe in some cases tunneling processes between degenerate states [25]. Such instantons may also be relevant for the problem of electric confinement [24] as we shall see later. It is therefore of interest to know that instantons in four dimensions and related objects called merons [26] have been found as solutions of non-abelian pure Yang-Mills theories [27]. Like the "monopole" instantons in three dimensions these four dimensional instantons carry topological charges which characterize different sectors in which the quantum mechanical functional integrals of exp iS are performed.

IV. MAGNETIC CONFINEMENT

1. The abelian model [9] [28]

The Dirac quantization condition is $eg = 2\pi n$ where the magnetic charge g is defined to be the magnetic flux Φ emitted by the Dirac monopole. If e is the charge of the particules forming the condensate of a superconductor this conditions is the same as the condition II.11) for the existence of quantized magnetic flux tube. This coïncidence is of course no accident and arises, as was already pointed out, from the fact that both flux tubes and Dirac monopoles are characterized by the same topological group, namely the first homotopy group $\pi_1(U_1)$ of the group U_1 which is isomorphic to the additive group of integers Z. It thus follows that a Dirac monopole plunged into a superconductor will radiate a total amount of flux which will be channelled into a flux tube by the Meissner effect. In a finite superconductor the axis of the flux tube would join the monopole to the surface along the shortest straight line possible; in an infinite superconductor, the introduction of a Dirac monopole would require infinite energy (part from the point singularity) as the flux tube will have infinite length and as the energy per unit length is a finite constant. This infinite energy can be absorbed by an anti-monopole ($n' = -n$); thus except for the end-point effects, monopole-antimonopole pairs in a superconductor represent the simplest realization of a confinement scheme which shares the qualitative aspect of a dual meson string. The monopoles play the role of quarks which cannot exist in isolation and the gauge field in an abelian "gluon" field. Of course the string has now a transverse section m_V^{-1} and a core of radius m_S^{-1}. For $n = \pm 1$, the equilibrium configuration of the tube axis for a fixed monopole-antimonopole pair would be a straight line; for $|n| = p > 1$ one expects from energetic consideration a splitting of the tube into p "elementary" tubes joining at the monopole ends. Thus a reasonable model for a "meson" should involve only "quarks" with magnetic charge $g = \pm 2\pi/e$.

To have a more quantitative picture one may introduce Dirac strings into the Lagrangian II.24) of the relativistic superconductor as well as monopole mass terms. A convenient description is obtained by using the unitary gauge for the tube as discussed in section (II.4). In this case indeed the gauge string coincidences with a Dirac string joining the two monopoles. It follows from the analysis in (II.4) that the Dirac constraint is automatically satisfied because the supercurrent never crosses the axis. One can also verify (see section V) that varying the extremity of the strings results in the correct relativistic equation of motion for the monopole. Thus, at least, in a first quantized theory, Dirac monopoles in a superconductor can be correctly described by an action principle.

We now give the explicit classical static solution for a confined monopole-antimonopole pair in a superconductor in the London limit. For sake of generality we keep a general shape for the axis of the tube joining the monopoles. From the above discussion the equation II.47) together with $\vec{\nabla} \cdot \vec{W} = 0$ for the potential $\vec{W}$ is valid in the unitary gauge provided the string terms in the right hand side extends only from the antimonopole a to the monopole b. The solution for $\vec{J}_S$ follows from II.45) and II.20) namely

$$\vec{J}_S(\vec{r}) = \frac{m_V^2}{4\pi} g \int_a^b d\vec{s}' \times \vec{\nabla} \left[\frac{-\exp(-m_V|\vec{r} - \vec{r}'|)}{|\vec{r} - \vec{r}'|} \right] \qquad \text{IV.1)}$$

As previously we evaluate $\vec{B}(\vec{r})$ from II.18) but now the second term in II.22) does contribute ; one thus obtains instead of II.23)

$$\vec{B} = \frac{g}{4\pi} \{ - m_V^2 \int_a^b d\vec{s}' \frac{\exp(-m_V|\vec{r} - \vec{r}'|)}{|\vec{r} - \vec{r}'|}$$

$$+ \vec{\nabla} \left[- \frac{\exp(-m_V|\vec{r} - \vec{r}_b|)}{|\vec{r} - \vec{r}_b|} \right]$$

$$- \vec{\nabla} \left[- \frac{\exp(-m_V|\vec{r} - \vec{r}_a|)}{|\vec{r} - \vec{r}_a|} \right] \qquad \text{IV.2)}$$

The resulting confined flux tube is pictured in Fig. 6; around each pole the flux emerges or recedes spherically up to a radial distance m_V^{-1} whereupon it becomes canalized into a flux tube of radius m_V^{-1} which flows between the poles.

As in absence of monopoles, the logarithmic divergence in IV.2) along the tube axis is removed when m_S is finite. Whether the infinite energies due to the monopole singularities could be removed by a consistent quantum mechanical procedure is not known; nor does one know if the theory admits a consistent second quantized version for "quarks". Whatever the issue on these points will turn out to be, it is in any case impossible to construct a "baryon" out of three basic "quarks" of the abelian model.

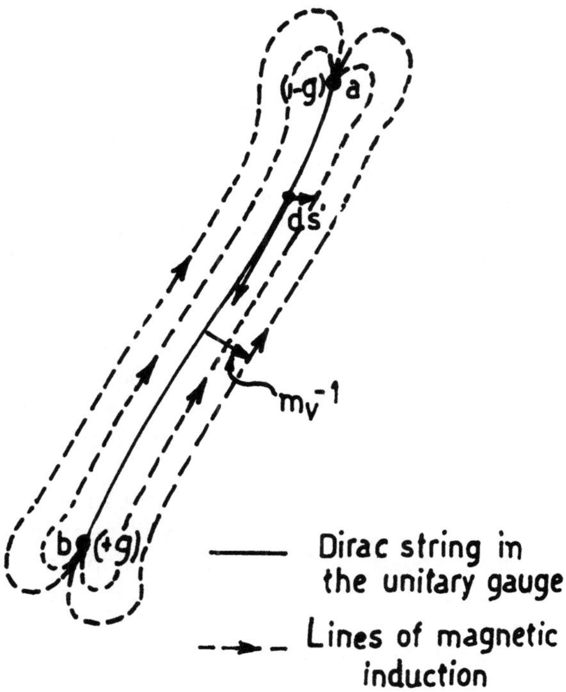

Fig. 6 : Confined Flux tube in a superconductor generated by a Dirac monopole-antimonopole pair (London limit)

2. Non-abelian models

A possible way to remove the monopole singularities of the abelian model in non abelian gauge field theories would lay in the use of regular monopoles as sources of magnetic flux tubes. To build such a regular confined solution for a "meson" let us consider the specific example when the gauge group is SU_2 or SO_3 [29]. Breaking the symmetry down to U_1 as in Chapter III it is possible to construct regular monopole-antimonopole solutions by first constructing a singular solution in the abelian gauge. This solution has a Dirac string joining the poles. The removal of string can then be performed by a gauge transformation $\Omega = \exp(i\phi T^3)\exp(-i\delta T^2)\exp(-i\phi T^3)$; δ is the angle from which the monopoles are seen. Introducing form factors the monopoles acquire finite energy and their location may be characterized by zero's of the scalar field. This finite energy solution has a field distribution of a dipole and hence is not confined. To achieve confinement it is therefore necessary to break the left over U_1 symmetry and this can be done for instance by using two or more scalar isovector fields. The result looks, in the abelian gauge, similar to the one described in the abelian model (Fig. 6) except that the classical solution has now finite energy and that the monopoles carry two units of flux. Such classical meson solutions would be unstable for two reasons. First the flux tube could split into two elementary flux tubes but this needs not be a dramatic effect. Second the whole structure is topologically unstable and may decay into the vacuum; this is obvious from the fact that monopole-antimonopole may annihilate but in fact the flux tube itself need not be stable as will be explained below. How much rigidity may be restored by quantum effects for tubes in motion is not known; but even if these effects would yield resonable meson-like structures, the construction of a baryon using only regular monopoles is certainly not a straightforward task.

There exists however a way to introduce baryon within the context of non abelian magnetic confinement which takes full advantage of the topology of group space. This was proposed by Mandelstam [30]; unfortunately his method requires the use of Dirac monopoles, at least at the first step of the construction. The crucial point is that if the group is completely broken, the abelian subgroup is again entirely "unlocked" because there is no long range residual gauge field and the monopole has no more the asymptotic behaviour III.12). Rather, when one switches on the mass of the gauge fields laying in the chosen subgroup t where a "bare" (Dirac or regular) monopole has been constructed, the flux is canalized into tubes in the same way as in the example described above. At the same time the solution can be moved continuously to another solution laying in a subgroup t'; the only topological characterization left of the monopole, as well as of the flux tube carried with it, is the class of the curve C" in group space into which an infinitesimal curve in space surrounding the axis of the flux tube in space is mapped. Flux tubes corresponding to topologically distinct curves C" cannot be transformed in a continuous way one into another. From the discussion at the end of section III it follows that for any compact simple Lie group $G = G^c/E$ one can construct exactly n-1 classes of topologically stable flux tubes where n is the finite order of E. Each such flux tube can be terminated by a single class of Dirac monopoles characterized by the same topological winding number. When flux tube meet they combine according to the group multiplication in E and flux tube corresponding to the unit element may be continuously deformed by a physical matter motion into the vacuum, for instance by formation of a regular monopole.

We illustrate the above consideration by choosing the gluon gauge group to be SU_3/Z_3 which is entirely broken. There are two distinct Dirac monopoles classes D_1 and D_2 and two corresponding topologically distinct flux tubes F_1 and F_2; they correspond to the elements $\exp(i2\pi/3)$ and $\exp(i4\pi/3)$ of Z_3. We assume that Dirac monopoles carry spin 1/2 as well as flavour indices. The monopoles are quarks and one may tentatively identify D_1 to a quark and D_2 to an antiquark. Mesons are then flux tubes joining monopoles D_1 and D_2 and baryons are built out of three D_1-quarks; the fluxes emitted by these three quarks join into a regular monopole. We may construct a classical solution by using Dirac monopoles and fluxes laying in the same abelian gauge $t \equiv (3,8)$ and characterized by the lowest magnetic charges $g = g^i H^i$ available in each class along the charge axis $T^3 + Y/2$. We get from III.26) and III.27)

$$g_{D_1} = 2\pi/e \quad ; \quad g_{D_2} = -2\pi/e \qquad \text{IV.3)}$$

and the flux tube structure of such a solution in the abelian gauge is shown in Fig. 7. The genuine Dirac strings are taken along the tube axis as in the unitary gauge of the abelian model; they are labelled type I or type 2 according to the class of the monopole on which they terminate. Type zero string is a pure gauge string which can be removed by going to the spherical gauge; this string ends up in Fig. 7 at the center of a regular monopole. The solutions pictures in Fig. 7 give only a qualitative idea of the hadron in a semi-classical limit because there are other configurations available which may be obtained from the one considered by continuous deformation. If fact were the group SU_3/Z_3 broken locally but globally unbroken the solution of Fig. 7 would be degenerate with respect to the choice of t. To get a correct description one should then know how to superpose quantum mechanically such solutions;

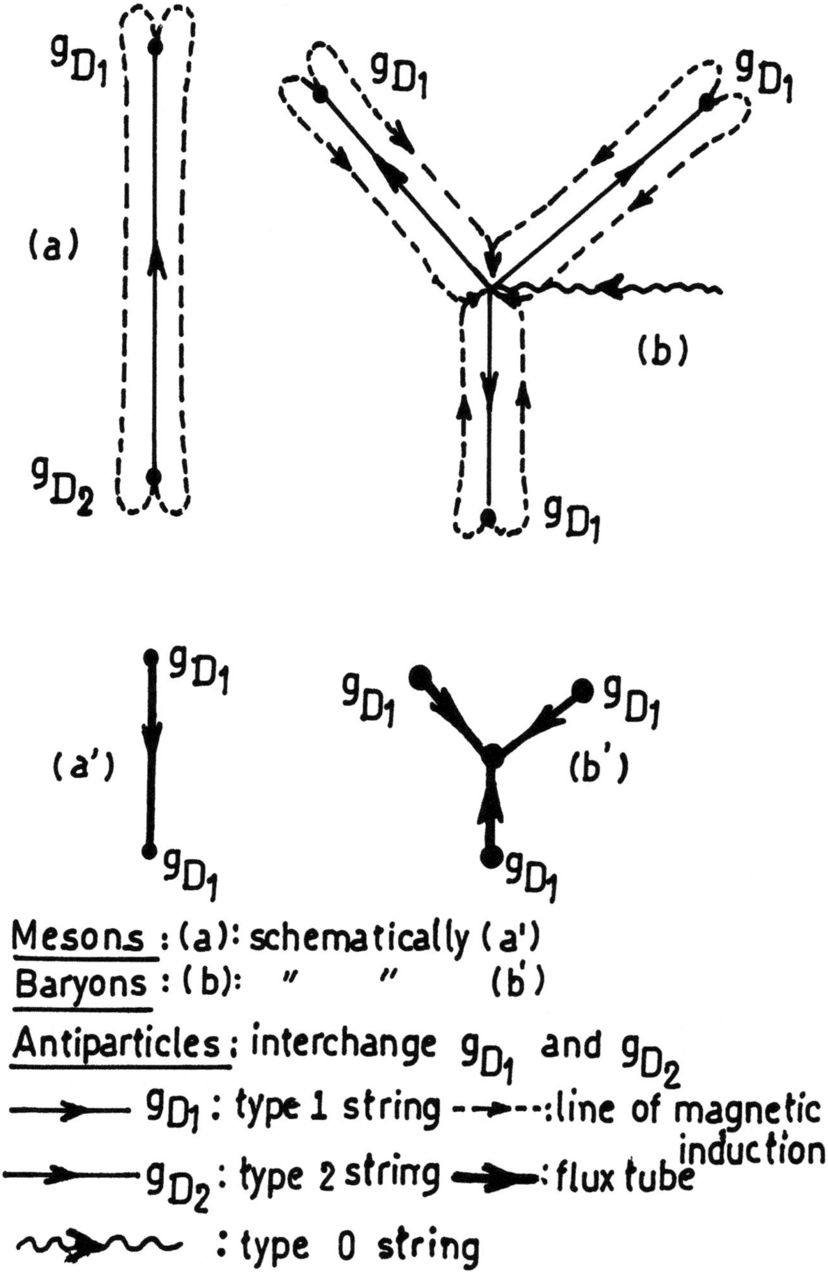

Fig. 7 : Approximate flux tube configurations for Mesons and baryons for the gluon gauge group SU_3/Z_3.

clearly however a better understanding of the Dirac monopole entering this scheme is required before further developments are envisaged. An interesting feature of the model is that the baryon wave function is antisymmetric in the interchange of two identical monopoles so that the quarks have the required symmetrical character in flavour space. Thus despite the fact that the gluon group SU_3/Z_3 is very different from the conventional colour group SU_3 used in quantum chromodynamics, it does provide the correct statistical properties of the quarks.

The shortcomings of non-abelian magnetic confinement are clear:

a) the recourse to Dirac monopoles casts strong doubts on the whole approach because of theoretical ambiguities. However one should keep in mind that Dirac strings in SU_3/Z_3 may be gauged away if SU_3/Z_3 would be imbedded in a larger group, say SU_4;

b) there is at present no hint in magnetic confinement models that the monopole-quarks exhibit quasi free behaviour at high momentum transfer. Wether this is indeed a critical test for rejecting these models depends on a better understanding of the quantum theory of magnetic charge.

To conclude this section we wish to point out that the magnetic models described here are based on a generalization of the Ginzburg-Landau Lagrangian. Because the latter Lagrangian in the theory of superconductivity does not take into account the motion of individual electrons, it may well be that we miss some essential physical ingredient. In particular the fact that the gap vanishes inside flux tube in superconductors is interesting in view of the fact that the relativistic analog of the gap is a fermionic mass. Therefore models in which tubes play the role of bags inside which quarks would have small masses may be of interest.

V. ELECTRIC CONFINEMENT

1. Introduction

The apparent or real shortcomings of magnetic confinement models motivate a search for alternative confinement schemes. The possibility arises that quark fields minimally coupled to the non-abelian gauge fields of a "colour" group, could bind quarks in colour singlet states. If such a mechanism is operative, we shall refer to it as to electric confinement. That such an electric confinement is possible is suggested from renormalization group arguments [31]. Consider the Lagrangian of quantum chromodynamics (Q.C.D.)

$$L = i \sum_j (\psi_j \gamma^\mu D_\mu \psi_j + m_j \psi_j \psi_j) - \frac{1}{4} \mathrm{tr} A_{\mu r} A^{\mu r} \qquad V.1)$$

$$D_\mu \psi_j = \partial_\mu \psi_j + e A^a_\mu T^a \psi_j \qquad V.2)$$

Here j is a flavour index, ψ_j is a vector belonging to the 3-representation of a colour group SU_3 generated by the matrix T^a; $A_{\mu r} = A^a_{\mu r} T^a$ is the corresponding SU_3 colour field tensor and e the bare colour charge. The effective charge felt by two quarks separated by a distance r is affected by the dielectric properties of the vacuum. While for r^{-1} greater than any quark mass quark -antiquark pairs tend to screen the charge, the Yang-Mills fields give rise to antiscreening; therefore as long as there are not to many quarks present, the net result is that the effective charge increases with distance at least at small distances. In terms of the renormalized charge $e(\lambda)$ where λ is a momentum of order r^{-1}, this means that $e(\lambda) \to 0$ when $\lambda \to \infty$ or that the theory is asymptotically free. If such a behaviour can be extrapolated to the low momentum region without leading to a broken colour symmetry it would imply an effective dielectric constant of the vacuum $\varepsilon(\lambda)$ which would tend to zero at large distances, that is $\varepsilon(\lambda) \to 0$ as $\lambda \to 0$. This suggests that the vacuum would exhibit an "electric" Meissner effect [32][33] because the displacement field $\vec{D}$ due to an "electric" colour source cannot radiate into the medium despite the fact that $\vec{\nabla} \cdot \vec{D} = 0$ outside the sources. This situation is, except for non-abelian effects, completely analogous to the magnetic confinement of monopoles in a relativistic superconductor with electric and magnetic fluxes interchanged. There the medium may be characterized by a vanishing magnetic permeability $\mu(\lambda) \simeq 0$ for $\lambda < m_V$; the zero of $\mu(\lambda)$ for $\lambda \to 0$ expresses that $\vec{B}$ cannot radiate into the medium and magnetic confinement then follows from $\vec{\nabla} \cdot \vec{B} = 0$ outside the magnetic poles.

Thus one expects that quarks are bound together by "electric" colour flux lines with a width which in the limit of vanishing quark mass m_j would now be determined by some typical hadron mass as a result of "dimensional transmutation". Due to the non-abelian character of the unbroken colour group one cannot, even in the semi classical limit, restrict oneself to abelian configurations : one must take into account in a symmetrical way all configurations where the quarks undergo transitions from one colour to another while emitting a coloured Yang-Mills field. As a consequence of the unbroken colour symmetry, electric confinement would then imply that the only objects which are confined and hence have finite energy are colour singlets and not only colour neutral objects. Note that at high momentum transfer, because of asymptotic freedom, the usual quasi-free quark picture inside hadrons is recovered.

The above argument suggests that electric confinement is possible but does not constitute a proof of such confinement. A more practical approach to the problem has been suggested by Wilson [34] and further developed by Kogut and Susskind [35] and others : fields are defined on a lattice of space-time points. The advantage of the procedure is that one can perform a perturbation expansion in $1/e(\lambda)$, and the unperturbed state in the limit $e(\lambda)\to\infty$ exhibits confinement as a consequence of the canonical communication relations [33]; this is true whether the gauge field is abelian or not. The lattice method does not by itself prove the existence of confinement but it provides a method for constructing low energy excitations if confinement is indeed correct.

Clearly electric confinement must be found in particular properties of the vacuum. Guided by the examples of magnetic confinement and by the analogy between electric and magnetic flux tubes, we shall construct simple models where electric confinement arises at a semi-classical level [36]. This will permit us to characterize electric confinement in general. We shall then understand how localized solutions of field theory, such as monopoles and instantons may lead to formation of electric flux tubes.

2. The superconductor revisited.

a) The axiovector potential.

The abelian relativistic superconductor exhibits magnetic confinement of Dirac monopoles. Maxwell equations are symmetrical with respect to electric charges e and magnetic charges g under the interchange of $F_{\mu\nu}$ and $F^{+}_{\mu\nu}$; one thus expects that a description of such a superconductor is available in which monopoles are minimally coupled to $F^{+}_{\mu\nu}$ through an axiovector potential C_μ. In the conventional description, confinement would be described by the gauge invariant mass m_V acquired by the vector field A_μ. What would be the characterization of magnetic confinement in terms of C_μ ? The answer to that question is of interest because in terms of C_μ a superconductor may be thought of as a model for electric confinement with g playing the role of an electric charge. Thus if a C_μ-field is indeed available we shall get both a criterium for electric confinement and a model which may suggest mechanisms for such a confinement in a more realistic situation.

We now show how to formulate the action for a relativistic abelian superconductor in term of the C_μ-field; our study is restricted to the London limit and will be valid in first quantized theory. The latter restriction is not new as the Dirac monopoles previously used were considered in that context. To introduce the C_μ-field we describe the charges e of the condensate by Dirac strings carrying fictitious electric fluxes; thus we pose

$$F^{+}_{\mu\nu} = C_{\mu\nu} - E^{+}_{\mu\nu} \qquad V.1)$$

$$C_{\mu\nu} = \partial_\mu C_\nu - \partial_\nu C_\mu \qquad V.2)$$

$$E_{\mu\nu}(x) = -\sum_i e \iint d\sigma_i d\tau_i [y_\mu, y_\nu]_i \delta^4(x-y) \qquad V.3)$$

The string term in V.3) is the same as the magnetic string $M_{\mu\nu}$ (see equation III.4) with g_i and z replaced by e and y. We then have

$$\partial_\mu E^{\mu\nu} = - j^\nu \qquad V.4)$$

$$j^\nu(x) = \sum_i e \int \delta^4(x-y_{(i)}) dy^\nu_{(i)} \qquad V.5)$$

and hence as V.1) implies $- F_{\mu\nu} = C^{+}_{\mu\nu} + E_{\mu\nu}$, the first Maxwell equation follows :

$$\partial_\mu F^{\mu\nu} = j^\nu \qquad V.6)$$

Introducing the magnetic current as before (equation III.7))

$$k^\nu(x) = \sum_i g_i \int \delta^4(x-z_{(i)}) dz^\nu_{(i)} \qquad V.7)$$

we may try to describe a superconductor in presence of monopoles by the action

$$S^q = S^0 + S^{int} + S^g \qquad V.8)$$

Here

$$S^0 = -\frac{1}{4}\int F^{+}_{\mu\nu} F^{+\mu\nu} d^4x \qquad V.9)$$

where $F^{+}_{\mu\nu}$ is defined by V.1), V.2) and V.3),

$$S^{int} = -\frac{1}{2m_V^2}\int \partial_\mu E^{\mu\sigma} \partial_\nu E^\nu_\sigma \, d^4x \qquad V.10)$$

$$S^g = -\sum_i m_i \int ds_{(i)} - \sum_i g_i \int C_\nu(z_{(i)}) dz^\nu_{(i)} \qquad V.11)$$

Note that the sign in V.9) which is required in order to treat C_ν as a vector field is opposite to the sign of the usual free electromagnetic action as $F_{\mu\nu}F^{\mu\nu} = -F^+_{\mu\nu}F^{+\mu\nu}$. The necessity of such a flip in sign is best understood if one notices that the Lagrangian density in V.9) is $(E^2 - B^2)/2$ which correctly reduces to the potential energy term $-B^2/2$ for static monopoles. We shall indeed see below that this sign yields the correct Lorentz force for the magnetic charges minimally coupled to the C_μ-field in S^g. The sign in V.11) is imposed as the kinetic energy of the monopoles has to be positive. The interaction action S^{int} is equal, from V.4), to minus the current-current interactions which is the gauge invariant manifestation of an induced meson mass term in II.24) in the London limit, as seen for instance from the relation II.45); here also the sign of the interaction term had to be changed for consistency. Note that no kinetic energy term was introduced for electric charges as the properties of the condensate should already be contained in $S^0 + S^{int}$.

If the action V.8) would yield the correct classical equations of motion for the relativistic superconductor, the quantized theory could be described by the vacuum fluctuations Z, which expressed in terms of functional integrals is, [37]

$$Z = \int e^{iS_q} \, \mathcal{D}(\chi) \qquad V.12)$$

Here the integration is on all fields C_μ and all distinguishable electric string configuration including all values of N^+ and N^-; $N^+(N^-)$ is the number of particles (antiparticles) of charge e but we shall restrict ourselves to the topological sector $N^+ = N^-$. As we are interested in the study of flux line formation we need not vary the number of monopoles.

We shall show in c) and d) below that while V.12) is indeed correct, the classical action V.8) is false. This apparant paradoxical situation arises because we have missed in V.8) a term necessary to insure the gauge invariance of the action in the C_μ-field language. As it will turn out that the missing term in S^q is a topological integral which takes for all string configurations values $2\pi n$ (n integer), it will not affect V.12) which is indeed gauge invariant. We now introduce this topological integral which will play an essential role in the subsequent analysis.

b) A topological integral.

We consider a magnetic describing a closed path in space-time, where the monopole moving backward in time may always be interpreted in terms of an antimonopole. We join the monopole-antimonopole pair by a Dirac string $M_{\mu\nu}$ as given in III.4). From by now standard application of Stokes' theorem, we have

$$g \oint C_\nu dz^\nu = \frac{1}{2} \int C_{\mu\nu} M^{\mu\nu} d^4x \tag{V.13)}$$

Writing $M_{\mu\nu} d^4x = - M_{\nu\mu} d^4x = g dS_{(\mu\nu)}$ we may write V.13) in the form

$$g \oint C_\nu dz^\nu = g \int F^+_{(\mu\nu)} dS^{(\mu\nu)} + \frac{1}{2} \int E^+_{\mu\nu} M^{\mu\nu} d^4x \tag{V.14)}$$

It follows from V.14) that the left hand side is not gauge invariant and hence the action V.8) is not admissible. We therefore replace V.8) by the gauge invariant action

$$S^c = S^q + S^t \tag{V.15)}$$

$$S^t = \frac{1}{2} \int E^+_{\mu\nu} M^{\mu\nu} d^4x \tag{V.16)}$$

where for sake of generality we may include in $M_{\mu\nu}$ not only strings ending up at each monopole entering S^g but also an arbitrary number of closed strings or strings of infinite length.The significance of such terms will become clear in next section.

The use of S^c instead of S^q in the quantum mechanical action apparently violates minimal coupling. To prove the quantum consistency of V.15) we must thus show that S^t does not contribute to the functional integral and that therefore V.12) is gauge invariant as it stands. This is indeed the case because S^t is for all configurations equal to $2\pi n$, owing to the Dirac quantization condition III.8). The proof is as follows.

From III.4) and V.3) one has

$$S^t = \sum_{ij} eg_j \int d\sigma_i d\tau_i d\sigma_j d\tau_j \, \varepsilon_{\mu\nu\rho\eta} \quad \times$$

$$\frac{\partial y^{\rho}}{\partial \sigma^{i}} \frac{\partial y^{\eta}}{\partial \tau^{i}} \frac{\partial z^{\mu}}{\partial \sigma^{j}} \frac{\partial z^{\nu}}{\partial \tau^{j}} \delta^{4}(y-z) \qquad \text{V.17)}$$

The contributions to V.17) may only occur when an electric string meets a magnetic string. If the two strings have a line element in common, the determinant in the integrand is zero because the four homogeneous linear equations $dy^{\mu} = dz^{\mu}$ admit a nontrivial solution in $d\sigma^{i}$, $d\tau^{i}$, $d\sigma^{j}$, $d\tau^{j}$. Therefore one has to consider only cases in which an electric string crosses a magnetic string at an isolated point. At this point the determinant cannot vanish and hence has an inverse. This permits to change variables to the y^{μ} (or z^{μ})and the integral over the δ-function is thus one. From Dirac quantization condition $eg_j = 2\pi n_j$ we thus have

$$S^{t} = \sum_{ij} 2\pi n_j p_{ij} \qquad \text{V.18)}$$

where p_{ij} is the number of isolated points where an electric string i crosses a magnetic string j. This proves our assertion and we shall therefore from now on use the total action S^c in the classical limit while in the quantum formulation we may equivalently use S^q or S^c.

C. The classical equations of motions.

The variational calculations are similar to those carried earlier with magnetic strings (see sections (II.4) and (III.2)).

Variation with respect to C_{μ} yields the second Maxwell equation

$$\partial_{\mu} F^{+\mu\nu} = k^{\nu} \qquad \text{V.19)}$$

Variations with respect to electric string variables yields the analog of equation II.44)

$$\frac{\partial O^{\mu\nu}}{\partial y^{\mu}} [y_{\nu}, y^{\rho}]_{i}^{+} = 0 \qquad \text{V.20)}$$

where now

$$O^{+}_{\mu\nu} = F_{\mu\nu} + \frac{1}{m_V^2}(\partial_\mu j_\nu - \partial_\nu j_\mu) - M^{+}_{\mu\nu} \qquad \text{V.21)}$$

and hence

$$\partial_\mu O^{\mu\nu} = -\partial_\mu F^{+\mu\nu} - \partial_\mu M^{\mu\nu} = 0 \qquad \text{V.22)}$$

Hence the electric string is not physical; the fact that we do not encounter any Dirac constraint is clearly traceable to the contribution $M^{+}_{\mu\nu}$ to V.21) which comes from the topological term V.16).

Variation with respect to magnetic string variables yields

$$\partial_\mu E^{\mu\nu}\,[z_\nu, z_\rho]^{+}_{i} = \jmath^{\nu}[z_\nu, z_\rho]^{+}_{i} = 0 \qquad \text{V.23)}$$

which is equation II.44) again and is correct only if the magnetic string is along the axis of a flux tube. That this is indeed the case follows from the variation of the ends of the electric strings; this yields indeed

$$O^{+}_{\mu\nu}\, y^{\nu}_{(i)} = 0 \qquad \text{V.24)}$$

as follows from Stokes' theorem applied to the last term of II.42) with $F_{\mu\nu}$ replaced by $O_{\mu\nu}$. Equation V.24) is the covariant generalization of the London equation II.18) and the magnetic string term in then just the source term $\vec{B}_S$. The magnetic strings in V.23) and V.24) which arise from the topological term are thus physical and follow the axis of confined magnetic tubes joining monopoles-antimonopoles pairs. Closed or infinite magnetic strings generate flux tubes which may exist in a superconductor even in absence of magnetic charges.

We still have to vary the extremity of magnetic string tubes. Thies yields the correct equation of motion for a magnetic charge

$$m_i \frac{d^2 z^{\mu}_{(i)}}{ds^2_{(i)}} = g\, F^{+\mu}{}_{\nu} \frac{dz^{\nu}_{(i)}}{ds_{(i)}} \qquad \text{V.25)}$$

the fact that this equation involves the complete $F^{+}_{\mu\nu}$ and not simply $C_{\mu\nu}$ comes from the additional term $-E^{+}_{\mu\nu}$ in the variation of S^{t}.

Thus we have proven that V.18) yields all the classical features of a relativistic superconductor in the London limit. Note that it is the introduction of the topological term S^{t} which ensures that the electric strings are free of any constraint; at the same time S^{t} provides the correct source terms in the London equation V.24).

We now show how the confinement is described classically in the C_{μ}-field language by considering a cylindrical flux tube $\Phi = g$ whose axis is taken along the z direction; this flux tube may be thought as originating from a monopole anti-monopoles pair located at $z = \pm a$; $a >> m_{V}^{-1}$. Far from the monopole ends we thus have, as $\partial_{\mu} j^{\mu} = 0$,

$$j^{1} = -\frac{x^{2}}{\rho} j(\rho), \quad j^{2} = \frac{x^{1}}{\rho} j(\rho); \; \rho = [(x^{1})^{2} + (x^{2})^{2}]^{1/2} \qquad \text{V.26)}$$

$$j^{0} = j^{3} = 0 \qquad \text{V.27)}$$

The current V.26) may be viewed as superpositions of circular currents generated by particle-antiparticle pairs created out of the condensate and moving around the magnetic string before they annihilate. We may thus choose a gauge in which the strings move in the x-y plane an join particles to antiparticles; then during the motion of the pair the electric string has to cross the magnetic string, thereby contributing a unit "bip" $2\pi n_{j}$ to S^{t}. In fact we shall see that in such a "bip-bip" gauge the entire contribution to the flux $F^{+}_{\mu\nu}$ in the tube arises from these "Bips", and C_{μ} is classically zero.

As the sytem is neutral, stationary in time and has cylindrical symmetry, the only non vanishing component of $E_{\mu\nu}$ classically averaged, should be in the "bip-bip" gauge $E^{12}(\rho) = E^{+03}(\rho) = \phi(\rho)$. We verify indeed that V.26) and V.27) satisfay $\partial_{\mu} E^{\mu\nu}$ with the identification

$$j(\rho) = -\frac{\partial\phi(\rho)}{\partial\rho} \qquad \text{V.28)}$$

The equation of motion for C_{μ} in the gauge $\partial_{\mu} C^{\mu} = 0$ is $\Box C^{\nu} = \partial_{\mu} E^{+\mu\nu}$ and as the right hand side is always zero we may choose $C_{\mu} = 0$ as announced. The only non vanishing component of the flux is thus $F^{+03} = - E^{+03}$; as $F^{+03} = F^{12} = - B_Z$ we thus find

$$B_z = \phi(\rho) \qquad V.29)$$

To evaluate B_Z, we have to solve the London equation V.24) which reduces to $0^{+12} = 0$. Recalling that $M^{+}_{12} = - \Phi\delta(x)\delta(y)$ (see equation II.31), this equation reduces to $-B_Z - (\partial_1 j^2 - \partial_2 j^1)(1/m_V^2) - \Phi\delta(x)\delta(y) = 0$; using V.26), V.28), V.29) this may be written as

$$\Delta B_x - m_v^2 B_z = \Phi\delta(x)\delta(y) \qquad V.30)$$

This is indeed the correct equation for the induction in the tube, far away from the monopole ends.

The fact that C_μ is zero is simply a consequence of our choice of gauge; had we chosen strings parallel to the z-axis we would have found that C_0 is non zero, massive and generates the entire flux tube; in such a gauge no contribution would arise from the topological term.

It follows from the example that in the language of the C_μ-field, it may appear inconvenient to characterize the confinement by a vector meson mass term. An alternative gauge invariant characterization is provided by the Wilson criterium 34) and we now discuss this point.

d. The Wilson criterium

Consider a superconductor in absence of monopoles. The creation and subsequent annnihilation after a time Δt of a monopole-antimonopole pair may be represented in the action S^q by a term S^g containing a single monopole of charge $g = \Phi$ describing closed loops. Such a process would cost an energy ΔE which is usually given if $\Delta t >> 1/\Delta E$ by

$$e^{-i\Delta E\Delta t} = \frac{\int e^{iS^q} \mathcal{D}(\chi)}{\int e^{iS_0^q} \mathcal{D}(\chi)} \qquad V.31)$$

where S_0^q is the action in absence of pairs. Selecting a particular loop C we write

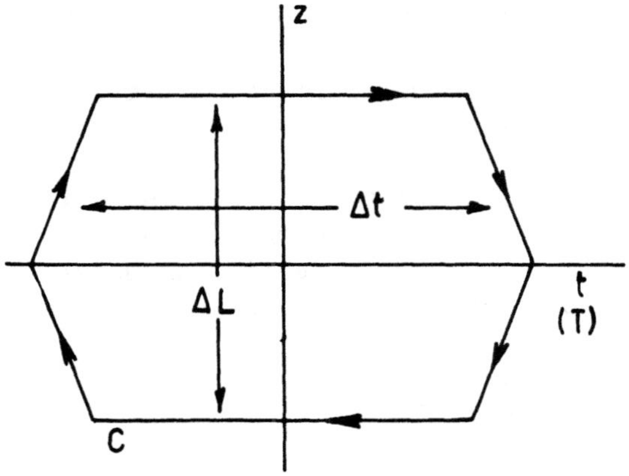

Fig. 8 : Closed loop described by a monopole-antimonopole pair in Minkowskian space (z,t) or in analytically continued Euclidian space (z, T).

$$e^{-iW_C} = \langle e^{-ig \oint_C C_\nu(z) dz^\nu} \rangle_0 \qquad \text{V.32)}$$

where the average is performed with the unperturbed action S^A. Consider the contour drawn in Fig. 8 when the monopole antimonopole pair is created along the z-axis.

For large loops, that is large time Δt and large monopole-antimonopole separation ΔL, one expects $W_C \simeq \beta \Delta L \Delta t$ where β is the energy per unit length of the flux tube flowing between the poles; thus W_C is expected to be proportional to the area of the loop drawn in Fig. 7 for large ΔL and Δt. W_C is conveniently characterized in general through analytic continuation to the Euclidian space $x_k \equiv (x^1, x^2, x^3, ix^0 = x_4 = T)$, $\overline{C}_k = (C_1, C_2, C_3, -iC_0)$ and one writes

$$e^{-W_C} = \langle e^{-ig \oint_C \overline{C}_k(x) dz_k} \rangle_0 \qquad \text{V.33)}$$

and for large loops $W_C \simeq \beta A$ where A is the Euclidian area of the smallest surface laying on the Euclidian contour chosen. Thus the correlation function V.33) is expected to fall off proportionally to the area A owing to flux tube formation. Such a criterium for confinement which is easily generalized for non-abelian gauge groups was proposed by Wilson[34].

We now display V.33) explicitely for the superconductor; our purpose is to exhibit and to understand the semi-classical nature of the confinement in a formulation in which this simple feature is hidden. To this effect we shall construct the mathematical tool required to remove spurious difficulties due to the quantum mechanical nature of V.32) or V.33)

The semi-classical evaluation of V.31) is defined by restricting oneself to the Gaussian approximation (that is the one-loop approximation in Feynman diagram language).
For any probability distribution, this is

$$< e^{-\zeta\alpha} >_o = e^{-\frac{\zeta^2}{2} < \alpha^2 >_o} \qquad \text{V.34)}$$

where α is a random variable with vanishing expectation value. This follows from the semi-invariant expansion

$$\log < e^{-\zeta\alpha} >_o = \sum_n \frac{(-\zeta)^n}{n!} M_n ;$$

$$M_1 = <\alpha>_o , \; M_2 = <\alpha^2>_o - <\alpha>_o^2 , \; \ldots \qquad \text{V.35)}$$

In the Gaussian approximation, M_n is put equal to zero for $n > 2$ so that V.34) follows. If the probability distribution is of the form $p(\eta) \sim \exp - S_o(\eta)$ where α and S_o are functions of η we have explicitely

$$< \alpha^n >_o = \frac{\int e^{-S_o(\eta)} \alpha^n(\eta) \, d\eta}{\int e^{-S_o(\eta)} \, d\eta} \qquad \text{V.36)}$$

The calculation of $<\alpha^2>_o$ can be performed by defining the perturbed "action"

$$S(\eta, \zeta) = S_o(\eta) + \zeta\alpha \qquad \text{V.37)}$$

and

$$< \alpha >_\zeta = \frac{\int e^{-S(\eta,\ \zeta)} \alpha(\eta) d\eta}{\int e^{-S(\eta,\ \zeta)} d\eta} \tag{V.38)}$$

From V.37) and V.38) one gets the "fluctuation-dissipation" theorem

$$- \left. \frac{d<\alpha>_\zeta}{d\zeta} \right|_{\zeta=0} = < \alpha^2 >_o - < \alpha >_o^2 \ . \tag{V.39)}$$

Thus V.34) is equivalent to $<\exp(-\zeta\alpha)>_o = \exp[-(\zeta^2/2) \frac{d<\alpha>_\zeta}{d\zeta}]$. Furthermore in first order in z, $<\alpha>_\zeta = \zeta \left. \frac{d<\alpha>_\zeta}{d\zeta} \right|_{\zeta=0}$ so that in the stated approximation,

$$< e^{-\zeta\alpha} >_o = e^{-\frac{1}{2}\zeta<\alpha>_\zeta} \tag{V.40)}$$

With the identification $\zeta = ig$ and $\alpha = \oint_C C_\nu dz^\nu$ V.40) would permit to express the Wilson integral V.31) in the semi-classical limit in terms of the quantity $<\oint C_\nu dz^\nu>$ obtained by solving the classical equation of motion; this method indeed would yield for the integral a result proportional to g thereby ensuring the validity of V.40) in the Gaussian limit. Such a straightforward procedure fails for two related reasons :

a. S^q is not gauge invariant and hence incorrect as a classical action; in particular the "one loop" approximation is not gauge invariant.

b. The validity of the fluctuation-dissipation theorem V.39) rests in the existence of the limit $\zeta \to 0$. When g varies however, e varies accordingly to ensure the Dirac quantization condition and when $g \to 0$, $e \to \infty$. The "unperturbed" action has thus an essential singularity when $\zeta \to 0$

To cure these two defects it suffices to replace S^q by the gauge invariant action $S^c = S^q + S^t$. One may then vary g and leave e constant, thereby relaxing the quantization condition. The fluctuation-dissipation theorem remains valid despite the fact that S^c is not equivalent to S^q for $q \neq 2\pi n/e$. Identifying now α with $\oint_c C_\nu dz^\nu - (1/2g) \int E^+_{\mu\nu} M^{\mu\nu} d^4x$ and using V.14), we now get from tne "fluctuation-dissipation" theorem

$$< e^{-ig \oint C_\nu dz^\nu} >_0 = e^{-i \frac{g}{2} \int F^+_{(\mu\nu),cl} dS^{(\mu\nu)}} \qquad V.41)$$

where $F^+_{(\mu\nu)cl}$ is the solution of the classical field equation. From section c) we see that the exponent of V.41) is, for the curve pictured in Fig. 7, proportional to the Euclidian area $\Delta L \Delta t$. Explicitely $W_c \approx \beta \Delta L \Delta t$ if the field is averaged over the transverse flux size; the divergence in β is due to the London limit and hence is not relevant.

Thus the Wilson criterium is indeed satisfied in the superconductor at the simi-classical level. The fact that this feature is not immediately apparent in terms of the quantum mechanical action S^q is due to the fact that S^q differs from the classical action S^t by a topological term S^t.

e. Electric confinement from monopoles : the magnetic superconductor [38]

Under the substitution $e \to g$, $C_\mu \to A_\mu$, $E^+_{\mu\nu} \to -M^+_{\mu\nu}$, the action V.8) becomes

$$\tilde{S}^q = - \frac{1}{4} \int (A_{\mu\nu} + M^+_{\mu\nu})(A^{\mu\nu} + M^{+\mu\nu}) d^4x$$
$$- \frac{1}{2m_V^2} \int \partial_\mu M^{\mu\sigma} \, \partial_\nu M^\nu_\sigma d^4x$$
$$- \sum_i m_i \int ds_{(i)} - \sum_i e_i \int A_\nu(y_{(i)}) dy^\nu_{(i)} \qquad V.42)$$

V.42) describes a "magnetic" superconductor which confines electric flux tubes; it would obey Wilson's criterium in its conventional form, that is with gC_μ replaced by cA_μ. Of course if a classical description of these effects is required one must again add the topological term V.16). The interest of V.42) is that it suggests that in some non-abelian generalization $M^+_{\mu\nu}$

may simply be an approximate description of a regular solution of the monopole type; the interaction term could then find its origin in the framework of non abelian gauge field theory. We shall see that such a situation indeed arises when the present model is taken over in 2 + 1 dimensions.

3. Instanton confinement mechanism.

A magnetic superconductor in 2 + 1 dimensions can be formally constructed form the action V.42) by projecting out one space dimension in the following way. Define the field tensor

$$F_{\mu\nu} \quad (g^{00} = +1, \; g^{11} = g^{22} = -1; \qquad \varepsilon^{012} = +1)$$

through

$$\begin{aligned} F^{\mu\nu} &= A^{\mu\nu} + \varepsilon^{\mu\nu\lambda} M_\lambda \\ A_{\mu\nu} &= \partial_\mu A_\nu - \partial_\nu A_\mu \end{aligned} \qquad \text{V.43)}$$

and

$$M^\sigma(x) = \sum_j g_j \int_{-\infty}^{j} \delta^3(x-z)\, dz^\sigma ; \quad g_h e = 2\pi n_j \quad (n_j \text{ integer}) \qquad \text{V.44)}$$

The string M^σ define a "monopole" located at the point z(j) whose charge is a δ function in time, indeed if

$$F^\sigma = \frac{1}{2} \varepsilon^{\sigma\mu\nu} F_{\mu\nu} \qquad \text{V.45)}$$

one gets

$$\partial_\sigma F^\sigma = -\partial_\sigma M^\sigma = + \sum_j g_j \delta^3(x - z_{(j)}) = k \qquad \text{V.46)}$$

Thus k plays the role of a magnetic current crossing the 1-2 plane from "outer space". To ensure electric confinement, it then suffices to postulate a "current-current" interaction term and to consider the "projected" action.

$$\mathcal{S}^q = -\frac{1}{4}\int (A_{\mu\nu}+\varepsilon_{\mu\nu\sigma}M^\sigma)(A^{\mu\nu}+\varepsilon^{\mu\nu\sigma}M_\sigma)d^3x$$
$$+\frac{1}{2m_v^2}\int \partial_\sigma M^\sigma\ \partial_\rho M^\rho d^3x$$
$$-\sum_i m_i\int ds_{(i)} - \sum_i e_i \int A_\nu(y_{(i)})dy^\nu_{(i)} \qquad \text{V.47)}$$

Introducing the topological term

$$\mathcal{S}^t = -\frac{1}{2}\int \varepsilon^{\mu\nu\sigma}E_{\mu\nu}M_\sigma d^3x = 2\pi N \qquad \text{(N integer)} \qquad \text{V.48)}$$

where $E_{\mu\nu}$ is the electric string tensor V.3) in 2 + 1 dimensions, one gets the gauge invariant classical action

$$\mathcal{S}^c = \mathcal{S}^q + \mathcal{S}^t \qquad \text{V.49)}$$

from which the classical equations of motion can be deduced. One finds by the same procedures as before

$$\partial_\mu F^{\mu\nu} = j^\nu \qquad \text{V.50)}$$

$$[F_\sigma + \frac{1}{2}\varepsilon_{\mu\nu\sigma}E^{\mu\nu} - \frac{1}{m_v^2}\partial_\sigma(\partial_\rho M^\rho)]dz^\sigma_{(i)} = 0 \qquad \text{V.51)}$$

as well as the Lorentz force for electric charges; no constraint appears for magnetic strings but no magnetic current may cross the electric string.

To exhibit the static electric flux tube engendered by two distant electric charges laying in the 2-axis, we select a "bip-bip" gauge such that only $M^1(x^1)$ is non vanishing. Then it follows from V.50) that $A_\mu = 0$ and that the only non vanishing component of the field is $F^{02} = M^1 = \phi(x^1)$. The "London equation" V.51) then shows that $\phi(x^1)$ falls off as exp $-\mu|x^1|$ from its maximum value on the x^2 axis. Thus electric charges are confined and the Wilson criterium follows. This is clear from the method used in the preceeding section. Again it is the use of a topological term V.48) which has permitted the description of an electric confinement in classical terms.

The action S^q is a formal construct as the "monopoles" V.44) are not really particles. However we can field an interesting interpretation for the analytic continuation of S^q in a 3-dimensional Euclidian space. Defining $x_k \equiv (x^1, x^2, ix^0)$, $\bar{A}_k \equiv (A_1, A_2, -iA_0)$,

$$\bar{M}_k = \sum_j g_j \int_{-\infty}^{j} \delta^3(x-z)\,dz_k \qquad \text{V.52)}$$

$$\bar{F}_{kl} = \bar{A}_{kl} + \varepsilon_{klm}\bar{M}_m$$

$$\bar{A}_{kl} = \partial_k\bar{A}_l - \partial_l\bar{A}_k \qquad \text{V.53)}$$

one has in absence of electric sources

$$\bar{S}^q_o = i\, S^q_o = -\frac{1}{4}\int (\bar{A}_{kl}+\varepsilon_{klm}\bar{M}_m)(\bar{A}_{kl}+\varepsilon_{klp}\bar{M}_p)\,d^3x$$

$$-\frac{1}{2m_v^2}\int (\partial_k\bar{M}_k)(\partial_l\bar{M}_l)\,d^3x \qquad \text{V.54)}$$

One recognizes in $\bar{M}_m$ the strings of singular instantons. It may well be that V.54) provides an approximate description of the regular instanton fluid which appears if the U_1 group considered here is the unbroken subgroup of a larger SO_3 group, the finite instanton size being described by the "interaction" $(-1/2m_v^2)\int k^2(x)\, d^3x$. The action V.54) is indeed quite similar to the one proposed by Polyakov to describe instantons in this circumstance [23) 24)]. In any case we know from the preceeding argument that, as in his treatment, instantons in V.54) do confine electric charges. Note that the classical approach used previoulsy to verify Wilson's criterium could have been carried in Euclidian space directly.

Thus instantons may lead to confinement in much the same way as particle do, this opens perspectives for confining quarks in non abelian "colour" gauge theories. Indeed in 3 + 1 dimensions, instantons (and merons) exist in non-abelian gauge theories without requiring a broken symmetry. The fact that in the above example the instanton confinement leads to something resembling to a "projected" magnetic superconductor is perhaps not an accident and may be more relevant than the statistical analogy with symmetry restauration : Yang-Mills fields coupled minimally to massless fermions are, at the classical level, invariant under conformal transformations which are often conveniently formulated in a projective space.

VI. PERSPECTIVES

An understanding of the confinment mechanism would have far reaching consequences. Not only would this provide the necessary justification for the use of a suitable field theoretical framework, say quantum chromodynamics, but it would permit to relate the "dual string" to more fundamental approaches. Hence the knowledge of the exact confinement mechanism,if is exists, should cure the defects of present dual string models.

Confinement should then yield the correct explanation of the rather mysterious relation that exists between the dual phenomenology and chiral symmetry. It is well known that P.C.A.C. and the smallness of the pion mass are best understood in the framework of spontaneously broken chiral invariance [3)]. This is turn yields, through the Adler consistency condition [39)], as applied to dual phenomenology relations between Regge intercepts which are well verified experimentally [40)]. In fact, nearly the whole dual phenomenology within the non-strange, non-charmed sector as well as the related meson spectrum, can be deduced from chiral symmetry supplemented by dynamical assumptions based on the absence of exotic resonances [41)]. Thus one is tempted to believe that if chiral symmetry can be understood in framework of confinement, the whole structure of the hadron resonance spectrum and the pattern of their interactions will be brought to light.

The crucial point is that, if confinement can occur in presence of massless quarks, then chiral symmetry <u>has</u> to be broken spontaneously because there will appear massive baryon states. The Nambu-Goldstone theorem then implies the existence of a massless pion in the context of a confinement and hence probably links this massless pion to dual phenomenology, as described by a "correct" string model. It is thus of great interest to inquire how the Nambu-Goldstone theorem is realized in a context which may be different from the original situation treated by Nambu and Jona-Lasinio. To this effect, one dimensional models which mimic the inside of a flux tube, whether electric or magnetic, may be relevant. As deep inside the flux tube the Yang-Mills fields behave as in the unconfined case one should focus one's attention to one-dimensional models where the gauge field remains massless. This is the case in the model proposed by 't Hooft [42] which indeed satisfies the Nambu-Goldstone theorem in that the lowest meson state is massless when quarks have zero bare mass; further studies of this model are therefore of great interest.

If indeed confinement would provide the necessary link between dual phenomenology and chiral symmetry, then the mass scale of the non-strange quarks might really be the same as the electron mass scale and the correspondance may perhaps be extended to other quarks and leptons including charmed quarks and heavy leptons. Unified models of quarks and leptons [43] should then be required to understand these facts as well as the quark-lepton proliferation. What would then be the unifying principle underlying such unified schemes ? We cannot give presently an answer to such a question, but the very fact that chiral symmetry would play a key role suggests that maybe all masses are dynamical quantities and this points towards the existence of higher symmetries.

REFERENCES

1. The "dual string" interpretation of dual phenomenology and in particular of rising Regge trajectories was introduced by Nambu, Nielsen, Susskind and Takabayashi in 1970. For a comprehensive study of "dual strings" and of their application to dual models sea.
C. Rebbi, Physics Reports 12C, n° 1 (1974).
S. Mandelstam, Physics Reports 13C, n°6 (1974).

2. J. Bardeen, L.N. Cooper and J.R. Schrieffer, Phys. Rev. 108, 1175 (1957). An approach to the B.C.S. theory closer to the field theoretical analysis is given by Y. Nambu Phys. Rev. 117, 648 (1959).

3. Y. Nambu and L. Jona-Lasinio, Phys. Rev. 122, 345 (1961). For applications of the Nambu-Goldstone theorem in statistical mechanics see R. Brout "Phase Transitions" W.A. Benjamin Inc. New York (1963).

4. F. Englert and R. Brout, Phys. Rev. Lett. 13, 321 (1964). F. Englet, R. Brout and M-F Thiry, Il N. Cimento, 43, 244 (1966).

5. V.L. Ginzburg and L.D. Landau, Zh. Eksp. i. Th. Fiz. 20, 1064 (1950). For the application of the Ginzburg-Landau formalism in type II superconductors and in particular to the flux tubes discovered by Abrikosov (A.A. Abrikoxov, Zh. Eksp. i. Th. Fiz. 32, 1442 (1957) see A.L. Fetter and P.C. Hohenberg "Theory of type II superconductors" in "Superconductivity" Vol. II pp. 817, Edited by R.D. Parks; M. Dekker Inc., New York (1969).

6. P.W. Higgs, Phys. Letters 12, 132 (1964); Phys. Rev. Letters 13, 508 (1964), Phys. Rev. 145, 1156 (1966).

7. One could equivalent restrict the domain of validity of $\vec{\nabla} \times \vec{\nabla}\chi$ outside the axis and give up Stokes' theorem. The alternate approach presented here makes contact with the Dirac string procedure (see chapter III), it was suggested to me by J. Nuyts. See also R. Brout "Flux Line and Monopole Mechanisms of Confinement" Symposium on Dynamical Broken Symmetry I.C.T.P. Trieste (1976).

8. B. Nielsen and P.O. Olesen, Nucl. Phys. B61, 45 (1973).

9. Y. Nambu, Phys. Rev. D10, 4262 (1974).

10. P.A.M. Dirac, Phys. Rev. 74, 817 (1948).

11. T.T. Wu and C.N. Yang, Phys. Rev. D12, 3845 (1975).

12. Z.F. Ezawa and H.C. Tze, Phys. Rev. D15, 1647 (1977).

13. S. Arafune, P.G.O. Freund and C.S. Goebel, J. Math. Phys. 16,

433 (1975).

14. T.T. Wu and C.N. Yang, "Properties of Matter under Unusual Conditions" p. 349 Interscience (1969).

15. F. Englert and P. Windey, Phys. Rev. D14, 2728 (1976).

16. See for instance D. Speiser in "Group Theoretical Concepts and Methods in Elementary Particle Physics" Edited by F. Gürsey - Gordon and Breach, New York (1964).

17. G. 't Hooft, Nucl. Phys. B79, 276 (1974).
A.M. Polyakov, Zh. Eksp. Theor. Fiz. 20, 403 (1974).
(J.E.T.P. Lett. 20, 199 (1974)).

18. M.K. Prasad and C.M. Sommerfield, Phys. Rev. Lett. 35, 760 (1975).

19. E.J. Weinberg and A.H. Guth, Phys. Rev. D14, 1660 (1976).

20. F.A. Bais and J.R. Primack, to be published in Nucl. Phys. B.

21. A general discussion of topological stability is given in S. Coleman. "Classical Lumps and their Quantum Descendants", Lectures at the 1975 Erice Summer School, Harvard preprint (December 1975).

22. N.H. Christ, Phys. Rev. Lett. 34, 355 (1975).

23. A.M. Polyakov, Phys. Lett. 59B, 82 (1975).

24. A.M. Polyakov, Nordita Preprint 76/33 (1976).

25. G. 't Hooft, Phys. Rev. Lett. 37, 8 (1976).
C.G. Callan, R.F. Dashen and D.J. Gross, Phys. Lett. 63B, 334 (1976).

26. C.G. Callan, R.F. Dashen and D.J. Gross, Phys. Lett. 66D, 375 (1977).

27. A.A. Belavin, A.M. Polyakov, A.S. Schwartz and Yu. S. Tyupkin, Phys. Lett. 59B, 85 (1975).

28. G. Parisi, Phys. Rev. D11, 970 (1975).

29. R. Brout, F. Englert and W. Fischler, Phys. Rev. Lett. 36, 649 (1976).

30. S. Mandelstam, Phys. Lett. 476 (1975).

31. G. 't Hooft, Marseille Conference on Gauge Theories, (1972) (unpublished); H.D. Politzer Phys. Rev. Lett. 30, 1346 (1973); D.G. Gross and F. Wilczek, Phys. Rev. Lett., 30, 1343 (1973).

32. J. Kogut and L. Susskind, Phys. Rev. D9, 3501 (1974).

33. L. Susskind, Bonn Summer School Lectures 1975. This paper contains also a review of the lattice gauge theories.

34. K.G. Wilson, Phys. Rev. D10, 2445 (1974).

35. J. Kogut and L. Susskind, Phys. Rev. D11, 395 (1975).

36. F. Englert and P. Windey : "Electric confinement and magnetic superconductors", to be published. For detailed discussions on these problems, see P. Windey, PHD. Thesis - Université Libre de Bruxelles, (1977).

37. In principle, gauge fixing terms should be included in the functional integrand V.12). We shall ignore this inessential complication as those terms do not affect the semi-classical results of the following section.

38. The suggestion that monopoles could lead to an electric Meissner effect was made by S. Mandelstam at the Paris Meeting on Extended Systems in Field Theories (1975).

39. S.L. Adler, Phys. Rev. 137B, 1022 (1965).

40. M. Ademollo, G. Veneziano and S.Weinberg, Phys. Rev. Lett. 22, 83 (1969).

41. R. Brout, F. Englert and C. Truffin, Phys. Lett. 29B 590 (1969); 29B 686 (1969); F. Englert, R. Brout and H. Stern, Il Nuovo Cimento, 66, 845 (1969); R. Brout, F. Englert and C. Truffin, Phys. Rev. D9, 2694 (1974).

42. G. 't Hooft, Nucl. Phys. B75, 461 (1974).

43. J.C. Pati and A. Salam, Phys. Rev. D8, 1240 (1973).

TRANSVERSE LATTICE THEORY OF QUANTUM CHROMODYNAMICS

William A. BARDEEN

Fermi National Accelerator Laboratory, Batavia,

Illinois 60510 and Max Planck Institute, Germany

A. INTRODUCTION.

Quantum chromodynamics (QCD) has been proposed as a complete Lagrangian field theory for the description of strongly interacting particles. The theory consists of colored quarks coupled to a color octet of Yang-Mills vector bosons. This theory can be solved in the short distance region using renormalization group methods as the effective-coupling constant vanished in this limit. This "asymptotic freedon" at short distance provides a fundamental basis for understanding the success of the quark-parton phenomenology.

Various methods have been employed to study the large distance behavior of this theory in attempt to interpret strong interaction physics in terms of confined quarks and gluons. Classes of perturbation theory diagrams have been analyzed in order to extract the leading infrared behavior of the theory. While renormalization group equation seem to be no applicable[1)], no self consistent solution to the infrared problem appears to exist at weak coupling in a manner analogous to the behavior at short distance. Another approach employs semiclassical methods[2)] to include effects missing in a perturbation theory analysis but present at small but non zero effective coupling. Instantons and other semiclassical objects appear to have important consequences for the chiral structure of QCD[3] but their possible role in the confinement mechanism is not yet evident. A third approach involves the reformation of QCD in terms of lattice gauge theories. Wilson[4)] has studied QCD using a four dimensional euclidean lattice wile Kogut and Susskind[5)] have constructed a hamiltonium version of QCD on a three dimensional spacial lattice. The lattice formulations permit a study of the strong coupling limit where confinement is a manifest consequence of gauge invariance. Whether these lattice versions of QCD truly reflect the confinement aspects of continuum QCD remains an open question at this time. Also the qualitative success of these theories with respect to strong interaction physics is certainly not excessively successful quantitatively particularly with respect to aspects involving chiral symmetry.

In this lecture I will discuss yet another approach to QCD based on a reformulation of the theory using a transverse lattice and infinite momentum frame techniques. This formulation of QCD has the unique advantage (disadvantage) that neither the weak coupling limit nor the strong coupling limit are trivial in the theory. This theory was developed in conjunction with R.B. Pearson [6]. The strong coupling aspects of the theory were analyzed by R.B. Pearson and E. Rabinovici [7].

In section B, I will briefly review certain aspects of the axial gauge description of QCD relevant to confinement. The transverse lattice theory is formulated in Section C with particular reference to the strong coupling (confinement) phase of QCD. The methods developed for study of hadronic bound states are described in Section D and Section E. In section F, the preliminary results of this program are discussed.

B. AXIAL GAUGE FORMULATION OF QUANTUM CHROMODYNAMICS.

The quantum mechanics of non-abelian gauge theories is complicated by the fact that only the transverse degrees of freedom are independent. The standard methods of quantization (convariant gauge, temporal gauge, or coulomb gauge) all involve the introduction of additional unphysical degrees of freedom or "ghost" contributions. Hence the physical aspects of the theory may be obsured. This is true for both continuum and lattice versions of the theory.

In the axial gauge only the two independent degrees of freedom associated with each vector potential are quantized with the other fields determined as dependent variables. The standard lagrangian for QCD is given by :

$$\mathcal{L} = -\frac{1}{4}\vec{G}_{\mu\nu}\cdot\vec{G}_{\mu\nu} + \bar{q}(i\not{\partial}-m)q, \qquad (1)$$

where $\vec{G}_{\mu\nu} = \partial_\mu\vec{A}_\nu - \partial_\nu\vec{A}_\mu + g\vec{A}_\mu\times\vec{A}_\nu,$

and

$$D_\mu = \partial_\mu + ig\vec{T} \cdot \vec{A}_\mu .$$

The rest fram axial gauge is defined by $\vec{A}_z = 0$. Standard canonical quantization results in the hamiltonian of Eq.(2) where the field $\vec{A}_0$ has been eliminated in favor of a "coulomb" potential,

$$H = \int dx_\perp \int dz \left\{ \frac{1}{2} \vec{P}_\alpha^2 + \frac{1}{2}(\partial_z \vec{A}\alpha)^2 + \frac{1}{4} G_{\alpha\beta}^2 + q^+ [\alpha_z \frac{1}{i} \nabla_z + \alpha_\beta \frac{1}{i} D_\beta + \beta m] q \right\} + \int dx_\perp \int dz \int dz' g^2 \vec{J}_0(z,x_\perp) \frac{1}{4} |z-z'| \cdot \vec{J}_0(z',x_\perp), \tag{2}$$

where

$$\vec{J}_0 = \frac{1}{g} \partial_\alpha P_\alpha + \vec{A}_\alpha x \vec{P}_\alpha - q^+ \vec{T} q .$$

The specification of $\vec{A}_z$ and $\vec{A}_0$ does not completely determine the gauge. The hamiltonian of Eq.(2) is invariant under the residual symmetry of transverse gauge transformatives which are local in $x_\perp$ but are global with respect to (z,t). The charges which generate these transformatives are given by

$$Q(\vec{\Lambda}) = \int dx\, \overrightarrow{P_\alpha} \cdot \overrightarrow{D_\alpha \Lambda} + \Lambda \cdot q^+ \vec{T} q, \tag{3}$$

with $\vec{\Lambda} = \vec{\Lambda}(x_\perp)$ independent of z. These charges satisfy the local algebra $[Q(\vec{\Lambda}), Q(\vec{\Lambda}')] = iQ(\vec{\Lambda} \times \vec{\Lambda}')$. Each term in the hamiltonian of Eq.(2) is seperately invariant under this residual symmetry.

The linear potential in Eq.(2) represents the focusing of the "coulomb" electric fields along the z direction in the axial gauge. If this theory is studied perturbatively, the usual quark-gluon dynamics is recovered but in a manner which is reminiscent of the spontaneous breakdown symmetry in a Higgs theory in the coulomb gauge. We may see this effect by first ignoring the "coulomb" term in Eq.(2). To lowest order, the gluons have two independent modes. The mode with transverse polarization, $k_{\perp\alpha}\, \varepsilon_{\perp\alpha} = 0$, has the usual dispersion $\omega_{\perp k}^2 = k_z^2 + \vec{k}_\perp^2$. The "longitudinal" mode, $\varepsilon_{1\alpha} = \hat{k}_{1\alpha}$, satisfies $\omega_{1k}^2 = k_z^2$. When we include the "coulomb" term the linear term in $\vec{J}_0$ implies a mixing between the longitudinal mode and the "coulomb" potential which results in the usual dispersion for the longitudinal mode $\omega^2{}_k = k_z^2 + k_\perp^2$ in perturbation theory. As in the Higgs theory, this effect may be interpreted as arising from the fact that there is a complete screening of the local color charge in the perturbation theory vacuum with only the global color charge as a residual symmetry. The local color screening may also be seen in the static coulomb potential where the liner (confining) "coulomb" potential of Eq.(2) is screened to become the usual

1/r potential.

At this point it is natural to speculate that color confinement would result in the true vacuum if the screening of the local color were incomplete. In this case the linear "coulomb" potential of Eq.(2) may reflect the true infrared structure of QCD. It is difficult to see how this "symmetrical" phase can naturally occur directly in the contiuum theory. In the next section, we formulate a cutoff version of QCD where the possible realization of a symmetrical phase is more evident.

C. TRANSVERSE LATTICE FORMULATION OF QUANTUM CHROMODYNAMICS.

A gauge theory may be separated into potentials and the independent physical degrees of freedom. The precise nature of this separation is of course gauge dependent. In the axial gauge, we have indicated the possibility of different realizations of the residual gauge symmetries than that indicated in perturbation theory. In order to discuss these symmetry properties, we introduce a cutoff version of QCD where all the gauge symmetries are preserved but where only the independent degrees of freedom are directly affected by the cutoff.

The cutoff procedure makes use of a discrete spacial lattice for the transverse directions (α, β = x, y) while the longitudinal directions (μ, ν = t, z) remain continuous. The transverse gauge degrees of freedom are associated with links of the transverse lattice. The longitudinal gauge potentials and quark fields are to be associated with sites of the lattice. The gauge fields are defined by

$$\begin{aligned} \vec{B}_{\vec{n}\mu}(t,z) &= a\vec{A}_{\mu}(\vec{n}a,t,z) & \mu &= t,z \\ M_{\vec{n}\hat{\alpha}}(t,z) &= \frac{1}{g}\, e^{ig\vec{T}\cdot\vec{A}_{\alpha}(\vec{n}a,t,z)}, & \alpha &= x,y \end{aligned} \tag{1}$$

where a is the lattice spacing and $\{\vec{T}\}$ are a matrix representation for the color generators. The link variables $M_{\vec{n}\hat{\alpha}}(t,z)$ are analogous to the U variables introduced by Wilson [4] except for their normalization and the fact that they are fields in the longitudinal variables.

Using these lattice fields, the lattice analogue of the action of Eq.(1) for the pure gauge theory becomes

$$S = \int dtdz \sum_{\vec{n}\mu\nu} \left\{ -\frac{1}{4}\vec{G}^{2}_{\vec{n}\mu\nu} \right\}$$

$$+\int dtdz \sum_{\vec{n}\mu\alpha} \mathrm{tr}\left\{ D_\mu M_{\vec{n}\hat{\alpha}} D_\mu M^+_{\vec{n}\hat{\alpha}} \right\}$$
$$+\int dtdz \sum_{\vec{n}\alpha\beta} \frac{H}{a^2} \mathrm{tr}\left\{ M_{\vec{n}\hat{\alpha}} M_{\vec{n}+\hat{\alpha}\hat{\beta}} M^+_{\vec{n}+\hat{\beta}\hat{\alpha}} M^+_{\vec{n}\hat{\beta}} \right\}, \tag{5}$$

where $\vec{G}_{\vec{n}\mu\nu} = \partial_\mu \vec{B}_{\vec{n}\nu} - \partial_\nu \vec{B}_{\vec{n}\mu} + \frac{g}{a} \vec{B}_{\vec{n}\mu} \times \vec{B}_{\vec{n}\nu}$

and $D_\mu M_{\vec{n}\hat{\alpha}} = \partial_\mu M_{\vec{n}\hat{\alpha}} + i \frac{g}{a} \vec{T} \cdot B_{n\mu} M_{n\hat{\alpha}} - i \frac{g}{a} M_{n\hat{\alpha}} \vec{T} \cdot \vec{B}_{n\mu}$.

The action of Eq.(5) is the simplest lattice action which reduces to continuum action in the naive continuum limit $a \to 0$ with g, $H = g^2$, and $\{\vec{A}_\mu(x_\mu)\}$ fixed. This action is not unique and other terms may be needed to obtain physical results for large lattice spacing.

This lattice action was constructed to be invariant under a full set of local gauge transformations. These transformations are associated with local color rotations at individual sites on the transverse lattice and may depend continuously on the longitudinal variables. The link fields, $M_{\vec{n}\hat{\alpha}}(z,t)$, are matrices which transform on the left according to color rotations at $(\vec{n},z,t)$ and on the right according to color rotations at $(\vec{n} + \hat{\alpha},z,t)$.

We may make use of this gauge freedom to study the theory in a longitudinal axial gauge. For the purpose of this lecture, we will use an infinite momentum axial gauge defined by $\vec{A}_{n-}(z,t)= 0$ where $\vec{A}_{\vec{n}\pm}= (\vec{A}_{\vec{n}t} \pm \vec{A}_{\vec{n}z})/\sqrt{2}$. The longitudinal potential $\vec{A}_{n+}$ may be eliminated in terms of the independent transverse fields, $M_{\vec{n}\hat{\alpha}}$, using the equations of motion. In this light cone gauge, the action of Eq.(5) becomes

$$S = \int dx_+ dx_- \sum_{n\mu\alpha} \mathrm{tr}(\partial_\mu M_{\vec{n}\hat{\alpha}} \partial_\mu M^+_{\vec{n}\hat{\alpha}})$$
$$+ \int dx_+ dx_- \sum_{n\alpha\beta} \frac{H}{a^2} \mathrm{tr}\,(M_{\vec{n}\hat{\alpha}} M_{\vec{n}+\hat{\alpha}} M^+_{\vec{n}+\hat{\beta}\hat{\alpha}} M^+_{\vec{n}\hat{\beta}}) \tag{6}$$
$$+ \int dx_- dx_+ dx'_+ \sum_{n} \frac{g^2}{4a^2} \vec{J}_{\vec{n}-}(x_- x_+) |x_+ - x'_+| \vec{J}_{\vec{n}-}(x_-, x'_+),$$

where the longitudinal charge density $\vec{J}_{\vec{n}-}(x_-,x_+)$ is given by

$$\vec{J}_{\vec{n}-}(x_-,x_+) = \sum_{\alpha} \mathrm{tr}\, |\vec{T}(M_{\vec{n}\hat{\alpha}} i \overleftrightarrow{\partial}_- M^+_{n\hat{\alpha}}) + T(M^+_{n-\hat{\alpha},\hat{\alpha}} i \overleftrightarrow{\partial}_- M_{n-\hat{\alpha},\hat{\alpha}})|$$

As was the case in Section B, the axial gauge leaves a residual gauge symmetry. The action of Eq.(6) is invariant color rotations at each site of the transverse lattice which are inde-

pendent of t and z.
The conserved charge is given by

$$\vec{Q}_{\vec{n}} = \int dx_+ \; \vec{J}_{\vec{n}-}(x_-, x_+).$$

A perturbative treatment of the gauge theory assumes that this residual symmetry is spontaneously broken as the link fields are expanded as $M_{\vec{n}\hat{\alpha}} = 1/g\ I + a\vec{T}\cdot\vec{A}_{\hat{\alpha}}(\vec{n}a,z,t) + --$. With this formulation, the possibility that this residual gauge symmetry is not dynamically broken can be investigated.

If we focus only on the first term in the action of Eq.(6), we recognize SU(N) x SU(N) sigma model in two dimensions in the case where the full theory is a SU(N) gauge theory. The nonlinear sigma model is asymptotically free and renormalizable in two dimensions. The large N analysis of similar theories [8] indicates that only the disordered, symmetrical-phase is expected to exist contrary to the result of perturbation theory.

We may also study another piece of the gauge theory if we also include terms from the coulomb potential and focus on a single link of the lattice. An analysis of a similar theory [9] indicated that there could exist a broken symmetry, perturbation phase for weak coupling while the symmetrical phase exists for strong coupling. By studying these elements of the full gauge theory we conclude that the longitudinal dynamics of the lattice theory is nontrivial with respect to the possible realizations of the residual transverse gauge symmetries. Which phase actually is realized in the transverse lattice gauge theory would appear to be a detailed dynamical question. However it should also be clear that confinement of color would result in the disordered, symmetrical phase. For the remainder of this lecture we shall assume this to be true.

The link fields, $M_{\vec{n}\hat{\alpha}}(z,t)$, used to define the action in Eq.(6) transform linearly under the transverse color rotations but are restricted by the nonlinear constraints, $MM^+=M^+M=I/g^2$ and $\det\{Mg\}=1$. The independent degrees of freedom have nonlinear transformation properties. <u>If</u> the gauge theory results in a symmetrical, disordered phase, then we expect linear realizations of the symmetry to provide a more appropriable description of the physics especially for large lattice spacing.

We may obtain a linear realization of the symmetry by relaxing the nonlinear constraints on the link fields. Each component of the matrix, $M_{\vec{n}\hat{\alpha}}$ becomes an independent field. The physics of the nonlinear theory may be preserved through the addition of a local potential function of $M_{\vec{n}\hat{\alpha}}$. The potential may be constructed such that the nonlinear theory results as a strong coupling limit of the linear theory. It may also be the case that the effective interactions in a large lattice approximation need not be strong. The nonlinear limit may then be combined with the continuum limit of the lattice theory.

In the linearized theory the full action may be written as the sum of action of Eq.(5) and a local potential, V. V Contains terms of the form

$$V = \int dtdz \sum_{n\alpha} \{mtrM_{\vec{n}\hat{\alpha}}M^{+}_{\vec{n}\hat{\alpha}} + \lambda_1 tr(M_{\vec{n}\hat{\alpha}}M^{+}_{\vec{n}\hat{\alpha}})^2 + \lambda_2[detM_{\vec{n}\hat{\alpha}} + detM^{+}_{\vec{n}\hat{\alpha}}] + \ldots\}, \tag{7}$$

where we have explicitly indicated those terms necessary to achieve the nonlinear theory as a strong coupling limit. In the light cone axial gauge, the linearized theory is obtained by adding the potential of Eq.(7) to the nonlocal terms in Eq.(6).

Since we will be interested primarily in the construction of the color singlet hadronic bound states of the theory, it is useful to consider the hamiltonian formulation of the theory. The infinite momentum frame hamiltonian is simply obtained from Eq.(6) and Eq.(7) with the result

$$\begin{aligned} H = &- \int dx_+ \sum_{n\alpha\beta} \frac{H}{a^2} tr(M_{\vec{n}\hat{\alpha}}M_{\vec{n}+\hat{\alpha}\hat{\beta}}M^{+}_{\vec{n}+\hat{\beta}\hat{\alpha}}M^{+}_{\vec{n}\hat{\beta}}) \\ &- \int dx_+dx'_+ \sum_{n} \frac{g^2}{4a^2} \vec{J}_{n-}(x_+)|x_+-x'_+|\vec{J}_{n-}(x'_+) \\ &+ \int dx_+ \sum_{n\alpha} \{mtr(M_{\vec{n}\hat{\alpha}}M^{+}_{\vec{n}\hat{\alpha}}) + \lambda tr(M_{\vec{n}\hat{\alpha}}M^{+}_{\vec{n}\hat{\alpha}})^2 + \ldots\}. \end{aligned}$$

In the symmetrical vacuum, terms arising from normal ordering the hamiltonian may be absorbed as a mass renormalization. In the following sections we show how this hamiltonian may be used to determine what might be considered the strong coupling limit for the hadronic bound states of the non abelian gauge theory on the transverse lattice.

D. LONGITUDINAL DYNAMICS.

In the previous section, a linearized hamiltonian formulation of QCD was defined on a tarnsverse lattice. The physical degrees of freedom are associated the transverse link fields, $M_{\vec{n}\hat{\alpha}}$, which create and destroy the transverse gluons, link mesons in the linearized theory. The link mesons are generally expected acquire mass as indicated in the effective local potential, Eq.(7). On the lattice, the linear coulomb potential associated with each site of the lattice implies that finite energy states are color singlet bound states of link mesons (and quarks). In addition to the coulomb interaction the magnetic term (nonlocal box) in Eq.(8) results in a nonlocal interaction between the link mesons.

In this section, we focus on the longitudinal dynamics responsible for the color confinement of this theory. For a given configuration of link mesons in a color singlet state, we may identify the direct coulomb interaction which does not change the link meson configuration. This part of the dynamics must be treated nonpertubatively as it is responsible for the binding of the link mesons.

In order to study the bound states we first expand the link meson fields in creation and destruction operators appropriate to the infinite momentum frame

$$M(x_+) = \int_0^\infty \frac{dk}{k} \{A_x f_k(x_+) + B_k^+ f_k^*(x_+)\}$$
$$[A_k, A_{k'}^+] = 2k\delta(k-k'), \quad [B_k, B_{k'}^+] = 2k\delta(k-k') \tag{9}$$

where the color indices have been suppressed. A color singlet two gluon state is easily constructed using these operators [6].

$$|M\bar{M}P> = \int_0^1 dx\Phi(x)\ (2x(1-x))^{-1/2} A_{xp}^+ B_{(1-x)p}^+ |0>$$
$$< M\bar{M}P'|M\bar{M}P> = 2P\delta(P-P'), \quad \int_0^1 dx|\Phi(x)|^2 = 1 \tag{10}$$

If we include only the mass term and the direct coulomb interaction in the hamiltonian, the equation of motion may be obtain simply by applying H to the state of Eq.(10). In this approximation, the twobody wavefunction, $\Phi(x)$, satisfies the wave equation

$$2P^+P^-\Phi(x) = m^2\left(\frac{1}{x} + \frac{1}{1-x}\right)\Phi(x)$$
$$- \frac{2g^2}{\pi a^2} C_N \,⨍_0^1 \frac{dy}{(x-y)^2}\,\Phi(y)\,\frac{(x+y)(2-x-y)}{4[y(1-y)x(1-x)]^{1/2}} \tag{11}$$

where C_N is the SU(N) Casimir in the singlet representation and we have used a principle value definition for the coulomb potential. This wave equation is analogous to the one considered by 't Hooft for the meson bound states in two dimensional QCD. It differs from his result by the "spin" factor which follows $\Phi(y)$ in Eq.(11). This wave equation can be solved numerically to obtain an approximately linear trajectory of states.

In a similar fashion, multimeson bound states may be studied by simply constructing the appropriate color singlet state parametrized by a wavefunction $\Phi(y_1 \ldots y_N)$ as in Eq.(10). The wave equation for Φ is simply derived by evaluating the matrix element of the hamiltonian in such a state. As an example the wave equation for a string which consists of a chain of mesons on different links bound together by the coulomb interactions at the connecting sites is given by

$$|P\Phi> = \sqrt{2}\int dy_1 \ldots \int dy_N \Phi(y_1 \ldots y_N)(2y_1 \ldots 2y_N)^{-1/2}$$
$$\delta(1-y_1 \ldots -y_N) A^+_{y_1 p} \ldots A^+_{y_N p}|0>$$
$$2P_+P_-\Phi(y_1 \ldots y_N) = m^2 \left(\frac{1}{y_1} + \ldots + \frac{1}{y_N}\right)\Phi(y_1 \ldots y_N) \tag{12}$$
$$- \frac{g^2}{\pi a^2} C_N \int_{y_1}^{y_2} \frac{dz}{z^2} \frac{(2y_1+z)(2y_2-z)}{\sqrt{2(y_1+z)2y_1 2(y_2-z)2y_2}} \Phi(y_1+z,\ y_2-z,$$
$$y_3 \ldots y_N) + \text{sum on cyclic permutations.}$$

A WKB solution to this wave equation leads to a linear spacing of eigenvalues for the energy $E^2 = 2P_+P_-$. Complicated configurations with more than one meson associated with each link lead to modified coulomb interactions due to the more complex color structure of the bound states.

While the wave equation such as exhibited in Eq.(12) seem somewhat complex, a systematic approximation scheme can be developed [7]. It is easy to check that the amplitude Φ must vanish at the endpoints $y_k=0$, with limiting behavior $\Phi(y_1 \ldots y_k \ldots y_N) \to y_k^\beta$ as $y_k \to 0$. The parameter β is determined from a self-consistency relation between the mass operator and the coulomb potential, $m^2 = 2\pi\beta \tan \pi\beta$. This condition differs from a similar condition for the quark bound states discussed by 't Hooft [10]. Contrary to the quark case, the "renormalized" meson mass can not be that of a tachyon. Since all binding energies are positive, this result implies that the gluon bound states should be heavier than the corresponding quark bound states.

Except for the end point behavior determined by β, the wave function Φ is expected to be a smooth function of $\{y_k\}$. We have found that good approximations to the wave functions can be obtained using the form $\Phi(y_1 \ldots y_N) = (y_1 \ldots y_N)^\beta \times$ Polynomials. This form for the wavefunctions is also convenient for evaluating matrix elements of hamiltonian as all of the integrals can be computed analytically for polynomial wavefunctions. A systematic truncation of the longitudinal dynamics results if only independent polynomials below a certain degree are retained. The truncation reduces the hamiltonian to matrix form. Since the matrix elements can be computed analytically in this basis, the numerical problem is greatly simplified.

As a result of these calculations, good numerical results for the spectrum of these bare hadron states are obtained as shown in Fig. 1.

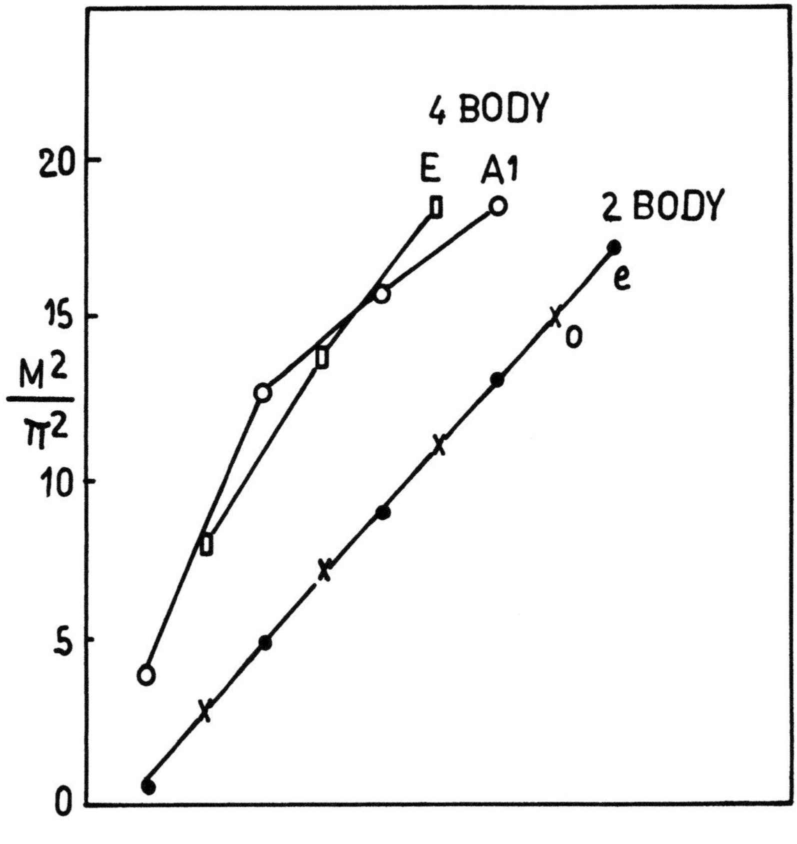

FIG. 1 : Two and Four body Bare Hadron Masses.

The bare hadrons are color singlet bound states of gluons (and quarks) and are identified with a specific configuration of link mesons on the transverse lattice. In the following section, transverse motion of these states is considered.

E. TRANSVERSE LATTICE DYNAMICS.

The bare hadron states are static configurations on the transverse lattice. The remaining operators in the full hamiltonian generate the transverse dynamics for the bare hadrons. The magnetic, coulomb production, etc... interactions couple neighboring bare hadron states. These interactions cause a mixing of static configurations and result in transverse motion when diagonalized. The physical hadrons consistof those linear combinations of the bare hadrons which move together on the transverse lattice.

For the low mass spectrum of states the few body bound states are expected to provide a good approximation to the physical states in the large lattice approximation. If only a few bare hadron states are kept the lattice problem becomes tractable. In Fig. 2, we show the complete set of two and four link

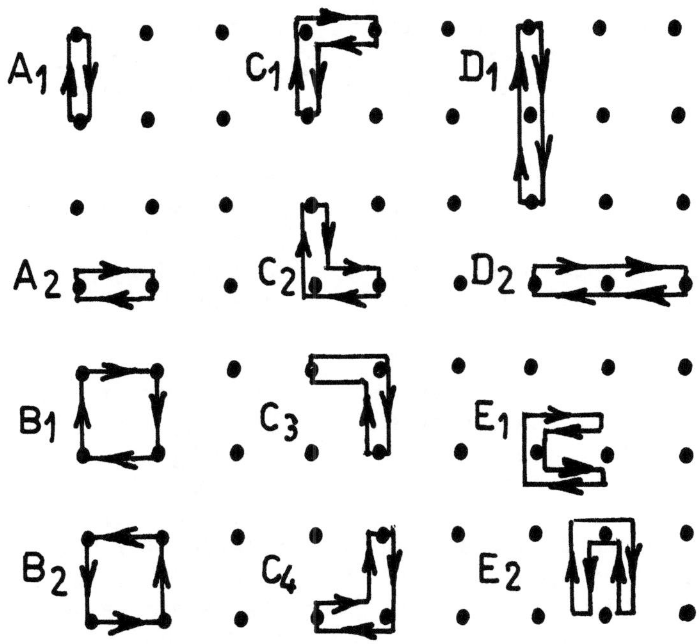

FIG. 2 : Two and Four Body Lattice Configurations.

meson bound state configurations on the lattice. Each configuration can be placed anywhere on the transverse lattice. The hamiltonian induces nearest neighbor type interactions between configurations.

The transverse motion is more easily discussed if the transverse momentum is diagonalized. For each value of transverse momentum, the energy, $E_\perp^2(k_\perp) = 2P_+P_-$, acts as a matrix operator on bare hadrons states labelled by the specific type of lattice configuration and by the state of longitudinal excitation. For a truncated set of states, this matrix may be diagonalized for each value of k . The results may be improved either by incorporating more complicated lattice configurations or more longitudinal states for each configuration.

F. PRELIMINARY RESULTS AND CONCLUSIONS.

The calculational method described in Section D and E was implemented by R.B. Pearson and E. Rabinovici [7] to compute the low mass gluonic states of QCD on the transverse lattice. For these calculations only the two and four link meson bound states shown in Fig. 2 were included. The longitudinal dynamics was approximated by wavefunctions involving low degree polynomials but consistant with the permutation symmetries of the two and four body wavefunctions. In this approximation eighty-two bare hadron states were used as the basis of the calculation. For each value of the transverse momentum, the eighty-two by eighty-two matrix for the transverse energy, $E_\perp^2(k_\perp)$, was constructed. The diagonalization of this matrix results in transverse energies for the eighty-two hadronic states.

The masses of the hadronic bound states are obtained by evaluating $E_\perp$ at zero transverse momentum. The mass spectrum is shown for the lightest states in Fig. 3. The labels, A1, B2, E, etc..., refer to states having definite rotational symmetry on the lattice at $k_\perp^2 = 0$. A1 is totally symmetric, E is a helicity one doublet, and so forth.

We may also consider states with having nonzero transverse momentum. The transverse energy should depend on the transverse momentum through the usual relation, $E_\perp^2 = m_\perp^2 + k_\perp^2$, for each bound state. On the lattice, the relation is modified but it should be a good approximation at least within the first brillouin zone. In Fig. 4, we show the dispersion relation for $E_\perp(k_\perp)$ for the lowest states of symmetry type A1, B2 and E. The two lowest states clearly behave as expected. These two states are largely two body states with sufficient mixing with four body states to have approriate transverse motion. However the E states and the other excited states do not appear to mix properly and have

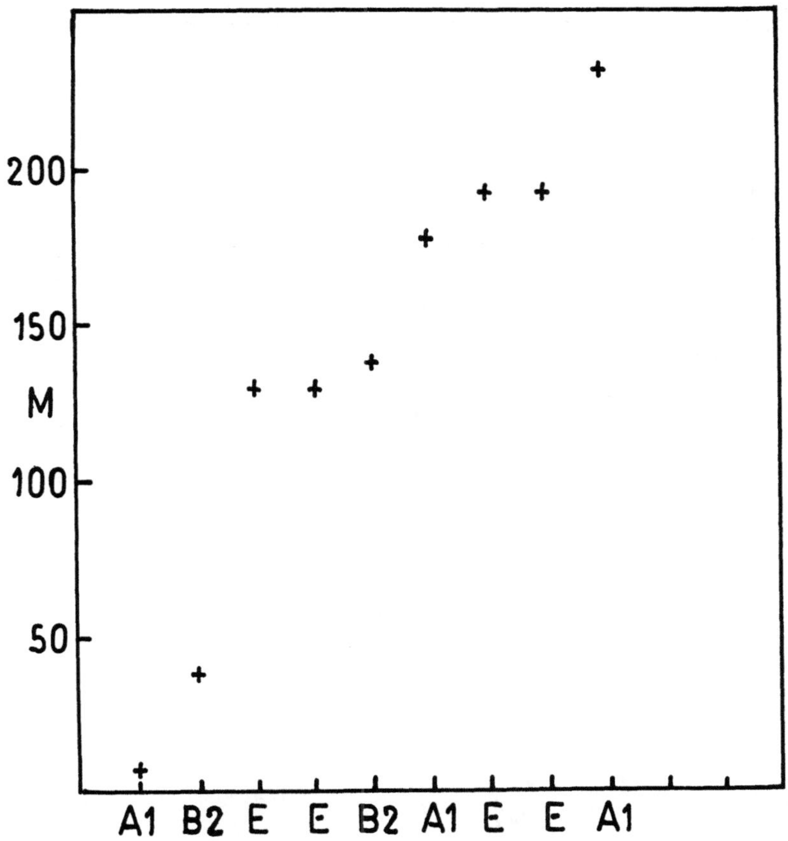

FIG. 3 : Low Mass Hadron Spectrum.

unphysical dependence on the transverse momentum. Whether these results reflect intrinsic problems or merely reflect the fact that the higher states have energies comparable to the ground state energies at the brillouin zone boundary is uncertain at this time.

In this lecture I have presented a systematic analysis of the bound state structure of a transverse lattice version of Quantum Chromodynamics. In particular, the equivalent of the strong coupling expansions of the other lattice approachs [4,5] requires a nontrivial but tractable analysis of the longitudinal and transverse dynamics. We obtain encouraging results for the behavior of the ground state hadrons but the behavior of the excited states indicates that we are far from the contiuum limit for these states. Although only bound states of the pure

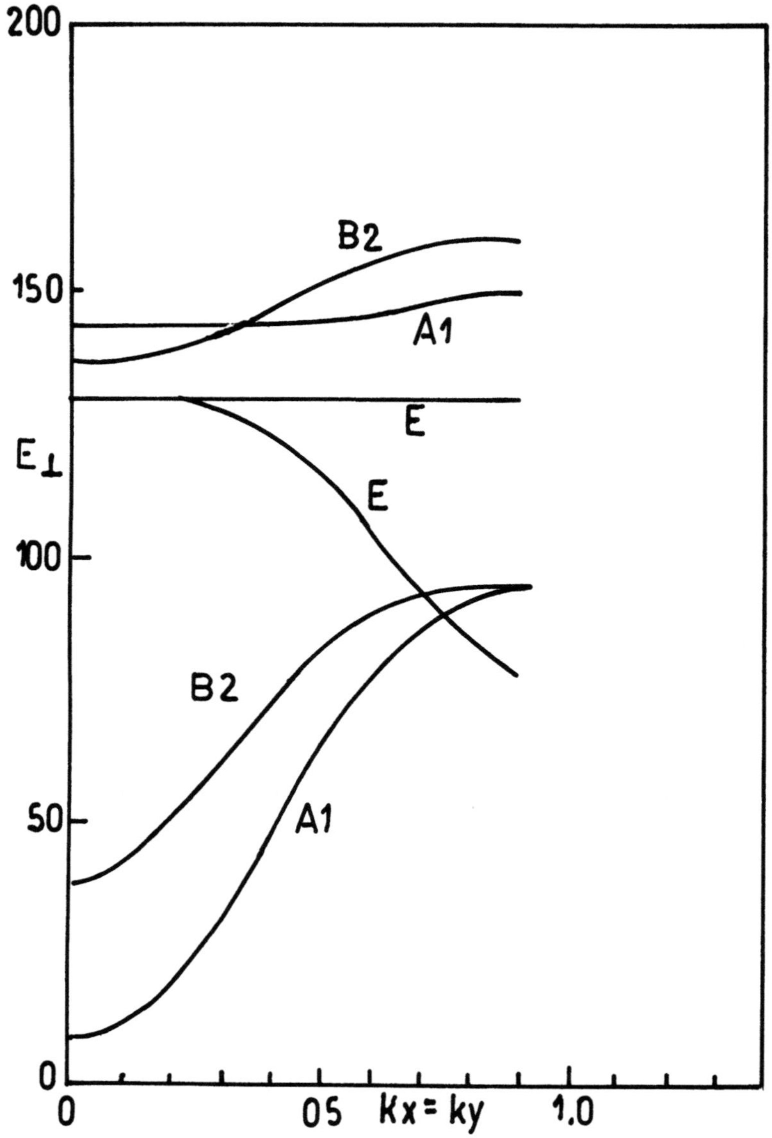

FIG. 4 : Transverse Energy vs. Transverse Momentum.

Yang-Mills theory have been considered, quarks can be included in the lattice formulation in a number of ways.

The transverse lattice, infinite momentum frame version of QCD represents the attempt to focus attention directly on the independent physical degrees of freedom of QCD while maintaining the full symmetry structure of the theory. The analysis of the longitudinal and transverse dynamics in this lecture represents the beginning to an understanding of the hadronic aspects of this theory.

ACKNOWLEDGMENT

The work reported in this lecture was performed in collaboration with R.B. Pearson and E. Rabinovici. I would like to thank the Max-Planck Institute for Physics and Astrophysics and the Alfred von Humboldt Foundation for making my visit to Europe possible.

BIBLIOGRAPHY

1. T. Appelquist, et al., Phys. Rev. Lett. 36, 768 (1976);
J.M. Cornwall and G. Tiktopoulos, Phys. Rev. D13, 3370(1976);
Akio Sugamoto, Phys. Rev. D16, 1065 (1977);
T. Kinoshita and A. Ukawa, Phys. Rev. 13, 1573 (1976).
2. A.A. Belavin, A.M. Polyakov, A.S. Schwartz and Y.S. Tyupkin, Phys. Rev. Lett. 59B, 85 (1975);
G. 't Hooft, Phys. Rev. D14, 3432 (1976).
G.G. Callan, R.F. Dashen, D.J. Gross, Princeton preprints.
3. D.G. Caldi, Phys. Rev. Lett. 39, 121 (1977) and Cargèse Lecture, 1977;
G.G. Callan, R.F. Dashen, D.J. Gross, Princeton preprints;
D.J. Gross, Cargèse Lecture, 1977.
4. K.G. Wilson, Phys. Rev. D10, 2445 (1974);
K.G. Wilson, Erice Lectures, 1976.
5. J. Kogut and L. Susskind, Phys. Rev. D11, 395 (1975);
T. Banks et al., Phys. Rev. D15, 1111(1977).
6. W.A. Bardeen and R.B. Pearson, Phys. Rev. D14, 547 (1976).
7. W.A. Bardeen, R.B. Pearson, E. Rabinovici, unpublished.
8. A. Polyakov, Phys. Lett. 59B, 79 (1975);
E. Brezin and J. Zinn-Justin, Phys. Rev. Lett. 36, 691 (1976) and Phys. Rev. B;
W.A. Bardeen, B.W. Lee and R.E. Shrock, Phys. Rev. D14, 985 (1976).
9. M.Bander and W.A. Bardeen, Phys. Rev. D14, 2117 (1976).
10. G. 't Hooft, Nucl. Phys. B75, 461 (1974).

A THEORY OF THE STRONG INTERACTIONS

David J. GROSS

The Institute for Advanced Study

Princeton, New Jersey 08540

I. INTRODUCTION.

The most promising candidate for a fundamental microscopic theory of the strong interactions is a gauge theory of colored quarks--Quantum Chromodynamics (QCD). There are many excellent reasons for believing in this theory. It embodies the broken symmetries, SU(3) and chiral SU(3)xSU(3), of the strong interactions and reflects the success of (albeit crude) quark models in explaining the spectrum of the observed hadrons. The hidden quantum number of color, necessary to account for the quantum numbers of the low lying hadrons, plays a fundamental role in this theory as the SU(3) color gauge vector 'gluons' are the mediators of the strong interactions. The absence of physical quark states can be 'explained' by the hypothesis of color confinement, i.e. that hadrons are permanently bound in color singlet bound states. Finally this theory is unique in being asymptotically free, thus accounting for the almost free field theory behavior of quarks observed at short distances.

QCD is easily described. The theory is an SU(3) color gauge theory whose dynamics is given by the Lagrangian

$$= \frac{1}{2g^2} \mathrm{Tr} F_{\mu\nu} F^{\mu\nu} + \sum_{i=1}^{F} \bar{\psi}_i (i\not{D} - m_i) \psi_i \qquad (1)$$

where $F_{\mu\nu} = \partial_\mu A_\nu - \partial_\nu A_\mu + [A_\mu, A_\nu]$ is the gauge field tensor, $A_\mu = i/2 \lambda^a A^a_\mu$ (a=1...8), ψ_i is the ith quark field of mass m_i where i runs over the quark flavors (up, down, strange, charmed, etc...) and $D_\mu = \partial_\mu - A_\mu$. The theory is thus parametrized by a single dimensionless coupling constant, g, and the quark mass parameters m_i. The light quark masses, m_{up}, m_{down}, $m_{strange}$ can, to the extent that Chiral SU(3)xSU(3) is a good symmetry, be neglected or treated as small perturbations. Furthermore heavy quarks can be ignored if we are only interested in hadrons composed of light quarks. Thus to a good approximation QCD is characterized by a single dimensionless parameter g. The actual number of quark flavors, the quark masses as well as the number of colors are not to be understood within the framework of QCD, but rather within a unified theory of the weak and strong interactions. However the number of quark flavors and the value of the masses of the heavy quarks has little bearing on the structure of the flavor blind strong interactions.

Since QCD contains no essential dimensional parameters the scale of lengths is determined by the arbitrary renormalization scale parameter μ, and all physical quantities will be functions, P(g,μ), of μ and the coupling g. The arbitrariness of μ is summarized by the renormalization group equation

$$\left(\mu \frac{\partial}{\partial\mu} + \beta(g) \frac{\partial}{\partial g}\right) P\,(g,\mu) \quad = \quad 0 \qquad (2)$$

consequently there are no arbitrary dimensionless physical parameters in QCD. The only relevent parameter is the mass scale, μ. All dimensionless couplings are calculable numbers. This situation is unprecedented in physics. It means that any small expansion parameter in QCD must be determined self consistently. It also means that the structure of physical states is fundamentally non-perturbative. Thus hadronic masses, using Eq. 2, must depend on g according to

$$m_H \sim \mu \exp - \int^{g} \frac{dx}{\beta(x)} \sim \mu \exp(- \frac{1}{g^2})$$

and cannot be calculated in perturbation theory. At short distances or large spacelike momenta perturbation theory can still be used since the effective coupling in QCD is momentum dependent and vanishes as $g^2/8\pi^2 \sim (11\ell n\, P\mu)^{-1}$, thus suppressing non-perturbative effects proportional to $\exp(-1/g^2)$ by powers of P. However as we proceed to larger distances g(P) increases leading presumably to the increased importance of the non-perturbative mechanisms which bind the quarks to form hadrons. We must therefore determine the non-perturbative mechanisms in QCD responsible for hadronic structure, determine the numerical value of the hadronic coupling (i.e. $g^2/8\pi^2$ at momenta of order hadronic masses) and hopefully discover self consistent small parameters. We now believe that the dynamics of QCD is due mainly to the nonperturbative structure of the vacuum, and that the hadronic coupling itself is always small ($g^2/8\pi^2 \sim 1/10 - 1/20$). This small soupling, which suppresses quantum fluctuations, is the basis for the semi-classical approximation to the structure of the vacuum.

Even though QCD has been remarkably successful in describing deep inelastic scattering and is compatible with the partially successful models (the quark, bag and string models) that have illuminated the structure of hadrons, we still do not possess a true theory of the strong interactions. Such a theory would, starting from first principles, yield a qualitative understanding of the major dynamical features of the strong interactions as well as a rough (say at the 10% level) quantitative calculation of hadronic masses, couplings and scattering amplitudes. Given the lack of experimental surprises in the realm of the strong interactions it is very likely that the time is ripe for such a comprehensive theory. In these lectures I shall outline an approach recently developed by C. Callan R. Dashen and myself, based on a new understanding of the QCD vacuum, which has already made substantial progress towards this goal [1]. In these lectures I shall only be able to give the barest of outlines of this approach. For details the reader is referred to the references cited, in particular Ref. 1.

II. Tunneling and Vacuum Structure.

The phenomenon of vacuum degeneracy, although somewhat novel in quantum field theory, is familiar in ordinary quantum mechanics. Consider a system with one degree of freedom in a periodic potential $V(x) = 1/g^2[1-\cos(gmx)] = V(x+2n\pi/gm)$. The classical minimal energy configurations (classical vacua) are clearly $x_n = 2n\pi/mg$, $n=0, \pm 1, \ldots$, all with zero energy. However the quantum mechanical ground state is a non-degenerate eigenstate of the translation operator $T = \exp i\, 2\pi P/mg$. In terms of the states one constructs perturbatively about $x = x_n, |n\rangle$, (Gaussian wave packets localized at $x = x_n$ to lowest order in g^2) the possible quantum ground states are the coherent superpositions

$$|\mathrm{vac},\theta\rangle = \sum_{n=-\infty}^{+\infty} e^{in\theta}|n\rangle, \quad T|\mathrm{vac},\theta\rangle = e^{i\theta}|\mathrm{vac},\theta\rangle \tag{3}$$

The structure of the θ-vacuum is easily discussed in the semi-classical (WKB) approximation. The path-integral formulation of this approximation is most appropriate when we wish to generalize to field theory. Here one is interested in the imaginary time propagator $\langle x_2|e^{-HT}|x_1\rangle$, where H is the Hamiltonian, which can be written as a sum over all path histories which proceed from x_1 to x_2 in (imaginary) time T, weighted by the exponential of the classical action for the path

$$\langle x_2|e^{-HT}|x_1\rangle = \text{const.}\int_{x(0)\,x_1}^{x(T)=x_2} Dx(t)\, \exp[-S_{cl}[x(t)]] \tag{4}$$

For large T we probe the vacuum wave function, since

$$< x_2|e^{-HT}|x_2 > \underset{T\to\infty}{\to} \psi_0^+(x_2) < vac|e^{-HT}|vac > \psi_0(x_1).$$

The relevant paths that contribute will be those of finite action. One of these is $x(t) = 0$, which contributes 1 to the amplitude. However there exist finite action paths $x(t)$ that interpolate between x_0 and any x_n.
For example, the minimum action path that interpolates between $x_0 = 0$ and $x_1 = 2\pi/gm$ is the solution of $\ddot{x}_I(t) = V'_I(x)$ where $x_I(-\infty) = x_0$, $x_I(+\infty) = x_1$ which has action equal to

$$S_0 = \int_0^{2\pi/gm} \sqrt{V(x)}dx = 2\sqrt{2}/mg^2.$$

The contribution of paths in the vicinity of this path can be estimated by saddle point techniques, which yield

$$\int_{x_0}^{x_1} Dx(t) \exp[-S_{cl}] \simeq \exp(-S_0)[\det[S''_{cl}(x_I(t))]]^{-1/2} = \frac{T}{T_0} \exp(-S_0) \tag{5}$$

This is recognizable as the WKB expression for the amplitude to tunnel between the two classically degenerate vacua. The imaginary time solution, $x_I(t)$, of the classical equations of motion which yields the minimum action interpolation path is known as an "instanton", and the existence of such solutions with finite Euclidean action is a sure sign of vacuum tunneling. The above semi-classical approximation is valid for sufficiently small g.

Similarly there exist paths which interpolate between x_0 and x_n and consist of superpositions of widely separated n_+ instantons and n_- anti-instantons ($n = n_+ - n_-$).

$$x(t) \simeq \sum_{i=1}^{n_+} x_I(t-t_i) + \sum_{j=1}^{n_-} x_I(t_j - t), \tag{6}$$

which are approximate solutions of the equation of motion when $|t_i - t_j| >> T_0$ with action equal to $(n_+ + n_-)S_0$. Paths near this path contribute a term (when g is small)
$(n_+!n_-!)^{-1}\,(T/T_0)^{n_+ + n_-} \exp[-(n_+ + n_-)S_0]$ to the vacuum to vacuum amplitude. The tunneling amplitude is proportional to $T^{n_+ + n_-}$ since each tunneling (instanton) can occur at any time $0 \leq t \leq T$.

Adding the contributions of all these configurations yields

$$< \mathrm{vac},\theta|e^{-HT}|\mathrm{vac},\theta > \sim \sum_{n_+,n_-} \frac{1}{n_+!}\frac{1}{n_-!}\left(\frac{T}{T_0}\right)^{n_++n_-} e^{-(n_++n_-)S_0} e^{i(n_+-n_-)\theta}$$

$$= \exp\left(2\frac{T}{T_0}\cos\theta\exp(-S_0)\right). \qquad (7)$$

Thus due to tunneling, the energy of the θ-vacua is lowered by the amount $2/T_0 \cos\theta\exp(-S_0)$.

One recognizes Eq. (7), for say θ=0, as the grand canonical partition function for a free gas of instantons and anti-instantons which may be located on the line $0 \leq t \leq T$, where $T_0^{-1}\exp(-S_0) = \exp(-\beta\mu)$ is the activity and the temperature is proportional to g^2. This is a very useful analogy. The density of instantons in the gas, or the tunneling amplitude, is proportional to $T_0^{-1}\exp(-S_0)$ and thus vanishes rapidly as $g^2 \to 0$. For this reason one was justified in assuming that $|t_i-t_j| >> T_0$ in Eq.(6). This is the dilute gas approximation, DGA. When the density increases one must take into account the fact that the action for such a configuration differs from $(n_++n_-)S_0$ by an amount $S_{int}(t_i,t_j)$ which can be interpreted as 1/β times an interaction energy of the analog gas. For small g^2, therefore small density, one can presumably treat S_{int} as a perturbation and derive an expression for the vacuum to vacuum amplitude in powers of the density, although a systematic procedure for carrying out such an expansion has not been developed yet.

The structure of the vacuum in a four dimensional nonabelian gauge theory bears many similarities with the above example. Classical vacua are pure gauge fields, say $A_0 = 0$, $A_i = U^+\nabla_i U$, $F_{\mu\nu} = 0$ where $A_\mu = A_\mu{}^a\lambda^a/2$. Such fields fall into discrete classes, which can be labelled (in $A_0 = 0$ gauge) by the number of times that $U(\vec{x})$ covers the gauge group as $\vec{x}$ runs over all of space. These classes (the homotopy classes, Π_3 (gauge group)) are labelled by an integer n - the winding number. Within each class any two gauge fields are continuously deformable into each other and thus equivalent. Standard perturbation theory constructs the vacuum by perturbing about one such class, say about the fields equivalent by continuous gauge transformations to A = 0. However, there exist imaginary time paths in function space that interpolate between different winding numbers. These time histories can be labelled by a topological charge

$$\frac{1}{8\pi^2}\int d^4x\, \mathrm{Tr}F_{\mu\nu}\tilde{F}^{\mu\nu} ,$$

which in $A_0 = 0$ gauge is simply equal to the change in winding number between t = -∞ and t = +∞. Therefore tunneling will occur with non-vanishing amplitude and the true vacuum will be a coherent superposition of winding number vacua. The minimal action path between winding number zero and one is the famous instanton solution of the Yang-Mills equations [2], which for $A_0 = 0$ has the property (for color SU(2))

$$A_i(x,t) \underset{t\to-\infty}{\to} 0, \quad A_i(x,t) \underset{t\to+\infty}{\to} U_1^+(x)\partial_i U_1(x),$$
$$U_1(x) = \exp \frac{\pi}{2} \frac{\vec{x}\cdot\vec{\tau}}{\sqrt{x^2+\rho^2}} . \qquad (8)$$

This tunneling path or instanton is labelled by its position (x,t), scale size ρ and 7 continuous parameters (for color SU(3)) that determine the group orientation. The classical action is finite and equal to $8\pi^2/g^2$.

The existence of these instantons implies that the true QCD vacuum will be a coherent superposition of winding number vacua

$$|vac,\theta> = \Sigma e^{i\theta n}|n> \qquad (9)$$

labelled by a continuous parameter [3,4,5]. The parameter $|\theta| < \pi$ labels superselection sectors of QCD and must be regarded as a free parameter. Since P and T are not symmetries of QCD for $\theta\neq 0$ (unless one of the quarks is strictly massless in which case θ can be always made to vanish) we must choose $\theta=0$. The vacuum to vacuum amplitude can be calculated, as in the quantum mechanical example, by a semiclassical (saddle point integration) treatment of the Euclidean path integral. The resulting tunneling amplitude, i.e., the contribution of one instanton to the vacuum to vacuum amplitude, is

$$.1 \int d^4x \int \frac{d\rho}{\rho^5} \left(\frac{8\pi^2}{g^{-2}(\rho)} \right)^6 \exp \left| - \frac{8\pi^2}{g^{-2}(\rho)} \right| .$$

This formula requires some explanation. The integration over x and ρ are due to the fact that the instanton can be located anywhere and have any scale size ρ. Thus the operator $S''[A_I(x,\rho)]$ will have 12 zero energy modes corresponding to translating, dilating or rotating in group space the instanton solution A_I. These modes are treated as collective coordinates and one replaces $\int dA_\mu$ by $\int dA'_\mu \int d^4x \int d\rho/\rho^5 d\Omega J$ ($d\Omega$ is an integration over the compact group volume) where J is a Jacobian and A'_μ is orthogonal to the zero energy modes. Since $J \sim \Pi_i(dA_I/dz_i)$ (z_i is the ith collective coordinate) and since $A_I \sim 1/g$ each zero energy mode yields a factor of $1/g$. This is the origin of the factor $(1/g)^{12}$ in the tunneling amplitude. Finally the determinant of $[S''(A_I)]^{-1/2}$(whose logarithm is simply the sum of connected one loop vacuum Feynman diagrams in the external gauge field) has ultraviolet divergences, proportional to $\ln\Lambda\rho$, which are cancelled by the standard coupling constant renormalization of the classical action $8\pi^2/g_0^2 \to 8\pi^2/g^2 + \ln \Lambda/\mu$. The net effect is to replace g^2 by $\bar{g}^2(\rho\mu)$, where $\bar{g}$ is the effective coupling whose ρ dependence is given by $d\bar{g}(\rho\mu) / d\ln\rho\mu = -\beta(\bar{g})$.

If we make the dilute gas approximation, i.e. perform the functional integral $\int DA_\mu\, e^{-S(A)}$ by saddlepoint integration about well separated multiple instanton anti-instanton configurations, we will obtain a free gas of instantons and anti-instantons with a density given by

$$\int \frac{d\rho}{\rho^5} \left(\frac{8\pi^2}{\bar{g}^2(\rho)} \right)^6 \exp \left| \frac{-8\pi^2}{\bar{g}^2(\rho)} \right| = \int \frac{d\rho}{5} D(\rho). \tag{10}$$

Note that the density of instantons of size ρ measured in units of ρ^{-4}, $D(\rho)$, depends strongly on ρ. For small ρ the density will vanish rapidly due to asymptotic freedom. In other words $\bar{g}^2/8\pi^2 \underset{\rho\to 0}{\sim} -11\ell n\rho\mu$ (for a pure SU(3) gauge theory) and thus $D(\rho) \underset{\rho\to 0}{\sim} \rho^{11}(\ell n\rho)^6$. However as ρ increases the density increases and can become very large. Thus in QCD there is no way of cranking down the density of instantons (or the tunneling probability) independent of the process one is considering. Instantons of large size have corresponding large couplings, which might lead to a breakdown of the semi-classical approximation. Thus one again we are faced in QCD with the problem of strong coupling. However, it is clear that we are not interested in instantons of arbitrarily large size. Thus an instanton whose radius is one mile will surely not affect physical processes that take place on a length scale of a Fermi. One would expect that the only instantons that are relevant to physical processes are those whose size is comparable with the relevant scale parameter of the process. Therefore when one calculates the effect of instantons on some physical process the ρ integration in Eq.(10) will be effectively cut off at some ρ_c. If ρ_c is sufficiently small that $\bar{g}^2(\rho_c)/8\pi^2 << 1$ one can still use the semi-classical approximation, and if ρ_c is small enough that the total instanton density is $\lesssim 1$ one can still use the DGA.

The actual value of ρ_c as well as the dynamical mechanisms that apply in the region of $\rho \geq \rho_c$ cannot be treated within the DGA. One must consider the interactions of the instantons and take into account other fluctuations about the vacuum (i.e. merons). I shall return to this point in Sections V and VI.

Finally we note that we have not made use of the exact multiple instanton solutions that have attracted so much attention. The existence of these solutions does tell us that instantons do not interact with other instantons classically (when quantum fluctuations are taken into account an instanton-instanton interaction $\sim (1/g^2)^0$ will be induced). However as we have seen the dominant configuration, even within the DGA, is a superposition

of instantons and anti-instantons with $< n_+ > = < n_- >$.
The probability of a configuration with topological charge $Q=n_+$ vanishes as

$$\exp \left| - \frac{VT}{V_0} e^{-8\pi^2/g^2} \right|_{VT\to\infty} \to 0.$$

These superpositions cannot be exact solutions of the equations of motion. However for small instanton densities they can be dealt with perturbatively.

III. The effect of Instantons.

The existence of degenerate vacua and the ensuing tunneling leads to new dynamical effects that would not be seen by perturbing about the naive vacuum. In addition to the standard Gaussian fluctuations about the naive vacua, which one calculates by standard perturbation theory (summing Feynman diagrams), there will exist instanton fluctuations. They are characterized by large fields ($A_\mu \sim 1/g$), however they are rare for small coupling and thus suppressed. In the DGA one can easily evaluate the contribution of instantons to any process. I shall show how this is done in general and then consider two potentially important applications -- the heavy quark potential [1,6] and the instanton contributions to e^+e^- annihilation at short distances [7].

Let us first consider calculating the vacuum expectation value of a gauge invariant operator, $O = O(A_\mu, \psi, \bar{\psi})$, constructed out of the gauge and quark fields. As an example take O to be a product of electromagnetic currents, $O = \bar{\psi}(x)\gamma_\mu Q\psi(x)\bar{\psi}(y)\gamma_\nu Q \psi(y)$. Then

$$\frac{< \mathrm{vac}|O|\mathrm{vac} >}{< \mathrm{vac}|\mathrm{vac} >} = \frac{\int d\psi d\bar{\psi} \mathcal{D}(A_\mu) \exp[-S(A,\psi,\bar{\psi})]\ O}{\int d\psi d\bar{\psi} \mathcal{D}(A_\mu) \exp[-S(A,\psi,\bar{\psi})]} \tag{11}$$

Ordinarily one perturbs about the naive vacuum, A=0, thereby deriving the standard Feynman rules for evaluating the expectation value of O. This we denote by $< O >_0$. Now there will be additional contributions to Eq. 11, arising from the instanton-anti-instanton saddle points. In the DGA one easily shows that

$$\frac{< \mathrm{vac}|O|\mathrm{vac} >}{< \mathrm{vac}|\mathrm{vac} >} = < O >_0 + \int \frac{d\rho}{\rho^5} D(\rho) \int d^4x_I \left[<O>_{A_{I(x,\rho)}} + <O>_{\bar{A}_{I(x,\rho)}} - 2 <O>_0 \right] + O[\mathrm{density}^2] \tag{12}$$

where$(\langle O\rangle_{A_I(x,\rho)})(\langle O\rangle_{\bar{A}_I(x,\rho)})$ is the expectation value in instanton (anti-instanton) field, located at x_I of size ρ In other words <O> is calculated using the ψ and A_μ propagators one calculates in a background instanton field.

The integration over instanton position, x_I, will in general converge since when the instanton is far away from the fields contained in O it will have no effect on O (the $F_{\mu\nu}$ due to the instanton falls off like $1/r^4$). Similarly if ρ is very large compared to the typical distance, d, associated with O(say $\rho >> |x-y|$ if $O = J(x)J(y)$), then again it will have little effect on <O> since it will act like a constant field in the vicinity of O that can be gauged away. Thus by letting d get small we suppress the effect of large instantons by factors of (d/ρ). This allows one to reduce the effects of instantons by going to small distances as well as improving the DGA. As d is increased both the instanton effects, proportional to the density of instanton $D(\rho)$ with $\rho \approx d$, and higher order perturbative effects, proportional to $\bar{g}^2(d)/8\pi^2$ increase. It is simply a numerical question as to which of these effects dominates.

It is possible to show that the effect of instantons in the DGA can be summarized by an effective non-local Lagrangian [7]. The effective Lagrangian contains all possible vertices, which can be systematically evaluated in powers of the instanton density. To lowest order, however, it is simplest to use Eq. 12 directly.

In ordinary quantum mechanics the perturbative corrections about the naive vacuum, proportional to g^2, are always more important than the tunneling effects, proportional to $1/g \exp(-1/g^2)$, unless g is quite large. Thus for small g one can safely ignore tunneling in calculating, say, the energy levels of the periodic potential. In QCD it appears that the situation is quite different. Due to the fact that there exist so many ways to tunnel from one winding number vacuum to another, the net tunneling amplitude can be very large event for small coupling. In other words the factor of $(8\pi^2/g^2)^6$, which is a measure of the volume in function space of tunneling paths, greatly enhances the total tunneling amplitude. In fact $D(x) = .1x^6e^{-x} \gtrsim 1/x$ $(x = 8\pi^2/g^2)$ for $x \lesssim 18$ or $g^2/8\pi^2 \gtrsim 1/18$ which is a very small coupling indeed. Thus we expect in general that as one proceeds to large distances in QCD the first thing that happens is that the non-perturbative instanton generated effects become large and that this occurs when the coupling is still very small. Thus these effects can be calculated using semi-classical approximations.

We have calculated a wide variety of instanton generated effects including coupling constant renormalization [1], massive quark-anti-quark interaction energy, [1,6] dynamical chiral symmetry breaking [1], and the short distance behavior of hadronic currents [7]. In all the above cases we find strong instanton effects in the region of small coupling. Furthermore, as we go out to large distances meron configurations, as I shall explain below, appear and seem to lead to quark confinement. This again occurs at distances where $g^2/8\pi^2$ is small ($\sim 1/10 - 1/20$). Thus it appears that in QCD, even at the hadronic size, the quantum coupling is always weak. Although strong non-perturbative effects are generated they can be treated semi-classically. Furthermore, the small value of $g^2/8\pi^2$ for all relevant momenta might explain precarious scaling [7] as well as the success of naive quark models.

To illustrate the dynamical effects of instantons let us consider the interaction energy, $\varepsilon(R)$, between a heavy quark anti-quark pair as a function of their separation R. This energy can be used as a potential in calculating heavy quark bound states (charmonium) and its asymptotic behavior as $R \to \infty$ provides a test of whether or not color is confined. Thus if $\varepsilon(R) \underset{R \to \infty}{\to} \infty$ (or is bounded) than color is (or is not) confined.

The interaction energy of infinitely massive quarks (or external color sources) can be calculated by evaluating the vacuum expectation value of the ordered exponential of the line integral of the gauge field about a Euclidean loop L, namely

$$< \text{vac}|P \; \text{TR}[\exp i \oint_L A^\mu \cdot dx_\mu]\text{vac}> \underset{T\to\infty}{\to} \exp(-\varepsilon(R)T) \qquad (13)$$

where L is a loop of spatial extent R (the separation between the quarks) and time extent T. This can most easily be seen in $A_0=0$ gauge, where only the spatial legs of the loop contribute. It is then clear that the above expression evaluates the Euclidean propagator, e^{-HT}, between gauge invariant states of the form

$$P\{\psi^+(0)\exp\int_0^R A_i \cdot dx_i \psi(R)\}|\text{vac}>$$

which consist of a quark (anti-quark) at x = 0 (R). As $T \to \infty$ we project out the state with the lowest value of the energy which contains a quark-anti quark pair separated by R.

It is possible to show that this expression is the first term in a systematic (Foldy-Wouthuysen) expansion of the quark-anti quark Hamiltonian in inverse powers of the quark mass. One can systematically calculate the $1/m^2$ spin-orbit and spin-spin corrections to $\varepsilon(R)$ [6]. Here however I shall only discuss $\varepsilon(R)$, which is the spin independent term in the non-relativistic quark anti-quark potential. If we were to calculate the loop by expanding about the ordinary vacuum ($A_\mu=0$) we would of course obtain from the ordinary vacuum fluctuations the standard Coulomb interaction energy $- 4/3\ (g^2/4\pi)\ 1/R$. Higher order corrections would renormalize g, replacing it by $\bar{g}(R)$. However in a θ-vacuum we must also consider multiple instanton and anti instanton configurations, which will produce non-perturbative corrections to the Coulomb potential.

In the DGA the instanton contribution to $\varepsilon(R)$ is easily calculated [1] in terms of the one instanton contribution to the loop integral. Thus in the DGA

$$\varepsilon(R) = -\lim_{T\to\infty} \frac{1}{T} \int \frac{d\rho}{\rho^5} 2D(\rho)\frac{1}{3}\int \mathrm{TR}\{\exp \oint_L A_I^\mu(x-x_1,\rho)dx_\mu - 1\}d^4x_I \tag{14}$$

This expression can easily be evaluated numerically. For small R ε will behave as :

$$\begin{aligned}\varepsilon(R)_{\widetilde{R\to 0}} & \frac{2\pi^3}{9}\left(\frac{\pi}{4} - 4J_1(2\pi)\right) \int \frac{d\rho}{\rho^3} D(\rho)R^2 \\ &= 11.27\ R^2 \int \frac{d\rho}{\rho^3} D(\rho) + O(R^4).\end{aligned} \tag{15}$$

Thus we see that the non perturbative corrections vanish rapidly as $R \to 0$, but can be quite substantial for large R. Unfortunately we also note that in order to determine the coefficient of the R^2 term we must have control over instantons of all sizes. The integration over instanton size is peaked in a region where the density, $D(\rho)$, is already quite large. Thus we are not truly able to make a self consistent DGA to $\varepsilon(R)$. However we can see that the instanton correction will be very important even for distances R which are small enough so that $g^2(R)/8\pi^2$ is still very small. To see this we cutoff the ρ integration at a size, ρ_c, corresponding to the point where the net fraction of space occupied by instantons and anti instantons is of order 1. This yields $x(\rho_c) \sim 18$, and $\varepsilon(R) \sim 8\mu^3R^2$. This is surely a gross underestimate of $\varepsilon(R)$, but even so it becomes comparable to the perturbative Coulomb interaction for $\mu R \sim .3$ or $x(R) \sim 14$.

For large values of R $\varepsilon(R)$ approaches a constant. This indicates that instantons alone, at least within the DGA will not confine quarks. The large R behavior of $\varepsilon(R)$ can be interpreted as a mass and coupling constant renormalization. Thus

$$\begin{aligned}\tilde{\varepsilon}(R)_{\overrightarrow{R\to\infty}} & - \frac{16\pi^3}{9}\left(J_1(\pi)+J_0(\pi)\right)\int\frac{d\rho}{\rho^2}D(\rho) - \frac{4\pi^3}{3R}\int\frac{d\rho}{\rho}D(\rho) \\ &= 37 \int \frac{d\rho}{\rho^2}D(\rho) - \frac{41.3}{R}\int \frac{d\rho}{\rho}D(\rho),\end{aligned} \tag{16}$$

the first term corresponding to a mass renormalization of the quarks, $\delta m = 18.5 \int d\rho/\rho^2\, D(\rho) \gtrsim 2.2\mu$ (if we use the same cutoff as above) whereas the second corresponds to a renormalization of the coupling :

$$\frac{g^2}{8\pi^2} \rightarrow \frac{g^2}{8\pi^2} + \frac{\pi^2}{2} \int \frac{d\rho}{\rho} D(\rho). \tag{17}$$

This question can also be derived by evaluating the gluon propagator as in ReF. 1.

Although the DGA to $\varepsilon(R)$ is not quantitatively reliable we can deduce the following qualitative features. First instantons do not confine. This is essentially because the instanton field strength, $F_{\mu\nu}$, falls off too rapidly for large x ($\sim 1/x^4$). In order for $< \exp i \oint_L A\cdot dx >$ to behave like Area (L) for large Euclidean loops it is necessary for $F_{\mu\nu}$ to behave, for large distances, as $1/x^2$. This we shall see is the case for meron configurations which are important in considering large loops and which will produce the confining part of $\varepsilon(R)$. Second the instanton contribution to the vacuum fluctuations yield a non-perturbative renormalization of the effective coupling. This, just like the standard perturbative renormalization, is equivalent to an anti-screening of color; i.e. $g^2/8\pi^2$ increases as R increases. When the correlations (interaction) between instantons and anti-instantons are taken into account we find that this effect is greatly enhanced [1].

As a second application let us consider the contribution of instantons in the DGA to the vacuum polarization tensor of the electromagnetic current.

$$\begin{aligned} \pi_{\mu\nu}(p) &= \int d^4x e^{ipx} < 0\{T|J_\mu(x)J_\nu(0)\}|0 > \\ &= (g_{\mu\nu}p^2 - p_\mu p_\nu)\Pi(p^2) \end{aligned} \tag{18}$$

which is related, for timelike p^2, to the total e^+e^- annihilation cross section to hadrons. Using the methods outlined above it is a simple matter to evaluate Π to lowest order in the density of instantons. One need only calculate $\Pi_{\mu\nu}$ using as a quark propagator the propagator in a single instanton background field in order to obtain $< J_\mu(x)J_\nu(0) >_{AI}$ [7]. The explicit form of this propagator has been derived for massless quarks [8].

One finds [7] that the instantons (plus anti-instanton) contribution (for large spacelike p) is for small quark masses

$$\sum_i Q_i^2 \frac{8\pi^2}{p^4} \int \frac{d\rho}{\rho^5} D(\rho)\Delta(m_i\rho) \tag{19}$$

compared to the perturbative expression $\Sigma Q_i^2 \ln p^2/\mu^2$, where Q_i is the charge of the ith quark, m_i its mass and $\Delta(m_i\rho)$ the value of the fermionic determinent, det[iD-m], in the background instanton field. As we shall see below Δ depends strongly on the mass of the quark, vanishing like $m\rho$ for small mass and approaching 1 as $m\rho \to \infty$. For the light quarks m is to be understood as consisting of a bare part and a dynamically generated momentum dependent mass. According to the picture developed below this dynamical mass turns on rapidly as the momentum is decreased. In $\Delta(m\rho)$ m should be evaluated at $p \sim 1/\rho$. We argue that the net effect is that $\Delta(m\rho)$ for the light quarks can be approximated by $\Theta(\rho-\rho_A)$, thus providing a lower cutoff on the instanton size [7]. Again we must introduce an upper cutoff into Eq. 19 to supress large, and thereby dense and interacting, instantons and merons.

We note that the instanton corrections fall rapidly with increasing p. For moderate values of p they are quite large. With reasonable cutoffs on the ρ integration we find that the magnitude of the instanton corrections becomes comparable to the perturbative term for values of p that correspond to $\bar{g}^2(p)/8\pi^2 \sim 1/20$! Thus we conclude that one cannot trust ordinary perturbative corrections to asymptotic freedom once the effective coupling is of order 1/20. Although at present one cannot reliably calculate $\pi(p)$ for such values of p, since the integrals in Eq. 19 are very sensitive to the upper cutoff and we have not included large instantons and merons it is clear that the effects are large and rapidly varying with momentum. They are of the form $(CM_H/p)^4$ where M_H is the hadronic mass scale and $c \approx 1$. Thus we expect that the corrections are negligible for $p \gtrsim CM_H$ and increase rapidly for $p \lesssim CM_H$. Since the effective coupling $g^2(p)/8\pi^2$ remains small down to $p \sim CM_H$ one expects a sharp transition from the region of asymptotic freedom, where there are small logarithmic corrections to free field behavior, to the region where there are non-perturbative effects. This could explain precocious scaling.

IV. Massless Fermions and Dynamical Symmetry Breaking.

The effect of massless fermions on the structure of the θ-vacuum is substantial as is the effect of vacuum tunneling on the chiral symmetry of the QCD. Indeed instantons provide a cure for the notorious U(1) problem [3,4,5] as well as a mechanism for the dynamical symmetry breaking of chiral U(2) [2,4]. This is due to the Adler-Bell-Jackiw anomaly in the divergence of the axial baryon number current $J_\mu = \bar{\Psi}\gamma_\mu\gamma_5\psi$. In the presence of this anomaly one can still define a conserved charge, Q_5, however this charge is not invariant under the discrete gauge transformations that transform one winding number vacuum into another. Therefore in the presence of massless fermions there can be no tunneling between vacua with different winding number since the Hamiltonian, and thus the time evolution operator, conserves axial baryon number. The coherent superposition of these vacua, the θ-vacuum, is still the correct ground state of the theory. However in the absence of tunneling all the θ-vacua are physically equivalent.

The U(1) problem disappears simply due to the fact that a θ-vacuum is not U(1) invariant. This does not lead to a Goldstone U(1) particle. However the θ-vacuum still possesses a chiral SU(N) symmetry if there are N massless quarks, which must be broken dynamically.

Although tunneling is suppressed by the existence of massless fermions, virtual tunneling can still occur. In other words a single instanton configuration cannot contribute to the vacuum to vacuum amplitude, but an instanton anti-instanton pair can. Such a pair now interacts strongly, via an attractive logarithmic interaction whose coupling is proportional to the number of massless fermions. This means that the analog instanton gas is similar to a dipole gas, consisting of tightly bound instanton anti-instanton pairs. The nonperturbative effects of such pairs will be quite small, since they do not differ much from standard vacuum configurations. Thus in order to obtain significant nonperturbative effects it is necessary to liberate the instantons. This will occur if the quarks develop a dynamical mass, and we argue that the instantons themselves are a source of dynamical quark mass generation [1].

The easiest way to see the source of dynamical quark mass generation is to consider the effective Lagrangian that summarizes the DGA for quark Greens functions. As I explained previously the vacuum expectation value of any product of fields is easily calculated in the DGA using Eq.(12). In the presence of fermions $D(\rho)$ must be multiplied by the determinant of the operator $(i\not\partial - m - \not A_I)$ as a consequence of integrating over the fermion fields in a background instanton field A_I. For N zero mass fermions this determinant vanishes as m^N, since the massless Dirac equation in a background instanton field possesses a zero energy mode. For this reason, unless $< O >_{A_I}$ is singular as $m \to 0$, there will be no modification to $< O >$ due to instantons in the DGA. For U(1) invariant Greens functions this is the case, which is just another way of saying that massless fermions suppress tunneling.

However if $O_D = \det\{\bar\psi_i(1\pm\gamma_5)\psi_j\}$, where i and j run over all flavors, then $< O >_{A_I} \sim 1/m^N$ as $m \to 0$. This is because the fermion propagator contains a term $\psi_0\psi_0^+/m$ (for small m), where ψ_0 is the above mentioned zero energy mode. The powers of m then cancel and such a Greens function does receive a contribution from a single instanton. This is to be expected since O_D changes the axial baryon number by 2N, and thus can connect states with winding number zero and one. Thus in the DGA one can represent the effect of instantons by an effective Lagrangian of the form

$$L_{eff} = \det_{i,j} \bar{\psi}_i (1+\gamma_5)\psi_j + (\gamma_5 \to -\gamma_5) \tag{20}$$

which is SU(N) chirally invariant, non-local (and thus renormalizable) and U(1) non-invariant. Thus we see the explicit breaking of axial baryon number symmetry.

If N=1 such an effective Lagrangian directly generates a quark mass, which we calculate to be [1]

$$m(p) = 32\pi^2 \int \frac{d\rho}{\rho^2} D(\rho)\, e^{-2p\rho}. \tag{21}$$

Numerical evaluation shows that m(p) (defined so that the quark propagator is $(\not{p}+m(p))^{-1}$ vanishes for rapidly large p(as $\sim 1/p^9$ but turns on very strongly for moderate values of P($\approx 15\mu$). What this means is that instantons of large enough size, ρ_A, will no longer be confined in dipole-like pairs; since a background instanton field of size $\gtrsim \rho_A$, the mean fermion momentum will be small enough so that the effective fermion mass will be substantial and the fermion determinent will no longer vanish. Thus the quark mass generation will liberate instantons of size $\gtrsim \rho_A$ which then can be treated as a dilute gas and which will give rise to the dynamical effects discussed previously.

In the case of 2 or more massless fermions the situation is much more complicated. The determinental effective interaction does not directly produce a quark mass, since it still is Su(N) chirally invariant. However we argue that it serves as a source for dynamical chiral symmetry breaking. In fact it has the precise (although non-local) form of the $(\bar{\psi}\psi)^2$ interactions which have historically served as models of dynamical symmetry breaking [10,11]. Now to actually construct a self consistent quark mass is a difficult task, one which is outside the domain of the DGA. However one can argue, as we do in the case of N/2, that the chirally symmetric θ-vacuum is unstable against chiral symmetry breaking, and that the true groundstate must have $\langle\bar{\psi}\psi\rangle \neq 0$. To see this we calculate the $\sigma = \bar{\psi}\psi$ propagator in the DGA, using the above effective vectex, and find that it develops a tachyon pole. Such a singularity is indicative of the instability of the theory, as if one is perturbing in a theory with a potential $V(\sigma) = -m_\sigma^2\sigma^2 + \lambda\sigma^4$ about the symmetric $\sigma = 0$ vacuum instead of the stable asymmetric $\sigma = m_\sigma 2/2\lambda$ vacuum.

Consequently the true vacuum of QCD, in the presence of quarks with zero bare mass, will be a Goldstone-θ vacuum in which $<\bar{\psi}\psi> \neq 0$ and the quarks will possess a dynamical mass m(p). From the analysis of the momentum dependence of the σ propagator we conclude that this mass will be large for rather large values of P, thus liberating instantons of rather small size. Furthermore since the effective interaction that produces the chiral symmetry breaking is U(1) noninvariant the theory will possess SU(N) Goldstone bosons (the π in the use of SU(2)) but no U(1) Goldstone boson (η').

V. Instanton Interactions.

In this section I shall discuss some features of the instanton gas in QCD. First we recall that the analog gas of instantons in the perfect gas approximation (i.e. neglecting all interactions) has a density of instantons of scale size between ρ and ρ+dρ

$$\frac{d\rho}{\rho^5} D(\rho) = \frac{d\rho}{\rho^5} \cdot 1 \left(\frac{8\pi^2}{\bar{g}^2(\rho)} \right)^6 e^{-\frac{8\pi^2}{\bar{g}^2(\rho)}} \tag{22}$$

where the effective coupling $8\pi^2/\bar{g}^2(\rho) = 8\pi^2/g_0^2 + 11 \ln 1/\rho\mu$. The gas of instantons can be consistently treated as dilute as long as the fraction, f(ρ), of space-time occupied by instantons of size less than ρ is less than unity. If we take the "volume" of an instanton of size ρ to be $\pi^2/2\ \rho^4$(i.e. that of a sphere of radius ρ) we find that

$$f(\rho) = \pi^2 \int_0^\rho \frac{d\rho'}{\rho'} D(\rho'). \tag{23}$$

For SU-3, using $dx / d\ln 1/\rho\mu = 11$, we have

$$f(x) = \frac{\pi^2}{11} \int_x^\infty dx \cdot 1\ x^6 e^{-x}.$$

This attains the critical value 1 at $x_c \approx 14$. We find that when f is less than one, say of order .1, instantons cause significant modifications of the vacuum properties, in spite of the smallness of the coupling in this region ($x \sim 20$). But when $f \gtrsim 1$, the DGA breakdown and some new vacuum physics must take over. Everything that we will find strongly suggests that this transition is associated with confinement, or at least the physics that sets the scale size of hadrons. Since the rise in f is so rapid as a function of x or ρ one gets a rather sharp definition of the hadronic coupling constant and scale size : $x_c \sim 16$, $\rho_c\mu \sim .25$.

The analysis of various effects arising from merons or instanton interactions will cause as to revise these numbers slightly but the basic point remains : the DGA reveals how and where dimensional transmutation occurs (i.e. $f[x(\rho_c\mu)]\sim 1$) and shows, most importantly, that the hadronic coupling is a small number. The value associated with this coupling ($\alpha_c \sim .4$) is not far from the numbers which have been extracted from optimistic studies of scaling in electroproduction.

I shall now discuss the effects that arise from the interaction of instantons within the DGA, i.e. when the density is still small but large enough for the instantons to influence each other. To study this problem consider imposing on the analog gas a weak slowly varying external field $F^{ext}_{\mu\nu}$, which in the normal vacuum contributes

$$\frac{1}{4g^2}\int dx^4 \sum_a (F^{ext}_{\mu\nu,a})^2$$

to the action. In fact A^{ext}_μ must be regarded as a perturbation on $[A_\mu]^0$, the multiple instanton configuration which dominates the DGA vacuum. The interaction of the external field with the instanton must be included in δS and we shall see that the net effect is just a coupling constant renormalization. We write $A_\mu = [A_\mu]^0 + \delta A_\mu$, and adjust δA_μ to approach the potential of $F^{ext}_{\mu\nu}$ ($\delta A_\mu = -\frac{1}{2} F^{ext}_{\mu\nu} x^\nu$ in Landau gauge) far away from the instanton locations in $[A_\mu]^0$; and solve the linearized equations of motion to find δA_μ near the instantons. The analysis shows [1] that a single instanton is a weak slowly varying external field may be assigned an interaction energy

$$S_{int} = -\frac{2\pi^2}{g^2}\rho^2 \bar{\eta}_{a\mu\nu} F^{ext}_{\mu\nu} \tag{24}$$

where F^{ext} is the background field at the instanton position. Note that if F^{ext} is self dual (say arises from another, far away, instanton) S_{int} vanishes since $\eta_{a\mu\nu}$ is anti-self dual. However, if F^{ext} arises from an anti-instanton of size $\bar{\rho}$ at a distance R we find

$$S_{int} = \frac{32\pi^2}{g^2}\frac{\rho^2\bar{\rho}^2}{R^4} R_{ab}\bar{\eta}_{a\mu\nu}\eta_{b\mu\nu}\hat{R}_\nu\hat{R}_\nu, \tag{25}$$

where $\hat{R}$ is the unit vector pointing from the instanton to the anti instanton and R_{ab} the matrix describing to relative group orientation. The maximum value of $-S_{int}$, for parallel orientation is

$$\frac{96\pi^2}{g^2}\frac{\rho^2\bar{\rho}^2}{R^4}$$

independent of $\hat{R}$, in agreement with Forster's calculation of the action of a pair constrained as to location.

This interaction energy is that of a dipole-dipole interaction. Instantons behave as four dimensional dipoles with a color magnetic dipole moment proportional to $\rho^2\bar{\eta}_{a\mu\nu}$. Thus we expect the dilute instanton gas will behave like a dilute gas of spins whose response to an external field will be described by a susceptibility, which will lead to a "classical coupling constant renormalization".

This can be calculated using a virial expansion to find the effect of an instanton medium on an external field. Given the interaction of a single instanton with F^{ext}, one computes $-S_{eff} = < e^{-S_{INT}} - 1 >$, averaged over orientations of the instanton, weighted with the appropriate instanton density and integrates over instantons position and scale size to get the full effective action. For weak F^{ext}

$$< e^{-S_{INT}} - 1 > = \left(\frac{F_{ext} - \tilde{F}_{ext}}{2} \right)^2 \frac{8\pi^2}{g^2} \frac{\pi^2 \rho^4}{8} + O(F_{ext}{}^4)$$

where $\tilde{F}_{ext}$ is the dual of F_{ext}. The appearance of $F_{ext} - \tilde{F}_{ext}$ reflects the lack of instanton-instanton interaction. Adding instantons and anti-instantons converts this to F^2_{ext}, so that the effect of the medium is to renormalize the original action by a multiplicative constant $1/k$, where

$$\frac{1}{k} = 1 - \int \frac{d\rho}{\rho^5} \frac{D(\rho)}{8} \frac{8\pi^2}{g^2(\rho)} \; 4\pi^2 \rho^4 \tag{26}$$

Since $k > 1$ the vacuum is behaving like a "paramagnetic" medium, i.e. the interaction energy between fixed external sources is increased. In other words the instantons <u>anti-screen</u> external fields. One finds that k becomes quite big even in the dilute gas region where $f << 1$.

This mechanism of instantons anti screening color fields can lead to a picture of confinement. Namely if k is big enough it may be energetically favorable in the presence of external colored sources (quarks) to form a flux tube (or bag !) in which the flux is expelled from the normal vacuum region (full of instantons and thus large K) and confined to a region of abnormal vacuum ($K \approx 1$) where no instantons are present. The vacuum energy that it costs to expel the instantons will make up for the lowered interaction energy between the quarks. This will lead to a linear increase in energy as one separates the quarks, and determine the confinement scale as that at which K departs significantly from one, which is not too different from our earlier criterion based on F. Absolute confinement will only obtain if K actually diverges. In the following section we shall see that the instanton dipoles of sufficiently large size dissociate into meron null-poles, and that this leads to a divergent susceptibility and thus confinement.

The above effect can be interpreted as a coupling constant renormalization, in that the action of an instanton is reduced by small instantons within it, crudely

$$\frac{8\pi^2}{\bar{g}^2(\rho)} = \frac{8\pi^2}{g^2(\rho)} - \int_0^\rho \frac{d\rho'}{\rho'} \frac{4\pi^2}{8} \left(\frac{8\pi^2}{g^2(\rho')} \right)^2 D(\rho') \qquad (27)$$

where $g(\rho)$ is the asymptotic freedom coupling and $\bar{g}(\rho)$ includes the "classical" renormalization. Replacing $x = 8\pi^2/g^2(\rho)$ under the integral by $\bar{x} = 8\pi^2/\bar{g}^2(\rho)$ one derives a self consistent equation for $\bar{x}$

$$\bar{x}(\rho) = x(\rho) - \int^\rho \frac{d\rho'}{\rho'} \frac{4\pi^2}{8} \bar{x}^{2}(\rho) \, D(\bar{x}(\rho))$$

or

$$\frac{d\bar{x}}{d\ln \frac{1}{\rho\mu}} = 11 + \frac{4\pi^2}{8} \bar{x}^{2} D(\bar{x}) \qquad (28)$$

Numerical integration shows [1] a rapid use in $\bar{g}$ for $\rho\mu \sim .15$, whereas at this point x is only ≈ 22, and $F \sim .1$. Again we find that the instanton effects once significant are dominant.

Since $\bar{x}$ decreases more rapidly with ρ then given by asymptotic freedom Eq. (28) should be modified, replacing 11 by

$$\frac{d\bar{x}}{d\ln \frac{1}{\rho\mu}} \,.$$

The effect is to reduce F(x) so that F(x) is always <1 for $x \gtrsim 10$. We conclude that once instanton interactions are included the density no longer rises so dramatically but there is still a well defined scale size and coupling constant at which the renormalized coupling (and vacuum susceptibility) rise rapidly.

We also note that the same "classical" coupling renormalization can be calculated by evaluating the gauge field propagator $D^{ab}_{\mu\nu}(x-y)$ in the DGA [1].

VI. Merons and Confinement.

We have seen above that instantons alone, at least within

the DGA, do not lead to quark confinement. This is essentially due to the fact that they act as four dimensional magnetic dipoles with a field strength that falls off as $1/r^4$ for large r whereas one can argue that for

$$< \exp i \oint_L A.dx > \approx \exp < -\int_L\int_L dx_1^\mu dx_2^\nu A_\mu(x_1)A_\nu(x_2) >$$

to grow as exp[-Area(L)] requires the configurations where $F_{\mu\nu} \sim 1/r^2$. This however implies infinite action for such configurations. I shall argue that infinite configurations can contribute to the vacuum wave function if the coupling constant is large enough and that there exists a class of configurations-merons- that will lead to quark confinement for reasonably small values of the effective coupling.

Alternatively one can think of the instantons as four dimensional dipoles, whose dipole moment is proportional to ρ^2, where ρ is the instanton scale size. The susceptibility of this medium is proportional, in the DGA, to the mean of ρ^4 (see Eq. 26). Thus the effect of the instantons can be increased by increasing their size and thereby their dipole moment. This however might not be the optimal way to increase the dipole moment. If we were to imagine that the instanton dipole was formed out of monopoles then separating the monopoles would also increase the dipole moment, as well as the phase space available to the distorted instanton. Non-abelian gauge fields can be classified in the same way as Abelian fields as long as $A_\mu r \to 0$ as $r\to\infty$, since then the Feld equation linearize asymptotically. An

$$A_\mu \sim \frac{1}{r^{1+\ell}} \quad , \quad F_{\mu\nu} \sim \frac{1}{r^{2+\ell}}$$

corresponds to a four dimensional ℓ-pole. Now an instanton dipole cannot be regarded as a formed from monopoles since one could not conserve its topological charge. (i.e. if $A \sim 1/r^2$ and is non-singular then the topological charge vanishes). Instead if the instanton can be regarded as being formed from null-poles, for which $A_\mu \sim 1/r$ each of which carries one-half the topological charge of the instanton. The existence of such null-poles, or merons, is special to a non-abelian gauge theory. They necessarily have an action which diverges logarithmically, but allow us to distort an instanton into a configuration consisting of a meron pair, separated by a distance R, which acts like a dipole of dipole moment R^2. As we shall argue below for large R the free energy of such a configuration is much greater than that of a large instanton of comparable size, and for large enough R the susceptibility will diverge leading to a phase transition where the instantons dissociate into merons.

Instantons contribute to the vacuum to vacuum amplitude for arbitrarily small values of the coupling since they occupy an infinite volume in function space. Namely the instanton contributions is proportional to V exp$(-S_0)$, where the factor of V = volume of space time overwhelms the action term no matter how small g^2 is. In the statistical mechanical analog we might say that $S_0 = 8\pi^2/g^2$ is proportional to the energy it costs to create an instanton (its chemical potential), $S_0 = \beta\mu$; whereas V is a measure of the entropy of the instanton, entropy = S = lnV. What is important in the vacuum to vacuum amplitude = partition function is not the energy, E, of a configuration, but its free energy, F=E-TS. For an instanton $F = 8\pi^2/g^2 - \ln V$, which is negative as $V\to\infty$ for any g. In terms of the analog gas this implies that we will be in the "instanton phase" with a finite density of instantons for arbitrarily small values of g^2=temperature.

Now let us consider other field configurations. As long as one considers configurations which are localizable in the appropriate sense the entropy will always behave as lnV. Therefore we could very well tolerate configurations whose energy=action diverges logarithmically as $V\to\infty$. A classical configuration with logarithmically divergent action will have a free energy equal to $c\ 8\pi^2/g^2 \ln V - \ln V$, which can vanish for large enough $g^2/8\pi^2$ ($\geq c$). Thus such configurations will appear, with finite density, once the coupling is greater than some value and must be taken into account in the functional integral. The only constraint on configurations with action that diverges as lnV is that they can be superimposed with independent entropy of position (i.e. that a superposition of N such configurations occupies a volume of function space $\sim V^N$) and that the coupling be large enough.

We therefore are interested in singular field configurations which are solutions of the Yang-Mills Euclidean equations. They should be solutions so that they can be combined, with a finite number of constraints to fix the collective coordinates of each configuration, to yield multiple field configurations with independent entropy of position for each one and can be singular as long as the action only diverges logarithmically. (It is a reasonable conjecture that the only finite action solutions are instantons). Such a solution has been discovered, it is the "meron" solution

$$A_\mu^a = \eta^a_{\mu\nu} \frac{x_\nu}{x^2} . \tag{29}$$

This configuration has topological charge 1/2 and its action diverges for large volume as $3\pi^3/g^2 \ln(V)^{1/4}$. The above solution is singular at x=0 and corresponds to a meron of zero size located at the origin. This ultraviolet singularity will be removed by the quantum fluctuations that, due to asymptotic freedom, will effectively smear out the meron and give it a finite size. According to the above argument the free energy of such a configuration will vanish for $g^2/8\pi^2 = 3/32$ and for such a coupling, which is still quite small, merons will contribute significantly to the functional integral.

The above form of the meron solution actually contains a second meron located at $x=\infty$. By a conformal mapping we obtain the configuration

$$A^a_\mu = \eta^a_{\mu\nu} \left\{ \frac{(x-x_1)^\nu}{(x-x_1)^2} + \frac{(x-x_2)^\nu}{(x-x_2)^2} \right\}, \tag{30}$$

which consists of a meron at x_1 and a meron at x_2 (both of zero size). A non-singular 2 meron configuration is then obtained by smearing out the singularities of $x=x_1$, x_2; i.e., by introducing constraints that force the topological charge within 2 spheres, of radii r_1 and r_2, to be equal to 1/2. One then obtains a non-singular configuration consisting of half instantons inside the spheres and the above meron field outside the spheres. The total action is then

$$S_M = \frac{4\pi^2}{g^2} + \frac{4\pi^2}{g^2} + \frac{6\pi^2}{g^2} \ln \left(\frac{R}{\sqrt{r_1 r_2}} \right),$$

where R is the separation of the merons. This consists of the energy to create each meron, $4\pi^2/g^2$, and the interaction energy of the pair

$$\frac{6\pi^2}{g^2} \ln \left(\frac{R}{\sqrt{r_1 r_2}} \right).$$

It is necessary to discuss the quantum fluctuations about such a configuration in order to prove that they actually contribute to the functional integral and to determine how the couplings that enter in S_M are renormalized. This we have done for the two-meron configuration [1]. We find that the couplings that appear in the self energy of a meron are renormalized at its core size,

$$\frac{4\pi^2}{g^2} \to \frac{4\pi^2}{\bar{g}^2(r_i)} ;$$

whereas the coupling that appears in the interaction energy is renormalized at a distance $\sqrt{Rr}$ (for equal size merons), i.e.

$$\frac{6\pi^2}{g^2} \ln \frac{R}{r} \rightarrow \frac{6\pi^2}{\bar{g}^2(\sqrt{Rr})} \ln \frac{R}{r} \, .$$

This has two important implications. First

$$\exp\left\{-\frac{4\ ^2}{\bar{g}^2(r)}\right\} \sim r^{11/2}$$

will make the density of zero size (singular) merons vanish. The dominant contribution in fact is obtained from finite size merons. Second the fact that as the separation between the merons increase the coefficient of lnR decreases suggests that the merons will effectively dissociate for large R ! Indeed the probability of finding a pair with separation R behaves as

$$R^4 \exp - \left| \frac{6\pi^2}{\bar{g}^2(\sqrt{Rr})} \ln \frac{R}{r} \right| \, .$$

In order to determine whether or not there exists a phase transitions at some meron size R we have considered the dynamical effects of merons [1]. In particular we consider the response of a meron pair to an external color field. We find that a meron pair behaves as a magnetic dipole, whose dipole moment is proportional to R^4. The paramagnetic susceptibility of a gas of meron pairs is proportional to $< R^4 >$ which, with the above probability for a pair of size R, leads to a divergent susceptibility once R is big enough that $\bar{g}^2(\sqrt{Rr})/8\pi^2 = 3/32$. This we believe signals the onset of a new phase, in which merons are effectively liberated. This (plasma ?) phase will confine quarks.

A quantitative treatment of the confinement due to merons can be carried out for a dilute gas of meron pairs. As in the DGA to the instanton gas the effects of meron pairs on the quark loop is easily evaluated. We find now a quark-anti quark potential, $\varepsilon(R)$ which increases with quark separation as

$$\varepsilon(R) \sim R^{7 - \frac{6\pi^2}{g^2(R)}} \qquad (31)$$

This starts confining once $g^2/8\pi^2 \geq 3/28$ and becomes linear at $g^2/8\pi^2 = 1/8$. This calculation is very reassuring since the value of the coupling at which the potential becomes linear is still relatively small. However we believe that there is a phase transition before this point, at $g^2/8\pi^2 = 3/32$, where $\varepsilon(R) \sim 1/R$, and thus we cannot rigorously penetrate the confinement domain until we understand the meron "plasma". It is presumably the properties of the plasma phase that cause the potential to saturate once $\varepsilon(R) \sim R$.

VII. Conclusions.

In this paper we have initiated a systematic exploration of the relevant degrees of freedom and the dynamics of quantum chromodynamics. We have discovered that to a large extent the dynamical properties of QCD are a consequence of the structure of the vacuum arising from the tunneling between degenerate, classically stable vacua; and that the relevant degrees of freedom can be taken to be the Euclidean path histories that can be used to calculate the tunneling amplitudes in the semiclassical approximation. We have seen that the non perturbative structure of the vacuum that arises due to this tunneling explains the major features of QCD, i.e. the dimensional transmutation which determines the size of the hadrons and the strong interaction coupling constant, the source of dynamical chiral symmetry breaking and the mechanism responsible for confinement.

Let us summarize the picture of QCD that has emerged from our investigation. The general structure of the θ vacuum as a superposition of n-vacua and the solution of the U(1) problem follow from general considerations. The detailed dynamics of QCD however depend crucially on the scale of phenomena under investigation. At very short distances asymptotic freedom guarantees that the effective coupling is small enough so that the theory is essentially free. The density of instantons of small size vanishes rapidly and at short distances the "perfect gas approximation" applies. Since the up and down quark masses are very light, tunneling is suppressed at short distances or, alternatively, instantons of small size are tightly bound to anti-instantons. We have shown that at distance ρ_A, corresponding to couplings or order $g^2/8\pi^2 \sim 1/20$ chiral symmetry occurs, via a Nambu-Hona-Lasinio type mechanism generated by the effective determinal interactions provided by instantons. At this distance therefore the quark dynamical mass will become substantial and tunneling will be restored. Alternatively, instantons of size greater than ρ_A will exist in isolation. At this distance scale the instanton gas is dilute and can be treated by mean field theory techniques. Their dynamical effects are large and significant. For example we have seen how they substantially modify the Coulomb potential between massive quarks. They are certainly much greater in importance than the standard perturbative corrections. Physically the reason that such tunneling effects are so large for small coupling is the existence of many distinct tunneling paths (degrees of freedom for instantons). The interactions in the dilute instanton gas, which behaves as a paramagnetic medium of magnetic dipoles, cause a renormalization of the effective coupling resulting in a rapid increase in its value at a sharply defined distance $\rho_C \sim .2\mu$. At this distance, where the effective coupling is of order $g^2/8\pi^2 \sim 1/10$ instantons dissociate into merons. The latter configuration are such as to provide a mechanism for confinement of quarks. In the confinement phase we conjecture that the vacuum is dominated by configurations that can be described as a plasma of merons, leading to a picture of electric confinement in ordinary space similar to that in a magnetic super conductor. We associate ρ_C with the confinement scale or the size of a hadron and $g^2/8\pi^2 \sim 1/10$ with the hadronic coupling constant, thereby concluding that in QCD the coupling is always relatively small and that the strong interactions can be treated by semiclassical methods.

Much work remains to be done. The quantitative details of the chiral phase transition require further investigation. A formalism for generating systematic improvements of the dilute gas approximation should be developed. Most important we must gain control over the meron configurations and the confinement phase where our understanding is still at a qualitative and conjectural stage. Finally we must proceed to construction of hadronic states, and the calculation of their masses, couplings and scattering amplitudes. This is a difficult problem in a theory such as QCD where it is not even clear what is the best way to begin the attack. Perhaps the most reasonable strategy is to construct a phenomenological model of hadrons which would play a role similar to that of the Landau-Ginzburg theory in superconductivity. Given our picture of confinement as described above the most likely candidate for such a model is a bag model.

In this model the quarks would be confined in a region of space-time void of instantons and merons, since normal vacuum would expel large scale color fields. In principle one could desire the parameters of such a bag (the volume energy of the bag, etc...) from first principles. It would differ in many respects from the MIT bag. Most importantly chiral symmetry breaking would be maintained, the instantons outside the bag would generate dynamical chiral symmetry healing and a Goldstone pion !

Clearly much work remains to be done.

REFERENCES

1. C. Callan, R. Dashen and D. Gross
Phys. Lett. 66B, 375 (1977).
I.A.S. preprint COO-2220-115 (1977).

2. A. Belavin, A. Polyakov, A. Schwartz and Y. Tyuphin
Phys. Lett. 59B (1975).

3. G. 't Hooft, Phys. Rev. Letters 37, 8(1976);
Phys. Rev. D14, 3432 (1976).

4. C. Callan, R. Dashen and D. Gross,
Phys. Lett. 63B, 334 (1976).

5. M. Tackin and C. Debbi,
Phys. Rev. Lett. 37, 1721 (1976).

6. C. Callan, R. Dashen, D. Gross, F. Wilczek and A. Zee
To be published.

7. N. Andrei and D. Gross,
Princeton preprint, to be published, Phys. Rev.

MAGNETIC MONOPOLES IN CLASSICAL FIELD THEORIES

Walter TROOST *

K.U.L., Leuven, Belgium

The subject of magnetic monopoles will be treated from a purely theoretical point of view. An exploration will be made into classical field theories to try and find solutions of classical field equations with a magnetic charge, without introducing this charge as a quantum number of the fields at the start. I hope to show that they are not as rare as is generally thought.

To begin with, I will introduce the theoretical apparatus I will use. It is of course important that this apparatus is not restricted in such a way as to make monopoles "unobservable". Therefore, it should allow for the description of magnetic charges in a very natural way, and in a general framework. The apparatus I have in mind was modeled by T.T.Wu and C.N.Yang[1] after some concepts in the mathematical theory of fibre bundles. I will not go into the mathematicl details, so the presentation will be very pragmatic. After that, I will describe the gauge-theory solution found by A.M. Polyakov and G.'t Hooft[2], which can be interpreted as containing a magnetic charge. I will give a description of this solution in a unitary gauge,where the use of the theoretical formalism described in the first part is adequate. I will try to picture the forces at work which stabilise this solution from a physical point of view. The recognition of the essential forces will then lead to the generalisations discussed in the next part. In so far the work described here is original (most of sec 2, and sec.3), it was done in collaboration with P. Vinciarelli[3].

* Aangesteld navorser van het NFWO, Belgium

1. FORMALISM

Let us first look at some indications for the inadequacy of the traditional treatment of monopoles.
The magnetic field of a magnetic pole with charge g is given by*:

$$\vec{B} = g\,\frac{\vec{r}}{r^3} \qquad (1)$$

Apart from the problem that this gives rise to an infinite magnetostatic self-energy, a description in terms of the usual electromagnetic potentials $A_\mu(x)$ also poses problems. A possible candidate is

$$A_0 = 0$$

$$\vec{A} = g(-1 + \cos\theta)\vec{\nabla}\phi \quad = g\,\frac{-1 + \cos\theta}{r\sin\theta}\,\vec{1}_\phi \qquad (2a)$$

Formally, the curl of this potential gives the desired $\vec{B}$ of eq.(1), but $\vec{A}$ is singular at $\theta = \pi$, i.e. the negative z-axis. Another candidate is :

$$\vec{A} = g(1 + \cos\theta)\vec{\nabla}\phi \quad = g\,\frac{1 + \cos\theta}{r\sin\theta}\,\vec{1}_\phi \qquad (2b)$$

with a curl formally equal to the same $\vec{B}$, but with singularities on the positive z-axis. It is obtainable from (2a) by a formal gauge transformation

$$\vec{A} \to \vec{A} + \vec{\nabla}\Lambda$$

where $\Lambda = -2g\phi$. This explains why it gives the same $\vec{B}$.
These singularities cannot be avoided : it is a mathematical impossibility to find a set of potentials $\vec{A}$ regular everywhere (except possibly at the origin) whose curl is $\vec{B}$ of eq.(1). Whatever the gauge, there will always be at least one line of singularities (the Dirac string) from the magnetic charge to infinity.

---

* : $\vec{r}$ is the 3-dimensional radial vector, r its magnitude. θ and ϕ are polar angles, $\vec{\nabla}$ means gradient, and $\vec{1}_\phi(\vec{1}_\theta)$ is in each point a unit vector in the direction of increasing ϕ (θ).

This problem is reminiscent of the problem of describing the points on the surface of a sphere with coordinates. It is impossible to find a coordinate system that has no singularities anywhere on the sphere. For example, in terms of usual longitude and latitude, a velocity component of a point passing through one of the poles is singular.

The solution of this problem is well known. One chooses different coordinate systems (maps) for different regions (at least two), so that these different regions taken together cover the whole surface. One can then choose the coordinate system in each of these regions in such a way that no coordinate singularities are introduced. In the overlap regions one given in addition formulae which connect the coordinate systems.

This procedure can be adapted rather straightforwardly to th description of monopole potentials,thus overcoming the Dirac string singularity problem, with the following correspondences :

surface of sphere	$\leftrightarrow$	world with a magnetic charge
coordinate system	$\leftrightarrow$	(electromagnetic) gauge
coordinate singularity	$\leftrightarrow$	Dirac string

The choice of a coordinate system corresponds to a choice of gauge: one needs it (or at least finds it convenient) for the description, but there is nothing physical to any particular choice. There is no reason why one should impose oneself the restriction of choosing a definite coordinate system valid over the whole sphere. Likewise, there is no reason to restrict oneself to a particular gauge valid for all space . In fact, if one choses to make this restriction, one finds that in certain physically very regular situations, singularities pop up (example, the motion through a point which happens to be the Northpole of a longitude - latitude coordinate system). Likewise, consider eq.(1) : the physical fields are very regular outside the origin. But if one wishes to choose one definite gauge, one finds oneself confronted with the Dirac string singularities.

Once the analogy of the problem is realised, a possibility for the solution is apparent. Let us divide the space (except the monopole position) into two parts according to the value of the polar angle θ :

$$\text{region a} \quad (R_a) \qquad 0 \leq \theta \leq \frac{3\pi}{4} \tag{3a}$$

$$\text{region b} \quad (R_b) \qquad \frac{\pi}{4} \leq \theta \leq \pi$$

Furthermore, in $R_a (R_b)$ we use the potentials given by eq. 2a(2b). It is then clear that we have, in R_a and R_b taken separately, a singularity-free description of the monopole potentials. Since R_a and R_b together cover the whole space (except the origin of

course) it only remains to be shown how to "connect" the two descriptions in the overlap region without introducing a singularity. In the case of coordinate systems on a sphere this is done by giving a regular coordinate transformation; in our case, to guarantee that the two choices of potentials (2a) and (2b) describe the same physics in the overlap region R_{ab} $(\pi/4 \leqslant \theta \leqslant 3\pi/4)$, we want the patch to be made by a regular gauge transformation. This is indeed the case for the potentials (2a)-(2b); as mentioned before, they are connected by a gauge transformation, $\vec{\nabla}\Lambda = 2g\vec{\nabla}\phi$. Note that this function is regular in R_{ab}.
Up to this point only the electromagnetic potentials were considered. We will also need other fields. These fields, when carrying electric charge, have functional forms that depend on the choice of gauge. Correspondingly, they will be represented by different functions in R_a and R_b (this pair of functions represents a "section" - see ref. (1)). The "patch" in R_{ab} will again be made by a regular gauge transformation, which is typically a multiplication by $S_{ab} = \exp(\, ie\, \Lambda(x))$, where e is the charge of the field being considered. For the gauge transformation at hand this becomes

$$S_{ab} = \exp\left(-2ieg\phi\right) \quad .$$

This looks regular enough in R_{ab}, but in general it is not:since $\phi \to 0^+$ and $\phi \to 2\pi^-$ correspond to the same point in space, S_{ab} will be discontinuous unless

$$2ieg.2\pi \ = \text{a multiple of } 2\pi$$

i.e. unless the Dirac quantisation condition

$$eg = \frac{1}{2}\, n$$

is satisfied.

All of this is an elementary exposition of what the use of Wu and Yang's formalism amounts to in our present simple case, viz the possibility of a U(1) monopole, i.e.a. monopole connected with the U(1) gauge transformations of electromagnetism. It can be extended to higher groups as well, but that will not concern us here. The general theory is adopted from the theory of connections on fibre bundles (a branch of modern geometry), and asserts that all "sections"belong to a definite gauge type, characterised by an integer, the "chain class number", which corresponds to the magnetic charge (in Dirac units).

2. THE 'T HOOFT-POLYAKOV SOLUTION

Consider the classical SU(2) gauge theory with a triplet scalar meson inducing "spontaneous symmetry breakdown" with Lagrangian density

$$\mathcal{L} = \frac{1}{4}\,(F^a_{\mu\nu})^2 + (D_\mu \Phi^a)^2 - \lambda((\Phi^a)^2 - c^2)^2 \qquad (4)$$

(standard notation has been used; $F^a_{\mu\nu}$ are the SU(2) gauge field strengths, Φ^a the scalar triplet).
The following Ansatz reduces the field equations to 2 equations for 2 unknown functions [2]

$$A^a_i = \varepsilon^{aij}\, x_j\, \frac{A(r)}{er^2} \qquad (5)$$

$$\Phi^a = \frac{x^a}{r}\, \Phi(r) \qquad (6)$$

(all other field components are 0)

A solution can be found by computer. The explicit form in the limit $\lambda \to 0$ is also known[4]. Its qualitative features are as in fig. (1). Its (classical) energy is finite. It is analytic in x,y,z everywhere ($A(r) \simeq O(r^2)$, $\Phi(r) \simeq O(r)$ in $r=0$).

To interpret this field configuration, let us first consider the vacuum (lowest energy configuration) and its small excitations. The vacuum is clearly given by $\Phi^a \Phi_a = c^2$, Φ^a in some fixed direction, and A^a_μ = a pure gauge. Let us take the fixed direction of Φ^a to be the 3-direction, $\phi^a = c\,\delta^{a3}$. There is a U(1) gauge invariance left for this configuration, viz. rotations around the 3-axis, which we will associate with electromagnetism. Small oscillations around this vacuum can then be associated with

a) the photon (A^3_μ)

b) two oppositely charged massive vector particles ($A^1_\mu \pm iA^2_\mu$)

c) a neutral scalar meson ($\Phi^3 - c$)

Now we turn to eq.(5)-(6). First of all, there is also a U(1) gauge invariance left : rotations of Φ^a around the "radial" direction. This U(1) gauge invariance is again identified with the electromagnetic gauge group. Its mathematical realisation is more complicated than usual, because this U(1) group is a subgroup of the original SU(2) group which differs from place to place. The field strenght tensor corresponding to it has been given by 't Hooft. Its value for the present solution is :

$$\vec{E} = 0$$

$$\vec{B} = g \frac{\vec{r}}{r^3} \qquad \text{with } eg = 1 \qquad (7)$$

Note that at spatial infinity the scalar field Φ^a has a "length" equal to c (see fig. 1) and is covariantly constant. A^a_μ becomes a pure gauge. It follows that, far away from the origin, the field configuration is the same as in the vacuum (up to a gauge transformation - which has no physical significance). In addition, there is a pointlike magnetic charge in the origin, as shown by the magnetic field of eq.(7), with a strength equal to 2 Dirac units.

This interpretation of the solution (5)-(6) is rather well known. One may not be completely satisfied and ask for more details on the content of (5)-(6) in terms of the physical "particle" fields, which we identified with the small oscillations of the vacuum (photon, charged vector and scalar) and on the mutual interactions of these configurations. Also, although a rather beautiful "topological" argument can be invoked to guarantee the existence of a stable configuration with a magnetic charge, it is by no means clear what the physical reasons are for the stability of this configuration.

Let me first indicate why the answer to these questions is not ridiculously trivial. For this it is enough to consider the expression for the electromagnetic field tensor :

$$F_{\mu\nu} = \frac{1}{|\Phi|} \Phi^a F^a_{\mu\nu} - \frac{1}{e|\Phi|^3} \varepsilon_{abc} \Phi^a (D_\mu \Phi)^b (D_\nu \Phi)^c \qquad (8)$$

proposed by 't Hooft [2)]. This expression is of course considerably more complicated than the corresponding familiar

$$F_{\mu\nu} = \partial_\mu A_\nu - \partial_\nu A_\mu$$

to which it reduces when $\Phi^a = \delta^{a3} c$ $(A_\mu = A^3_\mu)$.

The reason for the complication is that (5)-(6) imply the choice of a very unusual gauge (called "radial gauge"), in which there are no field variables corresponding explicitly to the photon fields A_μ . Likewise, there are no variables corresponding explicitly to the charged vector meson fields, and in fact a gauge invariant expression for them analogous to eq.8 has never

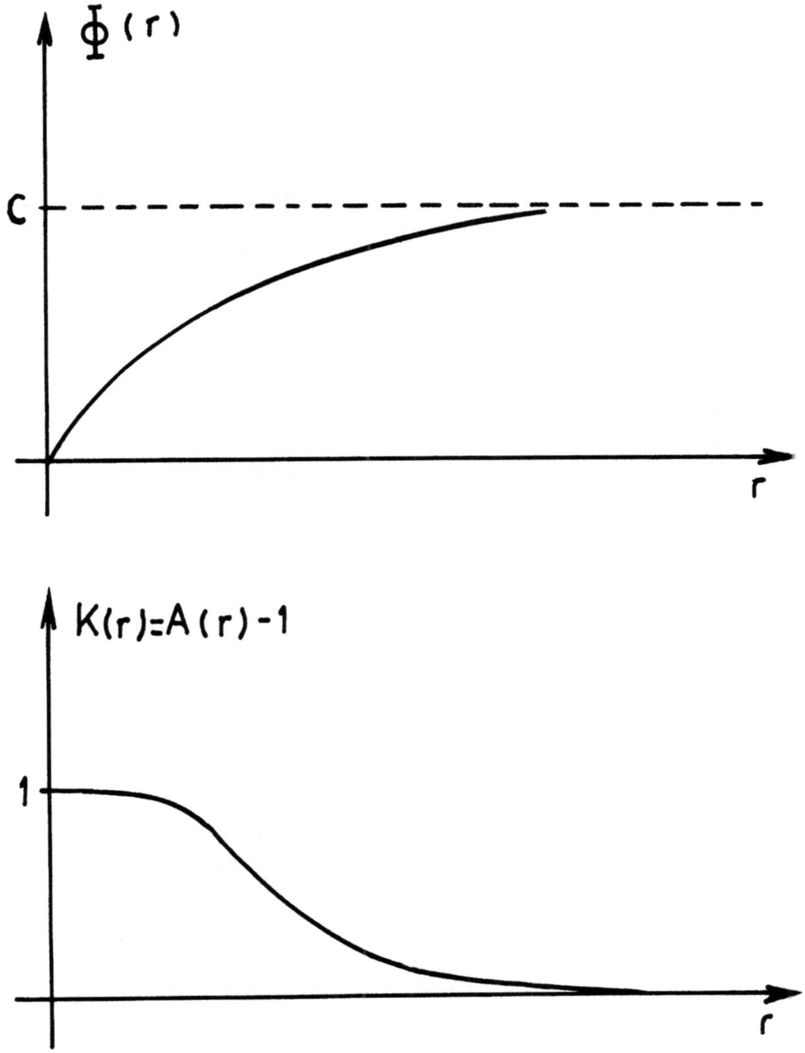

Figure 1. Behaviour of the monopole solution (scalar field ϕ, vector field A) in space.

been written down as far as I know.

To remedy this lack of transparence one would like to go to a more usual gauge, where electromagnetism corresponds to a fixed direction in the SU(2) group space. This can be accomplished by doing a local gauge rotation on the solution (5)-(6), such that the rotated Φ^a satisfy

$$\Phi^a = \delta^{a3}\ \Phi(r)$$

Then, clearly, the 3 direction corresponds to the photon, ($A_\mu = A^3_\mu$) as is the case in the vacuum. Φ^1 and Φ^2 are absent from the solution, also as in the vacuum, which is also satisfying since they correspond to unphysical ("gauge") degrees of freedom. The charged vector meson fields will of course be identified with

$$W^{\pm}_\mu = \frac{1}{\sqrt{2}}\,(A^1_\mu \pm i\, A^2_\mu)$$

Having spelled out the advantages, let us now actually do this local gauge rotation. Φ^a , eq. (6), has to be rotated to $\delta^{a3}\ \Phi(r)$. We focus our attention on the configurations in the x-z plane. A possibility is to rotate around the y-axis over an angle θ in the direction of the positive z-axis (see figure 2 a,b). Formally one obtains

$$\Phi^a = \delta^{a3}\ \phi(r)$$

$$A^3_i = \text{as in (2a)}$$

The rotated A^3_i become equal to the monopole potentials, with a singularity along the negative z-axis. This singularity is not unexpected : we see from fig. 2b that the rotation we performed is singular there. Of course we could have performed the rotation in a different manner. But as a corollary of the impossibility mentioned before, to write down a set of monopole potentials without a line of singularities, we cannot find a non singular local gauge rotation for all space which brings

$$\Phi^a \simeq \frac{r^a}{r}\ \Phi(r) \quad \text{to} \quad \Phi^a = \delta^{a3}\ \Phi(r)\ .$$

Luckily this is not really an obstacle anymore : look back to section 1 to see what the solution is. To repeat, there is no physical reason to insist on defining a single gauge rotation over all space. One can allow oneself to rotate in the upper region as described above, thus obtaining the singularity-free eq. 2a in R_a. In the lower region we can rotate clockwise everywhere, see fig. 2a-2c. (Doing this in the upper hemisphere, one would have a rotation that is singular along the positive z-axis).

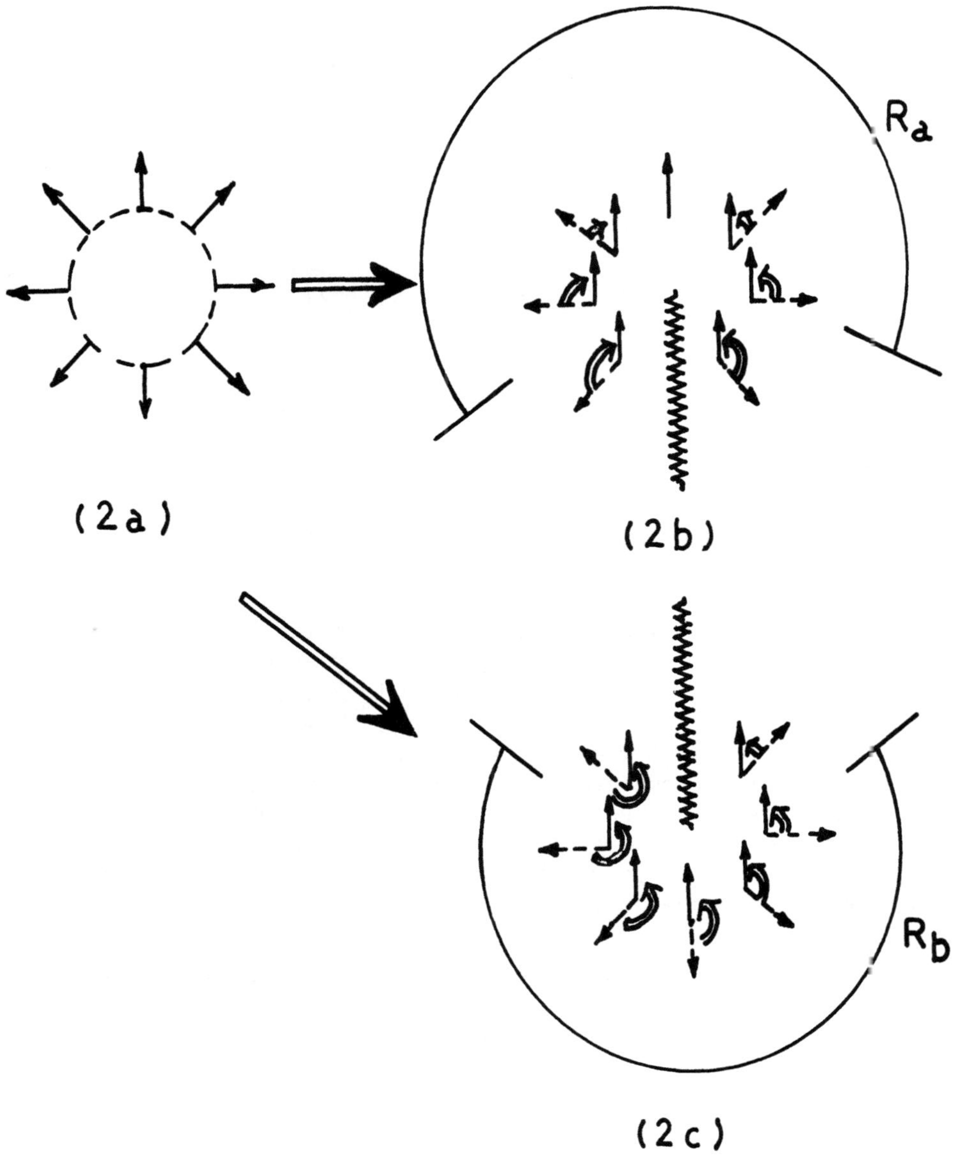

Figure 2. The gauge rotations connecting the radial and the unitary gauge used in the text.

One then obtains a set of potentials for Rb. Thus by applying the ideas of (1), we avoid the problem of a string singularity or of a gauge transformation with a line of singularities, and we can proceed to investigate the physical content of the 't Hooft-Polyakov solution.

The electromagnetic potentials and fields we discussed already. Now let us turn to the charged vector meson field configuration. Its explicit form :

$$\vec{W}^{+} = \frac{1}{\sqrt{2}}\ \frac{1}{e}\ \frac{K(r)}{r}\ (\vec{1}_\phi + i\vec{1}_\theta)e^{-i\phi} \qquad (9a)$$

$$\vec{W}^{+} = \frac{1}{\sqrt{2}}\ \frac{1}{e}\ \frac{K(r)}{r}\ (\vec{1}_\phi + i\vec{1}_\theta)e^{i\phi} \qquad (9b)$$

$$(K(r) = A(r) - 1,\quad W^{-}_{\mu} = (W^{+})^{+}_{\mu},\quad W^{+}_{o} = 0))$$

to be used in R_a and R_b respectively.

Note that, in accordance with the general requirements of sec. 1, (9b) and (9a) are connected by a nonsingular gauge transformation in R_{ab}, $S_{ba} = \exp(2i\phi)$. This form also indicates immediately that the magnetic charge for which (9a-b) is appropriate is 2 Dirac units. Finally, and also in accordance with the general requirements, (9a) is nonsingular in R_a, 9b in R_b. This is not very explicit from eq. 9a, since $\vec{1}_\phi, \vec{1}_\theta, e^{-i\phi}$ are not well defined on the positive z-axis. However, it is true since it is a nonsingular transform of a nonsingular set of functions, eq.(5) . Later we will meet the same functional forms as (9a), without being able to go back to a nonsingular gauge to check their analyticity. So let us check it explicitly. Expressing (9a) in components, one obtains

$$\vec{W}^{+} \sim \left| \begin{pmatrix} -\sin\phi \\ \cos\phi \\ 0 \end{pmatrix} + i \begin{pmatrix} \cos\theta\ \cos\phi \\ \cos\theta\ \sin\phi \\ -\sin\theta \end{pmatrix} \right| e^{-i\phi} \xrightarrow[\theta\to 0]{} \begin{pmatrix} -\sin\phi + i\cos\phi \\ \cos\phi + i\sin\phi \\ 0 \end{pmatrix} e^{-i\phi} = \begin{pmatrix} i \\ 1 \\ 0 \end{pmatrix}$$

So the combination $(\vec{1}_\phi + i\vec{1}_\theta)e^{-i\phi}$ is at least well-defined for $\theta = 0$. Showing that (9a) is actually analytic in x,y,z (inside Ra) requires only very little extra effort.

There is another singularity in (9a)-(9b) which has not been discussed so far : the singularity at r = 0, which is also present in $\vec{B}$. I would like to call it a "physical" singularity corresponding to the position of the pointlike magnetic charge. Two questions arise : whether this singularity is physically acceptable, and what happens to the equations of motion in the singular point. On the first question, I can only give my criterium for the acceptability; it is that the energy is finite, in spite of the singularity. This is the case for (9a),(9b), (1). On the second there can also be given a definite answer : the equations of motion are considered to be satisfied in a singular point, when the variation of the action is zero for field variations belonging to some test-function space. This is also the case for (9a)-(9b)-(1). (As can be shown by using some distribution theory). The situation is to be contrasted with the 1/r solution to $\Box V = 0$ ($V \neq 0$). First of all, the corresponding Lagrangian density being $(\partial_\mu V)^2$, its energy is not finite. And, even if one is willing to forget this problem by appealing to renormalisation theory, it still does not satisfy the second criterium : the action is not stationary with respect to variations of V that are test functions. Therefore, 1/r is not a solution of $\Box V = 0$ in r = 0. By contrast, (1), 9(a)(9b) constitute a field configuration which does satisfy the two criteria. (In section 3 we will encounter very similar functional forms, which also satisfy these criteria).

To close this interludium on singularities, let me go back for a moment to the radial gauge. This "physical" singularity is not clear from (5)-(6) : the radial-gauge variables show no singularity, but neither are they very physical (if one assumes that the local U(1) gauge invariance corresponds to electromagnetism). Rather, the definition of the physical quantities, like magnetic fields, can be singular in terms of the Φ^a, A^a_μ, as shown by the explicit form in eq. 8, when $|\Phi|$ is zero at some point.

Thus, if one rejects a solution which exhibits a singularity in a physical quantity, one rejects either the 't Hooft-Polyakov solution altogether (since $|\Phi| = 0$ in r = 0, and (8) is then singular), or at least the physical significance of the U(1) local gauge group. I adopt neither of these rebellious attitudes.

I now gather up the threads which temporarily slipped through my hands at contemplating eq. (9a)-(9b). The meaning of these eqs. is not very transparent yet, but let us compute the physical quantities associated with them. The computation is trivial and the results are :

(a) the charge density is zero
(b) the spin density $\vec{\Sigma}$ of the vector mesons is non zero and given by

$$\vec{\Sigma} = \frac{\vec{r}}{r}\ \Sigma(r) \quad , \qquad \Sigma(r) = g^2\ \left(\frac{K(r)}{r}\right)^2$$

$\Sigma(r)$ goes to zero exponentially at infinity, and is singular as $1/r^2$ in the origin.

In words : at each point in space there is an equal density of positively and negatively charged vector mesons. All the spins are aligned along the radial direction, parallel to the local magnetic field.

Finally, as far as the Φ^a fields are concerned, only the physical component Φ^3 -c remains. There is a density of these neutral mesons proportional to $|\Phi(r)-c|^2$, clustered around the origin. This density is nonsingular.

What are now the forces with which all these particle densities interact ? To see this most clearly, one can write down the form of the energy density for the field configuration at hand.

$$\varepsilon = \frac{1}{2}\ (\vec{B} - e\vec{\Sigma})^2 + \varepsilon_{kin}(W^{\pm}) + E_{kin+pot}(\Phi^3) \qquad (10)$$

(a) $\frac{1}{2}\ \vec{B}^2$ is the magnetostatic self-energy of the monopole field.

(b) $\frac{1}{2}\ e^2\ \vec{\Sigma}^2$ is a spin-spin coupling term, between the anomalous parts of the magnetic moments of the vector mesons. It induces an anti-aligning force for the spins.

(c) $e\vec{B}.\vec{\Sigma}$ is of the familiar magnetic moment-magnetic field coupling type. It induces a force that tends to align the magnetic moments with the magnetic field.

Each of these terms constributes on infinite term to the energy but in the sum their infinite contributions cancel.

(d) $\varepsilon_{kin}(W^{\pm})$ contains the kinetic ("distortion") energy and the masses of the W-mesons, as well as the minimal coupling terms. Again it can be split up into separate infinite parts, but the negative term from the magnetic moment (normal part) - magnetic field interaction again cancels the infinite "distortion" energy.

(e) $\varepsilon_{kin+pot}(\Phi)$ includes all energies involving the $(\Phi^3$-c) field. All its terms give finite contributions to the energy.

The forces at work should now be clear. The strongest forces are those in (a) - (d). The magnetic field attracts the magnetic moments to the places in space where it is strongest (term (c)). On the other hand, the repulsive forces between these spins (term (b)) prevents the conglomeration of the spins to become arbitrarily high. A balance is reached between these forces in the 't Hooft-Polyakov configuration : it would cost more energy to create an extra $W^{\pm}$ pair from the vacuum, or to have a pair disappearing into the vacuum. This balance is the physical reason for the stability of the configuration.

This intuitive reasoning is of course not meant to be a replacement for a mathematical proof of the stability. But neither is a mathematical proof a replacement for the physical insight. Indeed, guided by the recognition of the forces which play an essential part, generalisations have been found to other field theories, where monopole solutions are possible also (section 3).

Summarizing : it is perfectly possible to discuss the 't Hooft-Polyakov solution in a physical gauge, by making use of the formalism of Wu and Yang. In this gauge, the physical singularities occur explicitly in the physical field variables, rather than being hidden in a singular definition of these physical fields in terms of unphysical (including redundant) variables that are regular. This way, a physical picture of the solution in terms of the underlying fields is obtained.

3. EXTENSIONS

In the balancing mechanism (between forces and infinite contributions to the energy) described in the previous section there are several properties of the specific model at hand that play no essential role.

(a) The scalar degree of freedom. It breaks the symmetry and provides a mass scale, but on the classical level at which we are working this can be done in other ways. It also provides a weakly attractive potential for the charged vector mesons (once one has attributed a fixed mass $M \sim eC$ to them), but this seems inessential too, since there are stronger forces at work around the monopole position.

(b) The SU(2) symmetry. The magnetic charge is essentially connected to the unbroken U(1) gauge group, not with the SU(2) group. The fact that the anomalous magnetic moment of the vector field is equal to 1 is also a consequence of the SU(2) group. Whereas the presence of this anamalous magnetic moment played an important part, it is not clear why its value should be precisely 1.

(c) The vector mesons. Physically their role is that they provide spinning matter fields, with a magnetic moment, giving rise to a spin and magnetic moment density. Why would one need spin 1 fields for this ?

In fact, each of these specific points is amenable to a generalisation : first I will get rid of the scalar particle, and the SU(2) symmetry. Then I will also exhibit a theory with nonrelativistic spin 1/2 fields. In each of these theories, a stationary solution will exist to the field equations with a nonzero magnetic charge.

3a. A model with and without SU(2) symmetry.

$$\mathcal{L}_1 = \frac{1}{4}(F^a_{\mu\nu})^2 + (D_\mu \Phi^a)^2, \text{ with constraint } (\Phi^a)^2 - c^2 = 0$$

This model looks of course very similar to (4). It is therefore hardly a surprise that the ansatz

$$A^a_i = \varepsilon^{aij} x_j \frac{A(r)}{er^2} \qquad\qquad (12)$$

$$\Phi^a = \frac{x^a}{r} C$$

solves the field equation, provided $A(r) = 1 - K(r)$

$$K'' = \frac{K(K^2-1)}{r^2} + e^2c^2K \qquad (K'' = \frac{d^2}{(dr)^2} K)$$

There is a small difference with 't Hooft-Polyakov: there is (not even in this gauge) no analyticity in the origin. Nevertheless, the energy of the configuration is finite, and it is a stationary point of the action. (This statement will be valid for all solution I discuss and will not be repeated anymore).
No surprises, but note that the first extension has been achieved in this effortless manner : there is no longer a scalar degree of freedom. In loose terms, the mass of the scalar in the model of sec. 2 has gone to infinity, and the scalar is correspondingly "frozen". In superconductivity this is called the London limit.

Now let me rewrite $\mathcal{L}$ in the physical gauge. $\Phi^1 = \Phi^2 = 0$ and $\Phi^3 = c$ because of the constraint. Φ disappears from the Lagrangian, and writing out the remaining terms as functions of the photon field A_μ $(= A^3_\mu)$ and the charged vector fields $(w^\pm \sim A^1 \pm iA^2)$ we obtain :

$$\mathcal{L}_1 = \frac{1}{4}\left[F_{\mu\nu} - e \cdot \frac{1}{2}(W_\mu^+ W_\nu^- - W_\nu^+ W_\mu^-)\right]^2$$

$$+ \frac{1}{2}\;|(D^\mu W_\nu^+ - D^\nu W_\mu^+) \quad + M^2\;|W_\mu^+|^2 \tag{13}$$

where $D^\mu W_\nu^+ = \partial^\mu W_\nu^+ + ie\, A_\mu W_\nu^+$

$M^2 = e^2 C^2$

which one recognises as containing an explicit vector meson mass and a non minimal electromagnetic coupling. It is a special case of the model discussed next, viz with anomalous magnetic moment $\kappa = 1$. This more general model has κ arbitrary, thus losing the last remnant of the SU(2) symmetry :

$$\mathcal{L}_\kappa = \frac{1}{4}\left[F_{\mu\nu} - \frac{e}{2}\kappa\,(W_\mu^+ W_\nu^- - W_\nu^+ W_\mu^-)\right]^2$$

$$+ \frac{1}{2}\;(D^\mu W_\nu^+ - D_\nu W^{+\mu})|^2 + M^2|W_\mu^+|^2 \tag{14}$$

The following Ansatz is made in Ra :

$$\vec{W}^+ = \frac{1}{\sqrt{\kappa}}\;\frac{1}{e}\;\frac{1}{\sqrt{}}\;\frac{K(r)}{r}\;(\vec{1}_\phi + \vec{1}_\theta)e^{-i\phi} \tag{15a}$$

$$\vec{A} = \frac{1}{\rho}\;\frac{\cos\theta - 1}{r\sin\theta}\;\vec{1}_\phi$$

$$(\vec{W}^- = (\vec{W}^+)^+ \quad , A_0 = 0 = W_0^+)$$

and a similar equation in R_b.

The Ansatz is compatible with the field equations, and the equation for K reads :

$$K'' = \kappa \frac{K(K^2-1)}{r^2} + M^2 K \qquad (16)$$

The singularities have been discussed in section 2. In this case one can make no appeal to a radial gauge, since there is no such thing. Instead one is forced to check the analyticity properties required by the Wu-Yang formalism explicitly. As indicated in sec. 2,this presents no problem.

Eq. 16 can be solved by computer. The solution for different κ are exhibited in fig. 3. Only for $\kappa>0$ can we find a solution (which can also be seen from eq. (6)). The energies are finite and fig. 4 shows the behaviour of the monopole mass as a function of κ .

A final remark about ℓ_κ , eq. 14 is in order. It is essential that a spin-spin coupling is introduced in such a way that it forms a complete square when combined with the Maxwell Lagrangian and the anomalous coupling term. This may seem rather special but in fact it has been shown[5] that some very physical requirements on a theory of interacting vector mesons (with terms of dimension ≤ 4) constrains the Lagrangian very much. If one considers a theory with just 2 oppositely charged vector mesons and a local U(1) gauge invariance, $\mathcal{L}_\kappa$ is the most general Lagrangian allowed (with κ arbitrary). The only extra constraint one has to impose for the existence of a monopole solution is that κ be positive.

3b. A model with nonrelativistic spin 1/2 fields

$$H = \frac{1}{2m} \{\Phi^+ |\vec{\sigma}.(\frac{\vec{\nabla}}{i} - e\vec{A})^2|\Phi + \chi^+|\vec{\sigma}.(\frac{\vec{\nabla}}{i} + e\vec{A})^2|\chi\}$$

$$+ m (\Phi^+\Phi + \chi^+\chi)$$

$$+ \frac{1}{4} \left[F^{ij} - \frac{\kappa e}{2m} \varepsilon^{ijk} (\Phi^+ \sigma_k\Phi - \chi^+ \sigma_k\chi)\right]^2 \qquad (17)$$

(it has been anticipated that the solution we are interested in is time-independent).

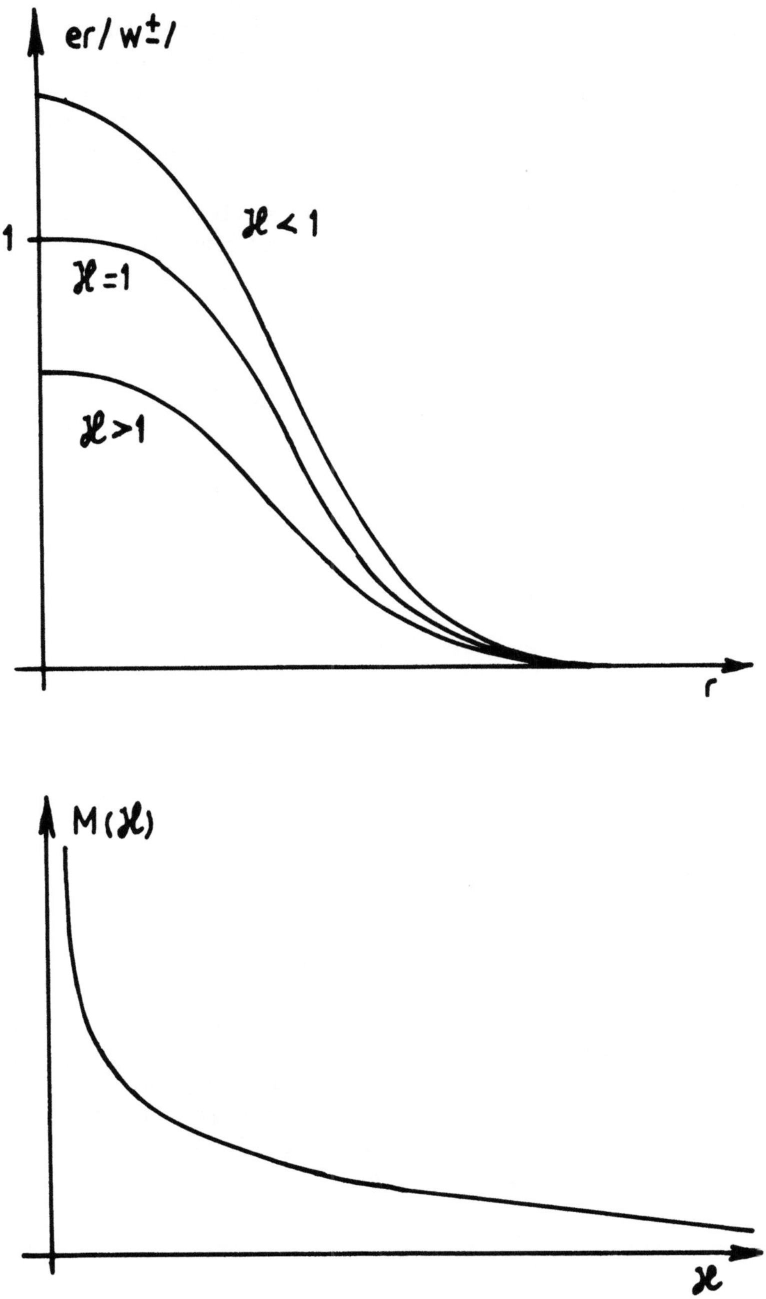

Figure 3. Behaviour of the charged vector fields (K) and the mass of the monopole solution for different values of κ .

Φ and χ are oppositely charged spin 1/2 fields with the same mass, interacting with the electromagnetic fields $\vec{A}$. The model is patterned after the nonrelativistic limit of the Dirac Hamiltonian, duplicated for particles and antiparticles, which are represented by the 2-component spinors Φ and χ. In addition to the "normal" magnetic moment, an anomalous coupling is added in the same way as in 14. (This makes the energy at least positive definite).
Motivated by the physical description in section 2, one is led to look for a solution with aligned magnetic moments,

$$\vec{r}.\vec{\sigma}\Phi = r\Phi \quad , \qquad \vec{r}.\vec{\sigma}\chi = -\vec{r}\chi$$

This helps in finding the following Ansatz :

$$R_a \qquad\qquad\qquad\qquad R_b$$

$$\vec{A} = g\,\frac{\cos\theta - 1}{r\sin\theta}\,\vec{1}_\phi \qquad\qquad \vec{A} = g\,\frac{\cos\theta + 1}{r\sin\theta}\,\vec{1}_\phi$$

$$\Phi = \begin{pmatrix}\cos\theta/2 \\ \sin\theta/2\; e^{i\phi}\end{pmatrix}\frac{K(r)}{er}\sqrt{\frac{m}{2\kappa}} \qquad \Phi_{(b)} = \Phi_{(a)}\, e^{-i\phi}$$

$$\chi = \begin{pmatrix}\sin\theta/2 \quad e^{-i\phi} \\ -\cos\theta/2\end{pmatrix}\frac{K(r)}{er}\sqrt{\frac{m}{2\kappa}} \qquad \chi_{(b)} = \chi_{(0)}\, e^{+i\phi} \qquad (18a,b)$$

Analyticity can be checked as before. Note that the gauge transformation connecting the fields in Region R_{ab} is $S_{ba} = e^{i\phi}$. (In sections 2 and 3a it was $e^{2i\phi}$). This corresponds to a single unit of magnetic charge, eg = 1/2. One finds indeed that the Ansatz (18) is only compatible with the field equations if eg = 1/2. There remains of course the equation for K,

$$K'' = \frac{\kappa}{2}\;\frac{K(K^2-1)}{r^2} + 2m^2K \qquad (19)$$

which is almost identical to (16). So again one finds a solution provided $\kappa > 0$. K(r) and the mass of the monopole behave as in sec. 3a.

4. CONCLUSIONS AND SPECULATION

From the physical appearance of the 't Hooft-Polyakov monopole, which was investigated in a physical gauge with the Wu-Yang formalism, inferences were drawn on the physical reasons for the stability of the monopole solution. The essential elements were kept in more general theories, in which we expected - and found - monopoles as well. The models discussed include a model with spin 1/2 fields. This invites an interesting speculation. Let us take a bold step and consider this model to be an approximation to QED, where the effect of the first radiative correction to the magnetic moment has been taken into account in the effective classical hamiltonian of eq. 17. Putting in the numbers for the electron mass, $\alpha \sim \frac{1}{137}$, $\kappa = \frac{\alpha}{2\pi}$, one obtaines for the energy of the monopole

$$M \simeq 43 \text{ GeV}$$

This number should be compared with $M \simeq 10^4$ GeV, arising by speculating on the applicability of the model of sec. 2, to weak interactions.

REFERENCES

1) T.T. Wu and C.N. Yang, Nucl.Phys. B107 (1976) 365 and references therein.
2) A.M. Polyakov, JETP Lett. 20 (1974) 194.
G. 't Hooft, Nucl.Phys. B79 (1974) 276.
3) W. Troost and P. Vinciarelli, CERN preprint TH2195 (1976) and unpublished calculations. See also P. Vinciarelli, Lectures at the 1976 Erice Summer School on subnuclear physics.
4) M.K. Prasad and C.M. Sommerfield, Phys.Rev.Lett. 35(1975) 760.
5) G. Velo, Nucl.Phys. B65 (1973) 427.

INSTANTONS AND QUARK-MASS GENERATION - THE HOW OF SPONTANEOUS CHIRAL SYMMETRY BREAKING

D.G. CALDI

Department of Physics, Rutgers University

Piscataway, New Jersey 08854

There is abundant evidence that one of the important characteristics of the physical world is that quarks, if they exist, appear to be massive and so chiral symmetry is spontaneously broken so that is is realized in the Nambu-Goldstone mode. [1] To mention just two pieces of evidence, we point to the absence of parity doubling and the success of PCAC. The leading candidate for a theory of the strong interactions, quantum-chromodynamics[2] (QCD), starts out with massless quarks and a chiral $SU(N) \otimes SU(N)$ symmetry, N being the number of flavors. How the quarks get a mass has been an open question, the importance of which has been stressed in the lecture of Professor Gross. [3] Recently, it was proposed by Callan, Dashen and Gross[4] (CDG) that the answer, at least in two space-time dimensions, is provided by instantons and their interactions with fermions. I have found that dynamical quark-mass generation via instantons does occur in four dimensions.[5]

Our starting point is the effective Lagrangian found by t'Hooft[6] for a color-SU(2) gauge theory instanton[7] interacting with N flavors of massless quarks,

$$\mathcal{L}_{eff} = (.64) 2^{2+3N} \pi^{2N-2} \left(\frac{8\pi^2}{g^2}\right) \int d\rho \, \rho^{-5+3N} \exp\left\{- \frac{8\pi^2}{g^2(\mu)} + \ln(\mu\rho)\left[\frac{22}{3} - \frac{2}{3} N\right] + 2N(0.146)\right\} \left< \prod_{s=1}^{N} (\bar{\psi}_s \omega)(\bar{\omega} \psi_s) \right> + \text{H.C.}, \quad (1)$$

where ρ is the instanton size, μ is the renormalization subtraction point, and

$$<\omega_\alpha \bar{\omega}_\beta> = \frac{1}{4} \delta_{\alpha\beta}(1 + \gamma_5) \qquad (2)$$

For the case of two flavors,

$$<\prod_{s=1}^{2} (\bar{\psi}_s \omega)(\bar{\omega} \psi_s) = \frac{1}{24} (2\delta_{\alpha_1}^{\beta_1} \delta_{\alpha_2}^{\beta_2} - \delta_{\alpha_1}^{\beta_2} \delta_{\alpha_2}^{\beta_1})$$

$$\varepsilon^{st} \bar{\psi}_1^{\alpha_1} (1+\gamma_5)\psi_s^{\beta_1} \psi_2^{\alpha_2} (1+\gamma_5)\psi_t^{\beta_2} \quad . \qquad (3)$$

Because the coefficient of $\ell n(\mu\rho)$ is that of the Callan-Symanzik coefficient for $g^2(\mu)$, we may to this one-loop order replace g by the running coupling constant $\bar{g}(\rho)$.

This $L_{eff}(1)$ was obtained by t'Hooft from a calculation of the vacuum-to-vacuum amplitude for QCD with N massless fermions by path integral techniques,

$$<0|0> = \int \mathcal{D} A \mathcal{D} \psi \exp \int d^4\chi\{ \mathcal{L} (A) + \mathcal{L}(A ,\psi) + \dots \} \qquad (4)$$

using the classical instanton solution,

$$A_\mu^a(\chi)^{cl} = \frac{2}{9} \frac{\eta_{a\,\mu\nu}(\chi-Z)^\nu}{(\chi-Z)^2+\rho^2} \qquad (5)$$

as the stationary phase point.

Since the 2N fermions interaction of this effective Lagrangian (1) has the chiral symmetry properties of the determinant of an N x N matrix, it is invariant under chiral SU(N) ⊗ SU(N). (Note that it is not invariant under U(N) ⊗ U(N), so that the apparent $U_A(1)$ invariance is actually not present. This is one way of seeing how instantons solve the $U_A(1)$ problem [6,8]. We therefore have a situation similar to that of the Nambu-Jona-Lasinio model[9], except that here the 2N fermion interaction is an induced one not appearing in the fundamental Lagrangian. The theory remains renormalizable and there is no need to introduce an extraneous cutoff since, as we shall see, the instanton size provides a natural cutoff. Given this induced 2N fermion interaction we may ask whether the fermions without a bare mass may nevertheless dynamically develop a mass due

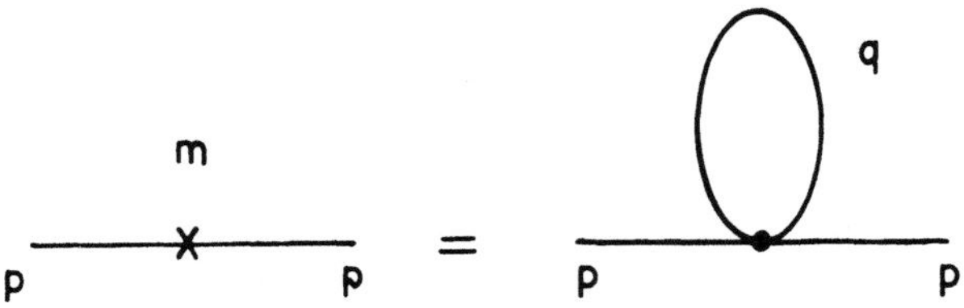

Fig. 1. Equation for the quark mass operator for N = 2.

to the self-energy from this interaction. To answer this we look at the self-consistent or Hartree-Fock equation for the quark mass operator.

For heuristic purposes I shall do the calculation in two different ways : the first is more transparent but is more approximate; the second justifies some of the assumptions and leads to additional results.

Let us consider the quark-mass operator equation for N = 2 and for an instanton of fixed size ρ . The integral equation is given graphically in Fig. 1 where on the left we have the mass operator and the right side shows the vertex of the instanton-induced effective four-quark interaction with the internal line being the full massive-quark propagator. The resulting equation is

$$m(p) = C\left(\frac{8\pi^2}{\bar{g}^2}\right) C^{-\frac{8\pi^2}{\bar{g}^2}} \rho^2 \int \frac{d^4q}{(2\pi)^4} \frac{m(q)F(p^2)F(q^2)}{q^2 + m^2(q)} \qquad (6)$$

where C = 2900 and we have introduced cutoff factors $F(p^2)F(q^2)$ in the definition of the vertex. This is done since L_{eff}(1) is only valid outside the instanton (which we have located at the origin). After some transformations[5], one finds the result of the Euclidean integration to be

$$\left|1 - \tilde{C}\left(\frac{\bar{g}^2}{8\pi^2}\right) C^{\frac{8\pi^2}{\bar{g}^2}}\right| = \rho^2 m^2 \ln\left(\frac{1}{m^2\rho^2} + 1\right) , \qquad (7)$$

where $\tilde{C}$ = .055. Since the right side of (7) is ≥ 0, there is a spontaneous generated quark mass for

$$0 < \tilde{C}\left(\frac{\bar{g}^2}{8\pi^2}\right)^4 C^{\frac{8\pi^2}{\bar{g}^2}} < 1 \quad . \qquad (8)$$

The result is, of course, non perturbative.

The question now is whether for reasonable values of $\bar{g}$ the inequality (8) is satisfied. Due to the large numerical value of C in (6), which comes mainly from the powers of 2 and π in $L_{eff}(1)$, it turns out that (8) is satisfied for $\frac{1}{13} \lesssim \frac{g^2}{8\pi^2} \lesssim 1.8$ and at least for the lower range of these limits the result of the calculation is credible. For example, if we pick the value

$$\frac{\bar{g}^2}{8\pi^2} = \frac{1}{8} \quad (\frac{\bar{g}^2}{4\pi} \simeq .8)$$

at which CDG predict[10)] for color SU(2) that hadronic processes occur (i.e. confinement) then $m = \frac{4}{\rho}$.
Taking ρ on the order of an inverse pion mass, the quark mass comes out to be several hundred MeV. It is also important that for this value of the coupling the density[10] of instantons, $.26\ (\frac{8\pi^2}{\bar{g}^2})^4 \exp(-\frac{8\pi^2}{\bar{g}^2}) = .36$, is small enough so that the dilute gas approximation[8] implicit in our calculation is reasonable. As for our neglect of the m = 0 solution to (6), one can show[9)] that the energy of the vacuum for massive fermions is lower than that for massless ones.

I would like to repeat the calculation now without having to use a cutoff, and also including the integration over all instanton sizes which I have thus far ignored. To do so we first make the observation that L_{eff} only approximately reproduces the exact massless fermion contribution to the vacuum-to-vacuum amplitude computed by t'Hooft. In order to keep the local vertex of L_{eff} and still obtain the exact vacuum-to-vacuum amplitude, we will employ a fermion propagator modified to take into account the presence of the instanton.[5)] For the massive fermion Euclidean propagator in the presence of an instanton of size ρ we have,

$$\tilde{S}_m(p) = \rho\left\{\frac{i\not{p}+m}{(p^2+m^2)^{1/2}} K_1(\rho(p^2+m^2)^{1/2}) + iK_0(\rho(p^2+m^2)^{1/2})\right\} \quad (9)$$

where $K_{1,0}$ are modified Bessel functions of integer order which asymptotically behave as

$$K_\nu(Z) \xrightarrow[Z \to \infty]{} (\frac{\Pi}{2Z})^{1/2} e^{-Z} \quad (10)$$

The self-consistent integral equation for the quark mass (Fig. 1) using the modified propagator $\tilde{S}_m(p)$ is

$$m = C\ (\frac{8\pi^2}{\bar{g}^2})^4\ e^{-\frac{8\pi^2}{\bar{g}^2}}\ \rho^2 \int \frac{d^4q}{(2\pi)^4}\ \frac{\rho m K_1(\rho(q^2+m^2)^{1/2})}{(q^2+m^2)^{1/2}}\ , \quad (11)$$

where for the moment we still keep ρ fixed. In (11) we no longer have cutoff factors as in (6), since $\tilde{S}_m(p)$ explicitly takes the instanton size into account, and indeed the integral is finite due to the exponential dumping provided by $K_1(\rho(q^2+m^2)^{1/2})$. The result of the integration is

$$1 = C\left(\frac{8\pi^2}{\bar{g}^2}\right) e^{-\frac{8\pi^2}{\bar{g}^2}} \frac{1}{4\pi^2} \rho\, m\, K_1(\rho m) \tag{12}$$

If we again use $\frac{\bar{g}^2}{8\pi^2} = \frac{1}{8}$, we find a result $m \sim \frac{6}{\rho}$, similar to that obtained from (7).

But now we can go further and perform the ρ integration. The integral is (to one-loop order in the expansion of $\bar{g}(\rho)$ as in (1))

$$\mu^6 m \int_0^\infty d\rho\, \rho^6\, K_1(\rho m). \tag{13}$$

One notes that the integrand goes to zero at both ends of the scale and that it has a fairly sharp maximum at $\rho m = 5.5$ (where as we have seen $\frac{\bar{g}^2}{8\pi^2} \simeq \frac{1}{8}$ to satisfy (12)). Hence it was a reasonable approximation to take a single instanton of fixed size. More importantly, the ir divergence which seemed to be present in L_{eff} as $\rho \to \infty$ does not appear here because of the exponential damping of $K_1(\rho m)$. (As $\rho \to 0$ the integral converges due to asymptotic freedom). What appears to be happening is that essentially one size (or at least a small range of sizes) of instanton is reponsible for interacting with fermions to generate a fermion mass.

Since QCD is a scale invariant theory, all dimensional quantities are ultimately expressed in terms of the only scale in the theory, the renormalization subtraction point μ. Hence if we actually perform the integration of (13) then m is expressed in terms of $g(\mu)$ and we must pick a renormalization subtraction point μ since the result is

$$1 = C \frac{1}{4\pi^2} \left(\frac{8\pi^2}{g^2(\mu)}\right)^4 e^{-\frac{8\pi^2}{g^2(\mu)}} 2^5\, \Gamma(4)\Gamma(3) \frac{\mu^6}{m^6}. \tag{14}$$

If we pick $\frac{g^2(\mu)}{8\pi^2} = \frac{1}{8}$ with $\mu = 100$ MeV, which is consistent with our choice above for $\frac{\bar{g}^2(\rho)}{8\pi^2}$, we find that again $m \simeq 550$ MeV.

This also confirms our observation that essentially one instanton size is involved.

Turning our attention to the question of the momentum dependence of m, one can argue that if it is generally true that only one instanton size contributes then the approach followed in the first calculation (6) is valid. In particular, the factorization of the cutoff factors, F(p) F(q), is allowed. However, I have shown that one instanton size contributes by performing the second calculation (11) only for constant m. While I believe that this approach is a reasonable approximation and that the result is correct, to establish it rigorously involves a complicated, highly non-linear integral equation for m(p) which has so far proved intractable.

One should also note that the calculations done here are to lowest order. The higher order self-energy graphs (e.g. Fig. 2) will give corrects to the quark mass, but whether or not these corrections are large I do not yet know. Estimating these corrections is complicated by the fact that it is just the iterated bubbles in this class of graphs which give rise to an important collective effect, namely the Nambu-Goldstone mesons, just as in the Nambu-Jona-Lasinio model.

I have also calculated the quark mass for the cases N= 3, 4. I find masses slightly smaller than that for N = 2, and they occur for smaller coupling and much lower density. For N = 3 the coupling is approximately $\frac{1}{11}$ and the density is 10^{-2}; for N = 4, the values are $\frac{1}{18}$ and 10^{-4}. Hence the results are even more trustworthy than the N = 2 case. This situation is in contrast to the two-dimensional work of CDG where there is a phase transition at N = 2 so that m = 0 for N > 2 and m is exceedingly small for N = 2. I find no evidence for such a phase transition in four dimensions, though there may well be one for high enough N.

All these calculations have been done for a color-SU(2) gauge theory. Of course, QCD has SU(3) for the color gauge group. I expect that the qualitative features of this work will also be found for color-SU(3), but the actual values of the predicted quark mass will probably be somewhat different. These are the values of course that should be compared with whatever one can deduce from experimental evidence. This extension to color-SU(3) is currently under investigation.

Subsequent to the work I have been describing, CDG have done a related calculation with similar results [11]. They looked at the so-called σ propagator, namely

$$\langle\theta|T(\bar{\psi}\psi,\ \bar{\psi}\psi)|\theta\rangle = D_\sigma(p), \tag{15}$$

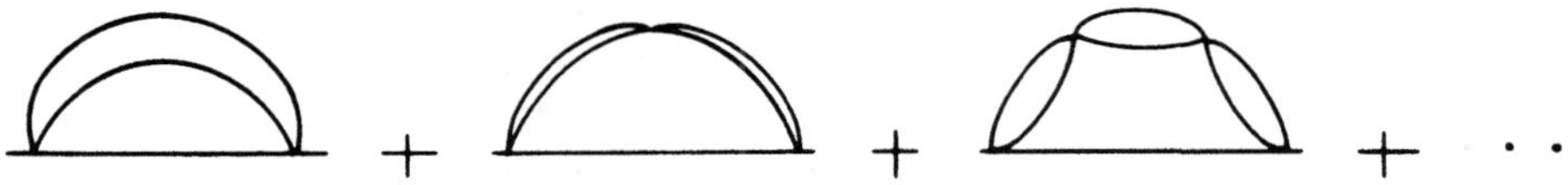

Fig. 2. A class of high order self-energy graphs.

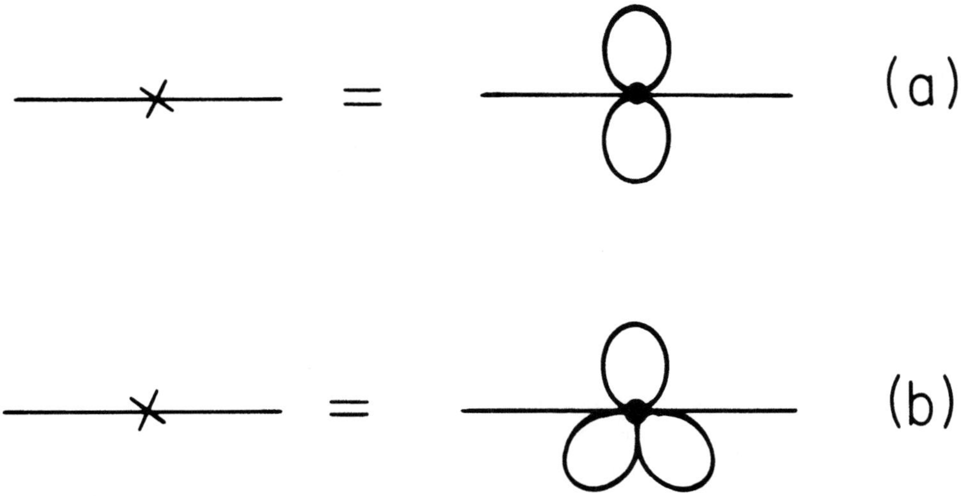

Fig. 3. Equation for the quark mass operator for (a) N = 3, (b) N = 4.

to see if a tachyon pole develops, since this would be a sign of vacuum instability and so at least indirect evidence of chiral symmetry breaking. They use an approach somewhat different from the one I have followed mainly in their use of a non-local vertex.

For details, consult the lectures of Professor Gross. Their result is that $D_\sigma(p)$ does have a tachyon pole at $p \sim 5\text{-}6\mu$. Hence the two independent calculations lead to quite similar effects.

The conclusions to be drawn are that instanton interactions with massless quarks have extensive consequences. Not only do they solve the $U_A(1)$ problem so that the apparent chiral U(N) ⊗ U(N) symmetry of the QCD Lagrangian is really only SU(N) ⊗ SU(N). More significantly, this quark-instanton interaction produces a spontaneous generated mass for the quark, thus dynamically breakking the chiral SU(N) ⊗ SU(N) symmetry. This, as usual, produces a multiplet of massless Nambu-Goldstone pseudoscalar mesons in the adjoint representation of SU(N) which arise from iterated bubble graphs. For the mesons to acquire (different) masses requires mechanisms for conventional breaking of SU(N) symmetry (hence explicitly breaking SU(N) O Su(N)) as treated, for example, in the Gell-Mann - Oakes - Renner model. [12] Perhaps this breaking of SU(N) comes from the weak interactions, but that must be left as the subject for a future lecture.

REFERENCES

1. For a review see H. Pagels, Phys. Rep. 16C, 219 (1975).
2. For a review see W. Marciano and H. Pagels, Phys. Rep. C, to be published.
3. D.J. Gross, these Proceedings.
4. C.G. Callan, R.F. Dashen, and D.J. Gross, IAS Princeton preprint COO-2220-110 (1977).
5. D.G. Caldi, Phys.Rev.Lett. 39, 121 (1977).
6. G. 't Hooft, Phys.Rev. D14, 3432 (1976); Phys.Rev.Lett. 37, 8 (1976).
7. A.A. Belavin, A.M. Polyakov, A.S. Schwartz, and Y.S. Tyupkin, Phys.Lett. 59B, 85 (1975).
8. C.G. Callan, R.F. Dashen, and D.J. Gross, Phys.Lett. 63B, 334 (1976); R. Jackiw and C. Rebbi, Phys.Rev.Lett. 37, 172 (1976).
9. Y. Nambu and G. Jona-Lasinio, Phys.Rev. 122, 345 (1961); ibid. 124, 246 (1961).
10. C.G. Callan, R.F. Dashen, and D.J. Gross, Phys.Lett. 66B, 375 (1977).
11. C.G. Callan, R.F. Dashen, and D.J. Gross, IAS Princeton preprint (1977).
12. M. Gell-Mann, R. Oakes and B. Renner, Phys.Rev. 175, 2195 (1968). See also S. Glashow and S. Weinberg, Phys.Rev.Lett. 20, 224 (1968).

ASYMPTOTIC ESTIMATE OF PERTURBATION THEORY AT LARGE ORDERS

Jean ZINN-JUSTIN

Princeton University, Department of Physics

Princeton, New Jersey 08540

INTRODUCTION

In these lectures we want to show how a new method, due to Lipatov, and based on a steepest descent evaluation of the euclidean path integral, allows one to estimate the coefficients of the perturbative expansion at large orders. For boson field theories one finds in general that the Kth order of perturbation theory behaves like :

$$K!a^K K^b c \left(1 + O\left(\frac{1}{K}\right)\right)$$

where a depends only on the interaction, where b and c depend on the particular Green's function one is calculating.

The present subject has already been partially covered by lecture notes written for the Salamanca School 1977 [1], which I do not want to duplicate. Therefore I shall often only give an outline of the method, referring the reader to these lecture notes for more details. On the other hand I want to emphasize the consequences of these large estimates for practical calculations.

The set-up of these lectures is the following :
In section I, I shall explain the method on the example of a one variable integral. In section II, I shall present the calculation of the large order behavior of the energy levels of the anharmonic oscillator. In section III, I shall briefly describe the same calculation in Φ Field Theory. In section IV I shall review the various generalizations of this method. In section V, I shall review some numerical examples, and the application to the problem of resumming the perturbation series.

I. THE SIMPLE INTEGRAL

Consider the integral :

$$Z(g) = \frac{1}{\sqrt{\pi}} \int_{-\infty}^{+\infty} dx \exp -(x^2 + gx^4) \tag{1}$$

For small g, the function Z(g) is given by an asymptotic series in the whole cut plane :

$$Z(g) = \sum_{k=0}^{\infty} Z_k g^k \tag{2}$$

with

$$Z_k = \frac{(-1)^k}{k!} \frac{1}{\sqrt{\pi}} \int_{-\infty}^{+\infty} e^{-x^2} x^{4k} dx \tag{3}$$

which yields after integration :

$$Z_k = \frac{1}{\sqrt{\pi}} \frac{(-1)^k}{k!} \Gamma(2k + \frac{1}{2}) \tag{4}$$

If we are only interested in the large order behavior of Z_k, we can obtain it directly from equation (3), by using steepest descent to evaluate it. The saddle point equation is :

$$-2 x_c + \frac{4k}{x_c} = 0$$

$$\text{or} \quad x_c = \pm\sqrt{2k} \tag{5}$$

After integrating the small fluctuations around the saddle point, we get at leading order :

$$Z_k \sim \frac{(-1)^k}{k!} \sqrt{2} \exp - 2k + 2k \ln 2k \tag{6}$$

or equivalently :

$$Z_k \simeq \frac{(-1)^k}{\pi \sqrt{2}} 4^k (k-1)! (1 + O(\frac{1}{k})) \tag{7}$$

An alternate way of proceeding, the motivation for which will become clearer in the next section, is to extract the Kth order of perturbation through a contour integral:

$$Z_k = \frac{1}{\sqrt{\pi}} \frac{1}{2i\pi} \int d_x \oint_c \frac{dg}{g^{k+1}} e^{-(x^2+gx^4)} \tag{8}$$

One can then calculate Z_k for k large through a steepest descent. We have now a saddle point in x and g :

$$\begin{aligned} 1 + 2gx^2 &= 0 \\ \frac{K}{g} + x^4 &= 0 \end{aligned} \tag{9}$$

so that :

$$\begin{aligned} x &= \pm \sqrt{2k} \\ g &= -\frac{1}{4k} \end{aligned} \tag{10}$$

Integrating then over the fluctuations around the saddle point, leads to the same result as before. The same calculation can be interpreted in a slightly different way. By integrating on x first, for g negative, we are calculating through steepest descent the discontinuity of Z(g) on its cut for g small.
But the quantity Z_k can be written :

$$Z_k = \frac{1}{\pi} \int_{-\infty}^{0} \frac{\text{Im } Z(g\ )}{g^{k+1}}\, dg \tag{11}$$

For k large, this integral is dominated by the small g values. Integrating on x at g fixed through steepest descent yields

$$1 + gx_c^2 = 0 \tag{12}$$

Then :

$$\text{Im}Z(g) = \frac{1}{\sqrt{2}} \, e^{\frac{1}{4g}} \qquad g \to 0_-$$

and from equation (11) :

$$Z_k \sim \frac{(-4)^k}{\pi\sqrt{2}} (k-1)! \tag{13}$$

We shall generalize this last method to quantum mechanics and field theory.

II. ENERGY LEVELS OF THE ANHARMONIC OSCILLATOR

We can use a completely similar method to evaluate the large order behavior of the ground state energy of the anharmonic oscillator. The large behavior of the perturbation series for this potential, had been obtained previously by Bender and Wu [2], using the W.K.B. method. This method is not easy to generalize to Field Theory. We present here the application to this potential problem of the generalization of the method used by Lipatov [3] for ϕ^4 theory in 4 dimension, as it can be found in ref. 4.

We shall start from the path integral representation for $e^{-\beta H}$:

$$\langle x | \exp - \beta H | x' \rangle = \int_{q(0)=x}^{q(\beta)=x'} [dq(t)] \exp - A[q(t)] \quad (14)$$

with the euclidean action $A[q(t)]$ given by :

$$A[q(t)] = \int_0^\beta dt \left[\frac{1}{2}\dot{q}^2(t) + \frac{1}{2}q^2(t) + \frac{g}{4}q^4(t)\right] \quad (15)$$

The ground state energy $E_0(g)$ is then given by the large β limit :

$$\ln \operatorname{tr} e^{-\beta H} \underset{\beta \to +\infty}{\sim} e^{-\beta E_0(g)} \quad (16)$$

The kth order of the perturbative expansion of $\exp -\beta H$ is given by

$$\{\langle x | \exp - \beta H | x' \rangle\}^{(k)} = \frac{1}{2i\pi} \int_{q(0)=x}^{q(\beta)=x'} [dq(t)] \frac{dg}{g^{k+1}} \exp -A[q] \quad (17)$$

The normalization of the path integral (14) is obtained by dividing the path integral by its value at g=0. We shall now calculate expression (17) for k large by steepest descent. The saddle point is now defined by a classical trajectory $q_c(t)$ satisfying the euclidean equation of motion

$$\ddot{q}_c(t) = q_c(t) + g_c q_c^3(t) \quad (18)$$

and by the value of the coupling constant g_c :

$$\frac{k}{g_c} + \frac{1}{4} \int_0^\beta q_c^4(t)dt = 0 \quad (19)$$

This last equation shows that again g_c is negative if there exists a real solution $q_c(t)$ to equation (18).

Setting :

$$q_c(t) = \frac{1}{\sqrt{-g_c}}\, y(t) \tag{20}$$

Leads to :

$$\begin{aligned} \ddot{y}(t) &= y(t) - y^3(t) \\ g_c &= -\frac{1}{4k}\int_0^\beta y^4(t) \end{aligned} \tag{21}$$

which show that again g_c is of order $\frac{1}{k}$ and $q_c(t)$ of order $\sqrt{k}$.
As we want to calculate only $\mathrm{tr}^k\ e^{-\beta H^{q_c}}$, we are interested in periodic trajectories.
Furthermore we are looking for a trajectory which gives a firite action in the large β limit.

Such a trajectory corresponds to an instanton solution describing the decay of the false ground state located in the local minimum of the unbounded potential $y^2/2 - y^4/4$. Actually the method used by Bender and Wu, consisted in estimating through the W.K.B. method the same barrier penetration coefficient.

The solution is :

$$y(t) = \frac{\sqrt{2}}{\mathrm{ch}\,(t-t_0)} \tag{22}$$

So that g_c is ;

$$g_c = -\frac{4}{3k} \tag{23}$$

From this alone we see that the perturbative expansion is a divergent series with terms growing like $(-3/4)^k\ k!$

To calculate the precise power of k and the constant multiplying $(-3/4)^k\ k!$, we have to integrate over the fluctuations around the saddle point. The integration over the fluctuations in q(t) space, would naively generate a factor $(\det M)^{-1/2}$, where M is an operator, second derivative of $A[q]$ taken at $q(t) = q_c(t)$.

$$M(t_1 t_2) = \left[-\left(\frac{d}{dt}\right)^2 + 1 + 3y^2(t_1)\right]\delta(t_1 - t_2) \tag{24}$$

There is one subtlety in this calculation.
There exists not only one classical trajectory but a family of trajectories obtained by taking one and translating it in time of a time interval comprised between 0 and β, giving all the same action.

One has therefore to sum on all these trajectories, taking the time translation[5] as one of the integration variables in the path integral. Furthermore, to conserve the number of degrees of freedom, one has then to integrate over fluctuations which are orthogonal to infinitesimal time translations, i.e. to $\dot{y}(t)$. This is fortunate because, as $A[q_c(t)]$ is independent of the time translation parameter, M has a zero eigenvalue associated to the eigenvector $\dot{y}(t)$. Therefore the real factor coming from the remaining integration is $J\{(\det M)_{\perp}\}^{-1/2}$ where J is the jacobian coming from the change of variable necessary to take time translation as a variable :

$$J = \|\dot{y}(t)\| \tag{25}$$

and $(\det.M)_{\perp}$ is the determinant of M in the subspace orthogonal to $y(t)$.

Although this is the method of calculation which can be generalized to Field Theory problems, we shall expose here a method special to completely integrable hamiltonian systems, like one dimensional systems. It will allow us to find an analytical form of the result for arbitrary potentials.

Calculation of $\langle x|\exp-\beta H|x'\rangle$ in the Classical Limit

We shall calculate at leading order the contribution of a saddle point to the path integral :

$$\langle x|\exp-\beta H|x'\rangle = \int_{q(0)=x}^{q(\beta)=x'} [dq(t)] \exp-A[q(t)] \tag{26}$$

where $A[q]$ is given by :

$$A[q] = \int_0^{\beta} \left[\frac{1}{2}\dot{q}^2(t) + V(q(t))\right] dt \tag{27}$$

with the condition :

$$V(q) = \frac{1}{2} q^2 + O(q^3) \tag{28}$$

The saddle point is given by the euclidean equation of motion :

$$\ddot{q}_c(t) = \frac{\partial V}{\partial q}|q_c(t)| \tag{28}$$

which gives :

$$\frac{1}{2}\dot{q}_c^2(t) = V\,|q_c(t)| + E \tag{30}$$

From this equation we calculate E as a function of x, x' and β :

$$\beta = \int_x^{x'} \frac{dq}{\left[2(V(q)+E)\right]^{1/2}} \tag{31}$$

In this calculation we shall assume $x < o < x'$.
It is straightforward to generalize these equations to other situations.
Setting :

$$q(t) = q_c(t) + \delta q(t) \tag{32}$$

and expanding the action $A[q]$ up to second order in δq we get :

$$\langle x|\exp - \beta H|x'\rangle \sim \{\exp - A\,[q_c(t)]\}G(x,x';\beta) \tag{33}$$

where we have defined :

$$G(x,x';\beta) = \int_{\delta q(0)=0}^{\delta q(\beta)=0} [d\delta q(t)]\ \exp - A^{(2)}[\delta q(t)] \tag{34}$$

$$A^{(2)}[\delta q(t)] = \frac{1}{2}\int_0^{\beta} \{[\delta\dot{q}(t)]^2 + [\delta q(t)]^2\ V''[q_c(t)]\}dt \tag{35}$$

It is a remarkable feature of completely integrable hamiltonian systems that this last path integral can be calculated analytically[6]. It yields the famous Van Vleck formula[7] :

$$\langle x|\exp - \beta H|x'\rangle \sim \frac{1}{\sqrt{2\pi}}\left\{-\frac{\partial^2 A_c}{\partial x\partial x'}\right\}^{1/2} \exp - A_c \tag{36}$$

This formula for n degrees of freedom becomes :

$$\langle x|\exp - \beta H\,|x'\rangle \sim \frac{1}{(2\pi)^{n/2}}\left\{\det - \frac{\partial^2 A_c}{\partial x_i \partial x_j}\right\}^{1/2} \exp - A_c \tag{37}$$

The quantity called A_c is just the euclidean action of the calssical trajectory.

Equation (37) can be rewritten in another form.
The quantity A_c is given by :

$$A_c = \int_0^\beta \left[\frac{1}{2}\dot{q}^2(t) + V\, q(t)\right] dt \tag{38}$$

$$A_c = \int_0^\beta \dot{q}_c^2(t)dt - E\beta \tag{39}$$

Using again equation (30) we finally obtain :

$$A_c = \int_x^{x'} \{2\,(V(q) + E)\}^{1/2}\, dq - E\beta \tag{40}$$

We can now calculate $\frac{\partial A_c}{\partial x}$

$$\frac{\partial A_c}{\partial x} = - \{2\,[V(x) + E]\}^{1/2} \tag{41}$$

Notice that :

$$\frac{\partial E}{\partial x'} = - \frac{\partial E}{\partial \beta} \frac{1}{[2(V(x')+E)]^{1/2}} \tag{42}$$

We have then another expression for $\frac{\partial^2 A_c}{\partial x \partial x'}$:

$$\frac{\partial^2 A_c}{\partial x \partial x'} = \frac{\partial E}{\partial \beta} \frac{1}{[2(V(x)+E)]^{1/2}[2(V(\dot{x})+E)]^{1/2}} \tag{43}$$

Equation (37) now becomes :

$$\langle x|\exp - \beta H|x'\rangle \sim \frac{1}{\sqrt{2\pi}} \frac{1}{[2(V(x)+E)]^{1/4}} \frac{1}{[2(V(x')+E)]^{1/4}} \left(- \frac{\partial E}{\partial \beta}\right)^{1/2} \exp -A_c \tag{44}$$

Application to the Anharmonic Oscillator

We now have :

$$\{e^{-\beta E_o}\}_k \sim \sum_{\substack{\text{leading}\\ \text{saddle}\\ \text{points}}} \frac{1}{\sqrt{2\pi}} \frac{1}{2i\pi} \oint \frac{dg}{g^{k+1}} \int \frac{dx}{\{2[V(x)+E]\}^{1/2}} \left(-\frac{\partial E}{\partial \beta}\right)^{\frac{1}{2}} \exp - A_c \tag{45}$$

As we have set x = x', E(β) and A_c are now independent of x. Furthermore the equation for E :

$$\int \frac{dx}{\{2[V(x)+E]\}^{1/2}} = \beta \tag{46}$$

allows us to integrate over x :

$$\{e^{-\beta E_o}\}_k \sim \sum_{\text{l.s.p.}} \frac{1}{\sqrt{2\pi}} \frac{1}{2i\pi} \beta \oint \frac{dg}{g^{k+1}} \left(-\frac{\partial E}{\partial \beta}\right)^{\frac{1}{2}} \exp - A_c \tag{47}$$

Now for β large, equation (46) shows that E(β) goes to zero. We need only the asymptotic form of E(β)

$$\begin{cases} \beta = 2 \displaystyle\int_{x_-}^{x_+} \frac{dx}{\{2[V(x)+E]\}^{1/2}} \\ V(x) = \frac{1}{2} x^2 + O(x^3) \end{cases}$$

The logarithmic divergence of the integral for E small is governed by the small x behavior of the potential. By subtracting and adding to the integrand $(x^2+2E)^{-1/2}$, we get :

$$E(\beta) \sim -\{ 2x_+^2 + \exp 2 \int_0^{x_+} \left(\frac{1}{\sqrt{2V(x)}} - \frac{1}{x}\right) dx \} e^{-\beta} \tag{48}$$

where x_+ in the small E limit is just the zero of V(x), end point of the classical trajectory.
We now get for $\{e^{-\beta E_o}\}_k$:

$$\{e^{-\beta E_o}\}_k \sim \sum_{\text{l.s.p.}} \frac{e^{-\beta/2}}{2\pi^{3/2}} \beta \int x_+ \exp \int_0^{x_+} \left(\frac{1}{\sqrt{2V(x)}} - \frac{1}{x}\right) dx$$

$$\frac{dg}{g^{k+1}} \quad \exp -2 \int_0^{x_+} \sqrt{2V(x)}\, dx \tag{49}$$

Replacing the potential by the anharmonic potential we then obtain:

$$\{e^{-\beta E_o}\}_k \sim 2\beta e^{-\beta/2} \left(\frac{2}{\pi^3}\right)^{1/2} (-1)^{k+1} \int_\infty^0 \frac{dg}{(-g)^{k+3/2}} \exp \frac{4}{3g} \tag{50}$$

Setting

$$E_o = \frac{1}{2} + \sum_{k=1}^{\infty} E_k\, g^k \tag{51}$$

and as E_k is of order $k!$:

$$\{e^{-\beta E_o}\}_k \sim -e^{-\beta/2} \beta\, E_k \tag{52}$$

So that we finally obtain :

$$E_k \sim (-1)^{k+1}\, 2 \left(\frac{2}{\pi^3}\right)^{1/2} \int_{-\infty}^{0} \frac{dg}{(-g)^{k+3/2}}\, e^{\frac{4}{3g}} \tag{53}$$

Or :

$$E_k \sim (-1)^{k+1} \left(\frac{6}{\pi^3}\right)^{1/2} \left(\frac{3}{4}\right)^k \Gamma\left(k + \frac{1}{2}\right) \tag{54}$$

It also shows that :

$$I_m E_o(g) \sim \left(\frac{-8}{\pi g}\right)^{1/2} e^{\frac{4}{3g}} \tag{55}$$

$$g \to 0$$

in agreement with Bender and Wu's calculation[2)].

It is also possible to obtain from the same calculation the behavior of the Nth level :

Notice that :

$$\frac{\partial A_c}{\partial \beta} = -E(\beta) \sim \frac{16}{g} e^{-\beta} \qquad (56)$$

So that:

$$A_c = -\frac{4}{3g} - \frac{16}{g} e^{-\beta} + O(\frac{e^{-2\beta}}{g}) \qquad (57)$$

If we expand exp $-A_c$ in powers of $e^{-\beta}$, the coefficient of $e^{-N\beta}$ corresponds to the Nth state above the ground state. Furthermore the leading term at large order comes from the largest negative power of g, therefore from the $e^{-\beta}$ contribution to A_c. Therefore :

$$E_{k,N} \sim (-1)^{k+N+1} (16)^N (\frac{8}{\pi^3})^{1/2} \frac{dg}{(-g)^{k+N+3/2}} e^{4/3g} \qquad (58)$$

Or :

$$E_{k,N} \sim (-1)^{k+N+1} (12)^N (\frac{6}{\pi^3})^{1/2} (\frac{3}{4})^k \Gamma(k+N+\frac{1}{2}) \qquad (59)$$

Again in agreement with Bender and Wu's result.[2)]

In addition as we have written formulae valid for any potential, it will be easy to generalize this result to an arbitrary potential.

A last remark is in order:

We have chosen a trajectory of period exactly β . We could also consider trajectories of period β/2, β/3 ... . But they give an action 2, 3 ... times larger and an exponentially small and therefore negligible contribution to the large order behavior.

III. THE ϕ^4 FIELD THEORY

We shall now study the ϕ^4 theory in order to show that, up to minor modifications, the same method applies. We shall restrict ourselves first to 2 and 3 dimensions because in 4 dimension the ϕ^4 is just renormalizable so that new features arise. We shall calculate the large order behavior of the N-point function :

$$\{Z^{(N)}(x,\ldots x_N)\}_{(K)} = \frac{1}{2i\pi}\oint \frac{dg}{g^{k+1}} \left[d\phi(x)\right]\phi(x_1)\ldots\phi(x_N)\exp - A(\phi) \tag{60}$$

with :

$$A[\phi] = \int d^d x \left[\frac{1}{2}(\partial_\mu \phi)^2 + \frac{1}{2} m^2\phi^2 + \frac{1}{4} g\, m^{4-d}\, \phi^4\right] \tag{61}$$

Again we look for the saddle point equations for $\phi(x)$ and g :

$$\frac{k}{g} + \frac{1}{4} m^{4-d} \int \phi_c^4 \, d^d x = 0 \tag{62}$$

$$\left[-\Delta + m^2\right] \phi(x) + g\, m^{4-d}\, \phi^3(x) = 0 \tag{63}$$

Setting :

$$\phi_c(x) = \frac{m^{\frac{d}{2}-1}}{\sqrt{-g}} \chi(mx) \tag{64}$$

We get :

$$\begin{cases} (-\Delta+1)\chi(x) - \chi^3(x) = 0 & (65) \\ Kg = -\frac{1}{4}\int \chi^4(x)\, d^d x & (66) \end{cases}$$

Therefore, provided there exists a finite action solution to the euclidean field equation, g at the saddle point will again be negative and of order 1/k.

It is easy to find spherically symmetric solutions to the field equation . Let us choose an arbitrary origin x_0 and call r :

$$r = |(x-x_0)| \tag{67}$$

Then the field equation becomes :

$$\left[-\left(\frac{d}{dr}\right)^2 - \left(\frac{d-1}{r}\right)\frac{d}{dr} + 1\right] \chi(r) - \chi^3(r) = 0 \qquad (68)$$

This equation has a solution even in r, finite at the origin, zero free, and decreasing as e^{-r} for large r. Sobolev inequalities, or a recent more general result proven by Coleman, Glaser and Martin[9)], show that this is the minimal action solution.

At the saddle in $\phi(x)$, the action is :

$$A_c[\phi_c] = -\frac{a(d)}{g} \qquad (69)$$

where a(d) is a positive number.
Again we have to calculate the determinant of small fluctuations around the saddle point. As the solution of the field equation depends on an arbitrary origin x_o^μ, we have to take x_p^μ as a collective variable and this generates a Jacobian. Let us introduce the operator M :

$$M = \frac{\delta^2 A}{\delta\phi_c(x)\delta\phi_c(x')} \qquad (70)$$

$$M(x,x') = \left[(-\Delta + m^2) - 3m^2 \chi^2(mx)\right] \delta(x-x') \qquad (71)$$

This operator has now d zero eigenvalues corresponding to translations, with eigenvectors $\partial_\mu\phi_c(x)$. As before the determinant of M that we need, is the determinant of M in the subspace orthogonal to $\partial_\mu\phi(x)$.

Combining all factors we get :

$$\{Z^{(N)}(x_1 \ldots x_N)\}_k \sim \frac{(-1)^{k+1}}{2i\pi} \left(\frac{a}{2\pi}\right)^{\frac{d}{2}} \{(\det M_\perp \; m^{2d}\}^{-1/2} \times$$

$$F_N(x_1 \ldots x_N) \int \frac{dg}{(-g)^{k+\frac{d}{2}+\frac{N}{2}+1}} e^{\frac{a}{g}} \qquad (72)$$

where F_N is defined as :

$$F_N(x_1 \ldots x_N) = m^{\frac{N}{2}(\frac{d}{2}-1)} \int d^d x_0 \, m^d \prod_{i=1}^{N} \chi[m(x_i - x_0)] \quad (73)$$

The Fourier transform of $F_N(x_1 \ldots x_N)$ is just the product of the Fourier transforms $\tilde{\chi}(\frac{p_i}{m})^N$ of the functions $\chi(mx_i)$.

Performing the last integrations on g leads to :

$$\{Z^{(N)}(x_1 \ldots x_N)\}_K \sim \frac{(-1)^k}{2i\pi} \frac{1}{(2\pi)^{d/2}} \{m^{2d}(\det M)_\perp\}^{-1/2} \times$$

$$F_N(x_1 \ldots x_N) \, a^{-k-\frac{N}{2}} \, \Gamma(k + \frac{d}{2} + \frac{N}{2}) \quad (74)$$

The factor i in front of the expression is just canceled by the negative eigenvalue of M. Up to now the calculation was not really different from the calculation of the large order behavior of the anharmonic oscillator. The only problem is that, as we shall see, $(\det M)_\perp$ is divergent.

Renormalization

We have mainly to calculate (up to the problem of zero eigenvalues) det M.

$$\ln \det M = \operatorname{tr} \ln M$$

$$\operatorname{tr} \ln M = \operatorname{tr} \ln (-\Delta + m^2) - \operatorname{tr} (-\Delta + m^2)^{-1} 3m^2 \chi(mx)$$

$$- \sum_{n=2}^{\infty} \frac{1}{n} \operatorname{tr} \{(-\Delta + m^2)^{-1} 3m^2 \chi(mx)\}^n \quad (75)$$

The first term in this expansion is removed by the normalization of the path integral. But the second is U.V. divergent. Our problem comes from the fact that we have forgotten the mass counterterms necessary to make the ϕ^4 theory finite in two and three dimensions.

We have to replace the action by the renormalizated action with a cut-off :

$$A_n[\phi] = \int d^dx \left\{ \frac{1}{2} \left[\phi(x)\left(-\Delta + \frac{\Delta^2}{\Lambda^2}\right) \phi(x)\right] + \frac{1}{2} m^2\phi^2 + \frac{1}{4} g\phi^4 + \frac{1}{2} \delta m^2(\Lambda)\phi^2\right\} \tag{76}$$

At cut-off large but fixed, the location of the saddle point is not changed up to corrections small for large Λ :
First the term $\int d^dx \left[\phi(x) \frac{\Delta^2}{\Lambda^2} \phi(x)\right]$ is of order $\frac{1}{\Lambda^2}$, and therefore negligible.
Second the counterterm $\delta m^2(\Lambda)$ is at least of order g, and therefore smaller by a factor 1/k than the classical action. The only effect at leading order comes from the one-loop counterterm which yields a constant multiplicative factor, and, as it is easy to verify, just cancels the divergence appearing in det M. Of course, as such, this argument only applies at Λ large but fixed, for k becoming infinite. We are interested instead in a situation in which k is large but fixed, and Λ becomes infinite. It is not clear that these limits commute.

Another assumption would lead to the same conclusion. If we assume that the function we are considering here is, for g small, only singular on the negative real axis, then we must calculate its imaginary part for g small and negative, and our whole procedure is justified.

In this way we see that a connection seems to exist, between the inversion of the large cut-off and the large order limit, and the possible appearance of new singularities.

Actually there is very good numerical evidence that no such singularities arise in two and three dimensions, where the ϕ^4 is superrenormalizable. On the other hand, in four dimension where a coupling constant renormalization is required, the situation, as we shall see, is more complicated.

The ϕ^4 Theory in Four Dimension

The Sobolev inequalities. Consider the functional $R(\phi)$:

$$R(\phi) = \frac{\left\{\int d^dx \left[(\partial_\mu\phi)^2 + m^2\phi^2\right]\right\}^2}{\int d^dx\phi^4(x)} \tag{77}$$

In any dimension $d \lesssim 4$ $R(\phi)$ is bounded from below by a strictly positive constant R.

$$\min_{\{\phi(x)\}} R(\phi) = R > 0 \tag{78}$$

Furthermore there exists for $d < 4$ a spherically symmetric and zero free function $\phi_c(x)$ solution of the differential equation

$$\frac{\delta R}{\delta \phi_c(x)} = 0 \tag{78}$$

such that

$$R(\phi_c) = R \tag{80}$$

It is easy to verify that the equation $\frac{\delta R}{\delta \phi(x)} = 0$ is the same as the equation $\frac{\delta A}{\delta \phi(x)} = 0$ up to a change in normalization in $\phi(x)$. Furthermore

$$R = - A\ (\phi_c(x),\ g=1) \tag{81}$$

For $d = 4$, a scaling argument (one replaces $\phi(x)$ by $\phi(\lambda x)$ and one minimizes on λ first) shows that the minimum of $R(\phi)$ is the same for the massive and massless theory :

$$\min R(\phi) = \min \frac{(\int (\partial\mu\phi)^2 d^4x)^2}{\int \phi^4 d^4x} \tag{82}$$

This shows that only the massless field equation can have a solution. The solution can now be found explicitly, due to the conformal invariance of the classical massless ϕ^4 theory in 4 dimensions :

$$m = 0 \qquad \phi_c(x) = \frac{1}{1 + x^2} \tag{83}$$

The path integral in the massive case is dominated by quasi-solutions of the form $\lambda\phi_c(\lambda x)$ with $x < a$, and λ very large.

Notice that the same scaling argument shows that for $d > 4$, R vanishes, so that if one modifies the free part to incorporate a cut-off :

$$R = \underset{\{\phi(x)\}}{\mathrm{Min}} \quad \{R(\phi) = \frac{\left[\int d^2x \{\left[\partial_\mu \phi (1 + \frac{\Delta}{\Lambda^2})\partial_\mu \phi\right] + m^2\phi^2\}^2\right]}{\int d^d x \; \phi^4(x)} \tag{84}$$

One finds :

$$R \sim \Lambda^{d-4} \tag{85}$$

which is not surprising since the theory is not renormalizable.

The massless theory. This example has been worked out first by Lipatov[3]. Due to the conformal invariance of the classical theory, all calculations can be done analytically. The main new feature of this model is that the solution of the euclidean equation depends on an arbitrary scale parameter, and at the same time, as the quantum theory is just renormalizable, a coupling constant renormalization is necessary.

The renormalized action is :

$$A_\Lambda(\phi) = \int d^4x \, \{\frac{1}{2}\left[\phi(-\Delta + \frac{\Delta^2}{\Lambda^2} - \frac{\Delta^3}{\Lambda^4})\phi\right] + \frac{g}{4}\phi^4$$

$$+ (Z-1)(\partial_\mu \phi)^2 + \delta m^2 \phi^2 + \frac{g}{4}(Z(4) - 1)\phi^4\} \tag{86}$$

And the classical solution is :

$$\phi_c(x) = \frac{2\sqrt{2}}{\sqrt{-g}} \quad \frac{\lambda}{1 + \lambda^2 (x-x_0)^2} \tag{87}$$

As this solution now breaks dilatation invariance we shall remain at the end of the calculation with an integral not only on x_0^p which restores translation invariance, but also only on λ .

The classical action $A(\lambda)$ at the saddle point is :

$$A(\lambda) = \frac{1}{g}\left(c_1 \frac{\lambda^2}{\Lambda^2} + c_2 \frac{\lambda^4}{\Lambda^4} - \frac{3}{8\pi^2}\right) \tag{88}$$

so that the integral on λ has the form :

$$\int_0^\infty \frac{d\lambda}{\lambda}\ \lambda^{N+4}\ \phi_c(\lambda x_1)\ldots\phi_c(\lambda x_N)\exp - \frac{1}{g(\lambda)}\ \left[c_1 \frac{\lambda^2}{\Lambda^2} + \frac{c_2\lambda}{\Lambda^4} - \frac{3}{8\pi^2}\right] \tag{89}$$

where $g(\lambda)$ is the effective couplant constant :

$$\frac{1}{g(\lambda)} = \frac{1}{g} - 8\ \pi^2\ \frac{(n+8)}{9}\ \ell n\ \lambda + O(g) \tag{90}$$

We see that the term $\int \phi\Delta^2\phi$ and $\int \phi\Delta^3\phi$ provide a cut-off to the λ integration.
If $g(\lambda)$ would not depend on λ, the integral on λ would just have the divergence expected from power counting, quadratic for the 2-pt function and logarithmic for the 4-pt function. But as for $g < 0$ the ϕ^4 theory is asymptotically free, $g(\lambda)$ behaves as $-\frac{9}{8\pi^2(n+8)}\ \frac{1}{\ell n\lambda}$, which yields an additional power of λ in the integral and makes it convergent. At this stage it is clear that the exchange in the large K, large Λ limit plays a crucial role. This additional power of λ is the result of this exchange of limit. Another somewhat related problem which comes from the same exchange of limits is the following : a natural interpretation of the large order behavior is that the factor K! comes just from the number of diagrams, and that the values of each individual diagram is close to $(\frac{a}{4})^K$. In four dimensions this interpretation is not compatible with the structure of the diagrams. Some diagrams are by themselves of order K!, as pointed out by Parisi and 't Hooft. (See also J. Zinn-Justin Cargèse lectures 1973).

Consider for instance the following contribution to the two point function :

$$D_K(g) \sim (-1)^K g^K \int \frac{d^4q}{(p-q)^2}\ I^K(q) \tag{91}$$

where $I(q)$ is the one-loop four point function

$$I(q) = \int^1 \frac{d^4k}{(k-q)^2 q^2} \sim \ell n\ \frac{\Lambda}{q} \tag{92}$$

After renormalization we obtain an integral which has a large momentum behavior of the form :

$$D_K(g) \sim (-1)^K g^K \int^{\infty} \frac{d^4q}{q^6} \left[-\ell n\, q\right]^K \tag{93}$$

Setting $q = e^t$ we get :

$$D_K(g) \sim g^K \int^{\infty} e^{-2t} t^K dt \sim g^K \frac{1}{2^K} K! \tag{94}$$

We see that this set of diagrams by itself gives a contribution of order K! without oscillations in sign. In our calculation we have neglected the weight factor of the diagram. A more careful calculation shows that the factor which multiplies K! is related to the small g behavior of the Callan Symanzik $\beta(g)$ function, and that this K! behaviour is a reflexion of the so-called "Landau ghost". So we see that the calculation of the large order behavior of a theory which is just renormalizable is more difficult to justify. On the other hand the numerical comparison between the large order estimate of the C.S. $\beta(g)$ function, and the actual calculation of the perturbative expansion up to order g^4 seems favorable (see Section V).

IV. GENERAL POTENTIALS

A) Quantum Mechanics

We shall now discuss the large order behaviour of the energy levels of a general potential[4)], calculated through a semi-classical or loopwise expansion.

We shall consider the following euclidean classical action :

$$A(q) = \int_0^{\beta} \left[\frac{1}{2} \dot{q}^2(t) + \frac{1}{\lambda^2} V(\lambda q(t))\right] dt \tag{95}$$

and the energy levels are expanded in powers of λ^2.
The potential V(q) is an entire function of q and has a minimum at the origin :

$$V(q) = \frac{1}{2} q^2 + O(q^3) \tag{96}$$

We calculate the energy levels from the large β behavior of $\operatorname{tr} \exp - \beta H$

$$\{\operatorname{tr} e^{-\beta H}\}_{(K)} = \frac{1}{2i\pi} \int [dq(t)] \oint \frac{d\lambda^2}{(\lambda^2)^{K+1}} \exp -A[q] \qquad q(0) = q(\beta) \tag{97}$$

In section II we have already given all the relevant expressions, needed in this large order behavior calculation provided we find a solution to the saddle point equations. Let us set :

$$\begin{cases} \lambda g(t) = x(t) \\ \lambda^2 = g \end{cases} \tag{98}$$

Then the saddle point is given by:

$$\ddot{x}(t) = \frac{\partial V}{\partial x} \tag{99}$$

$$gK = \int_0^\beta \left[\frac{1}{2} \dot{x}^2(t) + V(x(t))\right] dt \tag{100}$$

If we are able to find solutions of finite action in the large β limit to the equation of motion, we shall obtain a behavior analogous to that of the anharmonic oscillator.

A solution of finite action corresponds to the trajectory of a classical particle moving in a potential -V, leaving a point very close to the origin (for β large) at time 0, being reflected by a zero of the potential at time β/2 and coming back to the origin.

So the nature of such a solution depends of the nature of the zeros of the potential

$$V(x_+) = 0 \qquad x_+ \neq 0 \tag{101}$$

Given such trajectories, we obtain for the large order behavior :

$$\{tr^{-\beta H}\}_K \sim \frac{\beta e^{-\beta/2}}{2\pi^{3/2}} \sum_{\ell \leq p} x_+ \exp \int_0^{x_+} \left(\frac{1}{2\sqrt{2V(x)}} - \frac{1}{x}\right) dx$$

$$\int \frac{dg}{g^{K+3/2}} \exp - \frac{2}{g} \int_0^{x_+} \sqrt{2V(x)}\, dx \tag{102}$$

This leads to an expression for the large order behavior of the ground state energy $E_0(g)$:

$$E_{0,K} \underset{K \to +\infty}{\sim} \sum_{\ell \leq p} \frac{1}{2\pi^{3/2}} \{x_+ \exp \int_0^{x_+} \left[\frac{1}{\sqrt{2V(x)}} - \frac{1}{x}\right] dx\} \frac{\Gamma(K+\frac{1}{2})}{a^{K+\frac{1}{2}}} \tag{103}$$

where we have defined :

$$a = 2 \int_0^{x_+} \sqrt{2V(x)} dx \quad . \tag{104}$$

So we find a behavior indeed very similar to the large order behavior of the anharmonic oscillator.

Discussion

(i) If the origin is a relative minimum of the potential, then there exists real zeros of the potential, and therefore real trajectories.

Thus a the classical action is positive.

(ii) If the origin is an absolute minimum, but this minimum is degenerate, then again a is positive. In this case there are no solutions to the equation of motion satisfying the boundary conditions in the large β limit. One has instead to consider a family of quasi solutions corresponding to the superposition of a solution going from the origin to another minimum and its time reversed, in a configuration in which the points around which the action is concentrated are largely separated[9].

(iii) The origin is a unique absolute minimum of the potential. The zeros of the potential are now necessarily complex. The trajectories are also complex and a is not real positive.

As we shall see in the last section,these facts have direct implications on the summability of the perturbative expansion.

B. Scalar Boson Field Theories[4)]

The generalization to Field Theory is straightforward the results of what we already done. Consider a Field theory defined by the euclidean action :

$$A(\phi) = \int d^d x \left[(\partial_\mu \phi)^2 + \frac{1}{\lambda^2} V(\lambda\phi) \right] \qquad (105)$$

We calculate various correlation functions through loopwise expansion, i.e. in powers of λ^2 . If we can find real or complex solutions of the euclidean field equation :

$$\Delta\phi(x) = \frac{\delta V}{\delta\phi(x)} \qquad (106)$$

Then the calculation of the large order behavior follows very closely the lines of section III.

We again obtain a large behaviour of the form
$c_N(x_i)\, K^{N/2 + \delta/2 - 1}\, (a)^{-K}\, K!$
where δ is the number of symmetries of the classical action broken by the classical solution (translations, dilatation, internal symmetry ...). The constant a is the classical action and is positive if and only if the classical solution is real.

C. Other Field theories

Other field theories have been studied. We shall just list them :

1) QED with charged scalar bosons :

$$A(A_\mu, \phi) = \int d^4 x \left[\frac{1}{4} F_{\mu\nu}^2 + D_\mu \phi\, D_\mu \phi^\dagger \right] \qquad (107)$$

where $D_\mu \phi$ is the covariant derivative :

$$D_\mu \phi = \partial_\mu \phi + i\, e\, A_\mu \phi \quad . \tag{108}$$

Solutions to the euclidean field equations have been found for $e^2 < 0$ by Itzykson,Parisi,Zuber [10] on one side and Lipatov[11] on the other side. The main problem here is to choose an Ansatz for the vector field A_μ .

2) Non Abelian Gauge Theories in 1 dimension . Belavin[12] et al. have found real instantons in this case, so that the classical action is positive, and again all terms of perturbation series at large order are of the same sign.

3) Theories with fermions. This problem is discussed in the lectures of G. Parisi so we shall not discuss it again here.

V. NUMERICAL EXAMPLES AND BOREL SUMMATION

There have been a number of comparisons between large order behavior calculations and perturbative expansions.

A. Quantum Mechanics

The most systematic study has been made in Quantum Mechanics. Bender and Wu have calculated 150 terms of the series of the ground state of the anharmonic oscillator. We have studied as a function of the parameter γ the potential $V(x)$:

$$V(x) = \frac{1}{2}\left(x^2 - 2\,\gamma x^3 + x^4\right) \tag{109}$$

When γ varies, one goes from a situation in which the classical action is real to an other in which it is complex.

Extrapolation of the 50 first terms of the series has shown agreement with the large order estimate with an accuracy of 10^{-4} to 10^{-7}.

We have also studied some other potentials like $(1 - \cos x)$.

The agreement is so general and so accurate that there seems to be little doubt about the applicability of the method.

B. A Model of a Finite Field Theory

In the case of Field theory it is interesting to separate the problem of renormalization from the question of the number of degrees of freedom. Therefore a theory with a very convergent propagator has been studied by Bervillier et al.[13)], described by the following action :

$$A(\phi) = \int d^d x \left| \frac{1}{2} \phi(x) \left(\exp - \frac{\Delta}{\Lambda^2}\right) \phi(x) + \frac{g}{4!} \phi^4(x) \right| \qquad (110)$$

As the propagator is of the form e^{-p^2/Λ^2} in momentum space, all Feynman diagrams are both finite and easy to calculate. It has been possible to calculate up to order g^7 the zero, two and four functions.

Furhtermore the equation of motion can be solved analytically for any dimension

$$\left(\exp - \frac{\Delta}{\Lambda^2}\right) \phi(x) + \frac{g}{3!} \phi^3(x) = 0 \qquad (111)$$

It is easy to verify that a gaussian function $c_1 \exp - c_2 x^2$ satisfies the equation for an appropriate choice of the constants c_1 and c_2.

Furthermore to complete the calculation one has to evaluate the determinant of the operator $\frac{\delta^2 A}{\delta\phi_c(x)\delta\phi_c(x')}$. More precisely, taking into account the normalization of the path integral one has to calculate :

$$D(z) = \det \left[1 + z\, e^{\frac{\Delta}{\Lambda^2}} \frac{g}{2!} \phi_c^2(x)\right] \qquad (112)$$

The variable z is just there to allow us to cancel the zero eigenvalues of the determinant when $z = 1$.

Now the operator $e^{\frac{\Delta}{\Lambda^2}} \phi_c^2(x)$ is of the form $e^{\frac{\Delta}{\Lambda^2}} e^{-2ax^2}$. Using Baker-Hausdorf formula it is easy to verify that this operator is the exponential of a quadratic form in x_μ and ∂_μ, which is an harmonic oscillator hamiltonian.

So the large order behavior calculation can be performed analytically for any dimension and any correlation function. For $d = 0$ the results are those of the simple integral of section I.

Comparison between large order behavior and explicit calculation of diagrams, shows that the agreement which is very good at low dimensions deteriorates both when the dimension and the number of arguments of the correlation function increase . Table I gives an idea of what happens.

Table I

	D =	0	1	2	3	4
N =	0	0.935	0.954	0.794	0.460	0.140
	2	0.935	0.827	0.644	0.395	0.151
	4	0.847	0.515	0.280	0.120	0.0309

Ratios of the series and the asymptotic estimate of 7th order at D dimensions for the N point function in the ϕ^4 theory with gaussian propagator.

C. The $(\vec{\phi}^2)^2$ Field Theory

1. <u>The massless theory in four dimension</u>. A numerical comparison has been made between the large order behavior of the Callan-Symanzik renormalization group function $\beta(g)$, and the perturbation series at order g^4.

The results are as follows :

	n = 1	n = 2	n = 3
β_4 exact :	4.90	6.08	7.34
β_4 asymptotic :	5.33	5.28	4.84

One sees that the agreement is rather good in spite of the arguments which seem to suggest that the large order behavior calculation could be incomplete.

Of course this leaves open the possibility that the large order behavior calculation is correct for some well defined renormalization group functions, without being correct for the general correlation functions.

2. <u>The massive theory in 3 dimensions.</u> Nickel[14] has calculated the renormalization group functions up to order 7 or 6 for the massive $(\vec{\phi}^2)^2$ theory in 3 dimensions, renormalized at zero momentum.

The Saclay[15] group has calculated numerically the large order behavior in this theory. Therefore we have here an example of a superrenormalizable theory with mass renormalization, in which an interesting comparison is possible.

Table II shows the results for the R.G. $\beta(g)$ function. The agreement is extremely good, better than in the corresponding theory with gaussian propagator. This seems to result from the finite coupling constant renormalization. If we perform the expansion in terms of the bare coupling constant, the convergence to the large order behavior is slower.

Table II

	K=0	1	2	3	4	5	6	7
$(-1)^K$	β_K	1	1	0.42249	0.35107	0.37653	0.49555	0.74967
10^2	c_K	12.93	15.91	6.997	5.246	4.479	4.199	4.094

Coefficients β_K of the perturbative expansion of $\beta(g)$, as a function of the order K. The coefficients c_K are defined by :

$$c_K = \beta_K/(-a)^K\ \Gamma(K+\frac{g}{2}) \qquad \text{and should have a limit}$$

$$c_\infty \simeq 4.11\ 10^{-2} \qquad a = .14777423\ \dots .$$

D. The Problem of Resummation

Consider a function F(g) which is analytic, in the neighborhood of the origin, for

$$|\mathrm{Arg}\ (g)| < \alpha$$

and has an asymptotic series expansion in this angle for g small, with a bound on the error of the form :

$$\left|F(g) - \sum_{n=0}^{K} F_n\, g^n\right| < M\, A^K (K!)^\rho |g|^{K+1} \quad \text{for } |\mathrm{Arg}(g)|<\alpha \qquad (113)$$

At g fixed, the best estimate of F(g) from the series is obtained by truncating the series at an order K such that the error is minimum :

$$K \sim (A|g|)^{-1/\rho} \tag{114}$$

and the minimal error ε(g) is then :

$$\varepsilon(g) \sim \exp -\rho|A|g||^{-1/\rho} \tag{115}$$

If we now try to reconstruct the function F(g) from the series, the ambiguity in the solution of this problem will be of the form of functions analytic in the same region, and decreasing faster than ε(g) at the origin. From Phragmen Lindelöf theorem, we know that no such functions exist if the angle α is large enough :

$$\alpha > \rho \frac{\pi}{2} \tag{116}$$

If such a condition is fulfilled, the function F(g) is uniquely defined by its asymptotic series[16]. We see in particular that if ρ is larger than 2, the asymptotic series must also be asymptotic in a second sheet of the function.

Furthermore this unique solution can be obtained from a Borel transformation.

Let us call B(g) the Borel transform of F(g) :

$$B(g) = \sum_0^\infty \frac{F_n}{\Gamma(K\rho+1)} g^n \tag{117}$$

Then in the sense of formal power series we have the relation :

$$F(g) = \int_0^\infty e^{-t} B(gt^\rho)dt \tag{118}$$

From the bound on the error, we see B(g) is analytic in a circle of radius R :

$$R = \frac{\rho^\rho}{A}$$

Furthermore it can be shown that B(g) is also analytic in an angle θ :

$$\mathrm{Arg}(g) < \alpha - \frac{\pi}{2}\rho \qquad (120)$$

and does not increase too fast for $|t|$ large, so that the integral (118) converges.

Reciprocally assume that such an integral representation exists. It can be rewritten as a Laplace transform :

$$F(g) = g^{-\frac{1}{\rho}} \int_0^\infty e^{-\frac{t}{g^{1/\rho}}} B(t^\rho)\,dt \qquad (121)$$

Then, as noticed by Sokal[17], even if $B(t^\rho)$ is a distribution, if it exists for some value g_0 of g, it exists in the region

$$\mathrm{Re}\,\frac{1}{g^{1/\rho}} > \mathrm{Re}\,\frac{1}{(g_0)^{1/\rho}} \qquad (122)$$

which in the neighborhood of the origin implies :

$$|\mathrm{Arg}(g)| < \frac{\pi}{2}\rho \qquad (123)$$

A necessary condition for representation (121) to exist is therefore that $F(g)$ is analytic in the region defined by the condition (122).

A last problem remains to be solved. Assume the analytic domain of $B(g)$ and a few terms of the asymptotic series are known. We have still to continue $B(g)$ outside the circle of convergence of the Taylor series in a neighborhood of the real positive axis. A first possible method is to construct the Pade approximants of $B(g)$. In the absence of any information on the function $B(g)$, it is even one of the few methods available. The main defect of this method is its instability, which makes an estimation of the error difficult. If we know the analytic domain of $B(g)$ another method, first suggested by Loeffel[18] is available. One can perform a conformal mapping :

$$u = u(g)$$

with

$$u(0) = 0$$

such that the domain of analyticity of B(g) is mapped onto the circle : $|u| < 1$.
From the asymptotic series of F(g), we derive now a convergent expansion for B(g(u)) in power series of u :

$$B(g(u)) = \sum_{K=0}^{\infty} B_K u^K \qquad (124)$$

where we know as many coefficients B_K as F_K .
We then get an expansion for F(g) :

$$F(g) = \sum_{K=0}^{\infty} B_K \int_0^{\infty} e^{-t} u^K \left[gt^{\rho}\right] dt \qquad (125)$$

In this last method it is of course very easy to incorporate the information about the large order behavior.

If F(g) is a Green's function of some field theory, for which we know the large order behavior :

$$F_K \underset{K\to +\infty}{\sim} c\, K^b\, K!\, a^K \left(1 + O\left(\frac{1}{K}\right)\right) \qquad (126)$$

then we know the nature and the location of the nearest singularity of B(g). Furthermore, other subleading instanton contributions will probably correspond to the other singularities of B(g).

If there are no real instanton solutions (or quasi solutions), no singularity will appear on the real positive axis, and the Borel transformation will exist, provided B(g) does not increase too much for g large and positive. A too fast rise of B(g) for g large would be related to terms not increasing as fast as K! (as $\sqrt{K}!$ for example) and all of the same sign. These terms are not seen in the large order behavior analysis, but generate a singularity of $g = +\infty$. Up to now, no example of such a situation has been found in field theoretical models.

Let us now illustrate by a few numerical examples[20] this method. All examples will involve Borel transform analytic in a plane cut on the negative real axis[21]. Large order behavior tells us that the cut starts at $g = \frac{1}{a}$ where a is now negative. The conformal mapping is then :

$$u(g) = \frac{\sqrt{1 - ag} - 1}{\sqrt{1 - ag} + 1} \qquad (127)$$

Furthermore, at it is useful to vary the singularity of B(g) at g = 1/a, we used in practice a Borel-Leroy[19] transformation :

$$B(g) = \sum_{0}^{\infty} \frac{F_K}{\Gamma(K + \beta + 1)} g^K \tag{128}$$

so that :

$$F(g) = \int_{0}^{\infty} e^{-t} t^{\beta} B(gt)dt \tag{129}$$

The singularity of B(g) at $g = \frac{1}{a}$ is then :

$$B(g) \sim (1 - ag)^{\beta-b-1} \tag{130}$$

The parameter β was varied in neighborhood of b. When β-b is a positive integer, the previous formula has to be multiplied by $\ell n(1-ag)$.
A last practical remark. Let us set :

$$F(g) = F_0 + gF_1(g) \tag{131}$$

One can of course calculate the Borel transform of $F_1(g)$ instead. This changes the singularity of B(g) for g large.

Table III shows the convergence for the integral :

$$F(g) = \int_{0}^{\infty} x \exp - \frac{x^2}{2} - g \frac{x^4}{4} dx \tag{132}$$

Table IV shows the convergence of the anharmonic oscillator. Notice that for the value -150 of the product ag, the accuracy obtained with 20 terms of the series is still better than 10^{-2}.

Table V shows the convergence of the infra-red stable zero of the renormalization group function of massive ϕ_3^4 . Table VI shows the convergence of some combination of the anomalous dimensions of the ϕ and ϕ^2 fields for the fixed point coupling constant.

Table III

	K=1	2	3	4	5	10
β= 1	$-6.65\ 10^{-2}$	$1.57\ 10^{-2}$	$-4.63\ 10^{-3}$	$1.56\ 10^{-3}$	$-5.76\ 10^{-4}$	$8.47\ 10^{-6}$
= 1.5	0	0	0	0	0	0
= 2.0	$5.29\ 10^{-2}$	$-3.59\ 10^{-3}$	$5.72\ 10^{-4}$	$-1.26\ 10^{-4}$	$3.38\ 10^{-5}$	$-1.85\ 10^{-7}$

Error for the simple integral for the product ag = -2. , as a function of the number of terms of the series K, and β the parameter of the Borel transform. The exact value is 0.65567954 ... .

Table IV

K=	1	2	3	4	5	6
β= 0.5	$-4.8\ 10^{-2}$	$-3.2\ 10^{-2}$	$-2.26\ 10^{-3}$	$-4.2\ 10^{-3}$	$-1.18\ 10^{-3}$	$1.22\ 10^{-5}$

Error for the ground state energy of the anharmonic oscillator as a function of the order K, for the value 0.5 of the parameter β of the Borel Leroy transform, and the value ag = -1.5. The exact result is 1.3923516 ...

Table V

K =	2	3	4	5	6	7
g^* =	1.0	1.5202	1.4256	1.4301	1.41928	1.4163

Zero of the renormalization group function $\beta(g)$ for ϕ^4_3 as a function of the order K. The parameter a is in

Table VI

K =	0	1	2	3	4	5	6
γ =	1.0	1.2	1.2439	1.2414	1.2429	1.2411	1.2412

The critical exponent γ as a function of the order as calculated from the I.R. fixed point of ϕ^4_3.

REFERENCES

1) J. Zinn-Justin, Lecture notes for the GIFT School 1977, SALAMANCA, to be published in "Lecture Notes in Physics" (Springer Verlag)
2) C. Bender and T.T. Wu, Phys.Rev.Lett. 27, 461 (1971) Phys.Rev. D7, 1620 (1973).
3) L.N. Lipatov, Zh. Eksp.Teor.Fiz., Pis'ma Red 25, 116 (1977) and Leningrad Institute of Nuclear Physics Reports 253 and 255 (1976)
4) E.Brezin, J.C. Le Guillou and J. Zinn-Justin, Phys.Rev. D15, 1544 and 1558 (1977)
5) The collective coordinate method has been discussed in : J.S. Langer, Ann.Phys (NY) 41, 108(1967); J.L. Gervais, B. Sakita, Phys.Rev. D11, 2943 (1975); V.E. Korepin, P.P. Kulish and L.D. Faddeev, J.E.T.P. Lett. 21, 138 (1975); See also N.H. Christ and T.D. Lee, Phys.Rev. D12, 1606 (1975)
6) R. Dashen, B. Hasslacher and A. Neveu, Phys.Rev. D10, 4114 (1974)
7) J.H. Van Vleck, Proceedings National Academy of Sciences 178, 14(1928)
8) S. Coleman, V.Glaser and A. Martin, CERN preprint TH 2364
9) E. Brezin, G. Parisi and J. Zinn-Justin, Phys.Rev. D16, 408 (1977)
10) C. Itzykson, G. Parisi and J.B. Zuber, Phys.Rev.Lett. 38, 306 (1977)
11) L.N. Lipatov, Leningrad Nuclear Physics Institute Preprint (1977)
12) A.A. Belavin, A.M. Polyakov, A.S. Schwartz and Yu.S.Tyupkin, Phys.Lett. 59B, 85 (1975)
13) C. Bervillier, J.M. Drouffe, C. Godrèche and J. Zinn-Justin, Saclay Preprint DPhT/77/101 (1977)
14) B.G. Nickel to be published. See also : G.A. Baker, B.G.Nickel and D.T. Meiron, Saclay preprint DPhT/77/39 (1977)
15) E. Brezin, G. Parisi and J. Zinn-Justin, in preparation
16) E. Borel, Leçon sur les séries divergentes (Paris 1928) G.H. Hardy, Divergent series (Oxford Univ.Press, Oxford 1948)

17) A. Sokel, private communication see ref. 16)
18) J.J. Loeffel, Proceedings of the 1976 Workshop on Padé approximants, D. Bessis, J. Gilewicz and P. Mery ed. CEA publication
19) E. Leroy, Ann.Fac.Sci.Toulouse 2, 317 (1900)
20) J.C. Le Guillou and J. Zinn-Justin, Phys.Rev.Lett. 39, 95 (1977) and in preparation.
21) For anharmonic oscillators, the Borel summability has been shown by S. Graffi, V. Grecchi and B. Simon, Phys.Lett. 32B, 631 (1970).
Analyticity properties and convergence of diagonal Padé approximants to the actuel energy levels have been shown by J.J. Loeffel, A. Martin, B. Simon and A.S. Wightman ibid. 30B, 656 (1969).
In quantum field theory the Borel-summability has been shown for ϕ^4_3theory by J.P. Eckmann, J. Magnen and R. Sénéor, Comm.Math.Phys. 39, 251 (1975) and also by J.S. Feldman and K. Osterwalder, Ann.Phys. (N.Y.) 97, 80 (1976).

THE PHYSICAL BASIS OF THE ASYMPTOTIC ESTIMATES IN PERTURBATION THEORY

G. PARISI [+]

Laboratoire de Physique Théorique de l'Ecole Normale Supérieure [++], 24 rue Lhomond 75231 Paris Cedex 05 France

I. INTRODUCTION

During last year, the asymptotic behaviour in K of the Kth order of the perturbative expansion has been widely investigated with remarkable success[1-3]. If the interaction is superrenormalizable, the problem has been solved for both interacting bosons[4,5] and fermions[6-8] : of course technical details may change from theory to theory. When the interaction is renormalizable new difficulties arise[9,10]; they are not completely understood at the present moment.

The aim of these lectures is to present a general review of the subject and to discuss the motivations of these investigations. The asymptotic estimates obtained are quite interesting : they are an essential step toward the use of the perturbative expansion outside the weak coupling regime. Indeed, if the perturbative expansions were convergent, by computing a high enough number of Feynman diagrams it would be possible to obtain numerical results with arbitrarily small accuracy also in the strong coupling region. Unfortunately, the perturbative series have a zero radius of convergence. We can obtain reasonable answers for strong coupling only if we use resummation techniques which cope with non convergent series. The success of these techniques is not guaranteed : it depends on the properties of the asymptotic estimates.

[+] On leave of absence from INFN (Frascati-Roma) Italy.

[++] Laboratoire Propre du C.N.R.S. associé à l'Ecole Normale Supérieure et à l'Université de Paris Sud.

To be concrete, let us consider a typical example. The function A(g) (g being the coupling constant) has a formal expansion in powers of g

$$A(g) = \sum_{0}^{\infty}{}_K A_K g^K \qquad (I.1)$$

Our aim is to get an expression of A(g) in terms of the A_K. The solution is not unique; however, if the function A(g) is analytic in the domain $\mathscr{D}$ ($|g|< R^{-1}, |\arg(g)|<\delta + \frac{\pi}{2}$, $\delta>0$) and the bounds $|\frac{d^K}{dg^K} A(g)|<(K!)^2 R^K$ are valid inside $\mathscr{D}$, (this happens in many field theories[11-13]; e.g. super-renormalizable $g\phi^4$ and Yukawa theories) A(g) can be unequivocally written as :

$$A(g) = \int_0^{\infty} B(bg)\exp(-b)db = g^{-1}\int_0^{\infty}\exp(-b/g)B(b)db \qquad (I.2)$$

The function B(b) (the Borel transform of A(g)) is defined by the convergent expansion :

$$B(b) = \sum_{0}^{\infty}{}_K A_K b^K/K! \qquad (I.3)$$

inside its radius of convergence and by analytic continuation on the positive real axis; the conditions we have imposed on the function A(g) imply that B(b) is analytic in a sector containing the positive real axis ($|\arg b|<\delta$); the following representation holds

$$B(b) = (2\pi i)^{-1}\oint_{\varphi} G(g)/g \exp(b/g)dg \qquad (I.4)$$

φ being the border of the domain .

The Borel transform is a quite naturel concept in quantum field theory : in a rough sense it has the meaning of the contribution to A(g) of those field configurations having an action equal to b/g, as it can be seen from the functional integral representation.

The asymptotic estimates for the A_K are connected to the singularities of the Borel transform : if

$$A_K \underset{K\to\infty}{\to} K!\, R^K \cos(\theta K)\, K^{\alpha}\left[C + O(K^{-1})\right] \qquad (1.5)$$

the nearest singularity to the origine is at $b = R^{-1}\exp(\pm i\theta)$.

Although the Borel sum (eqs. (I.2)-(I.3)) gives a unique definition of the function A(g), at first sight, it seems that the knowledge of the first few A_K is not sufficient to obtain convergent sequence of approximants to A(g) : if we restrict the sum in the definition of B(b) to a finite number of terms, eqs. (I.2)-(I.3) reproduce the standard non convergent perturbative expansion. However, we can take advantage of the analyticity properties of the Borel transform to map conformally the real positive line on the interval 0-1 [14-16]

$$A(g) = g^{-1} \int_0^1 \exp(-b(z)/g)\, F(z) dz$$

$$F(z) = B(b(z)) \frac{db}{dz} \qquad \text{(I.6)}$$

$$b(0) = 0; \quad b(1) = \infty$$

The conformal map must be chosen such as not to map the singularities of the Borel transform inside the unity circule. A convergent sequence of approximants is given by :

$$A^N(g) = g^{-1} \int_0^1 \exp\left[-b(z)/g\right] F^N(z) dz$$

$$F(z) = \sum_0^\infty{}_K F_K z_K \qquad \text{(I.7)}$$

$$F^N(z) = \sum_0^N{}_K F_K z_K$$

where the F_K are linear combinations of the A_K .

In theories in which representations like (I.6) can be established, we have an algorithm based on the Feynman diagrammatical expansion which allows us to compute physically interesting quantities for strong coupling constants with arbitrarily small accuracy. This technique has been succesfully applied to superrenormalizable $g\phi^4$ interactions, also for very large values of the coupling constant[14-16].

In the general case, a disaster may happen which ruins our beautiful dreams of computing physical quantities using the perturbative expansion in the strong coupling region. If the angle θ in eq. (I.5) is equal to zero, the Borel transform has a singularity on the real axis and the perturbation expansion cannot be

Borel summed. However, a disaster of such a magnitude (we do not know how to define the theory for strong coupling) must be a punishment for a wrong doing. Indeed the presence of singularities on the real line in the Borel transform is the signal of existence of non perturbative effects which are ignored in the perturbative expansion. The study of the singularities of the Borel transform (i.e. the asymptotic estimates) aims to find and to compute these non perturbative effects. Let us discuss a few examples.

1) The interacting Hamiltonian is unbounded from below : no ground state exists and the perturbative vacuum is a metastable state (a resonance). Its energy has an imaginary part which is exponentially small for a small coupling constant. This phenomenon is typical of theories in which the coupling constant has the wrong sign.

2) The interacting Hamiltonian is bounded from below. However there are more than one classical vacua and the tunneling between these vacua is exponentially small but non zero. There is a non perturbative effect due to the mixing between the different vacua. This usually happens if we study the perturbation expansion around a classical vacuum which breaks an unbreakable symmetry (i.e. a discrete symmetry in one dimension or any symmetry in a finite volume).

3) The short (or the large) momentum behavior of the Green functions is different in the free and in the interacting theories. If P_c (the cross over momentum between the free and the interacting behaviour) is a power of the coupling constant (as when the interaction is non renormalizable or super-renormalizable but massless[17]) the perturbative expansion is undefined because of ultraviolet or infrared divergences. If P_c is exponentially small or large as in a renormalizable theory, the perturbative expansion is well defined, however, a remnant of this phenomenon appears as a singularity of the Borel transform.

In the following sections of these lectures, we will see how to compute explicitly these non perturbative effects and consequently how to derive asymptotic estimates for the high orders of perturbation theory. Semi-classical methods will be used to study cases 1 and 2 while renormalization group arguments will be the main tools in the discussion of case 3. Unfortunately, we are not able to reach definite conclusions in this last case. Finally, in the last section we will discuss our understanding of Yang-Mills theories.

II. THE UNSTABLE VACUUM

In this section we study the asymptotic estimates connected to the metastability of the vacuum (case 1). To be definite let us consider the $g\phi^4$ interaction in dimension D < 4. The Lagrargian density is

$$\mathcal{L}(x) = \frac{1}{2} (\partial_\mu \phi)^2 + \frac{1}{2} \phi^2 + \frac{1}{4} g\phi^4 \tag{II.1}$$

We concentrate our attention on the "vacuum energy" E(g) i.e. the sum of all the vacuum to vacuum connected diagrams. We suppose that E(g) is analytic on the cut complex g plane and the disconti- nuity on the negative real axis is related to the mean life of the vacuum for negative values of g. The interaction among particles is repulsive for positive g, but attractive for negative g; an attractive interaction among bosons leads to a collapse. Let us see why. Denoting by $E_N(g)$ the energy of a state of N particles with overlapping wave functions, we roughly find :

$$E_N(g) \simeq Nm + gN^2 \tag{II.2}$$

For negative g there are collapsed states of negative energy. The vacuum decays in one or more of these objects with probability roughly proportional to $\exp(-E_B)$; E_B being the potential barrier, i.e. the maximum of $E_N(g)$ which is proportional to 1/g. The imaginary part of the energy is proportional to the mean life and it is of order $\exp((-R|g|)^{-1})$; R being an appropriate constant.

The asymptotic estimate

$$\begin{aligned} E_K &\sim K!(-R)^K \\ E(g) &= \sum_0^\infty{}_K E_K g^K \end{aligned} \tag{II.3}$$

can be easily derived from the result

$$\mathrm{Im}\, E(g) = \exp(-|Rg|^{-\alpha}) \qquad g < 0 \qquad \Longrightarrow$$

$$E_K \sim \Gamma(\alpha^{-1}K)(-R)^K \tag{II.4}$$

The particles in the collapsed state will be in a coherent state. They will be described by a classical field satisfying the classical equation of motion. Semi-classical methods may be applied to compute the constant R. For technical reasons it is convenient to study the problem in the Euclidean version of field theory. We have to find the non trivial solution φ_c of the classical equa- tions of motion having a minimal action. In this case, the classi- cal equations of motion have non trivial solutions only for negati- ive g.

It has been shown that if $\mathcal{A}_c/|g|$ is the action corresponding to φ_c, the imaginary part of $E(g)$ is $\exp(-\mathcal{A}_c/|g|)$ and the nearest singularity of the Borel transform of $E(g)$ is located at $b = -\mathcal{A}_c$. The existence of this singularity of the Borel transform for negative b forbids us to define the theory for negative g. These results can be obtained[18] starting from the functional integral representation for E(g)

$$E(g) = \lim_{V\to\infty} -V^{-1} \ell n \int d|\varphi|_V \exp - \mathcal{A}(\varphi)$$

$$\mathcal{A}(\varphi) = \int d^D x \mathcal{L}(x) \tag{II.5}$$

and integrating on those field near to φ_c. The final result is

$$\mathrm{Im}\, E(g) \simeq |\det M|^{-1/2} \exp - \mathcal{A}_c/|g| \implies$$

$$E_K \sim |\det M|^{-1/2} (K-1)! \, (-\mathcal{A}_c)^{-K} \tag{II.6}$$

M being defined as

$$\mathcal{A}(\varphi_c + \tilde{\varphi}) = \mathcal{A}(\varphi_c) + \frac{1}{2} (\tilde{\varphi}, M\tilde{\varphi}) + O(\tilde{\varphi}^3) \tag{II.7}$$

Notice that any solution of the classical equation of motion (in particular φ_c) is a stationary point of the integral and, in the limit of small g, the saddle point method of integration yields the correct result. Using this technique and taking care of the first corrections to the saddle point method, the terms of $O(K^{-1})$ in eq. (II.3) can also be computed.

It is believed that using the same method all the singularities of the Borel transform can be computed; however, there is no numerical evidence of this conjecture.

For simplicity, in this sketchy derivation of the asymptotic estimates, we have neglected all problems connected to the symmetries of the Lagrangian, in particular the need of collective coordinates to take care of translational invariance[18] has not been discussed.

If we substitute bosons by fermions there is only a crucial difference[6] : fermions satisfy the Pauli exclusion principle and they cannot be in a coherent state : a description based on a classical field is consequently forbidden. The correct results can be obtained bu using the same intuitive argument as before. The energy of a system of N interacting fermions with overlapping wave functions is

$$E_N(g) \sim N^{1+1/d} + gN^2 \qquad \text{(II.8)}$$

$d = D-1$ being the number of spatial dimensions. Fermi statistics implies that the kinetical energy term is no more proportional to N : at high density at each point of the space we must fill the Fermi sphere up to a momentum proportional to $\rho(x)^{1/d}$, $\rho(x)$ being the local density.

Using the same arguments as before we find that

$$E_B \sim g^{-\frac{d+1}{d-1}} \qquad \text{Im } E(g) \simeq \exp\left[-(Rg)^{-D/(D-2)}\right] \qquad \text{(II.9)}$$

The reader will recognize in this argument the familiar results of the Thomas Fermi approximation[19]. In order to obtain quantitative results we must extend the Thomas Fermi approximation to field theory. In the standard derivation of the TF approximation (e.g. in atomic physics), one obtains an approximate form for the fermionic energy in terms of the field in which the fermions are imbedded. This expression is valid for strong external field and it is obtained without having to discuss the fermionic wave functions. The same strategy can also be applied to field theory. The generating functional for a Yukawa theory can be written as

$$Z_V(g) = \int d[\psi]\, d[\bar\psi]\, d[\sigma] \exp\left[-\int_V \mathcal{L}(x)\, d^D x\right]$$

$$\mathcal{L}(x) = 1/2\,((\partial_\nu \sigma)^2 + \mu^2\sigma^2) + \bar\psi\,(\not\partial + m)\psi + \lambda\bar\psi\psi\sigma \qquad \text{(II.10)}$$

$$g = \lambda^2$$

Integrating over the fermionic fields, we find

$$Z_V(g) = \int d[\sigma] \exp\left[-\int_V \mathcal{L}_F(\sigma)\, d^D x\right] D(\lambda\sigma)$$

$$\mathcal{L}_F(\sigma) = \frac{1}{2}\,(\partial_\nu\sigma)^2 + \mu^2\sigma^2 \;; \qquad D(\lambda\sigma) = \det(\not\partial + m + \lambda\sigma) \qquad \text{(II.11)}$$

The determinant which is an entire function of λ , appears at a positive power because of Fermi statistics. In the bosonic case, it appears to a negative power and its zeros are responsible for the leading part of the discontinuity along the cut in the complex g plane. In the fermionic case non analytic terms in the coupling constant may only arise from the integration in the region where $\lambda\sigma$ is large. In this region, the determinant is asymptotically proportional to

$$\det\left[\not{\partial} + m + \lambda\sigma\right] \sim \exp - S_D \left[\int (\lambda\sigma(x))^D \, d^D x\right]$$

$$S_D = 2\,\Gamma(-D/2)(4\pi)^{-D/2} \qquad \text{(II.12)}$$

This result on the large field behaviour of the determinant is obtained[7] by using the same arguments as those needed to derive the standard TF approximation[19]. It is now a simple job to derive the asymptotic estimates

$$Z_K \sim \Gamma\left[(1 - \frac{2}{D})K\right] R^K \cos (2\pi K/D)$$

$$Z(g) = \sum_0 {}_K \, Z_K g^K \qquad \text{(II.13)}$$

The Borel transform is now an entire function : it may be convenient to introduce the Borel transform of order $\alpha = 1 - 2/D$ defined by

$$Z(g) = \int db \, \exp(-b^{1/\alpha}) \; B_\alpha(bg) \qquad \text{(II.14)}$$

The asymptotic estimates give informations on the nearest singularity to the origine of $B_\alpha(b)$.

The introduction of fermions does not change the basic principles of the approach, although rather delicate estimates[8] are needed to control the large field behaviour of det $(\not{\partial} + eA)$ which is relevant in the study of the asymptotic estimates in QED.

III. THE WRONG VACUUM

In this section we study the asymptotic estimates for a theory in which the perturbation expansion is done around the wrong vacuum, (case 2 of the introduction). This happens in a one dimensional scalar theory (quantum mechanics) in which the potential has two degenerate minima. The simplest example is given by the Lagrangian

$$\mathcal{L}(\phi) = \frac{1}{2} (\partial_\mu \phi)^2 + \frac{1}{2} \phi^2 (1 - \lambda\phi)^2$$

$$\lambda^2 = g \qquad \text{(III.1)}$$

The sum of all the vacuum to vacuum connected diagrams yields formally the ground state energy of the Hamiltonian

$$H = \frac{1}{2} p^2 + \frac{1}{2} (1 - \lambda x)^2 x^2$$

$$E(g) = \sum_0^\infty {}_K E_K g^K = -\lim_{T\to\infty} T^{-1} \ell n \int d[p] \exp \left[- \int_0^T \mathcal{L}(\phi) dx \right] \qquad \text{(III.2)}$$

For small coupling constant, the Hamiltonian (III.2) has two nearly degenerate levels, whose wave functions are the symmetric and anti-symmetric combinations of those of two harmonic oscillators centered respectively at 0 and λ^{-1}. The splitting between the levels can be computed by semi classical methods[20] and it is proportional to $\exp(-1/(6g))$. Of course, the standard perturbative expansion is obtained by perturbing around the classical vacuum $\phi = 0$ (or equivalently $\phi = \lambda^{-1}$). The asymptotic estimates can be obtained by a simple modification of the approach described in the previous sections.
The Kth order of the perturbative expansion can be formally written as

$$E_K = (2\pi i)^{-1} \oint dg/g^{K+1} E(g) \qquad \text{(III.3)}$$

In the limit K going to infinity, E_K is evaluated by inserting eq. (III.2) into eq. (III.3) and considering a simultaneous saddle point in the variables ϕ and g.

The saddle point equations for ϕ are the classical equations of motion. The existence of solutions for real values of the coupling constant, (the so called instantons[21-23]), having an action equal to $+\mathcal{A}_c/g$, implies the asymptotic estimates

$$E_K \sim K! \, (2\mathcal{A}_c)^{-K} \qquad \text{(III.4)}$$

Consequently, the Borel transform $B(b)$ of $E(g)$ has a singularity on the real line at $b = 2 \, \mathcal{A}_c = \frac{1}{3}$.

The derivation of the asymptotic estimates would be straigthforward except for the following technical point which is the origine of the factor 2^{-K} in eq. (III.4). The solution to the classical Euclidean equation of motion which governs the large order behaviour has to satisfy periodic boundary conditions. In the case of degenerate minima such a solution does not exist. In particular, the well known instanton which interpolates between the two minima (its action being 1/(6g)) and corresponds to the quantum mechanical tunneling, obviously does not satisfy periodic boundary conditions. The correct result can be obtained by expanding around an interacting instanton-anti-instanton path configuration and integrating over a parameter describing the separation of the instanton-anti-instanton pair[24]. The instanton anti-instanton path configurations have, asymptotically, twice the action of a single instanton 1/3g; they are not solutions of the equation of motion, but the derivative of the action with respect to the path vanishes exponentially when the separation between the instantons goes to infinity. They form a family of quasi solutions of the equation of motion.

Using this technique, the correct asymptotic estimates have been obtained, and they agree with the computed first 73 orders of the perturbative expansion.

Unfortunately, the perturbation theory is no more Borel summable, the Borel transform being singular on the real axis, so that the problem remains : how to extract the true ground state energy from the perturbation series ?

In order to discuss this question we must sharpen our concepts and introduce a few definitions.

Given a function A(g) having the asymptotic expansion

$$A(g) = \sum_{0}^{\infty}{}_K A_K g^K \qquad |A_K| < K! \; R^K \tag{III.5}$$

we define the perturbative Borel transform by

$$B^P(b) = \sum_{0}^{\infty}{}_K A_K b^K/K! \tag{III.6}$$

for $|bR| < 1$ and by analytic continuation elsewhere.

The true Borel transform is defined by the representation

$$A(g) = \frac{1}{g}\int_0^{\infty} \exp(-b/g)\, B(b)\, db \tag{III.7}$$

where B(b) may also be a distribution. The asymptotic expansion

is Borel summable if $B^p(b) = B(b)$ and it is analytic on the real positive axis.
We can also introduce the concept of asymptotic Borel transform if

$$A(g) = \frac{1}{g}\int_0^T \exp(-b/g)B(b)db + O\left(\exp(-T/g)\right) \qquad \text{(III.8)}$$

and we do not require the convergence of the integral in the limit $T \to \infty$.

The perturbative expansion fixes $B^p(b)$; how can we compute $B(b)$? The situation is not simple : the presence of exponentially small terms due to level splitting implies that $B(b)$ contains a non analytic term proportional to $\theta(b - 1/6)$, while the nearest singularity of $B^p(b)$ is located at $b = 1/3$. Moreover, both the symmetric level and the anti-symmetric level have the same perturbation expansion and consequently, the same perturbative Borel transform, their true Borel transform being obviously different. The best that can be hoped is a relation of the kind

$$\bar{B}(b) = \mathrm{Re}\, B^p(b)$$

$\bar{B}(b)$ being the half sum of the Borel transform for the symmetric and anti-symmetric levels. Eq. (III.9) is a suggestive possibility which is quite likely wrong.

In principle, the function $B(b)$ may be fixed studying the perturbative expansion around the instanton and the instanton-anti-instanton path configurations, however, this problem remains unsolved, especially for what concerns the many instanton contribution.

There is still an additional difficulty. The representation (III.7) implies that $E(g)$ is analytic at least in the sector

$$\mathrm{Re}(g^{-1}) > \delta \qquad \text{(III.10)}$$

The WKB approximation strongly suggests that for complex g in the right plane $E(g)$ is singular, due to level crossing[25]. This phenomenon produces a sequence of cuts in the right plane accumulating toward the origine. Only the sector $|g| < r$ $|\arg(g)| < \theta$, $\theta \sim 10^{-2}$ is singularity free. The analytic properties of $E(g)$ clash with those implied by eq. (III.7) : $B(b)$ may be the Borel transform of $E(g)$ only in the asymptotic sense (III.8).

How to reconstruct the function $E(g)$ from the knowledge of its asymptotic Borel transform ? The solution is not unique. The simplest possibility is given by a two step Borel transform.

In the previous chapter we have seen that we can introduce the Borel transform of order α . The validity of such a representation implies that the function E(g) is analytic in the sector $\mathrm{Re}(g^{-\alpha}) > \delta$. We can pick an α compatible with the analyticity properties of E(g) and write it as

$$E(g) = \int_0^\infty \exp(-\beta^{1/\alpha}) B_\alpha(g\beta)\, \beta^{(1/\alpha\, -1)}\, d\beta \tag{III.11}$$

The function $B_\alpha(\beta)$ must be computed not from the non-convergent Taylor expansion around the origine, but as an integral over the asymptotic Borel transform B(b). A simple example is given if $\alpha = 1/2$.

$$E(g) = \int_0^\infty d\beta\, \exp(-\beta^2)\, \beta B_{1/2}(g\beta)$$

$$B_{1/2}(\beta) = 4/\pi^{1/2} \int_0^\infty db\, \exp(-b^2) B(2b\beta) \tag{III.12}$$

In the case in which there is no sector free of singularities we can generalize the two step procedure by introducing a "Borel transform of infinitesimal order"

$$E(g) = \frac{1}{g} \int_0^\infty F(s\, \exp(-1/g))\, \exp(s) ds \tag{III.13}$$

The function F(s) can be formally written as

$$F(s) = \int_0^\infty B(b)/\Gamma(b+1)\, s^b\, db \tag{III.14}$$

This last construction introduced by 't Hooft[9] may be quite useful if the function B(b) increases as $\Gamma(b)$ at infinity and has strong oscillations.

IV. THE LANDAU GHOST

At least in the massless case, there is no difficulty in extending the arguments of the previous section to a renormalizable interaction. In the massive case, the Euclidean equation of motion may have no solutions. However, a family of quasi solutions (in the sense of the previous section) can be constructed. Ultraviolet divergences are present in the intermediate steps of the calculations; however, they cancel at the end giving finite estimates for the high orders of the perturbative expansion in the renormalized coupling constant. In other words, the formation of collapsed states does not show any new physically interesting feature when the interaction becomes renormalizable.

Unfortunately, semi-classical methods are well adapted to control the configuration to the functional integral coming from one field configuration (a finite dimensional manifold of field configurations) which is essentially different from zero, but the study of the contributions of an infinite dimensional manifold of field configurations (each of them being infinitesimally near to zero) is outside the range of application of semi-classical methods. This last region of integration becomes important when the theory is renormalizable and it is the origine of the ultraviolet divergences. The estimates for the singularities of the Borel transform obtained in the previous sections are undoubtedly correct, however, there may be other singularities which correspond to non perturbative effects connected to the renormalization procedure.

For simplicity, let us consider the massive $g\,\phi^4$ interaction in 4 dimensions. It is well known that in the leading logarithmic approximation one finds a Landau ghost[26] in the large momentum region, i.e. a pole of the 4 point amputated Green function $\Gamma_4(P|g)$, in the Euclidean region at momentum $P_G \sim \exp(1/\beta' g)$; β' is defined by

$$\beta(g) = \beta' g^2 + O(g^3)$$

$\beta(g)$ being the Callan-Symanzik beta function. The theory we are considering is not asymptotically free and β' is positive.

Let us assume for the time being that this feature of the leading logarithmic approximation is shared by the exact theory. The Landau ghost is present in any theory in which there are no ultraviolet stable fixed point, i.e. the function $\beta(g)$ has no zero for positive g and $\int_\varepsilon^\infty \beta^{-1}(g)dg < \infty$. If we insert a 4 point function with a Landau ghost in the D-S (Dyson-Schwinger) renormalized equations of motion, the region of integration near the ghost will give an imaginary contribution proportional to $P_G^{-\alpha}$. Consequently, the Borel transform has a singularity on the positive line at $b^{-1} = (\alpha/\beta')^{-1}$, corresponding to an additional contribution to the asymptotic estimates proportional to

$$G_K \sim K!(\beta'/\alpha)^{-K} \qquad \alpha > 0 \qquad \text{(IV.2)}$$

This new singularity is quite interesting and it cannot be studied with semi-classical methods. If the interaction is super-renormalizable, each diagram contributing to the Kth order is bounded by $Cnst^K$. The K! behaviour is due to the increase in the number of diagrams. The asymptotic estimates may be obtained by computing the contribution of the generical diagram using statistical methods[27,28]. If the interaction is renormalizable only a tiny minority of the diagrams of the Kth order has a number n of subdivergent diagrams of the same order as K. At K-n fixed, their number increases as $Cnst^K$. Contrary to the first appearance they cannot be neglected in the asymptotic estimates. Each of them gives a contribution proportional to K! [9,10]. These diagrams are at the origine of the asymptotic estimates (IV.2) : they have a structure rather different from those contributing to the estimates (II.3).

The existence of singularities of the Borel transform connected to the large momentum behaviour seems to be independent from the existence of a Landau ghost for real g. Also, if we assume the existence of an ultraviolet stable fixed point for real values of g, under a simplifying hypothesis, a slight modification of arguments due to Khuri[29] and based on the renormalization group shows the existence of a Landau ghost for complex values of g, i.e. the function $\Gamma_4(P|g)$ has a singularity on the curve $\mathrm{Im}\, g = O(\mathrm{Re}\, g^2)$ when $P \sim \exp\left[1/(\beta'\mathrm{Re}(g))\right]$.

If we denote by r(p) the position of the nearest singularity to the origin of $\Gamma_4(P|g)$ in the right complex plane, we find

$$r(p) \sim \frac{1}{\beta' \ln P} + O\left(\frac{\ln \ln P}{\ln^2 P}\right) \qquad \text{(IV.3)}$$

Eq. (IV.3) suggests that the large momentum behaviour of the Borel transform $G_4(P|b)$ of the 4 point Green function $\Gamma_4(P|b)$ is given by

$$C_4(P|b) \simeq P^{b\beta'} \qquad \text{(IV.4)}$$

Eq. (IV.4) can be derived starting from the representation (I.4) for the Borel transform, by shifting part of the integration path to the right, picking the contribution of the Landau ghost and neglecting the rest. Eq. (IV.4) becomes exact in the leading log.approximation.

The increase in the large momentum region of C_4 is the origine of the new singularities of the Borel transform. The argument is quite general. The D-S equations for the Green functions induce similar equations for the Borel transform. Indeed, if we slightly change our notation and write

$$\Gamma(g) = \int \exp(-b/g)C(b)db \tag{IV.5}$$

the following convolution theorem holds

$$\mathcal{B}\left|f(g)h(g)\right|(b) =$$

$$\int_0^b \int_0^b db_1\, db_2\, \delta(b-b_1-b_2)\,\mathcal{B}\left[f(g)\right](b_1)\,\mathcal{B}\left[h(g)\right](b_2) \tag{IV.6}$$

where β stands for "Borel transform of".

The Borel transformed D-S equation can be obtained by simple use of the convolution theorem (IV.6). Let us see an example. A typical contribution to $\Gamma_6(0|g)$ (i.e. the six point function at zero momentum) is given by

$$\Gamma_6(0|g) = \int \frac{d^4p}{(p^2+m^2)^3} \left[\Gamma_4(P,0|g)\right]^3 \tag{IV.7}$$

$\Gamma_4(P, 0|g)$ being a short notation for $\Gamma_4(P,-P\ 0,0|g)$.
The Borel transform statement is

$$C_6(0|b) = \int \frac{d^4p}{(p^2+m^2)^3} I(P|b)$$

$$I(P|b) = \iiint_0^b C_4(P,0|b_1)C_4(P,0|b_2)C_4(P,0|b_3)\delta(b_1+b_2+b_3-b)\, db_1 db_2 db_3 \tag{IV.8}$$

If the large momentum behaviour of $C_4(P\ 0|b)$ is given by eq. (IV.4), the integral in eq. (IV.8) becomes divergent for $b = 2\beta'$. This argument shows that the Borel transforms of the Green functions have a sequel of singularities at the points $b = 2\,\beta' n$, n being a positive integer. In this derivation we have neglected the difference between asymptotic behaviour when all the momenta are large and when only a subset of the momenta is large; it is possible that some of the found singularities are absent.

A two line derivation of this singularities of the Borel may be interesting. Let us consider the Callan Symanzik equation [30,31] for $\Gamma^6(0\ g)$ in which the anomalous dimension has been taken equal to zero and terms of order g^3 have been neglected in the $\beta(g)$ function (these approximations are harmless for our aims)

$$\left[\beta' g^2 \frac{d}{dg} + 2\right] \Gamma^6(0|b) = \Delta\Gamma^6(0|b) \qquad \text{(IV.9)}$$

The Borel transformed statement is

$$\left[2-\beta' b\right] C^6(0|b) = \Delta C^6(0|b) \qquad \text{(IV.10)}$$

This equation may be interpreted either as a pole of $C^6(0|b)$ at $b = 2/\beta'$, or as a sum rule $(\Delta C^6(0|\ 2/\beta') = 0)$ for ΔC^6. If we disregard the second interpretation we find the wanted result. This argument may be dangerous as far as the precise position of the singularity is concerned, but it should be considered as an indication for the existence of singularities of the Borel transform on the real positive axis.

It may be useful to study a similar problem. Let us consider the $g\phi^4$ interaction in less than 4 dimensions $(D = 4-\varepsilon)$. The renormalized coupling constant spans the range $0-g_c(\varepsilon)$ $(\beta_\varepsilon(g_c(\varepsilon)) = 0;\ g_c(\varepsilon) = O(\varepsilon))$. Arguments similar to the previous ones [32,33] suggest that when $g \to g_c(\varepsilon)$, $\Gamma_6(g)$ behaves as $(g_c(\varepsilon)-g)^{2(1-\varepsilon)/\beta'_c}$ $\left(\frac{d}{dg}\beta_\varepsilon(g)\Big|_{g=g_c(\varepsilon)} = \beta'_c = O(\varepsilon)\right)$.
A more refined analysis shows that a singularity is present, but it is slightly weaker (the coefficient of the term linear in ε in the exponent is wrong)[34].

A singularity when $g = O(\varepsilon)$ implies that the Kth order of the perturbative expansion increases as $(1/\varepsilon)^K$ at large K. This result may seem to be in variance with the well known fact that all the coefficients of the renormalized perturbative expansion have a finite limit when $\varepsilon \to 0$. This contradiction is removed if there is a cross over region for $\varepsilon K \sim O(1)$ such that

$$G_K \sim K! \qquad (\varepsilon K) \ll 1$$

$$G_K \sim (1/\varepsilon)^K \qquad (\varepsilon K) \gg 1 \qquad \text{(IV.11)}$$

In the limit $\varepsilon \to 0$ only the first region survives and we reproduce asymptotic estimates similar to (IV.2).

We have seen explicitly the existence of a relation between the singularities of the Borel transform for positive b and the behaviour of the Green function in the critical region ($g \sim g_c(\varepsilon)$). We note en passant that it is very likely that the Borel transform of the ε expansion for the critical exponents is free of these singularities and the results of ref. 2 should be correct. In principle, the critical exponents can be computed by using the conformally invariant self consistency conditions[35-39], in which there is no trace of a possible Landau ghost.

Concluding this review of the asymptotic estimates for renormalizable interactions, we note that there are three main problems which wait for a solution.

1) To compute the position, nature and strength of these singularities of the Borel transform in the right complex b plane as it has been done for the singularities in the left plane.
2) To find a relation between the true Borel transform and the perturbative one (the presence of singularities on the real positive axis destroys the Borel summability).
3) To understand if a two step Borel transform may work in this case (the presence of a Landau ghost in the complex g plane implies that the Borel transform can be only asymptotic).

V. GAUGE THEORIES

It is believed that hadrons are composed by a multiplet of coloured quarks interacting via coloured gluon exchange. It would be quite interesting to extend the results of the previous section to gauge theories. Let us consider for simplicity a pure Yang-Mills theory. The existence of the instanton, i.e. a solution of the classical Euclidean equations of motion, implies the presence of tunneling between different vacua which are classified by a topological winding number[40,41]. There are non perturbative effects proportional to $\exp{-\mathcal{A}_I/g^2}$, ($\mathcal{A}_I/g^2$ is the instanton action); they reflect in singularities of the Borel transform with respect to g^2 having the form $\theta(b-\mathcal{A}_I)$.

As explained in section III only those solutions of the classical equations of motion which satisfy periodic boundary conditions are relevant for the estimates of the high orders of the perturbative expansion. In gauge theories, the instanton field goes to zero at infinity and it is not clear in which sense it does or does not satisfy periodic boundary conditions. We suggest that a solution A(x) of the classical equation of motion contributes to the asymtptotic estimates only if we can find a sequence of functions $A_K(x)$, having compact support, such that

$$\lim_{K\to\infty} A_K(x) = A(x)$$

$$\lim_{K\to\infty} \mathcal{A}(A_K) = \mathcal{A}(A) \tag{V.1}$$

$\mathcal{A}[A]$ being the action functional. We think that condition (V.1) reflects the fact that at any finite order in the perturbative expansion we can always compute the diagrams in a finite volume and perform the infinite volume limit only at the end.

Using the inequality $|F\tilde{F}| \lesssim F^2$ it is possible to prove that the single instanton field does not satisfy condition (V.1) and consequently it does not contribute to the asymptotic estimates. The relevant fields are the instanton-anti-instanton configurations. They are not solutions of the equations of motion but in the limit of large distances they form a family of quasi solutions. The nearest singulairity of the perturbative Borel transform is at $b = 2\mathcal{A}_I$ provided that in the zero topological charge section there are no true solutions of the classical equation of motion having an action smaller than $2\mathcal{A}_I/g^2$. Borel summability is lost. We are not arrived at the end of the troubels. Gauge theories describe the renormalizable interaction of massless particles; in the leading logarithmic approximation we find a Landau ghost in the ultraviolet region when g^2 is negative (as implied by asymptotic freedom) and in the infrared region when g^2 is positive. The arguments of the previous section suggest the presence of singularities of the Borel transform having the form of cuts starting from the points $b = \pm 2\, n/\beta'$. Of course some of these cuts may be absent.

the meaning of these singularities is rather transparent in this case. In the approximation in which we retain only the first non zero contribution in the $\beta(g^2)$ function and we neglect anomalous dimensions, we find

$$G(P,g) = 1/p^2\; G(\mu,\bar{g}^{2}(g^2,P/\mu)) \tag{V.2}$$

where

$$\bar{g}^{-2}(g^2,P/\mu) = g^{-2} - \beta'\,\ell n\, P/\mu \tag{V.3}$$

For simplicity, we are considering the case of a two point function

$$G(P,g) = \int_0^\infty \frac{1}{p^2+m^2}\, P(m^2,g)\,dm^2 \tag{V.4}$$

We denote by $B(P,b)$ the Borel transform of $G(P,g)$

$$G(P,g) = \int_0^\infty db\, \exp -b/g\; B(P,b) \tag{V.5}$$

Using eq. (V.3) we find

$$G(P,g) = \frac{1}{p^2} \int_0^\infty d\alpha \, \exp(-2\alpha/\beta' g) B(\mu, 2\alpha/\beta')(P^2/\mu^2)^{-\alpha} \qquad (V.6)$$

Singularities at integer α correspond to negative integer powers of p^2 in the asymptotic region. These terms look like mass corrections. They are welcome if we think that gluons are confined and a mass gap is automatically generated as in the two dimensional non linear σ model[46].

As in the previous section the Borel transform may be only asymptotic. Only a Borel representation of infinitesimal order (eqs. (III.13-14) may be valid. The singularities in the complex g plane accumulate toward the origine, leaving no singularity free sector. As suggested by 't Hooft[9] we can write a two step Borel representation

$$F(P,s) = \int_0^\infty \frac{db \, s^{2b/\beta'}}{\Gamma(2b/\beta'+1)} B(P,b)$$

$$G(P,g) = \int_0^\infty ds \, e^{-s} F(P, s \exp(-2/\beta' g)) \qquad (V.7)$$

Using the approximate relation (V.3), we find

$$G(P,g) = \frac{1}{\mu^2} \int_0^\infty ds \, F(\mu, s \exp(-2/\beta' g)) \exp(-sK^2/\mu^2)$$

$$F(\mu, s \exp(-2/\beta' g)) = \int_0^\infty dm^2 \, \rho(m^2, g) \exp{-sm^2/\mu^2} \qquad (V.8)$$

As it can be seen from eq. (V.7), the function has a non convergent expansion in powers of $1/\ell n \, s$ around $s = 0$. The singularities of $F(\mu,s)$ are quite interesting. A sequence of narrow resonance equidistant in M^2 produces a singularity of the function $F(\mu,s)$ on the imaginary s axis. Oscillations in the hadronic spectrum reflect in singularities of $F(\mu,s)$. In order to obtain WKB type approximations[47] to the hadronic masses, we need to know the singularities of $F(\mu,s)$. Unfortunately, the representation(V,7)is useless as long as we do not know how to compute the Borel transform $B(\mu,b)$. We only know the perturbative Borel transform $B^p(\mu,b)$ and it is not evident if the singularities of $F^p(\mu,s) \equiv \int_0^\infty \frac{db \, s^{2b/\beta'}}{\Gamma(2b/\beta'+1)} \mathrm{Re}\, B^p(\mu,b)$ coincide with those of $F(\mu,s)$.

This lengthy analysis is necessary if we want to use the perturbative expansion to obtain numerical and reliable results for strong interactions. This aim may seem hopeless. However, when the physics is quite clear and many orders of the perturbation expansion have been computed, (as in the theory of the critical exponents[47,48]) extremely good results have been obtained also for strong coupling[15,49]. The difficulties we face in gauge theories are connected to our lack of physical comprehension of the problem.

The reader may think that in spite of our efforts, the situation is still a mess. That is true. However, in these last few years a progress has been made: we do not yet know how to get correct answers for a field theory of strong interactions, but we begin to understand which are the right questions to ask.

REFERENCES

1. L.N. Lipatov, JEPT 72 (1977) 411
2. E. Brezin, J.C. Le Guillou, J. Zinn-Justin, Phys.Rev. D15 (1977) 1544
3. G. Parisi, Phys.Lett. 66B (1977) 167
4. E. Brezin, J.C. Le Guillou, J. Zinn-Justin, Phys.Rev. D15 (1977) 1558
5. C. Itzykson, G. Parisi, J.B. Zuber, Phys.Rev.Letters 38 (1977) 306
6. G. Parisi, Phys.Lett. 66B (1977) 382
7. C. Itzykson, G. Parisi, J.B. Zuber, Saclay preprint DPh/T 77/27 (Phys.Rev. D to be published)
8. R. Balian, C. Itzykson, G. Parisi, J.B. Zuber, Saclay preprint DPh/T 77/92
9. G. 't Hooft, Lectures given at Erice, July 1977
10. B. Lautrup, Phys.Lett. 69B (1977) 109
11. S. Graffi, V. Grecchi, B. Simon, Phys.Lett. 32B (1970) 631
12. J.D. Eckmann, J. Magnen, R. Seneor, Comm. Math. Phys. 39 (1975) 251
13. P. Renuard, preprint Ecole Polytechnique Palaiseau, n°A247 1076 (1976)
14. J.J. Loeffel "Workshop on Pade Approximants", Ed. D. Bessis, J. Gilewiczond, P. Merry CEA (1976)
15. J.C. Le Guillou, J. Zinn-Justin, Phys.Rev.Lett. 39 (1977)95
16. G. Parisi, Phys.Lett. 65 B (1977) 329
17. G. Parisi, Lectures given at the Cargèse Summer Institute (1976)

18. J.S. Langer, Ann. of Phys. 41 (1967) 108
19. H. Lieb, Rev. Mod. Phys. 49 (1977) 137
20. See for example L. Landau, E. Lifschitz, "Mécanique Quantique", Ed. Mir, Moscou (1970)
21. A.A. Belavin, A.M. Poliakov, A.S. Schwartz, Yu S. Tyupkin, Phys.Lett. 59B (1977) 85
22. G. 't Hooft, Phys.Rev.Lett. 37 (1976) 8
23. A.M. Poliakov, Nucl. Phys. B120 (1977) 429
24. E. Brezin, G. Parisi, J. Zinn-Justin, Saclay Preprint DPh/T 77/16 (Phys.Rev. D to be published)
25. T. Banks, C.M. Bender, T.T. Wu, Phys.Rev. D8 (1973) 3346
26. L.D. Landau, A.A. Abrikosov, I.M. Khalatnikow, Doklady Acad. Nauk URSS 95 (1954) 773
27. C.M. Bender, T.T. Wu, Phys.Rev.Lett. 37 (1976) 117
28. G. Parisi, Phys.Lett. 68B (1977) 361
29. N.N. Khury, Phys.Rev. D12 (1975) 2298 and Rockfeller Univ. Preprint COO 2232B 131
30. C.G. Callan, Phys.Rev. D2 (1970) 1541
31. K. Symanzik, Comm.Math.Phys. 18 (1970) 227
32. G. Parisi, Lectures given at the Cargèse Summer School (1973), Columbia University Preprint
33. G. Parisi, Nuovo Cimento, 21A (1974) 179
34. K. Symanzik, DESY preprint 77/05
35. A.A. Migdal, Sov.Phys. JEPT 28 (1969) 1036
36. A.M. Poliakov, Sov.Phys. JEPT 20 (1969) 533
37. A.M. Poliakov, JEPT Letters 12 (1970) 381
38. G. Parisi, L. Peliti, Lett.Nuov. Cimento 2 (1971) 627, Phys.Lett. 41A (1972) 331
39. G. Mack, L. Todorov, Phys.Rev. D8 (1973) 1764
40. C. Callan, R. Dashen, D. Gross, Phys.Lett. 63B (1976) 334
41. R. Jackiw, C. Rebbi, Phys.Rev.Lett. 37 (1976) 172
42. K. Symanzik, Nuovo.Cim.Lett. 6(1973) 420
43. G. 't Hooft, Marseille Conference June 1972 (unpublished)
44. H.D. Politzer, Phys.Rev.Lett. 30 (1973) 1346
45. D.J. Gross, F. Wilczek, Phys.Rev.Lett. 30 (1973) 1343
46. E. Brezin, J. Zinn-Justin, Phys.Rev. 36 (1976) 691
47. R. Balian, C. Bloch, Annals of Phys. 85 (1974) 514
48. K.G. Wilson, J. Kogut, Phys.Reports 12C (1974) 75
49. C.A. Baker, B.C. Nikel, M.S. Green, D.I. Meiron, Phys.Rev. Lett. 36 (1976) 1351.

STRONG INTERACTION MODELS IN 6-ε DIMENSIONS

D.J. WALLACE

Department of Physics, The University

SOUTHAMPTON SO9 5NH - ENGLAND

ABSTRACT

A unique theory of the strong interactions can arise as a critical phenomenon with respect to some fundamental length scale. We discuss models which give a concrete realisation of this idea through ε-expansions in 6-ε dimensions. Illustrative calculations and problems in this approach are reviewed.

1. INTRODUCTION

This talk is about models in which one can realise Wilson's idea [1] that strong interactions of elementary particles might be describable as a critical phenomenon.

In order to motivate this approach, let me remind you that critical phenomena are characterised by two very different length scales. The first is the microscopic length scale appearing in the Hamiltonian describing the interactions of the microscopic degrees of freedom of the system. The second length scale is the range of co-operative behaviour of these degrees of freedom; in a critical system it is much larger than the microscopic length scale scale, and it controls the macroscopic behaviour of the system. A prototype example is given by a magnetic system. The microscopic length scale is the lattice spacing, a, and the second length scale is the range of co-operative behaviour, the correlation length ξ; near the phase transition point $\xi \gg a$.

There are two basic problems when one studies critical phenomena : (a) one frequently does not have detailed knowledge of the form of the microscopic Hamiltonian.

(b) Even if the Hamiltonian is known it is difficult to calculate the co-operative many-body effects associated with the large correlation length ξ.

The remarkable developments of the last few years [1,2] have shown how the renormalization group (RG) can solve both of these problems. Essentially unique critical behaviour, independent of the detailed nature of the microscopic Hamiltonian, is obtained provided that there exists a fixed point of the RG which is suitably infra-red stable. The success of this approach in phase transitions and other critical phenomena is well-established [1,2].

The idea that one might obtain a unique theory of strong interactions as critical phenomena was considered by Wilson [1]. To believe in this approach one has to imagine the existence of a meaningful bare Lagrangian containing a microscopic length scale i.e. a cut-off Λ in momentum space ($\Lambda \sim a^{-1}$). The idea that gravity might provide some kind of cut-off is not new [3]; the microscopic scale is then the Planck mass, $\Lambda \sim 10^{19}$ GeV. Qualitatively one may argue that in order to obtain Green functions correctly at <u>all</u> momenta scales including $q \gtrsim 10^{19}$ GeV one must explicitly include the gravitational field $g_{\mu\nu}$. Then Green functions of matter fields ψ look like

$$<T\{\psi(x_1) \ldots \psi(x_N)\}> = \int D\psi\ \psi(x_1) \ldots (x_N) \int Dg \exp i\, A(g,\psi) \tag{1}$$

where A is some general-relativistic invariant action. One can then obtain in principle an effective action for the ψ fields

$$<T\{\psi(x_1) \ldots \psi(x_N)\}> = \int D\psi\ \psi(x_1) \ldots \psi(x_N) \exp i\, A_{eff}(\psi) \tag{2}$$

by averaging on $g_{\mu\nu}$:

$$\exp i\, A_{eff}(\psi) = \int Dg \exp i\, A\,(g,\psi) \tag{3}$$

If one now makes the simple postulate that all dimensional couplings in $A_{eff}(\psi)$ have a natural momentum scale governed by the gravitational interaction from which they are obtained (i.e. the Planck mass Λ) then high energy physics indeed becomes a critical phenomenon; there is a meaningful bare Lagragian with a cut-off Λ and the physics we are interested in occurs in the infra-red region of small masses and momenta, $(m,q) << \Lambda$.

Other mass scales such as super heavy vector bosons in Grand Unified Models may play the same role [4]. My aim is not to explore such possibilities directly but simply to use them to motivate a search for infra-red stable fixed points in four-dimensional space-time.

As a first candidate one may consider a one-component boson field theory with interaction $\lambda\phi^4$. This certainly has an infra-red stable fixed point $\lambda^*=0$. This is clearly no good for our purposes however, since it implies zero effective coupling at one Fermi in this picture; this is not a good starting point for a theory of strong interactions. Attempts (Wilson and Kogut, [1], and ref. 5) to find non-trivial fixed points in this model have been unsuccessful so far.

Clearly it is sensible to look for non-trivial infra-red behaviour in models with stronger infra-red behaviour than ϕ^4. There are two obvious candidates; non abelian gauge theories and models whose symmetry permits ϕ^3 coupling. In this talk I propose to explore the latter [6,7,8].

The outline of the talk is as follows. In the second section I shall discuss how some ϕ^3 models with desirable symmetries have non-trivial fixed points in 6-ε space-time dimensions. In the third section I shall use these models to illustrate the benefits of this approach; we shall see the possibility of calculable symmetry breaking effects, scaling violations in deep inelastic scattering and dynamical symmetry generation. In the final section we discuss the diseases of ε-expansions in 6-ε dimensions.

2. ε-EXPANSIONS IN 6-ε DIMENSIONS.

Our starting point is a bare Action of the form

$$A = \int d^dx\{\frac{1}{2}(\nabla\phi)^2 + \frac{1}{2}m_o{}^2\phi^2 + \frac{1}{3!}(K_d)^{-\frac{1}{2}}g_o\Lambda^{(6-d)/2}d_{ijk}\phi_i\phi_j\phi_k + \text{O.T.}\} \tag{4}$$

Here $\phi_i (i = 1,2, \ldots)$ transforms according to some (irreducible) representation of some symmetry group; $\phi^2 = \sum_{i=1}^{n} \phi_i\phi_i$ and d_{ijk} is an invariant tensor of the symmetry group. The Λ-dependence of the ϕ^3 interaction comes from naive dimensional analysis; g_o is the dimensionless bare coupling constant, with some generic value of order 1. (Sometimes we shall have more than one ϕ^3 interaction) The factor involving $K_d = S_d/(2\pi)^d$.

Where S_d is the surface of the unit sphere in d dimensions conveniently absorbs the results of angular integrations, etc... Similarly, m_o^2 behaves like Λ^2. A cut-off, e.g. $1/2\Lambda^{-2}(\nabla^2\phi)^2$ may be understood. This assumed Λ-dependence is the mathematical statement that $(q,m_{phys}) << \Lambda$ is a critical region. "O.T." in eq. (4) represent interactions with higher powers of fields and derivations; such couplings are neglected for the time being since the lowest powers of fields are expected to dominate critical behaviour.

It is technically most convenient to use renormalized perturbation theory, by introducing a momentum scale μ with renormalized coupling constant $g(\mu)$. If we take $q, m \ll \mu \ll \Lambda$ then the cut-off Λ can be removed in the renormalized vertex function since $\mu^2/\Lambda^2 \ll 1$, and the physics is still in the infra-red region with respect to μ (with $\mu \ll \Lambda$, $g(\mu)$ would already be very close to an infra-red stable fixed point if one existed). The following calculations then involve standard RG calculations in renormalized perturbation theory.

The RGβ function for the model (4) has the form

$$\beta(g) = -\frac{(6-d)}{2} g + Ag^3 + Bg^5 + \ldots \qquad (5)$$

The first term comes from the naive dimension of the ϕ^3 interaction and the coefficient A comes from a one loop calculation, etc... We see that if A > 0, there exists a non-trivial fixed point g^* in $d = 6-\varepsilon$ dimensions [9] :

$$g^{*2} = \varepsilon/2A + O(\varepsilon^2) \qquad (6)$$

The coefficient B in (5) gives terms of order ε^2. To first order in ε this fixed point is automatically infrared stable :

$$\beta'(g^*) = \varepsilon + O(\varepsilon^2) > 0 \qquad (7)$$

Therefore the aim is to find those models which have $A > 0$ and to calculate perturbatively infra-red behaviour as a power series in ε. The picture is that strong interactions weaken as one approaches six dimensions. For "physics", one must set $\varepsilon = 2$ (for $d = 4$); we return to this point in section 4. There are no free coupling cons tants in this approach since the effective infra-red coupling is fixed at the value (6); corrections due to finite cut-off and $g \neq g^*$ are of the form $(g^2 - g^{*2})$ $[(q/\Lambda)^{\beta'(g^*)}, (m/\Lambda)^{\beta'(g^*)}]$ and are vanishingly small if g^* is infra-red stable ($\beta'(g^*) > 0$).

(a) One-Component Scalar Field

This theory is known [10] to be asymptotically free in six dimensions; A is negative,

$$A = -3/4.$$

Therefore it has no infra-red stable fixed point in $(6-\varepsilon)$ dimensions.

(b) Adjoint Representation of Su(n).

Here ϕ_i is an (n^2-1) component scalar field and d_{ijk} is the usual symmetric tensor (n > 3). Explicit calculation gives [11]

$$A = \frac{20-n^2}{4n} > 0 \quad \text{for } n = 3,4 \text{ only.}$$

It is remarkable that this approach specifically picks out SU(3) and SU(4) as the only models in this class which have fixed points in 6-ε dimensions.

(c) $(3,\overline{3}) + (\overline{3},3)$ Representation of SU(3) x SU(3).

This model involves nine scalar fields σ_α and nine pseudoscalar fields π_α which are conveniently written in terms of the complex 3 x 3 matrix $\phi = (\sigma_\alpha + i\pi_\alpha)\lambda^\alpha$ using the Gell-Mann λ matrices (α = 0,1, ... 8). The invariant interaction is det ϕ and, with the normalisation 1/3! $d_{ijk}\phi_i\phi_j\phi_k \equiv \det\phi$, one finds

$$A = 8$$

This model therefore has an ε-expansion in 6-ε dimensions.

The combinatorial coefficients of many graphs in this model are zero and even the three-loop calculations of the β function and the anomalous dimension of the field are not too tedious [7].

3. SOME ILLUSTRATIVE CALCULATIONS.

The above calculations show that there are models with desirable symmetry properties which permit one to realise concretely Wilson's idea of strong interactions as critical phenomena. These models enable one to illustrate the attractive possibilities in this approach, as the following calculations show.

(a) Calculable Symmetry Breaking Effects [6].

Consider the model of eight scalar fields with SU(3) symmetry discussed in 2(b) above. Let us add by hand a symmetry breaking term $c\phi_8$ to the Lagrangian. This will split the masses in the multiplet. First order in SU(3) breaking gives the usual Gell-Mann-Okubo mass formula

$$4M_\kappa^2 - M_\delta^2 - 3M_\sigma^2 = 0$$

where κ, δ and σ have isospin 1/2, 1 and 0 respectively. The same result is also obtained in tree diagram approximation. Since there exists an ε-expansion for this model, Green functions can be calculated as a power series in ε. In particular one can find the (Minkowski) values of q^2 at which the Green functions for the κ, δ and σ particles have poles. A nice exercise in renormalization of a broken symmetry yields

$$4M_K^2 - M_\delta^2 - 3M_\sigma^2 = \frac{(80\pi/\sqrt{3}-119)}{33} \cdot \frac{(M_K^2-M_\delta^2)^2}{2M_K^2+M_\delta^2} \cdot \varepsilon$$

$$+ O\,(\varepsilon^2(M_K^2-M_\delta^2)^2,\quad \varepsilon(M_K^2-M_\delta^2)^3) \tag{8}$$

$$\simeq 0.8\ (M_K^2-M_\delta^2)^2\quad \varepsilon/(2M_K^2+M_\delta^2)$$

Note the uniqueness of the result (because g^2 is fixed at $g^{*2}=6\xi/11$) and the fact that the right hand side is indeed of second order in SU(3) breaking. The $(3,\bar{3}) + (\bar{3},3)$ model of SU(3) x SU(3) provides a possible vehicle for similar calculations of f_K/f_π, $K_{\ell 3}$ form factors, etc..., although rather complicated symmetry breaking is probably required.

(b) Deep Inelastic Scattering.

The main point of this calculation is to stress that even very high energies ($\sim 10^3$ GeV) are still in the infra-red region ($q << \Lambda$) in this approach and still governed by the infra-red stable fixed point. For example structure functions in deep inelastic scattering will exhibit power-law scaling violations determined by anomalous dimensions of operators. This is to be contrasted with the asymptotic freedom of non-abelian gauge theories [12]; there one has a generic coupling constant for $q \sim m$ and uniquely calculable high q behaviour because $g^* = 0$ is an ultraviolet stable fixed point. For example, the operator

$$P_\ell = 1/2\ \phi_i\ \{\partial_{\alpha_1}\ \dots\ \partial_{\alpha_\ell}\}\ \phi_i$$

where $\{\partial_{\alpha_1}\ \dots\ \partial_{\alpha_\ell}\}$ is traceless will appear in the operator product expansion for deep inelastic scattering. The anomalous dimension of P_ℓ is calculable;

$$\eta_\ell = -\ g^{*2}\ \left(\frac{d_{ijk}d_{ijk}}{n}\right)\ \frac{2}{(\ell+1)(\ell+2)} + O(\varepsilon^2) \tag{9}$$

where n is the number of field components. We have not attempted to make a phenomenological fit to data with such power-law scaling violations; one suspects that the existence of reliable data supporting the fermionic point-like structure of hadrons at short distances will rule out such a fit.

(c) Dynamical Symmetry Generation.

In this formalism there is a rather naturel framework for the dynamical generation of the symmetries of the strong interactions. The fundamental reason is that the interactions at "infra-red" distances have the symmetry of the fixed point irrespective of the possible absence of any symmetry in the bare Lagrangian.

A concrete example involving the O(n) symmetric fixed point in 4-ε dimensions is discussed by Brézin, Le Guillou and Zinn-Justin [13]. An example in 6-ε dimensions is the extension of the SU(3) model of eight scalars σ_i to include eight pseudoscalars π_i with interactions

$$\frac{1}{3!} g_1 d_{ijk} \sigma_i \sigma_j \sigma_k + \frac{1}{2} g_2 d_{ijk} \sigma_i \pi_j \pi_k$$

The case $g_1 = g_2$ corresponds to an invariant line for the running coupling constant since then the interaction is SU(3) x SU(3) invariant with σ and π transforming according to the (1,8) + (8,1) representation. It turns out [8] that there is a fixed point $g_1^* = g_2^* = \sqrt{3\varepsilon + O(\xi^2) / 11}$ which is infra-red stable. Thus theories with generic bare couplings in the domain of stability of this fixed point have effective interactions which are SU(3) x SU(3) symmetric in the infra-red.

Following the lines of ref. (13), one may also ask if the SU(3) symmetry of the scalar model 2(b) can be dynamically generated starting from eight fields with arbitrary trilinear interactions. There are 120 independent interactions which are classified according to their SU(3) symmetry transformation properties; in the symmetrized Kronecker cube we have

$$\{8 \times 8 \times 8\}_{\mathcal{S}} = 1 \oplus 8 \oplus 10 \oplus \overline{10} \oplus 27 \oplus 64$$

The remarks in §2(b) show that the SU(3) fixed point is stable to perturbation by the singlet operator ("1" $\equiv d_{ijk}\, \phi_i \phi_j \phi_k$). If all four other anomalous dimensions $\lambda^{(8)}$, $\lambda^{(10+\overline{10})}$, $\lambda^{(27)}$ and $\lambda^{(64)}$ are also positive then the SU(3) is dynamically generated. One finds [8]

$$\lambda^{(8)} = -\frac{49}{33}\varepsilon + O(\varepsilon^2)$$

$$\lambda^{(10+\overline{10})} = -\frac{9}{11}\varepsilon + O(\varepsilon^2)$$

$$\lambda^{(27)} = \frac{1}{33}\varepsilon + O(\varepsilon^2)$$

$$\lambda^{(64)} = 0 + O(\varepsilon^2)$$

The negative nature of these calculations is contained in the general result [8] : for any fixed point in 6-ε dimensions there is always at least one unstable trilinear perturbation (i.e. at least one negative iegenvalue of the stability matrix) if the number of field components $\leq$ 135. For $n \leq 135$, inequalities required for the stability matrix to be positive definite are violated.

4. DISEASES IN 6-ε DIMENSIONS.

There are three related questions one should ask in order to justify the above calculations :

(i) What is the form of the effective potential, and is one perturbing about the correct ground state ?

(ii) What is the nature of the ε-expansion ?

(iii) Do the O.T. in eq. (4) indeed correspond to irrelevant interactions whose insertion does not change the critical behaviour ?

As regards the effective potential, it is well known that the conventional ϕ^3 theory with interaction $gm^{(6-d)/2}d_{ijk}\phi_i\phi_j\phi_k$ where m is a physical mass, is fatally diseased. It is asymptotically free for $d < 6$; the effective potential at short distances is given by tree diagrams and is claerly not bounded below. By contrast the action (4), with coupling $g\ \Lambda^{(6-d)/2}$ is not asymptotically free and at short distance ϕ^4 interactions are present in O.T. and may provide stability. However in 6-ε dimensions the effective potential can be calculated in any given model; its absolute minimum, if it has one, is presumably not at $\phi = 0$.

The structure of the ground state is also important for the nature of the ε-expansion, in particular high order behaviour. As discussed in the lectures of Zinn-Justin and Parisi perturbation about a metastable ground-state is characterised by non-oscillatory behaviour at high orders in the perturbation expansion. In this case it is straightforward to do a calculation for a theory with arbitrary ϕ^3 interactions in six dimensions. One finds a growth at order g^{2K} of the usual from $K!a^K K^b c(1+O(1/K))$ with $a = 5(192p.\ p\pi^3)^{-1}$ where $p_i = d_{ijk}p_j p_k$ arises in the solution of the saddle point equations. Clearly $a > 0$, and the above ε expansions are about a metastable ground state and are invalidated by the neglected tunneling out of that state.

It is amusing that the RG does appear to sense this disease when one looks at the third question of the relevance or otherwise of the O.T. in eq. (4). Consider ϕ^4 operators which appear in the Action (4) in the form $v\ \Lambda^{4-d}\int O(\phi^4)d^dx$. Naively these are soft infra-red insertions and hence irrelevant for critical behaviour if the power of Λ is negative, i.e. $2 - \varepsilon > 0$ ($\varepsilon = 6-d$). In the presence of ϕ^3 interactions the condition ofr irrelevance is

$$2 - \varepsilon + \lambda > 0,$$

for all anomalous dimensions λ of ϕ^4 operators (and of all others which mix with ϕ^4 because of renormalization). In four dimensions we therefore require $\lambda > 0$ for all such operators. However λ can be calculated as a power series in ε and one can show that at order ε the ϕ^4 anomalous dimension matrix is not positive definite; the quadratic from from it corresponding to $(\phi^2)^2$ interactions is not positive definite. This indicates that $(\phi^2)^2$ becomes a relevant perturbation as d decreases towards four, i.e. that interaction becomes important which is precisely required to stabilise the ϕ^3 theory.

The conclusion of this talk is that ϕ^3 models in 6-ε dimensions provide a framework in which Wilson's ideas can be concretely realised in perturbation theory. However this perturbation theory neglects the existence of relevant $(\phi^2)^2$ interactions which may stabilise the vacuum. To study this question further requires techniques beyond the simple perturbation theory discussed here. Nevertheless the above exercises serve as a useful reminder of this approach to elementary particle interactions.

REFERENCES

1. These ideas and some calculations in 4-ε dimensional space-time are reviewed in K.G. Wilson and J. Kogut, Phys. Reports, 12C, (1974) 75.
   See also S. Weinberg in Proceedings of the International School of Sub-Nuclear Physics, Ettore Majorana, 1976, and S.S. Shei and T. Yan, Phys. Rev. D8, (1973) 2457.
2. For reviews of the renormalization group in critical phenomena see for example, S.K. Ma, Modern Theory of Critical Phenomena, (W.A. Benjamin, Inc., Reading, Mass., 1976); E. Brézin, J.C. Le Guillou and J. Zinn-Justin in Phase Transitions and Critical Phenomena, Vol. 6, eds. C. Domb and M.S. Green (Academic Press, 1976), and other articles in that volume; M.E. Fisher, Rev. Mod. Phys. 46, (1974) 597; K.G. Wilson, Rev. Mod. Phys. 47, (1975) 773.
3. For example the non-polynomial Lagrangian approach is discussed in C.J. Isham, A. Salam and J. Strathdee, Phys. Rev. D5, (1972) 2548.
4. H. Georgi, H.R. Quinn and S. Weinberg, Phys. Rev. Letters, 33, (1974) 451, give an example of renormalization effects at momentum scales infra-red with respect to super-heavy vector boson masses.
5. J.C. Le Guillou and J. Zinn-Justin, Phys. Rev. Letters 39 (1977) 95.
6. A.J. McKane, D.J. Wallace and R.K.P. Zia, Phys. Letters 65B, (1976) 171.
7. A.J. McKane, J. Phys. G.
8. A.J. McKane and D.J. Wallace, to be published.

9. Expansions in 6-ε dimensions are studied in many critical phenomena. See e.g., A.B. Harris, T.C. Lubensky, W.K. Holcomb and C. Dasgupta, Phys. Rev. Letters 35 (1975) 327, E 1937; R.G. Priest and T.C. Lubensky Phys. Rev. B13, (1976) 4159 and Erratum; D.J. Amit, J. Phys. A9 (1976) 1441; A.B. Harris, T.C. Lubensky and J.H. Chen, Phys. Rev. Letters 36 (1976) 415. Many of these problems involve n → 0 limits which can evade the diseases discussed in section 4.

10. G. Mack in Proceedings of the International Summer Institute on Theoretical Physics, Kaiserslautern, 1972, ed. J. Ehlers, K. Hepp and H.A. Weidenmuller (Springer Verlag, Berlin, 1973), p. 300.

11. This number was also calculated, for different reasons, by S. Panfil, Contributed Paper, 1976 International Conference on High Energy Physics, Tbilisi.

12. D.J. Gross and H. Wilczek, Phys. Rev. Letters 30 (1973) 1343; H.D. Politzer, Phys. Rev. Letters 30 (1973) 1346; S. Coleman and D.J. Gross, Phys. Rev. Letters 31 (1973) 1851. For a recent phenomenological fit and further references see the lectures by C.H. Llewellyn-Smith.

13. E. Brézin, J.C. Le Guillou and J. Zinn-Justin, Phys. Rev. B10, (1974) 892.

SUMMING ALL THE EIKONAL GRAPHS

H.M. FRIED

Physique Théorique, Université de Nice, France
and
Physics Department, Brown University, Providence, R.I.,USA

During the past year there has been some progress in al old problem of eikonal physics, that of estimating the high-energy contributions of all eikonal pionization graphs (tower plus non-planar) in some relevant Lagrangian field theory. Since so much material has already been presented concerning non-Abelian physics at this Summer Institute, and partly to remind us that much high-energy phenomenology is still couched in the language of non-non-Abelian theories, I will try to present a very brief description of this recent work [1]. There is also a personal reason for this choice : at Cargèse, five years ago, I presented a critique of the then-fashionable Cheng-Wu "explanation" of rising total cross sections[2]; and now it is most satisfying to be more constructive, to point out in the context of an approximate, semi-classical type of computation that sums over all eikonal graphs, which features of their calculation remain and which must be changed.

It must be emphasized, however, that this is an exercise in phenomenology. It is not here attempted to sum all the graphs of a given interaction, in a specified order of perturbation theory, and then to sum over all perturbation orders. Rather, what is desired is a representation of certain selected sets of processes to all orders in perturbation theory; and it is extremely convenient to be able to perform such partial sums via the techniques of functional field theory. For simplicity, all interacting Pomerons are omitted, along with effects of diffractive disassociation; what is retained are the basic pionization graphs out of which multiple Reggeized structures may be formed. Alternatively, one might say that selected, and relatively simple classes of graphs in a Hybrid theory (introduced below) are used to model the effects of quite complicated closed-fermion-loop graphs in QED.

At the outset, the eikonal idea assumes that all momenta can be divided into two classes, the "large" (e.g., incident $p_{1,2}$ and final $p'_{1,2}$ nucleon momenta) and the "small" (just about everything else, e.g. all exchanged meson momenta q_i); with the exception of the not-too-large diffractive disassociation processes, at current energies this seems to be an assumption in reasonable agreement with experiment. The result of such approximations will be the replacement of nucleon propagators $G_c(x,y|A)$, defined in the presence of ficticious (but convenient!) c-number fields $A_\mu(z)$, by "no-recoil" or Bloch-Nordsieck (BN) propagators [3] $G_c^{BN}(x,y|A)$, in effect an expression of the judicious neglect of any $q_i \cdot q_j$ combination in comparison to terms of form $p.q_i$. And then, because the explicit source dependence of $G_c^{BN}[A]$ is usually simple, one may carry through almost all functional computations in a straight-forward way.

It will be useful to set the stage for what is to come by considering the eikonal forms relevant to a "hybrid" theory, one resembling both ϕ^3-theory and (massive photon) QED. Here, one treats nucleons (ψ) scattering by the exchange of many NVMs(A_μ), which may themselves exchange arbitrary numbers of "scalar pions" π ,

$$\mathcal{L}' = ig\, \bar{\psi}\gamma.A\psi - \frac{\lambda}{2}\pi\, A^2 \quad .$$

The "natural language" to use in representing the scattering amplitude

$$T_{EIK}(s,t) = i\,\frac{s}{2}\int d^2b\; e^{i\vec{q}.\vec{b}}\left[1 - e^{i\chi(s,b)}\right] ,$$

with normalization given by $d\sigma_{EL}/dt = |T|^2/\pi s^2$ and $\sigma_{TOT} = \frac{4}{s}\,\mathrm{Im}T(s,0)$, is the language of functional methods, wherein all normalization factors, phases, etc..., are automatically included. Neglecting radiative corrections corresponding to NVM self-linkages on the same nucleon line, and neglecting closed nucleon and closed NVM loops (e.g., discarding A- and π-self-energies), it is easy to see that [4]

$$e^{i\chi} = \exp\left[-\frac{i}{2}\int \frac{\delta}{\delta\pi}\, D_c\, \frac{\delta}{\delta\pi}\right].\,\exp\left[ig^2\int \mathcal{F}_1\, \bar{\Delta}_c(\pi)\, \mathcal{F}_2\right]\Big|_{\pi\to 0} \qquad (3)$$

or

$$i\chi = \left(\exp\left[-\frac{i}{2}\int \frac{\delta}{\delta\pi}\, D_c\, \frac{\delta}{\delta\pi}\right].\,\exp\left[ig^2\int \mathcal{F}_1\, \bar{\Delta}_c(\pi)\, \mathcal{F}_2\right]\Big|_{\pi\to 0}\right)_{conn} - 1 , \qquad (4)$$

where the $\mathcal{J}^{\mu}_{1,2}(u) = P^{\mu}_{1,2} \int_{-\infty}^{+\infty} d\xi\ \delta(u - Z_{1,2} + \xi p_{1,2})$ denote classical currents of the interacting nucleons, with moment $p_{1,2}$ and spatial coordinates $z_{1,2}$; the impact parameters is $\vec{b} = (\vec{z}_1 - \vec{z}_2)_{\perp}$; D_c and $\delta_{\mu\nu}\Delta_c$ denote pion and NVM propagators of masses μ and m, respectively $\bar{\Delta}_c(\pi) = \Delta_c(1+ \lambda\pi\Delta_c)^{-1}$ represents the NVM propagator in the presence of a ficticious, c-number field $\pi(x)$; and the subscript "conn" specifies the retention of connected graphs only, with at least one pion linkage between any NVM and the rest of that graph. S-channel unitarity then requires that $\sigma_{TOT} = \sigma_{EL} + \sigma_{IN}$, where

$$\sigma_{TOT} = 2\ \mathrm{Re} \int d^2b\ \lfloor 1 - e^{i\chi} \rfloor\ , \tag{5a}$$

$$\sigma_{EL} = \int d^2b\ |1 - e^{i\chi}|^2\ , \tag{5b}$$

$$\sigma_{IN} = \int d^2b\ \lfloor 1 - e^{-2\mathrm{Im}\chi} \rfloor . \tag{5c}$$

Expansion of the g^2-dependent exponential of (4) provides a representation of the conventional eikonal approximations. Omitting all self-linkages (all pion radiative corrections to the same NVM line), to order g^2 one obtains just the Lévy-Sucher result [5], $i\chi_1(s,b) \simeq - \frac{g^2}{2\pi}\ \gamma(s)\ K_o(mb)$, where $\gamma(s) \to 1$ as $s/m^2 \to \infty$, in this NVM theory. (Where the NVM replaced by scalar mesons A, then $\gamma(s) \to -m^2/s$ as $s/m^2 \to \infty$). The g^4-term of this expansion contains all the ladder and crossed pion linkages between a pair of NVM lines; and upon extracting the leading rapidity dependence from the set of ladder graphs, one reproduces an absorptive eikonal $i\chi_2 \sim -a_1\ s^{a_2} / \ell ns.\ \exp\lfloor - a_3\ b^2/\ell ns \rfloor$, equivalent in physical content to the original Cheng-Wu/Chang Yan calculations [2]; here, the a_i are constants, and $a_2 \sim \lambda^2/m^2$. (In a scalar A-meson theory, $a_2 \sim \lambda^2/m^2 - 2$); and the phenomenological choice $a_2 = 0$ produces an effective, unitarized Pomeron, complete with the correct cuts). Estimates of the remaining $i\chi_n$, $n \geqslant 3$ are exceedingly difficult to obtain; and when obtained are difficult to believe [4].

In a recent paper [6], an attempt was made to incorporate the effects of all the χ_n, in an approximate way, by performing a semi-classical calculation for an averaged, "self-consistent" pion field, whose properties would reflect the probable importance of the higher eikonal terms, in the limit of ultra-high energies and in the region of small impact parameters. There, a formalism was set-up to perform such an "averaged" estimate, by first converting the functional differential operations of (3) and (4) to an equivalent function integral -- along with the corresponding functional representations of all the pion inelastic cross sections -- and then approximating these functional integrals by a stationary phase method, with the condition for stationary phase providing the "semi-classical" Euler equation for the "averaged" field. At very high energies, one may expect that every pion exchanged between a pair of NVMs will, in a leading-rapidity, ladder graph approximation, generate a factor of Y essentially corresponding to the available phase space for the inelastic production of a pion with limited transverse momentum; and hence one is faced with an effective strong-coupling problem, in which it is not acceptable to neglect any particular sets of eikonal graphs. Mathematically, one evaluates an integral in such a strong-coupling limit by a method such as stationary phase; physically, one is shifting the description from one of overly-many quanta to one of a single, averaged field.

This type of calculation has one distinct advantage over other, more properly classical computations, wherein one merely hopes that quantum fluctuations are small, and will not disturb the classical effects obtained. Because one is here dealing with an eikonal representation of quantum effects (which are then to be approximated by a semi-classical technique), one has the added feature of s-channel unitarity, relating inelastic pion emission at every impact parameter to the imaginary part of the elastic eikonal. After the transition to an averaged field has been made, this requirement of unitarity translated into a restriction on the possible solutions of the Euler equations. Hence an averaged field that satisfies this unitarity condition provides an approximate representation of all the complicated, non-linear, internal dynamics specified by the original Lagrangian interaction; and one which is also consistent with the quantum mechanical restrictions of probability for scattering and all production processes.

One may add that it is precisely because the Euler equations are non-linear that the method works at all. That is, solutions to a particular non-linear Euler equation have no reason to satisfy another non-linear relation, that of unitarity; one cannot expect miracles. But because the semi-classical equations are non-linear, they involve the magnitude, phase and branch of their solutions -- three parameters, or functions, rather than just two. In essense, unitarity determines the third quantity, for example, the branch; and in this sense, the non-linearity of the problem is essential.

The equation for the averaged field $\varphi_0(x)$ was obtained in ref.[6] as :

$$K \varphi_0(x) = - i \left. \frac{\delta \mathcal{F}}{\delta \varphi(x)} \right|_{\varphi=\varphi_0} , \tag{10}$$

with

$$\mathcal{J}(y) = i\, g^2 \int \mathcal{J}_1\, \bar{\Delta}_c\, [\varphi]\, \mathcal{J}_2\ , \tag{11}$$

and $K = \mu^2 - \partial^2 = D_c^{-1}$. Eq. (10) is "self-consistent" in the sense that each "quasi-pion" φ_0 emitted by a NVM is defined by the behavior of many (an infinite number of such) quasi-pions along that line. Inelastic emission then turns out to be described by the simplest of all unitary mechanisms, as if no more than a single particle is emitted by each multiperipheral chain. It is most convenient to introduce a specific form for $\bar{\Delta}_c(\varphi)$, the use of which is equivalent to the extraction of the leading rapidity dependence of all ladder-graph pions exchanged between any pair of NVMs,

$$\bar{\Delta}_c(Z|\varphi) = \int \frac{d^4q}{(2\pi)^4} \cdot \frac{e^{iq.Z}}{q^2+m^2} \cdot \exp\left[\lambda \int d^4k \quad \frac{\tilde{\varphi}(k)}{(q-k)_\perp^2 + m^2}\right],$$

together with the subsequent replacement

$$\int \frac{d^4K}{k^2+\mu^2} \rightarrow i\pi\, Y \cdot \int d^2k \quad , \qquad Y \sim \ln\left(\frac{s}{\mu^2}\right) \quad .$$

In terms of possible solutions φ_0 to (8), the eikonal is given by

$$\chi = -i\mathcal{F}(\varphi_0) - \frac{1}{2}\int \varphi_0\, K\, \varphi_0 + \frac{i}{2}\,\mathrm{Tr}\,\ln\left[1 + i\, D_c\, \frac{\delta^2 \mathcal{F}}{\delta\varphi\,\delta\varphi}\bigg|_{\varphi=\varphi_0}\right]. \tag{12}$$

We do take seriously the trace-log contributions of (12), which might be thought of as a normalization correction to the eikonal, and physically corresponds to a sum of quasi-pion linkages between the NVMs such that no more than two virtual particles are emitted or absorbed by any NVM line. However, due to the non-linearity imposed by unitarity, inelastic emissions still have the form of just one particle ejected per chain.

One point concerning self-linkages along every NVM line should be made, for the formalism of (3) and (4) will automatically include such dependence, even though the propagator $\bar{\Delta}_c(\varphi)$ has really been designed to reproduce those pionization effects delivered by cross-linkages between two different NVMs. But one may expect that this difficulty is relatively unimportant. In the context of the stationary phase approximation, such unwanted self-linkages may be removed by subtracting from χ the first-order expansion of the trace-log of (12),

$$\chi \to \chi' = \chi - \frac{i}{2} \mathrm{Tr} \left| i D_c \frac{\delta^2 \mathcal{L}}{\delta\varphi . \delta\varphi} \right|_{\varphi=\varphi_0} , \tag{13}$$

and this form will be used in what follows. For small b this has essentially no effect; but in the region of larger impact parameters close to $b_{max} \sim \frac{1}{m} Y$, it ensures the Froissart bound in its usual form, rather than $b_{max} \sim \frac{1}{m} [Y \ell n Y]^{1/2}$.

In momentum space, (10) reads

$$\tilde{\varphi}_0(k) = - \frac{g^2 \lambda}{(2\pi)^2} \cdot \frac{1}{k^2 + \mu^2} \cdot \int \frac{d^2 q \, e^{i\vec{q}.\vec{b}}}{q^2 + m^2} \frac{e^{R(q/m)}}{(q-k)^2 + m^2} \tag{14}$$

with

$$R(q/m) \equiv \lambda \int d^4 p \, \tilde{\varphi}_0(p) \, [m^2 + (q - p_\perp)^2]^{-1}$$

where we reject, at the outset, any additional mass-shell contribution to φ_0. It is immediately clear that finding solutions to (14) is a difficult task. To simplify the analysis, we look for solutions under the assumption $mb \gtrsim 1$, which suggests that the only significant values of q, in the integrand of (14), will be limited by $q \lesssim m$. Hence the R(q/m) factor may be approximated as

$$R(q/m) \simeq R_0 + i\vec{q}.\vec{R}_1 - q_i q_j R_2^{ij} \tag{15}$$

retaining only (the integrable) quadratic q dependence. If the $|R_2^{ij}|$ are of size m^{-2}, this manipulation is equivalent to a cut-off of $\int d^2q$ at $q_{max} \sim m$; but non-linear effects can serve to significantly decrease q_{max}. For example, if $m^2 |R_2^{ij}| \sim Y$, or larger (and here we are assuming ultra-high energies, where $Y >> 1$), then $q_{max}^2 << m^2$. It is solutions of this form, corresponding to sizable φ_0, that we have in mind.

The substituion of (15) into (14), together with the use of the pionization form of $\bar{\Delta}_c(\varphi)$ and the assumption that $q_{max}^2 << m^2$, generates

$$\vec{R}_1 = 0 \qquad \text{(by symmetry)},$$

$$R_2^{ij} = R_2 \, \delta_{ij}, \quad R_2 = R_0/6m^2, \tag{16}$$

and

$$R^2 = -i\xi \exp(R_o - \frac{\eta}{R_o}) ,$$

with $\xi = 3(g\lambda)^2\pi Y/2m^2$, $\eta = \frac{3}{2}(mb)^2$. By the use of these approximations, which follows naturally for large R_2, or R_o, all the non-linear complexity of the original eq.(14) has been transformed into a single relation for a complex R_o, (17). If solutions to (17) can be found which have the property

$$Re(R_o) > 0 , \tag{18}$$

necessary for the convergence of the original q-integral of (14), then φ_o may be written in the form

$$\tilde{\varphi}_o(k) = -i\left(\frac{m^2R_o}{\lambda\pi^2Y}\right)\frac{1}{(k^2+\mu^2)(k_\perp^2+m^2)} . \tag{19}$$

It is then straightforward to substitute (19) into (13), and obtain :

$$i\chi = +\left(\frac{m}{\lambda\pi}\right)^2 \frac{1}{Y} R_o - \left(\frac{m}{\lambda\pi}\right)^2 \frac{1}{2Y} R_o^2 - \frac{1}{2}\ln(1-R_o) - \frac{1}{2} R_o . \tag{20}$$

Evaluation of the trace-log terms is performed by an approximate evaluation of its nth iterate, using the same approximation methods followed in reaching (19); and then the sum over all iterates (each is finite!) has been put into logarithmic form by a simple continuation argument.

With (20) one can now supply a specific input to the s-channel unitarity relation of ref. 10, which demands, for any such eikonal, that

$$Im\ \chi = \frac{1}{2}\int(K\varphi_o)^{*}.\ D_{(+)}.(K\varphi_o) ,$$

where $\tilde{D}_{(+)}$ represents the (positive definite) pion's phase space function, $\tilde{D}_{(+)}(k) = (2\pi)^{-3}.\theta(k_o).\delta(k^2+\mu^2)$. In terms of the explicit solution (19), this becomes :

$$Im\ \chi = \frac{1}{2Y}\left(\frac{m}{\lambda\pi}\right)^2 . |R_o|^2 , \tag{22}$$

where the LHS of (22) is given by the imaginary part of (20). We shall insist that any solution for R_o to (17) and (18) must also satisfy (22).

Approximate solutions to the Euler equations-plus-unitarity have been found which depend on b and (very large !) Y in the following way : $R_o = \rho \exp |i(\theta + 2\pi n)|$.

(i) For relatively small b, $mb \lesssim (Y \ell n Y)^{1/2}$:

$$\rho \sim aY \ell n Y, \qquad \theta \sim \pi/2$$

Here, the branck n is large, $n \sim \rho$, while the eikonal is properly absorptive,

$$i\chi \sim -\frac{1}{8} \ell n Y - ia \left(\frac{1}{6} Y \ell n Y\right)^{1/2}, \qquad a = \left(\frac{\pi\lambda}{2m}\right)^2 .$$

(ii) As the Froissart limit is approached, $(Y\ell n Y)^{1/2} \lesssim mb \lesssim Y$:

$$\rho \sim \alpha Y + \beta \ell n Y , \quad \theta \rightarrow \theta_o < \pi/2 ,$$

with $\alpha = 2a \cos \theta_o .(\sin\theta_o)^{-2}$, $\beta^{-1} = \cos \theta_o$. Here $n \sim \rho$ again; and hence we are on a higher branch. In this region, $Re(i\chi) \sim -Y$.

(iii) As b continues to increase, $\theta \rightarrow 0$, and there occurs a sharp drop in ρ such that $\eta > \xi > \rho \sim O(1)$. The, for larger b, $mb > Y$,

$$\rho \sim Y^{-1} \cos \theta \sim Y^{1/2} \exp \left(-\frac{a}{4\pi} n Y\right),$$

so that $\theta \rightarrow \frac{\pi}{2}$ and $n \sim \frac{n}{\rho}$. Here, ρ falls off very rapidly as Yb^2 increases, and is effectively zero for $b > b_{max} \sim \frac{1}{m} Y$. This behavior provides the Froissart bound for our inclusive and total cross sections, $\sigma_{TOT} \sim Y^2$.

Inclusive cross sections are obtained by calculating the functional derivatives of

$$\sigma_{IN} = \int d^2b \, e^{-2 \, Im\chi} \left(e^{\int K\varphi_o^{*} . D_{(+)} . K\varphi_o} - 1\right)$$

with respect to $\tilde{D}_{(+)}(k)$, and produce non-correlated inclusive distributions. The one-particle inclusive cross section, for example, is given by

$$\omega \frac{d\sigma}{d^3k} = \pi \int d^2b \; |K \tilde{\varphi}_o(k)|^2 ,$$

with an integrand that cuts off sharply at $b \geqslant b_{max}$. Using the larger b forms throughout, this gives approximately

$$\omega \frac{d\sigma}{d^3k} \sim b^2_{max} \cdot \left(\frac{m^2}{m^2+k^2_\perp} \right)^2 \quad . \tag{23}$$

The limited k-distribution of (23) is not significant, since it was effectively assumed at the beginning; but the growth of such an inclusive cross section with Y^2, and the flat plateau in the rapidity version of this result are specific predictions of the model. The other n-particle inclusive cross sections show a similar rise with Y^2, and contain no correlations between any such inclusive emissions. Except for the flat plateau, these results are similar to those recently found in a strong-coupling, super-critical Pomeron computation [7].

Within the context of this averaged, or semi-classical calculation, the present computation thus provides an answer to the long-standing eikonal question : How important are all the $i\chi_n$, $n \geqslant 2$? We do not find the almost complete cancellation suggested in ref.2; rather there is an effective saturation of the Froissart bound, which would have been given by the pionization model of $i\chi_2$, except for a slight decrease in σ_{TOT} and the inclusive cross sections coming from the region of small b, $mb < \left(Y \ln Y\right)^{1/2}$. One difference is that here the ratio of total one-particle inclusive cross section to σ_{TOT} produces a multiplicity $<n> \sim Y$, in contrast to the Cheng-Wu/ Chang-Yan behavior $<n> \sim S^{a/1+2a}$. Except for the complete absence of correlations, these results are not in overt disagreement with existing data, and may be looked upon as a prediction of future, higher-energy experiments.

REFERENCES

1) H. MORENO and H.M. FRIED, Nice Preprint NTH 77/5.
2) H. CHENG and T.T.WU, Phys.Rev.Letters 24, 1455 (1970); S-J CHANG and T-M YAN, Phys.Rev. D4, 537 (1971).
3) Rather extensive discussions and applications of the fundamental Bloch-Nordsieck approximation may be found in : H.M. FRIED, "Functional Methods and Models in Quantum Field Theory", M.I.T. Press, Cambridge (1972).
4) R. BLANKENBECLER and H.M. FRIED, Phys.Rev. D8, 678 (1973).
5) M. LEVY and J. SUCHER, Phys. Rev. D2, 1716 (1970).
6) H. MORENO and H.M. FRIED, Phys.Rev. D12, 2031 (1975).
7) D. AMATI, M. LE BELLAC, G. MARCHESINI and M. CIAFALONI, Nucl.Phys. B112 (1976) 107.

PARITY VIOLATION EFFECT IN ATOMIC PHYSICS AND THE STRUCTURE OF NEUTRAL CURRENTS IN GAUGE THEORIES *)

T.C. YANG

Deutsches Elektronen-Synchrotron DESY, Hamburg and

Department of Physics and Astronomy, University of Massachusetts, Amherst, Mass. **)

We study here three models, namely, the SU(2) ⊗ U(1) model, the $SU_L(2) \otimes SU_R(2) \otimes U(1)$ model and the $SU_L(2) \otimes SU_L(2) \otimes U(1)$ model, which seem to be the most natural extensions of the Weinberg-Salam model in order to accomodate the recent atomic Bismuth experimental results. The differences between the model predictions in neutrino reactions are small except for the elastic $\nu_\mu e$ and $\bar{\nu}_\mu e$ cross sections. We remark on the other specific experiments which could provide meaningful checks between these models. We also comment on the "naturalness" of each model.

I. INTRODUCTION

Several lectures were given in this Summer Institute, which discussed the Weingerg-Salam model[1] with GIM mechanism[2], and its predictions as compared with the present experimental data. The general conclusions are that Weinberg-Salam model is in good agreement with the neutrino data,[3] but seems to predict too big a parity violation effect as compared with the recent atomic Bismuth Experiments.[4] The good feature of the Weinberg-Salam model lies in its simple and elegant structure of the neutral current, and therefore it seems nontrivial that the Weinberg-Salam model is in good agreement with the present neutrino data.

---

*) Research work supported by the National Science Foundation

**) Present address

The atomic Bismuth experiments measured the parity violation effect induced by the weak neutral currents and found an effect about ten times smaller than what was predicted by using the atomic calculations[5] for the Weinberg-Salam model.[6] (The data is also consistent with no effect within the error bars). In contrast with what seems to be the success of the Weinberg-Salam model in neutrino reactions, the atomic Bismuth results remain as a puzzle. It is possible that the atomic calculations might have overestimated the parity violation effect in heavy atoms. It is, however, generally believed that the atomic physics calculation is probably reasonable within a factor of 2, which is still not large enough to bring the theory and the data into agreement. Thus insofar as the atomic physics calculation is believable, the atomic Bismuth results imply modification of the basic Weinberg-Salam model. We note that postulating more left-handed quarks does not change the weak neutral currents in gauge theories.

It is theoretically a challenging problem and perhaps a ncessity to study whether the Weinberg-Salam model can be extended to incorporate the atomic Bismuth results and still agree with the present neutrino data. I would like to review here three models which seem to be the natural choices to study and which are consistent with the present data - these are the $SU(2) \otimes U(1)$ model,[7] the $SU_L(2) \otimes SU_R(2) \otimes U(1)$ model[8] and the $SU_L(2) \otimes SU_L(2) \otimes U(1)$ model (or the $SU_L(4) \otimes U(1)$ model[9]). One of our aims is to study what types of experiments are most useful in distinguishing between these models. These involve the following processes:

(1) Cross sections and the y distributions of elastic $\nu_\mu e$ and $\bar{\nu}_\mu e$ scattering.

(2) Parity violation in light atoms.

(3) Forward-backward asymetry in the angular distributions of $e^+e^- \to \mu^+\mu^-$ and $e^+e^- \to hX$ at say, PETRA and PEP energies.

(4) I=0 and I=1 interference in inclusive or semi-inclusive neutral-current neutrino reactions.

For convenience, we shall always compare the various model predictions with that of the Weinberg-Salam model and state only the differences.

II. THE $SU(2) \otimes U(1)$ MODEL[7]

The difference between this model and the Weinberg-Salam model is in the representations of the right-handed fermions. Both the left-handed and the right-handed electrons are assigned in doublets. Thus the electron neutral current is pure vector and one does not expect parity violation effect in atomic Bismuth experiments.

In order to cancel the triangle anomaly induced by the right-handed lepton currents, we need to introduce right-handed quark currents or right-handed heavy lepton currents. The second alternative seems ad hoc but is certainly a possibility. As we have heard from the "naturalness" argument[10] (namely, $O(G_F^2)$ suppression of flavor-changing neutral currents for all values of the parameters of the theory) the only model which has right-handed quark currents and which is natural is the vector model (if the quark charges are only 2/3 and -1/3), but the vector model is inconsistent with the neutral current neutrino data. The $SU(2) \otimes U(1)$ model appears to be "unnatural".

This difficulty can be avoided, if we impose on the Lagrangian an additional discrete symmetry R, which changes the signs of the right-handed quark doublets and the left-handed quark singlets, but leaves the left-handed quark doublets, the right-handed quark singlets, the leptons, the Higgs scalars and the gauge bosons invariant. As long as R symmetry is not broken, the quarks which are odd or even under R do not mix with each other - the R symmetry does not allow Yukawa coupling between quarks having different R-parity quantum numbers. We identify u,d,s,c quarks as even and t,b,g,h quarks as odd under the R symmetry. Since the (t,b) or (g,h) quarks do not mix with (u,d) or (c,s) quarks in masses by R-parity conservation, the flavor-changing neutral currents can be naturally suppressed. The representation of the quarks and the leptons are then given by

$$\begin{pmatrix} u \\ d(\theta_c) \end{pmatrix}_L, \quad \begin{pmatrix} c \\ s(\theta_c) \end{pmatrix}_L, \quad \begin{pmatrix} t \\ b \end{pmatrix}_R, \quad \begin{pmatrix} g \\ h \end{pmatrix}_R; \quad u_R, d_R, c_R, s_R; t_L, b_L; g_L, h_L,$$

$$\begin{pmatrix} \nu_e \\ e \end{pmatrix}_L \quad \begin{pmatrix} E^\circ \\ e \end{pmatrix}_R, \quad \begin{pmatrix} \nu_\mu \\ \mu \end{pmatrix}_L \quad \begin{pmatrix} M^\circ \\ \mu \end{pmatrix}_R; \quad E_L^\circ, M_L^\circ \ldots$$

where mixings between quarks of the same R-parity are allowed. The price one has paid to maintain the "naturalness" is 4 more heavy quarks

The R symmetry forbids right-handed coupling between the u and b quarks. Thus this model predicts no y anomaly in anti-neutrino reactions which is consistent with the recent data.[3]

We have seen the data of a new resonance of mass ∿ 9.5 GeV decaying into μ pairs in hadronic collisions[11] which could suggest more heavy quarks beyond charm. We could identify 9.5 GeV resonances as bound states of t,b,g or h.

Experimental implications. Since the u,d,s,c, quarks have the same representations as in the Weinberg-Salam model, this model has the same predictions for the inclusive neutral-current neutrino reactions and for $\nu_\mu p$, $\bar{\nu}_\mu p$ elastic scatterings as the Weinberg-Salam model. The only difference is the weak electron-neutral-currents in processes such as $e^+e^- \to \mu^+\mu^-$, hX at high energies, the deep inelastic electron scattering at high energies and neutrino-electron elastic scatterings.

(1) Elastic $\nu_\mu e$ and $\bar{\nu}_\mu e$ scatterings seem to provide the most critical tests of this model which are experimentally readily accessible to us. Since the electron current is pure vector in this model, elastic $\nu_\mu e$ and $\bar{\nu}_\mu e$ scatterings should have equal cross sections and the same y distribution. The present data, although inconclusive at present, disfavors the vector current hypothesis but is consistent with V+A or the Weinberg-Salam model[12]. With improved statistics, it seems that the vector current hypothesis could definitely be tested in the near future.

(2) $e^+e^- \to \mu^+\mu^-$. The electron and muon currents being pure vector in this model, there is no forward-backward asymmetry (defined below) in $e^+e^- \to \mu^+\mu^-$. In contrast, asymmetry in the muon angular distribution is predicted by the Weinberg-Salam model. The muon angular distribution can be written as

$$\frac{d\sigma}{d\Omega}\left(\frac{\alpha^2}{4q^2}\right)^{-1} = F_1(q^2)(1+\cos^2\theta) + F_2(q^2)2\cos\theta$$

where θ is the angle between $\mu^{\mp}$ and $e^{\mp}$. For C.M. energy $\sqrt{q^2} << M_Z$ one finds[13]

$$F_1(q^2) \simeq 1 + 2R(q^2)g_V g_V'$$

$$F_2(q^2) \simeq 2R(q^2)g_A g_A'$$

$$R(q^2) = \frac{g^2}{e^2}\,\frac{q^2}{q^2-M_Z^2} \sim -\frac{g^2}{e^2}\,\frac{q^2}{M_Z^2}$$

where g_V, g_V' and g_A, g_A' are the vector and axial-vector coupling constants for the electron and muon neutral currents respectively. For the Weinberg-Salam model, one finds[13]

$$\frac{d\sigma}{d\Omega} \sim (1 + \cos^2\theta) + 4\varepsilon \cos\theta$$

$$\varepsilon \simeq \frac{-q^2}{m_Z^2} \frac{1}{4\sin^2 2\theta_W} = \frac{-q^2}{M_P^2} \left(\frac{G_F M_P}{8\pi\alpha\sqrt{2}}\right) = -0.4 \times 10^{-4} \frac{q^2}{M_P^2} .$$

The forward-backward asymmetry is given by

$$\alpha = \frac{\sigma_f - \sigma_b}{\sigma_f + \sigma_b} \simeq \frac{3}{8} 4\varepsilon = -0.6 \times 10^{-4} \frac{q^2}{M_P^2}$$

where σ_f and σ_b are the forward and backward cross sections for $0 < \theta < \pi/2$ and $\pi/2 < \theta < \pi$ respectively.

(3) $e^+e^- \to h(p)X$. For inclusive production of hadrons in e^+e^- annihilation, the forward-backward asymmetry (in the hadron angular distribution) is given by[14] $(q^2 << m_Z^2)$

$$\alpha \simeq \frac{3Eg_A R(q^2)\, d|p|\, |p|^3\, W_3^{weak}(q^2,\nu)}{\int d|p|\, |p|^2 \{3m^2 W_1^{em}(q^2,\nu) + |p|^2 W_2^{em}(q^2,\nu)\}}$$

where $E = E_{beam}$; $g_A, R(q^2)$ defined above; $|p|$ is the momentum of the (detected) hadron; W_1, W_2, W_3 are the standard structure functions of electromagnetic and weak neutral currents. Since W_3^{weak} is the vector and axial-vector interference term, there will be no forward-backward asymmetry for $e^+e^- \to g(p)X$ in this model in contrast with the Weinberg-Salam model.

III. MORE W BOSONS ?

If the weak electron neutral currents are not pure vector, in order to accomodate the atomic Bismuth results, one needs to enlarge the gauge symmetry of the weak and electromagnetic interactions beyond SU(2) ⊗ U(1). We will consider here two models which seem to be the simplest choices to be considered.
They are the $SU_L(2) \otimes SU_R(2) \otimes U(1)$ model and $SU_L(2) \otimes SU_L(2) \otimes U(1)$ model. As far as the neutral current sector is concerned, we are making use of the third "isospin" component of the $SU_R(2)$ and $SU_L(2)$ groups, which means effectively an additional U(1) degree of freedom. We dismiss the general SU(2) ⊗ U(1) ⊗ U(1) model since in the general case the phase of one of the U(1) group is arbitrary-

the model has too many parameters and thus seems uninteresting. When embedded in a larger group, the new piece of neutral current is constrainted by the gauge transformations. The structure of the neutral currents shows significant deviations from the Weinberg-Salam model as discussed below.

(II.1) The $SU_L(2) \otimes SU_R(2) \otimes U(1)$ Model

The $SU_L(2) \otimes SU_R(2) \otimes U(1)$ models have been considered by Mohapatra and Sidhu; by De Rujula, Georgi and Glashow; and by others.[8] The neutral currents structure is sufficiently illustrated by the four quark model of Mohapatra and Sidhu[8], to which we turn our attention next. Models with more than four quarks which predict the y-anomaly are phenomenologically not interesting in view of the recent neutrino data. The quarks and the leptons are all in doublets.

$$\begin{pmatrix} u \\ d(\theta) \end{pmatrix}_L, \quad \begin{pmatrix} c \\ s(\theta) \end{pmatrix}_L, \quad \begin{pmatrix} \nu_e \\ e \end{pmatrix}_L, \quad \begin{pmatrix} \nu_\mu \\ \mu \end{pmatrix}_L$$

$$\begin{pmatrix} u \\ d(\theta') \end{pmatrix}_R, \quad \begin{pmatrix} c \\ s(\theta') \end{pmatrix}_R, \quad \begin{pmatrix} \nu_e \\ e \end{pmatrix}_R \quad \begin{pmatrix} \nu_\mu \\ \mu \end{pmatrix}_R$$

where θ, θ' denote the mixing angles. The charged right-handed currents are assumed heavy and not relevant for low energy weak interactions - experimentally, we have no evidence for right-handed charged currents. Thus effectively, the difference between this model and the Weinberg-Salam model is in the new neutral currents generated by $(T_3)_R$ of the $SU_R(2)$ group.

If one imposes a reflection symmetry which changes L with R, parity is conserved at the Lagrangian level and the $SU_L(2)$, $SU_R(2)$ gauge coupling constant are equal. Two I=1 triplets of Higgs scalars are assumed to give the heavy masses only to $W_R^{\pm}$, as a result parity is violated in low energy charged current interactions g_L and g_R will differ by finite renormalization.

The reflection symmetry is responsible for the absence of parity violation in atomic experiments. Because of the reflection symmetry, the left-handed and right-handed neutral gauge bosons are allowed to mix equally with the U(1) gauge boson. As a result, the neutral gauge boson eigenstates couple to either the vector currents or the axial-vector currents but not both. In other words, at the tree level the neutral current interactions conserve parity (note that neutrinos are massive in this model). The fact that parity is violated in neutrino reactions is due to the fact that the neutrino beam in laboratories are dominantly left-handed. The left-hand neutrinos can couple to both of the neutral gauge bosons through the vector and axial-vector components. The model thus avoids the constraint of the vector-like models. The neutral current interactions can be written as

$$(m_{Z_1}^2 \simeq m_{W_L}^2 (1+\varepsilon)\ , \quad m_Z^2 \simeq m_{W_L}^2 \csc^2\theta(1-\varepsilon))$$

$$\frac{e}{2\cos\theta} \{-\bar{e}\gamma_\mu\gamma_5 e - \frac{1}{2}\bar{\nu}\gamma_\mu(1-\gamma_5)\nu + \bar{u}\gamma_\mu\gamma_5 u - \bar{d}\gamma_\mu\gamma_5 d\} Z_1^\mu$$

$$- \frac{e}{\sin 2\theta} \{\cos 2\theta\, \bar{e}\gamma_\mu e + \frac{1}{2}\bar{\nu}\gamma_\mu(1-\gamma_5)\nu - \frac{1}{3}(1 - 4\sin^2\theta))\bar{u}\gamma_\mu u$$

$$- \frac{1}{3}(1 + 2\sin^2\theta)\,\bar{d}\gamma_\mu d\} Z_2^\mu$$

where $\sin\theta$ is the mixing angle and ε another parameter.

Experimental implications. For $\varepsilon = 0$, this model has identical predictions for all the low energy neutrino reactions as the Weinberg-Salam model ($\cos^2\theta = 2\sin^2\theta_W$). The ε dependence allows more flexibility, but the difference is not expected to be large. Because of the nature of the two neutral gauge bosons and their mass relations, this model has different predictions for the following processes as compared with the Weinberg-Salam model.

(1) Parity violation in light atoms. This model predicts no parity violation in any atom, light or heavy, to order G_F, whereas the Weinberg-Salam model, would predict parity violation in light atoms. If one releases the left and right reflection symmetry of the Lagrangian, then this unique prediction is lost.

(2) $e^+e^- \to \mu^+\mu^-$. The forward-backward angular symmetry is enhanced in this model by

$$\left(\frac{m_Z^2}{m_{Z_1}^2}\right)^2 \simeq \left(\frac{m_Z^2}{m_{W_L^\pm}^2}\right)^2 \simeq \left(\frac{1}{\cos^2 W}\right)^2 \simeq 2$$

as compared with the Weinberg-Salam model for $q^2 \ll m_Z^2$.

(3) $e^+e^- \rightarrow h(p)X$. The forward-backward angular symmetry is enhanced by $m_Z^2/m_{Z_1}^2 \sim 1.4$ as compared with the Weinberg-Salam model for $q^2 \ll m_Z^2$.

(4) The neutrinos are massive in this model, thus it is in principle possible to produce right-handed neutrinos at high energies. For e^+e^- annihilation one would find

$$\sigma(e^+e^- \rightarrow \nu_\mu \bar{\nu}_\mu) = 2\,(e^+e^- \rightarrow \nu_{\mu_L} \bar{\nu}_{\mu_L}) \quad .$$

The cross section increases linearly with q^2 for $q^2 \ll m_Z^2$.

(III.2) The $SU_L(2) \otimes SU_L(2) \otimes U(1)$ Model

The first $SU_L(2)$ is the weak isospin group, the second SU(2), which commutes with the first $SU_L(2)$, is actually isosinglet, but describes an isospin-like symmetry group which will be specified below. The neutral current structure of the $SU_L(2) \otimes SU_L(2) \otimes U(1)$ model has been treated in detail in ref. (9) where $SU_L(2) \otimes SU_L(2) \otimes U(1)$ is imbedded in a larger group, namely, $SU_L(4) \otimes U(1)$. We'll illustrate the neutral currents here without referring to the $SU_L(4) \otimes U(1)$ model. To lowest order in coupling squared, the effective strength of the "fragrance" interaction is given by $g^2/m_{W'}^2$, where the coupling constant and $m_{W'}$ refer to the second SU(2) group. For convenience, we can set the two SU(2) coupling constants equal (as if they are imbedded in $SU_L(4)$) by adjusting $m_{W'}$.

If we write the four quarks in a matrix

$$\begin{pmatrix} u & c \\ d & s \end{pmatrix}_L , \qquad \text{Model (A)}$$

the $SU_L(2) \otimes SU_L(2)$ symmetry couple the quarks in columns and rows respectively. We could arrange the u,d and c,s quarks in different matrices by postulating more quarks.

$$\begin{pmatrix} u & t \\ d & b \end{pmatrix}_L , \qquad \begin{pmatrix} c & g \\ s & h \end{pmatrix}_L \qquad \text{Model (B)}$$

where mixings between the quarks are not explicitly shown. Under SU(4) they all transform like quartets.

Model (A) is economical. See ref. (9) for some of the clean predictions in this model. As far as the neutral currents interactions are concerned, both models give the same result.

Model (B) has more quarks. One of them might be the constituent of the 9.5 resonance about which we have heard in this school.[11] Let us use Model (B) for the following discussions.

To cancel triangle anomalies in Model (B), we need 8 leptons,[15] whose representations are being identified with known and unknown leptons as follows :

$$\begin{pmatrix} \nu_\tau & e \\ \tau^- & e^- \end{pmatrix}_L, \qquad \begin{pmatrix} M^\circ & \nu_\mu \\ M^- & \mu \end{pmatrix}_L$$

Thus the τ heavy leptons is identified as an excited electron, but because of the available final states the decays of τ lepton are not very different from a sequential heavy lepton[16].

We denote the vertical degrees of freedom of the quarks as "flavor", and the horizontal degrees of freedom (of the above quark matrix) as "fragrance". Together with the "color" degrees of freedom, they form the basis of the weak, semi-weak and strong interactions. (The effective strenght of the "fragrance" interactions turns out to be about a quarter of G_F.)

The (t,b) quarks differ from the (u,d) quarks by the "fragrance" quantum numbers - they have the same properties as far as "flavor" (defined specifically above) is concerned. The third components of the "fragrance" symmetry are conserved, thus (u,d) or (c,s) can not mix with (t,b) or (g,h). This means strangeness-changing neutral currents are "naturally" suppressed. (For the Higgs sector see below). Similarly neutral currents coupling b with h or t with g quarks are suppressed in order $G_F\alpha$. Note that the neutral currents coupling d to b quarks, s to h quarks are fragrance-changing (associated with raising and lowering operators of the second SU(2) group) and are allowed.

Before we can write down the neutral currents, we need to specify the representations of the Higgs scalars. We assume two Higgs scalars, transforming as (1/2, 0, 1) and (0, 1/2, 0) under $SU_L(2) \otimes SU_L(2) \otimes U(1)$ respectively. The first Higgs scalar is the same as the one in the Weinberg-Salam model. The second Higgs scalar is neutral under the first SU(2) group. We find that the eigenstates of the photon and the Z boson, and the mass relations between Z and W are given as in the Weinberg-Salam model. The second Higgs scalars are electrically neutral and do not couple to the U(1) gauge boson. After spontaneous symmetry breaking, they give equal masses to $W'_{1\pm i2}$, W'_3, where prime denotes the second SU(2) (fragrance) group. Thus we have

$$A_\mu = -\sin\theta\, W^3_\mu + \cos\theta\, B_\mu$$

$$Z_\mu = \cos\theta\, W^3_\mu + \sin\theta\, B_\mu\,, \qquad m_Z^2 = \frac{m_W^2}{\cos^2\theta_W}$$

and

$$m^2_{W'_3} = m^2_{W'_\pm}$$

the neutral currents interactions can now be written down as

$$L_{neut} = \frac{g}{\cos\theta_W}\{j^3_\mu - \sin^2\theta_W j^{em}_\mu\}\, Z^\mu$$

$$+ g\{j'^3\}\, W'^3_\mu$$

where j^3_μ, j'^3_μ are the "isospin" currents of the two SU(2) groups given by

$$j^3_\mu = \frac{1}{2}\left[\bar{u}\gamma_{\mu L}u - \bar{d}\gamma_{\mu L}d + t\gamma_{\mu L}t - b\gamma_{\mu L}b\right.$$

$$+ \bar{c}\gamma_{\mu L}c - \bar{s}\gamma_{\mu L}s + \bar{g}\gamma_{\mu L}g - \bar{h}\gamma_{\mu L}h$$

$$+ \bar{\nu}_e\gamma_{\mu L}\nu_e - \bar{e}\gamma_{\mu L}e + \bar{\nu}_\tau\gamma_{\mu L}\nu_\tau - \bar{\tau}\gamma_{\mu L}\tau$$

$$\left.+ \bar{\nu}_\mu\gamma_{\mu L}\nu_\mu - \bar{\mu}\gamma_{\mu L}\mu + \bar{M}_o\gamma_{\mu L}M_o - \bar{M}^-\gamma_{\mu L}M^-\right]$$

$$j'^{13}_\mu = \frac{1}{2}\left[\bar{u}\gamma_{\mu L} + \bar{d}\gamma_{\mu L}d - \bar{t}\gamma_{\mu L}t - \bar{b}\gamma_{\mu L}b\right.$$

$$+ \bar{c}\gamma_{\mu L}c + \bar{s}\gamma_{\mu L}s - \bar{g}\gamma_{\mu L}g - \bar{h}\gamma_{\mu L}h$$

$$- \bar{\nu}_e\gamma_{\mu L}\nu_e - \bar{e}\gamma_{\mu L}e + \bar{\nu}_\tau\gamma_{\mu L}\nu_\tau + \bar{\tau}\gamma_{\mu L}\tau$$

$$\left.- \bar{\nu}_\mu\gamma_{\mu L}\nu_\mu - \bar{\mu}\gamma_{\mu L}\mu + \bar{M}_o\gamma_{\mu L}M_o + \bar{M}^-\gamma_{\mu L}M^-\right]$$

where $\gamma_{\mu L} \equiv \gamma_\mu \frac{1-\gamma_5}{2}$. We see that the neutral currents have the structure of the Weinberg-Salam model plus an additional piece which in the ordinary quark sector is pure isoscalar and purely left-handed. In contrast, the $SU_L(2) \otimes SU_R(2) \otimes U(1)$ model introduces I=1 and right-handed components to the neutral currents.

Phenomenology of neutral currents. From the above interactions, one obtains the following effective Lagrangrian for lepton-hadron interactions

$$\frac{G_W}{\sqrt{2}} \{ [\bar{u}\gamma_{\mu L} u - \bar{d}\gamma_{\mu L} d - 2X_W (\frac{2}{3}\bar{u}\gamma_\mu u - \frac{1}{3}\bar{d}\gamma_\mu d)] - (\frac{m_W^2}{m_{W'}^2}) [\bar{u}\gamma_{\mu L} u + \bar{d}\gamma_{\mu L} d] \} \bar{\nu}_\mu \gamma^\mu (1-\gamma_5)\nu_\mu$$

$$\frac{G_W}{\sqrt{2}} \{ [\bar{u}\gamma_{\mu L} u - \bar{d}\gamma_{\mu L} d - 2X_W(\frac{2}{3}\bar{u}\gamma_\mu u - \frac{1}{3}\bar{d}\gamma_\mu d)] \bar{e}\gamma^\mu (1-4X_W-\gamma_5)e + (\frac{m_W^2}{m_{W'}^2}) (\bar{u}\gamma_{\mu L} u + \bar{d}\gamma_{\mu L} d \quad \bar{e}\gamma^\mu (1-\gamma_5)e \}$$

and

$$\frac{G_W}{2\sqrt{2}} [-\bar{e}\gamma_\mu (1-4X_W-\gamma_5)e + (\frac{m_W^2}{m_{W'}^2}) \bar{e}\gamma_\mu (1-\gamma_5)e] \bar{\nu}_\mu \gamma^\mu (1-\gamma_5)\nu_\mu$$

where $X_W = \sin^2\theta_W$. For the atomic Bismuth experiment (heavy atoms in general), the time-component of the vector hadronic current is enhanced by the atomic numbers and is given by

$$Q^W = Z(1-4X_W) - N + (\frac{m_W^2}{m_{W'}^2})\, 3\,(Z+N)$$

$$\sim -129 + (\frac{m_W^2}{m_{W'}^2})627 \qquad \text{for } X_W \sim 0.26$$

For the absence of large parity violation effect in Bismuth experiment, we are able to determine $m_W^2/m_{W'}^2$. We find that

$$m_W^2/m_{W'}^2 \sim \frac{1}{4} , \qquad \text{or} \qquad \frac{g^2}{8m_{W'}^2} \sim \frac{1}{4} (\frac{g^2}{8m_W^2}) \sim \frac{1}{4} \quad \frac{G_F}{\sqrt{2}} .$$

The "fragrance" interaction is semi-weak.

The inclusive neutral current cross sections in neutrino reactions can be readily calculated from the effective Lagrangian given above. Neglecting the sea contribution, one has

$$R^{\nu} = \frac{\sigma^{\nu}_{NC}}{\sigma^{\nu}_{CC}} = \frac{1}{4}\{(1 - \frac{4}{3}X_W - \frac{1}{\rho})^2 + (-1 + \frac{2}{3}X_W - \frac{1}{\rho})^2 + \frac{1}{3}\left[(\frac{4}{3}X_W)^2 + (\frac{2}{3}X_W)^2\right]\}$$

$$= R^{\nu}(W\text{-}S) - \frac{1}{2\rho}(\frac{1}{\rho} - \frac{2}{3}X_W)$$

$$R^{\bar{\nu}} = \frac{\sigma^{\bar{\nu}}_{NC}}{\sigma^{\bar{\nu}}_{CC}} = \frac{1}{4}\{(1 - \frac{4}{3}X_W - \frac{1}{\rho})^2 + (-1 + \frac{2}{3}X_W - \frac{1}{\rho})^2 + 3\left[(\frac{4}{3}X_W)^2 + (\frac{2}{3}X_W)^2\right]$$

$$= R^{\bar{\nu}}(W\text{-}S) - \frac{1}{2\rho}(\frac{1}{\rho} - \frac{2}{3}X_W)$$

where $\frac{1}{\rho} \equiv \frac{M_W^2}{M_{W'}^2} \sim \frac{1}{4}$. For $X_W \sim$ 0.2 - 0.4, we see that the corrections to the Weinberg-Salam (W-S) model are at the level of a few percent. The Weinberg-Salam model is in good agreement with the data[3] - the ratios of the neutral-current to charged-current cross sections for both the neutrino and anti-neutrino reactions are consistent with one mixing angle and the scale set by $m_Z^2 = m_W^2/\cos^2\theta_W$. These features are preserved in this model. The reason is partly because the "fragrance" contribution (being of the order $\frac{g^2}{m_{W'}^2} \sim \frac{1}{4}\frac{g^2}{m_W^2}$) is small compared with ordinary weak interaction. But in heavy atoms like Bismuth, even though the "fragrance" contribution is semi-weak, its contribution is greatly enhanced by the atomic numbers as we noted above and cannot be neglected. This enhancement mechanism is why the parity violation effect is small in heavy atoms like Bismuth. For hydrogen atom, to the first approximation of neglecting form factors, this model predicts opposite sign of the parity violation as compared with the Weinberg-Salam model.

For elastic $\nu_\mu e$ and $\bar{\nu}_\mu e$ cross sections, we have plotted this model's predictions in Fig. 1 together with those of the Weinberg-Salam model. For large $\sin^2\theta_W$, ($\geq$0.4), the difference is negligible (since $m_W^2/m_{W'}^2 \sim \frac{1}{4}$). For small $\sin^2\theta_W \sim 0.26$, the difference

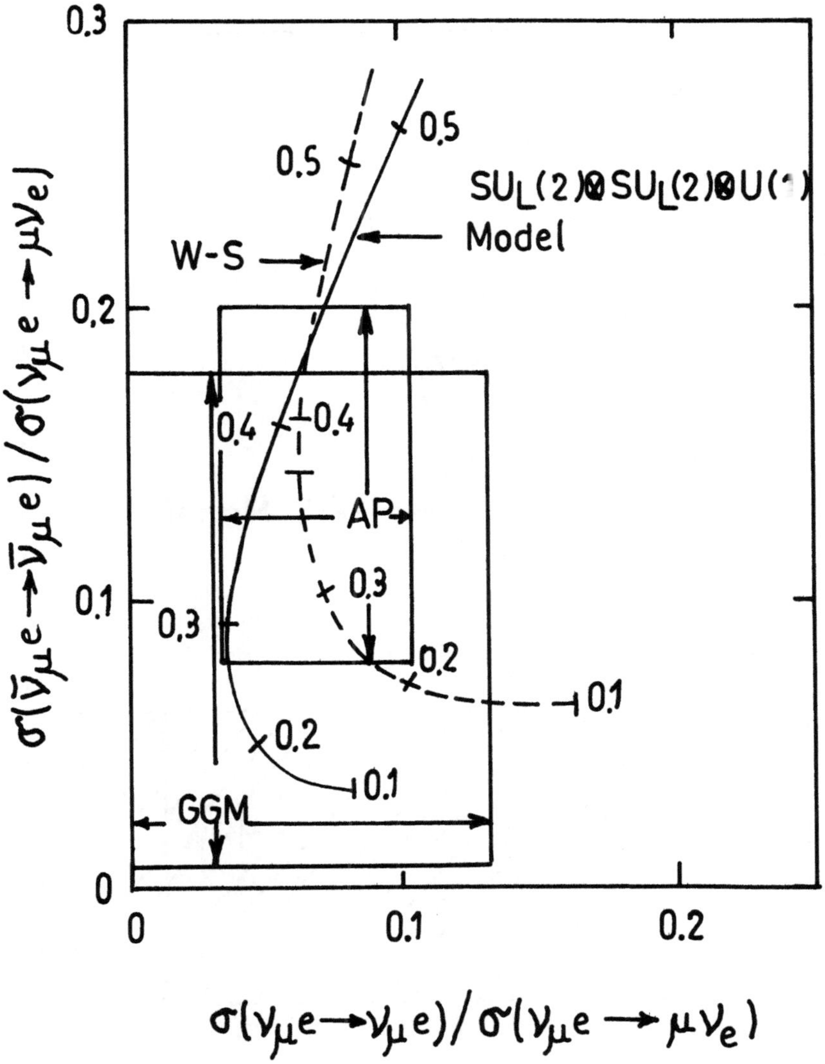

Fig.1. The elastic $\nu_\mu e$ and $\bar{\nu}_\mu e$ cross sections as functions of $\sin^2\theta_W$. The data are from ref. 12.

GGM : $\sigma(\nu_\mu e) \leqslant 2.3 \times 10^{-42}\ E_\nu\ \mathrm{cm}^2/\mathrm{GeV}$

$\sigma(\bar{\nu}_\mu e) = (1.0 {}^{+\,2.1}_{-\,0.9}) \times 10^{-42}\ E_\nu\ \mathrm{cm}^2/\mathrm{GeV}$

AC-P : $\sigma(\nu_\mu e) = (1.2 \pm 0.6) \times 10^{-42}\ E_\nu\ \mathrm{cm}^2/\mathrm{GeV}$

$\sigma(\bar{\nu}_\mu e) = (2.4 \pm 1.1) \times 10^{-42}\ E_\nu\ \mathrm{cm}^2/\mathrm{GeV}$

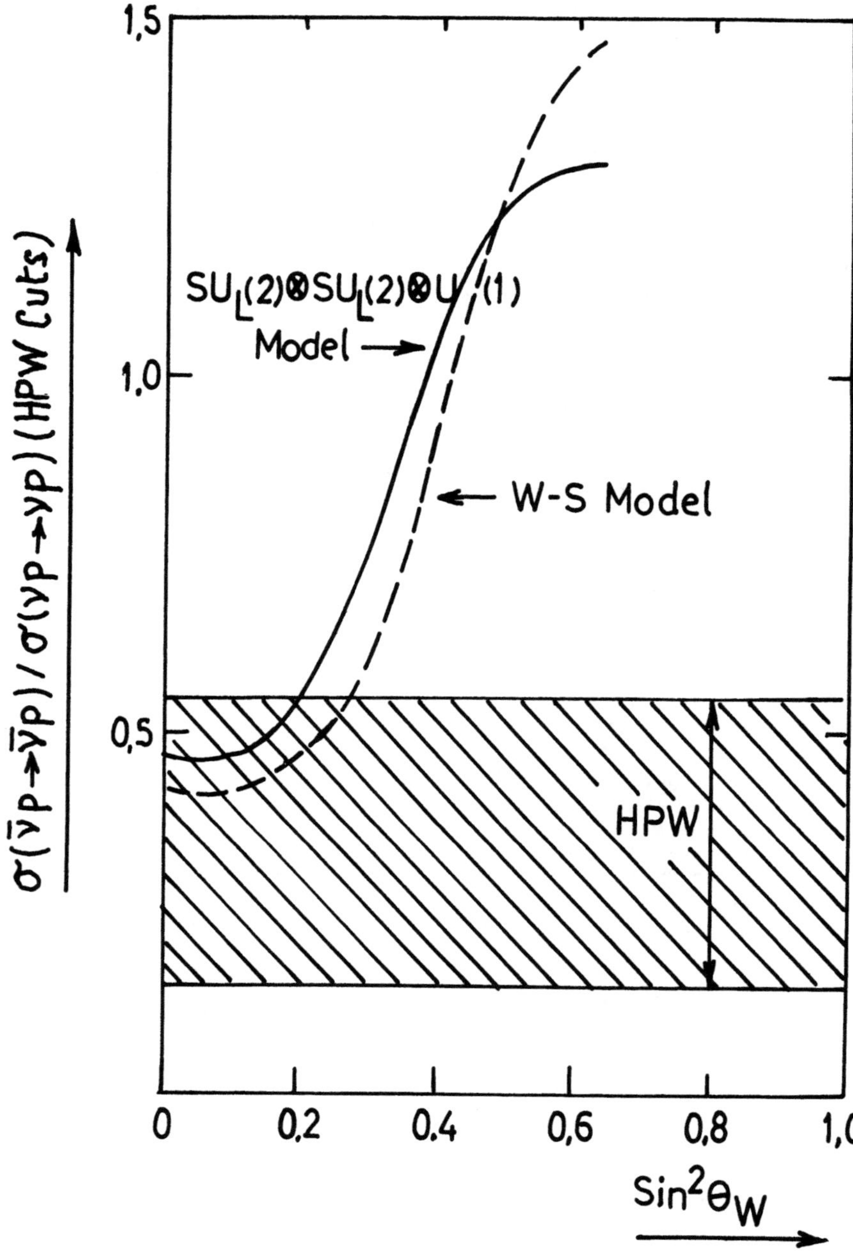

Fig. 2 (a) The ratio of elastic $\bar{\nu}_\mu$ p versus ν_μp cross sections as a function of $\sin^2\theta_W^\mu$.

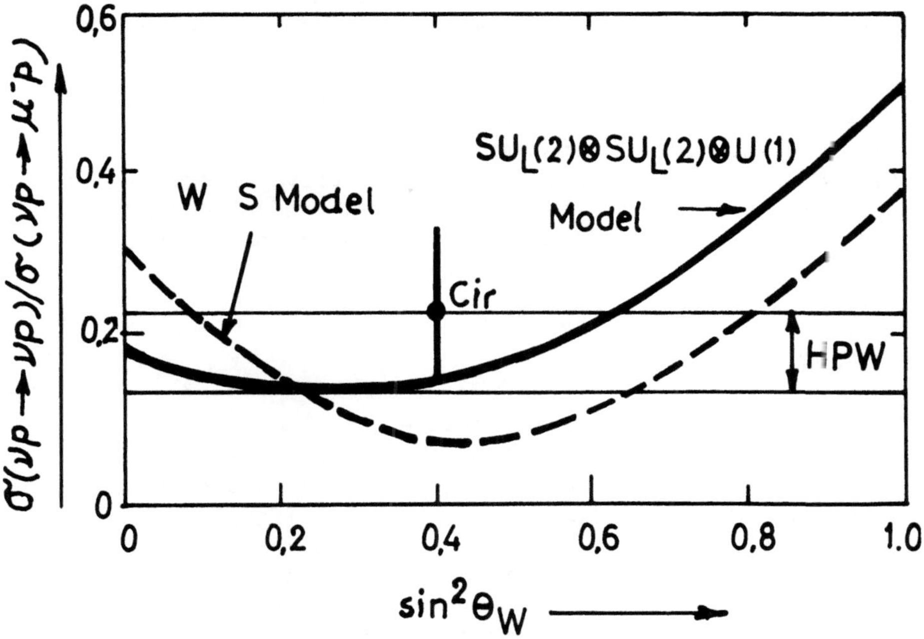

Fig. 2 (b) The elastic $\nu_\mu p$ cross section as a function of $\sin^2\theta_W$. The data are from W. Lee et al., Phys. Rev. Lett. 37, 186 (1976); D. Cline et al., ibid., 37, 252 (1976).

becomes noticeable. We conclude that measuring the cross section of $\nu_\mu e$ elastic cross section will be a crucial test of this model.

In Fig. 2 we have shown the cross sections of elastic ν_μ p and $\bar{\nu}_\mu$ p scattering. The curves have assumed dipole form factors with $m_V^2 = m_A^2 = 0.7$ GeV^2 and the same ratio for the I=0/I=1 axial-vector matrix elements as that of the magnetic contribution's to the e.m. form factors.

The isospin coupling constants of the hadronic weak neutral currents are given in Table 1 to compare with the Weinberg-Salam model. We note that both models have identical I=1 components but differ in the magnitudes of the I=0 components. Hence the isospin interference effects are quite different between these two models and could be checked experimentally. The isospin interference can be measured experimentally through either the inclusive neutral current cross sections on proton and neutron targets[17] or the pion multiplicities in inclusive or semi-inclusive neutral current processes.[18] The model predictions are given in Table 1 in terms of the differences of the neutral current cross sections according to ref. 17.

IV. CONCLUSIONS

As the experimental data accumulated and the experiments improved with the time, we find that many gauge models become obsolete. The Weinberg-Salam model still survives with respect to the neutrino data but there seems to be a difficulty with the atomic Bismuth experimental results. Whether that problem will also go away with the time remains to be seen. For the maintime, we have tried in this note to study three models which appear to be the simplest extensions of the Weinberg-Salam model which can accomodate the atomic Bismuth results.

The next thing we have done here is to study the different phenomenological predictions of the models. Our conclusions are best summarized by the following table.

Acknowledgement

I would like to thank my colleagues at DESY and University of Massachusetts, Amherst for their comments and A. Swift for reading the manuscript.

	Parity Violation in heavy atoms	Parity Violation in light atoms	$\sigma(\nu_\mu e)$, $\sigma(\bar{\nu}_\mu e)$	Angular Asymmetry in $e^+e^- \to \mu^+\mu^-$, hX	Isospin Interference
W-S Model	yes	yes	See Fig.1	yes	see Table 1
$SU(2) \otimes U(1)$ Model	very small	yes	$\sigma(\nu_\mu e)=\sigma(\bar{\nu}_\mu e)$	as (W-S)	as (W-S)
$SU_L(2) \otimes SU_R(2) \otimes U(1)$ Model	no	no	as (W-S)	$\sim$ 2(W-S)	as (W-S)
$SU_L(2) \otimes SU_L(2) \otimes U(1)$ Model	small	yes	different from (W-S) for $\sin^2\theta < 0.4$ (see fig. 1)	$\sim$ (W-S)	$\sim$ 3(W-S)

where (W-S) stands for the Weinberg-Salam model predictions.

$C_V^{I=1}$	$C_V^{I=0}$	$C_A^{I=1}$	$C_A^{I=0}$	$\frac{(d\sigma_{NC}^{\nu p}+d\sigma_{NC}^{\bar{\nu}p})-(d\sigma_{NC}^{\nu n}+d\sigma_{NC}^{\bar{\nu}n})}{(d\sigma_{CC}^{\nu p}+d\sigma_{CC}^{\bar{\nu}n})-(d\sigma_{CC}^{\nu n}+d\sigma_{CC}^{\bar{\nu}p})}$ $= -\frac{1}{2}(C_V^{I=1}\ C_V^{I=0}+C_A^{I=1}\ C_A^{I=0}$	$\frac{(d\sigma_{NC}^{\nu p}-d\sigma_{NC}^{\bar{\nu}p})-(d\sigma_{NC}^{\nu n}-d\sigma_{NC}^{\bar{\nu}n})}{(d\sigma_{NC}^{\nu p}-d\sigma_{CC}^{\bar{\nu}n})-(d\sigma_{CC}^{\nu n}\ d\sigma_{CC}^{\bar{\nu}p})}$ $= \frac{-1}{2}(C_V^{I=1}C_A^{I=0}+C_V^{I=0}\ C_A^{I=1})$	
$1-2X_W$	$-\frac{2}{3}X_W$	+1	0	$\frac{X_W}{3}(1-2X_W)$ ~ 0.04, for $X_W \sim 0.3$	$\frac{X_W}{3}$ $\sim$ 0.1, for $X_W \sim 0.3$	Weinberg-Salam model
$1-2X_W$	$-\frac{2}{3}X_W - \frac{1}{\rho}$	+1	$-\frac{1}{\rho}$	$\left((1-2X_W)(\frac{X_W}{3}+\frac{1}{2\rho})+\frac{1}{2\rho}\right)$ $\sim$ 0.21, for $X_W \sim 0.3$ and $\frac{1}{\rho} \sim 0.25$	$\left((\frac{X_W}{3}+\frac{1}{2\rho})+\frac{1-2X_W}{2\rho}\right)$ $\sim$ 0.27, for $X_W \sim 0.3$ and $\frac{1}{\rho} \sim 0.25$	$SU(2)\otimes SU_L(2) \otimes U(1)$ Model

Table I. The I=0, I=1 vector and axial-vector coupling constants of the Weinberg-Salam model and the $SU_L(2) \otimes SU_L(2) \otimes U(1)$ Model as defined by

$$\frac{G_F}{\sqrt{2}} \{C_V^{I=0}\frac{1}{2}(\bar{u}\gamma_\mu u + \bar{d}\gamma_\mu d) + C_V^{I=1}\frac{1}{2}(\bar{u}\gamma_\mu u - \bar{d}\gamma_\mu d) + C_A^{I=0}\frac{1}{2}(\bar{u}\gamma_\mu\gamma_5+\bar{d}\gamma_\mu\gamma_5 d)$$
$$+ C_A^{I=1}\ \frac{1}{2}(\bar{u}\gamma_\mu\gamma_5 u - \bar{d}\gamma_\mu\gamma_5 d)\}$$

. The expressions for the cross sections are derived with parton model assumptions, see ref.17.

REFERENCES AND FOOTNOTES

1. S. Weinberg, Phys.Rev.Lett. 19, 1264 (1967).
Abdus Salam, in Elementary Particle Physics, edited by N. Svartholm (Almquist and Wikskells, Stockholm, 1968), p.367.
2. S.L. Glashow, J. Iliopoulos, and L. Maiani, Phys.Rev.D2, 1285 (1970).
3. See the lectures given by J. Steinberger.
4. See the lectures given by V. Telegdi. One should bear in mind that in heavy atoms like Bismuth, the time component of the hadronic neutral currents are enhanced by the atomic numbers. What the present experiments have measured is thus the parity violation effect induced by the vector hadronic neutral currents and the axial-vector electron currents.
5. E.M. Henley and L. Wilets, Phys.Rev. A14, 1411 (1976); M. Brimicombe, C.E. Loving and P.G.H. Sandars, J. Phys. (London), B9, L237 (1976); I.B. Khriplovich, JETP Lett., Vol. 20, 315 (1974).
6. The latest results reported by P.G.H. Sanders at International Symposium on Lepton and Photon Interactions at High Energies, Hamburg (1977) are
$R_{876nm} = (-0.7 \pm 3.2) \times 10^{-8}$ (U. of Washington experiment) and
$R_{648nm} = (2.7 \pm 4.7) \times 10^{-8}$ (Oxford experiment).
The Weinberg-Salam model with the atomic physics calculation predicts $R_{876nm} \sim -23 \times 10^{-8}$ and $R_{648nm} = -30 \times 10^{-8}$.
7. P. Fayet, Nucl.Phys. B78, 14 (1974); T.P. Cheng and L.-F.Li, Phys.Rev. Lett. 38, 381 (1977). Fayet first proposed that the weak electron current should be pure vector. This model was noted by Cheng and Li with regard to $\mu \to e\gamma$ decay.
8. R.N. Mohapatra and D.P. Sidhu, Phys.Rev.Lett. 38, 667 (1977) and to be published; A. De Rujula, H. Georgi and S.L. Glashow, to be published; J.C. Pati, S. Rajpout and A. Salam, ICTP preprint ICTP/76/11; J.C. Pati and A. Salam, Phys.Rev. D10, 275 (1974); R.N. Mohapatra and J.C. Pati, Phys.Rev. D11, 566, 2558 (1975); H. Fritzch and P. Minkowski, Nucl.Phys.B103, 61 (1976); E. Ma, Oregon preprint OITS-68 (to be published); S. Weinberg, Phys.Rev.Lett. 29, 388 (1972).
9. T.C.Yang, DESY 77/39, to be published and DESY preprint 77/51.
10. See the lectures given by J. Ellis. Naturalness condition was first proposed by S.L. Glashow and S. Weinberg, Phys.Rev. D15, 1958 (1977).
11. See the lectures given by L. Lederman.
12. For recent results, see H. Reithler reported at International Symposium on Lepton and Photon Interactions at High Energies, Hamburg (1977).
13. A. McDonald, Nucl.Phys. B75, 343 (1974) and references therein.
14. R. Gatto and G. Preparata, Nuovo Cimento Letters 7, 89 (1973) and ref. 13.

15. 8 quarks and 8 leptons were previously proposed in a different context (see T.C. Yang, Phys.Lett. 65B, 358 (1976)). We received a recent preprint by M.A.B. Bég, R.N. Mohapatra, A. Si lin and H.-S. Tsao who have discussed 8 quarks and 24 leptons in a $SU_L(4) \otimes SU_R(4) \otimes U(1)$ model. Note that under SU(4) symmetry, 8 quarks would require 24 leptons in order to cancel anomalies.
16. The main point is that t, b, g, h quarks are heavier than τ lepton. The τ decay via "flavor" SU(2) interactions is just like the case of a sequential heavy lepton. The only place where the "fragrance" interaction contributes is in $\tau \to e\bar{\nu}_e\nu_\tau$. We thus find that $\frac{BR(\tau\to\nu_\tau e\bar{\nu}_e)}{BR(\tau\to\nu_\tau\mu\bar{\nu}_\mu)} \neq 1$.
17. P.Q. Hung and J.J. Sakurai, Phys.Lett. 63B, 295 (1976).
18. L.M. Sehgal, Nucl. Phys. B90, 471 (1975) and references therein.

QUARK CONFINEMENT AND TIME SPACE STRUCTURE

David FINKELSTEIN

Yeshiva University

New York, 10033 - U.S.A.

ABSTRACT

A program for the reconstruction of the time space continuum and the linear space of quantum states, from a more basic theory of discrete diagrams, it outlined in which quantum complementarity occurs because the observer is part of the "entire situation" described by the state vector according to Bohr, and quark confinement occurs because there is a finite elementary time space unit and quark lines in the same hadron diagram are a small number of units apart.

I. INTRODUCTION

Quantum chromodynamics, as others here have pointed out, is too good to ignore, too ugly to be true. (Somewhat similar gauge theories of weak electromagnetic interactions in which I have taken part are reported by M. Tavel [1]). Since that work I have been trying more fundamental approaches to the particle problem. The newest and most radical of these renounces (and finally reconstructs) both the common state vector or Ψ space and the common Minkowski or time space.

Imperatives for the renunciation of the time space continuum include such syntactic inconsistencies as the ultraviolet infinities, which I do not discuss, and such semantic inconsistencies as the non-observability of time space points, which I do.

Bohr and Rosenfeld [2] measure a field at a "point" of time space of size L by means of a test particle of size L and arbitrarily large mass/charge ratio (approaching infinity). For $L \sim 1$ millimeter, pith balls serve well, because they have a large range of variation for this ratio (say, 10^{10} - 10^{20} amu/e, a range of about 10^{10}). At the molecular scale the range of this ratio is smaller (say 10^3), at the atomic scale, smaller still (10^2) and at the particle scale it hardly varies at all, let alone, approaching infinity, contrary to the needs of the measurement.

Adhering to the operational spirit of quantum mechanics, I take it as obvious therefore that ordinary time space concepts and at the particle scale of size and time (10^{-13}cm, $\tau \sim 10^{-23}$ sec) not the Planck scale (10^{20} smaller). I suppose quark confinement is another of Mother Nature's marvelously tactful evasions of a question she knows to be stupid : here, "Is common geometry valid inside a hadron ?" The renunciation of time space is a natural next step in the delinerizations of physics in this century [3]. There have been attempts to make a physics without benefit of a time space continuum, adhering to the common Ψ space, with a small success [4].

The imperative for a renunciation of Ψ space is mentioned less often. It is a semantic difficulty, not a syntactic one. The Ψ space of quantum theory represents the transition probability prob A)B from (pure, sharp) quantum effectors A), e.g. emitters, to quantum receptors (B, e.g. detectors . Indeed the linear space of quantum theory (the coefficient field and dimension, not the metric) are already completely defined by the possibility relation poss A)B, defined to be 0 (false) if the transition probability prob A)B is zero, and 1 (true) otherwise. The continuity of this space therefore represents only the continuity of our effectors and receptors. For example, the continuity of the two-dimensional Hilbert space of photon spin represents the continuity of rotation of a polarizer about the beam axis and the continuity of thickness of wave plates in the beam.

Therefore it is as naive to extrapolate Ψ space into arbitrarily small regions as time space, and I now consider only theories in which common Ψ space and time space are constructed together. It makes no physical sense to construct such a theory by quantizing a Lagrangian theory, however heuristic that procedure has been. I seek a concept of theory quantum from the start, with common quantum mechanics as a limit of large scale ($\tau \to 0$) and classical mechanics are a limit of large action ($\hbar \to 0$). The renunciation of Ψ space must be the last step in the delinearization of physics. Here is the kind of reconstruction I seek.

II. COMPLEMENTARITY REPRESENTS PARTICIPATION

The prime clue for this search is Bohr's dictum that Ψ describes the "entire situation". This dictum is a deep truth in the sense of Bohr :

In classical physics the descriptive symbols (p's and q's) do not involve the reference system in one way that Ψ does. A Ψ or Dirac ket |A> is best read as a sentence of type subject-verb-object. The bar | is the subject, "I". The label A is the verb, says what I do. The terminal bracket > is the object, the quantum I prepare. In the dual bra symbol written <A| by Dirac, I and the quantum change positions; I prefer the notation >A|, therefore. Bra and Ket are descriptions of the quantum before and after the fact of the observer. In classical physics the effect of the observer is negligible and the distinction between vector and covector is otiose. Not in quantum theory. For example processes of position and momentum preparation, |x'> and |p'>, are incompatible, but position preparation and momentum detection, |x'> and >p'|, are not. The distinction between preparation and detection is the added extent by which Ψ describes the entire situation and classical coordinates do not.

The negation of Bohr's dictum is also a deep truth :

The Ψ says as little about me as it can. It expresses the existence and action of the subject and the temporal relation of these to the rest of the situation. This cannot maximally describe any but the simplest systems, none complex enough actually to carry out the action named.

This descriptive omission of the common quantum theory leads me to ask for a maximal description of the entire situation subject-plus-object-plus-surroundings, which I call the metasystem. This departs from the usual position of physics, which I call objective, that so nearly deletes the subject. Such fuller theories may be called holistic.

The point of this paper is that a certain degree of complementarity is natural to holistic theories. I show how participation can give rise to some degree of complementarity, and postulate that all complementarity has that origin : complementarity represents participation. The central position of participation in quantum theory has been emphasized by Wheeler [5].

The starting point and motivation for this postulate is also one of the origins of Bohr's own conception of complementarity in the existential philosophy and the psychology of his youth, such as that of William James. The effect of observation on the quantum is likened to that of introspection on the stream of consciousness. Bohr and James discuss a complementarity of modes of though (love and judgement, introspection and stream of consciousness, etc...) that inevitably arises even in a classical model just because such modes compete to operate in the same mind. Likewise, I postulate that complementarity arises between quantum effector processes because they compete for place in the same world or metasystem. From the early days of quantum theory complementarity has been related to such operational incompatibility by Bohr and Heisenberg; I try to strengthen the relation and derive the Ψ space formalism from it.

III. HOLISTIC DESCRIPTION

A holistic language that describes the metasystem can be much simpler than an objective one for the system alone. Many of the properties attributed to the object in objective theories are relations between subject and object in holistic theories. In particular all the familiar invariance groups of physics need not act on the basic elements of a holistic theory. Insofar as such invariances are exact they are unobservable; insofar as they are observable they are inexact.

For example, there need be no description of translational invariance. If the translation meant is that of an entire particle experiment, accelerator, physicist, universe and all, that is not an actual process that can take place, and is not even supposed observable by those two consider it meaningful. If the translation is that of the accelerator relative to the physicist's frame of reference, invariance fails when his/her nose enters the particle beam. Similarly for the other important invariances of physics. In a holistic theory they are all approximate, asymptotic, or dynamical invariances, and must be derived, not postulated.

For definiteness I assume a language of (discrete finite) diagrams, somewhat like Feynman's or those of [4], but using neither time space nor Ψ space. I cannot give the syntax of this language : Is the diagram labelled ? Are its lines oriented ? (I suppose not, since arrow reversal would then be expected to be an invariance transformation or automorphism, and I have said there are none). Is there a graph connection [4] ? I will not pin the theory down yet. I designate a typical diagram by p, regard it as a description of the metasystem, and suppose that each Ψ vector of the system goes with a set P = {p} of metasystem diagrams, each dual Ψ' vector with a set P' = {p'} of diagrams :

$$P = \text{dia}\ \Psi, \quad P' = \text{dia}\ \Psi'$$

The key element of the holistic theory, though it need not be described simply in the theory, is an exactness or null relation pop' defined at least for the diagrams p, p' representing effectors and receptors, and meaning that the receptor (process described by the diagram) p' is not actuated by the effector p. (Why call it exactness ? Because the sequence of maps

$$\underset{\sim}{0} \to \underset{\sim}{1} \xrightarrow{\Psi} L \xrightarrow{\Psi'} 1 \to 0$$

is exact, in which $\underset{\sim}{0}$ and $\underset{\sim}{1}$ stand for linear spaces of those dimensions, L is the linear space of the Ψ vectors, the maps $0 \to 1$ and $1 \to 0$ are the obvious constant maps, the map Ψ carries a complex number c into cΨ, and the map Ψ' carries a vector Ψ into the functional value $\Psi'(\Psi)$, a complex number. Why call it null ? Because the transition probability prob $\Psi)\Psi'$ is then null, zero. The pretty symmetry of the exactness condition is lost in the limiting case of classical set theory, where the relation "o" becomes simply "$\neq$", the inequality of points .

We use a construction due to Galois to make a lattice from the o relation, and a construction due to von Staudt to make a linear space from the lattice. The relation o on which Galois based his construction is a fixed-point relation in which pop' means that p' is an element of a number field, p is an automorphism of the number field, and p leaves p' fixed. It is not hard to modify the concept of measurement so that the null case $\Psi'(\Psi) = 0$ likewise corresponds physically to a case in which the quantum produced by the first process is left unchanged by the second. For this one uses not 3-state devices like von Neumann's measuring instruments (ready, yes, no), but 2-state devices (ready, yes), which I call determination rather than measurements. For common quantum mechanics, measurement lagebra is invariant under the unitary group, determination algebra under the linear group. I return later to the question : How is the relation pop' computed from the diagrams p, p' ?

IV. THE GALOIS CONSTRUCTION

First form the Galois connection [5] of the null relation pop' as follows.

The Galois connection consists of two lattice Λ, Λ' and a dual isomorphism $\Lambda \to \Lambda'$. Here Λ is a lattice of physical sets of effector diagrams, Λ' of receptor diagrams.

The elements of Λ are sets of diagrams p; of Λ', p' dually. A set P of diagrams p belongs to Λ if and only if there exists a set P' of diagrams p' such that

$$P = \{p \mid \text{for all } p' \varepsilon P',\ pop'\}.$$

A set P' of diagrams p' belongs to Λ' if and only if there exists a set P of diagrams p such that

$$P' = \{p' \mid \text{for all } p \varepsilon P,\ pop'\}.$$

These express the idea that for a set of effectif diagrams to be physical there must be a receptor that distinguishes those diagrams from others, and dually. Verify that Λ, Λ' are lattices under set inclusion.

There is accordingly no need for the Birkhoff-von Neumann [5] postulation of the lattice property.

The dual isomorphism P → p' associates with each set P in the set in Λ'

$$P' = \{p' \mid \text{for all } p \text{ in } P,\ pop'\}.$$

Verify that if the p's are vectors in a linear space of a quantum system and the p' covectors in the dual space, with

$$pop' \text{ if and only if } pp' = 0$$

(vanishing inner product or contraction of the vector p and co-vector p'), then Λ is the lattice of subspaces, the "logic" of the quantum system.

This construction does not yet refer to a particular subject. For this the pop' in the above construction should be appropriately limited.

V. THE VON STAUDT CONSTRUCTION

Then test the lattice Λ of the Galois connection as follows [6] :

1. Does the Galois connection have polar symmetry : i.e. is there an isoporphism $*$: Λ → Λ' such that pop^* never holds. Such an isomorphism is called a polarity of the connection, and of the relation o.

2. Is Λ modular ? (If not, does it become modular in a suitable limit of large situations ?).

3. Is it irreducible ? (If not, reduce it).

The most critical of these tests is modularity.

If these tests are passed then von Staudt has shown how to construct a number field F such that Λ is the projective geometry of the F-linear space of a suitable dimension [6]. I take this as the linear space of the quantum theory flowing from the diagram theory.

VI. ORIGIN OF QUARK CONFINEMENT

Having programmed the reconstruction of the linear space of the quantum theory, we turn now to the microstructure of time space.

The procedures given elsewhere [3] for constructing the time space from the diagram theory are being reconsidered within the present more basic program. The basic idea, that infinitesimal coordinate differentials are replaced by finite but small spin-like operators acting on the connections of the diagram, is still tenable for there is a natural correspondence already assumed between some diagrams at least and kets, and changes of connection in the diagram still represent, therefore, linear operators (quantities) of the quantum theory thus reconstructed.

While we see a time space continuum between us, particles or quarks need not. The seeming continuity of the time space coordinates of an event relative to a macroscopic observer arises entirely, in this view, from the huge number of vertices in the diagram for the observer, and in particular from averaging over the multiplicity of parallel paths. Quark confinement, the limited range of the relative coordinates of one quark with respect to another in the same hadron, may therefore be attributed to the small number of vertices between the quark events in the diagrams of the hadron process, each vertex contributing an operator of order τ to the interval.

The first steps in recovering a gauge field theory like chromodynamics or that of [1] from a diagram theory like that assumed here are taken in [4] and seem to remain meaningful in spite of the renunciation of the quantum state vector space since that work.

The main problem now is the definition of the relation pop'. In view of the complexity of the conditions IV. 1-3 that o must satisfy, I expect o itself to be complex, the result of a recursive construction that takes the place of the differential equations of common quantum field theory.

REFERENCES

1. M. Tavel, Weak and Electromagnetic Interactions in Quaternion Quantum Mechanics", Ph. D. Thesis, Yeshiva University, New York, 1964.

2. N. Bohr and L. Rosenfeld, Danske Vid. Selskab Mat-Fys Med. XII, 8, Copenhagen, 1933.

3. D. Finkelstein, "The Delinearization of Physics", in Symposium on the Foundations of Modern Physics, Loma-Koli, Finland, August 1977, to be published.

4. D. Finkelstein and G. McCollum, "The Concept of Particle in Quantum Mechanics", in Quantum Theory and Structure of Time and Space, L. Castell, M. Drieschner and C.C. von Weizsaecker eds., Vol. 2, Carl Hanser Verlag, Munich, Germany, 1977.

5. G. Birkhoff, Lattice Theory, Amer. Math. Soc. Publ., 1948.

6. G. Birkhoff and J. von Neumann, "Logic of Quantum Mechanics", Annals of Math. 37, 823 (1936).

INDEX

GPSR Compliance

The European Union's (EU) General Product Safety Regulation (GPSR) is a set of rules that requires consumer products to be safe and our obligations to ensure this.

If you have any concerns about our products, you can contact us on ProductSafety@springernature.com

In case Publisher is established outside the EU, the EU authorized representative is:

Springer Nature Customer Service Center GmbH
Europaplatz 3
69115 Heidelberg, Germany

Batch number: 10683525

Printed by Printforce, the Netherlands